Sperm Chromatin

Armand Zini • Ashok Agarwal
Editors

Sperm Chromatin

Biological and Clinical Applications in Male Infertility and Assisted Reproduction

Springer

Editors
Armand Zini, MD
Department of Surgery
Division of Urology
McGill University
St. Mary's Hospital Center
Montreal, QC, Canada
ziniarmand@yahoo.com

Ashok Agarwal, PhD, HCLD (ABB)
Center for Reproductive Medicine
Glickman Urological and Kidney Institute
OB-GYN and Women's Health Institute
Cleveland Clinic, Cleveland, OH, USA
agarwaa@ccf.org

ISBN 978-1-4419-1781-2 e-ISBN 978-1-4419-6857-9
DOI 10.1007/978-1-4419-6857-9
Springer New York Dordrecht Heidelberg London

Library of Congress Control Number: 2011932679

Printed on acid-free paper

Springer is part of Springer Science+Business Media (www.springer.com)

Foreword

The composition, stability, and integrity of sperm chromatin have major bearings on the fertilizing potential of male gametes and their capacity to support normal embryonic development. This assertion is supported by clinical data generated over the past 10–15 years, indicating the existence of significant correlations between DNA damage in spermatozoa and a variety of adverse reproductive outcomes including reduced conception rates, a high frequency of miscarriage and an escalation in the incidence of birth defects. Despite this wealth of correlative data, uncertainties remain with respect to such key questions as the precise nature of the DNA damage, the relative importance of genetic, epigenetic and environmental factors in its etiology, and the most effective method for detecting its presence. Answers to these questions are critical if we are to develop rational strategies for alleviating or, ultimately, preventing DNA damage in spermatozoa. Furthermore, answers to these questions are needed if clinicians are to provide patients with informed advice about the significance of their DNA damage results. In *Sperm Chromatin: Biological and Clinical Application in Male Infertility and Assisted Reproduction* Ashok Agarwal and Armand Zini have assembled contributions from a panel of world experts in this area and, in so doing, created an invaluable text that brings us up-to-date with recent advances in this field.

The remodeling of sperm chromatin during spermiogenesis is as remarkable as it is complex. It involves a dramatic morphological and biochemical transformation of chromatin structure through the coordinated movement of proteins in and out of the nucleus to generate a unique, highly compacted matrix. Furthermore, this extraordinary metamorphosis takes place in the absence of *de novo* gene transcription. Indeed, spermiogenesis is an object lesson in how biological processes can be controlled through the regulated translation of pre-existing mRNA species. If we could only replicate this process in vitro, research into the underlying control mechanisms would be greatly facilitated. Unfortunately, this is not possible at the present time and, as a result, progress in this area has been painfully slow. Nevertheless, as the pioneer of sperm chromatin research, Rod Balhorn, indicates in Chap. 1, we might anticipate more rapid progress in this area in the future because we now have an array of sophisticated tools to investigate the chromatin remodeling process.

Thus, our new found ability to generate detailed inventories of proteins and mRNA species in selected cells and tissues is one of the driving forces behind the modern systems approach to biological research, and the analysis of sperm chromatin will be one of the major beneficiaries of these technical developments. In the first section of this book, we see beautifully illustrated reviews of spermatogenesis (Chap. 2) and chromatin organization (Chaps. 4 and 5) followed by a number of authoritative reviews on aspects of the genome (Chap. 6), epigenome (Chap. 7), transcriptome (Chap. 8), and proteome (Chap. 3) from a spermatozoon's perspective. The detailed chemical analysis of sperm chromatin using the tools of the *-omics* revolution will certainly provide important clues as to the formation and ultimate function of this material and act as a major stimulus for increased understanding in this area.

The impetus to study the composition and integrity of sperm chromatin from a clinical perspective can be traced back to the pioneering studies of Don Evenson, who not only initiated research in this area long before it became fashionable but also pioneered one of the major analytical techniques used in the assessment of sperm chromatin, the Sperm Chromatin Structure Assay (SCSA) (Chap. 9). This assay has now become the industry standard against all other techniques. In the second section of this book Agarwal and Zini have gathered together a very impressive list of authors, all of whom share an interest in the laboratory evaluation of sperm chromatin composition and integrity. The list of techniques is extensive and comprises, in addition to SCSA, methods that have been borrowed from research in toxicology (Comet; Chap. 15) or somatic cell apoptosis (TUNEL; Chap. 14) as well as a range of protocols that are more specific to the male germ line including the Sperm Chromatin Dispersion assay (Chap. 10), the acridine orange test (Chap. 13), the isolation, purification and quantification of protamines (Chap. 16), the aniline blue test for histone retention (Chap. 12), gene expression profiling (Chap. 18), and analyses of epigenetic modifications to the haploid paternal genome (Chap. 17).

Additional probes such as toluidine blue and chromomycin 3 (CMA3) are also important because they provide indirect evidence of the efficiency of chromatin packaging (Chap. 11). In this context, Denny Sakkas deserves special mention for his work on the development of CMA3 as an extremely valuable diagnostic probe, which has consistently been found to generate significant information about sperm chromatin status in infertile men.

The clinical and environmental factors that contribute to the etiology of DNA damage are reviewed in the third section of this book. The range of possible contributory factors is again extensive and includes oxidative stress (Chap. 19), apoptosis (Chap. 20), and defective repair of physiological DNA nicks (Chap. 21). The primary drivers for the oxidative stress, abortive apoptosis, and defective DNA repair include infertility (Chap. 22), age (Chap. 23), cancer (Chap. 24), environmental pollutants (Chap. 25), infection (Chap. 26), and cryopreservation (Chap. 27). If we understood more about the chemical nature of the DNA damage seen in human spermatozoa, we might be able to reduce this long list of potential causative factors down to the major culprits. However, at present, the only clue we have to the types of DNA damage present in human spermatozoa is the high prevalence of oxidative base lesions detected in the patient population. The mechanism by which such stress is generated is still something of an unresolved mystery.

If we do not understand the etiology of DNA damage, then there is little we can do to treat this condition or prevent it from arising. However, we can try to develop strategies for limiting the impact of such damage on human embryos conceived in vitro including antioxidant therapy (Chap. 30), electrophoretic sperm isolation (Chap. 29), and the selective binding properties of hyaluronic acid polymers (Chap. 28), all of which help select non-DNA damaged spermatozoa for the insemination of oocytes.

The importance of developing strategies for avoiding conception with DNA damaged spermatozoa is generally supported by the clinical data, although the results are not always as clean cut as one would like (Chaps. 31 and 32). A major problem with such clinical studies is that pregnancy is a very bad test of sperm function because it depends on so many confounding variables. By contrast, the animal data are incontrovertible in demonstrating that DNA damage in the male germ line has profound, lasting effects on the viability of pregnancy and the health of the offspring (Chap. 33). In light of such certainty, it is critical that the information available on this topic is assembled and presented for careful consideration. Agarwal and Zini have done a remarkable job of pulling together the protagonists in this field and creating a compendium of knowledge that will be of intense interest to clinicians and scientists who share an appreciation of the significant contributions made by the male gamete to a healthy start to life.

Callaghan, NSW — R. John Aitken, PhD, ScD, FRSE

Preface

The evaluation of sperm DNA and chromatin abnormalities has gained significant importance in the past several years, largely as a result of the recent advances in assisted reproductive technologies (ARTs). In vitro fertilization (IVF) and intracytoplasmic sperm injection (ICSI) have revolutionized the treatment of male-factor infertility. However, we have come to realize that in the context of these ARTs, the genetic integrity of the sperm is a key aspect of the paternal contribution to the offspring. With the growing concerns about the long-term safety of ARTs (especially ICSI), we have seen an increasing number of studies on the male genome's influence on reproductive outcomes. These studies now shed some light on the influence of sperm chromatin and DNA abnormalities on reproductive outcomes. Along with these clinical studies, we also have made real advances in our understanding of the basic aspects of sperm chromatin and DNA integrity. We are now starting to better understand the unique organization of the sperm chromatin, as well as the nature and etiology of sperm DNA damage.

We assembled this textbook with the idea of bringing together the key fundamental and practical elements of this rapidly evolving field. The 34 chapters in our book and four supplementary sections covering test protocols and guidelines are written by contributors from 16 countries. These authors were carefully selected based on their expertise and proven track record of high-quality research in the field. Our book is intended for researchers and clinicians alike and is meant to bridge the gap between our basic and clinical knowledge on sperm chromatin and DNA integrity. For the basic scientist, this textbook will serve as a sound foundation for any further studies in this field. For the clinician, this book will help guide clinical practice in this area.

We would like to thank Richard Lansing, executive editor, for his support and advice and Margaret Burns, developmental editor, for her tireless efforts in reviewing and editing each of the manuscripts. Furthermore, we would like to thank all of the outstanding contributors for sharing their knowledge and for submitting their manuscripts on time. Finally, we are indebted to our families who have endured many long nights when we were working late on this book.

Montreal, QC — Armand Zini
Cleveland, OH — Ashok Agarwal

Editor Biographies

Dr. Armand Zini is Associate Professor of Surgery and Director of the Andrology Fellowship program at McGill University. Dr. Zini received his Medical degree and completed his urologic training at McGill University in Montreal. He then completed a fellowship in Male Infertility at the New York Hospital-Cornell Medical Centre and The Population Council in New York, working with Drs. Marc Goldstein and Peter Schlegel. Dr. Zini's main expertise is in clinical male infertility. Over the past 10 years, he has focused his research activity on the study of human sperm chromatin and DNA integrity, and, he has published numerous important papers on the influence of sperm DNA damage on reproductive outcomes. In 2005, he gave the John Collins lecture entitled "Sperm DNA damage and Male Infertility" at the annual meeting of the American Society for Reproductive Medicine. In 2006, he was invited to present on the "Tests of sperm DNA damage" at the Canadian Fertility and Andrology Society (CFAS) annual meeting and in 2008 was invited to present on the "Clinical importance of sperm DNA damage" at both the Canadian Fertility and Andrology Society (CFAS) and the American Society of Andrology (ASA) annual meetings. Dr. Zini has recently presented on the "Role of antioxidants and sperm DNA damage" (Sperm DNA Symposium in Rome, Italy, March 2009) and at the 2009 European Society for Human Reproduction and Embryology (ESHRE) consensus workshop on sperm DNA testing in Sweden. Dr. Zini is currently funded for studies on sperm physiology and the epigenetic effects of vitamin supplements.

Dr. Ashok Agarwal is the Director of Research at the Center for Reproductive Medicine at Cleveland Clinic Foundation and a Professor at the Lerner College of Medicine of Case Western Reserve University. His current research interests include studies on molecular markers of oxidative stress, DNA integrity, apoptosis in the pathophysiology of male and female reproduction, and effect of radio frequency radiation on fertility and fertility preservation in patients with cancer. Dr. Agarwal has published over 500 scientific articles and reviews in peer reviewed scientific journals, authored over 50 book chapters, and presented over 700 papers at scientific meetings. He is on the editorial board of over a dozen scientific journals. His laboratory has trained more than 100 basic scientists and clinical researchers from the United States and abroad. He is the Program Director of the highly successful Summer Internship Course in Reproductive Medicine. In the last 4 years, over 100 premed and medical students from across the United States and overseas have graduated from this highly competitive program. Dr. Agarwal has been invited as a guest speaker to over 20 countries for important international meetings. He has directed more than a dozen Andrology Laboratory and ART Workshops in recent years.

Contents

Contributors

Ashok Agarwal, PhD, HCLD (ABB) Center for Reproductive Medicine, Glickman Urological and Kidney Institute, OB-GYN and Women's Health Institute, Cleveland Clinic, Cleveland, OH, USA

R. John Aitken, PhD, ScD, FRSE Discipline of Biological Sciences, School of Environmental and Life Sciences, University of Newcastle, University Drive, Callaghan, NSW, Australia

ARC Centre of Excellence in Biotechnology and Development, Priority Research Centre in Reproductive Science, University of Newcastle, Callaghan, NSW, Australia

Naif Al-Hathal, MD Department of Surgery, Division of Urology, Royal Victoria Hospital, McGill University, Montreal, QC, Canada

Chris D.R. Arca, BS Institute for Biogenesis Research (IBR), John A. Burns School of Medicine, University of Hawaii at Manoa, Honolulu, HI, USA

Rod Balhorn, PhD Department of Applied Science, University of California, Davis, CA, USA

Alberto Barros, MD, PhD Centre for Reproductive Genetics, Porto, Portugal

Davide Bizzaro, PhD Department of Biochemistry, Biology and Genetics, Polytechnic University of Marche, Ancona, Italy

Guylain Boissonneault, PhD Department of Biochemistry, Université de Sherbrooke, Sherbrooke, QC, Canada

Jason Matthew Boman, MD, FRCS(C) Department of Surgery, Division of Urology, Centre Hospitalier Régional du Suroît, Montreal, QC, Canada

Mona Bungum, MSc, Med Dr, PhD Reproductive Medicine Centre, Skåne University Hospital Malmö, Lund University, Sodra Forstadsgatan, Malmö, Sweden

Alexandra Calle, BS Department of Animal Reproduction and Animal Genetic Resources Conservation, National Research Institute for Agriculture and Food Technology (INIA), Madrid, Spain

Aldo E. Calogero, MD Section of Endocrinology, Andrology and Internal Medicine, and Master in Andrological, Human Reproduction and Biotechnology Sciences, Department of Internal Medicine and Systemic Diseases, University of Catania, Policlinico "G. Rodolico", Catania, Italy

Serge Carreau, PhD Department of Biochemistry, University of Caen, Caen, France

Douglas T. Carrell, PhD, HCLD Andrology and IVF Laboratories, Department of Surgery, University of Utah, Salt Lake City, UT, USA

Judit Castillo, MS Human Genetics Research Group, IDIBAPS, Faculty of Medicine, University of Barcelona, Barcelona, Spain

Biochemistry and Molecular Genetics Service, Hospital Clínic i Provincial, Barcelona, Spain

Donovan Chan, BSc Departments of Pharmacology and Therapeutics, Pediatrics, and Human Genetics, McGill University and Montreal Children's Hospital of the McGill University Health Centre Montreal, QC, Canada

Peter Chan, MD Department of Surgery, McGill University and the Research Institute of the MUHC, Montréal, QC, Canada

Wai-Yee Chan, BSc, PhD Reproduction, Development and Endocrinology Program, School of Biomedical Sciences, Chinese University of Hong Kong, Shatin, Hong Kong, SAR, China

Albert Hoi-Hung Cheung, PhD Reproduction, Development and Endocrinology Program, School of Biomedical Sciences, The Chinese University of Hong Kong, Shatin, Hong Kong, SAR, China

Rosita A. Condorelli, MD Section of Endocrinology, Andrology and Internal Medicine, Human Reproduction and Biotechnology Sciences, Department of Biomedical Sciences, University of Catania, Catania, Italy

Jean-Pierre Dadoune, MD, PhD Department of Histology, Biology of Reproduction, Hôpital Tenon, Paris, France

Rosario D'Agata, MD Section of Endocrinology, Andrology and Internal Medicine, Human Reproduction and Biotechnology Sciences, Department of Biomedical Sciences, University of Catania, Catania, Italy

Geoffry N. De Iuliis, PhD, BSc Department of Biological Sciences, ARC Centre of Excellence in Biotechnology and Development, Priority Research Centre in Reproductive Science, University of Newcastle, Newcastle, NSW, Australia

Kenneth Dominguez, MS Institute for Biogenesis Research (IBR), John A. Burns School of Medicine, University of Hawaii at Manoa, Honolulu, HI, USA

Hasan M. El-Fakahany, MD Department of Dermatology, STD's and Andrology, Al-Minya University, Al-Minya, Egypt

Benjamin R. Emery, M Phil Andrology and IVF Laboratories, Department of Surgery, University of Utah School of Medicine, Salt Lake City, UT, USA

Juris Erenpreiss, MD, PhD Andrology Laboratory, Riga Stradins University, Riga, Latvia

Donald P. Evenson, PhD, HCLD SCSA Diagnostics, Po Box 107, Volga, SD, USA

Emeritus, South Dakota State University, Brookings, SD, USA

Department of Obstetrics and Gynecology, Sanford Medical School, University of South Dakota, Sioux Falls, SD, USA

José Luís Fernández, MD, PhD Genetics Unit, INIBIC-A Coruña University Hospital, As Xubias, Coruña, Spain

Molecular Genetics and Radiobiology Laboratory, Centro Oncológico de Galicia, Coruña, Spain

Raúl Fernández-González, BS Department of Animal Reproduction and Animal Genetic Resources Conservation, National Research Institute for Agriculture and Food Technology (INIA), Madrid, Spain

C. Fischer-Hammadeh, MD Department of Obstetrics and Gynecology, Assisted Reproduction Technology Unit, University of Saarland, Homburg/Saar, Germany

Steven Fleming, BSc (Hons), MSc, PhD Greenslopes Private Hospital, Greenslopes, QLD, Australia

Isabelle Galeraud-Denis, PhD Section of Biology of Reproduction, Université de Caen Basse-Normandie, Caen, France

Aleksander Giwercman, MD, PhD Reproductive Medicine Centre, Skåne University Hospital Malmö, Lund University, Malmö, Sweden

Jaime Gosálvez, PhD, BSc School of Biological Sciences, Madrid Autonoma University, Cantoblanco, Madrid, Spain

Marie-Chantal Grégoire, MSc Department of Biochemistry, University of Sherbrooke, Sherbrooke, QC, Canada

Alfonso Gutiérrez-Adán, PhD Department of Animal Reproduction and Animal Genetic Resources Conservation, National Research Institute for Agriculture and Food Technology (INIA), Madrid, Spain

M.E. Hammadeh, MD Department of Obstetrics and Gynecology, Assisted Reproduction Technology Unit, University of Saarland, Homburg/Saar, Germany

Gabor Huszar, MD Director, Sperm Physiology Laboratory and Male Fertility Program, Department of Obstetrics, Gynecology and Reproductive Sciences, Yale University, School of Medicine, New Haven, CT, USA

Timothy G. Jenkins, BS Andrology and IVF Laboratories, Department of Andrology Development, University of Utah School of Medicine, Salt Lake City, UT, USA

Sandro La Vignera, MD Section of Endocrinology, Andrology and Internal Medicine, Human Reproduction and Biotechnology Sciences, Department of Biomedical Sciences, University of Catania, Catania, Italy

Frédéric Leduc, MSc Department of Biochemistry, Université de Sherbrooke, Sherbrooke, Canada

Tin-Lap Lee, PhD Reproduction, Development and Endocrinology Program, School of Biomedical Sciences, The Chinese University of Hong Kong, Shatin, Hong Kong SAR, China

Laboratory of Clinical and Developmental Genomics, Eunice Kennedy Shriver National Institute of Child Health and Human Development, National Institutes of Health, Bethesda, MD, USA

Sheena E.M. Lewis, PhD Centre for Public Health, Queen's University of Belfast, Room 208, Institute of Clinical Science, Belfast, Northern Ireland, UK

Carmen López-Fernández, PhD Genetics Unit, Department of Biology, Universidad Autónoma de Madrid, Madrid, Spain

Gian Carlo Manicardi, PhD Department of Agricultural and Food Science, University of Modena and Reggio Emilia, Reggio Emilia, Italy

Cristina Joana Marques, PhD Genetics Department, Faculty of Medicine, University of Porto, Portugal, UK

Olga Mudrak, PhD The Jones Institute for Reproductive Medicine, Eastern Virginia Medical School, Norfolk, VA, USA

Rafael Oliva, MD, PhD Human Genetics Research Group, IDIBAPS, Faculty of Medicine, University of Barcelona, Barcelona, Spain

Biochemistry and Molecular Genetics Service, Hospital Clínic i Provincial, Barcelona, Spain

Eleonora B. Pasqualotto, MD, PhD Department of Gynecology, University of Caxias do Sul, Caxias do Sul, RS, Brazil

Fábio F. Pasqualotto, MD, PhD Department of Urology, University of Caxias do Sul, Rua Pinheiro Machado, Bairro São Pelegrino, Caxias do Sul, RS, Brazil

Miriam Pérez-Crespo, PhD Department of Animal Reproduction and Animal Genetic Resources Conservation, National Research Institute for Agriculture and Food Technology (INIA), Madrid, Spain

Eva Pericuesta, PhD Department of Animal Reproduction and Animal Genetic Resources Conservation, National Research Institute for Agriculture and Food Technology (INIA), Madrid, Spain

Miguel Ángel Ramírez, PhD Department of Animal Reproduction and Animal Genetic Resources Conservation, National Research Institute for Agriculture and Food Technology (INIA), Madrid, Spain

Owen M. Rennert, MD Laboratory of Clinical Genomics, Eunice Kennedy Shriver National Institute of Child Health and Human Development, National Institutes of Health, Bethesda, MD, USA

Bernard Robaire, PhD Department Pharmacology and Therapeutics, Department of Obstetrics and Gynecology, McGill University, Montréal, QC, Canada

Denny Sakkas, PhD Department of Obstetrics, Gynecology and Reproductive Sciences, Yale University School of Medicine, New Haven, CT, USA

Maria San Gabriel, BS Division of Urology, Department of Surgery, Royal Victoria Hospital, McGill University, Montreal, QC, Canada

Leyla Sati, MSc Department of Histology and Embryology, School of Medicine, Akdeniz University, Antalya, Turkey

Rakesh Sharma, PhD Andrology Laboratory and Center for Reproductive Medicine, Glickman Urological and Kidney Institute, OB-GYN and Women's Health Institute, Cleveland Clinic, Cleveland, OH, USA

Luke Simon, MSc, PhD Centre for Public Health, Queens University Belfast, Institute of Clinical Sciences, Belfast, UK

Mário Sousa, MD, PhD Departments of Laboratory Cell Biology and Microscopy, Institute of Biomedical Sciences Abel Salazar, University of Porto, Porto, Portugal

Marcello Spanò, PhD Laboratory of Toxicology, Unit of Radiation Biology and Human Health, Italian National Agency for New Technologies, Engery, and Sustainable Economic Development, Casaccia Research Centre, Casaccia, Rome, Italy

Bianca St. John, MD The Mitochondrial Genetics Group, Centre for Reproduction & Development, Monash Institute of Medical Research, Monash University, Clayton, VIC, Australia

Justin C. St. John, MD The Mitochondrial Genetics Group, Centre for Reproduction & Development, Monash Institute of Medical Research, Monash University, Clayton, VIC, Australia

Jacquetta Trasler, MD, PhD Departments of Pharmacology and Therapeutics, Pediatrics, and Human Genetics, McGill University and Montreal Children's Hospital of the McGill University Health Centre, Montreal, QC, Canada

Igor Tsarev, MD Andrology Laboratory, Riga Stradins University, Riga, Latvia

Alex C. Varghese, PhD Fertility Clinic and IVF Division, AMRI Medical Centre (A Unit of AMRI Hospitals), Kolkata, India

Enzo Vicari, MD Department of Internal Medicine and Systemic Diseases, University of Catania, Catania, Italy

W. Steven Ward, PhD Institute for Biogenesis Research (IBR), John A. Burns School of Medicine, University of Hawaii at Manoa, Honolulu, HI, USA

Dan Yu, MD Centre for Public Health, Queens University Belfast, Belfast, UK

Irina Zalenskaya, PhD CONRAD, Department of Obstetrics and Gynecology, Eastern Virginia Medical School, Norfolk, VA, USA

Andrei Zalensky, PhD, DSci The Jones Institute for Reproductive Medicine, Eastern Virginia Medical School, Norfolk, VA, USA

Armand Zini, MD Department of Surgery, Division of Urology, McGill University, St. Mary's Hospital Center, Montreal, QC, Canada

Part I

Human Sperm Chromatin: Structure and Function

Sperm Chromatin: An Overview

1

Rod Balhorn

Abstract

The dramatic changes in the structure and function of sperm chromatin that occur during spermatogenesis have continued to intrigue researchers for more than a century. In addition to wanting to understand how these changes in chromatin organization affect genome function, many of the studies conducted in placental mammals have been driven by a desire to understand the relationship between sperm chromatin organization and sperm function (fertility) or dysfunction (subfertility or infertility). While we have learned a great deal, many important questions still remain unanswered. Major technological advances in imaging techniques, transgenic animal production, gene function disruption, molecular and compositional analysis at the single cell and subcellular level as well as the development of many new molecular probes now make it possible to design and carry out studies that examine structure and function at the level of the individual cell in ways that have not been previously possible.

Keywords

Sperm chromatin • Spermatogenesis • Chromatin remodeling • DNA–protamine complex • Chromatin reorganization

Origins of Sperm Chromatin Research

The first research conducted on sperm chromatin, which dates back almost 150 years, began with the discovery of its two primary molecular components – DNA and protamine. Only a year after Gregor Mendel reported his work on the laws of heredity in 1865 [1], Ernst Haeckel suggested that the nuclei of cells must contain the material responsible for the transmission of genetic traits [2]. Friedrich Miescher, working in Felix Hoppe Seyler's laboratory in Germany, had become intrigued by cells and began conducting experiments to determine their chemical composition. Working initially with lymphocytes obtained from blood and later enriched populations

R. Balhorn (✉)
Department of Applied Science,
University of California, Davis, CA, USA
e-mail: rodbalhorn@hughes.net

A. Zini and A. Agarwal (eds.), *Sperm Chromatin: Biological and Clinical Applications in Male Infertility and Assisted Reproduction*, DOI 10.1007/978-1-4419-6857-9_1,

of leukocytes he obtained from hospital bandages, Miescher noticed a precipitate that formed when he added acid to extracts of cells he was using to isolate proteins [3]. While he and the rest of the scientific community were unaware that this material, which he called nuclein, was the genetic material Mendel and Haeckel had referred to, he became fascinated by and continued to study its properties [4]. Walther Flemming's work over the next decade introduced the scientific community to the cellular substructures called chromosomes and the concept of mitosis, and Flemming was the first to introduce the term *chromatin* [5]. It took the next 30 years, however, before cellular biologists began to realize the importance of individual chromosomes as the carriers of genetic information.

Miescher, who began his research career isolating and characterizing proteins, spent the majority of his time investigating nuclein (DNA). When he discovered he could not obtain enough of the nuclein from human cells to properly examine its properties, he turned to working with fish sperm. Salmon provided an abundance of sperm, and the sperm cells were considered ideal because they had almost no cytoplasm to contaminate his nuclear preparations with protein. In addition to being the first to isolate DNA, Miescher was also the first to isolate protamine, which he called protamin, and to discover its highly basic nature [6]. He discovered that nuclein and protamin made up the majority of the mass of the sperm head, and he also provided the first insight into the fundamental interaction that bound these two components together inside the sperm nucleus – that nuclein was bound in a salt-like state to protamin. As the interest in DNA and protamine grew, other researchers began to examine the molecules present in sperm. The majority of the initial work characterizing the composition of protamine molecules was carried out by Kossel and his group, not Miescher, over several decades spanning from about 1890–1920 [7–10]. The proteins bound to DNA in sperm were distinguished from those found in other cells very early on, but the real significance of this difference was not appreciated until almost half a century later when more detailed studies of spermatogenesis and spermiogenesis revealed significant differences in DNA packaging and sperm chromatin compaction. Up until this time, sperm chromatin was considered by many to be similar to the chromatin found in somatic cells.

Spermatogenesis: A Special Form of Terminal Differentiation

In species that reproduce sexually, testicular cells undergo a radical transformation as they progress through a process of differentiation called spermatogenesis. Diploid somatic cells that contain two complements of the genome divide in meiosis to produce haploid cells containing only a single copy of each chromosome. The nuclei and chromatin inside these haploid cells also undergo a series of structural and functional changes. In mammals, specific genes within the male genome are imprinted to identify their "parent of origin" [11, 12], and the chromatin is transformed from a highly functional, genetically active state characteristic of the somatic testis cell it was derived from to a quiescent or completely inactive state found in the fully mature sperm cell.

One might think of this transformation as the testicular cell embarking on a path of terminal differentiation similar to the differentiation of a stem cell into a liver, kidney or brain cell. The final cell not only differs structurally from the stem cell but also performs very different functions. Unlike the genome in most stem cells, however, the genome of most maturing vertebrate spermatids undergo an additional step in the process, a transient stage in which the entire genome is deprogrammed and shut down. This genome-wide inactivation bears some similarity to processes of heterochromatinization that have been observed to occur with one X-chromosome in vertebrates [13, 14], the entire genome in avian erythrocytes [15], and one set of chromosomes in mealy bugs [16]. These changes, which are induced by modifying or replacing the proteins that bind to and package DNA, enable the male genome of the sperm to be deprogrammed and maintained in a quiescent state until it enters the oocyte and is ready to be combined with the genome of the female to create a diploid embryonic cell. The process provides a mechanism by which the genes contributed by the male can be reactivated in the proper combinations to ensure the first cells function as embryonic stem cells, subpopulations of which later redifferentiate into the other types of cells that are required for the development of a fully functional organism.

Variability in the Composition of Sperm Chromatin

Both Miescher's and Kossel's studies of sperm focused on the morphological and compositional differences they observed between sperm and other cells. Kossel examined the proteins found in the sperm head, using the properties and composition of the proteins as indicators of the differences or similarities that might distinguish these cells in different species. The majority of the fish protamines analyzed by Kossel and others were found to be small proteins with unusually high contents of the two amino acids arginine and lysine. While these two amino acids were known to be present in all proteins at a low level (typically ~5%), the arginine-rich fish protamines were found to contain 50–90% arginine and the lysine-rich fish protamines contained as much as 28% lysine. Because the fish protamines appeared to be comprised mostly of arginine and lysine, Kossel proposed that the protamines might be one of the simplest proteins.

As researchers began examining the sperm chromatin proteins of other species, it became clear that there was a great deal of variability in the types of proteins used to package DNA in sperm. Sea urchins also proved to be an easy source from which sperm could be obtained in large numbers, and analyses of sea urchin sperm revealed that protamines were not present in the sperm chromatin of this organism. Instead, the DNA was found to be packaged by histones [17, 18]. Each of the five histones is larger (by a factor of two) than protamines and significantly less basic. In contrast to the protamines, the histones contain a great deal less arginine (2–10% of the total amino acids) and more lysine (13–28%). Subsequent analyses of sperm chromatin proteins isolated from the sperm of other invertebrates and vertebrates have shown that the size and amino acid sequences of the proteins used to package sperm DNA vary considerably [19]. Many of these proteins are smaller and substantially more basic than the histones and larger and less basic than protamines.

Amphibian and fish sperm provide one of the best examples of this variability. Sperm produced by frogs in the genus *Rana*, for example, have their DNA packaged entirely by histones [20]. Both histones and protamine-like intermediate proteins are found in the sperm chromatin of the clawed African frog (*Xenopus*) [21], while histones and protamines package the DNA in toad (*Bufo*) sperm [22]. Similar observations have been made in studies of fish sperm. Different species of fish, even within the same order, have been shown to use histones, protamine-like proteins, or protamines to condense their sperm chromatin, demonstrating that these differences do not correspond strictly with phylogeny. In addition, the particular type of protein used to package sperm DNA does not appear to be linked to mode of fertilization, as had been suggested based on the studies conducted with amphibian sperm. While several internally fertilizing fish such as *Xiphophorus helleri guentheri* (swordtail), *Xiphophorus maculatus* (platyfish), *Poecilia reticulata* (guppy), *Poecilia picta* (guppy), and *Cymatogaster aggregata* (shiner perch) all produce sperm containing protamines [23], several externally fertilizing species such as the grass carp (*Ctenopharyngodon idella*) [24], tub gurnard (*Trigla lucerna*) [25], and sea bream (*Sparus aurata*) [26] produce sperm containing DNA packaged by histones. However, this relationship between the mode of fertilization and type of protein used to package DNA in sperm does not extend to all species of fish. The sperm produced by salmon, herring, and many other species of fish that spawn and fertilize externally contain DNA that is packaged by protamines.

What these studies and those of chromatin in the sperm of other vertebrates and invertebrates have demonstrated is an evolutionary pattern in which the sperm chromatin proteins transition from histones to protamine-like proteins to protamines [27]. The variation observed in amphibians show that sporadic reversions are possible [28], and the fish studies [29] are consistent with this idea and provide additional examples that show the change from protamine to histone (or alternatively histone to protamine) has occurred independently several times during evolution.

Spermatid Differentiation and Chromatin Remodeling

Prior to meiosis, the chromatin in the spermatocyte nucleus is diffusely organized and appears structurally similar to that found in the nuclei of all other somatic cells. The predominant chromatin proteins are the somatic histones and a wide variety of other proteins that interact with DNA to regulate gene activity, anchor the genome to the nuclear matrix, and contribute to chromatin function. As the cell proceeds through meiosis and enters the early stages of spermiogenesis, several new DNA-binding proteins are synthesized that bind to DNA and initiate a series of subtle transformations in the organization and activity of the spermatid's chromatin. The nature of these proteins and their impact on chromatin organization and function differ widely among species.

The changes that have been characterized in the greatest detail are those that occur in placental mammals. The first new proteins to appear are four histone variants that replace some or the majority of their somatic H2B, H3, H2A, and H1 histone counterparts [30]. These proteins were originally referred to as testis specific histones with a "T" designation being added to the histone's name. More recently, the same histone variants have been referred to as sperm-specific histones because they are frequently retained at some level in mature sperm. TH3 histone appears very early in spermatogenesis in spermatogonia. TH2B and TH2A histone variants are synthesized and integrated into the chromatin of pachytene spermatocytes just prior to meiosis, and a new H1 histone variant, H1t or TH1, appears near the end of meiotic prophase. Up to 90% of H2B is replaced by TH2B. The proportion of replacement for H3 and H2A is unknown. Seven H1 variants or subtypes have been identified in mice and men. In the case of the spermatid H1 variant, H1t, it replaces approximately half of the other H1 subtypes. However, some of these subtypes, such as H1a, actually increase in abundance and are not replaced. While these sperm histone variants are thought to play some role in altering the functionality of the chromatin, the basic structural subunit of chromatin organization, the nucleosome, is retained.

Electron microscopy studies have shown that the first noticeable change in chromatin structure occurs when the sperm specific histone H1t variant is deposited in spermatid chromatin. Prior to H1t deposition, the chromatin appears more diffuse and contains regions that are more clumped than others. When H1t appears, the chromatin is transformed into a more uniform and granular state. H1t remains bound to DNA for a relatively short period of time and then begins to disappear in elongating spermatids. Following its loss, the chromatin takes on a more filamentous organization [31].

In mammals, the majority of the histones are replaced after meiosis by three smaller, more basic proteins that have been designated "transition proteins" because they only remain associated with DNA for a relatively short period of time. The mammalian transition proteins TP1, TP2, and TP4 appear in the chromatin of mid-stage spermatids at the same time the majority of the histones are removed from the chromatin. Studies in human and rat spermatids have shown that TP2 synthesis and deposition in spermatid chromatin precedes that of TP1 [32, 33]. With the appearance of TP1 and TP2, the chromatin begins to condense somewhat with condensation progressing in the nucleus from an apical to caudal direction [31, 34]. Very little is currently known about TP4. While a great deal remains to be learned about the function of these proteins, it is clear that they play important roles in replacing histones (TP1 has been reported to destabilize nucleosomes by preventing DNA bending [35]), initiating the termination of gene transcription by TP2 binding to CpG sites [35], enabling or facilitating the repair of DNA strand breaks [36], and contributing to chromatin condensation. By the time TP1, TP2, and TP4 deposition are completed, the chromatin becomes uniformly condensed and no longer appears to retain the subunit structure characteristic of nucleosomes. A fourth protein, TP3, was also considered to be a member of this group when it was first observed in spermatid chromatin. Once the protein was sequenced, however, TP3 was identified to be the precursor form of protamine 2 [37]. Instead of being displaced from late-spermatid DNA, the protein is simply processed to a smaller form (protamine 2) that remains bound to DNA throughout the remainder of spermiogenesis.

These transition proteins are replaced by a set of positively charged proteins called protamine in late-step spermatids as the chromatin is reorganized one final time before the sperm becomes fully mature. The mammalian protamines are small proteins rich in cysteine and the basic amino acids arginine, lysine, and histidine. Considerable variation in amino acid sequence has been observed within the protamines of mammals [38–41], but all the proteins examined fall into one of two protamine families, protamine P1 or protamine P2. The nature of protamine binding to DNA and the consequences of the synthesis and incorporation of the protamines into spermatid chromatin suggest that these proteins may perform a number of functions. These include protecting the DNA from physical and chemical damage while the chromatin is in a state in which it cannot repair DNA damage and compacting the genomic material to produce a smaller, more hydrodynamically shaped cell. The compaction of the genome that occurs when protamine binds to DNA also ensures the entire genome is retained in a genetically inactive state until fertilization, and it may even aid in the shaping of the sperm head by generating the forces needed to shape the nucleus from within [42].

Higher Ordered Organization of Chromatin in Mature Sperm

In contrast to the variability that has been observed in the composition of sperm chromatin in many vertebrates and invertebrates, there appears to be remarkably little variation in the final modes of DNA packaging that have been observed in sperm produced by different species of mammals. The sperm of all mammals examined to date, including monotremes, marsupials, and placental mammals, use protamines to package the majority of their DNA into the sperm head. In several mammalian species, a small fraction of the sperm genome has been observed to retain its histone packaging. This histone-containing fraction, which is currently thought to be present in all mammalian sperm, is small, comprising not more than a fraction to 1% of the genome. In human sperm, however, the fraction of DNA bound by histones is significantly larger, possibly as high as 10–15% [43–47].

Recent studies have identified a number of DNA sequences or genes that remain associated with histones in mammalian sperm. These include telomeric DNA [48], genes for epsilon and gamma globin [49], a paternally imprinted IGF-2 gene [50], microRNA clusters, the promoters of a number of genes expressing signaling proteins important for early embryonic development, and genes that produce transcription factors such as those in the Hox family [51]. Based on the types of genes that have been identified in histone associated sperm chromatin, it has been suggested that one function for the retention of these histones may be to maintain a subset of genes contributed by the male in a quiescent but accessible state so they can be activated immediately after fertilization and prior to the removal of the protamines. The histone-associated genes were also found to be highly enriched in a variety of imprinted genes, indicating another function of these histones may also be to play a role in epigenetic programming.

The chromatin in monotreme and marsupial spermatids is condensed during spermiogenesis in a fashion similar to that observed in other species that use only protamines to package their DNA, but the nature of the nuclear protein–DNA interactions that lead to this condensation in monotreme sperm have not yet been characterized. Chromatin condensation in platypus sperm is initiated by the formation of a layer of electron dense chromatin granules under the nucleolemma [52]. As the spermatids continue to mature, foci of condensing chromatin are observed throughout the nucleus. These studies have not, however, provided much information about either the organization or subunit structure of mature sperm chromatin in monotremes. A combination of EM and AFM studies of sperm chromatin in two marsupials, the fat tailed dunnart (*Sminthopsis crassicaudata*) and brush-tailed possum (*Trichosurus vulpecula*), has indicated the DNA is organized in nodular subunits [53]. Those regions of the chromatin that appear to be packaged by protamines have nodules with diameters of 50–80 nm, while other regions believed to contain histones bound to DNA contained much larger clusters (120–160 nm) of smaller nodules.

Chromatin reorganization and compaction occurs in a similar manner in placental mammals. The chromatin is transformed from the diffuse, genetically active state to a highly electron dense, compact form of chromatin that is completely inactive. Both electron and atomic force microscopy studies of spermatid chromatin and partially decondensed sperm chromatin have provided insight into the higher ordered structure of sperm chromatin in placental mammals. EM images of the chromatin in differentiating late-step spermatids have shown that the DNA starts off organized with features characteristic of somatic chromatin (~11 nm nodules and 30 nm fibers [54]), which are subsequently transformed into nodular structures or fibers with diameters (50–100 nm) much larger than individual nucleosomes. As chromatin condensation progresses, these nodules coalesce into increasingly larger masses or fibers that eventually become so electron dense and tightly packed that they can no longer be distinguished.

Similar structural information has been derived from high resolution microscopy studies of sperm chromatin that has been partially decondensed by treatment with polyanions, reducing agents, or high ionic strength or by partial digestion by nucleases [55–62]. Analyses of partially decondensed sperm chromatin by electron microscopy have shown that at least two different sized structural units are present, small nodules similar in size to nucleosomes and much larger globular structures. Atomic force microscopy images of decondensed human sperm also revealed the presence of two types of structures: small subunits similar in diameter (~10 nm) and thickness (~5 nm) to somatic nucleosomes and lifesaver shaped larger structures approximately 60–100 nm in diameter and 20 nm thick with a hole or depression in the center [56]. Toroids with lifesaver-like features and similar dimensions have also been generated in vitro when protamine or other polycations were added to dilute solutions of DNA or to individual DNA molecules [63–65]. These toroids, which contain approximately 50,000 bp of DNA complexed with protamine, are spontaneously generated when protamine binds to and neutralize the phosphodiester backbone of double-stranded DNA [56, 66]. Closely packed beads with diameters similar to these toroids were found by Koehler to comprise the lamellar sheets of chromatin packed inside rat, rabbit, bull, and human sperm [59, 60, 67].

Mammalian Protamines

While the unusually high arginine content of protamine was recognized by both Miescher and Kossel to be a unique feature of fish sperm nuclear proteins more than 100 years ago, it took more than 50 years for researchers to begin to understand and appreciate the structural and functional differences between the protamines and histones. Structurally, the two families of DNA-binding proteins are very different. The four core histones interact with each other to form a well-defined octamer core of protein around which almost two turns of DNA are wrapped [68]. The DNA bound to the histones remains accessible to polymerases and other proteins and the genes packaged by histone remain active or can be readily activated. By marked contrast, the protamines contain so many positively charged amino-acid side chains that when protamine binds to DNA, it wraps around the DNA helix, neutralizing the negatively charged phosphodiester backbone of DNA and creating a maximally compact form of chromatin [56]. This prevents the genes packaged by protamines from being accessed by other proteins and modified, transcribed or repaired.

Two different types of protamines package DNA in mammalian sperm, P1 and P2. The smaller protein, protamine P1, is found in the sperm of all mammals [69]. The P1 protamine of placental mammals is a single peptide chain containing only 50 amino acids [70]. The one known exception is stallion P1, which contains 51 amino acids. The P1 protamines in marsupials and monotremes are larger (57–70 residues). The platypus and echidna protamines also differ from the P1 protamines of placental mammals in that they do not contain any cysteine residues [71]. This is also the case for most marsupial protamines [41]. One exception has been reported, however, in the family of Dasyuridae. Shrew-like marsupials in the genus *Planigales* produce protamines that containing 5–6 cysteines [72], a number similar to the number of cysteines that are typically found in the P1 protamines of placental mammals.

The P1 protamine of placental mammals is unstructured in solution and only adopts a specific conformation when bound to DNA [73]. Protamine P1 sequences are typically divided into three small domains, a central DNA-binding domain comprised of a series of $(Arg)_n$ DNA-binding domains interspersed with one or two uncharged amino acids and two short N- and C-terminal peptide domains that do not bind to DNA [70, 74]. Only the DNA-binding domain appears to be present in monotreme and marsupial P1 molecules [41, 71]. The two short terminal peptide domains in placental mammal P1 molecules contain serine and threonine residues that are phosphorylated shortly after the protein is synthesized, and this modification is thought to facilitate the protein's binding correctly to DNA. Similar phosphorylatable residues appear to be distributed throughout the monotreme and marsupial P1 sequences. These domains in placental mammal P1 molecules also contain multiple cysteine residues that form inter- and intraprotamine disulfide bonds and link each protamine molecule to its neighbor when the maturing spermatid passes through the epididymis [74].

Protamine P2, which is slightly larger than P1 (63 amino acids in mouse) is only expressed in the differentiating spermatids of a subset of placental mammals. These include primates, most rodents, lagomorphs, and perissodactyls [69]. Unlike protamine P1, P2 is synthesized as a larger precursor protein (106 residues in mouse) that is deposited onto DNA and subsequently shortened over a period of several days [75]. This processing of the precursor protein occurs by progressive and sequential cleavage of short peptide fragments from the amino terminus of the precursor [76–78]. The function of this processing remains unknown. P2 also appears to be phosphorylated transiently. How the final processed form of P2 interacts with DNA has not yet been determined, but studies of P1 and P2 in several species suggest the majority of the length of the P2 molecule binds to DNA. The "footprint" of P1 when bound to DNA is 10–11 base pairs, or one full turn of DNA, while the "footprint" of P2 appears to be larger (15 bp) [43]. The final processed form of P2 also appears to use a series of $(Arg)_n$ anchoring peptide segments to bind to DNA. These segments are shorter and less well defined than those found in the DNA-binding domain of P1, and they are distributed throughout the entire length of the P2 sequence. P2 also contains multiple cysteine residues that participate in the formation of the disulfide bonds that interconnect all the protamines late in spermiogenesis.

Structure of the DNA–Protamine Complex

While the relative proportion of the two protamines in sperm chromatin varies widely between mammalian genera, the proportion appears to be conserved among the species within a genus [69]. P2 is believed to bind to DNA in a manner similar to P1, but the evidence for this is limited and primarily circumstantial. Beyond the knowledge that both protamines P1 and P2 bind along the DNA in some manner that allows the two proteins to be cross-linked together by disulfide bridges during the final stage of sperm maturation, very little is known about the details of P2 binding to DNA or the distribution of the two protamines along a segment of DNA.

Because it has not been possible to determine the structure of a native or artificial protamine–DNA complex by X-ray crystallography or NMR spectroscopy, most of the information that has been learned about how the protamines interact with DNA has been determined using lower resolution techniques. Low-angle X-ray scattering experiments performed on intact sperm heads confirmed the close packing of the DNA within sperm chromatin, showing the center to center distance between adjacent DNA molecules is approximately 2.7 nm [79]. To achieve this tight packing, the molecules must be organized in a hexagonal arrangement with only 7 Å distance of separation between the surfaces of adjacent molecules. High-resolution EM studies of individual toroidal subunits [80] have shown that the individual DNA molecules coiled into the toroid are tightly packed in a hexagonal arrangement, consistent with what has been observed by low-angle X-ray scattering. Such a packing arrangement for DNA is also consistent with the microscopy data obtained from stallion sperm heads [81], particularly if the toroidal structures are stacked tightly together as lifesavers and organized in layers similar to the lamellae reported by Koehler [59, 60, 67].

At the molecular level, the protamines bind to duplex DNA in a manner that is independent of base sequence [66, 82]. The primary interactions are electrostatic and involve the binding of the positively charged guanidinium groups in the arginine residues present in the DNA anchoring domains of protamine to the negatively charged phosphates that comprise the DNA phosphodiester backbone. The high affinity of binding is derived from two aspects of these interactions, the formation of a salt bridge and hydrogen bond between the guanidinium group and the phosphate and the binding of every arginine residue in the DNA-binding domain of protamine to every phosphate group in one turn of DNA. Both computer modeling and X-ray scattering and other experimental studies [73, 83–85] have shown that the DNA-binding domain of protamine P1 wraps in an extended conformation around the DNA helix, partially filling the major groove. By interacting in this way, adjacent arginine residues in the $(Arg)_n$ anchoring domains would be expected to bind to phosphates on opposite strands of the duplex DNA molecule, interlocking the relative positions of the bases together and preventing strand separation or changes in DNA conformation throughout the period that the protamines remain bound to DNA. This would result in the production of a neutral, highly insoluble complex that allows the DNA strands to be packed tightly together without charge repulsion.

Chromosome Territories, Loop Domains, and Matrix Attachment Regions

Three important structural features of somatic chromatin organization appear to be retained by mammalian sperm chromatin even after all the nuclear protein transitions and condensation have been completed. Confocal microscopy of somatic cells hybridized to fluorochrome-tagged DNA probes have shown that the DNA of individual chromosomes are not randomly distributed throughout the nucleus, but each is confined to a specific domain or territory inside the interphase nucleus [86–90]. Not only is there evidence that the chromosomal DNA molecules occupy a reproducible position, but there is also evidence that the domains are folded into shapes characteristic of a particular chromosome [91]. Similar observations have been made regarding the distribution of chromosomal DNA in mammalian sperm nuclei. Fluorescence in situ hybridization has been used to demonstrate that the DNA of individual chromosomes are also localized to specific domains inside the heads of human, bull, mouse, echidna, and platypus sperm [48, 91–94].

While these studies have not provided strong evidence that the chromosomes are arranged in any particular order relative to each other in the sperm heads of placental mammals, there is some evidence for a particular arrangement in echidna and platypus sperm.

Two other organizational features that are retained in sperm cell nuclei are the chromatin loop domains and the attachment of the chromatin to a nuclear protein scaffold or nuclear matrix [95–98]. The protein content of the nuclear matrix changes as the spermatid differentiates [95], but the DNA remains bound to the matrix at a very large number of sites (~50,000). This matrix appears by EM to be a network of dense protein filaments filling the interior of the head of the spermatid and sperm bounded by a peripheral structure, the lamina. The DNA in between the sites of attachment to the matrix appears to retain the loop organization present in somatic cells [99, 100]. These loops, which contain 40,000–50,000 bp of DNA in both the somatic and sperm nucleus, are anchored to a matrix through specific chromatin domains, called nuclear scaffold/matrix attachment regions (SARs/MARs). The retention of the matrix and its associations with DNA in sperm are important to maintain because their presence would facilitate and speed up the process of genome reactivation following fertilization and the initiation of the first cycle of DNA replication in the male pronucleus [101, 102]. The loop domains are believed to play an essential role in transcriptional regulation, DNA replication, and chromosome organization both prior to spermiogenesis and after fertilization. In sperm, these loops may also aid in the packing of the DNA by protamines into toroids, which also contain ~50,000 bp of DNA.

The retention of these particular features of chromosome and chromatin organization appears to preserve important genome organizational information critical to both germinal and somatic cell function. Clearly, the primary function of spermiogenesis is to produce a package of genomic information, the sperm cell, that will facilitate the transport of one complement of the male's chromosomes to and into the oocyte for the purpose of generating an embryo containing genomic contributions from both the male and female of the species. Once this is accomplished, the genome must be quickly reactivated so that it can begin functioning as a somatic cell, with subsets of genes being turned on and off as the cells are transformed from embryonic stem cells into the cells of the various tissues and organs.

Reorganization of Sperm Chromatin Following Fertilization

The formation of the male pronucleus and other processes associated with early embryonic development that occur immediately after fertilization have been well characterized by light microscopy. However, remarkably little is known at the molecular level about the early events that contribute to the unpackaging of sperm chromatin following fertilization. The current hypothesis is that the protamines are actively removed from the DNA by a histone chaperone similar to the nucleoplasmin first identified in frogs [103–105]. This protein chaperone has been shown to bind and carry core histones and, in the presence of DNA, is able to load the histones onto the DNA and generate nucleosomes. Sequence analyses of the frog and related mammalian proteins have shown that these proteins contain a series of polyglutamic acid sequences. Experiments conducted with sperm chromatin have also shown that the protein is able to remove protamine from the DNA prior to loading it with histones [106]. One possible mechanism of protamine removal may involve these segments of polyglutamic acid. The polyglutamic acid regions in nucleoplasmin-like proteins could form a series of salt bridges with the $(Arg)_n$ DNA-binding domains of the protamines and remove the protamines from DNA intact prior to depositing the histones and reestablishing the nucleosomal organization required to reactivate the new embryo's genome.

Another early event associated with the unpacking of the sperm chromatin that occurs almost immediately after removing the protamines is the initiation of a period of DNA synthesis associated with DNA damage repair [107–110]. This repair synthesis is required to repair DNA strand breaks and remove DNA adducts or other damage that is acquired during spermiogenesis and epididymal transit and storage when repair activities could not be performed due to the packaging of the genome by protamines. Studies have shown that the majority of the damage brought into the oocyte by the sperm is repaired during this period of DNA synthesis, and this process is considered to be critical for maintaining the integrity of the male genome and for ensuring normal embryonic development.

Consequences of Disrupting Sperm Chromatin Remodeling

Several changes associated with the reorganization of spermatid chromatin have been shown to be important for male fertility. One involves the removal of the majority of the histones and their replacement by protamines. Numerous studies have suggested that there is a positive correlation between male subfertility or infertility and elevated levels of histone in mature human sperm [77, 111–117]. It is not known, however, whether the problems encountered relate to the lack of removal of somatic histones from genes that need to be packaged by protamines, deficiencies in expression and incorporation of the sperm specific histone variants into subsets of nucleosomes, or errors in imprinting that may involve histone packaging.

Alterations in the expression and/or translation of the protamine genes have also been linked to infertility. Changes in the proportion of the P1 or P2 proteins present in sperm chromatin have been shown to not only be linked to infertility [118–124] but also adversely impact in vitro fertilization outcome and early embryonic development [125–129]. The observed differences in protamine content ranged from having very little protamine, to having too little protamine P1 or too little protamine P2. By contrast, analyses of sperm obtained from fertile human males have shown repeatedly that the sperm contain a specific proportion (1:1) of P1 and P2 [118–120, 130]. The primary cause for the observed changes in sperm protamine content appears to involve errors in gene expression, although incomplete processing of the P2 precursor may also contribute to decreased levels of the mature P2 protein.

Other studies have shown that the timely formation of the protamine disulfide cross-links that occur during the final stages of sperm maturation are important. In mammals, both protamines P1 and P2 contain multiple cysteine residues. The thiol groups of these cysteines are in the reduced form (free thiols) when the protamines are synthesized and deposited onto DNA, and they remain reduced until the final stage of spermiogenesis when they participate in the formation of both inter-and intramolecular protamine disulfides as the sperm pass through the epididymis [74, 131–134]. Cases of human, stallion, and bull infertility have been correlated with what appear to be errors in disulfide cross-linking among the protamines. What role these disulfide bonds play is still not known, but one theory is that the formation of interprotamine disulfide bonding stabilizes the chromatin and protects it from physical damage. An equally feasible possibility is that these disulfide bonds not only stabilize the chromatin but also prevent the thiol groups from being oxidized or alkylated during the long period of time required for spermatid maturation and sperm storage prior to fertilization. This might be important if the cysteine residues in mammalian protamine also play some other role in sperm chromatin, such as participating in protamine removal from DNA after fertilization. If the thiols were required for efficient protamine removal, the oxidation or alkylation of even a few cysteines could potentially complicate or prevent the efficient removal of the modified protamine from the male genome, and its retention would block the gene it was bound to from being transcribed or replicated later in development. Mice exposed to alkylating agents such as methyl methanesulfonate and ethylene oxide at a time prior to protamine disulfide bond formation have been shown to produce sperm with alkylated protamine thiols [135–137]. Matings conducted with the treated males resulted in the production of embryos that died early in development from dominant lethal mutations [136]. The sperm containing the protamines with alkylated cysteines succeeded in fertilizing oocytes and inducing embryonic development, but at some point after fertilization the embryo died when a key gene could not be turned on.

Male infertility has also been linked to deficiencies in sperm chromatin-associated zinc. Zinc is known to be essential for several aspects of sperm development, ranging from contributions to structural elements in the tail to roles in chromatin organization and protamine structure and function [138]. A deficiency in zinc can affect the developing sperm directly, or it can impact the function of other testicular cells that contribute to or play a role in spermatid maturation, such as sertoli cells. Because zinc plays multiple roles in spermatogenesis and testicular function, it has been difficult to decipher how sperm chromatin bound zinc impacts the functionality of the sperm cell. Chromatin associated zinc is almost exclusively bound to protamine P2 in mammals [139]. In human, bull, mouse, and hamster sperm, a single zinc atom is bound to each P2 molecule. Zinc does not appear to bind to protamine P1. Zinc ion coordination by P2 occurs sometime after the synthesis of P2 and its deposition onto DNA, long before the sperm cell enters the seminal fluid and the sperm chromatin can be impacted by seminal fluid zinc. Where the zinc binds in P2 has not been determined, but the amino acids in protamine P2 that coordinate the zinc appear to change during sperm maturation. In sonication resistant spermatids, the zinc is coordinated only by cysteines, while in mature sperm, both histidine and cysteine residues participate in the coordination (unpublished results). The function of this P2 bound zinc is not known, but it has been suggested that the coordination of the zinc by protamine may influence the binding of the protamine to DNA [140, 141] or to other protamines [138]. An alternative possibility is that zinc coordination by cysteine residues in protamine might also protect the thiol groups and prevent their oxidation until it is time for the cysteines to form inter- and intramolecular disulfide bonds. Several studies have also suggested that exposures to other metals, such as copper and lead, may result in these metals binding to the cysteines in protamine in place of zinc (or prior to disulfide bond formation) and their being transported into the oocyte upon fertilization [133, 142, 143]. In addition to potentially disrupting the function of sperm by altering chromatin decondensation or protamine P2 function, the delivery of these and other toxic metals into the oocyte would also be expected to have an adverse impact on early embryonic development.

Future Research and Practical Applications

The dramatic changes in the structure and function of sperm chromatin that occur during spermatogenesis have continued to intrigue researchers for more than a century. In addition to wanting to understand how these changes in chromatin organization affect genome function, many of the studies conducted in placental mammals have been driven by a desire to understand the relationship between sperm chromatin organization and sperm function (fertility) or dysfunction (subfertility or infertility). While we have learned a great deal, many important questions still remain unanswered. Major technological advances in imaging techniques, transgenic animal production, gene function disruption, molecular and compositional analysis at the single cell and sub-cellular level as well as the development of many new molecular probes now make it possible to design and carry out studies that examine structure and function at the level of the individual cell in ways that have not been previously possible. Studies to be conducted in the next decade using these tools should advance our understanding of sperm chromatin structure and function quickly while providing new information that can be used to diagnose and treat male infertility, develop new male contraceptives, and contribute to other unrelated areas of research such as improving the efficiency of creating transgenic animals or targeted genome silencing for cancer therapy.

References

1. Mendel G. Experiment in plant hybribization. Paper presented at: Brunn Natural History Society; March, 1865, 1865; Brunn, Czechoslovakia.
2. Haeckel E. Generelle Morphologie der Organismen. Berlin: Reimer; 1866.
3. Miescher F. Letter I to Wilhelm His; Tubingen, February 26th, 1869. In: His W, ed. Die Histochemischen und Physiologischen Arbeiten von Friedrich Miescher – Aus dem sissenschaft – lichen Briefwechsel von F. Miescher. Vol 1. Liepzig: F. C. W. Vogel; 1869:pp. 33–8.
4. Miescher F. Uber die chemische Zusammensetzung der Eiter – zellen. Med Chem Unters. 1871;4: 441–60.
5. Flemming W. Uber das Verhalten des Kern bei der Zellltheilung und uber dei Bedeutung mekrkerniger Zellen. Arch Pathol Anat Physiol. 1879;77:1–29.
6. Miescher F. Das Protamin – Eine neue organishe Basis aus den Samenssden des Rheinlachses. Ber Dtesch Chem Ges. 1874;7:376.
7. Kossel A. Ueber die Constitution der einfachsten Eiweissstoffe. Z Pysiologische Chemie. 1898;25: 165–89.
8. Kossel A, Dakin HD. Uber Salmin und Clupein. Z Pysiologische Chemie. 1904;41:407–15.
9. Kossel A, Dakin HD. Weitere Beitrage zum System der einfachsten Eiweisskorper. Z Pysiologische Chemie. 1905;44:342–6.
10. Kossel A, Edlbacher F. Uber einige Spaltungsprodukte des Thynnins und Pereins. Z Pysiologische Chemie. 1913;88:186–9.
11. Reik W, Walter J. Genomic imprinting: parental influence on the genome. Nat Rev Genet. 2001;2(1): 21–32.
12. Solter D. Differential imprinting and expression of maternal and paternal genomes. Annu Rev Genet. 1988;22:127–46.
13. Gartler SM, Goldman MA. X-chromosome inactivation, Encyclopedia of life. New York: Wiley Interscience; 2005. p. 1–6.
14. Heard E, Clerc P, Avner P. X-chromosome inactivation in mammals. Annu Rev Genet. 1997;31: 571–610.
15. Ney PA. Gene expression during terminal erythroid differentiation. Curr Opin Hematol. 2006;13(4): 203–8.
16. Berlowitz L. Chromosomal inactivation and reactivation in mealy bugs. Genetics. 1974;78(1):311–22.
17. Bloch D. Handbook of Genetics, vol. 5. New York: Plenum Press; 1976.
18. Palau J, Ruiz-Carrillo A, Subirana JA. Histones from sperm of the sea urchin *Arbacia lixula*. Eur J Biochem. 1969;7(2):209–13.
19. Eirin-Lopez JM, Ausio J. Origin and evolution of chromosomal sperm proteins. Bioessays. 2009; 31(10):1062–70.
20. Kasinsky HE, Huang SY, Mann M, Roca J, Subirana JA. On the diversity of sperm histones in the vertebrates: IV. Cytochemical and amino acid analysis in Anura. J Exp Zool. 1985;234(1):33–46.
21. Mann M, Risley MS, Eckhardt RA, Kasinsky HE. Characterization of spermatid/sperm basic chromosomal proteins in the genus Xenopus (Anura, Pipidae). J Exp Zool. 1982;222(2):173–86.
22. Takamune K, Nishida H, Takai M, Katagiri C. Primary structure of toad sperm protamines and nucleotide sequence of their cDNAs. Eur J Biochem. 1991;196(2):401–6.
23. Su H. Characterization of nuclear basic proteins in sperm and erythrocytes of vertebrates. Vancouver: Department of Zoology, University of British Columbia; 2004.
24. Kadura SN, Khrapunov SN, Chabanny VN, Berdyshev GD. Changes in chromatin basic proteins during male gametogenesis of grass carp. Comp Biochem Physiol B. 1983;74(2):343–50.
25. Saperas N, Lloris D, Chiva M. Sporadic appearance of histones, histone-like proteins, and protamines in sperm chromatin of bony fish. J Exp Zool. 2005; 265(5):575–86.
26. Kurtz K, Saperas N, Ausio J, Chiva M. Spermiogenic nuclear protein transitions and chromatin condensation. Proposal for an ancestral model of nuclear spermiogenesis. J Exp Zool B Mol Dev Evol. 2009;312B(3):149–63.
27. Ausio J. Histone H1 and evolution of sperm nuclear basic proteins. J Biol Chem. 1999;274(44):31115–8.
28. Kasinsky HE, Gutovich L, Kulak D, et al. Protamine-like sperm nuclear basic proteins in the primitive frog *Ascaphus truei* and histone reversions among more advanced frogs. J Exp Zool. 1999;284(7):717–28.
29. Saperas N, Chiva M, Pfeiffer DC, Kasinsky HE, Ausio J. Sperm nuclear basic proteins (SNBPs) of agnathans and chondrichthyans: variability and evolution of sperm proteins in fish. J Mol Evol. 1997; 44(4):422–31.
30. Churikov D, Zalenskaya IA, Zalensky AO. Male germline-specific histones in mouse and man. Cytogenet Genome Res. 2004;105(2–4):203–14.
31. Oko RJ, Jando V, Wagner CL, Kistler WS, Hermo LS. Chromatin reorganization in rat spermatids during the disappearance of testis-specific histone, H1t, and the appearance of transition proteins TP1 and TP2. Biol Reprod. 1996;54(5):1141–57.
32. Kistler WS, Henriksen K, Mali P, Parvinen M. Sequential expression of nucleoproteins during rat spermiogenesis. Exp Cell Res. 1996;225(2):374–81.
33. Steger K, Klonisch T, Gavenis K, Drabent B, Doenecke D, Bergmann M. Expression of mRNA and protein of nucleoproteins during human spermiogenesis. Mol Hum Reprod. 1998;4(10):939–45.
34. Alfonso P, Kistler WS. Immunohistochemical localization of spermatid nuclear transition protein 2 in the testes of rats and mice. Biol Reprod. 1993;48(3): 522–9.

35. Pradeepa MM, Rao MR. Chromatin remodeling during mammalian spermatogenesis: role of testis specific histone variants and transition proteins. Soc Reprod Fertil Suppl. 2007;63:1–10.
36. Caron N, Veilleux S, Boissonneault G. Stimulation of DNA repair by the spermatidal TP1 protein. Mol Reprod Dev. 2001;58(4):437–43.
37. Unni E, Zhang Y, Meistrich ML, Balhorn R. Rat spermatid basic nuclear protein Tp3 is the precursor of protamine 2. Exp Cell Res. 1994;210(1):39–45.
38. Queralt R, Adroer R, Oliva R, Winkfein RJ, Retief JD, Dixon GH. Evolution of protamine P1 genes in mammals. J Mol Evol. 1995;40(6):601–7.
39. Retief JD, Dixon GH. Evolution of pro-protamine P2 genes in primates. Eur J Biochem. 1993;214(2): 609–15.
40. Retief JD, Krajewski C, Westerman M, Dixon GH. The evolution of protamine P1 genes in dasyurid marsupials. J Mol Evol. 1995;41(5):549–55.
41. Retief JD, Krajewski C, Westerman M, Winkfein RJ, Dixon GH. Molecular phylogeny and evolution of marsupial protamine P1 genes. Proc Biol Sci. 1995; 259(1354):7–14.
42. Cree LH, Balhorn R, Brewer LR. Single molecule studies of DNA-protamine interactions. Protein Pept Lett. 2011;18(8):802–10.
43. Bench GS, Friz AM, Corzett MH, Morse DH, Balhorn R. DNA and total protamine masses in individual sperm from fertile mammalian subjects. Cytometry. 1996;23(4):263–71.
44. Gatewood JM, Cook GR, Balhorn R, Schmid CW, Bradbury EM. Isolation of four core histones from human sperm chromatin representing a minor subset of somatic histones. J Biol Chem. 1990;265(33): 20662–6.
45. Gusse M, Sautière P, Bélaiche D, et al. Purification and characterization of nuclear basic proteins of human sperm. Biochim Biophys Acta. 1986;884(1): 124–34.
46. Tanphaichitr N, Sobhon P, Taluppeth N, Chalermisarachai P. Basic nuclear proteins in testicular cells and ejaculated spermatozoa in man. Exp Cell Res. 1978;117(2):347–56.
47. Wykes SM, Krawetz SA. The structural organization of sperm chromatin. J Biol Chem. 2003;278(32): 29471–7.
48. Zalenskaya IA, Zalensky AO. Non-random positioning of chromosomes in human sperm nuclei. Chromosome Res. 2004;12(2):163–73.
49. Gardiner-Garden M, Ballesteros M, Gordon M, Tam PP. Histone- and protamine-DNA association: conservation of different patterns within the beta-globin domain in human sperm. Mol Cell Biol. 1998; 18(6):3350–6.
50. Banerjee S, Smallwood A. Chromatin modification of imprinted H19 gene in mammalian spermatozoa. Mol Reprod Dev. 1998;50(4):474–84.
51. Hammoud SS, Nix DA, Zhang H, Purwar J, Carrell DT, Cairns BR. Distinctive chromatin in human sperm packages genes for embryo development. Nature. 2009;460(7254):473–8.
52. Lin M, Jones RC. Spermiogenesis and spermiation in a monotreme mammal, the platypus, Ornithorhynchus anatinus. J Anat. 2000;196(Pt 2):217–32.
53. Soon LL, Bottema C, Breed WG. Atomic force microscopy and cytochemistry of chromatin from marsupial spermatozoa with special reference to *Sminthopsis crassicaudata*. Mol Reprod Dev. 1997;48(3):367–74.
54. Horowitz RA, Agard DA, Sedat JW, Woodcock CL. The three-dimensional architecture of chromatin in situ: electron tomography reveals fibers composed of a continuously variable zig-zag nucleosomal ribbon. J Cell Biol. 1994;125(1):1–10.
55. Allen MJ, Lee C, Lee JDt, et al. Atomic force microscopy of mammalian sperm chromatin. Chromosoma. 1993;102(9):623–30.
56. Balhorn R, Cosman M, Thornton K, et al. Protamine mediated condensation of DNA in mammalian sperm. In: Gagnon C, editor. The male gamete: from basic knowledge to clinical applications: Proceedings of the 8th International Symposium of Spermatology. Vienna, IL: Cache River; 1999.
57. Evenson DP, Witkin SS, de Harven E, Bendich A. Ultrastructure of partially decondensed human spermatozoal chromatin. J Ultrastruct Res. 1978; 63(2):178–87.
58. Koehler JK. Fine structure observations in frozen-etched bovine spermatozoa. J Ultrastruct Res. 1966;16(3):359–75.
59. Koehler JK. A freeze-etching study of rabbit spermatozoa with particular reference to head structures. J Ultrastruct Res. 1970;33(5):598–614.
60. Koehler JK, Wurschmidt U, Larsen MP. Nuclear and chromatin structure in rat spermatozoa. Gamate Res. 1983;8:357–77.
61. Sobhon P, Chutatape C, Chalermisarachai P, Vongpayabal P, Tanphaichitr N. Transmission and scanning electron microscopic studies of the human sperm chromatin decondensed by micrococcal nuclease and salt. J Exp Zool. 1982;221(1): 61–79.
62. Wagner TE, Yun JS. Fine structure of human sperm chromatin. Arch Androl. 1979;2(4):291–4.
63. Allen MJ, Bradbury EM, Balhorn R. AFM analysis of DNA-protamine complexes bound to mica. Nucleic Acids Res. 1997;25(11):2221–6.
64. Bloomfield VA. Condensation of DNA by multivalent cations: considerations on mechanism. Biopolymers. 1991;31(13):1471–81.
65. Marquet R, Wyart A, Houssier C. Influence of DNA length on spermine-induced condensation. Importance of the bending and stiffening of DNA. Biochim Biophys Acta. 1987;909(3):165–72.
66. Brewer LR, Corzett M, Balhorn R. Protamine-induced condensation and decondensation of the same DNA molecule. Science. 1999;286(5437): 120–3.

67. Koehler JK. Human sperm head ultrastructure: a freeze-etching study. J Ultrastruct Res. 1972;39(5): 520–39.
68. Finch JT, Lutter LC, Rhodes D, et al. Structure of nucleosome core particles of chromatin. Nature. 1977;269(5623):29–36.
69. Corzett M, Mazrimas J, Balhorn R. Protamine 1: protamine 2 stoichiometry in the sperm of eutherian mammals. Mol Reprod Dev. 2002;61(4):519–27.
70. Balhorn R. The protamine family of sperm nuclear proteins. Genome Biol. 2007;8(9):227.
71. Retief JD, Winkfein RJ, Dixon GH. Evolution of the monotremes. The sequences of the protamine P1 genes of platypus and echidna. Eur J Biochem. 1993;218(2):457–61.
72. Retief JD, Rees JS, Westerman M, Dixon GH. Convergent evolution of cysteine residues in sperm protamines of one genus of marsupials, the Planigales. Mol Biol Evol. 1995;12(4):708–12.
73. Hud NV, Milanovich FP, Balhorn R. Evidence of novel secondary structure in DNA-bound protamine is revealed by raman spectroscopy. Biochemistry. 1994;33(24):7528–35.
74. Balhorn R. Mammalian protamines: structure and molecular interactions. In: Adolph KW, editor. Molecular biology of chromosome function. New York: Springer; 1989. p. 366–95.
75. Yelick PC, Balhorn R, Johnson PA, et al. Mouse protamine 2 is synthesized as a precursor whereas mouse protamine 1 is not. Mol Cell Biol. 1987; 7(6):2173–9.
76. Carré-Eusèbe D, Lederer F, Lê KH, Elsevier SM. Processing of the precursor of protamine P2 in mouse. Peptide mapping and N-terminal sequence analysis of intermediates. Biochem J. 1991;277 (Pt 1):39–45.
77. Chauviere M, Martinage A, Debarle M, Sautiere P, Chevaillier P. Molecular characterization of six intermediate proteins in the processing of mouse protamine P2 precursor. Eur J Biochem. 1992; 204(2):759–65.
78. Elsevier SM, Noiran J, Carre-Eusebe D. Processing of the precursor of protamine P2 in mouse. Identification of intermediates by their insolubility in the presence of sodium dodecyl sulfate. Eur J Biochem. 1991;196(1):167–75.
79. Schellman JA, Parthasarathy N. X-ray diffraction studies on cation-collapsed DNA. J Mol Biol. 1984; 175(3):313–29.
80. Hud NV, Vilfan ID. Toroidal DNA condensates: unraveling the fine structure and the role of nucleation in determining size. Annu Rev Biophys Biomol Struct. 2005;34:295–318.
81. Livolant F. Cholesteric organization of DNA in the stallion sperm head. Tissue Cell. 1984;16(4): 535–55.
82. Bianchi F, Rousseaux-Prevost R, Bailly C, Rousseaux J. Interaction of human P1 and P2 protamines with DNA. Biochem Biophys Res Commun. 1994; 201(3):1197–204.
83. Feughelman M, Langridge R, Seeds WE, et al. Molecular structure of deoxyriboncleic acid and nucleoprotein. Nature. 1955;175:834–8.
84. Prieto MC, Maki AH, Balhorn R. Analysis of DNA-protamine interactions by optical detection of magnetic resonance. Biochemistry. 1997;36(39): 11944–51.
85. Wilkins MFH. Physical studies of the molecular structure of deoxyribonucleic acid and nucleoprotein. Cold Spring Harb Symp Quant Biol. 1956; 21:75–90.
86. Cremer T, Cremer C. Chromosome territories, nuclear architecture and gene regulation in mammalian cells. Nat Rev Genet. 2001;2(4):292–301.
87. Lichter P, Cremer T, Borden J, Manuelidis L, Ward DC. Delineation of individual human chromosomes in metaphase and interphase cells by in situ suppression hybridization using recombinant DNA libraries. Hum Genet. 1988;80(3):224–34.
88. Savage JR. Interchange and intra-nuclear architecture. Environ Mol Mutagen. 1993;22(4):234–44.
89. Schardin M, Cremer T, Hager HD, Lang M. Specific staining of human chromosomes in Chinese hamster × man hybrid cell lines demonstrates interphase chromosome territories. Hum Genet. 1985;71(4): 281–7.
90. Weierich C, Brero A, Stein S, et al. Three-dimensional arrangements of centromeres and telomeres in nuclei of human and murine lymphocytes. Chromosome Res. 2003;11(5):485–502.
91. Manuelidis L. Individual interphase chromosome domains revealed by in situ hybridization. Hum Genet. 1985;71(4):288–93.
92. Manvelyan M, Hunstig F, Bhatt S, et al. Chromosome distribution in human sperm – a 3D multicolor banding-study. Mol Cytogenet. 2008;1:25.
93. Mudrak O, Tomilin N, Zalensky A. Chromosome architecture in the decondensing human sperm nucleus. J Cell Sci. 2005;118(Pt 19):4541–50.
94. Zalensky A, Zalenskaya I. Organization of chromosomes in spermatozoa: an additional layer of epigenetic information? Biochem Soc Trans. 2007;35 (Pt 3):609–11.
95. Chen JL, Guo SH, Gao FH. Nuclear matrix in developing rat spermatogenic cells. Mol Reprod Dev. 2001;59(3):314–21.
96. Santi S, Rubbini S, Cinti C, et al. Ultrastructural organization of the sperm nuclear matrix. Ital J Anat Embryol. 1995;100 Suppl 1:39–46.
97. Ward WS, Coffey DS. DNA packaging and organization in mammalian spermatozoa: comparison with somatic cells. Biol Reprod. 1991;44(4): 569–74.
98. Yaron Y, Kramer JA, Gyi K, et al. Centromere sequences localize to the nuclear halo of human spermatozoa. Int J Androl. 1998;21(1):13–8.
99. Heng HH, Goetze S, Ye CJ, et al. Chromatin loops are selectively anchored using scaffold/matrix-attachment regions. J Cell Sci. 2004;117(Pt 7): 999–1008.

100. Heng HH, Krawetz SA, Lu W, Bremer S, Liu G, Ye CJ. Re-defining the chromatin loop domain. Cytogenet Cell Genet. 2001;93(3–4):155–61.
101. Shaman JA, Yamauchi Y, Ward WS. Function of the sperm nuclear matrix. Arch Androl. 2007;53(3):135–40.
102. Shaman JA, Yamauchi Y, Ward WS. The sperm nuclear matrix is required for paternal DNA replication. J Cell Biochem. 2007;102(3):680–8.
103. Frehlick LJ, Eirin-Lopez JM, Jeffery ED, Hunt DF, Ausio J. The characterization of amphibian nucleoplasmins yields new insight into their role in sperm chromatin remodeling. BMC Genomics. 2006;7:99.
104. McLay DW, Clarke HJ. Remodelling the paternal chromatin at fertilization in mammals. Reproduction. 2003;125(5):625–33.
105. Philpott A, Leno GH. Nucleoplasmin remodels sperm chromatin in Xenopus egg extracts. Cell. 1992;69(5):759–67.
106. Katagiri C, Ohsumi K. Remodeling of sperm chromatin induced in egg extracts of amphibians. Int J Dev Biol. 1994;38(2):209–16.
107. Derijck A, van der Heijden G, Giele M, Philippens M, de Boer P. DNA double-strand break repair in parental chromatin of mouse zygotes, the first cell cycle as an origin of de novo mutation. Hum Mol Genet. 2008;17(13):1922–37.
108. Generoso WM, Cain KT, Krishna M, Huff SW. Genetic lesions induced by chemicals in spermatozoa and spermatids of mice are repaired in the egg. Proc Natl Acad Sci USA. 1979;76(1):435–7.
109. Matsuda Y, Seki N, Utsugi-Takeuchi T, Tobari I. Changes in X-ray sensitivity of mouse eggs from fertilization to the early pronuclear stage, and their repair capacity. Int J Radiat Biol. 1989;55(2):233–56.
110. Matsuda Y, Yamada T, Tobari I. Studies on chromosome aberrations in the eggs of mice fertilized in vitro after irradiation. I. Chromosome aberrations induced in sperm after X-irradiation. Mutat Res. 1985;148(1–2):113–7.
111. Blanchard Y, Lescoat D, Le Lannou D. Anomalous distribution of nuclear basic proteins in round-headed human spermatozoa. Andrologia. 1990;22(6):549–55.
112. de Yebra L, Ballesca JL, Vanrell JA, Bassas L, Oliva R. Complete selective absence of protamine-P2 in humans. J Biol Chem. 1993;268(14):10553–7.
113. Foresta C, Zorzi M, Rossato M, Varotto A. Sperm nuclear instability and staining with aniline blue: abnormal persistence of histones in spermatozoa in infertile men. Int J Androl. 1992;15(4):330–7.
114. Hofmann N, Hilscher B. Use of aniline blue to assess chromatin condensation in morphologically normal spermatozoa in normal and infertile men. Hum Reprod. 1991;6(7):979–82.
115. Terquem A, Dadoune J. Aniline bule staining of human spermatozoa chromatin: evaluation of nuclear maturation. The Hague: Martinus Nijhoff; 1983.
116. van Roijen HJ, Ooms MP, Spaargaren MC, et al. Immunoexpression of testis-specific histone 2B in human spermatozoa and testis tissue. Hum Reprod. 1998;13(6):1559–66.
117. Zhang X, SanGabriel M, Zini A. Sperm nuclear histone to protamine ratio in fertile and infertile men: evidence of heterogeneous subpopulations of spermatozoa in the ejaculate. J Androl. 2006;27(3):414–20.
118. Aoki VW, Liu L, Carrell DT. Identification and evaluation of a novel sperm protamine abnormality in a population of infertile males. Hum Reprod. 2005;20(5):1298–306.
119. Balhorn R, Reed S, Tanphaichitr N. Aberrant protamine 1/protamine 2 ratios in sperm of infertile human males. Experientia. 1988;44(1):52–5.
120. Belokopytova IA, Kostyleva EI, Tomilin AN, Vorobev VI. Human male infertility may be due to a decrease of the protamine-P2 content in sperm chromatin. Mol Reprod Dev. 1993;34(1):53–7.
121. Carrell DT, Emery BR, Hammoud S. Altered protamine expression and diminished spermatogenesis: what is the link? Hum Reprod Update. 2007;13(3):313–27.
122. Carrell DT, Liu L. Altered protamine 2 expression is uncommon in donors of known fertility, but common among men with poor fertilizing capacity, and may reflect other abnormalities of spermiogenesis. J Androl. 2001;22(4):604–10.
123. Chevaillier P, Mauro N, Feneux D, Jouannet P, David G. Anomalous protein complement of sperm nuclei in some infertile men. Lancet. 1987;2(8562):806–7.
124. Oliva R. Protamines and male infertility. Hum Reprod Update. 2006;12(4):417–35.
125. Aoki VW, Christensen GL, Atkins JF, Carrell DT. Identification of novel polymorphisms in the nuclear protein genes and their relationship with human sperm protamine deficiency and severe male infertility. Fertil Steril. 2006;86(5):1416–22.
126. Aoki VW, Emery BR, Liu L, Carrell DT. Protamine levels vary between individual sperm cells of infertile human males and correlate with viability and DNA integrity. J Androl. 2006;27(6):890–8.
127. Aoki VW, Liu L, Jones KP, et al. Sperm protamine 1/protamine 2 ratios are related to in vitro fertilization pregnancy rates and predictive of fertilization ability. Fertil Steril. 2006;86(5):1408–15.
128. Cho C, Jung-Ha H, Willis WD, et al. Protamine 2 deficiency leads to sperm DNA damage and embryo death in mice. Biol Reprod. 2003;69(1):211–7.
129. Depa-Martynow M, Kempisty B, Lianeri M, Jagodzinski PP, Jedrzejczak P. Association between fertilin beta, protamines 1 and 2 and spermatid-specific linker histone H1-like protein mRNA levels, fertilization ability of human spermatozoa, and quality of preimplantation embryos. Folia Histochem Cytobiol. 2007;45 Suppl 1:S79–85.
130. Mengual L, Ballesca JL, Ascaso C, Oliva R. Marked differences in protamine content and P1/P2 ratios in sperm cells from percoll fractions between patients and controls. J Androl. 2003;24(3):438–47.

131. Bedford JM, Calvin HI. The occurrence and possible functional significance of -S-S- crosslinks in sperm heads, with particular reference to eutherian mammals. J Exp Zool. 1974;188(2):137–55.
132. Calvin HI, Bedford JM. Formation of disulphide bonds in the nucleus and accessory structures of mammalian spermatozoa during maturation in the epididymis. J Reprod Fertil Suppl. 1971;13 Suppl 13:65–75.
133. Calvin HI, Yu CC, Bedford JM. Effects of epididymal maturation, zinc (II) and copper (II) on the reactive sulfhydryl content of structural elements in rat spermatozoa. Exp Cell Res. 1973;81(2):333–41.
134. Saowaros W, Panyim S. The formation of disulfide bonds in human protamines during sperm maturation. Experientia. 1979;35(2):191–2.
135. Sega GA, Generoso EE. Measurement of DNA breakage in spermiogenic germ-cell stages of mice exposed to ethylene oxide, using an alkaline elution procedure. Mutat Res. 1988;197(1):93–9.
136. Sega GA, Owens JG. Methylation of DNA and protamine by methyl methanesulfonate in the germ cells of male mice. Mutat Res. 1983;111(2): 227–44.
137. Sega GA, Owens JG. Binding of ethylene oxide in spermiogenic germ cell stages of the mouse after low-level inhalation exposure. Environ Mol Mutagen. 1987;10(2):119–27.
138. Bjorndahl L, Kvist U. Human sperm chromatin stabilization: a proposed model including zinc bridges. Mol Hum Reprod. 2010;16(1):23–9.
139. Bench G, Corzett MH, Kramer CE, Grant PG, Balhorn R. Zinc is sufficiently abundant within mammalian sperm nuclei to bind stoichiometrically with protamine 2. Mol Reprod Dev. 2000;56(4):512–9.
140. Bianchi F, Rousseaux-Prevost R, Sautiere P, Rousseaux J. P2 protamines from human sperm are zinc -finger proteins with one CYS2/HIS2 motif. Biochem Biophys Res Commun. 1992;182(2): 540–7.
141. Gatewood JM, Schroth GP, Schmid CW, Bradbury EM. Zinc-induced secondary structure transitions in human sperm protamines. J Biol Chem. 1990; 265(33):20667–72.
142. Hernandez-Ochoa I, Sanchez-Gutierrez M, Solis-Heredia MJ, Quintanilla-Vega B. Spermatozoa nucleus takes up lead during the epididymal maturation altering chromatin condensation. Reprod Toxicol. 2006;21(2):171–8.
143. Johansson L, Pellicciari CE. Lead-induced changes in the stabilization of the mouse sperm chromatin. Toxicology. 1988;51(1):11–24.

Spermatogenesis: An Overview

2

Rakesh Sharma and Ashok Agarwal

Abstract

The purpose of this chapter is to provide a comprehensive overview of spermatogenesis and the various steps involved in the development of the male gamete, including cellular processes and nuclear transformations that occur during spermatogenesis, to provide a clear understanding of one of the most complex cellular metamorphosis that occurs in the human body.

Spermatogenesis is a highly complex temporal event during which a relatively undifferentiated diploid cell called spermatogonium slowly evolves into a highly specialized haploid cell called spermatozoon. The goal of spermatogenesis is to produce a genetically unique male gamete that can fertilize an ovum and produce offspring. It involves a series of intricate, cellular, proliferative, and developmental phases. Spermatogenesis is initiated through the neurological axis by the hypothalamus, which releases gonadotropin-releasing hormone, which in turn signals follicle-stimulating hormone (FSH) and luteinizing hormone (LH) to be transmitted to the reproductive tract. LH interacts with the Leydig cells to produce testosterone, and FSH interacts with the Sertoli cells that provide support and nutrition for sperm proliferation and development.

Spermatogenesis involves a series of cell phases and divisions by which the diploid spermatogonial cells develop into primary spermatocytes via mitosis. Primary spermatocytes in the basal compartment of Sertoli cells undergo meiosis to produce haploid secondary spermatocytes in the adluminal compartment of Sertoli cells in a process called spermatocytogenesis. This process gives the cells a unique genetic identity within the

A. Agarwal (✉)
Professor and Director, Center for Reproductive Medicine, Glickman Urological and Kidney Institute, OB-GYN and Women's Health Institute,
Cleveland Clinic, 9500 Euclid Avenue, Desk A19.1,
Cleveland, OH 44195, USA
e-mail: agarwaa@ccf.org

A. Zini and A. Agarwal (eds.), *Sperm Chromatin: Biological and Clinical Applications in Male Infertility and Assisted Reproduction*, DOI 10.1007/978-1-4419-6857-9_2,

population of secondary spermatocytes and subsequent developing cells. After spermatocytogenesis, spermatids elongate to form spermatozoa by spermiogenesis, a morphological development phase in which the nuclear transformations involving chromatin remodeling and compaction occur. Spermatozoa then leave the Sertoli cells through the lumen of the seminiferous tubules, exit through the rete testis, and enter the epididymis for final maturation. This is where spermatozoa acquire motility and acrosomal function. Spermatogenesis in the human male takes about 74 days.

Spermatogenesis is regulated by intrinsic and extrinsic factors. Not all spermatogonia mature into spermatozoa – most are eliminated and phagocytosed in a process called apoptosis. The overall goals of spermatogenesis are (1) to enable the male to transfer genetically recombined DNA by contributing to half of the offspring's genome and (2) to equip the spermatozoa to effectively navigate through the female reproductive tract and deliver the genetic material to the ovum. In the following sections, the complex transformation of the simple single diploid cell into a fully functional haploid cell is described.

Keywords
Spermatogenesis • Male gamete • Neurological pathways in spermatogenesis • Spermiogenesis • Meiosis and mitosis

Neurological Pathways

Spermatogenesis is initiated through hormonal controls in the hypothalamus (Fig. 2.1). The hypothalamus secretes gonadotropin-releasing hormone (GnRH), triggering the release of luteinizing hormone (LH) and follicle-stimulating hormone (FSH) from the adenohypophysis or anterior lobe of the pituitary. LH assists with steroidogenesis by stimulating the Leydig cells of the interstitium, and FSH stimulates the Sertoli cells to aid with the proliferative and developmental stages of spermatogenesis. In addition to LH and FSH, the adenohypophysis also secretes adrenocorticotropic hormone, prolactin, growth hormone, and thyroid-stimulating hormone – all of these hormones play important roles throughout spermatogenesis. The primary hormones are responsible for initiating spermatogenesis inside the testes, which is the central organ of the reproductive axis. GnRH stimulations are regulated through three types of rhythmicity: (1) seasonal – peak GnRH production occurs during the spring (2) circardian – daily regulator with the highest output during the early morning and (3) pulsatile – highest output occurring on average every 90–120 min.

Steroid Hormone Interaction and Neurological Axis

Androgens are an integral part of spermatogenesis. Dihydrotestosterone is formed by metabolizing testosterone with 5 alpha-reductase. Both testosterone and dihydrotestosterone regulate various genes and the various developmental stages during gestation [1]. Estrogen is necessary for proper spermatogenesis [2, 3]. During Sertoli cell differentiation, estrogen levels drop to minimum levels. During the prepubescent years, estrogen shuts off androgen production by the Leydig cells. When puberty begins, estrogen levels fall to enable androgen production by Leydig cells and initiate spermatogenesis. Thyroid hormones play a key role in spermatogenesis involving Sertoli cell proliferation and development. All of these hormones interact with one another

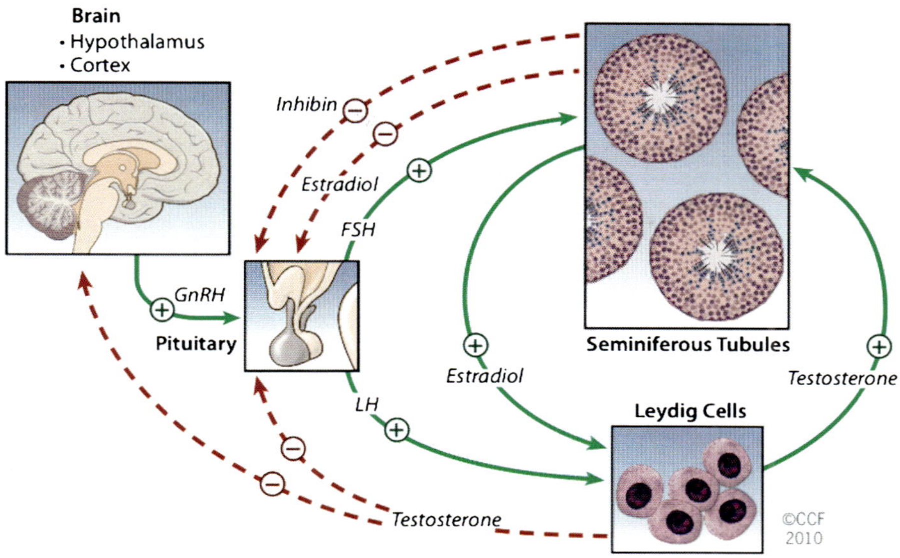

Fig. 2.1 Schematic representation of the hypothalamic pituitary axis and the hormonal feedback system (reprinted with permission, Cleveland Clinic Center for Medical Art & Photography © 2010. All Rights Reserved)

in the testicular axis in both the interstitial region and the Sertoli cells to enable spermatogenesis. In addition to the hormones, growth factors secreted directly by the Sertoli cells also play an important role in spermatogenesis. Transforming growth factor (alpha and beta), insulin-like growth factor, and beta fibroblast growth factor facilitate germ cell migration during embryonic development, proliferation, and regulation of meiosis and cellular differentiation.

Organization of the Testis

The testes are ellipsoid in shape, measuring of 4.5–5.1 cm in length [4, 5], 2.5×4 cm in width [6] and have a volume of 15–25 mL [7]. They are engulfed by a strong connective tissues capsule (tunica albuginea) [6] and are the only organs in humans that are located outside the body. Spermatogenesis occurs at temperatures that are optimally 2–4° lower than that the temperature of main body [8]. The testis is loosely connected along its posterior border to the epididymis, which gives rise to the vas deferens at its lower pole [9]. The testis has two main functions: to produce hormones, in particular testosterone, and to produce male gametes – the spermatozoa (Fig. 2.2).

Supporting Cells: Leydig Cells

The Leydig cells are irregularly shaped cells that have granular cytoplasm present individually or more often in groups within the connective tissue. They contribute to about 5–12% of the testicular volume [10–12]. Leydig cells are the prime source of the male sex hormone testosterone [13–15]. LH acts on Leydig cells to stimulate the production of testosterone. This acts as a negative "feedback" on the pituitary to suppress or modulate further LH secretion [15]. The intratesticular concentration of testosterone is significantly higher than the concentration in the blood. Some of the key functions of testosterone are as follows: (1) Activation of the hypophyseal-testicular axis, (2) Masculation of the brain and sexual behaviors, (3) Initiation and maintenance of spermatogenesis, (4) Differentiation of the male genital organs, and (5) Acquisition of secondary sex characteristics.

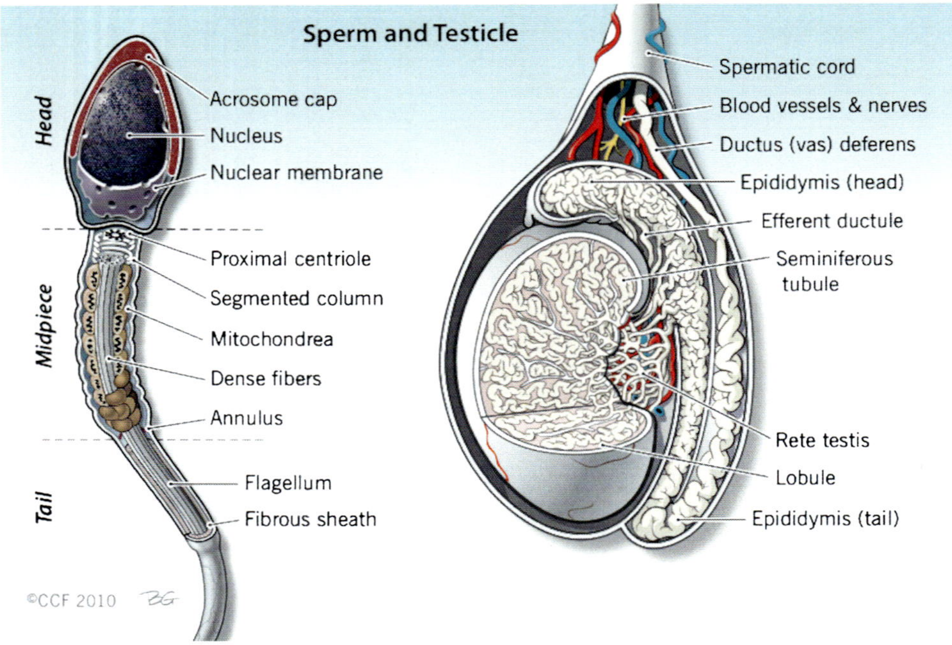

Fig. 2.2 The human testis and the epididymis. The testis shows the tunica vaginalis and tunica albuginea, seminiferous tubule septae, rete testis, and the overlying head, body, and tail of the epididymis. To the *left* is a diagrammatic representation of a fully mature spermatozoon (reprinted with permission, Cleveland Clinic Center for Medical Art & Photography © 2010. All Rights Reserved)

Seminiferous Tubules and Sertoli Cells

Most of the volume of the testis is made up of seminiferous tubules, which are packed in connective tissue within the confines of the fibrous septa. The testis is incompletely divided into a series of about 370 lobules or fibrous septae consisting of the seminiferous tubules and the intertubular tissue. The seminiferous tubules are a series of convoluted tubules within the testes. Spermatogenesis takes place in these tubules, scattered into many different proliferating and developing pockets (Fig. 2.3). The seminiferous tubules are looped or blind-ended and separated by groups of Leydig cells, blood vessels, lymphatics, and nerves. Each seminiferous tubule is about 180 μm in diameter. The height of the germinal epithelium measures 80 μm and the thickness of the peritubular tissue is about 8 μm [16].

Seminiferous tubules consist of three layers of peritubular tissue: (1) the outer adventitial layer of fibrocytes that originates from primitive connective tissue from the interstitium, (2) the middle layer composed of myoid cells that are distributed next to the connective tissue lamellae, and (3) the peritubular layer, a thick, inner lamella that mainly consists of collagen. The seminiferous tubule space is divided into basal (basement membrane) and adluminal (lumen) compartments by strong intercellular junctional complexes called "tight junctions." The seminiferous tubules are lined with highly specialized Sertoli cells that rest on the tubular basement membrane and extend into the lumen with a complex ramification of cytoplasm. They encourage Sertoli cell proliferation and development during the gestational period. Both ends of the seminiferous tubules open into the spaces of the rete testis [17]. The fluid secreted by the seminiferous tubules is collected in the rete testis and delivered into the excurrent ductal system of the epididymis.

Approximately 40% of the seminiferous tubules consist of Sertoli cells, and roughly 40%

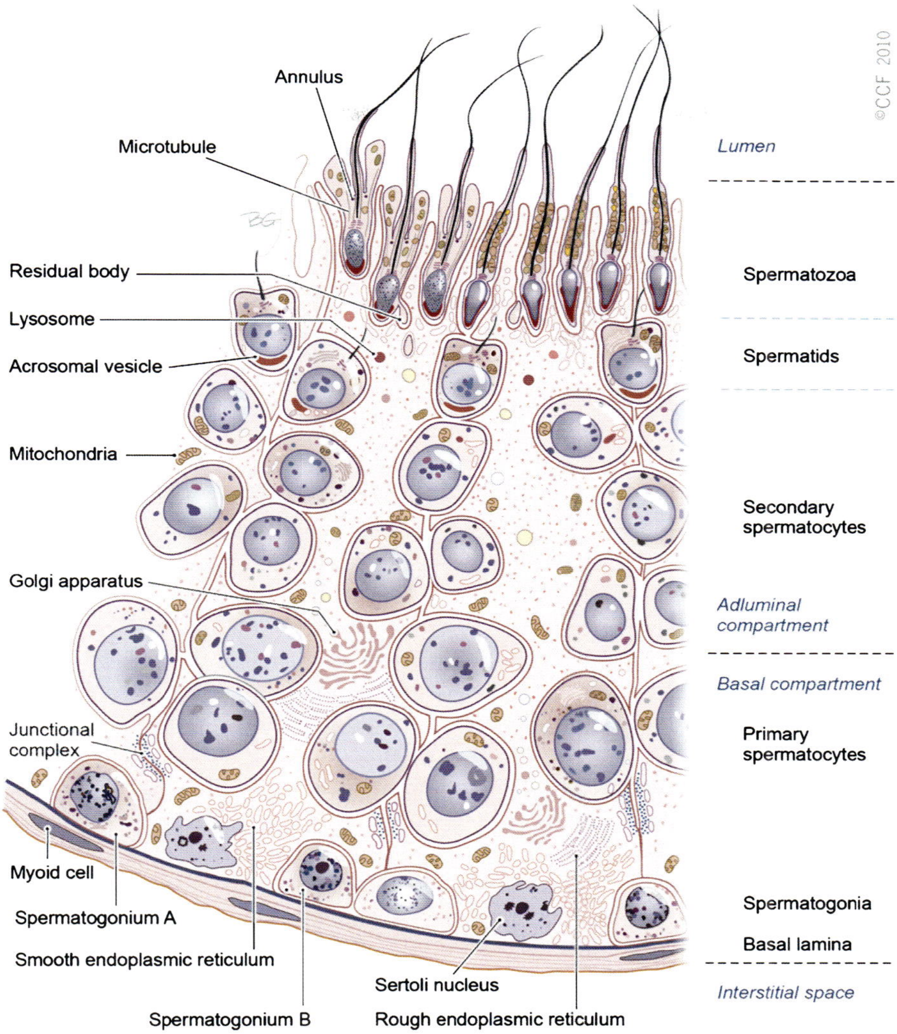

Fig. 2.3 Section of the germinal epithelium in the seminiferous tubule. Sertoli cells divide the germinal epithelium into a basal and adluminal compartment, via the Sertoli cell. Spermatozoa are released into the lumen (reprinted with permission, Cleveland Clinic Center for Medical Art & Photography © 2010. All Rights Reserved)

of the Sertoli cells are occupied with elongated spermatids [18, 19]. Sertoli cells have larger nuclei than most cells, ranging from 250 to 850 cm^3 [18]. Each Sertoli cell makes contact with five other Sertoli cells and about 40–50 germ cells at various stages of development and differentiation. The Sertoli cells provide structural, functional, and metabolic support to germ cells. Functionally and endocrinologically competent Sertoli cells are necessary for optimal spermatogenesis. During spermatogenesis, the earlier germinal cells rest toward the epithelium region of the seminiferous tubules in order to develop and mature while the more developed germinal cells move toward the lumen of the seminiferous tubules in order to exit the seminiferous tubule system and continue with the final phases of spermatogenesis.

Sertoli cells function as “nurse” cells for spermatogenesis, nourishing germ cells as they develop and participating in germ cell phagocytosis. Multiple sites of communication exist between Sertoli cells and developing germ cells for the maintenance of spermatogenesis within an appropriate hormonal milieu. FSH binds to the high-affinity FSH receptors found on Sertoli cells, signaling the secretion of androgen-binding protein (ABP). ABP allows androgens such as testosterone and dihydrotestosterone to bind and increase their concentrations to initiate and/ or continue the process of spermatogenesis. Sertoli cells also release anti-Müllerian hormone that allows for the embryonic development of the male by reducing the growth of the Müllerian ducts [20, 21]. Sertoli cells also secrete inhibin – a key macromolecule participating in pituitary FSH regulation.

Spermatozoa are produced at puberty but are not recognized by the immune system that develops during the first year of life. The blood–testis barrier provides a microenvironment for spermatogenesis to occur in an immunologically privileged site. The blood–testis barrier is divided into two regions: a basal region located near the seminiferous epithelium and an adluminal region that is positioned toward the lumen region of the seminiferous tubules. The basal region is the spermatogenic site for spermatogonial and primary spermatocyte development, while the adluminal region serves as the site for secondary spermatocyte and spermatid development. The blood–testis barrier has three different levels: (1) tight junctions between Sertoli cells, which helps separate premeiotic spermatogonia from the rest of the germ cells, (2) the endothelial cells in both the capillaries and (3) peritubular myoid cells.

Some of the main functions of the Sertoli cells are as follows:

1. Maintenance of integrity of seminiferous epithelium
2. Compartmentalization of seminiferous epithelium
3. Secretion of fluid to form tubular lumen to transport sperm within the duct
4. Participation in spermiation
5. Phagocytosis and elimination of cytoplasm
6. Delivery of nutrients to germ cells
7. Steroidogenesis and steroid metabolism
8. Movement of cells within the epithelium
9. Secretion of inhibin and ABP
10. Regulation of spermatogenic cycle
11. Provide a target for LH, FSH, and testosterone receptors present on Sertoli cells

Spermatogenesis

The process of differentiation of a simple diploid spermatogonium into a spermatid is known as spermatogenesis [17]. It is a complex, temporal event whereby primitive, totipotent stem cells divide to either renew them or produce daughter cells that are transformed into a specialized testicular spermatozoon (Fig. 2.4). It involves both mitotic and meiotic divisions and extensive cellular remodeling. Spermatogenesis can be divided into three phases: (1) proliferation and differentiation of spermatogonia, (2) meiosis, and (3) spermiogenesis, a complex process that transforms round spermatids after meiosis into a complex structure called the spermatozoon. In humans, the process of spermatogenesis starts at puberty and continues throughout the entire life span of the individual. Once the gonocytes have differentiated into fetal spermatogonia, an active process of mitotic replication begins very early in the embryonic development.

Within the seminiferous tubule, germ cells are arranged in a highly ordered sequence from the basement membrane to the lumen. Spermatogonia lie directly on the basement membrane, followed by primary spermatocytes, secondary spermatocytes, and spermatids as they progress toward the tubule lumen. The tight junction barrier supports spermatogonia and early spermatocytes within the basal compartment and all subsequent germ cells within the adluminal compartment.

Types of Spermatogonia

Fetal spermatogonia become transitional spermatogonia and later spermatogonia type Ad (dark). Spermatogonial stem cells undergo proliferative events and produce a population of cells that have distinct nuclear appearance that can be seen with hematoxylin and eosin staining. Spermatogonia

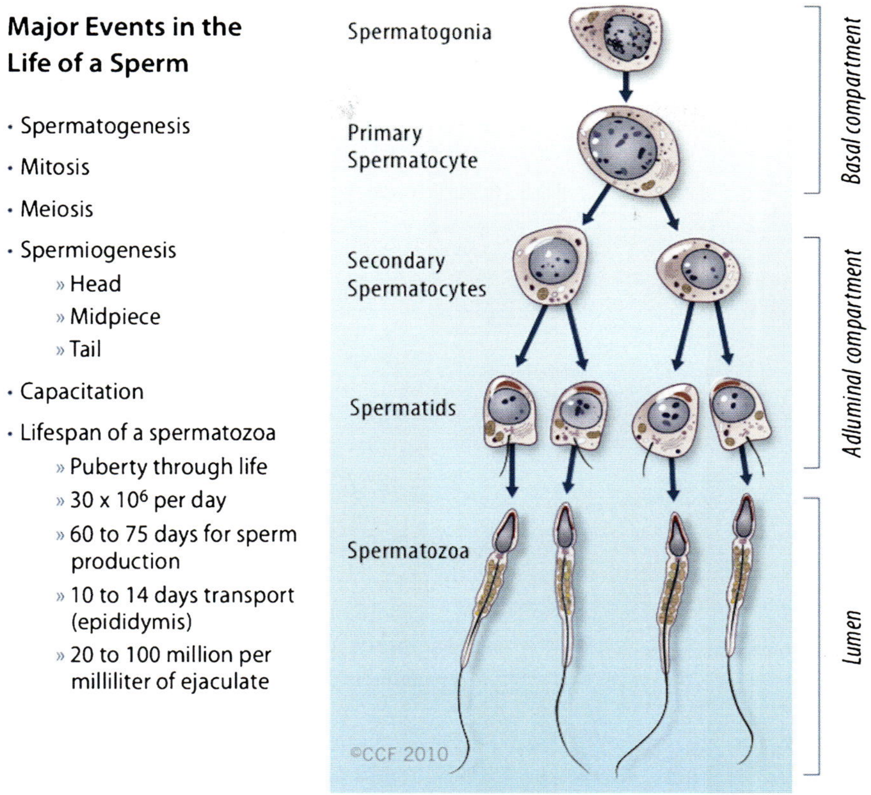

Fig. 2.4 A diagrammatic representation of major events in the life of a sperm involving spermatogenesis, spermiogenesis, and spermiation during which the developing germ cells undergo mitotic and meiotic division to reduce the chromosome content (reprinted with permission, Cleveland Clinic Center for Medical Art & Photography © 2010. All Rights Reserved)

can be categorized into three types: (1) Dark Type A, (2) Pale type A, and (3) Type B spermatogonia (Fig. 2.5).

Dark type A spermatogonia are stem cells of the seminiferous tubules that have an intensely stained dark ovoid nucleus containing fine granular chromatin. These cells divide by mitosis to generate Dark Type A and Pale Type A spermatogonia. Pale Type A spermatogonia have pale staining and fine granular chromatin in the ovoid nucleus. Other proliferative spermatogonia include A_{paired} (A_{pr}), resulting from dividing $A_{isolated}$, and subsequently dividing to form $A_{aligned}$ (A_{al}). Further differentiation of spermatogonia includes Type A1, A2, A3, A4, Intermediate, and Type B, each a result of the cellular division of the previous type. In humans, four spermatogonial cell types have been identified: A_{long}, A_{dark}, A_{pale}, and Type B [22–24]. In the rat, Type $A_{isolated}$ (A_{is}) is believed to be the stem cell [25, 26], whereas in humans, it is unclear which Type A spermatogonia is the stem cell. Type B spermatogonia are characterized by large clumps of condensed chromatin under the nuclear membrane of an ovoid nucleus. Type B spermatogonia divide mitotically to produce primary spermatocytes (preleptotene, leptotene, zygotene, and pachytene), secondary spermatocytes, and spermatids (Sa, Sb, Sc, Sd_1, and Sd_2), [22] (Fig. 2.6). Spermatogonia do not separate completely after meiosis but remain joined by intercellular bridges, which persist throughout all stages of spermatogenesis. This facilitates biochemical interactions and synchronizes germ cell maturation [27].

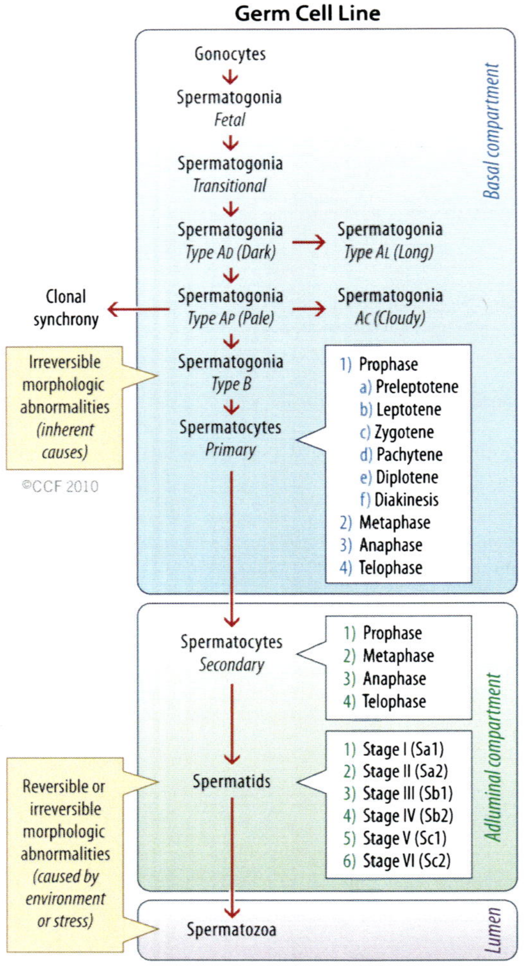

Fig. 2.5 Schematic representation of the development of a diploid undifferentiated germ cell into a fully functional haploid spermatozoon along the basal to the adluminal compartment and final release into the lumen. Different steps in the development of primary, secondary, and spermatid stages are also shown and the irreversible and reversible morphological abnormalities that may occur during various stages of spermatogenesis (reprinted with permission, Cleveland Clinic Center for Medical Art & Photography © 2010. All Rights Reserved)

Spermatocytogenesis

Spermatocytogenesis consists of the meiotic phase in which primary spermatocytes undergo meiosis I and meiosis II to give rise to haploid spermatids. This takes place in the basal compartment. Primary spermatocytes enter the first meiotic division to form secondary spermatocytes. The prophase of the first meiotic division is very long. Primary spermatocytes have the longest life span. Secondary spermatocytes undergo the second meiotic division to produce spermatids. Secondary spermatocytes are short-lived (1.1–1.7 days).

Mitosis

Mitosis involves the proliferation and maintenance of spermatogonia. It is a precise, well-orchestrated sequence of events in which the genetic material (chromosomes) is duplicated, with breakdown of the nuclear envelope and formation of two daughter cells as a result of equal division of the chromosomes and cytoplasm [28] DNA is organized into loop domains on which specific regulatory proteins interact [29–33]. The mitotic phase involves spermatogonia (types A and B) and primary spermatocytes (spermatocytes I). Primary spermatocytes are produced by developing germ cells interconnected by intracellular bridges through a series of mitotic divisions. Once the baseline number of spermatogonia is established after puberty, the mitotic component proceeds to provide precursor cells and initiate the process of differentiation and maturation.

Meiosis

The meiotic phase involves primary spermatocytes until spermatids are formed, and during this process, chromosome pairing, crossover, and genetic exchange take place until a new genome is determined. Meiosis consists of two successive divisions to yield four haploid spermatids from one diploid primary spermatocyte. After the first meiotic division (reduction division), each daughter cell contains one partner of the homologous chromosome pair, and they are called secondary spermatocytes (2n).

Meiosis is characterized by prophase, metaphase, anaphase, and telophase. The process starts when type B spermatogonia lose contact with the basement membrane and form

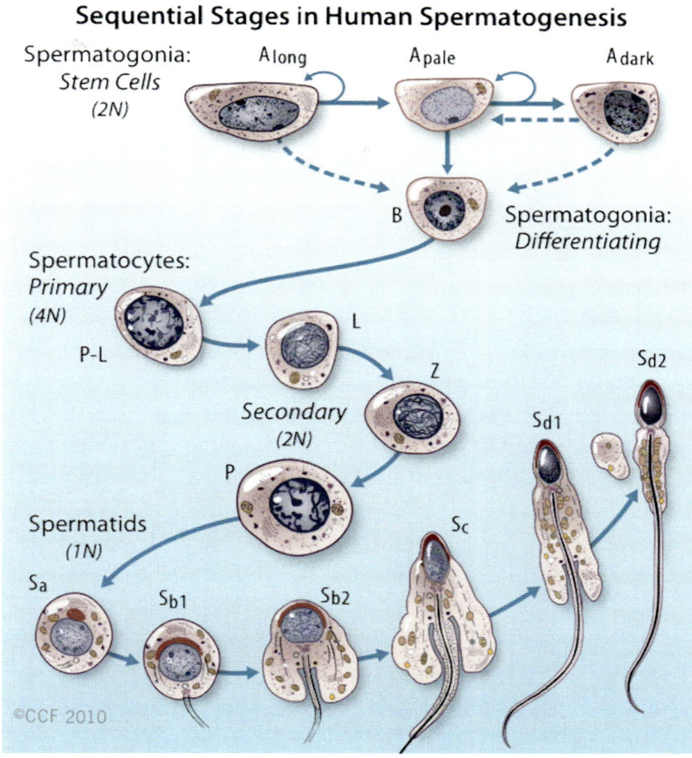

Fig. 2.6 Differentiation of a human diploid germ cell into a fully functional spermatozoon (reprinted with permission, Cleveland Clinic Center for Medical Art & Photography © 2010. All Rights Reserved)

preleptotene primary spermatocytes. During the leptotene stage of prophase, the chromosomes are arranged as long filaments. During the zygotene stage, the homologous chromosomes called tetrads are arranged linearly by a process known as synapsis and form synaptonemal complexes. Crossing over takes place during this phase, and the chromosomes shorten in the pachytene stage. The homologous chromosomes condense and separate from sites of crossing over during diakinesis. This random sorting is important to maintain genetic diversity in sperm. At the end of prophase, the nuclear envelope breaks down, and in metaphase, chromosomes are arranged in the equatorial plate. At anaphase, each chromosome consists of two chromatids migrating to opposite poles. In telophase, cell division occurs with the formation of secondary spermatocytes having half the number of chromosomes. Thus, each primary spermatocyte can theoretically yield four spermatids, although fewer actually result, as the complexity of meiosis is associated with a loss of some germ cells. The primary spermatocytes are the largest germ cells of the germinal epithelium.

The prophase of the second meiotic division is very short, and in this phase, the DNA content is reduced to half as the two chromatids of each chromosome separate and move to the opposite poles. At the end of telophase, the spermatids do not separate completely but remain interconnected by fine bridges for synchronous development. These spermatids are haploid with (22, X) or (22, Y) chromosome and undergo complete differentiation/morphogenesis known as spermiogenesis.

Spermiogenesis

Spermiogenesis is the process of differentiation of the spermatids into spermatozoa with fully compacted chromatin. During this process, morphological changes occur once the process of meiosis is completed. In humans, six different stages have been described in the process of spermatid maturation; these are termed as S_{a-1} and S_{a-2}, S_{b-1} and S_{b-2}, and S_{c-1} and S_{c-2} (Fig. 2.6). Each stage can be identified by morphological characteristics. During the S_{a-1} stage, both the Golgi complex and mitochondria are well developed and differentiated. In addition, the acrosomal vesicle appears, the chromatoid body develops in one pole of the cell opposite from the acrosomal vesicle, and proximal centriole and axial filament appear. During the S_{b-1} and S_{b-2} stages, acrosome formation is completed, the intermediate piece is formed and the tail develops. This process is completed during the Sc stages. During the post-meiotic phase, progressive condensation of the nucleus occurs with inactivation of the genome. The histones are converted into transitional proteins, and finally, protamines are converted into well-developed disulfide bonds.

Spermiation

A mature spermatid frees itself from the Sertoli cell and enters the lumen of the tubule as a spermatozoon in a process called spermiation. Spermatids that originate from the same spermatogonia remain connected by bridges to facilitate the transport of cytoplasmic products. Sertoli cells actively participate in spermiation, which may also involve the actual movement of the cells as the spermatids advance toward the lumen of the seminiferous tubules [18]. The mature spermatids close their intracellular bridges, disconnect their contact with the germinal epithelium, and become free cells called spermatozoa. Portions of the cytoplasm in the Sertoli cell known as the cytoplasmic droplet are completely eliminated, or at times, they may be retained in the immature spermatozoon during the process of spermiation [34].

The Cycle or Wave of Seminiferous Epithelium

A cycle of spermatogenesis involves the division of primitive spermatogonial stem cells into subsequent germ cell types through the process of meiosis. Type A spermatogonial divisions occur at a shorter time interval than the entire process of spermatogenesis. Therefore, at any given time, several cycles of spermatogenesis coexist within the germinal epithelium. Spermatogenesis is not a random but well orchestrated series of well-defined events in the seminiferous epithelium. Germ cells are localized in spatial units referred as stages. Each stage is recognized by development of the acrosome; meiotic divisions and shape of the nucleus and release of the sperm into lumen of the seminiferous tubule. A stage is designated by Roman numerals. Each cell type of the stage is morphologically integrated with the others in its development process. Each stage has a defined morphological entity of spermatid development called a step, which is designated by an Arabic number. Several steps occur together to form a stage, and several stages are necessary to form a mature sperm from immature stem cells [35, 36]. In rodent spermatogenesis, only one stage can be found in a cross section of seminiferous tubule.

Within any given cross section of the seminiferous tubule, there are four to five layers of germ cells. Cells in each layer comprise a generation or a cohort of cells that develop as a synchronous group. Each group has a similar appearance and function. Stages I–III have four generations comprising Type A spermatogonia, two primary spermatocytes, and an immature spermatid. Stages IV–VIII have five generations: Type A spermatogonia, one generation of primary spermatocyte, one generation of secondary spermatocytes, and one generation of spermatids. Thus, a position in the tubule that is occupied by cells comprising stage I will become stage II, followed by stage III, until the cycle repeats. The cycle of spermatogenesis can be identified for each species, but the duration of the cycle varies for each species [22].

The stages of spermatogenesis are sequentially arranged along the length of the tubule in such a way that it results in a "wave of spermatogenesis." Although it appears that the spatial organization is lacking or is poor in the human seminiferous tubule, these stages are tightly organized in an intricate helicine pattern [37]. In addition to the steps being organized spatially within the seminiferous tubule, the stages are organized in time. Spermatozoa are released only in certain cross sections along the length of the seminiferous tubule. In rat, all stages are involved in spermatogenesis, but spermatozoa are released only in stage VIII. In humans, this wave appears to be a spiral cellular arrangement as they progress down the tubule. This spatial arrangement probably exists to ensure that sperm production is a continuous and not a pulsatile process. The spermatocyte takes 25.3 days to mature. Spermiogenesis occurs in 21.6 days, and the duration of the cycle is 16 days. The progression from spermatogonia to spermatozoa or spermatogenesis is 74 days or 4½ cycles of the seminiferous cycle.

Chromatin Remodeling/Alterations During Sperm Differentiation

Mammalian sperm chromatin is unique in that it is highly organized, condensed, and compacted. This feature protects the paternal genome during transport through the male and the female reproductive tracts and helps ensure that it is delivered to the ova in good condition. Mammalian sperm DNA is the most tightly compacted eukaryotic DNA [38]. This feature is in sharp contrast to the DNA structure in somatic cell nuclei. Somatic cell nuclear DNA is wrapped around an octamer of histones and packaged into a solenoid structure [39]. This type of packaging adds histones, which increase the chromatin volume. The sperm nucleus does not have this type of packaging, and the volume is highly compacted. Chromatin changes occur in the testis during meiosis in which copies of the genome are partitioned into haploid spermatid cells and during spermiogenesis in which spermatids elongate to form sperm with fully compacted chromatin. These events are largely controlled by posttranslational events for transcription. Translation greatly subsides as DNA becomes compacted and the cytoplasm is jettisoned during spermiogenesis [40, 41]. After meiosis, sperm DNA experiences extreme chromosome compaction during spermiogenesis.

Chromatin modeling is accompanied by changes in the nuclear shape, conversion of negatively supercoiled nucleosomal DNA into a nonsupercoiled state [42], induction of transient DNA breaks [43], and chromatin condensation. It is mediated by drastic changes at the most fundamental level of DNA packaging where a nucleosomal architecture shifts to a toroidal structure [44]. This change is implemented by sperm nuclear basic proteins (SNBs) that include variants of histone subunits, transition proteins, and protamine proteins [45, 46]. Chromatin proteins do not act exclusively to compact sperm DNA. This transition occurs in a stepwise manner, replacing somatic histones with testis-expressed histone variants, transition proteins, and finally protamines [47]. Histone localization and posttranslational modification of histones encode epigenetic information that may regulate transcription important for sperm development [48]. They may also serve to mark the heterochromatin state of specific regions of the genome that may be important after fertilization, when somatic histones are incorporated back into paternal chromatin or during subsequent zygotic development [49]. Male infertility can result from deficits of SNBs [50–52].

Histone and Basic Nuclear Protein Transitions in Spermatogenesis

During spermatogenesis, histone proteins in developing sperm are replaced by testis-specific histone variants that are important for fertility [53]. The cells depend on posttranslational modifications to implement subsequent stages of sperm formation, maturation, and activation as de novo transcription in postmeiotic sperm is largely silenced [54]. During spermiogenesis, sperm chromatin undergoes a series of modifications in which histones are lost and replaced with transition proteins and subsequently with protamines [54–56]. Approximately 15% of the histones are retained in human sperm chromatin, subsequently making chromatin less tightly compacted [57, 58]. Chromatin remodeling is facilitated by the coordinated loosening of the chromatin by histone hyperacetylation and by the DNA topoisomerase II (topo II), which produce temporary nicks in the sperm DNA to relieve torsional stress that results from supercoiling [43, 59–61]. The same enzyme Topo II

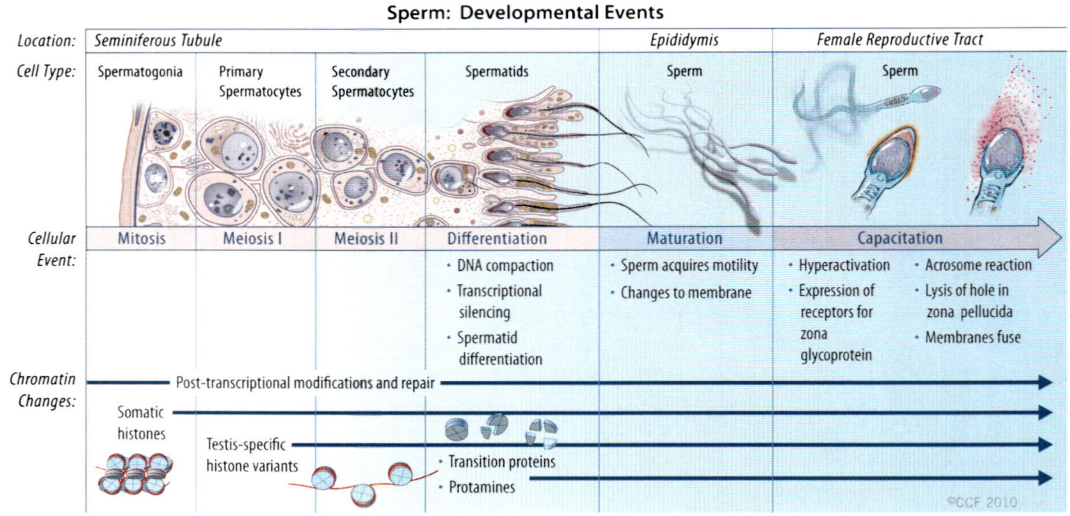

Fig. 2.7 Diagrammatic representation of the series of cellular and chromatin changes during the development of the germ cell into a spermatozoon and its subsequent release and storage into the epididymis and its journey into the female reproductive tract (reprinted with permission, Cleveland Clinic Center for Medical Art & Photography © 2010. All Rights Reserved)

normally repairs these temporary nicks prior to completion of spermiogenesis and ejaculation. However, if these nicks are not repaired, DNA fragmented sperm may be present in the ejaculate [62].

Role of Transition Proteins

The histone-to-protamine transition is important in the formation of spermatozoa [63]. This occurs in two steps in mammals: replacement of histones by transition nuclear proteins (TPs) – TP1 and TP2 – and replacement of TPs by protamines (protamine 1 and protamine 2). TPs are required for normal chromatin condensation, for reducing the number of DNA breaks and for preventing the formation of secondary defects in spermatozoa and the eventual loss of genomic integrity and sterility. TP1 is a 6.2-kDa, highly basic (about 20% each of arginine and lysine) protein with evenly distributed basic residues [64, 65], whereas TP2 is a 13-kDa basic (10% each of arginine and lysine) protein with distinct structural domains. The only similarity between the two is their high basicity, exon–intron genomic patterns, and developmental expression [66].

The transition nuclear proteins are localized exclusively to the nuclei of elongating and condensing spermatids [67] and were first detected in step 10–11 spermatids [68, 69] (Figs. 2.7 and 2.8). The maximum levels of TPs are acquired during steps 12–13, during which they constitute 90% of the chromatin basic protein, with the levels of TP1 being about 2.5 times those of TP2 [51]. They are not detected in the nucleus after the early part of step 15 [68, 69].

Some of the possible roles of TPs are as follows:

1. TP1 can destabilize nucleosomes and prevent binding of the DNA, both of which could contribute to displacement of histones [70, 71]
2. The zinc fingers of TP2 selectively bind to CpG sites and may be responsible for global expression of RNA synthesis [72]
3. Both TPs may play a role as alignment factors for DNA strand breaks, and TP1 is involved in the repair of strand breaks [73, 74]
4. Both TP1 and TP2 can condense DNA, and TP2 is more effective [70, 71, 75]. TP2 is not a critical factor for shaping of the sperm nucleus, histone displacement, initiation of chromatin condensation, binding of protamines to DNA, or fertility, but it is necessary for maintaining the normal processing of P2 and consequently the completion of chromatin condensation [52]

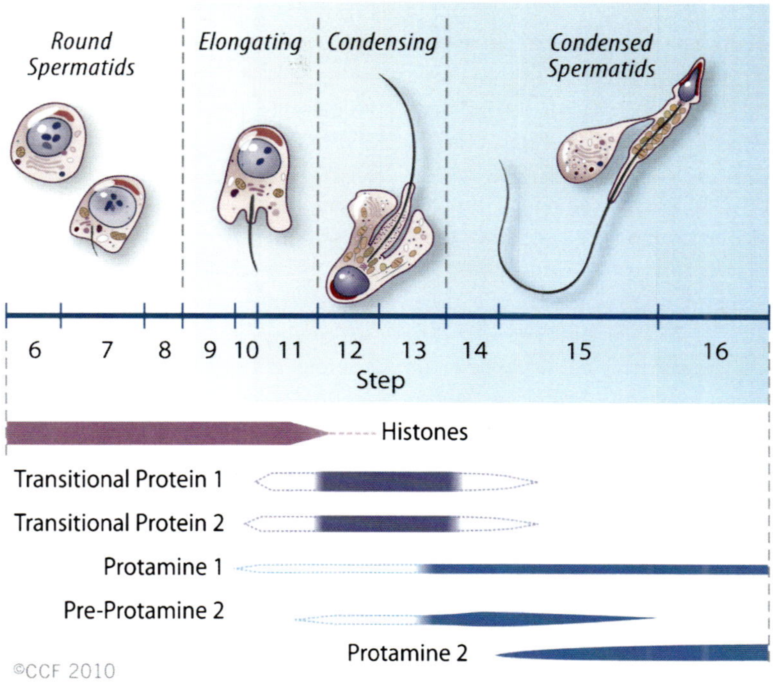

Fig. 2.8 Diagrammatic representation of the steps where the histones are replaced with the transition proteins and protamines in the round spermatid progresses into a condensed spermatid just before it is released into the lumen (reprinted with permission, Cleveland Clinic Center for Medical Art & Photography © 2010. All Rights Reserved)

Mice lacking either TP1 or TP2 alone had normal numbers of sperm with only minor abnormalities and were fertile, indicating either that the TPs were not essential or that the individual TPs complement each other [51, 52, 76]. Protamine 2 processing defects do not inhibit postfertilization processes because late spermatids containing unprocessed protamine 2 are able to initiate normal development [77]. Defective protamine 2 processing is correlated with infertility in humans [78] and mouse mutants [51, 52] and could be due solely to the secondary cytoplasmic effects on sperm development resulting in a reduced ability to penetrate the egg.

Protamines as Checkpoints of Spermatogenesis

Human sperm chromatin undergoes a complex transition during the elongating spermatid stage of spermiogenesis, in which histones are extensively replaced by protamines. Humans express equal quantities of two protamines: protamine 1 and protamine 2 [79–81]. Protamines are approximately half the size of histones [82]. They are highly basic sperm-specific nuclear proteins that are characterized by an arginine-rich core and cysteine residues [83, 84]. The high level of arginine causes a net positive charge, thereby facilitating strong DNA binding [85]. Cysteine residues facilitate the formation of multiple inter- and intraprotamine disulfide bonds essential for high-order chromatin packaging, which is necessary for normal sperm function [86–90]. P2 protamines contain fewer cysteine groups and thus contain fewer disulfide cross links [81]. This, theoretically, leaves the DNA more susceptible to damage. Altered P2 expression is common in men with infertility [77].

During spermiogenesis, protamines progressively replace somatic histones in a stepwise manner [83]. First, somatic histones are replaced by testis-specific histone variants, which are replaced by transition proteins (TP1a and TP2) in a process involving extensive DNA rearrangement and remodeling [42]. During the elongating spermatid stage, the transition proteins are replaced in the condensing chromatin by protamines. In humans, ~85% of the histones are replaced by protamines [54, 91–94]. This sequential process facilitates molecular remodeling of the male genome within the differentiating spermatid [40]. In human sperm, the mean P1/P2 ratio is approximately 1.0 [77, 80, 95]. Sperm from infertile men show an altered P1/P2 ratio and/ or no detectable P2 in mature sperm. Protamine abnormalities in sperm from fertile men are extremely rare [78, 95–98].

Two links are proposed between abnormal protamine expression and aberrant spermatogenesis: (1) abnormal protamine expression is indicative of a general abnormality of spermatogenesis, possibly due to abnormal function of the transcription or translational regulator (2) protamines act as checkpoint regulators of spermatogenesis, and abnormal protamine expression leads to induction of an apoptotic process and severely diminished sperm quality [99].

Protamines condense the DNA strands and form the basic packaging unit of sperm chromatin called a toroid. Intramolecular and intermolecular disulfide cross-links between cysteine residues present in protamines result in further compaction of the toroids [100]. Protamines confer a higher order of DNA packaging in sperm than that found in somatic cells. All of these levels of compaction and organization help protect sperm chromatin during transport through the male and female reproductive tract. This also ensures delivery of the paternal genome in a form that allows developing embryo to accurately express genetic information [58, 75, 80, 101]. Protamine replacement may also be necessary for silencing the paternal genome and reprogramming the imprinting pattern of the gamete [102]. Abnormal protamine expression is associated with low sperm count, decreased sperm motility and morphology, diminished fertilization ability, and increased sperm chromatin damage [77, 98, 103]. Infertile men are reported to have a higher histone-to-protamine ratio in their sperm chromatin [95, 104].

DNA Methylation During Spermatogenesis

Nucleohistones are present in human and rat sperm and are absent in mouse sperm. About ~15% of the histones are retained in the mature human spermatozoa [58]. The distribution of these histones within the sperm nucleus may have an important function. Chromatin associated with histones corresponds to specific sequences [58], suggesting that heterogeneity in the sperm nucleus may be the basis for male genetic information [105–107]. There are widespread differences in methylation of specific sequences during oogenesis and spermatogenesis. Maintenance methylases can stably preserve DNA methylation at cytosine residues through rounds of replication [108] and may have a role in gene regulation [109]. Methylation can also provide a mechanism for imprinting the maternal and paternal genomes as seen by the gametic differences in DNA methylation. This results in differential regulation of the paternal genomes during early development [110].The sequences that are highly methylated in pachytene spermatocytes are also highly methylated in spermatids and epididymal sperm, indicating that this state persists throughout spermatogenesis [111].

DNA methylation may be involved in genomic imprinting in mammals and is one of the major epigenetic marks established during spermatogenesis [112]. Mature sperms show a more unique DNA methylation profile than somatic cells [113]. The level of DNA methylation does not correlate with fertilization but with pregnancy rate after IVF [114].

Sperm Nuclear DNA Strand Breaks

Mammalian spermiogenesis involves important changes in the cytoarchitecture and dramatic remodeling of the somatic chromatin; most of the nucleosomal DNA supercoiling is eliminated [115, 116]. This modification in chromatin structure occurs in elongating spermatids and is an important contributor to the nuclear integrity and acquisition of full fertilization potential of the male gamete [117]. DNA damage involves (1) abortive apoptosis initiated post meiotically when the ability to drive this process to completion is in decline (2) unresolved strand breaks created during spermiogenesis to relieve torsional stress associated with chromatin remodeling and (3) oxidative stress as a result of reactive oxygen species. Three major mechanisms for the creation of DNA damage in the male germ line have been proposed: chromatin remodeling by topoisomerase, oxidative stress, and abortive apoptosis. DNA damage could arise due to a combination of all the three mechanisms. Furthermore, a two-step hypothesis has been proposed [117, 118]. According to this hypothesis, the first step in the DNA damage cascade has its origin in spermiogenesis during which DNA is remodeled prior to condensation. Defects in the chromatin remodeling process result in the production of spermatozoa that are characterized by reduction in the efficiency of protamination, abnormal protamine 1 to protamine 2 ratio, and relatively high nucleohistone content [101, 119, 120]. These defects in chromatin modeling create a state of vulnerability whereby spermatozoa become increasingly susceptible to oxidative damage. In the second step of this DNA cascade, reactive oxygen species attack chromatin.

One of the first hypothesis concerning the origins of DNA damage in the male germ line focused on the physiological strand breaks created by topoisomerase during spermiogenesis as a means of relieving the torsional stresses created as DNA is condensed and packaged into the sperm head [60, 101]. Normally, these strand breaks are marked by a histone phosphorylation event and are fully resolved by topoisomerase before spermatozoa are released from germinal epithelium during spermiogenesis [121].

Sperm chromatin compaction is believed to play an important role in protecting the male genome from insult. This specific chromatin structure of the sperm essential for proper fertility and is in part due to the proteins that are bound to the DNA, including the protamines, histones, and components of the nuclear matrix [122, 123]. The cascade of events leading to DNA damage involves an error in chromatin remodeling during spermiogenesis. This leads to generation of spermatozoa with poorly protaminated nuclear DNA that is increasingly susceptible to oxidative attack [118].

Efficiency of chromatin remodeling during spermiogenesis has been studied employing DNA sensitive fluorochrome chromomycin (CM3). Chromomycin competes with the nucleoproteins for binding sites in the minor groove of GC-rich DNA and serves as a marker for the efficiency of DNA protamination during spermiogenesis. Staining with this probe is positively related to the presence of nuclear histones [124] and poor chromatin compaction [125] and negatively related with presence of protamines [126]. Impaired chromatin remodeling during spermiogenesis is a consistent feature of defective human spermatozoa possessing fragmented DNA [127–131]. DNA damage depends on fundamental errors that occur during spermatogenesis and may explain the correlation of pathology with sperm count [132].

Sperm Apoptosis

Apoptosis in sperm is different from somatic apoptosis in many ways: (1) spermatozoa are transcriptionally and translationally silent, and therefore, cannot undergo programmed cell death or "regulated cell death," (2) sperm chromatin has a reduced nucleosome content due to extensive protamination and, therefore, lacks the characteristic DNA laddering seen in somatic cells, and (3) endonucleases that are activated in the cytoplasm or released from the mitochondria are prevented from physically accessing the DNA due to the inherent physical architecture of the spermatozoa. However, spermatozoa do exhibit some of the hallmarks of apoptosis including caspase activation and phosphatidylserine exposure on the surface of the cells [133].

Sertoli cells can support only a limited number of germ cells in the testis. In the testis, apoptosis normally occurs to prevent the overproduction of germ cells and to selectively remove injured germ cells [134]. Clonal expansion of the germ cells in the testis occurs at very high levels, and thus, apoptosis is necessary to limit the size of the germ cell population to one which the Sertoli cell is able to support [135]. Fas Ligand (FasL) is secreted by Sertoli cells. Fas is a protein located on the germ cell surface. Evidence of germ cell apoptosis has been demonstrated in FasL-defective mice [136]. Men with poor seminal parameters often display a large percentage of Fas-expressing sperm in their ejaculate [101]. Some of these sperms with DNA damage and Fas expression may have undergone "abortive apoptosis" in which they started but subsequently escaped the apoptotic pathway [137]. However, other studies have failed to find a correlation between DNA damage and Fas expression and other markers of apoptosis [62]. Recent studies examining loss of function have indicated that DNA damage checkpoints occur during spermatogenesis and may also involve excision repair genes, mismatch repair genes, and p53 [138].

Oxidative Stress in the Testis

Sertoli cells provide nutritional support to the differentiating germ cells in the testis. They are protected from oxidative stress as these cells pass through meiosis and emerge as haploid cells known as round spermatids. At this stage of development, these cells are transcriptionally silent. Even in the absence of any regulated gene transcription, they are able to undergo cellular transformation into fully differentiated, highly specialized cells – the spermatozoa. This is accomplished through a highly orchestrated differential translation of preexisting mRNA species through a process called spermiogenesis. Cells are sensitive to oxidative stress during spermiogenesis. Throughout this phase, they are highly dependent on the nurturing Sertoli cells, which possess antioxidants such as superoxide dismutase, glutathione reductase, transferase, and peroxidase [139]. Isolated spermatozoa have a limited capacity for DNA repair [140].

Spermiogenesis and Etiology of DNA Damage

Spermiogenesis, the process by which haploid round spermatids differentiate into spermatozoa, is a key event in the etiology of DNA damage in the male germ line. During spermiogenesis, the chromatin undergoes extensive remodeling, which enables the entire haploid genome to be compacted into a sperm head measuring 5×2.5 μm. This occurs as physiological DNA strand breaks are introduced by topoisomerase to relieve the torsional stresses involved in DNA packaging during sperm differentiation. These strand breaks are corrected by a complex process involving H2Ax expression, formation of poly(ADP-ribose) by nuclear poly (ADP-ribose) polymersases (PARP) and topoisomerase [141]. If the spermiogenesis process is disrupted for any reason, restoration of the cleavage sites is impaired, and defective spermatozoa with unresolved physiological strand breaks are released from the germinal epithelium. The "transition" proteins play a key role in maintaining DNA integrity during spermiogenesis as they move into the sperm nucleus between the removal of histones and the entry of protamines. Functional deletion of these proteins results in the production of spermatozoa with poor fertilizing ability, poor chromatin compaction, and high levels of DNA fragmentation [63]. DNA damage in human spermatozoa is associated with the disruption and poor chromatin remodeling during spermiogenesis [120, 128].

The efficiency of spermatogenesis is reflected by conventional semen characteristics such as sperm count and morphology and the correlation with DNA damage [132, 142]. Poor protamination results in spermatozoa that possess nucleohistone-rich regions of chromatin, which are vulnerable to oxidative attack [117]. Oxidative stress is a major determinant of the quality of spermiogenesis. When this process is disrupted, spermatozoa are produced that are vulnerable to oxidative stress, 8OHdg formation, and ultimately DNA fragmentation as a consequence of apoptosis [120, 143, 144].

Efficiency of Spermatogenesis

The efficiency of spermatogenesis varies between different species; it appears to be relatively constant in man. The time needed for a spermatogonium to differentiate into a mature spermatid is estimated to be 70 ± 4 days [145].

In comparison to animals, the spermatogenetic efficiency in man is poor, and the daily rate of spermatozoa production is about 3–4 million/g of testicular tissue [146]. Although a much higher sperm count should be expected in the ejaculate than the 20 million/mL described by WHO manual [147], this is not the case. This is largely because most developed cells (>75%) are eliminated as a result of apoptosis. In the remaining cells, more than half are abnormal. Therefore, only about 12% of the spermatogenetic potential is available for reproduction [148]. Furthermore, daily sperm production in men also declines with age; this is associated with a loss of Sertoli cells, an increase in germ cell degeneration during prophase of meiosis, or loss of primary spermatocytes along with a reduction in the number of Leydig cells, non-Leydig interstitial cells, and myoid cells.

Postspermiation Events

The process of spermiation and the journey of a sperm through the excurrent duct of the testis to a site where it can be included in the ejaculate take an additional 10–14 days. The nucleus progressively elongates as its chromatin condenses; the head is characterized by a flattened and pointed paddle shape, which is specific to each species, and involves the Golgi phase where the centrioles migrate from the cytoplasm to the base of the nucleus and proximal centriole becomes the implantation apparatus to anchor flagellum to the nucleus and distal centriole becomes the axoneme. In the cap phase, the acrosome forms a distinct cap over the nucleus covering about 30–50% of the nuclear surface [149]. The acrosome contains the hydrolytic enzymes necessary for fertilization. The manchette is formed, and the spermatids are embedded in Sertoli cells. During the maturation phase, mitochondria migrate toward the segment of the growing tail to form the mitochondrial sheath and dense outer fibers. A fibrous sheath is formed to complete the assembly of the tail. Most of the spermatid cytoplasm is discarded as a residual body, and the spermatid moves toward the lumen of the seminiferous tubule. Once elongation of the spermatid is complete, Sertoli cell cytoplasm retracts around the developing sperm, and all unnecessary cytoplasm is stripped. The spermatozoon is finally released it into the tubule lumen. The mature spermatozoon is an elaborate, highly specialized cell produced in large numbers – about 300 per gram of testis per second.

Spermatozoa

Spermatozoa are highly specialized and condensed cells that do not grow or divide. A spermatozoon consists of a head containing the paternal material (DNA) and the tail, which provides motility. The spermatozoon is endowed with a large nucleus but lacks a large cytoplasm, which is characteristic of most body cells. The heterogeneity of the ejaculate is a characteristic feature in men [150–152].

Head

The head is oval in shape, measuring about 4.0–5.5 μm in length and 2.5–3.5 μm in width. The normal length-to-width ratio is about 1.50–1.70 [153]. Under bright-field illumination, the most commonly observed aberrations include head shape/size defects (including large, small, tapering, pyriform, amorphous, and vacuolated (>20% of the head surface occupied by unstained vacuolar areas)) and double heads, or any combination thereof [154].

Acrosome

The acrosome is represented by the Golgi complex and covers about two thirds or about 70% of the anterior head area [151, 152]. When observed under the scanning electron microscope, the sperm head is unequally divided into the acrosomal and postacrosomal regions. Under the electron microscope, the sperm head is a flattened ovoid structure consisting primarily of the nucleus. The acrosome contains several hydrolytic enzymes, including hyaluronidase and proacrosin, which are necessary for fertilization [150]. During fertilization of the egg, the fusion of the outer acrosomal membrane with the plasma membrane at multiple sites releases the acrosomal enzymes at the time of acrosome reaction. The anterior half of the head is covered only by the inner acrosomal membrane, while the posterior region of the sperm head is covered by a single membrane called the postnuclear cap. The overlap of the acrosome and the postnuclear cap results in an equatorial segment. The equatorial segment does not participate in the acrosome reaction. The nucleus comprises 65% of the head and is composed of DNA conjugated with protein. The chromatin is tightly packaged, and no distinct chromosomes are visible. The genetic information, including the sex determining X or Y chromosome, is "coded" and stored in the DNA [150].

Neck

This forms a junction between the head and tail. It is fragile, and a common abnormality is the presence of a decapitated spermatozoon.

Tail

The sperm tail arises at the spermatid stage. During spermatogenesis, the centriole is differentiated into midpiece, principal piece, and endpiece. The mitochondria reorganize around the midpiece. An axial core composed of two central fibrils surrounded by a concentric ring of nine double fibrils continues to the end of the tail. An additional outer ring is composed of nine coarse fibrils. The main piece is comprised of 9 coarse outer fibrils that diminish in thickness until only the inner 11 fibrils of the axial core surrounded by a fibrous sheath remain. The mitochondrial sheath of the midpiece is relatively short but slightly longer than the combined length of the head and neck [150].

Endpiece

The endpiece is not distinctly visible by light microscopy. Both the tail sheath and coarse filaments are absent. The tail, which contains all the motility apparatus, is 40–50 μm long and arises from the spermatid centriole. It propels the sperm body via waves generated in the neck region. These waves pass distally along like a whiplash.

Under bright-field illumination, common neck and midpiece defects include bent tails, distended or irregular/bent midpieces, abnormally thin midpieces (no mitochondrial sheath), the absence of the neck or midpiece, or any of these combinations [154]. Tail defects include short, multiple hairpin broken tails, irregular widths, coiled tails with terminal droplets, or a combinations of these defects [154]. Cytoplasmic droplets greater than one third the area of a normal sperm head are considered abnormal. They are usually located in the neck/midpiece region of the tail [152].

Under scanning electron microscopy, the tail can be subdivided into three distinct parts, i.e., midpiece, principal piece, and endpiece. In the midpiece, the mitochondrial spirals can be clearly visualized. The midpiece narrows toward the posterior end. The short endpiece has a small diameter due to the absence of the outer fibers [150]. Under transmission electron microscopy, the midpiece possesses a cytoplasmic portion and a lipid-rich mitochondrial sheath that consists of several spiral mitochondria surrounding the axial filament in a helical fashion. The midpiece provides the sperm with the energy necessary for motility. An additional outer ring of 9 coarser fibrils surrounds the central core of 11 fibrils. Individual mitochondria are wrapped around these fibrils in a spiral manner to form the mitochondrial sheath, which contains the enzymes needed in the oxidative metabolism of the sperm. The mitochondrial sheath of the midpiece is relatively short and slightly longer than the combined length of the head and neck [150].

The principal or mainpiece is the longest part of the tail, and it provides most of the propellant machinery. The coarse nine fibrils of the outer ring diminish in thickness and finally disappear, leaving only the inner fibrils in the axial core for most of the length of the principal piece [155]. The tail terminates in the endpiece with a length of 4–10 μm and a diameter of <1 μm due to the absence of the outer fibrous sheath and distal fading of the microtubules.

Regulation of Spermatogenesis

Both intrinsic and extrinsic regulations influence spermatogenic process.

Intrinsic Regulation

Testosterone, neurotransmitters (neuroendocrine substances), and growth factors are secreted by Leydig cells to neighboring Leydig cells, blood vessels, the lamina propria of the seminiferous tubules and Sertoli cells [12, 148, 156] Leydig cells help maintain the nutrition of the Sertoli cells, and the cells of the peritubular tissue influence the contractility of myofibroblasts and regulate the peristaltic movements of seminiferous tubules and transportation of the spermatozoa. Leydig cells also help regulate blood flow in the intertubular microvasculature [6]. Sertoli cells deliver different growth factors, and various germ cells participate in the development and regulation of germ cells. These factors represent an independent intratesticular regulation of spermatogenesis.

Extrinsic Influences

The hypothalamus and hypophysis control local regulation of spermatogenesis by pulsatile secretion of GnRH and release of LH. Leydig cells produce testosterone, which influences spermatogenesis and provides feedback to the hypophysis, which regulates the secretory activity of Leydig cells. FSH action on the Sertoli cells is necessary for maturation of the germ cells. Both FSH and LH are necessary for complete spermatogenesis. Testicular function is determined by interaction between the endocrine and paracrine mechanisms [157–159]. Sertoli cells secrete inhibin, which functions in the feedback mechanism directed to the hypophysis. Thus, both growth and differentiation of testicular germ cells involve a series of complex interactions between somatic and germinal elements [157–159].

Immune Status of the Testis

The spermatozoa, late pachytene spermatocytes, and spermatids express unique antigens that are not formed until puberty, and therefore, immune tolerance is not developed. The blood–testis barrier develops as these autoantigens develop. The testis is considered to be an immune privileged site, i.e., transplanted foreign tissue can survive for a period of time without immunological rejection. An immune surveillance is present in the testis and the epididymis, which shows an active immunoregulation to prevent autoimmune disease [160, 161].

Disturbances of Spermatogenesis

Disturbances in both proliferation and differentiation of the male germ cells and the intratesticular and extratesticular mechanisms regulating spermatogenesis can occur as a result of environmental influences or as a result of diseases that directly or indirectly affect spermatogenesis [162, 163]. In addition, nutrition, therapeutic drugs, hormones and their metabolites, increased scrotal temperature, toxic substances, and radiation can reduce or completely inhibit spermatogenesis.

Sperm Transport in the Epididymis, Storage, and Capacitation

The epididymis lies along the dorsolateral border of each testis. It comprises the vasa efferentia, which emanates from the rete testis and the epididymal ducts. The primary function of the epididymis is posttesticular maturation and storage of spermatozoa during their passage from the testis to the vas deferens. The epididymal epithelium is androgen-dependent and has both absorptive and secretory functions. The epididymis is divided into three functionally distinct regions: the head, body, and tail, otherwise known as the caput epididymis, corpus epididymis, and cauda epididymis, respectively. Much of the testicular fluid that transports spermatozoa from the seminiferous tubules is reabsorbed in the caput, thereby increasing the concentration of the spermatozoa by 10- to 100-fold. As the newly developed spermatozoa pass through these regions of the epididymis, many changes occur including alterations in net surface charge, membrane protein composition, immunoreactivity, phospholipid and fatty acid content, and adenylate cyclase activity.

Epididymal Sperm Storage

As many as half of the spermatozoa released from the testis die and disintegrate within the epididymis and are reabsorbed by the epididymal epithelium. The remaining mature spermatozoa are stored in the cauda epididymis, and this provides a capacity for repetitive fertile ejaculations. The capacity for sperm storage decreases distally, and the spermatozoa in the vas deferens may only be motile for a few days. After prolonged sexual activity, caudal spermatozoa first lose their fertilizing ability, followed by their motility and then their vitality. They ultimately disintegrate. Older, senescent spermatozoa must be eliminated from the male tract at regular intervals. Otherwise, their relative contribution to the next ejaculate(s) increases, reducing semen quality, even though such ejaculates do have a high sperm concentration. The vas deferens is not a physiological site of sperm storage and contains only about 2% of the total spermatozoa in the male tract. Sperms transit through the fine tubules of the epididymis in approximately 10–15 days in humans.

Sperms mature outside the testis. The spermatozoa within the testis have very limited motility, or none at all, and are incapable of fertilizing an egg. Both epididymal maturation and capacitation are necessary before fertilization. Capacitation – the final step required for fertilization – may be an evolutionary consequence of the development of a storage system for inactive sperm in the caudal epididymis. Preservation of optimal sperm function during this period of storage requires adequate testosterone levels in the circulation.

Sperm Entry into Cervical Mucus

At the moment of ejaculation, spermatozoa from the cauda epididymis are mixed with secretions of the various accessory glands in a specific sequence and deposited around the external cervical os and in the posterior fornix of the vagina. The spermatozoa in the first fraction of the ejaculate have significantly better motility and survival than the later fractions. Most of the spermatozoa penetrate the cervical mucus within 15–20 min of ejaculation [164, 165]. Spermatozoa enter the uterine cavity from the internal "cervical os" by virtue of their own motility [166]. From here, the spermatozoa traverse to the site of fertilization in the ampulla of the Fallopian tube or the oviduct.

Capacitation and Acrosome Reaction

Capacitation is a series of cellular or physiological changes that spermatozoa must undergo in order to fertilize an egg [167, 168]. It is characterized by the ability to undergo the acrosome reaction, bind to the zona pellucida, and acquire hypermotility. Capacitation per se does not involve any morphological changes, even at the ultrastructural level. It does, however, represent a change in the molecular organization of the intact sperm plasmalemma, which gives spermatozoa the ability to undergo the acrosome reaction in response to the induction of the stimulus. During capacitation, the seminal plasma factors that coat the surface of the sperm are removed, and the surface charge is modified along with the sperm membrane, sterols, lipids, and glycoproteins, and the outer acrosomal membrane lying immediately under it. Levels of intracellular free calcium also increase [169, 170].

The acrosome reaction enables sperm to penetrate the zona pellucida and also spurs the fusogenic state in the plasmalemma overlying the nonreactive equatorial segment, which is needed for interaction with the oolemma. The changes termed as "acrosome reaction" prepare the sperm to fuse with the egg membrane. The removal of cholesterol from the surface membrane prepares the sperm membrane for the acrosome reaction [171, 172]. In addition, d-mannose binding lectins are also involved in the binding of human sperm to the zona pellucida [173, 174]. Thus, all these series of changes are necessary to transform the stem cells into fully mature, functional spermatozoa equipped to fertilize an egg (Fig. 2.7).

Conclusion

The testis is an immune privileged site. The blood–testis barrier provides a microenvironment for spermatogenesis to occur. The seminiferous tubules are the site of sperm production. The process of differentiation of a spermatogonium into a spermatid is known as spermatogenesis. It involves both mitotic and meiotic proliferation as well as extensive cell remodeling. In humans, the process of spermatogenesis starts at puberty and continues throughout life. Spermatogenesis produces genetic material necessary for the replication of the species. Meiosis assures genetic diversity. Along the length of the seminiferous tubule, there are only certain cross sections where spermatozoa are released. Sperm production is a continuous and not a pulsatile process. Spermatozoa are highly specialized cells that do not grow or divide. The spermatogenic process is maintained by different intrinsic and extrinsic influences. Spermatozoa have to undergo a series of cellular or physiological changes such as capacitation and acrosome reaction before they can fertilize. The epididymis is limited to a storage role. Nutrition, therapeutic drugs, hormones and their metabolites, increased scrotal temperature, toxic substances, or radiation can reduce or entirely inhibit spermatogenesis.

References

1. Wilson JD. Syndromes of androgen resistance. Biol Reprod. 1992;46:168–73.
2. Lubahn DB, Moyer JS, Golding TS, Couse JF, Korach KS, Smithies O. Alteration of reproductive function but not prenatal sexual development after insertional disruption of the mouse estrogen receptor gene. Proc Natl Acad Sci USA. 1993;90:11162–6.
3. Smith EP, Boyd J, Frank GR, Takahashi H, Cohen RM, Specker B, et al. Estrogen resistance caused by a mutation in the estrogen-receptor gene in a man. N Engl J Med. 1994;331:1056–61.
4. Tishler PV. Diameter of testicles. N Engl J Med. 1971;285:1489.
5. Winter JS, Faiman C. Pituitary-gonadal relations in male children and adolescents. Pediatr Res. 1972; 6:126–35.
6. Middendorff R, Müller D, Mewe M, Mukhopadhyay AK, Holstein AF, Davidoff MS. The tunica albuginea of the human testis is characterized by complex contraction and relaxation activities regulated by cyclic GMP. J Clin Endocrinol Metab. 2002;87: 3486–99.
7. Prader A. Testicular size: assessment and clinical importance. Triangle. 1966;7:240–3.
8. Agger P. Scrotal and testicular temperature: its relation to sperm count before and after operation for varicocele. Fertil Steril. 1971;22:286–97.
9. de Kretser DM, Temple-Smith PD, Kerr JB. Anatomical and functional aspects of the male reproductive organs. In: Bandhauer K, Fricks J, editors. Handbook of urology, vol. XVI. Berlin: Springer; 1982. p. 1–131.
10. Christensen AK. Leydig cells. In: Hamilton DW, Greep RO, editors. Handbook of physiology. Baltimore: Williams and Wilkins; 1975. p. 57–94.
11. Kaler LW, Neaves WB. Attrition of the human Leydig cell population with advancing age. Anat Rec. 1978;192:513–8.
12. DeKretser DM, Kerr JB. The cytology of the testis. In: Knobill E, Neil JD, editors. The physiology of reproduction. New York: Raven; 1994. p. 1177–290.
13. Payne AH, Wong KL, Vega MM. Differential effects of single and repeated administrations of gonadotropins on luteinizing hormone receptors and testosterone synthesis in two populations of Leydig cells. J Biol Chem. 1980;255:7118–22.
14. Glover TD, Barratt CLR, Tyler JJP, Hennessey JF. Human male fertility. London: Academic; 1980. p. 247.
15. Ewing LL, Keeney DS. Leydig cells: structure and function. In: Desjardins C, Ewin LL, editors. Cell and molecular biology of the testis. New York: Oxford University Press; 1993.
16. Davidoff MS, Breucker H, Holstein AF, Seidel K. Cellular architecture of the lamina propria of human tubules. Cell Tissue Res. 1990;262:253–61.
17. Roosen-Runge EC, Holstein A. The human rete testis. Cell Tissue Res. 1978;189:409–33.
18. Russell LD, Griswold MD, editors. The Sertoli cell. Clearwater: Cache Press; 1993.
19. de França LR, Ghosh S, Ye SJ, Russell LD. Surface and surface-to-volume relationships of the Sertoli cell during the cycle of the seminiferous epithelium in the rat. Biol Reprod. 1993;49:1215–28.

20. Behringer RR. The müllerian inhibitor and mammalian sexual development. Philos Trans R Soc Lond B Biol Sci. 1995;350:285–8.
21. Josso N, di Clemente N, Gouédard L. Anti-Müllerian hormone and its receptors. Mol Cell Endocrinol. 2001;179:25–32.
22. Clermont Y. Kinetics of spermatogenesis in mammals: seminiferous epithelium cycle and spermatogonial renewal. Physiol Rev. 1972;52:198–236.
23. Clermont Y. The cycle of the seminiferous epithelium in man. Am J Anat. 1963;112:35–51.
24. Schulze C. Morphological characteristics of the spermatogonial stem cells in man. Cell Tissue Res. 1974;198:191–9.
25. Clermont Y, Bustos-Obregon E. Re-examination of spermatogonial renewal in the rat by means of seminiferous tubules mounted "in toto". Am J Anat. 1968;122:237–47.
26. Huckins C. The spermatogonial stem cell population in adult rats. I. Their morphology, proliferation and maturation. Anat Rec. 1971;169:533–57.
27. Dym M, Fawcett DW. Further observations on the numbers of spermatogonia, spermatocytes, and spermatids connected by intercellular bridges in the mammalian testis. Biol Reprod. 1971;4:195–215.
28. Berezney R, Coffey DS. Nuclear matrix. Isolation and characterization of a framework structure from rat liver nuclei. J Cell Biol. 1977;73:616–37.
29. Mirkovitch J, Mirault ME, Laemmli UK. Organization of the higher-order chromatin loop: specific DNA attachment sites on nuclear scaffold. Cell. 1984;39:223–32.
30. Gasse S. Studies on scaffold attachment sites and their relation to genome function. Int Rev Cytol. 1989;119:57.
31. Izaurralde E, Kas E, Laemmli UK. Highly preferential nucleation of histone H1 assembly on scaffold-associated regions. J Mol Biol. 1989;210:573–85.
32. Adachi Y, Kas E, Laemmli UK. Preferential cooperative binding of DNA topoisomerase II to scaffold-associated regions. EMBO J. 1989;13:3997.
33. Dickinson LA, Joh T, Kohwi Y, Kohwi-Shigematsu T. A tissue-specific MAR/SAR DNA-binding protein with unusual binding site recognition. Cell. 1992;70:631–45.
34. Breucker H, Schäfer E, Holstein AF. Morphogenesis and fate of the residual body in human spermiogenesis. Cell Tissue Res. 1985;240:303–9.
35. Leblond CP, Clermont Y. Definition of the stages of the cycle of the seminiferous epithelium in the rat. Ann N Y Acad Sci. 1952;55:548–73.
36. Clermont Y, Perey B. The stages of the cycle of the seminiferous epithelium of the rat: practical definitions in PA-Schiff-hematoxylin and hematoxylin-eosin stained sections. Rev Can Biol. 1957;16:451–62.
37. Schulze W, Rehder U. Organization and morphogenesis of the human seminiferous epithelium. Cell Tissue Res. 1984;237:395–407.
38. Ward WS, Coffey DS. DNA packaging and organization in mammalian spermatozoa: comparison with somatic cells. Biol Reprod. 1991;44:569–74.
39. McGhee JD, Felsenfeld G, Eisenberg H. Nucleosome structure and conformational changes. Biophys J. 1980;32:261–70.
40. Sassone-Corsi P. Unique chromatin remodeling and transcriptional regulation in spermatogenesis. Science. 2002;296:2176–8.
41. Dadoune JP, Siffroi JP, Alfonsi MF. Transcription in haploid male germ cells. Int Rev Cytol. 2004;237:1–56.
42. Ward WS, Partin AW, Coffey DS. DNA loop domains in mammalian spermatozoa. Chromosoma. 1989;98:153–9.
43. McPherson S, Longo FJ. Chromatin structure-function alterations during mammalian spermatogenesis: DNA nicking and repair in elongating spermatids. Eur J Histochem. 1993;37:109–28.
44. Allen MJ, Lee C, Lee IV JD, Pogany GC, Balooch M, Siekhaus WJ, et al. Atomic force microscopy of mammalian sperm chromatin. Chromosoma. 1993;102:623–30.
45. Lewis JD, Abbott DW, Ausió J. A haploid affair: core histone transitions during spermatogenesis. Biochem Cell Biol. 2003;81:131–40.
46. Lewis JD, Song Y, de Jong ME, Bagha SM, Ausió J. A walk though vertebrate and invertebrate protamines. Chromosoma. 2003;111:473–82.
47. Braun RE. Packaging paternal chromosomes with protamine. Nat Genet. 2001;28:10–2.
48. Wu TF, Chu DS. Sperm chromatin: fertile grounds for proteomic discovery of clinical tools. Mol Cell Proteomics. 2008;7:1876–86.
49. Ooi SL, Henikoff S. Germline histone dynamics and epigenetics. Curr Opin Cell Biol. 2007;19:257–65.
50. Cho C, Willis WD, Goulding EH, Jung-Ha H, Choi YC, Hecht NB, et al. Haploinsufficiency of protamine-1 or -2 causes infertility in mice. Nat Genet. 2001;28:82–6.
51. Yu YE, Zhang Y, Unni E, Shirley CR, Deng JM, Russell LD, et al. Abnormal spermatogenesis and reduced fertility in transition nuclear protein 1-deficient mice. Proc Natl Acad Sci USA. 2000;97:4683–8.
52. Zhao M, Shirley CR, Yu YE, Mohapatra B, Zhang Y, Unni E, et al. Targeted disruption of the transition protein 2 gene affects sperm chromatin structure and reduces fertility in mice. Mol Cell Biol. 2001;21:7243–55.
53. Churikov D, Zalenskaya IA, Zalensky AO. Male germline-specific histones in mouse and man. Cytogenet Genome Res. 2004;105:203–14.
54. Dadoune JP. The nuclear status of human sperm cells. Micron. 1995;26:323–45.
55. Kierszenbaum AL. Transition nuclear proteins during spermiogenesis: unrepaired DNA breaks not allowed. Mol Reprod Dev. 2001;58:357–8.
56. Lee CH, Cho YH. Aspects of mammalian spermatogenesis: electrophoretical analysis of protamines in mammalian species. Mol Cells. 1999;9:556–9.
57. Bench GS, Friz AM, Corzett MH, Morse DH, Balhorn R. DNA and total protamine masses in individual sperm from fertile mammalian subjects. Cytometry. 1996;23:263–71.

58. Gatewood JM, Cook GR, Balhorn R, Bradbury EM, Schmid CW. Sequence-specific packaging of DNA in human sperm chromatin. Science. 1987;236: 962–4.
59. Laberge RM, Boissonneault G. On the nature and origin of DNA strand breaks in elongating spermatids. Biol Reprod. 2005;73:289–96.
60. Marcon L, Boissonneault G. Transient DNA strand breaks during mouse and human spermiogenesis new insights in stage specificity and link to chromatin remodeling. Biol Reprod. 2004;70:910–8.
61. McPherson SM, Longo FJ. Nicking of rat spermatid and spermatozoa DNA: possible involvement of DNA topoisomerase II. Dev Biol. 1993;158: 122–30.
62. Muratori M, Marchiani S, Maggi M, Forti G, Baldi E. Origin and biological significance of DNA fragmentation in human spermatozoa. Front Biosci. 2006;11: 1491–9.
63. Zhao M, Shirley CR, Mounsey S, Meistrich ML. Nucleoprotein transitions during spermiogenesis in mice with transition nuclear protein Tnp1 and Tnp2 mutations. Biol Reprod. 2004;71:1016–25.
64. Kistler WS, Noyes C, Hsu R, Heinrikson RL. The amino acid sequence of a testis-specific basic protein that is associated with spermatogenesis. J Biol Chem. 1975;250:1847–53.
65. Kleene KC, Borzorgzadeh A, Flynn JF, Yelick PC, Hecht NB. Nucleotide sequence of a cDNA clone encoding mouse transition protein 1. Biochim Biophys Acta. 1988;950:215–20.
66. Schlüter G, Celik A, Obata R, Schlicker M, Hofferbert S, Schlung A, et al. Sequence analysis of the conserved protamine gene cluster shows that it contains a fourth expressed gene. Mol Reprod Dev. 1996;43: 1–6.
67. Meistrich ML. Calculation of the incidence of infertility in human populations from sperm measures using the two-distribution model. Prog Clin Biol Res. 1989;302:275–85.
68. Alfonso PJ, Kistler WS. Immunohistochemical localization of spermatid nuclear transition protein 2 in the testes of rats and mice. Biol Reprod. 1993; 48:522–9.
69. Heidaran MA, Showman RM, Kistler WS. A cytochemical study of the transcriptional and translational regulation of nuclear transition protein 1 (TP1), a major chromosomal protein of mammalian spermatids. J Cell Biol. 1988;106:1427–33.
70. Baskaran R, Rao MR. Interaction of spermatid-specific protein TP2 with nucleic acids, in vitro. A comparative study with TP1. J Biol Chem. 1990;265: 21039–47.
71. Lévesque D, Veilleux S, Caron N, Boissonneault G. Architectural DNA-binding properties of the spermatidal transition proteins 1 and 2. Biochem Biophys Res Commun. 1998;252:602–9.
72. Kundu TK, Rao MR. Zinc dependent recognition of a human CpG island sequence by the mammalian spermatidal protein TP2. Biochemistry. 1996;35: 15626–32.
73. Boissonneault G. Chromatin remodeling during spermiogenesis: a possible role for the transition proteins in DNA strand break repair. FEBS Lett. 2002; 514:111–4.
74. Caron N, Veilleux S, Boissonneault G. Stimulation of DNA repair by the spermatidal TP1 protein. Mol Reprod Dev. 2001;58:437–43.
75. Brewer L, Corzett M, Balhorn R. Condensation of DNA by spermatid basic nuclear proteins. J Biol Chem. 2002;277:38895–900.
76. Adham IM, Nayernia K, Burkhardt-Göttges E, Topaloglu O, Dixkens C, Holstein AF, et al. Teratozoospermia in mice lacking the transition protein 2 (Tnp2). Mol Hum Reprod. 2001;7: 513–20.
77. Carrell DT, Liu L. Altered protamine 2 expression is uncommon in donors of known fertility, but common among men with poor fertilizing capacity, and may reflect other abnormalities of spermiogenesis. J Androl. 2001;22:604–10.
78. de Yebra L, Ballescá JL, Vanrell JA, Corzett M, Balhorn R, Oliva R. Detection of P2 precursors in the sperm cells of infertile patients who have reduced protamine P2 levels. Fertil Steril. 1998;69: 755–9.
79. Balhorn R, Corzett M, Mazrimas JA. Formation of intraprotamine disulfides in vitro. Arch Biochem Biophys. 1992;296:384–93.
80. Balhorn R, Cosman M, Thornton K, Krishnan VV, Corzett M, Bench G, et al. Protamine-mediated condensation of DNA in mammalian sperm. In: Gagnon C, editor. The male gamete: from basic science to clinical applications. Vienna: Cache River Press; 1999.
81. Corzett M, Mazrimas J, Balhorn R. Protamine 1: protamine 2 stoichiometry in the sperm of eutherian mammals. Mol Reprod Dev. 2002;61:519–27.
82. Fuentes-Mascorro G, Serrano H, Rosado A. Sperm chromatin. Arch Androl. 2000;45:215–25.
83. Dixon GH, Aiken JM, Jankowski JM, McKenzie D, Moir R, States JC, et al. Organization and evolution of protamine gene of salmoind fishes. In: Reeck GR, Goodwin GH, Puigdomenech P, editors. Chromosomal proteins and gene expression. New York: Plenum; 1986.
84. Krawetz SA, Dixon GH. Sequence similarities of the protamine genes: implications for regulation and evolution. J Mol Evol. 1988;27:291–7.
85. Balhorn R, Brewer L, Corzett M. DNA condensation by protamine and arginine-rich peptides: analysis of toroid stability using single DNA molecules. Mol Reprod Dev. 2000;56:230–4.
86. Courtens JL, Loir M. Ultrastructural detection of basic nucleoproteins: alcoholic phosphotungstic acid does not bind to arginine residues. J Ultrastruct Res. 1981;74:322–6.
87. Loir M, Lanneau M. Structural function of the basic nuclear proteins in ram spermatids. J Ultrastruct Res. 1984;86:262–72.
88. Singh J, Rao MR. Interaction of rat testis protein, TP, with nucleosome core particle. Biochem Int. 1988; 17:701–10.

89. Le Lannic G, Arkhis A, Vendrely E, Chevaillier P, Dadoune JP. Production, characterization, and immunocytochemical applications of monoclonal antibodies to human sperm protamines. Mol Reprod Dev. 1993;36:106–12.
90. Szczygiel MA, Ward WS. Combination of dithiothreitol and detergent treatment of spermatozoa causes paternal chromosomal damage. Biol Reprod. 2002;67:1532–7.
91. Hecht NB. Post-meiotic gene expression during spermatogenesis. Prog Clin Biol Res. 1988;267: 291–313.
92. Hecht NB. Regulation of 'haploid expressed genes' in male germ cells. J Reprod Fertil. 1990;88:679–93.
93. Oliva R, Dixon GH. Vertebrate protamine gene evolution I. Sequence alignments and gene structure. J Mol Evol. 1990;30:333–46.
94. Steger K. Transcriptional and translational regulation of gene expression in haploid spermatids. Anat Embryol (Berl). 1999;199:471–87.
95. Oliva R. Protamines and male infertility. Hum Reprod Update. 2006;12:417–35.
96. Chevaillier P, Mauro N, Feneux D, Jouannet P, David G. Anomalous protein complement of sperm nuclei in some infertile men. Lancet. 1987;2:806–7.
97. Balhorn R, Reed S, Tanphaichitr N. Aberrant protamine 1/protamine 2 ratios in sperm of infertile human males. Experientia. 1988;44:52–5.
98. Aoki VW, Moskovtsev SI, Willis J, Liu L, Mullen JB, Carrell DT. DNA integrity is compromised in protamine-deficient human sperm. J Androl. 2005; 26:741–8.
99. Carrell DT, Emery BR, Hammoud S. Altered protamine expression and diminished spermatogenesis: what is the link? Hum Reprod Update. 2007; 13:313–27.
100. Kosower NS, Katayose H, Yanagimachi R. Thiol-disulfide status and acridine orange fluorescence of mammalian sperm nuclei. J Androl. 1992;13:342–8.
101. Sakkas D, Mariethoz E, Manicardi G, et al. Origin of DNA damage in ejaculated human spermatozoa. Rev Reprod. 1999;4:31–7.
102. Aoki VW, Carrell DT. Human protamines and the developing spermatid: their structure, function, expression and relationship with male infertility. Asian J Androl. 2003;5:315–24.
103. Mengual L, Ballescá JL, Ascaso C, Oliva R. Marked differences in protamine content and P1/P2 ratios in sperm cells from percoll fractions between patients and controls. J Androl. 2003;24:438–47.
104. Steger K, Pauls K, Klonisch T, Franke FE, Bergmann M. Expression of protamine-1 and -2 mRNA during human spermiogenesis. Mol Hum Reprod. 2000; 6:219–25.
105. Rousseaux S, Caron C, Govin J, Lestrat C, Faure AK, Khochbin S. Establishment of male-specific epigenetic information. Gene. 2005;345:139–53.
106. Arpanahi A, Brinkworth M, Iles D, Krawetz SA, Paradowska A, Platts AE, et al. Endonuclease-sensitive regions of human spermatozoal chromatin are highly enriched in promoter and CTCF binding sequences. Genome Res. 2009;19:1338–49.
107. Hammoud SS, Purwar J, Pflueger C, Cairns BR, Carrell DT. Alterations in sperm DNA methylation patterns at imprinted loci in two classes of infertility. Fertil Steril. 2010;94:1728–33.
108. Razin A, Riggs AD. DNA methylation and gene function. Science. 1980;210:604–10.
109. Cedar H. DNA methylation and gene expression. In: Razin A, Cedar H, Riggs AD, editors. DNA methylation: biochemistry and biological significance. New York: Springer; 1985.
110. Sanford JP, Clark HJ, Chapman VM, Rossant J. Differences in DNA methylation during oogenesis and spermatogenesis and their persistence during early embryogenesis in the mouse. Genes Dev. 1987; 1:1039–46.
111. Rahe B, Erickson RP, Quinto M. Methylation of unique sequence DNA during spermatogenesis in mice. Nucleic Acids Res. 1983;11:7947–59.
112. Trasler JM. Epigenetics in spermatogenesis. Mol Cell Endocrinol. 2009;306:33–6.
113. Oakes CC, La Salle S, Smiraglia DJ, Robaire B, Trasler JM. Developmental acquisition of genome-wide DNA methylation occurs prior to meiosis in male germ cells. Dev Biol. 2007;307:368–79.
114. Benchaib M, Braun V, Lornage J, et al. Sperm DNA fragmentation decreases the pregnancy rate in an assisted reproductive technique. Hum Reprod. 2003;18:1023–8.
115. Ward WS. The structure of the sleeping genome: implications of sperm DNA organization for somatic cells. J Cell Biochem. 1994;55:77–82.
116. Risley MS, Einheber S, Bumcrot DA. Changes in DNA topology during spermatogenesis. Chromosoma. 1986;94:217–27.
117. Aitken RJ, De Iuliis GN. On the possible origins of DNA damage in human spermatozoa. Mol Hum Reprod. 2010;16:3–13.
118. Aitken RJ, De Iuliis GN, McLachlan RI. Biological and clinical significance of DNA damage in the male germ line. Int J Androl. 2009;32:46–56.
119. Carrell DT, Emery BR, Hammoud S. The aetiology of sperm protamine abnormalities and their potential impact on the sperm epigenome. Int J Androl. 2008;31:537–45.
120. De Iuliis GN, Thomson LK, Mitchell LA, Finnie JM, Koppers AJ, Hedges A, et al. DNA damage in human spermatozoa is highly correlated with the efficiency of chromatin remodeling and the formation of 8-hydroxy-2′, -deoxyguanosine, a marker of oxidative stress. Biol Reprod. 2009;81: 517–24.
121. Leduc F, Maquennehan V, Nkoma GB, Boissonneault G. DNA damage response during chromatin remodeling in elongating spermatids of mice. Biol Reprod. 2008;78:324–32.
122. Kramer JA, Krawetz SA. Nuclear matrix interactions within the sperm genome. J Biol Chem. 1996; 271:11619–22.

123. Ward WS, Kimura Y, Yanagimachi R. An intact sperm nuclear matrix may be necessary for the mouse paternal genome to participate in embryonic development. Biol Reprod. 1999;60:702–6.
124. Singleton S, Zalensky A, Doncel GF, Morshedi M, Zalenskaya IA. Testis/sperm-specific histone 2B in the sperm of donors and subfertile patients: variability and relation to chromatin packaging. Hum Reprod. 2007;22:743–50.
125. Iranpour FG, Nasr-Esfahani MH, Valojerdi MR, al-Taraihi TM. Chromomycin A3 staining as a useful tool for evaluation of male fertility. J Assist Reprod Genet. 2000;17:60–6.
126. Bizzaro D, Manicardi GC, Bianchi PG, Bianchi U, Mariethoz E, Sakkas D. In-situ competition between protamine and fluorochromes for sperm DNA. Mol Hum Reprod. 1998;4:127–32.
127. Manicardi GC, Bianchi PG, Pantano S, Azzoni P, Bizzaro D, Bianchi U, et al. Presence of endogenous nicks in DNA of ejaculated human spermatozoa and its relationship to chromomycin A3 accessibility. Biol Reprod. 1995;52:864–7.
128. Bianchi PG, Manicardi GC, Bizzaro D, Bianchi U, Sakkas D. Effect of deoxyribonucleic acid protamination on fluorochrome staining and in situ nick-translation of murine and human mature spermatozoa. Biol Reprod. 1993;49:1083–8.
129. Zini A, Gabriel MS, Zhang X. The histone to protamine ratio in human spermatozoa: comparative study of whole and processed semen. Fertil Steril. 2007;87:217–9.
130. Aoki VW, Emery BR, Liu L, Carrell DT. Protamine levels vary between individual sperm cells of infertile human males and correlate with viability and DNA integrity. J Androl. 2006;27:890–8.
131. Carrell DT, De Jonge C, Lamb DJ. The genetics of male infertility: a field of study whose time is now. Arch Androl. 2006;52:269–74.
132. Irvine DS, Twigg JP, Gordon EL, Fulton N, Milne PA, Aitken RJ. DNA integrity in human spermatozoa: relationships with semen quality. J Androl. 2000;21:33–44.
133. Weng SL, Taylor SL, Morshedi M, Schuffner A, Duran EH, Beebe S, et al. Caspase activity and apoptotic markers in ejaculated human sperm. Mol Hum Reprod. 2002;8:984–91.
134. Sinha Hikim AP, Swerdloff RS. Hormonal and genetic control of germ cell apoptosis in the testis. Rev Reprod. 1999;4:38–47.
135. Rodriguez I, Ody C, Araki K, Garcia I, Vassalli P. An early and massive wave of germinal cell apoptosis is required for the development of functional spermatogenesis. EMBO J. 1997;16:2262–70.
136. Hikim AP, Lue Y, Yamamoto CM, Vera Y, Rodriguez S, Yen PH, et al. Key apoptotic pathways for heat-induced programmed germ cell death in the testis. Endocrinology. 2003;144:3167–75.
137. Sakkas D, Seli E, Bizzaro D, Tarozzi N, Manicardi GC. Abnormal spermatozoa in the ejaculate: abortive apoptosis and faulty nuclear remodelling during spermatogenesis. Reprod Biomed Online. 2003;7:428–32.
138. Paul C, Povey JE, Lawrence NJ, Selfridge J, Melton DW, Saunders PT. Deletion of genes implicated in protecting the integrity of male germ cells has differential effects on the incidence of DNA breaks and germ cell loss. PLoS One. 2007;3:e989.
139. Bauché F, Fouchard MH, Jégou B. Antioxidant system in rat testicular cells. FEBS Lett. 1994;349:392–6.
140. Fraga CG, Motchnik PA, Wyrobek AJ, Rempel DM, Ames BN. Smoking and low antioxidant levels increase oxidative damage to sperm DNA. Mutat Res. 1996;351:199–203.
141. Meyer-Ficca ML, Lonchar J, Credidio C, Ihara M, Li Y, Wang ZQ, et al. Disruption of poly(ADP-ribose) homeostasis affects spermiogenesis and sperm chromatin integrity in mice. Biol Reprod. 2009;81:46–55.
142. Aitken RJ, Gordon E, Harkiss D, Twigg JP, Milne P, Jennings Z, et al. Relative impact of oxidative stress on the functional competence and genomic integrity of human spermatozoa. Biol Reprod. 1998;59:1037–46.
143. Piña-Guzmán B, Solís-Heredia MJ, Rojas-García AE, Urióstegui-Acosta M, Quintanilla-Vega B. Genetic damage caused by methyl-parathion in mouse spermatozoa is related to oxidative stress. Toxicol Appl Pharmacol. 2006;216:216–24.
144. Zubkova EV, Robaire B. Effects of ageing on spermatozoal chromatin and its sensitivity to in vivo and in vitro oxidative challenge in the Brown Norway rat. Hum Reprod. 2006;11:2901–10.
145. Heller C, Clermont Y. Kinetics of the germinal epithelium in man. Recent Prog Horm Res. 1964;20:545–75.
146. Sculze W, Salzbrunn A. Spatial and quantitative aspects of spermatogenetic tissue in primates. In: Neischlag E, Habenicht U, editors. Spermatogenesis-fertilization-contraception. Berlin: Springer; 1992. p. 267–83.
147. Rowe PJ, Comhaire F, Hargreave TB, Mellows HJ, editors. WHO manual for the standardized investigation and diagnosis of the infertile couple. Cambridge: Cambridge University Press; 1993.
148. Sharpe RM. Regulation of spermatogenesis. In: Knobill E, Neil JD, editors. The physiology of reproduction. New York: Raven; 1994. p. 1363–434.
149. De Kretser DM. Ultrastructural features of human spermiogenesis. Z Zellforsch Mikrosk Anat. 1969;98:477–505.
150. Hafez ES. The human semen and fertility regulation in the male. J Reprod Med. 1976;16:91–6.
151. Kruger TF, Menkveld R, Stander FS, Lombard CJ, Van der Merwe JP, van Zyl JA, et al. Sperm morphologic features as a prognostic factor in in vitro fertilization. Fertil Steril. 1986;46:1118–23.
152. Menkveld R, Stander FS, Kotze TJ, Kruger TF, van Zyl JA. The evaluation of morphological characteristics

of human spermatozoa according to stricter criteria. Hum Reprod. 1990;5:586–92.
153. Katz DF, Overstreet JW, Samuels SJ, Niswander PW, Bloom TD, Lewis EL. Morphometric analysis of spermatozoa in the assessment of human male fertility. J Androl. 1986;7:203–10.
154. World Health Organization. World Health Organization laboratory manual for the examination of human semen and sperm-cervical mucus interaction. 4th ed. Cambridge: Cambridge University Press; 1999.
155. White IG. Mammalian sperm. In: Hafez ESE, editor. Reproduction of farm animals. Philadelphia: Lea & Febiger; 1974.
156. Jegou B. The Sertoli cell. Baillières Clin Endocrinol Metab. 1992;6:273–311.
157. Bellve AR, Zheng W. Growth factors as autocrine and paracrine modulators of male gonadal functions. J Reprod Fertil. 1989;85:771–93.
158. Sharpe T. Intratesticular control of steroidogenesis. Clin Endocrinol. 1990;33:787–807.
159. Sharpe RM. Monitoring of spermatogenesis in man-measurement of Sertoli cell- or germ cell-secreted proteins in semen or blood. Int J Androl. 1992;15: 201–10.
160. Mahi-Brown CA, Yule TD, Tung KS. Evidence for active immunological regulation in prevention of testicular autoimmune disease independent of the blood-testis barrier. Am J Reprod Immunol Microbiol. 1988;16:165–70.
161. Barratt CL, Bolton AE, Cooke ID. Functional significance of white blood cells in the male and female reproductive tract. Hum Reprod. 1990;5:639–48.
162. Holstein AF, Schulze W, Breucker H. Histopathology of human testicular and epididymal tissue. In: Hargreave TB, editor. Male infertility. London: Springer; 1994. p. 105–48.
163. Nieschlag E, Behre H. Andrology. Male reproductive health and dysfunction. Berlin: Springer; 2001.
164. Tredway DR, Settlage DS, Nakamura RM, Motoshima M, Umezaki CU, Mishell Jr DR. Significance of timing for the postcoital evaluation of cervical mucus. Am J Obstet Gynecol. 1975;121:387–93.
165. Tredway DR, Buchanan GC, Drake TS. Comparison of the fractional postcoital test and semen analysis. Am J Obstet Gynecol. 1978;130:647–52.
166. Settlage DSF, Motoshima M, Tredway DR. Sperm transport from the external cervical os to the fallopian tubes in women: a time and quantitation study. In: Hafez ESE, Thibault CG, editors. Sperm transport, survival and fertilizing ability in vertebrates, vol. 26. Paris: INSERM; 1974. p. 201–17.
167. Eddy EM, O'Brien DA. The spermatozoon. In: Knobill EO, NO'Nneill JD, editors. The physiology of reproduction. New York: Raven; 1994.
168. Yanagamachi R. Mammalian fertilization. In: Knobill E, O'Brien NJ, editors. The physiology of reproduction. New York: Raven; 1994.
169. Mahanes MS, Ochs DL, Eng LA. Cell calcium of ejaculated rabbit spermatozoa before and following in vitro capacitation. Biochem Biophys Res Commun. 1986;134:664–70.
170. Thomas P, Meizel S. Phosphatidylinositol 4,5--bisphosphate hydrolysis in human sperm stimulated with follicular fluid or progesterone is dependent upon Ca^{2+} influx. Biochem J. 1989;264:539–46.
171. Parks JE, Ehrenwalt E. Cholesterol efflux from mammalian sperm and its potential role in capacitation. In: Bavister BD, Cummins J, Raldon E, editors. Fertilization in mammals. Norwell: Serono Symposia; 1990.
172. Ravnik SE, Zarutskie PW, Muller CH. Purification and characterization of a human follicular fluid lipid transfer protein that stimulates human sperm capacitation. Biol Reprod. 1992;47:1126–33.
173. Benoff S, Cooper GW, Hurley I, Mandel FS, Rosenfeld DL. Antisperm antibody binding to human sperm inhibits capacitation induced changes in the levels of plasma membrane sterols. Am J Reprod Immunol. 1993;30:113–30.
174. Benoff S, Hurley I, Cooper GW, Mandel FS, Hershlag A, Scholl GM, et al. Fertilization potential in vitro is correlated with head-specific mannose-ligand receptor expression, acrosome status and membrane cholesterol content. Hum Reprod. 1993; 8:2155–66.

Sperm Nucleoproteins

3

Rafael Oliva and Judit Castillo

Abstract

Protamines are the most abundant nuclear proteins packaging the mammalian male genome in the sperm nucleus. The main proposed functions of these proteins are condensation and streamlining of the sperm cell, protection of the genetic message, and contribution to the epigenetic organisation of the paternal genome. Different studies have demonstrated the presence of an altered expression of protamines in some infertile patients and a link to decreased DNA integrity. However, in addition to protamines, about 5–15% of the male sperm genome is also complexed with histones and histone variants. Furthermore, present proteomic approaches based on mass spectrometry are also identifying additional chromatin-associated proteins. Of importance, it has been demonstrated that there is a differential distribution of genes into the histone- and protamine-associated regions of the sperm nucleus, which supports the idea of an epigenetic marking of the sperm nucleoprotein, with potential relevance in early embryonic development.

Keywords

Spermatozoa • Proteome • Protamine • Epigenetic • Imprinting

Protamines: The Major Components of the Sperm Nucleus

Protamines were discovered and named by Friedrich Miescher more than a century ago [1]. Miescher identified a nitrogenous base from the sperm of salmon that he called protamine and found that this base was coupled to what he called nuclein, which later was to become known as DNA [1, 2]. Subsequent studies established

R. Oliva (✉)
Human Genetics Research Group, IDIBAPS, Faculty of Medicine, University of Barcelona, Barcelona, Spain

Biochemistry and Molecular Genetics Service, Hospital Clínic i Provincial, Barcelona, Spain
e-mail: roliva@ub.edu

A. Zini and A. Agarwal (eds.), *Sperm Chromatin: Biological and Clinical Applications in Male Infertility and Assisted Reproduction*, DOI 10.1007/978-1-4419-6857-9_3,

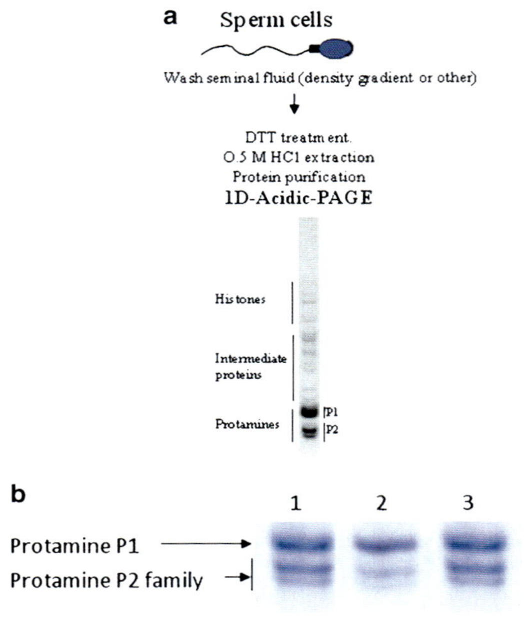

Fig. 3.1 Extraction and electrophoretic separation of protamines from human sperm. (**a**) A typical extraction of protamines from sperm cells involves reduction of the disulphide bonds of the protamines using DTT, followed by 0.5 M HCl extraction, protein precipitation and purification and separation using acidic polyacrylamide gel electrophoresis. Two major groups of bands can be visualised corresponding to the protamine 1(P1) and to the family of protamine 2 proteins (P2). (**b**) Protamines from three independent infertile patients. A reduction in protamine 2 in relation to protamine 1 can be observed in patient 2

the polypeptide nature of the protamines [3–6]. Protamines are the most abundant sperm nuclear proteins in many species and are involved in packaging the paternal genome [6–12].

A typical extraction of human protamines from mature sperm cells and its separation using electrophoresis in an acidic gel and visualisation using Coomassie blue staining is shown in Fig. 3.1. The most intense protein bands that can be visualised are the protamines (Fig. 3.1). The two major bands correspond, respectively, to the two types of protamines known to be present in mammals: the P1 protamine and the family of P2 proteins (Fig. 3.1). The content of protamine P1 in the human sperm nucleus is similar to the content of protamine P2 (P1–P2 ratio of approximately 1) [12–21]. The P1 protamine is present in all species of mammals studied [6, 22–27]. The protamine P2 is formed by the P2, P3 and P4 components, and it is only present in some mammalian species including human and mouse [6, 22–25, 27–30]. The genes encoding both protamines are closely linked in the genome and are subject to coordinate expression [31–36]. Another difference between the two protamines is that the protamine P1 is synthesised as a mature protein, whereas the components of the P2 family are generated by proteolysis from a precursor encoded by a single gene [23, 26, 37–41]. The components of the P2 family (P2, P3 and P4) differ only by the N-terminal extension of one to four residues, although the P2 component is the most abundant [6, 22, 23, 27, 30, 38, 42–44] (Fig. 3.2).

One of the most important characteristics of protamines is the high content of positively charged amino acids and specially arginine (48% in human protamines; Fig. 3.2). Indeed, protamines are proteins that have evolved to increase the number of positively charged residues in evolution, allowing the formation of a highly condensed complex with the paternal genomic DNA that has a strong negative charge [6, 40, 45–48]. In addition to a high arginine content, the protamines of different species also incorporate cysteines in their sequence, allowing the formation of disulphide bonds between adjacent protamine molecules, therefore strongly stabilising the nucleoprotamine complex [48–51] (Fig. 3.2). Related to the disulphide bonds and chromatin stabilisation it is also important the content of zinc and the formation of zinc bridges [52]. It is clear that the presence of protamines in the sperm nucleus results in a more compact nucleoprotamine structure. However, the question of the function of this higher compact structure remains unsolved. Several hypotheses have been proposed [6]:

1. Generation of a condensed paternal genome with a more compact and hydrodynamic nucleus.
2. Protection of the paternal genetic message delivered by the spermatozoa by making it inaccessible to nucleases or mutagens potentially present in the internal or external media.
3. Competition and removal of transcription factors and other proteins from the spermatid, resulting in a blank paternal genetic message devoid of epigenetic information, therefore allowing its reprogramming by the oocyte.
4. Involvement in the imprinting of the paternal genome during spermatogenesis. Also,

a

Protamine 1

P1 ARYRCCRSQSRSRYYRQRQRSRRRRRRSCQTRRRAMRCCRPRYRPRCRRH
** * ** *

Protamine 2 family

P2 RTHGQSHYRRRHCSRRRLHRIHRRQHRSCRRRKRRSCRHRRRHRRGCRTRKRTCRRH

P3 GQSHYRRRHCSRRRLHRIHRRQHRSCRRRKRRSCRHRRRHRRGCRTRKRTCRRH

P4 ERTHGQSHYRRRHCSRRRLHRIHRRQHRSCRRRKRRSCRHRRRHRRGCRTRKRTCRRH
* * * * *

b

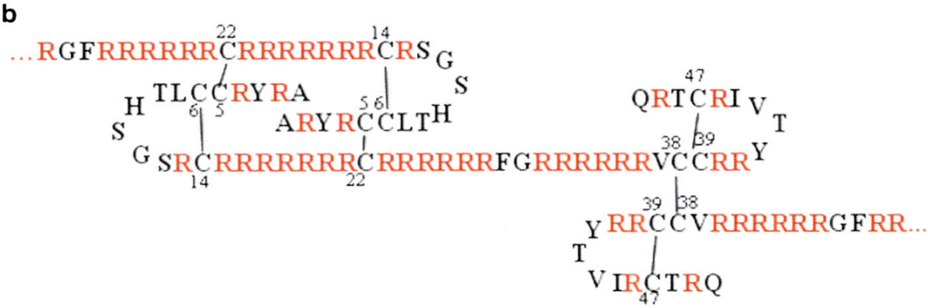

Fig. 3.2 Protamine amino-acid sequences. (**a**) Human protamine amino acid sequences. The positively charged arginine residues are shown in *red*. Note the high content in arginine and their distribution in clusters. The cysteines are indicated by an *asterisk*. Cysteines can form disulphide bonds intramolecularly or intermolecularly with other adjacent protamine molecules. (**b**) A model for the cross-linking of the cysteine residues of bull protamine P1 based on the in vivo and in vitro mapping of cross-linked cysteines [50, 51, 113]. Note that the situation in human is likely to the more complex than in bull. The position of some of the cysteines is changed in the human P1 sequences as compared to the bull P1 sequence. In addition, bull contains only protamine P1 in the sperm nucleus, whereas human contains approximately equal portions of protamine P1 and P2. Since protamine P2 also contains cysteines, which can also form disulphide bonds with other P1 and P2 molecules, the model shown here for the bull protamine P1 sequence may not be completely applicable to the human sperm

protamines themselves could confer an epigenetic mark on some regions of the sperm genome, affecting its reactivation upon fertilisation.

Another aspect that characterises the protamines is that they are among the proteins with one of the highest rates of evolutionary variation [6, 47, 48, 53, 54]. It has been proposed that one cause for this rapid evolution rate could be a positive Darwinian selection [55–57]. This proposal is supported by the observation, when comparing the sequence of protamines from different species, that the relation between non-synonymous substitutions (the nucleotide changes resulting in a change of amino acid) per residue and the synonymous substitutions per residue is superior to 1, and also that the protamine exons evolve faster than the protamine intron [55, 57]. However, a closer examination revealed an unusual form of purifying selection where the overall number of arginine residues is maintained at about 50% in mammals, but the total number of amino acids and the positions of arginine residues have changed considerably [58]. Concerning the origin of the protamines, the evidence indicates that they may have evolved from histone H1 ancestors [53, 54].

A critical issue in understanding the function of the protamines is to understand what has been the driving force that has directed its evolution. In addition to the DNA-binding function of protamines resulting in a more compact sperm nucleus, it has also been proposed a function in the oocyte through the interaction and strong activation of egg creatine kinase II by protamine [55, 59]. While the evolution of protamines is providing important clues towards understanding the function of protamines this aspect is not further covered in depth here, so the reader is also referred to other reviews and articles [6, 34, 40, 45–48, 53, 54, 56, 57, 60, 61].

The Nucleohistone to Nucleoprotamine Transition

Protamines are incorporated into the sperm cell at the final stages of spermatogenesis where the nucleosomal structure is progressively disassembled and replaced first by transition proteins and finally by protamines [6, 11, 35, 39, 62–69] (Fig. 3.3). This transition is preceded by extremely marked changes in many chromatin activities [6, 11, 41, 66, 68, 70–75]. One of the initial chromatin changes is the incorporation of histone variants [75–81]. Another important early event is histone hyperacetylation, which occurs during spermiogenesis prior to the nucleosome disassembly in vivo [72, 82–86] (Fig. 3.3). It was initially postulated that histone hyperacetylation and rapid turnover of acetyl groups could rapidly and reversibly expose binding sites in chromatin for subsequent binding of chromosomal proteins [72]. Subsequently, it was shown in vitro that histone hyperacetylation facilitated nucleosome disassembly and histone displacement by protamines [32, 87, 88]. However, in addition to the neutralisation of the positively charged lysine residues of the histone tails, histone acetylation has an even more powerful and specific role acting as highly specific marks (histone code) that determine the condensation state of the chromatin, binding of other proteins and chromatin

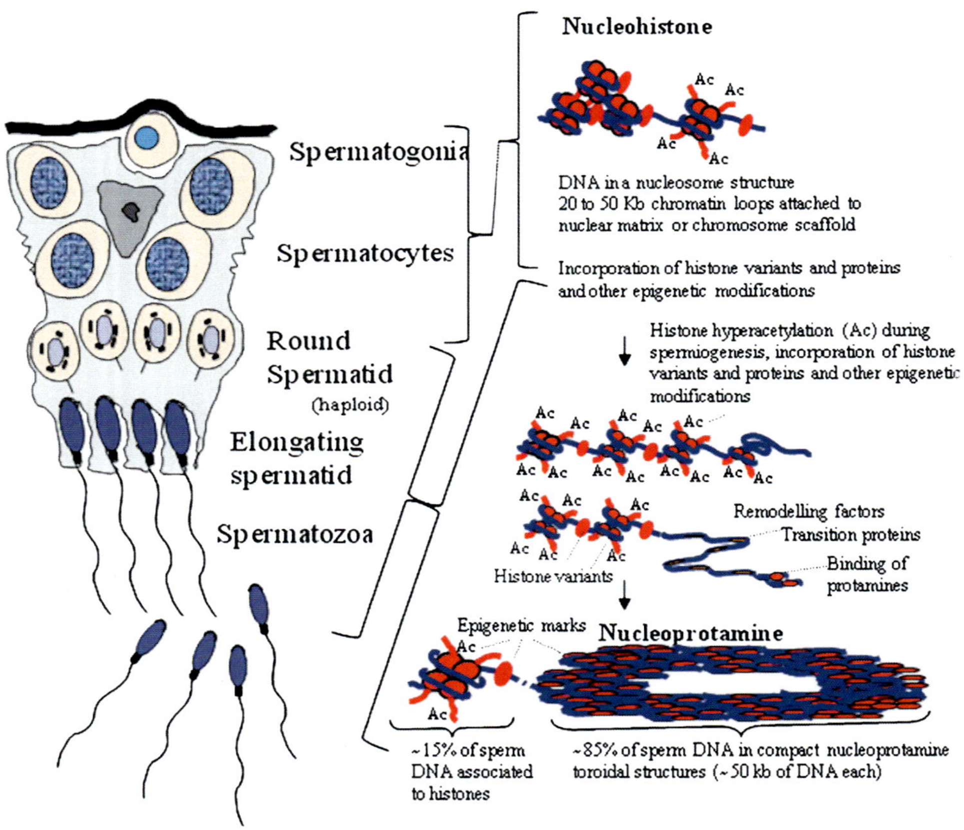

Fig. 3.3 Schematic representation of the major cellular and chromatin changes occurring during spermatogenesis. The *left side* of the figure represent a section of a spermatogenic tubule indicating the location of spermatogonia, spermatocytes and spermatids, and the liberation of spermatozoa to the tubular lumen. The *right side* of this figure represent the basic chromatin changes taking place during the nucleohistone to nucleoprotamine transition in spermiogenesis. The cellular changes in the *left side* of this figure are intended to correspond roughly to the chromatin structures and activities indicated in the *right side*. Histones are represented in *red* colour and DNA is drawn as *blue lines*. The indicated histone retention in approximately 5–15% of the sperm DNA corresponds to the situation in humans

remodelers and associated chromatin activities [89–94]. More recently, it has been demonstrated that the testis-specific bromodomain containing protein (BRDT) binds to hyperacetylated histone 4 (H4) triggering a reorganisation of the chromatin [95]. Impaired histone H4 hyperacetylation has been detected in infertile patients [96, 97]. In addition to histone acetylation, other types of chromatin modifications are also important for the correct nucleohistone to nucleoprotamine transition [11, 98–105].

Concomitant to nucleosome disassembly, the sperm DNA is extensively complexed with transition proteins [67, 106]. Transition proteins are then finally replaced by protamines to form a highly compact nucleoprotamine complex (Fig. 3.3). It is known that protamines are phosphorylated before binding to DNA and that a substantial dephosphorylation takes place concomitant to nucleoprotamine maturation [6, 107–109]. The dynamics of binding of the protamines to DNA has also been studied [110–112]. After binding to DNA ,the formation of inter-disulphide bonds between protamines further stabilises the nucleoprotamine complex [50, 51, 113] (Figs. 3.2 and 3.3). Different models for the structure of the nucleoprotamine have been proposed [51, 113–120]. A proposed model for the protamine P1 cross-linking in the bull sperm is shown in Fig. 3.2b. It has been demonstrated, using atomic force microscopy in vivo and in vitro, that the compact nucleoprotamine is formed by the presence of toroidal structures [50, 113]. Scale measurements indicate that each of the toroidal nucleoprotein structures would contain approximately 50 kb of highly packaged DNA [50, 113, 114, 120] (Fig. 3.3).

Organisation of the DNA in the Mature Sperm Nucleus

While the majority of the human sperm genome (about 85–95%) is tightly packaged by protamines into toroidal structures [6, 120] (Fig. 3.3), it is also important to take into account that about 5–15% of the sperm DNA is organised by histones, many of which are sperm-specific variants [121–123] (Fig. 3.3). The toroidal structures of the nucleoprotamine contain each about 50 kb of DNA, and it has been proposed that they could be attached through their linker region DNA to the sperm nuclear matrix [120, 124]. There is extensive evidence that the distribution of genes in the genomic regions organised by protamine and in the genomic regions organised by histones is not random [6, 121, 125–132]. This organisation of the sperm genes into the nucleoprotamine and nucleohistone compartments has been recently further demonstrated by two independent groups with the application of microarrays and deep genome sequencing technologies, respectively [131, 132].

In the first report, the authors used two different strategies to fractionate sperm human and the mouse sperm chromatin into the histone and protamine regions [131]. One of these strategies was based on the differential extraction of the histones using 0.65 M NaCl and subsequent digestion of the “free DNA” with a combination of BamHI and EcoRI to liberate the histone associated genomic domains following previously described procedures [126, 129]. The other strategy was based on the differential digestion of the nucleosome-associated regions of the sperm nucleus using micrococcal nuclease also following the previously described methods [127]. With the different chromatin fractions, the authors then use human and mouse whole chromosome microarray CGH to determine the differential distribution of genes. The basic conclusion of this work is that the regions of increased endonuclease sensitivity are closely associated with gene regulatory regions and that a similar differential packaging was observed in both mouse and man, implying the existence of epigenetic marks distinguishing gene regulatory regions in male germ with a potential role for subsequent embryonic development [131].

In the second study [132], the authors also used the differential digestion of the nucleosome-associated regions of the sperm nucleus using micrococcal nuclease [127]. The fractionated chromatin was then analysed by deep genome sequencing using the Illumina GAII sequencing. In this study, DNA methylation and the differential distribution of sequences were also investigated. The basic conclusions of this work are that retained sperm nucleosome-associated regions are significantly enriched at loci of developmental importance, including imprinted genes, microRNAs, HOX genes and the promoters of developmental transcription and signalling factors [132]. In addition, they demonstrated that histone modifications (H3K4me2, H3K27me3) localise to particular developmental loci and that developmental promoters are generally DNA hypomethylated in sperm, but acquire methylation during differentiation. Altogether, the results were interpreted in the sense that epigenetic marking in sperm is extensive and correlated with developmental regulators [132, 133].

In addition to these potential epigenetic marks encoded by the differential distribution of the genes in the histone- and protamine-associated nucleoprotein domains, there are other types of epigenetic information potentially transmitted by the sperm nuclei. One of these is the well-known and contrasted DNA methylation imprints set during gametogenesis [134]. More recently, the identification of sperm RNAs and the demonstration of their transfer to the ova have recently provided substrate for the potential involvement of the sperm RNAs upon fertilisation [135]. Of high potential importance, another source of epigenetic information can be the presence of other proteins in addition of histones and protamines.

One of the initial indications of sperm proteins crucial for embryo development was the finding that in humans and most mammals (with the exception of mouse) the centrosome is paternally inherited (see [136] for a review). It has also been demonstrated that sperm-derived histone variants contribute to zygotic chromatin in humans [137]. Thus, epigenetic processes implemented during spermatogenesis distinguish the paternal pronucleus in the embryo [138, 139]. There is also some evidence that alterations in some of the proteins present in the spermatozoa may be related to subsequent embryo development. This evidence has come from the observation that topoisomerase II-mediated breaks in spermatozoa cause the specific degradation of paternal DNA in fertilised oocytes [140]. Proteasomal proteins have also been detected in sperm cells [141, 142]. An important role for sperm proteasomes in zygotic development has recently been suggested based on the observation that the release of a functional sperm centriole that acts as a zygote microtubule-organising center relies on selective proteasomal proteolysis [143].

More recently, the analysis of the proteins identified in the different mature sperm proteomic projects has provided some unexpected results. For example, many transcription factors, DNA binding proteins and proteins involved in chromatin metabolism have been identified [141, 142, 144–147]. The catalogues corresponding to the sperm proteomes from human [141, 142, 144], bull [148], mouse [149] and rat [150] are now available. The presence of proteins such as histone acetyltransferase and deacetylase, histone methyltransferase, DNA methyltransferase, topoisomerase, helicase, transcription factor, zinc finger, leucine zipper, homoeobox proteins, chromodomain, centrosomal proteins and telomerase in cells that are transcriptionally inert and that have at least 85% of its DNA tightly packaged with protamine is remarkable [151, 152]. The proportion of these proteins identified most likely represent an underestimation since they have been identified in whole sperm proteomic analysis [141, 142, 144, 148–150]. A crucial issue is whether these newly identified transcription factors and nuclear proteins represent leftovers from the spermatogenic process or instead they are marking some regions of the male genome and have an epigenetic function [35, 151–153].

Protamine Anomalies in the Sperm Cells of Infertile Patients

Anomalies in the protamine content in infertile patients were already described more than 20 years ago [12, 154–156]. Subsequently, studies further confirmed the association of abnormal protamine content with abnormal seminal parameters and male infertility [14, 15, 21, 35, 157–165]. The type of protamine anomalies identified indicated a reduction in protamines relative to other proteins and an alteration of the P1–P2 ratio [35]. A clue to one of the potential causes of the abnormal P1–P2 ratio found in some infertile patients was found with the identification of an abnormal processing of protamine 2 and increase in protamine precursors in a subset of infertile patients [16, 17]. The reduction in protamine content in patients was consistent with the results of the analysis of the phosphorus and sulphur contents in individual spermatozoa by particle-induced X-ray emission (PIXE [17]). Thus, a potential cause for abnormal protamine P2 content in some infertile patients can be the presence of abnormal protamine P2 processing. However, it should be noted that small amounts of detectable levels of P2 precursors are also present in the normal mature sperm nucleus in human, mouse and rat [161, 166, 167]. An important question is whether the anomalies in protamine content found in some infertile patients are uniformly present in the different sperms in a sample or instead there are subpopulations within a single ejaculate different in protamine content. This subject has been studied both by measuring the effect of gradient centrifugation of spermatozoa on protamine content and through immunocytochemistry, indicating some degree of protamine heterogeneity within the cells of single samples [20, 168, 169].

In addition to the above studies in infertile patients, the expression of protamines has also been determined in response to thermal stress in normal testicles [170, 171]. Thermal stress in stallion testicle is associated to decreased formation of disulphide bridges in protamines [170]. This aspect has also been studied in a patient who just finished an episode of influenza detecting the appearance of protamine P2 precursors and a raise in the ratio of histones to protamines between 33 and 39 days post hyperthermia [171]. The expression of the gene corresponding to the protamine P2 also has been found altered concomitant to induced thermal stress in the mouse testicle [172]. It is also interesting to note that variation over time of protein and DNA contents in sperm from an infertile human male possessing protamine defects has been described [17].

Indirect detection methods to tentatively asses the amount of protamines or measuring chromatin structure based on different staining procedures or fluorochromes have also been used. For example, in-situ competition between protamine and chromomycin A3 (CMA3) indicated that CMA3 staining is inversely correlated with the protamination state of spermatozoa [173]. The CMA3 test has also been correlated to the extent of nicked DNA [174]. In the evaluation of the CMA3 staining sperm cells that bright yellow are CMA3-positive cells and those with dull yellow stain are CMA3-negative cells [175]. Interestingly, CMA3 staining has been shown to be increased in the sperm cells of infertile patients [176–181].

Another indirect approach to investigate the status of the sperm chromatin has been the use of aniline blue staining to detect the presence of histones and, therefore, indirectly infer the presence of lower amounts of protamines in the sperm nucleus [155, 182]. An increase in the percentage of aniline blue cells was found in asthenozoospermic samples as compared to normozoospermic ones [182]. Acidic aniline blue was also correlated with differences in sperm nuclear morphology in sperm donors and in infertile patients [183]. A decreased resistance to chromatin decondensation by treatment with sodium dodecyl sulphate (SDS) and dithiotreitol (DTT) in abnormal sperm as compared to normal sperm has also been taken as evidence for lower protamine S–S stability and chromatin packaging [184–186]. The accessibility additional fluorescent dyes to DNA have also been used as indirect methods to detect aberrant protamination [187, 188].

In addition to the protamine content, the disulphide bonds cross-linking status between cysteines has also been studied in infertile patients [52, 189–194]. There is many data indicating that the sperm protein thiols are oxidised upon passage from caput to the cauda epididymis [189]. When comparing the thiol labelling patterns, oligospermic or infertile samples were found to have higher SH content (less disulphide bonds) as compared to the normozoospermic ones [189–191]. After thiol-specific fluorochrome monobromobimane (mBBr)-flow cytometry, spermatozoa from subfertile patients with oligoasthenoteratozoospermia (the OAT syndrome) were characterised by a biphasic distribution reflecting both over oxidation and incomplete thiol oxidation and possibly a reduced protamine content [193]. Animal models also support a correlation between disulphide bond formation and integrity of the DNA [195–198].

As indicated previously, one of the hypotheses of the function of protamines is that they could be involved in the protection of the genetic message delivered by the spermatozoa [6, 20, 35]. Incomplete protamination could render the spermatozoa more vulnerable to attack by endogenous or exogenous agents such as nucleases [199, 200], free radicals [201, 202] or mutagens. Therefore, this issue has been assessed by different groups using a variety of direct or indirect approaches. Extensive evidence links decreased DNA integrity with poor reproductive outcomes [203–209]. A negative significant correlation between fertilisation rate and CMA3 staining has been reported [179]. Comet parameters also correlate with embryo cleavage score and with CMA3 staining, suggesting that DNA fragmentation is more frequent in protamine-deficient spermatozoa [181, 210–213]. A quite good direct proof that DNA integrity is compromised in protamine-deficient human sperm has been obtained by direct measuring protamines by electrophoresis [214, 215].

The correlation between protamines and integrity of the DNA has been further studied in vitro and in animal models. In vitro protamine-induced DNA compaction has been shown to result in radioprotection against double-strand breaks [216]. Using transgenic knockout mice for transition proteins it has been demonstrated that sperm fertility declines during epididymal passage, as revealed by ICSI, while genomic integrity deteriorates [198]. This loss of genomic integrity during passage from the caput to the cauda epididymis in these mice has been related to abnormalities in the protection of the DNA by protamines [37, 195, 198]. Furthermore, in these mice, the developmental defects appeared at implantation similarly as it has been described in clinical reports from infertile patients with decreased DNA integrity [198, 217]. Also, protamine P2 deficiency in mice has been shown to lead to sperm DNA damage and embryo death [218]. In humans, the use of ICSI with testicular sperm has demonstrated to improve pregnancy rates in patients with poor pregnancy rates and decreased DNA integrity of ejaculated spermatozoa [219]. Thus, a reasonable explanation could be that incomplete or abnormal protamination, as it has been observed in many studies, could lead to incomplete disulphide bond formation and incomplete DNA protection during epididymal passage in these patients.

All the above observations have led to the proposal of a two-step hypothesis for the generation of damaged DNA [220–223]. Abnormal protamination of the sperm cell, set during abnormal spermatogenesis, would leave the sperm genome more prone to be damaged by oxidative stress. Subsequently, free radicals would result in the attack of the sperm DNA, resulting in DNA damage. This hypothesis would explain the correlations detected between abnormal protamine content through gel electrophoresis [14, 20, 161, 224, 225] or indirectly with CMA3 staining [226, 227] and decreased DNA integrity [214].

If protamine alterations are present in infertile patients and are also associated with abnormalities in the DNA integrity, it is obvious to also consider whether protamines are related to the assisted reproduction outcome. One of the initial observations linking protamines and in vitro fertilisation capacity came from the observation of a limited number of patients with an altered P1–P2 ratio with a reduced fertilisation index [228]. Radical differences in protamine content in two siblings associated to different ICSI outcomes were also reported [229]. Also, a reduction in protamine P2 and the sperm penetration assay was reported [230]. More recently, it has been described that spermatozoa staining with CMA3, which indirectly indicates a possible deficiency in protamines, have a percentage of in vitro fertilisation of 36.8%, which is below the index reached (64,6%) with the negative spermatozoa after using this dye [180]. Subsequent work using this approach demonstrated the presence of increased DNA fragmentation in presumably protamine-deficient spermatozoa [181]. This group also measured directly the protamines

P1 and P2 by gel electrophoresis and found a negative significant correlation of the fertilisation rate with the protamine deficiency and the P1–P2 ratio [180, 181].

The expression of the genes encoding protamines P1 and P2 in testicular spermatids of azoospermic patients biopsied and treated by ICSI has also been studied [231, 232]. Using this approach a lower expression of the mRNA corresponding to the protamine P1 gene in couples that did not reach a pregnancy was found in comparison with the couples that reached a pregnancy. At the protein level it has been reported that a reduction of the P1–P2 ratio results in a marked reduction of the in vitro fertilisation index in comparison with the patients with a normal or an increased P1–P2 ratio [21]. Furthermore, the sperm P1–P2 ratios are related to in vitro fertilisation pregnancy rates and predictive of fertilisation ability [215]. These observations have been confirmed in independent laboratories including ours [225]. In this study, a significant decrease in fertilisation rate in the low P1–P2 group of patients was detected when using IVF, but not when using ICSI. But even in the ICSI group, a subsequent reduction in the pregnancy rate was detected [225]. Perhaps this result could be related to the findings of a series of in vitro fertilisation experiments using spermatozoa injured with dithiothreitol (DTT), where the binding and penetration of the oocytes in the hamster assay was markedly reduced, except if ICSI was used, where the DTT injured spermatozoa reach an even higher rate of pronuclear formation and decondensation of the sperm head of the spermatozoa [233, 234]. It is interesting to note that most of the above studies considered only the P1–P2 ratio, but this ratio provides limited information. For instance, it does not indicate whether the abnormal ratio is due to a change in P1, in P2 or in both. It does not provide information either on the distribution of the protamines along the genome. Thus, it will be interesting in future studies to consider also the protamine to DNA ratio and the distribution of the protamines related to the assisted reproduction results.

In addition to protamine alterations and the DNA integrity, it is also important to consider other protein abnormalities present in infertile patients. In fact, if protamine alterations are a consequence of abnormal spermiogenesis, then a concomitant presence of other abnormalities can also be expected. Thus, increased transition proteins have been detected in the sperm cells of some infertile men [235]. Also, abnormalities in histone retention have been described [14, 163, 169].

Of interest, recent proteomic approaches have led to the detection of additional altered proteins in infertile patients [141, 142]. Thus, it will be interesting in the future to further identify which of the additional proteins being identified through proteomics are related to the reproductive outcomes.

Acknowledgements Supported by grants from the Ministerio de Educación y Ciencia (BFU2009-07118), fondos FEDER to RO.

References

1. Miescher F. Das Protamin – Eine neue organische Basis aus denSamenf – den des Rheinlachses. Ber Dtsch Chem Ges. 1874;7:376.
2. Dahm R. Friedrich Miescher and the discovery of DNA. Dev Biol. 2005;278(2):274–88.
3. Kossel A. The protamines and histones. London: Longmans Green; 1928.
4. Felix K. Protamines. Adv Protein Chem. 1960;15:1–56.
5. Dixon GH, Smith M. Nucleic acids and protamine in salmon testes. Prog Nucleic Acid Res Mol Biol. 1968;8:9–34.
6. Oliva R, Dixon GH. Vertebrate protamine genes and the histone-to-protamine replacement reaction. Prog Nucleic Acid Res Mol Biol. 1991;40:25–94.
7. Bloch DP. A catalog of sperm histones. Genetics. 1969;61(1):Suppl:93-111.
8. Ando T, Yamasaki M, Suzuki K. Protamines. Isolation, characterization, structure and function. Mol Biol Biochem Biophys. 1973;12:1–114.
9. Calvin HI. Comparative analysis of the nuclear basic proteins in rat, human, guinea pig, mouse and rabbit spermatozoa. Biochim Biophys Acta. 1976;434(2):377–89.
10. Subirana JA. Proceedings of the forth international symposium of spermatogly. In: Andre J, editor. The sperm cell. The Netherlands: Martinus Nijhoff Pub; 1983. p. 197–213.
11. Mezquita C. Chromatin composition, structure and function in spermatogenesis. Revis Biol Celular. 1985;5:V–XIV, 1–124.
12. Balhorn R, Reed S, Tanphaichitr N. Aberrant protamine 1/protamine 2 ratios in sperm of infertile human males. Experientia. 1988;44(1):52–5.

13. Ammer H, Henschen A, Lee CH. Isolation and amino-acid sequence analysis of human sperm protamines P1 and P2. Occurrence of two forms of protamine P2. Biol Chem Hoppe Seyler. 1986;367(6):515–22.
14. de Yebra L, Ballesca JL, Vanrell JA, Bassas L, Oliva R. Complete selective absence of protamine P2 in humans. J Biol Chem. 1993;268(14):10553–7.
15. de Yebra L, Oliva R. Rapid analysis of mammalian sperm nuclear proteins. Anal Biochem. 1993;209(1):201–3.
16. de Yebra L, Ballesca JL, Vanrell JA, Corzett M, Balhorn R, Oliva R. Detection of P2 precursors in the sperm cells of infertile patients who have reduced protamine P2 levels. Fertil Steril. 1998;69(4):755–9.
17. Bench G, Corzett MH, De Yebra L, Oliva R, Balhorn R. Protein and DNA contents in sperm from an infertile human male possessing protamine defects that vary over time. Mol Reprod Dev. 1998;50(3):345–53.
18. Corzett M, Mazrimas J, Balhorn R. Protamine 1: protamine 2 stoichiometry in the sperm of eutherian mammals. Mol Reprod Dev. 2002;61(4):519–27.
19. Aoki VW, Carrell DT. Human protamines and the developing spermatid: their structure, function, expression and relationship with male infertility. Asian J Androl. 2003;5(4):315–24.
20. Mengual L, Ballesca JL, Ascaso C, Oliva R. Marked differences in protamine content and P1/P2 ratios in sperm cells from percoll fractions between patients and controls. J Androl. 2003;24(3):438–47.
21. Aoki VW, Liu L, Carrell DT. Identification and evaluation of a novel sperm protamine abnormality in a population of infertile males. Hum Reprod. 2005;20(5):1298–306.
22. Gusse M, Sautiere P, Belaiche D, et al. Purification and characterization of nuclear basic proteins of human sperm. Biochim Biophys Acta. 1986;884(1):124–34.
23. McKay DJ, Renaux BS, Dixon GH. Human sperm protamines. Amino-acid sequences of two forms of protamine P2. Eur J Biochem. 1986;156(1):5–8.
24. Balhorn R, Corzett M, Mazrimas J, Stanker LH, Wyrobek A. High-performance liquid chromatographic separation and partial characterization of human protamines 1, 2, and 3. Biotechnol Appl Biochem. 1987;9(1):82–8.
25. Bellve AR, McKay DJ, Renaux BS, Dixon GH. Purification and characterization of mouse protamines P1 and P2. Amino acid sequence of P2. Biochemistry. 1988;27(8):2890–7.
26. Chauviere M, Martinage A, Debarle M, Sautiere P, Chevaillier P. Molecular characterization of six intermediate proteins in the processing of mouse protamine P2 precursor. Eur J Biochem. 1992;204(2):759–65.
27. Yoshii T, Kuji N, Komatsu S, et al. Fine resolution of human sperm nucleoproteins by two-dimensional electrophoresis. Mol Hum Reprod. 2005;11(9):677–81.
28. Belaiche D, Loir M, Kruggle W, Sautiere P. Isolation and characterization of two protamines St1 and St2 from stallion spermatozoa, and amino-acid sequence of the major protamine St1. Biochim Biophys Acta. 1987;913(2):145–9.
29. Bower PA, Yelick PC, Hecht NB. Both P1 and P2 protamine genes are expressed in mouse, hamster, and rat. Biol Reprod. 1987;37(2):479–88.
30. Bianchi F, Rousseaux-Prevost R, Sautiere P, Rousseaux J. P2 protamines from human sperm are zinc -finger proteins with one CYS2/HIS2 motif. Biochem Biophys Res Commun. 1992;182(2):540–7.
31. Oliva R, Mezquita J, Mezquita C, Dixon GH. Haploid expression of the rooster protamine mRNA in the postmeiotic stages of spermatogenesis. Dev Biol. 1988;125(2):332–40.
32. Oliva R, Bazett-Jones DP, Locklear L, Dixon GH. Histone hyperacetylation can induce unfolding of the nucleosome core particle. Nucleic Acids Res. 1990;18(9):2739–47.
33. Queralt R, de Fabregues-Boixar O, Adroer R, et al. Direct sequencing of the human protamine P1 gene and application in forensic medicine. J Forensic Sci. 1993;38(6):1491–501.
34. Queralt R, Oliva R. Identification of conserved potential regulatory sequences of the protamine-encoding P1 genes from ten different mammals. Gene. 1993;133(2):197–204.
35. Oliva R. Protamines and male infertility. Hum Reprod Update. 2006;12(4):417–35.
36. Martins RP, Krawetz SA. Nuclear organization of the protamine locus. Soc Reprod Fertil Suppl. 2007;64:1–12.
37. Yelick PC, Balhorn R, Johnson PA, et al. Mouse protamine 2 is synthesized as a precursor whereas mouse protamine 1 is not. Mol Cell Biol. 1987;7(6):2173–9.
38. Sautiere P, Martinage A, Belaiche D, Arkhis A, Chevaillier P. Comparison of the amino acid sequences of human protamines HP2 and HP3 and of intermediate basic nuclear proteins HPS1 and HPS2. Structural evidence that HPS1 and HPS2 are pro-protamines. J Biol Chem. 1988;263(23):11059–62.
39. Green GR, Balhorn R, Poccia DL, Hecht NB. Synthesis and processing of mammalian protamines and transition proteins. Mol Reprod Dev. 1994;37(3):255–63. doi:10.1002/mrd.1080370303.
40. Queralt R, Adroer R, Oliva R, Winkfein RJ, Retief JD, Dixon GH. Evolution of protamine P1 genes in mammals. J Mol Evol. 1995;40(6):601–7.
41. Wouters-Tyrou D, Martinage A, Chevaillier P, Sautiere P. Nuclear basic proteins in spermiogenesis. Biochimie. 1998;80(2):117–28.
42. Martinage A, Arkhis A, Alimi E, Sautiere P, Chevaillier P. Molecular characterization of nuclear basic protein HPI1, a putative precursor of human sperm protamines HP2 and HP3. Eur J Biochem. 1990;191(2):449–51.
43. Arkhis A, Martinage A, Sautiere P, Chevaillier P. Molecular structure of human protamine P4 (HP4), a minor basic protein of human sperm nuclei. Eur J Biochem. 1991;200(2):387–92.

44. Alimi E, Martinage A, Arkhis A, Belaiche D, Sautiere P, Chevaillier P. Amino acid sequence of the human intermediate basic protein 2 (HPI2) from sperm nuclei. Structural relationship with protamine P2. Eur J Biochem. 1993;214(2):445–50.
45. Oliva R, Dixon GH. Vertebrate protamine gene evolution I. Sequence alignments and gene structure. J Mol Evol. 1990;30(4):333–46.
46. Retief JD, Winkfein RJ, Dixon GH, et al. Evolution of protamine P1 genes in primates. J Mol Evol. 1993; 37(4):426–34.
47. Oliva R. Sequence, evolution and transcriptional regulation of mammalian P1 type protamines. In: Jamieson BGM, editor. Advances in spermatozoal phylogeny and taxonomy. Paris: Museum National d'Histoire Naturelle; 1995.
48. Lewis JD, Song Y, de Jong ME, Bagha SM, Ausio J. A walk though vertebrate and invertebrate protamines. Chromosoma. 2003;111(8):473–82.
49. Saowaros W, Panyim S. The formation of disulfide bonds in human protamines during sperm maturation. Experientia. 1979;35(2):191–2.
50. Balhorn R, Corzett M, Mazrimas JA. Formation of intraprotamine disulfides in vitro. Arch Biochem Biophys. 1992;296(2):384–93.
51. Vilfan ID, Conwell CC, Hud NV. Formation of native-like mammalian sperm cell chromatin with folded bull protamine. J Biol Chem. 2004;279(19):20088–95.
52. Bjorndahl L, Kvist U. Human sperm chromatin stabilization: a proposed model including zinc bridges. Mol Hum Reprod. 2010;16(1):23–9.
53. Eirin-Lopez JM, Frehlick LJ, Ausio J. Protamines, in the footsteps of linker histone evolution. J Biol Chem. 2006;281(1):1–4.
54. Eirin-Lopez JM, Ausio J. Origin and evolution of chromosomal sperm proteins. Bioessays. 2009;31(10): 1062–70.
55. Rooney AP, Zhang J. Rapid evolution of a primate sperm protein: relaxation of functional constraint or positive Darwinian selection? Mol Biol Evol. 1999; 16(5):706–10.
56. Clark AG, Civetta A. Evolutionary biology. Protamine wars. Nature. 2000;403(6767):261, 263.
57. Wyckoff GJ, Wang W, Wu CI. Rapid evolution of male reproductive genes in the descent of man. Nature. 2000;403(6767):304–9.
58. Rooney AP, Zhang J, Nei M. An unusual form of purifying selection in a sperm protein. Mol Biol Evol. 2000;17(2):278–83.
59. Ohtsuki K, Nishikawa Y, Saito H, Munakata H, Kato T. DNA-binding sperm proteins with oligo-arginine clusters function as potent activators for egg CK-II. FEBS Lett. 1996;378(2):115–20.
60. Torgerson DG, Kulathinal RJ, Singh RS. Mammalian sperm proteins are rapidly evolving: evidence of positive selection in functionally diverse genes. Mol Biol Evol. 2002;19(11):1973–80.
61. Lewis JD, Ausio J. Protamine-like proteins: evidence for a novel chromatin structure. Biochem Cell Biol. 2002;80(3):353–61.
62. Poccia D. Remodeling of nucleoproteins during gametogenesis, fertilization, and early development. Int Rev Cytol. 1986;105:1–65.
63. Hecht NB. Gene expression during male germ cell development. In: Desjardins C, Ewing Ll, editors. Cell and molecular biology of the testis. New York: Oxford University Press; 1993.
64. Grootegoed JA, Siep M, Baarends WM. Molecular and cellular mechanisms in spermatogenesis. Baillières Best Pract Res Clin Endocrinol Metab. 2000;14(3):331–43.
65. Sassone-Corsi P. Unique chromatin remodeling and transcriptional regulation in spermatogenesis. Science. 2002;296(5576):2176–8.
66. Dadoune JP. Expression of mammalian spermatozoal nucleoproteins. Microsc Res Tech. 2003;61(1):56–75.
67. Meistrich ML, Mohapatra B, Shirley CR, Zhao M. Roles of transition nuclear proteins in spermiogenesis. Chromosoma. 2003;111(8):483–8.
68. Kierszenbaum AL, Tres LL. The acrosome-acroplaxome-manchette complex and the shaping of the spermatid head. Arch Histol Cytol. 2004;67(4):271–84.
69. Rousseaux S, Caron C, Govin J, Lestrat C, Faure AK, Khochbin S. Establishment of male-specific epigenetic information. Gene. 2005;345(2):139–53.
70. Puwaravutipanich T, Panyim S. The nuclear basic proteins of human testes and ejaculated spermatozoa. Exp Cell Res. 1975;90(1):153–8.
71. Oliva R, Vidal S, Mezquita C. Cellular content and biosynthesis of polyamines during rooster spermatogenesis. Biochem J. 1982;208(2):269–73.
72. Oliva R, Mezquita C. Histone H4 hyperacetylation and rapid turnover of its acetyl groups in transcriptionally inactive rooster testis spermatids. Nucleic Acids Res. 1982;10(24):8049–59.
73. Fuentes-Mascorro G, Serrano H, Rosado A. Sperm chromatin. Arch Androl. 2000;45(3):215–25.
74. Braun RE. Packaging paternal chromosomes with protamine. Nat Genet. 2001;28(1):10–2.
75. Govin J, Caron C, Lestrat C, Rousseaux S, Khochbin S. The role of histones in chromatin remodelling during mammalian spermiogenesis. Eur J Biochem. 2004;271(17):3459–69.
76. Prigent Y, Muller S, Dadoune JP. Immunoelectron microscopical distribution of histones H2B and H3 and protamines during human spermiogenesis. Mol Hum Reprod. 1996;2(12):929–35.
77. Prigent Y, Troalen F, Dadoune JP. Immunoelectron microscopic visualization of intermediate basic proteins HPI1 and HPI2 in human spermatids and spermatozoa. Reprod Nutr Dev. 1998;38(4):417–27.
78. Churikov D, Zalenskaya IA, Zalensky AO. Male germline-specific histones in mouse and man. Cytogenet Genome Res. 2004;105(2–4):203–14.
79. Tanaka H, Iguchi N, Isotani A, et al. HANP1/H1T2, a novel histone H1-like protein involved in nuclear formation and sperm fertility. Mol Cell Biol. 2005;25(16):7107–19.
80. Loppin B, Bonnefoy E, Anselme C, Laurencon A, Karr TL, Couble P. The histone H3.3 chaperone HIRA

is essential for chromatin assembly in the male pronucleus. Nature. 2005;437(7063):1386–90.
81. Ishibashi T, Li A, Eirín-López JM, Zhao M, Missiaen K, Abbott DW, et al. H2A.Bbd: an X-chromosome-encoded histone involved in mammalian spermiogenesis. Nucleic Acids Res. 2010;8(6):1780–9.
82. Candido EP, Dixon GH. Trout testis cells. 3. Acetylation of histones in different cell types from developing trout testis. J Biol Chem. 1972;247(17):5506–10.
83. Grimes Jr SR, Henderson N. Hyperacetylation of histone H4 in rat testis spermatids. Exp Cell Res. 1984;152(1):91–7.
84. Meistrich ML, Trostle-Weige PK, Lin R, Bhatnagar YM, Allis CD. Highly acetylated H4 is associated with histone displacement in rat spermatids. Mol Reprod Dev. 1992;31(3):170–81.
85. Hazzouri M, Rousseaux S, Mongelard F, et al. Genome organization in the human sperm nucleus studied by FISH and confocal microscopy. Mol Reprod Dev. 2000;55(3):307–15.
86. Marcon L, Boissonneault G. Transient DNA strand breaks during mouse and human spermiogenesis new insights in stage specificity and link to chromatin remodeling. Biol Reprod. 2004;70(4):910–8.
87. Oliva R, Mezquita C. Marked differences in the ability of distinct protamines to disassemble nucleosomal core particles in vitro. Biochemistry. 1986;25(21):6508–11.
88. Oliva R, Bazett-Jones D, Mezquita C, Dixon GH. Factors affecting nucleosome disassembly by protamines in vitro. Histone hyperacetylation and chromatin structure, time dependence, and the size of the sperm nuclear proteins. J Biol Chem. 1987;262(35):17016–25.
89. Cheung P, Allis CD, Sassone-Corsi P. Signaling to chromatin through histone modifications. Cell. 2000;103(2):263–71.
90. Strahl BD, Allis CD. The language of covalent histone modifications. Nature. 2000;403(6765):41–5.
91. Agalioti T, Chen G, Thanos D. Deciphering the transcriptional histone acetylation code for a human gene. Cell. 2002;111(3):381–92.
92. Peterson CL, Laniel MA. Histones and histone modifications. Curr Biol. 2004;14(14):R546–51.
93. Kimmins S, Sassone-Corsi P. Chromatin remodelling and epigenetic features of germ cells. Nature. 2005;434(7033):583–9.
94. Munshi A, Shafi G, Aliya N, Jyothy A. Histone modifications dictate specific biological readouts. J Genet Genomics. 2009;36(2):75–88.
95. Pivot-Pajot C, Caron C, Govin J, Vion A, Rousseaux S, Khochbin S. Acetylation-dependent chromatin reorganization by BRDT, a testis-specific bromodomain-containing protein. Mol Cell Biol. 2003;23(15):5354–65.
96. Sonnack V, Failing K, Bergmann M, Steger K. Expression of hyperacetylated histone H4 during normal and impaired human spermatogenesis. Andrologia. 2002;34(6):384–90.
97. Faure AK, Pivot-Pajot C, Kerjean A, et al. Misregulation of histone acetylation in Sertoli cell-only syndrome and testicular cancer. Mol Hum Reprod. 2003;9(12):757–63.
98. Chiva M, Mezquita C. Quantitative changes of high mobility group non-histone chromosomal proteins HMG1 and HMG2 during rooster spermatogenesis. FEBS Lett. 1983;162(2):324–8.
99. Corominas M, Mezquita C. Poly(ADP-ribosylation) at successive stages of rooster spermatogenesis. Levels of polymeric ADP-ribose in vivo and poly(ADP-ribose) polymerase activity and turnover of ADP-ribosyl residues in vitro. J Biol Chem. 1985;260(30):16269–73.
100. Agell N, Mezquita C. Cellular content of ubiquitin and formation of ubiquitin conjugates during chicken spermatogenesis. Biochem J. 1988;250(3):883–9.
101. Roca J, Mezquita C. DNA topoisomerase II activity in nonreplicating, transcriptionally inactive, chicken late spermatids. EMBO J. 1989;8(6):1855–60.
102. Meyer-Ficca ML, Lonchar J, Credidio C, et al. Disruption of poly(ADP-ribose) homeostasis affects spermiogenesis and sperm chromatin integrity in mice. Biol Reprod. 2009;81(1):46–55.
103. Liu Z, Zhou S, Liao L, Chen X, Meistrich M, Xu J. Jmjd1a demethylase-regulated histone modification is essential for cAMP-response element modulator-regulated gene expression and spermatogenesis. J Biol Chem. 2010;285(4):2758–70.
104. Lu LY, Wu J, Ye L, Gavrilina GB, Saunders TL, Yu X. RNF8-dependent histone modifications regulate nucleosome removal during spermatogenesis. Dev Cell. 2010;18(3):371–84.
105. Okada Y, Tateishi K, Zhang Y. Histone demethylase JHDM2A is involved in male infertility and obesity. J Androl. 2010;31(1):75–8.
106. Kierszenbaum AL. Transition nuclear proteins during spermiogenesis: unrepaired DNA breaks not allowed. Mol Reprod Dev. 2001;58(4):357–8.
107. Ingles CJ, Dixon GH. Phosphorylation of protamine during spermatogenesis in trout testis. Proc Natl Acad Sci USA. 1967;58(3):1011–8.
108. Marushige Y, Marushige K. Phosphorylation of sperm histone during spermiogenesis in mammals. Biochim Biophys Acta. 1978;518(3):440–9.
109. Papoutsopoulou S, Nikolakaki E, Chalepakis G, Kruft V, Chevaillier P, Giannakouros T. SR protein-specific kinase 1 is highly expressed in testis and phosphorylates protamine 1. Nucleic Acids Res. 1999;27(14):2972–80.
110. Prieto MC, Maki AH, Balhorn R. Analysis of DNA-protamine interactions by optical detection of magnetic resonance. Biochemistry. 1997;36(39):11944–51.
111. Brewer L, Corzett M, Lau EY, Balhorn R. Dynamics of protamine 1 binding to single DNA molecules. J Biol Chem. 2003;278(43):42403–8.
112. Brewer LR, Corzett M, Balhorn R. Protamine-induced condensation and decondensation of the same DNA molecule. Science. 1999;286(5437):120–3.

113. Balhorn R, Corzett M, Mazrimas J, Watkins B. Identification of bull protamine disulfides. Biochemistry. 1991;30(1):175–81.
114. Balhorn R. A model for the structure of chromatin in mammalian sperm. J Cell Biol. 1982;93(2): 298–305.
115. Hud NV, Allen MJ, Downing KH, Lee J, Balhorn R. Identification of the elemental packing unit of DNA in mammalian sperm cells by atomic force microscopy. Biochem Biophys Res Commun. 1993;193(3): 1347–54.
116. Allen MJ, Lee C, Lee IV JD, et al. Atomic force microscopy of mammalian sperm chromatin. Chromosoma. 1993;102(9):623–30.
117. Allen MJ, Bradbury EM, Balhorn R. AFM analysis of DNA-protamine complexes bound to mica. Nucleic Acids Res. 1997;25(11):2221–6.
118. Raukas E, Mikelsaar RH. Are there molecules of nucleoprotamine? Bioessays. 1999;21(5):440–8.
119. Biegeleisen K. The probable structure of the protamine-DNA complex. J Theor Biol. 2006;241(3): 533–40.
120. Balhorn R. The protamine family of sperm nuclear proteins. Genome Biol. 2007;8(9):227.
121. Gatewood JM, Cook GR, Balhorn R, Bradbury EM, Schmid CW. Sequence-specific packaging of DNA in human sperm chromatin. Science. 1987;236(4804): 962–4.
122. Zalensky AO, Siino JS, Gineitis AA, et al. Human testis/sperm-specific histone H2B (hTSH2B). Molecular cloning and characterization. Molecular cloning and characterization. J Biol Chem. 2002;277(45):43474–80.
123. Singleton S, Zalensky A, Doncel GF, Morshedi M, Zalenskaya IA. Testis/sperm-specific histone 2B in the sperm of donors and subfertile patients: variability and relation to chromatin packaging. Hum Reprod. 2007;22(3):743–50.
124. Shaman JA, Yamauchi Y, Ward WS. Function of the sperm nuclear matrix. Arch Androl. 2007;53(3): 135–40.
125. Zalensky AO, Allen MJ, Kobayashi A, Zalenskaya IA, Balhorn R, Bradbury EM. Well-defined genome architecture in the human sperm nucleus. Chromosoma. 1995;103(9):577–90.
126. Gardiner-Garden M, Ballesteros M, Gordon M, Tam PP. Histone- and protamine-DNA association: conservation of different patterns within the beta-globin domain in human sperm. Mol Cell Biol. 1998;18(6): 3350–6.
127. Zalenskaya IA, Bradbury EM, Zalensky AO. Chromatin structure of telomere domain in human sperm. Biochem Biophys Res Commun. 2000;279(1): 213–8.
128. Zalenskaya IA, Zalensky AO. Telomeres in mammalian male germline cells. Int Rev Cytol. 2002;218: 37–67.
129. Wykes SM, Krawetz SA. The structural organization of sperm chromatin. J Biol Chem. 2003;278(32): 29471–7.
130. Li Y, Lalancette C, Miller D, Krawetz SA. Characterization of nucleohistone and nucleoprotamine components in the mature human sperm nucleus. Asian J Androl. 2008;10(4):535–41.
131. Arpanahi A, Brinkworth M, Iles D, et al. Endonuclease-sensitive regions of human spermatozoal chromatin are highly enriched in promoter and CTCF binding sequences. Genome Res. 2009;19(8): 1338–49.
132. Hammoud SS, Nix DA, Zhang H, Purwar J, Carrell DT, Cairns BR. Distinctive chromatin in human sperm packages genes for embryo development. Nature. 2009;460(7254):473–8.
133. Miller D, Brinkworth M, Iles D. Paternal DNA packaging in spermatozoa: more than the sum of its parts? DNA, histones, protamines and epigenetics. Reproduction. 2010;139(2):287–301.
134. Reik W, Dean W, Walter J. Epigenetic reprogramming in mammalian development. Science. 2001;293(5532):1089–93.
135. Ostermeier GC, Miller D, Huntriss JD, Diamond MP, Krawetz SA. Reproductive biology: delivering spermatozoan RNA to the oocyte. Nature. 2004; 429(6988):154.
136. Chatzimeletiou K, Morrison EE, Prapas N, Prapas Y, Handyside AH. The centrosome and early embryogenesis: clinical insights. Reprod Biomed Online. 2008;16(4):485–91.
137. van der Heijden GW, Ramos L, Baart EB, et al. Sperm-derived histones contribute to zygotic chromatin in humans. BMC Dev Biol. 2008;8:34.
138. Biermann K, Steger K. Epigenetics in male germ cells. J Androl. 2007;28(4):466–80.
139. Wu TF, Chu DS. Sperm chromatin: fertile grounds for proteomic discovery of clinical tools. Mol Cell Proteomics. 2008;7(10):1876–86.
140. Yamauchi Y, Shaman JA, Ward WS. Topoisomerase II-mediated breaks in spermatozoa cause the specific degradation of paternal DNA in fertilized oocytes. Biol Reprod. 2007;76(4):666–72.
141. Martinez-Heredia J, Estanyol JM, Ballesca JL, Oliva R. Proteomic identification of human sperm proteins. Proteomics. 2006;6(15):4356–69.
142. de Mateo S, Martinez-Heredia J, Estanyol JM, et al. Marked correlations in protein expression identified by proteomic analysis of human spermatozoa. Proteomics. 2007;7(23):4264–77.
143. Rawe VY, Diaz ES, Abdelmassih R, et al. The role of sperm proteasomes during sperm aster formation and early zygote development: implications for fertilization failure in humans. Hum Reprod. 2008;23(3):573–80.
144. Baker MA, Reeves G, Hetherington L, Muller J, Baur I, Aitken RJ. Identification of gene products present in Triton X-100 soluble and insoluble fractions of human spermatozoa lysates using LC-MS/MS analysis. Proteomics Clin Appl. 2007;1: 524–32.
145. Codrington AM, Hales BF, Robaire B. Exposure of male rats to cyclophosphamide alters the chromatin

structure and basic proteome in spermatozoa. Hum Reprod. 2007;22(5):1431–42.
146. Lefievre L, Chen Y, Conner SJ, et al. Human spermatozoa contain multiple targets for protein S-nitrosylation: an alternative mechanism of the modulation of sperm function by nitric oxide? Proteomics. 2007;7(17):3066–84.
147. Martinez-Heredia J, de Mateo S, Vidal-Taboada JM, Ballesca JL, Oliva R. Identification of proteomic differences in asthenozoospermic sperm samples. Hum Reprod. 2009;23(4):783–91.
148. Peddinti D, Nanduri B, Kaya A, Feugang JM, Burgess SC, Memili E. Comprehensive proteomic analysis of bovine spermatozoa of varying fertility rates and identification of biomarkers associated with fertility. BMC Syst Biol. 2008;2:19.
149. Baker MA, Hetherington L, Reeves GM, Aitken RJ. The mouse sperm proteome characterized via IPG strip prefractionation and LC-MS/MS identification. Proteomics. 2008;8(8):1720–30.
150. Baker MA, Hetherington L, Reeves G, Muller J, Aitken RJ. The rat sperm proteome characterized via IPG strip prefractionation and LC-MS/MS identification. Proteomics. 2008;8(11):2312–21.
151. Oliva R, Martinez-Heredia J, Estanyol JM. Proteomics in the study of the sperm cell composition, differentiation and function. Syst Biol Reprod Med. 2008;54(1):23–36.
152. Oliva R, de Mateo S, Estanyol JM. Sperm cell proteomics. Proteomics. 2009;9(4):1004–17.
153. Rousseaux S, Reynoird N, Escoffier E, Thevenon J, Caron C, Khochbin S. Epigenetic reprogramming of the male genome during gametogenesis and in the zygote. Reprod Biomed Online. 2008;16(4): 492–503.
154. Silvestroni L, Frajese G, Fabrizio M. Histones instead of protamines in terminal germ cells of infertile, oligospermic men. Fertil Steril. 1976;27(12): 1428–37.
155. Chevaillier P, Mauro N, Feneux D, Jouannet P, David G. Anomalous protein complement of sperm nuclei in some infertile men. Lancet. 1987;2(8562):806–7.
156. Lescoat D, Colleu D, Boujard D, Le Lannou D. Electrophoretic characteristics of nuclear proteins from human spermatozoa. Arch Androl. 1988;20(1): 35–40.
157. Bach O, Glander HJ, Scholz G, Schwarz J. Electrophoretic patterns of spermatozoal nucleoproteins (NP) in fertile men and infertility patients and comparison with NP of somatic cells. Andrologia. 1990;22(3):217–24.
158. Blanchard Y, Lescoat D, Le Lannou D. Anomalous distribution of nuclear basic proteins in round-headed human spermatozoa. Andrologia. 1990;22(6): 549–55.
159. Belokopytova IA, Kostyleva EI, Tomilin AN, Vorob'ev VI. Human male infertility may be due to a decrease of the protamine P2 content in sperm chromatin. Mol Reprod Dev. 1993;34(1):53–7.
160. Chen S, Cao J, Fei RR, Mao QZ, Li HZ. Analysis of protamine content in patients with asthenozoospermia. Zhonghua Nan Ke Xue. 2005; 11(8):587–9. 593.
161. Torregrosa N, Dominguez-Fandos D, Camejo MI, et al. Protamine 2 precursors, protamine 1/protamine 2 ratio, DNA integrity and other sperm parameters in infertile patients. Hum Reprod. 2006;21(8):2084–9.
162. Carrell DT, Emery BR, Hammoud S. Altered protamine expression and diminished spermatogenesis: what is the link? Hum Reprod Update. 2007;13(3):313–27.
163. Zini A, Gabriel MS, Zhang X. The histone to protamine ratio in human spermatozoa: comparative study of whole and processed semen. Fertil Steril. 2007;87(1):217–9.
164. Carrell DT. Contributions of spermatozoa to embryogenesis: assays to evaluate their genetic and epigenetic fitness. Reprod Biomed Online. 2008;16(4): 474–84.
165. Carrell DT, Emery BR. Hammoud S. Int J Androl: The aetiology of sperm protamine abnormalities and their potential impact on the sperm epigenome; 2008.
166. Stanker LH, McKeown C, Balhorn R, et al. Immunological evidence for a P2 protamine precursor in mature rat sperm. Mol Reprod Dev. 1992;33(4): 481–8.
167. Debarle M, Martinage A, Sautiere P, Chevaillier P. Persistence of protamine precursors in mature sperm nuclei of the mouse. Mol Reprod Dev. 1995;40(1): 84–90.
168. Colleu D, Lescoat D, Gouranton J. Nuclear maturity of human spermatozoa selected by swim-up or by Percoll gradient centrifugation procedures. Fertil Steril. 1996;65(1):160–4.
169. Hammoud S, Liu L, Carrell DT. Protamine ratio and the level of histone retention in sperm selected from a density gradient preparation. Andrologia. 2009;41(2):88–94.
170. Love CC, Kenney RM. Scrotal heat stress induces altered sperm chromatin structure associated with a decrease in protamine disulfide bonding in the stallion. Biol Reprod. 1999;60(3):615–20.
171. Evenson DP, Jost LK, Corzett M, Balhorn R. Characteristics of human sperm chromatin structure following an episode of influenza and high fever: a case study. J Androl. 2000;21(5):739–46.
172. Iuchi Y, Kaneko T, Matsuki S, Sasagawa I, Fujii J. Concerted changes in the YB2/RYB-a protein and protamine 2 messenger RNA in the mouse testis under heat stress. Biol Reprod. 2003;68(1):129–35.
173. Bizarro D, Manicardi GC, Bianchi PG, Bianchi U, Mariethoz E, Sakkas D. In-situ competition between protamine and fluorochromes for sperm DNA. Mol Hum Reprod. 1998;4:127–32.
174. Manicardi GC, Bianchi PG, Pantano S, et al. Presence of endogenous nicks in DNA of ejaculated human spermatozoa and its relationship to chromomycin A3 accessibility. Biol Reprod. 1995; 52(4):864–7.
175. Nasr-Esfahani MH, Razavi S, Mardani M. Relation between different human sperm nuclear maturity

tests and in vitro fertilization. J Assist Reprod Genet. 2001;18(4):219–25.
176. Lolis D, Georgiou I, Syrrou M, Zikopoulos K, Konstantelli M, Messinis I. Chromomycin A3-staining as an indicator of protamine deficiency and fertilization. Int J Androl. 1996;19(1):23–7.
177. Franken DR, Franken CJ, de la Guerre H, de Villiers A. Normal sperm morphology and chromatin packaging: comparison between aniline blue and chromomycin A3 staining. Andrologia. 1999;31(6): 361–6.
178. Razavi S, Nasr-Esfahani MH, Mardani M, Mafi A, Moghdam A. Effect of human sperm chromatin anomalies on fertilization outcome post-ICSI. Andrologia. 2003;35(4):238–43.
179. Nasr-Esfahani MH, Razavi S, Mozdarani H, Mardani M, Azvagi H. Relationship between protamine deficiency with fertilization rate and incidence of sperm premature chromosomal condensation post-ICSI. Andrologia. 2004;36(3):95–100.
180. Nasr-Esfahani MH, Salehi M, Razavi S, et al. Effect of protamine-2 deficiency on ICSI outcome. Reprod Biomed Online. 2004;9(6):652–8.
181. Nasr-Esfahani MH, Salehi M, Razavi S, et al. Effect of sperm DNA damage and sperm protamine deficiency on fertilization and embryo development post-ICSI. Reprod Biomed Online. 2005;11(2):198–205.
182. Colleu D, Lescoat D, Boujard D, Le Lannou D. Human spermatozoal nuclear maturity in normozoospermia and asthenozoospermia. Arch Androl. 1988;21(3):155–62.
183. Auger J, Mesbah M, Huber C, Dadoune JP. Aniline blue staining as a marker of sperm chromatin defects associated with different semen characteristics discriminates between proven fertile and suspected infertile men. Int J Androl. 1990;13(6):452–62.
184. Bustos-Obregon E, Leiva S. Chromatin packing in normal and teratozoospermic human ejaculated spermatozoa. Andrologia. 1983;15(5):468–78.
185. Le Lannou D, Colleu D, Boujard D, Le Couteux A, Lescoat D, Segalen J. Effect of duration of abstinence on maturity of human spermatozoa nucleus. Arch Androl. 1986;17(1):35–8.
186. Jager S. Sperm nuclear stability and male infertility. Arch Androl. 1990;25(3):253–9.
187. Filatov MV, Semenova EV, Vorob'eva OA, Leont'eva OA, Drobchenko EA. Relationship between abnormal sperm chromatin packing and IVF results. Mol Hum Reprod. 1999;5(9):825–30.
188. Katayose H, Yanagida K, Hashimoto S, Yamada H, Sato A. Use of diamide-acridine orange fluorescence staining to detect aberrant protamination of human-ejaculated sperm nuclei. Fertil Steril. 2003;79 Suppl 1:670–6.
189. Rufas O, Fisch B, Seligman J, Tadir Y, Ovadia J, Shalgi R. Thiol status in human sperm. Mol Reprod Dev. 1991;29(3):282–8.
190. Lewis SE, Sterling ES, Young IS, Thompson W. Comparison of individual antioxidants of sperm and seminal plasma in fertile and infertile men. Fertil Steril. 1997;67(1):142–7.
191. Zini A, Bielecki R, Phang D, Zenzes MT. Correlations between two markers of sperm DNA integrity, DNA denaturation and DNA fragmentation, in fertile and infertile men. Fertil Steril. 2001;75(4):674–7.
192. Zini A, Kamal KM, Phang D. Free thiols in human spermatozoa: correlation with sperm DNA integrity. Urology. 2001;58(1):80–4.
193. Ramos L, van der Heijden GW, Derijck A, et al. Incomplete nuclear transformation of human spermatozoa in oligo-astheno-teratospermia: characterization by indirect immunofluorescence of chromatin and thiol status. Hum Reprod. 2008;23(2):259–70.
194. Omu AE, Al-Azemi MK, Kehinde EO, Anim JT, Oriowo MA, Mathew TC. Indications of the mechanisms involved in improved sperm parameters by zinc therapy. Med Princ Pract. 2008;17(2):108–16.
195. Shirley CR, Hayashi S, Mounsey S, Yanagimachi R, Meistrich ML. Abnormalities and reduced reproductive potential of sperm from Tnp1- and Tnp2-null double mutant mice. Biol Reprod. 2004;71(4):1220–9.
196. Conrad M, Moreno SG, Sinowatz F, et al. The nuclear form of phospholipid hydroperoxide glutathione peroxidase is a protein thiol peroxidase contributing to sperm chromatin stability. Mol Cell Biol. 2005;25(17):7637–44.
197. Zubkova EV, Wade M, Robaire B. Changes in spermatozoal chromatin packaging and susceptibility to oxidative challenge during aging. Fertil Steril. 2005;84 Suppl 2:1191–8.
198. Suganuma R, Yanagimachi R, Meistrich ML. Decline in fertility of mouse sperm with abnormal chromatin during epididymal passage as revealed by ICSI. Hum Reprod. 2005;20(11):3101–8.
199. Szczygiel MA, Ward WS. Combination of dithiothreitol and detergent treatment of spermatozoa causes paternal chromosomal damage. Biol Reprod. 2002;67(5):1532–7.
200. Sotolongo B, Lino E, Ward WS. Ability of hamster spermatozoa to digest their own DNA. Biol Reprod. 2003;69(6):2029–35.
201. Irvine DS, Twigg JP, Gordon EL, Fulton N, Milne PA, Aitken RJ. DNA integrity in human spermatozoa: relationships with semen quality. J Androl. 2000;21(1):33–44.
202. Alvarez JG, Sharma RK, Ollero M, et al. Increased DNA damage in sperm from leukocytospermic semen samples as determined by the sperm chromatin structure assay. Fertil Steril. 2002;78(2):319–29.
203. Evenson DP, Darzynkiewicz Z, Melamed MR. Comparison of human and mouse sperm chromatin structure by flow cytometry. Chromosoma. 1980;78(2):225–38.
204. Evenson DP, Wixon R. Clinical aspects of sperm DNA fragmentation detection and male infertility. Theriogenology. 2006;65(5):979–91.
205. Angelopoulou R, Plastira K, Msaouel P. Spermatozoal sensitive biomarkers to defective protaminosis and fragmented DNA. Reprod Biol Endocrinol. 2007;5:36.
206. Lewis SE, Agbaje I, Alvarez J. Sperm DNA tests as useful adjuncts to semen analysis. Syst Biol Reprod Med. 2008;54(3):111–25.

207. Agarwal A, Varghese AC, Sharma RK. Markers of oxidative stress and sperm chromatin integrity. Methods Mol Biol. 2009;590:377–402.
208. Barratt CL, Aitken RJ, Bjorndahl L, et al. Sperm DNA: organization, protection and vulnerability: from basic science to clinical applications – a position report. Hum Reprod. 2010;25(4):824–38.
209. Sakkas D, Alvarez JG. Sperm DNA fragmentation: mechanisms of origin, impact on reproductive outcome, and analysis. Fertil Steril. 2010;93(4):1027–36.
210. Zini A, Boman JM, Belzile E, Ciampi A. Sperm DNA damage is associated with an increased risk of pregnancy loss after IVF and ICSI: systematic review and meta-analysis. Hum Reprod. 2008;23(12): 2663–8.
211. Nili HA, Mozdarani H, Aleyasin A. Correlation of sperm DNA damage with protamine deficiency in Iranian subfertile men. Reprod Biomed Online. 2009;18(4):479–85.
212. Tavalaee M, Razavi S. Nasr-Esfahani MH. Fertil Steril: Influence of sperm chromatin anomalies on assisted reproductive technology outcome; 2008.
213. Zini A, Sigman M. Are tests of sperm DNA damage clinically useful? Pros and cons. J Androl. 2009; 30(3):219–29.
214. Aoki VW, Moskovtsev SI, Willis J, Liu L, Mullen JB, Carrell DT. DNA integrity is compromised in protamine-deficient human sperm. J Androl. 2005; 26(6):741–8.
215. Aoki VW, Emery BR, Liu L, Carrell DT. Protamine levels vary between individual sperm cells of infertile human males and correlate with viability and DNA integrity. J Androl. 2006;27(6):890–8.
216. Suzuki M, Crozatier C, Yoshikawa K, Toshiaki M, Yoshikawa Y. Protamine-induced DNA compaction but not aggregation shows effective radioprotection against double-strand breaks. Chem Phys Lett. 2009;480:113–7.
217. Tesarik J, Greco E, Mendoza C. Late, but not early, paternal effect on human embryo development is related to sperm DNA fragmentation. Hum Reprod. 2004;19(3):611–5.
218. Cho C, Jung-Ha H, Willis WD, et al. Protamine 2 deficiency leads to sperm DNA damage and embryo death in mice. Biol Reprod. 2003;69(1):211–7.
219. Greco E, Scarselli F, Iacobelli M, et al. Efficient treatment of infertility due to sperm DNA damage by ICSI with testicular spermatozoa. Hum Reprod. 2005;20(1):226–30. doi:10.1093/humrep/deh590.
220. Aitken RJ, De Iuliis GN. Origins and consequences of DNA damage in male germ cells. Reprod Biomed Online. 2007;14(6):727–33.
221. Leduc F, Nkoma GB, Boissonneault G. Spermiogenesis and DNA repair: a possible etiology of human infertility and genetic disorders. Syst Biol Reprod Med. 2008;54(1):3–10.
222. Aitken RJ, De Iuliis GN, McLachlan RI. Biological and clinical significance of DNA damage in the male germ line. Int J Androl. 2009;32(1):46–56.
223. Aitken RJ, De Iuliis GN. On the possible origins of DNA damage in human spermatozoa. Mol Hum Reprod. 2010;16(1):3–13.
224. Dominguez-Fandos D, Camejo MI, Ballesca JL, Oliva R. Human sperm DNA fragmentation: correlation of TUNEL results as assessed by flow cytometry and optical microscopy. Cytom A. 2007;71(12): 1011–8.
225. de Mateo S, Gazquez C, Guimera M, et al. Protamine 2 precursors (Pre-P2), protamine 1 to protamine 2 ratio (P1/P2), and assisted reproduction outcome. Fertil Steril. 2008;91(3):715–22.
226. Tarozzi N, Nadalini M, Stronati A, et al. Anomalies in sperm chromatin packaging: implications for assisted reproduction techniques. Reprod Biomed Online. 2009;18(4):486–95.
227. Chiamchanya C, Kaewnoonual N, Visutakul P, Manochantr S, Chaiya J. Comparative study of the effects of three semen preparation media on semen analysis, DNA damage and protamine deficiency, and the correlation between DNA integrity and sperm parameters. Asian J Androl. 2010;12(2):271–7.
228. Khara KK, Vlad M, Griffiths M, Kennedy CR. Human protamines and male infertility. J Assist Reprod Genet. 1997;14(5):282–90.
229. Carrell DT, Emery BR, Liu L. Characterization of aneuploidy rates, protamine levels, ultrastructure, and functional ability of round-headed sperm from two siblings and implications for intracytoplasmic sperm injection. Fertil Steril. 1999;71(3): 511–6.
230. Carrell DT, Liu L. Altered protamine 2 expression is uncommon in donors of known fertility, but common among men with poor fertilizing capacity, and may reflect other abnormalities of spermiogenesis. J Androl. 2001;22(4):604–10.
231. Steger K, Fink L, Failing K, et al. Decreased protamine-1 transcript levels in testes from infertile men. Mol Hum Reprod. 2003;9(6):331–6.
232. Mitchell V, Steger K, Marchetti C, Herbaut JC, Devos P, Rigot JM. Cellular expression of protamine 1 and 2 transcripts in testicular spermatids from azoospermic men submitted to TESE-ICSI. Mol Hum Reprod. 2005;11(5):373–9.
233. Ahmadi A, Ng SC. Destruction of protamine in human sperm inhibits sperm binding and penetration in the zona-free hamster penetration test but increases sperm head decondensation and male pronuclear formation in the hamster-ICSI assay. J Assist Reprod Genet. 1999;16(3):128–32.
234. Ahmadi A, Ng SC. Influence of sperm plasma membrane destruction on human sperm head decondensation and pronuclear formation. Arch Androl. 1999;42(1):1–7.
235. Becker S, Soffer Y, Lewin LM, Yogev L, Shochat L, Golan R. Spermiogenesis defects in human: detection of transition proteins in semen from some infertile men. Andrologia. 2008;40(4):203–8.

The Relationship Between Chromatin Structure and DNA Damage in Mammalian Spermatozoa

4

Kenneth Dominguez, Chris D.R. Arca, and W. Steven Ward

Abstract

At least two aspects of sperm chromatin seem to be vulnerable to DNA damage. The first is inherent within the structure of the chromatin, which predicts that the histone-bound segments are more susceptible to any type of DNA damaging agent than the protamine-bound DNA. The data support the conclusion that protamines do protect DNA from exogenous insults. The second aspect is the enzymes that are present in the sperm cell that can modify DNA, and activation of these segments can disrupt fertilization and/or embryonic development. We have reviewed only two of these, but it is possible that others exist that have not yet been documented. The idea that the highly condensed sperm chromatin retains some active enzymatic elements is an important consideration in the future development of assisted reproduction techniques (ARTs) that require the mechanical manipulation and storage of human sperm cells.

Mammalian sperm DNA is one among the most compact chromatins in nature. It is condensed to a pseudocrystalline state and in this respect differs markedly from organization of the same DNA in the next stage of its development, the prereplicative paternal pronucleus of the one-cell embryo. However, even the highly condensed sperm chromatin is organized according to functions that the spermatozoon must perform. It has become clear in the past decade that certain structural elements of sperm chromatin packaging are essential for proper embryogenesis, while others function almost entirely to protect the paternal genome during transit. Both types of sperm DNA packaging have implications for clinical infertility. The stability of this highly condensed sperm DNA provides a degree of assurance that the male donor's genome can withstand mechanical and biochemical manipulations

W. Steven Ward (✉)
Professor and Director,
Institute for Biogenesis Research (IBR),
John A. Burns School of Medicine,
University of Hawaii at Manoa, Honolulu, HI, USA
e-mail: wward@hawaii.edu

A. Zini and A. Agarwal (eds.), *Sperm Chromatin: Biological and Clinical Applications in Male Infertility and Assisted Reproduction*, DOI 10.1007/978-1-4419-6857-9_4,

required for ICSI and IVF. Biologically dead spermatozoa can still be used for successful ICSI if the DNA remains intact. However, recent evidence combined with new models for sperm chromatin structure also suggest that sperm DNA may be more sensitive to particular insults than others. Here, we review the current models of sperm chromatin structure with particular emphasis on its stability and unique vulnerability.

Keywords
Chromatin structure • DNA damage in mammalian spermatozoa • Spermatozoa and DNA damage • Sperm enzymes

Sperm Chromatin Structure Overview

Several recent reviews have provided updates on current knowledge and hypotheses about human sperm chromatin structure and its unique condensation of DNA in preparation for fertility [1–3]. We have recently described a new model for mammalian sperm chromatin structure that incorporates several recent findings in the sperm nucleus [4] (Fig. 4.1). While there is evidence to support each aspect of the model, it should be emphasized that it is by no means proven, and it is likely that as further data are accumulated, the model will be modified. The model does, however, provide a useful tool for thinking about the implications of chromatin structure for human fertility treatments.

Protamine Condensation of the Sperm DNA

The most unique aspect of mammalian sperm chromatin structure is that most of the DNA, 90–95% in human spermatozoa [5–7], is condensed into very tightly packed toroids by protamines [8–10] (Fig. 4.1b). Protamine toroids include roughly 50 kb of DNA, and evidence suggests that each protamine toroid is also one DNA loop domain [11] (see below). Each protamine toroid is linked by a short, region that we have termed the protamine linker (Fig. 4.1c). These toroid linkers represent a point in the mature chromatin structure that is particularly sensitive to DNA-damaging agents. For example, treatment of sperm chromatin with Dnase 1 easily digests the toroid linker regions, but does not harm the DNA within the protamine toroid, itself. Because these Dnase-1-sensitive linker regions are spaced approximately every 50 kb, destruction of these elements prevents embryogenesis beyond the one-cell stage [12, 13]. The toroids, themselves, may be packaged into a stacked "lifesaver" model (Fig. 4.1d), and evidence for this has recently been provided by Mudrak et al. [14]. But this is still uncertain, and the secondary packaging of protamine toroids in sperm chromatin remains an unsolved mystery.

Histone-Bound Sperm Chromatin

In the compaction of sperm DNA during spermiogenesis, the histones are replaced with protamines. However, in humans, about 4–10% of the histones remain attached to the chromatin. Two different laboratories have recently reported genome-wide mapping of histone-bound DNA in human spermatozoa [5, 7]. The results suggest that histone-bound chromatin is divided into two regions in human spermatozoa. First, there are large domains of 10 kb or larger that might constitute whole DNA loop domains. In our model, these are depicted as folded solenoids (Fig. 4.1a, d). However, there is not yet strong evidence that these histone-bound chromatin fragments are individual loop domains, so this particular point is speculative. Second, there are multiple, smaller histone-bound fragments whose properties are consistent with toroid linkers (Fig. 4.1c). The most interesting aspect of histone-bound sperm chromatin is that these

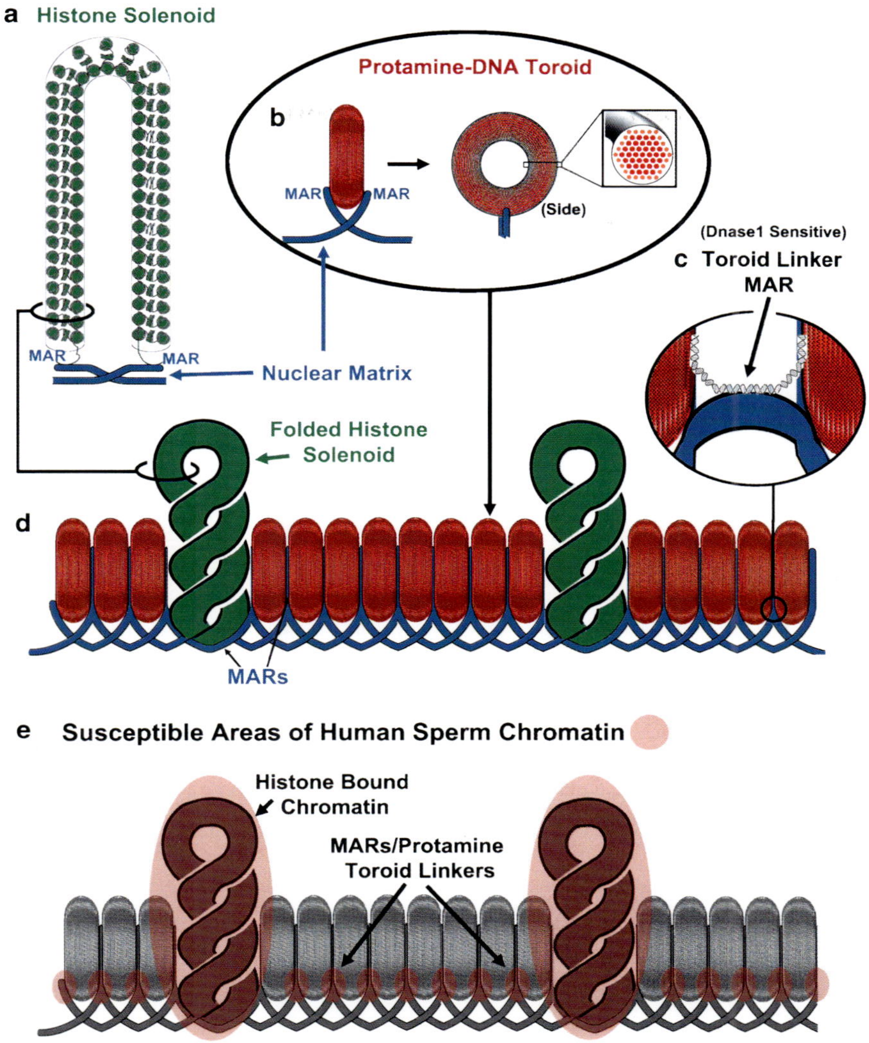

Fig. 4.1 Model for human sperm chromatin structure. We have recently proposed a model for human sperm chromatin [4] structure based on several recent publications (see text). Sperm DNA is organized into loop domains of about 50 kb, attached at their bases to a proteinaceous sperm nuclear matrix. While some of these loop domains are associated with histones, as in somatic cells (**a**), the majority of sperm chromatin is condensed into toroids by protamines, with each toroid representing a single loop domain (**b**). The toroids are linked by nuclease-sensitive segments of DNA called toroid linkers (**c**). Protamine toroids may be stacked side to side in a "lifesaver" model which also stabilizes the histone-bound loops (**d**). There are two areas of the sperm chromatin that are particularly sensitive to external DNA damaging agents, the histone-bound DNA and the protamine toroid linker regions (**e**). (Adapted from Ward [4], with permission)

regions may not require replacement by oocyte histones after fertilization. Evidence suggests that some histones are inherited by the embryo's paternal pronucleus [15, 16], suggesting that these chromatin structures may play a direct role in embryo function. These histone-bound chromatin segments represent another area of the sperm chromatin structure that are relatively more sensitive to all types of DNA damage than the protamine-bound elements.

Organization of DNA Loop Domains

We and others have also demonstrated that sperm DNA is organized into loop domains attached at their bases to a sperm nuclear matrix [17–21]. These loop domains are approximately 50 kb in length, although there is a wide variety in size. As mentioned above, our evidence supports a model in which each protamine toroid is one DNA loop domain [11]. More recently, we have shown that this organization of DNA into loop domains is probably inherited by the embryo in the organization of the paternal pronucleus [22]. Proper sperm loop structure is required for the replication of the paternal genome after fertilization. Moreover, these attachment sites also serve as the initiation of topoisomerase-mediated DNA degradation in mature spermatozoa [12]. The loop attachment sites are located with the DNA sequences that also serve as the protamine linkers and, therefore, as mentioned above represent a sensitive component of the sperm chromatin structure. Collectively, these data support a model in which the sperm nuclear matrix serves functional roles that are similar to DNA replication and apoptotic degradation in somatic cells.

Ability of Protamine Condensation to Protect Sperm DNA from Damage

There are at least four potential sources of the evolutionary pressure to evolve such tightly compact DNA in the mammalian sperm cell. They are as follows: (1) to protect the DNA during the transit of fertilization, (2) to make the DNA more compact for more efficient motility of the sperm cell, (3) to transcriptionally silence most of the genome in the mature sperm cell, and (4) to minimize cross species fertilization, which may be impacted by the presence and ratio of one or two different protamines (R. Balhorn, personal communication). For the purposes of human assisted reproduction techniques (ART), a knowledge of sperm chromatin structure is particularly important for the first reason, the protection of the sperm genome. This is because most ART procedures include mechanical and biomechanical manipulation of the sperm genome through micromanipulation and cryopreservation.

What is the evidence, then, that sperm chromatin is less susceptible to DNA damage by external factors than other cells? This question is not as simple to address as it may appear for two reasons. The first is that sperm the consequences of DNA damage are much more devastating to its biological function, fertilization and participation in proper embryonic development of the embryo, than are those for somatic cells. A somatic cell can grow and divide with some chromosomal aberrations, but even a single, unrepaired DNA strand break may disrupt the embryo's sensitive developmental program. The second reason is that spermatozoa have no known DNA repair mechanisms. This makes it difficult to compare, for example, the level of radiation-induced DNA breaks in spermatozoa as compared to somatic cells because the somatic cells can repair the breaks in some cases before the damage is assessed [23]. Some breaks in the sperm DNA can be repaired by the oocyte after fertilization, but it is clear that sperm can enter the oocyte with too many breaks to be repaired [24–28]. The embryo can, in fact, be induced to degrade all the paternal DNA at the onset of DNA synthesis if the damage in the sperm cell is too extensive [12].

While the relative resistance of the condensed mammalian sperm chromatin to radiation induced damage is difficult to assess, there does exist evidence that shows that DNA condensed into protamine toroids is more resistant to other insults than somatic cells. For example, it is clear that protamines protect DNA from degradation by a variety of nucleases [11, 29, 30]. When the degradation of hamster sperm DNA was compared directly to that of spleen cell nuclei by exogenous Dnase 1, the portion of the sperm DNA that was bound to protamines withstood very high concentrations of nuclease that completely degraded the histone-bound spleen DNA [11]. The structure of the protamine toroid (Fig. 4.1b) suggests a simple mechanism for this protection: the physical exclusion of the nuclease from most of the DNA. However, many parts of the condensed sperm chromatin are susceptible to nuclease digestion, including the toroid linker regions [11] and all the histone-bound segments of the sperm chromatin [5, 21]. The condensed sperm chromatin may protect the entire paternal genome from another potential DNA damaging agent – mechanical disruption. When mouse spermatozoa are briefly sonicated, they are still capable of fertilizing an oocyte that results in the production of viable offspring [31]. This suggests that the condensation of sperm DNA by protamines has a neighbor effect on the histone-bound chromatin, protecting the entire genome from the mechanical stresses it encounters during the vigorous cell motility of fertilization.

Finally, there is recent evidence to suggest that protamine condensation even protects the sperm DNA from degradation by reactive oxygen species (ROS). As recently noted by De Iuliis and Aitken [32], spermatozoa are very sensitive to oxidative stress [33–35]. These authors recently provided evidence that ROS attack DNA directly forming 8-hydroxy-20-deoxyguanosine (8OHdG) [32]. They also demonstrated that 8OHdG adduct formation was directly associated with CMA3 binding to sperm chromatin, an indication of aberrant protamine binding. This suggests that protamines may protect, to some degree, sperm DNA from ROS damage.

The current view of sperm chromatin structure, then, is one of a structurally extremely stable genome that has very little ability to correct any damage that does occur. Thus, the evolutionary pressure to condense sperm DNA, probably driven by a reproductive advantage found in protecting the paternal genome, also sacrificed much of the enzymatic machinery that a normal cell uses to repair damage.

Active Sperm Chromatin

The packaging of most of the sperm chromatin into an almost crystalline state (Fig. 4.1b) would seem to preclude any of the normal activities of chromatin in this transcriptionally silent, nonreplicating cell. However, at least two lines of evidence suggest that some DNA modification enzymatic machinery is still active. The first comes from two reports from the same laboratory that H2AX can be phoshorylated to gamma-H2AX in response to mutagenic agents or peroxide treatment [36, 37]. As Aitken and De Iuliis noted, "[t]hat a transcriptionally and translationally silent spermatozoon with such tightly compacted, histone-depleted chromatin, possesses the capacity to detect and mark DNA strand breaks for repair by phosphorylating H2AX is fascinating and deserves further attention. At face value such a concept runs contrary to the widely held belief that the chromatin with these cells is inert and once damaged has to wait until fertilization for repair to be effected by the embryo during a post fertilization round of DNA repair that unequivocally does involve activation of the gamma-H2AX signaling pathway" [1].

Another set of experiments from our laboratory suggests that an endogenous topoisomerase 2 in the sperm nucleus can be induced to fragment the entire genome into loop-sized domains. Epididymal mouse spermatozoa incubated with divalent cations can be induced to fragment all the DNA to about 50 kb fragments, mediated by topoisomerase 2 [38, 39]. These breaks can be reversed by treatment with EDTA, a hallmark of topoisomerase 2 DNA fragmentation. Spermatozoa from the vas deferens digest their DNA further, suggesting the involvement of an additional nuclease. Our evidence so far suggests the unexpected possibility that this topoisomerase-2-associated nuclease enters the sperm cell from the vas deferens luminal fluid. When spermatozoa that have been induced to cleave their DNA in this manner are injected into oocytes, the paternal pronucleus forms normally, but degrades all its DNA at the onset of the first round of DNA replication in the one-cell embryo [12, 13]. This topoisomerase-2-mediated DNA degradation is similar to that of somatic cells undergoing apoptosis [40–42]. Thus, the condensed chromatin of the sperm nucleus retains some of the enzymatic machinery that the cell uses to degrade its DNA in apoptosis, and activating this has severe consequences for embryonic development.

These two lines of evidence indicate that this silent, "sleeping genome" [43] may, in fact, retain some enzymatic activities associated with chromatin modification. One can speculate that these may represent important checkpoints that monitor the paternal genome during its difficult journey in fertilization [44]. It is also possible that the active chromatin modification enzymes may be residual components of the chromatin that were required for the intensive remodeling in spermiogenesis. A third possibility is that these sperm chromatin enzymes are actually required for the initiation of chromatin condensation that occurs after fertilization. Regardless of the evolutionary etiology for their existence, these studies indicate that the mature mammalian sperm cell does have the ability to manipulate its DNA to some degree.

Conclusions

At least two aspects of sperm chromatin seem to be vulnerable to DNA damage. The first is inherent within the structure of the chromatin, which predicts that the histone-bound segments are more susceptible to any type of DNA damaging agent than the protamine-bound DNA. As reviewed above, the data support the conclusion that protamines do protect DNA from exogenous insults. The second aspect is the enzymes that are present in the sperm cell that can modify DNA, and activation of these segments can disrupt fertilization and/or embryonic development. We have reviewed only two of these, but it is possible that others exist that have not yet been documented. The idea that the highly condensed sperm chromatin retains some active enzymatic elements is an important consideration in the future development of ART technologies that require the mechanical manipulation and storage of human sperm cells. The data suggest that for human sperm cells, two major component of chromatin structure are the most susceptible to DNA damage, the histone-bound segments and the matrix attachment regions (MARs) that are also the protamine linker regions (Fig. 4.1e).

There is much more work to be done before a firm model of sperm chromatin structure can be established. The model shown in Fig. 4.1 is consistent with the data so far described, but still lacks definitive proof for the lifesaver toroid stacking (Fig. 4.1d), or for any prediction of how this level of chromatin folding affects the overall structure of the sperm chromosome. One additional component that was not reviewed in this chapter, but discussed in two other chapters of this volume, is the presence of numerous RNA molecules in the sperm nucleus. The functions of these nucleic acids have not yet been identified. In somatic cells, siRNAs participate in the condensation of chromatin to inactivate genes [45], and it is possible that some of the sperm RNAs also contribute to the maintenance or modification of chromatin structure.

However uncertain the actual structure and related functional implications of sperm chromatin remain, it is clear that this highly condensed DNA does retain functional properties. Some, such as loop domain organization and histone binding, are essential for the initial processes of embryonic development. There are also areas that remain sensitive to external insults as somatic cell chromatin. Increased understanding of the packaging of sperm DNA will undoubtedly bear fruits in improved methods for storing and manipulating human spermatozoa for ART.

References

1. Aitken RJ, De Iuliis GN. On the possible origins of DNA damage in human spermatozoa. Mol Hum Reprod. 2010;16:3–13.
2. Barratt CL, Aitken RJ, Bjorndahl L, et al. Sperm DNA: organization, protection and vulnerability: from basic science to clinical applications – a position report. Hum Reprod. 2010;25(4):824–38.
3. Delbes G, Hales BF, Robaire B. Toxicants and human sperm chromatin integrity. Mol Hum Reprod. 2010;16:14–22.
4. Ward WS. Function of sperm chromatin structural elements in fertilization and development. Mol Hum Reprod. 2010;16:30–6.
5. Hammoud SS, Nix DA, Zhang H, et al. Distinctive chromatin in human sperm packages genes for embryo development. Nature. 2009;460:473–8.
6. Churikov D, Siino J, Svetlova M, et al. Novel human testis-specific histone H2B encoded by the interrupted gene on the X chromosome. Genomics. 2004;84: 745–56.
7. Arpanahi A, Brinkworth M, Iles D, et al. Endonuclease-sensitive regions of human spermatozoal chromatin are highly enriched in promoter and CTCF binding sequences. Genome Res. 2009;19:1338–49.
8. Hud NV, Allen MJ, Downing KH, et al. Identification of the elemental packing unit of DNA in mammalian sperm cells by atomic force microscopy. Biochem Biophys Res Commun. 1993;193:1347–54.

9. Hud NV, Downing KH, Balhorn R. A constant radius of curvature model for the organization of DNA in toroidal condensates. Proc Natl Acad Sci USA. 1995;92:3581–5.
10. Brewer L, Corzett M, Lau EY, et al. Dynamics of protamine 1 binding to single DNA molecules. J Biol Chem. 2003;278:42403–8.
11. Sotolongo B, Lino E, Ward WS. Ability of hamster spermatozoa to digest their own DNA. Biol Reprod. 2003;69:2029–35.
12. Yamauchi Y, Shaman JA, Boaz SM, et al. Paternal pronuclear DNA degradation is functionally linked to DNA replication in mouse oocytes. Biol Reprod. 2007;77:407–15.
13. Yamauchi Y, Shaman JA, Ward WS. Topoisomerase II mediated breaks in spermatozoa cause the specific degradation of paternal DNA in fertilized oocytes. Biol Reprod. 2007;76:666–72.
14. Mudrak O, Chandra R, Jones E, et al. Reorganisation of human sperm nuclear architecture during formation of pronuclei in a model system. Reprod Fertil Dev. 2009;21:665–71.
15. van der Heijden GW, Derijck AA, Ramos L, et al. Transmission of modified nucleosomes from the mouse male germline to the zygote and subsequent remodeling of paternal chromatin. Dev Biol. 2006;298: 458–69.
16. van der Heijden GW, Ramos L, Baart EB, et al. Sperm-derived histones contribute to zygotic chromatin in humans. BMC Dev Biol. 2008;8:34.
17. Nadel B, de Lara J, Finkernagel SW, et al. Cell-specific organization of the 5S ribosomal RNA gene cluster DNA loop domains in spermatozoa and somatic cells. Biol Reprod. 1995;53:1222–8.
18. Schmid C, Heng HH, Rubin C, et al. Sperm nuclear matrix association of the PRM1->PRM2->TNP2 domain is independent of Alu methylation. Mol Hum Reprod. 2001;7:903–11.
19. Martins RP, Krawetz SA. Nuclear organization of the protamine locus. Soc Reprod Fertil Suppl. 2007;64: 1–12.
20. Shaman JA, Yamauchi Y, Ward WS. Function of the sperm nuclear matrix. Arch Androl. 2007;53: 135–40.
21. Linnemann AK, Platts AE, Krawetz SA. Differential nuclear scaffold/matrix attachment marks expressed genes. Hum Mol Genet. 2009;18:645–54.
22. Shaman JA, Yamauchi Y, Ward WS. The sperm nuclear matrix is required for paternal DNA replication. J Cell Biochem. 2007;102:680–8.
23. Kamiguchi Y, Tateno H. Radiation- and chemical-induced structural chromosome aberrations in human spermatozoa. Mutat Res. 2002;504:183–91.
24. Ashwood-Smith MJ, Edwards RG. DNA repair by oocytes. Mol Hum Reprod. 1996;2:46–51.
25. Fernandez-Gonzalez R, Moreira PN, Perez-Crespo M, et al. Long-term effects of mouse intracytoplasmic sperm injection with DNA-fragmented sperm on health and behavior of adult offspring. Biol Reprod. 2008;78:761–72.
26. Genesca A, Caballin MR, Miro R, et al. Repair of human sperm chromosome aberrations in the hamster egg. Hum Genet. 1992;89:181–6.
27. Matsuda Y, Tobari I. Repair capacity of fertilized mouse eggs for X-ray damage induced in sperm and mature oocytes. Mutat Res. 1989;210:35–47.
28. Derijck A, van der Heijden G, Giele M, et al. DNA double-strand break repair in parental chromatin of mouse zygotes, the first cell cycle as an origin of de novo mutation. Hum Mol Genet. 2008;17: 1922–37.
29. Schmidt G, Cohen MP. The action of staphylococcal nuclease (EC-number 3. 1. 4. 7.) on thymus nucleohistone (TNH) and on some nucleoprotamines. Mol Cell Biochem. 1975;6:185–94.
30. Goscin LP, Guild WR. Enhancement of pneumococcal transfection by protamine sulfate. J Bacteriol. 1982;152:765–72.
31. Kuretake S, Kimura Y, Hoshi K, et al. Fertilization and development of mouse oocytes injected with isolated sperm heads. Biol Reprod. 1996;55:789–95.
32. De Iuliis GN, Thomson LK, Mitchell LA, et al. DNA damage in human spermatozoa is highly correlated with the efficiency of chromatin remodeling and the formation of 8-hydroxy-2'-deoxyguanosine, a marker of oxidative stress. Biol Reprod. 2009;81:517–24.
33. Aitken RJ, Baker MA. Oxidative stress, sperm survival and fertility control. Mol Cell Endocrinol. 2006;250:66–9.
34. Jones R, Mann T, Sherins R. Peroxidative breakdown of phospholipids in human spermatozoa, spermicidal properties of fatty acid peroxides, and protective action of seminal plasma. Fertil Steril. 1979;31: 531–7.
35. Alvarez JG, Touchstone JC, Blasco L, et al. Spontaneous lipid peroxidation and production of hydrogen peroxide and superoxide in human spermatozoa. Superoxide dismutase as major enzyme protectant against oxygen toxicity. J Androl. 1987;8:338–48.
36. Li Z, Yang J, Huang H. Oxidative stress induces H2AX phosphorylation in human spermatozoa. FEBS Lett. 2006;580:6161–8.
37. Li ZX, Wang TT, Wu YT, et al. Adriamycin induces H2AX phosphorylation in human spermatozoa. Asian J Androl. 2008;10:749–57.
38. Shaman JA, Prisztoka R, Ward WS. Topoisomerase IIB and an extracellular nuclease interact to digest sperm DNA in an apoptotic-like manner. Biol Reprod. 2006;75:741–8.
39. Boaz SM, Dominguez KM, Shaman JA, et al. Mouse spermatozoa contain a nuclease that is activated by pretreatment with EGTA and subsequent calcium incubation. J Cell Biochem. 2008;103:1636–45.
40. Widlak P, Garrard WT. Discovery, regulation, and action of the major apoptotic nucleases DFF40/CAD and endonuclease G. J Cell Biochem. 2005;94:1078–87.
41. Li TK, Chen AY, Yu C, et al. Activation of topoisomerase II-mediated excision of chromosomal DNA loops during oxidative stress. Genes Dev. 1999;13:1553–60.

42. Lagarkova MA, Iarovaia OV, Razin SV. Large-scale fragmentation of mammalian DNA in the course of apoptosis proceeds via excision of chromosomal DNA loops and their oligomers. J Biol Chem. 1995;270:20239–41.
43. Ward WS. The structure of the sleeping genome: implications of sperm DNA organization for somatic cells. J Cell Biochem. 1994;55:77–82.
44. Ward MA, Ward WS. A model for the function of sperm DNA degradation. Reprod Fertil Dev. 2004;16:547–54.
45. Iida T, Nakayama J, Moazed D. siRNA-mediated heterochromatin establishment requires HP1 and is associated with antisense transcription. Mol Cell. 2008;31:178–89.

Chromosome Positioning in Spermatozoa 5

Andrei Zalensky, Olga Mudrak, and Irina Zalenskaya

Abstract

Similar to interphase chromosomes, mammalian supercompact and genetically inactive sperm chromatin is arranged in non-overlapping chromosome territories. Recent works demonstrate that chromosome territories in sperm have defined and nonrandom intranuclear positioning. It is hypothesized that preferred localization of chromosomes in sperm together with well-organized chromosome architecture provide epigenetic signature to genome, which might be important at fertilization and early development. Here, we describe relevant experimental data with primary attention to the studies related to human spermatozoa. Possible implications of sperm chromosome positioning for modern reproductive technologies are discussed.

Keywords

Chromosome positioning • Spermatozoa • Sperm chromatin • Intranuclear positioning in chromosomes • Chromosome architecture

Overview of Chromosome Positioning in Interphase Cells

It is experimentally established and now commonly recognized that human genome is well-organized within nuclear volume. Progress in this field was possible due to success of the Genome Project, development of methods of multicolor fluorescence in situ hybridization (FISH) and the increasingly sophisticated microscopy technologies, including 3D imaging and reconstruction. The main feature of chromosome packaging in the interphase nucleus is their territorial organization [1, 2]. Culmination of studies in this direction was establishing 3D map for all 46 human chromosome territories (CT) in intact cell nuclei [3].

Importantly, individual CTs are characterized by their preferred and nonrandom intranuclear positions [4–7]. In the spherical interphase nuclei, only radial (e.g., preferably central or preferably peripheral) CT positioning may be determined by FISH localization of chromosome-specific painting

A. Zalensky (✉)
The Jones Institute for Reproductive Medicine,
Eastern Virginia Medical School,
601 Colley Avenue, Norfolk, VA 23507, USA
e-mail: ZalensAO@EVMS.EDU

A. Zini and A. Agarwal (eds.), *Sperm Chromatin: Biological and Clinical Applications in Male Infertility and Assisted Reproduction*, DOI 10.1007/978-1-4419-6857-9_5,

DNA probes. In most tissues, the gene-poor chromosomes are found to be located close to the nuclear edge, while the gene-rich are more central [8, 9]. In some cells, the distribution of CT correlates with chromosome size, larger chromosomes being located more peripherally than smaller ones [10].

It is suggested that distribution of CT within nuclear space has functional relevance to the regulation of gene expression [11, 12]. In fact, gene activation or silencing is often associated with repositioning of chromatin domains or whole CT [12–14].

So far, little is known about what determines preferred chromosome positioning in interphase nuclei. Cook and Marenduzzo [15] applied Monte-Carlo simulations to study the role of the nonspecific (entropic) forces acting to position and shape self-avoiding polymers within a confining sphere. In this computer simulation, long and flexible polymers (representing gene-rich chromosomes) were driven to the nuclear interior, while compact polymers (representing heterochromatic gene-poor chromosomes) were found at the sphere periphery. Authors conclude that self-organization may warrant nonrandom position of chromosomes within nuclei. On the other hand, using artificial introduction of human chromosomes into a mouse nucleus, Sengupta et al. [16] demonstrated conservation of the "donor-specific" chromosome positioning in the host cells. Authors propose the existence of a chromosomal determinant of the preferred intranuclear positioning. Using molecular approaches, it was shown that interphase position is influenced by proteins of the nuclear lamina. Lamin B1 is required to anchor chromosome 18 at the nuclear periphery; disruption of this interaction results in diffusion of the chromosome from its original location [17]. Changes in chromosome positioning with typical peripheral localization were also observed in haemopoietic lineage cells lacking A-type lamins and in primary fibroblast cell lines carrying mutations in lamin A [18].

While chromatin organization in spermatozoa differs significantly from that of somatic cells, spermatozoa preserve territorial organization of chromosomes and preferential localization of individual chromosomes within the nuclear volume, reviewed in [19]. The latter is the subject of the current review.

Methods of Determining Chromosome Nuclear Localization

Majority of methods of chromosomes localization depend on FISH of the labeled DNA probe specific to individual chromosome to the target DNA in the nucleus [20], Fig. 5.1 provides outline of procedure. Vast selection of available FISH probes complemented with sensitive hybridization and microscopic techniques encouraged rapid progress in localization of CT and selected chromosomal domains within the nuclear volume [20–23].

The hybridization procedure demands target DNA denaturation under harsh conditions, whereas truthful signal localization strongly depends on preservation of nuclear morphology. Therefore, methods of cell/nuclei fixation become very important. Most common are treatments with methanol/acetic acid or buffered formaldehyde, resulting in so-called 3D or 2D FISH, respectively [24, 25]. Finest preservation of 3D nuclear organization in interphase cells is achieved with the second approach. In the case of mammalian sperm, where DNA is tightly packed with protamines, [26] an additional step to liberate nuclear DNA before denaturation is needed. The prerequsite sperm decondensation may be achieved by treatment with either 1–3 M NaOH, or isolecithin/heparin [27], or lithium-3,5-diiodosalicylic acid [28–30]. Controlled decondensation was achieved using increasing concentrations of Heparin in the presence of DTT [28].

Determination of CT localization in interphase cells is hindered by the existence of chromosome homologues, the spherical shape of nuclei in many cell types, and the absence of internal spatial beacons. Therefore, only radial positioning of a chromosome can be measured. Two types of FISH probes are used to establish intranuclear chromosome positioning: "whole chromosome paints" – DNA probes, which allow the specific staining of individual chromosomes [30, 31], and

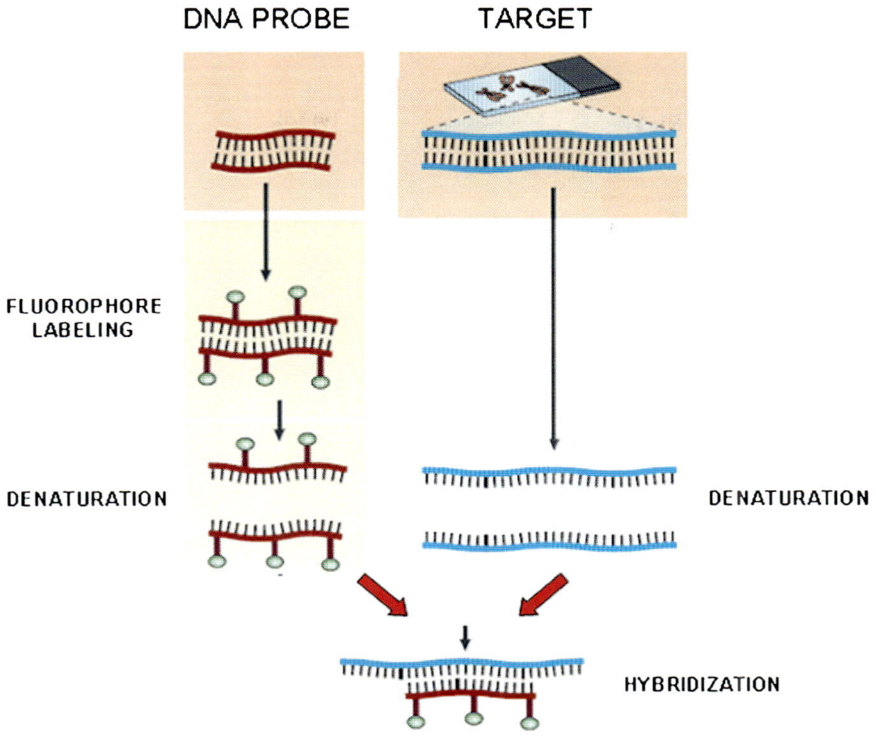

Fig. 5.1 Simplified outline of fluorescence in situ hybridization (FISH). Prior to hybridization, DNA probe (oligonucleotide, cloned sequence, microdissected DNA) is directly labeled with fluorophore (*left*). Alternative techniques for probe tagging include PCR, chemical labeling and others. Indirect labeling strategy, for example labeling with Biotin fluorescent tagging at later steps may be used. Probe DNA and target chromosomal DNA within cells (*right*) are denatured, then mixed and annealed. After completion of hybridization and washing of nonspecifically bound probe the signals are detected by fluorescence microscopy (modified from Speicher and Carter [20], with permission)

chromosome-specific centromere probes [31]. Due to the small size of the centromere-specific signals, determination of coordinates of the centromere positions does not present significant technical difficulties. FISH signals obtained using whole chromosome paints are comparatively large (Fig. 5.2a), and therefore, the intranuclear coordinates of a chromosome may be determined only roughly, for example, by the position of its CT geometrical center relative to the nearest nuclear edge [32]. In a more sophisticated approach [33], 3D radial distributions of chromosomes were evaluated utilizing special software [34], which measures the shortest distance of each voxel (3D equivalent of a pixel) within previously segmented CT to the border of the segmented nucleus followed by normalization of values collected from the set of the imaged nuclei.

Sperm cells of many species are asymmetrical. In humans, sperm nuclei are of ellipsoid shape, flattened, and have a fixed spatial marker–tail attachment point. Thus, the relative positions of CT can be much more easily defined than in the somatic cells. In addition, due to extended form of the sperm nucleus, the chromosome position can be assessed both in the radial and the longitudinal (along the anterior–posterior axis) direction. Figures 5.2b, c and 5.3a provide examples of chromosome painting and centromere FISH in human sperm cells.

In the absence of standardized methods, several approaches for the determination of the longitudal position in sperm have been used: (1) the nucleus was divided into sectors and the number of FISH signals found in each sector was calculated for each chromosome [32–38]; (2) normalized

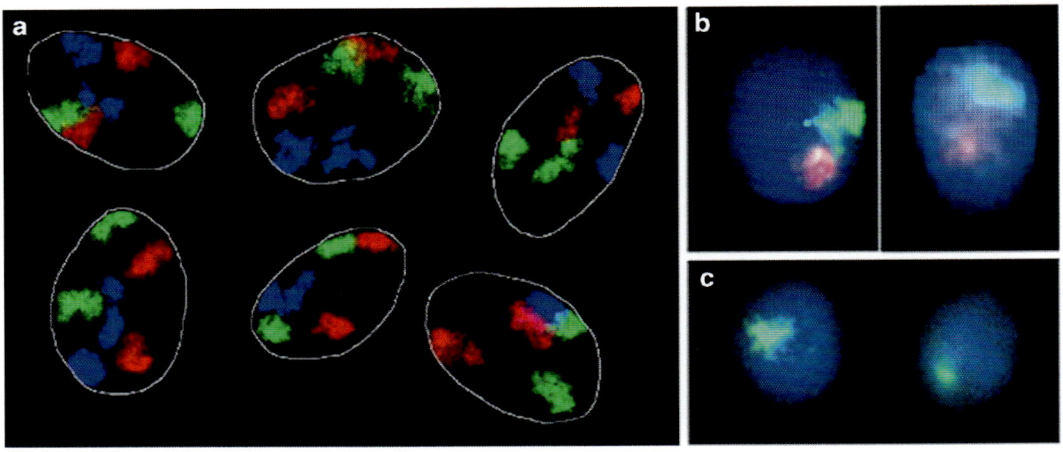

Fig. 5.2 FISH using chromosome painting probes. (**a**) Imaging of HSA3 (*green*), HSA5 (*blue*) and HSA11 (*red*) in HeLa cell nuclei (from Foster and Bridger [82], with permission). (**b**, **c**) Visualization of chromosomes in human spermatozoa. (**b**) HSA17 (*green*) and HSA19 (*red*); (**c**) HSA6. Total nuclear DNA is counterstained with DAPI (*blue*)

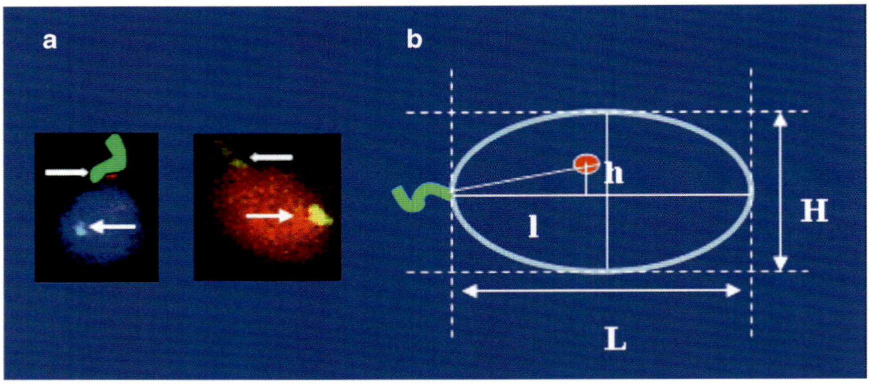

Fig. 5.3 Determination of human sperm chromosomes intranuclear position using FISH with centromere-specific probes. (**a**) Typical images after FISH using HSAY (*left*, *green*) and HSA17 (*right*, *yellow*). Total DNA is counterstained with DAPI (*blue*) or PI (*red*). (**b**) Schematic view of the sperm nucleus illustrating distances measured for the determination of hybridization signal (*red*) coordinates

distances between the tail attachment point and the position of chromosome-specific FISH centromere signals along the long nuclear axis (l/L in Fig. 5.3b) were measured and served as indicators of the longitude localization [36]. For assessment of radial positioning, number of FISH signals fallen into concentric radial "shells" [39] or "central and peripheral" zones [40] subdividing sperm nuclei were scored. Alternatively, distances from the CT centers to the nearest peripheral edge [32] or normalized distances between centromere signals and the long nuclear axis (h/H in Fig. 5.3) were measured [36].

Many different programs allow automatically analyze localization of FISH signals. Commonly used and user-friendly ImageJ program is available for free download [41]. More sophisticated software have been used to determine chromosome positioning as well [34, 42, 43]. At the final step, statistical analysis is applied using standard approaches and software.

Positioning of Chromosomes in Spermatozoa

Early studies of the chromosome arrangement in human sperm indicated the existence of their nonrandom localization [27, 44, 45]. The essence of these preliminary data has been supported by later works, and currently, we have base information on the nuclear positioning of all 23 chromosomes in haploid human male gamete. The story is still incomplete since data coming from different laboratories need to be reassessed to obtain the ultimate 2D or 3D map. Studies in this direction are of primary importance because sperm chromosome positioning bears elements of epigenetic information, and thus, is supposed to be critical for the correct arrangement and activation of chromosomes in zygote [19, 46].

Sperm Chromocenter

Localization of all human centromeres by FISH with α-satellite centromeric DNA or immunofluorescence of histone H3 centromere-specific variant (CENP-A) showed pronounced clustering of these chromosomal domains in mature sperm [28, 47]. Compact sperm chromocenter is buried within the nucleus interior as was demonstrated by confocal microscopy [47].

Existence of the chromocenter has been supported in several publications that followed. Gurevitch et al. were interested to know if there is a specific physical connection between the centromeres of the acrocentric chromosomes (HSA13, 14, 15, 21, and 22) within the human sperm nucleus [36, 48]. These chromosomes carry genes for the ribosomal RNA, and in somatic cells, are clustered in the nucleolus. Sperm cell nucleus has no observed nucleolus; nevertheless, the authors observed nonrandom proximity of acrocentric chromosomes within the chromocenter. Analysis of the absolute intranuclear position of seven centromeres (specific to HSA2, 6, 7, 16, 17, X, and Y) showed that they are located within a restricted space, which is located centrally and shifted toward the apical side of the sperm nuclei [36].

It has been proposed [19, 28] that sperm chromocenter is a structure playing the lead role in the formation of well-organized architecture of chromosomes in mature sperm and may be involved in establishing an ordered chromosome positioning [36].

Longitude and Radial Positioning

As discussed above, asymmetry in sperm nuclear shape makes it possible to establish the longitude and radial positioning of chromosomes after FISH. First indications of the preferential localization of the human sex chromosome HSAX in the anterior half of the nucleus were obtained using FISH with painting [45] and centromere [27] probes. Small numbers of sperm cells were analyzed in both works. Sbracia et al. [35] analyzed >36,000 sperm cells and showed that centromeres of both sex chromosomes were preferentially located in subacrosomal region (43–53% of cells, compare with 8–9% demonstrating basal localization). Apical positioning of the HSAX was confirmed later [36]. In disagreement, Manvelyan et al. [40] reported that only ~18% of HSAX positioned in the apical part and 40% in basal. The latter work provides data on localization of all human chromosomes using an advance technique of multicolor banding FISH followed by confocal microscopy. The fact that only 30 cells from one donor were studied and that different methods of sperm preparation for FISH and image analysis were used may account for the discrepancies in the results.

While the existence of chromosome radial arrangement is accepted in general, data for human sperm accumulated so far are few, and to some extent, inconsistent. For example, according to our data, [36] location of HSA6 is peripheral, and that of HSAY and HSAX, internal. Opposite, reverse positioning of HSAY, HSAX (peripheral), and HSA6 (central) was shown by others [40]. Both groups agree on positioning of HSA7 (most peripheral), HSA18 (peripheral), and HSA19 (internal) ([36, 40]; Mudrak et al., unpublished).

Ordered and nonrandom spatial localization of chromosomes has been observed in spermatozoa of other mammals. Most striking is the case of the monotreme mammals, which have highly

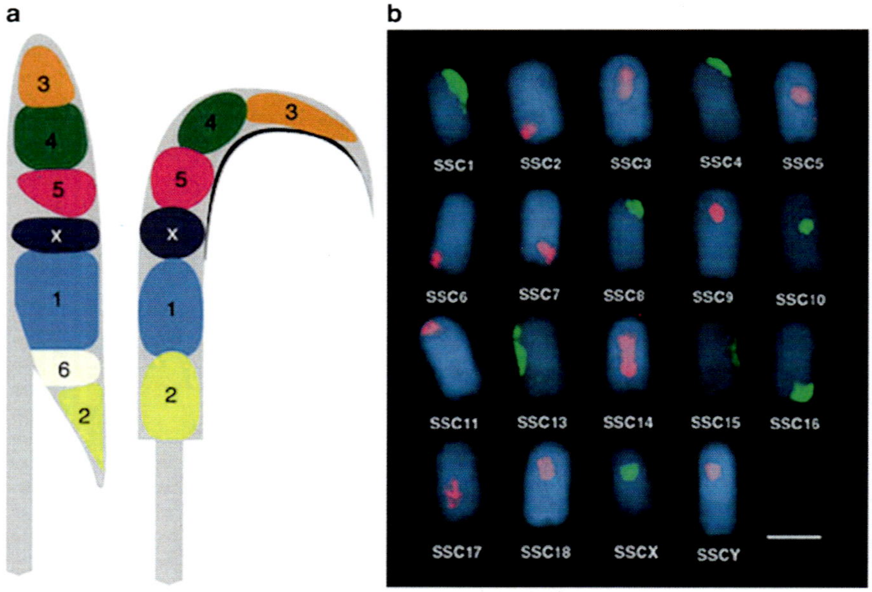

Fig. 5.4 Chromosome positioning in mammalian sperm. (**a**) Comparison of the homologous chromosomes positions in dunnart (*left*) and wombat (*right*) sperm. Scheme based on FISH using whole chromosome painting probes (from Greaves et al. [46], with permission). (**b**) Representative FISH images showing 19 CT (*green* or *red*) within porcine spermatozoa using chromosome-specific painting probes. The sperm nuclei are counterstained with DAPI (*blue*) (from Foster et al. [32], with permission)

asymmetrical, fibrillar sperm heads. Chromosome territories in platypus and echidna are arranged in telomere-to-telomere tandems along the narrow sperm nuclei [49]. Comparative analysis of the specific genes positions suggests a consistent chromosome order, which is conserved between these species [46, 49]. Strictly fixed and identical longitude positioning of chromosomes was demonstrated in sperm of two Australian marsupials ([50], Fig. 5.4a). Preferred localization of CT has been also observed in rat [51], mice [52], porcine [32], and bovine (Mudrak et al., unpublished) sperm. Most detailed study of the porcine spermatozoa showed that all chromosomes have preferential positions in two dimensions: radial and longitude [32] (Fig. 5.4b).

Importantly, similar to somatic cells, the radial intranuclear localization of sperm chromosomes in human ([40]; Mudrak et al., unpublished) and porcine [32] seems to be driven by gene density: the gene-rich chromosomes occupy more central positions, while the gene-poor chromosomes, a more peripheral one. At the same time, chromosomes distribution in human ([36]; Mudrak et al., unpublished) and porcine [32] sperm does not correlate with their size. Opposite result has been reported by [40].

Chromosome Movement during Spermatogenesis

Repositioning of whole chromosome territories in somatic cells during changes of genome activity or differentiation has been known for more than 20 years. One of the first examples was description of the HSAX move during epileptic seizure [53]. Later chromosome relocations were shown during differentiation of T-cell [54] and embryonic stem cells [55]; other examples are reviewed in [56].

Spatial arrangement of CT in sperm is much more ordered than that observed in somatic cell nuclei [19, 47, 57]. When and how is localization of chromosomes characteristic of mature spermatozoa established? A complex pattern of telomere and centromere repositioning at different stages of spermatogenesis has been observed, reviewed in [19, 58–61]. By using FISH on testicular preparations from normal fertile human males, it has been demonstrated [62] that XY pairing and "sex vesicle" formation comprise a complex series of spatial movements. During rat spermiogenesis, repositioning of RNO2 and RNO12 chromosomes were shown [51]. In the most detailed research of porcine spermatogenesis [32], noticeable CT movements were recorded that were "different in direction" for different chromosomes. As cells differentiated from spermatocytes through spermatids to mature sperm sex chromosome SSCX shifted from the nuclear periphery to interior locations, autosome SSC13 became more peripheral, while SSC5 did not demonstrate noticeable repositioning. Established new locations were preserved through the late stages of spermiogenesis and are characteristic to the mature sperm.

Mechanisms responsible for the specific intranuclear positioning of CT in sperm are totally unknown. For somatic cells, the hypothetical possibilities include activity-driven self-organization, specific interactions with nuclear membrane, associations between heterochromatic domains, and involvement of hypothetical protein complexes [63–65]. Recent study using mouse–human hybrid nuclei demonstrated conservation of human CT positions, which suggests existence of the conserved mechanism determining the nonrandom 3D CT placement in interphase nucleus [16].

Deviant Chromosome Positioning in Sperm of Subfertile Males

Existence of the preferred CT positioning in the human spermatozoa implies that deviation from the regular localization may be deleterious for proper fertilization and development. FISH using HSA1 painting probe revealed that while in 90% of donor sperm cells this chromosome was located in the apical half of the nucleus, its positioning in sperm of infertile individuals was noticeably less confined [66]. Similar observations were made by Finch et al. [39]. They compared sperm from 9 chromosomally normal men with the cohort of 15 infertile men undergoing male factor IVF treatment and having sperm samples with different types of abnormalities, including teratozoospermic. In the control group, the centromeres of chromosomes HSAX, HSAY, and HSA18 all demonstrated a central nuclear location while in the infertile men they were distributed over the nuclei in a random pattern.

Longitudinal and spatial localization of HSA7, HSA9, HSAX, and HSAY centromeres were compared in sperm nuclei of four control males with normal karyotypes and six carriers of reciprocal chromosome translocations [38]. This study revealed that chromosomes with translocations have shifted toward the intranuclear positions and that translocations studied might influence the localization of other chromosomes in sperm. This group also investigated localization of chromosome HSA15, HSA18, HSAX, and HSAY in fertile individuals and infertile patients with an increased level of aneuploidy [67]. In disomic sperm cells, chromosomes HSA15,15 and HSAY,Y were shifted toward the medial area, while chromosomes HSA18,18, toward the basal area of sperm nuclei. In hyperhaploidic sperm, slight changes of chromosome radial positioning were noted.

Potential connection between sperm chromosome positioning and fertility is an intriguing possibility, although it is challenging to achieve an unambiguous proof. One of the obstacles is a largely unknown degree of a populational variability in CT positioning. For example, interindividual differences were found in 25% of males [38]. More promising may be systematic study of teratozoospermic cases. Although aberration in chromosome positioning in sperm nuclei has been observed in a small number of infertility studies, there is a long way to go before the development of a valuable diagnostic test.

Possible Significance of Sperm Chromosome Positioning for Fertilization and Early Development

It has been proposed that different male chromosomes and chromosomal domains could be exposed to ooplasm components at different times after sperm penetration, resulting in diverse timing of chromatin remodeling preceding transcriptional activation and replication [19, 68, 69]. Nonrandom placement of sperm chromosomes may influence each of these events. In addition, potentially important may be the existence of a distinct chromosome neighborhood. In somatic cells, specific interchromosomal contacts, determined by their vicinity, play an important role in the regulation of gene expression [13, 70]. This may be also relevant to human sperm since some chromosomes are preferentially located in proximity of each other [36, 71] and can participate in programmed activation of male pronuclei in duo.

Does Positioning of Sex Chromosomes Have Functional Importance?

The attention of several groups was attracted by conserved location of sex chromosomes in mammalian sperm. Noticeably, in human [36] and porcine [32] spermatozoa, both sex chromosomes are localized most internally. The same is true for bovine X (Mudrak et al., unpublished).

As to the placement along the long sperm nuclear axis, sex chromosomes were found in the posterior subacrosomal position in humans ([27, 36, 45]; Mudrak, unpublished), monotremes and marsupial mammals [46, 49]. In porcine [32] and bovine (Mudrak, unpublished) sperm, the sex chromosomes are "shifted" to anterior-medial borderline, and thus, apical localization is not so explicit. Apparent conservation of the longitude sex chromosome position in the sperm of the monotremes, marsupials, and some eutherians implies important functions since these major mammal groups diverged 70 million years ago [46, 50].

Two opposite hypotheses concerning the implication of the sperm sex chromosome positioning for early embryonic development were put forward by [32] and [46]. Although both acknowledge the importance of chromosome placement within sperm nuclei, they however have diametrically opposite views on what is more important in fertilization: their radial or longitude localization. Foster et al. [32] suggest that peripheral regions of chromatin are affected first by the maternal cellular environment, and accordingly, chromosomes deep in the interior, such as X and Y, would respond to the signals from the oocyte the last. Greaves et al. [46, 50] hypothesize that apically located sex chromosomes enter the egg early during fertilization, and therefore, would be remodeled by the ooplasm one of the first. Indeed, in humans and monotremes, initial contact of gametes is by the anterior edge of the sperm head, reviewed by Bedford [72]. It should be noted that the point at which the sperm touches the egg may be irrelevant and delayed reorganization of subacrosomal chromatin may be the norm [73]. In summary, detailed studies of sperm chromosome remodeling during pronuclei formation are required to understand if and which chromosome location is essential – radial, longitude, or both.

Potential Role of Sperm Chromosome Positioning for Introcytoplasmic Sperm Injection

ICSI technique to treat some problems with male fertility became very popular in recent years because it avoids fertilization dependence on defects in some functional features of the spermatozoon [74]. In this procedure, a single spermatozoon is injected into oocyte, thus overcoming natural sperm selection during fertilization. Importantly, unlike spermatozoa naturally penetrated into an egg, the injected ones preserve intact plasma membrane and acrosome. Animal studies performed in rhesus monkeys demonstrated that preservation of acrosome appears to be associated with an abnormal pattern of chromatin decondensation during the formation of the male pronucleus [75]. During ICSI, nuclear decondensation was delayed in the sperm apical region compared to the basal region in monkey [76, 77], porcine [78], mouse [73], and in heterologous human-hamster ICSI [79, 80]. As discussed above, apical part of spermatozoa is a preferential "habitat" of sex chromosomes. Consequently, chromatin remodeling and replication of chromosomes located in subacrosomal part of the nucleus (e.g., X and Y) may be late in comparison with natural fertilization. According to initial observation using a small number of hamster oocytes injected with human spermatozoa, sex chromosomes remained condensed at 6 h post ICSI [79]. Apparent holdup of DNA replication in the male pronuclei in comparison with female pronuclei was also detected [79]. Greaves et al. [46] suggested that delayed DNA replication of sex chromosomes may lead to the increased frequency of their loss from human embryos after ICSI [81]. According to another view, which is based on a comparative study of the chromatin remodeling in mice IVF and ICSI, the delay in decondensation of the apical region of sperm head may be a normal stage of the pronuclei development [73].

Conclusion and Perspectives

Preferred nonrandom intranuclear localization of chromosomes in sperm is an element of emerging unique 3D structure of human genome, which is specific for the mature male germ cells. Together with the highly organized and conserved components of sperm nuclear architecture, it provides a mechanism for differential exposure of CT and chromatin domains to the ooplasm factors at fertilization. Consequently, CT positioning may be considered as a part of sperm-specific epigenetic code that will be deciphered in the descendant cells [82]. Further experiments are essential to fill numerous gaps in our understanding of sperm chromosome positioning. Some essential directions are: (1) establishment of a complete spatial map of sperm chromosomes; (2) identification of molecular mechanisms directing CT localization; (3) determination of CT positioning during spermatogenesis and in the developing male pronuclei; (4) analysis of the intra-population variability; and (5) ascertainment if gross deviations from "standard" CT localization exist in some cases of male infertility.

Acknowledgments This work has been supported by the Jones Foundation grant, and in part by National Institutes of Health Grant HD-042748 to A. Z.

References

1. Cremer T, Kurz A, Zirbel R, et al. Role of chromosome territories in the functional compartmentalization of the cell nucleus. Cold Spring Harb Symp Quant Biol. 1993;58:777–92.
2. Ramirez MJ, Surralles J. Laser confocal microscopy analysis of human interphase nuclei by three-dimensional FISH reveals dynamic perinucleolar clustering of telomeres. Cytogenet Genome Res. 2008;122(3–4):237–42.
3. Bolzer A, Kreth G, Solovei I, et al. Three-dimensional maps of all chromosomes in human male fibroblast nuclei and prometaphase rosettes. PLoS Biol. 2005;3(5):e157.
4. Cremer T, Kupper K, Dietzel S, Fakan S. Higher order chromatin architecture in the cell nucleus: on the way from structure to function. Biol Cell. 2004;96(8): 555–67.
5. Fedorova E, Zink D. Nuclear genome organization: common themes and individual patterns. Curr Opin Genet Dev. 2009;19(2):166–71.
6. Misteli T. Spatial positioning; a new dimension in genome function. Cell. 2004;119(2):153–6.
7. Verschure PJ. Positioning the genome within the nucleus. Biol Cell. 2004;96(8):569–77.
8. Parada LA, Roix JJ, Misteli T. An uncertainty principle in chromosome positioning. Trends Cell Biol. 2003;13(8):393–6.
9. Bickmore WA, Chubb JR. Dispatch. Chromosome position: now, where was I? Curr Biol. 2003;13(9):R357–9.
10. Sun HB, Shen J, Yokota H. Size-dependent positioning of human chromosomes in interphase nuclei. Biophys J. 2000;79(1):184–90.
11. Takizawa T, Meaburn KJ, Misteli T. The meaning of gene positioning. Cell. 2008;135(1):9–13.
12. Elcock LS, Bridger JM. Exploring the relationship between interphase gene positioning, transcriptional regulation and the nuclear matrix. Biochem Soc Trans. 2010;38(Pt 1):263–7.
13. Lanctot C, Cheutin T, Cremer M, Cavalli G, Cremer T. Dynamic genome architecture in the nuclear space: regulation of gene expression in three dimensions. Nat Rev Genet. 2007;8(2):104–15.
14. Kumaran RI, Thakar R, Spector DL. Chromatin dynamics and gene positioning. Cell. 2008;132(6): 929–34.
15. Cook PR, Marenduzzo D. Entropic organization of interphase chromosomes. J Cell Biol. 2009;186(6): 825–34.
16. Sengupta K, Camps J, Mathews P, et al. Position of human chromosomes is conserved in mouse nuclei indicating a species-independent mechanism for maintaining genome organization. Chromosoma. 2008;117(5):499–509.
17. Malhas A, Lee CF, Sanders R, Saunders NJ, Vaux DJ. Defects in lamin B1 expression or processing affect interphase chromosome position and gene expression. J Cell Biol. 2007;176(5):593–603.

18. Bridger JM, Foeger N, Kill IR, Herrmann H. The nuclear lamina. Both a structural framework and a platform for genome organization. FEBS J. 2007; 274(6):1354–61.
19. Zalensky A, Zalenskaya I. Organization of chromosomes in spermatozoa: an additional layer of epigenetic information? Biochem Soc Trans. 2007;35 (Pt 3):609–11.
20. Speicher MR, Carter NP. The new cytogenetics: blurring the boundaries with molecular biology. Nat Rev Genet. 2005;6(10):782–92.
21. Dernburg AF, Sedat JW. Mapping three-dimensional chromosome architecture in situ. Methods Cell Biol. 1998;53:187–233.
22. Fraser P, Bickmore W. Nuclear organization of the genome and the potential for gene regulation. Nature. 2007;447(7143):413–7.
23. Walter J, Joffe B, Bolzer A, et al. Towards many colors in FISH on 3D-preserved interphase nuclei. Cytogenet Genome Res. 2006;114(3–4):367–78.
24. Hepperger C, Otten S, von Hase J, Dietzel S. Preservation of large-scale chromatin structure in FISH experiments. Chromosoma. 2007;116(2): 117–33.
25. Goetze S, Mateos-Langerak J, Gierman HJ, et al. The three-dimensional structure of human interphase chromosomes is related to the transcriptome map. Mol Cell Biol. 2007;27(12):4475–87.
26. Balhorn R. The protamine family of sperm nuclear proteins. Genome Biol. 2007;8(9):227.
27. Hazzouri M, Rousseaux S, Mongelard F, et al. Genome organization in the human sperm nucleus studied by FISH and confocal microscopy. Mol Reprod Dev. 2000;55(3):307–15.
28. Zalensky AO, Breneman JW, Zalenskaya IA, Brinkley BR, Bradbury EM. Organization of centromeres in the decondensed nuclei of mature human sperm. Chromosoma. 1993;102(8):509–18.
29. Wyrobek AJ, Alhborn T, Balhorn R, Stanker L, Pinkel D. Fluorescence in situ hybridization to Y chromosomes in decondensed human sperm nuclei. Mol Reprod Dev. 1990;27(3):200–8.
30. Pinkel D, Landegent J, Collins C, et al. Fluorescence in situ hybridization with human chromosome-specific libraries: detection of trisomy 21 and translocations of chromosome 4. Proc Natl Acad Sci USA. 1988; 85(23):9138–42.
31. Sun HB, Yokota H. Correlated positioning of homologous chromosomes in daughter fibroblast cells. Chromosome Res. 1999;7(8):603–10.
32. Foster HA, Abeydeera LR, Griffin DK, Bridger JM. Non-random chromosome positioning in mammalian sperm nuclei, with migration of the sex chromosomes during late spermatogenesis. J Cell Sci. 2005;118 (Pt 9):1811–20.
33. Koehler D, Zakhartchenko V, Froenicke L, et al. Changes of higher order chromatin arrangements during major genome activation in bovine preimplantation embryos. Exp Cell Res. 2009; 315(12):2053–63.
34. Kupper K, Kolbl A, Biener D, et al. Radial chromatin positioning is shaped by local gene density, not by gene expression. Chromosoma. 2007;116(3):285–306.
35. Sbracia M, Baldi M, Cao D, et al. Preferential location of sex chromosomes, their aneuploidy in human sperm, and their role in determining sex chromosome aneuploidy in embryos after ICSI. Hum Reprod. 2002;17(2):320–4.
36. Zalenskaya IA, Zalensky AO. Non-random positioning of chromosomes in human sperm nuclei. Chromosome Res. 2004;12(2):163–73.
37. Mudrak O, Tomilin N, Zalensky A. Chromosome architecture in the decondensing human sperm nucleus. J Cell Sci. 2005;118(Pt 19):4541–50.
38. Wiland E, Zegalo M, Kurpisz M. Interindividual differences and alterations in the topology of chromosomes in human sperm nuclei of fertile donors and carriers of reciprocal translocations. Chromosome Res. 2008;16(2):291–305.
39. Finch KA, Fonseka KG, Abogrein A, et al. Nuclear organization in human sperm: preliminary evidence for altered sex chromosome centromere position in infertile males. Hum Reprod. 2008; 23(6):1263–70.
40. Manvelyan M, Hunstig F, Bhatt S, et al. Chromosome distribution in human sperm – a 3D multicolor banding-study. Mol Cytogenet. 2008;1:25.
41. Collins TJ. ImageJ for microscopy. Biotechniques. 2007;43(1 Suppl):25–30.
42. Gue M, Sun JS, Boudier T. Simultaneous localization of MLL, AF4 and ENL genes in interphase nuclei by 3D-FISH: MLL translocation revisited. BMC Cancer. 2006;6:20.
43. Iannuccelli E, Mompart F, Gellin J, Lahbib-Mansais Y, Yerle M, Boudier T. NEMO: a tool for analyzing gene and chromosome territory distributions from 3D-FISH experiments. Bioinformatics. 2010;26(5):696–7.
44. Geraedts JP, Pearson PL. Spatial distribution of chromosomes 1 and Y in human spermatozoa. J Reprod Fertil. 1975;45(3):515–7.
45. Luetjens CM, Payne C, Schatten G. Non-random chromosome positioning in human sperm and sex chromosome anomalies following intracytoplasmic sperm injection. Lancet. 1999;353(9160):1240.
46. Greaves IK, Rens W, Ferguson-Smith MA, Griffin D, Marshall Graves JA. Conservation of chromosome arrangement and position of the X in mammalian sperm suggests functional significance. Chromosome Res. 2003;11(5):503–12.
47. Zalensky AO, Allen MJ, Kobayashi A, Zalenskaya IA, Balhorn R, Bradbury EM. Well-defined genome architecture in the human sperm nucleus. Chromosoma. 1995;103(9):577–90.
48. Gurevitch M, Amiel A, Ben-Zion M, Fejgin M, Bartoov B. Acrocentric centromere organization within the chromocenter of the human sperm nucleus. Mol Reprod Dev. 2001;60(4):507–16.
49. Watson JM, Meyne J, Graves JA. Ordered tandem arrangement of chromosomes in the sperm heads of

monotreme mammals. Proc Natl Acad Sci USA. 1996;93(19):10200–5.
50. Greaves IK, Svartman M, Wakefield M, et al. Chromosomal painting detects non-random chromosome arrangement in dasyurid marsupial sperm. Chromosome Res. 2001;9(3):251–9.
51. Meyer-Ficca M, Muller-Navia J, Scherthan H. Clustering of pericentromeres initiates in step 9 of spermiogenesis of the rat (*Rattus norvegicus*) and contributes to a well defined genome architecture in the sperm nucleus. J Cell Sci. 1998;111(Pt 10): 1363–70.
52. Garagna S, Zuccotti M, Thornhill A, et al. Alteration of nuclear architecture in male germ cells of chromosomally derived subfertile mice. J Cell Sci. 2001;114(Pt 24):4429–34.
53. Borden J, Manuelidis L. Movement of the X chromosome in epilepsy. Science. 1988;242(4886):1687–91.
54. Kim SH, McQueen PG, Lichtman MK, Shevach EM, Parada LA, Misteli T. Spatial genome organization during T-cell differentiation. Cytogenet Genome Res. 2004;105(2–4):292–301.
55. Morey C, Da Silva NR, Perry P, Bickmore WA. Nuclear reorganisation and chromatin decondensation are conserved, but distinct, mechanisms linked to Hox gene activation. Development. 2007;134(5):909–19.
56. Bartova E, Kozubek S. Nuclear architecture in the light of gene expression and cell differentiation studies. Biol Cell. 2006;98(6):323–36.
57. Zalensky AO. Genome architecture. In: Verma RS, editor. Advances in genome biology, vol. 24. London: JAI Press; 1998. p. 179–210.
58. Scherthan H. A bouquet makes ends meet. Nat Rev Mol Cell Biol. 2001;2(8):621–7.
59. Scherthan H, Wang H, Adelfalk C, et al. Chromosome mobility during meiotic prophase in *Saccharomyces cerevisiae*. Proc Natl Acad Sci USA. 2007;104(43): 16934–9.
60. Zalenskaya IA, Bradbury EM, Zalensky AO. Chromatin structure of telomere domain in human sperm. Biochem Biophys Res Commun. 2000;279(1): 213–8.
61. Zalensky AO, Tomilin NV, Zalenskaya IA, Teplitz RL, Bradbury EM. Telomere-telomere interactions and candidate telomere binding protein(s) in mammalian sperm cells. Exp Cell Res. 1997;232(1):29–41.
62. Armstrong SJ, Kirkham AJ, Hulten MA. XY chromosome behaviour in the germ-line of the human male: a FISH analysis of spatial orientation, chromatin condensation and pairing. Chromosome Res. 1994;2(6): 445–52.
63. Parada LA, McQueen PG, Misteli T. Tissue-specific spatial organization of genomes. Genome Biol. 2004;5(7):R44.
64. Parada LA, Sotiriou S, Misteli T. Spatial genome organization. Exp Cell Res. 2004;296(1):64–70.
65. Mehta IS, Elcock LS, Amira M, Kill IR, Bridger JM. Nuclear motors and nuclear structures containing A-type lamins and emerin: is there a functional link? Biochem Soc Trans. 2008;36(Pt 6):1384–8.
66. Mudrak O, Zalensky A. Genome architecture in human sperm cells: possible implications for male infertility. In: Kruger TF, Oehninger S, editors. Diagnosis and treatment of male infertility. Informa Healthcare; 2006.
67. Olszewska M, Wiland E, Kurpisz M. Positioning of chromosome 15, 18, X and Y centromeres in sperm cells of fertile individuals and infertile patients with increased level of aneuploidy. Chromosome Res. 2008;16(6):875–90.
68. McLay DW, Clarke HJ. Remodelling the paternal chromatin at fertilization in mammals. Reproduction. 2003;125(5):625–33.
69. Adenot PG, Szollosi MS, Geze M, Renard JP, Debey P. Dynamics of paternal chromatin changes in live one-cell mouse embryo after natural fertilization. Mol Reprod Dev. 1991;28(1):23–34.
70. Cavalli G. Chromosome kissing. Curr Opin Genet Dev. 2007;17(5):443–50.
71. Tilgen N, Guttenbach M, Schmid M. Heterochromatin is not an adequate explanation for close proximity of interphase chromosomes 1–Y, 9–Y, and 16–Y in human spermatozoa. Exp Cell Res. 2001;265(2): 283–7.
72. Bedford JM. The co evolution of mammalian gametes. In: Dunbar BS, O'Rand MG, editors. A comparative overview of mammalian fertilization. New York: Plenum; 1991. p. 3–28.
73. Ajduk A, Yamauchi Y, Ward MA. Sperm chromatin remodeling after intracytoplasmic sperm injection differs from that of in vitro fertilization. Biol Reprod. 2006;75(3):442–51.
74. Palermo GD, Neri QV, Takeuchi T, Rosenwaks Z. ICSI: where we have been and where we are going. Semin Reprod Med. 2009;27(2):191–201.
75. Sutovsky P, Hewitson L, Simerly CR, et al. Intracytoplasmic sperm injection for Rhesus monkey fertilization results in unusual chromatin, cytoskeletal, and membrane events, but eventually leads to pronuclear development and sperm aster assembly. Hum Reprod. 1996;11(8):1703–12.
76. Hewitson L, Dominko T, Takahashi D, et al. Unique checkpoints during the first cell cycle of fertilization after intracytoplasmic sperm injection in rhesus monkeys. Nat Med. 1999;5(4):431–3.
77. Ramalho-Santos J, Sutovsky P, Simerly C, et al. ICSI choreography: fate of sperm structures after monospermic rhesus ICSI and first cell cycle implications. Hum Reprod. 2000;15(12):2610–20.
78. Katayama M, Koshida M, Miyake M. Fate of the acrosome in ooplasm in pigs after IVF and ICSI. Hum Reprod. 2002;17(10):2657–64.
79. Terada Y, Luetjens CM, Sutovsky P, Schatten G. Atypical decondensation of the sperm nucleus, delayed replication of the male genome, and sex chromosome positioning following intracytoplasmic human sperm injection (ICSI) into golden hamster eggs: does ICSI itself introduce chromosomal anomalies? Fertil Steril. 2000;74(3): 454–60.

80. Jones EL, Mudrak O, Zalensky AO. Kinetics of human male pronuclear development in a heterologous ICSI model. J Assist Reprod Genet. 2010;27(6):277–83.
81. In't Veld PA, van Opstal D, Van den Berg C, et al. Increased incidence of cytogenetic abnormalities in chorionic villus samples from pregnancies established by in vitro fertilization and embryo transfer (IVF-ET). Prenat Diagn. 1995;15(10):975–80.
82. Foster HA, Bridger JM. The genome and the nucleus: a marriage made by evolution. Genome organisation and nuclear architecture. Chromosoma. 2005;114(4):212–29.

Sperm Mitochondrial DNA

6

Justin C. St. John and Bianca St. John

Abstract

Human mitochondrial DNA (mtDNA) is 16.6 kb in size and resides in the mitochondrion. It encodes 13 of the subunits of the electron transfer chain that generates the vast majority of cellular ATP through the process of oxidative phosphorylation (OXPHOS). The importance of OXPHOS to sperm motility and function has been controversial. However, we present a case for the importance of OXPHOS in sperm function based on the effects that pathogenic mtDNA mutations and deletions have on sperm motility and function and how they are descriptive of certain forms of male subfertility. We also describe patterns of inheritance for the mitochondrial genome and how the elimination of sperm mtDNA in mammals prevents the transmission of mutant/deleted mtDNA to subsequent generations but when there is leakage it leads to a severe phenotype. This is also portrayed in the light of how mtDNA copy is reduced during the later stages of spermatogenesis and how reduced mtDNA copy number in the mature spermatozoa is indicative of good-quality, not poor-quality spermatozoa.

Keywords

Mitochondrial DNA • Replication • Oxidative phosphorylation • Polymerase gamma • Spermatozoa

What is mtDNA?

The human mitochondrial DNA (mtDNA) genome is approximately 16.6 kb in size [1] (Fig. 6.1) and is located in the inner membrane of the mitochondrion. It consists of a heavy (H) strand and a light (L) strand, which encode a total

J.C. St. John (✉)
The Mitochondrial Genetics Group, Centre for Reproduction & Development, Monash Institute of Medical Research, Monash University, 27-31 Wright Street, Clayton, VIC 3168, Australia
e-mail: Justin.StJohn@monash.edu

A. Zini and A. Agarwal (eds.), *Sperm Chromatin: Biological and Clinical Applications in Male Infertility and Assisted Reproduction*, DOI 10.1007/978-1-4419-6857-9_6,

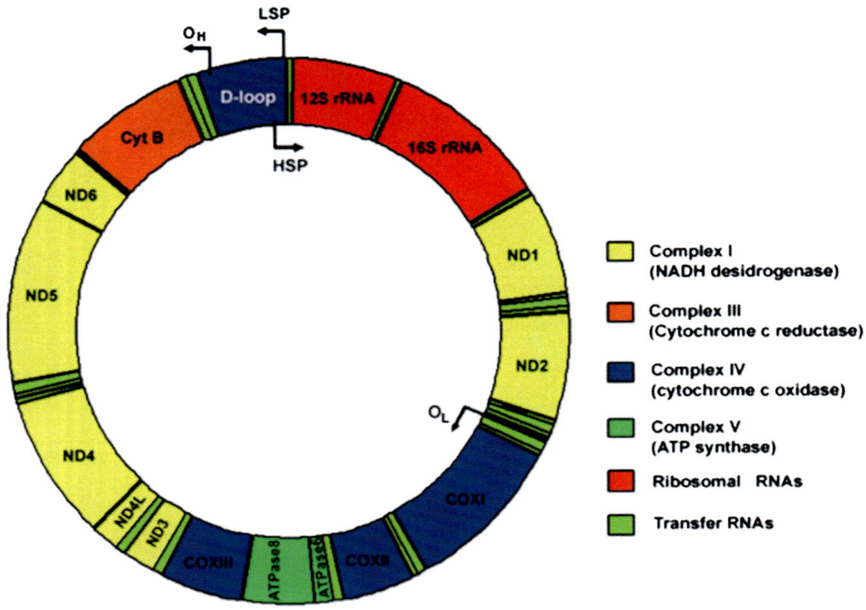

Fig. 6.1 The human mitochondrial genome. mtDNA encodes 13 of the subunits residing in four of the complexes of the ETC. It comprises a heavy (H) strand, which encodes 12 of these subunits along with 14 tRNAs and the 2 rRNAs, and a light (L) strand, which encodes one subunit (ND 6) and 8 tRNAs. The D-loop houses the H-strand origin of replication (O_H), the H- and L-strand promoters (LSP) and conserved sequence boxes. The D-loop is the only region of mtDNA that is not transcribed. However, it is the location of two hypervariable regions that can identify individuals from the same maternal lineage through molecular fingerprinting. O_L L-strand origin of replication

of 13 proteins associated with the subunits of Complexes I, III, IV and V of the electron transfer chain (ETC), the biochemical process that generates the vast majority of cellular ATP [2] (Fig. 6.2). The mitochondrial genome also encodes 22 tRNAs and 2 rRNAs (Fig. 6.1), thus contributing some, but not all of the transcription and translational machinery that is required for transcription and protein synthesis (Fig. 6.3). This demonstrates the importance of the symbiotic relationship between the cell and the mitochondria. The tRNAS are interspersed between most of the coding genes, while the coding regions for ATPase 6, ATPase 8 and ND4, and NDL4 overlap [1]. Furthermore, some of the genes do not have sequences for termination codons, which are thus generated through post-transcriptional polyadenylation [3].

There is one non-coding region of 1,121 bp, known as the displacement (D)-Loop (Fig. 6.1). This multifunctional control region is the site for interaction with the nuclear-encoded transcription and replication factors, which ensure efficient transcription and replication of this genome [4, 5]. Within the D-Loop (Fig. 6.1), there are two hypervariable (HV) regions, HV1 and HV2 [1] which contain specific sequences that distinguish distinct maternal lineages from one another. These regions are used by forensic scientists to determine perpetrators of crime [6] and to identify unidentified remains [7]. HV1 and 2 are also used to determine patterns of mtDNA transmission in offspring derived through fertilisation protocols and a range of assisted reproductive technologies including cytoplasmic transfer [8] and nuclear transfer [9–11].

Why is mtDNA Important?

The 13 subunits of the ETC encoded by the mtDNA genome are key components contributing to the process of oxidative phosphorylation (OXPHOS; Fig. 6.2). OXPHOS generates

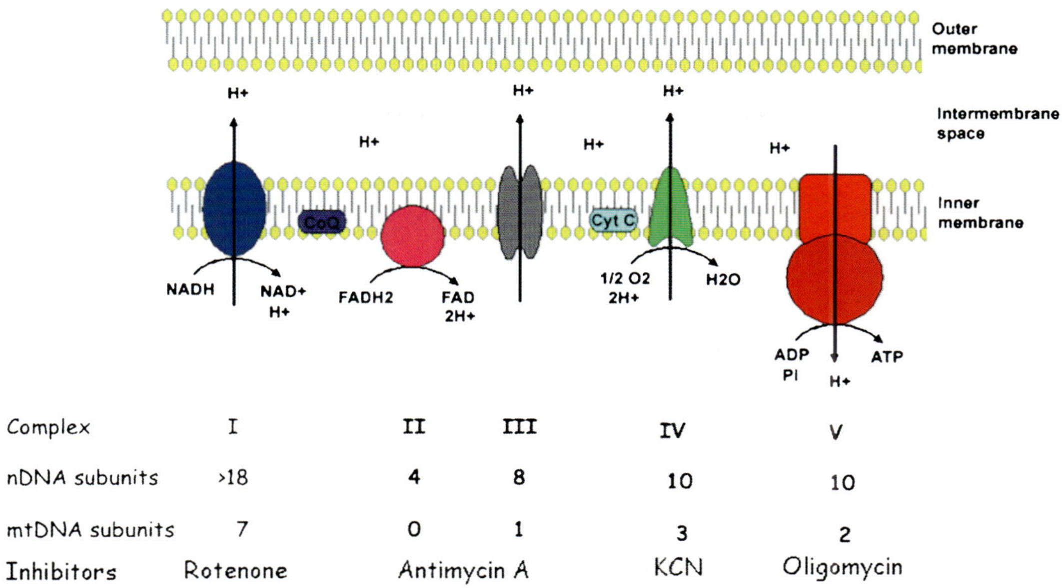

Fig. 6.2 The electron transfer chain. The subunits for each of the complexes of the ETC, except for Complex II, are encoded by both the mitochondrial and chromosomal genomes. ATP is generated by electrons passing along each of the complexes. Protons are pumped across the inner mitochondrial membrane to establish an electrochemical gradient whilst molecular oxygen reacts with protons to generate H_2O. This process generates sufficient energy to support ATP synthesis. *nDNA* nuclear DNA; *mtDNA* mitochondrial DNA; *KCN* potassium cyanide

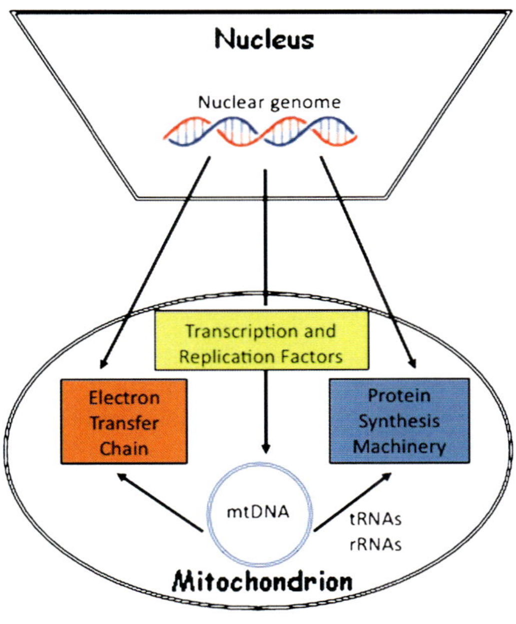

Fig. 6.3 Nucleo-mitochondrial interactions. MtDNA is reliant on nuclear-encoded transcription, replication and translation factors to generate proteins for the ETC. Likewise, the nucleus is dependent on the mitochondrial genes to contribute proteins to the ETC. This symbiotic relationship ensures that there is sufficient cellular energy so that the cell can perform its specific functions

32 molecules of ATP to every 2 produced through glycolysis but is highly dependent on substrates generated through the other anaerobic biochemical processes, such as the Krebs cycle and β-oxidation, and utilises these fuels in an O_2-mediated process [2]. This form of metabolism is especially essential for cells with high aerobic energy requirements, such as neurons and skeletal muscle [12]. The remaining 70+ genes of Complexes I, III, IV and V and all of the genes of Complex II are encoded by the chromosomal genome (Figs. 6.2 and 6.3), which translocate to the mitochondrion through a variety of import and chaperone proteins [13]. This again highlights the symbiotic nature of the mitochondria and the cell.

Until recently, it has been purported that all copies of mtDNA within an organism are identical and thus homoplasmic [14, 15]. However, the recent advances in sequencing technologies, and specifically deep sequencing, have demonstrated that many individuals have variable levels of polymorphic variants that contribute to wide-ranging levels of heteroplasmy [16]. Some of these mutations contribute to the genetic basis of hereditary mitochondrial disorders. These include mitochondrial myopathy, encephalopathy, lactic acidosis and stroke (MELAS) syndrome [17], neuropathy, ataxia and retinitis pigmentosa (NARP; [18]), Leber's hereditary optic neuropathy (LHON; Wallace et al. [19]) and myoclonic epilepsy and ragged-red fibre (MERRF) syndrome [20]. Single point mutations in the mtDNA coding regions have been identified in all these disorders except MERRF syndrome [20], which results from an A→G substitution in the mitochondrial tRNA(Lys) gene. Furthermore, a single large-scale deletion of 4,977 bp is indicative of Kearns–Sayre syndrome [21], whilst multiple deletions ranging from 2 bp to >10 kb have been observed with ageing [22].

Generally, the phenotype for each of these diseases is determined by the degree of mutant to wild-type (WT) loading within the affected tissue, except in the case of LHON, where the mutation is usually homoplasmic and other factors, such as the sex of the individual, modify the phenotype, suggesting a role for *trans*-acting nuclear genetic factors in this disease [23]. Nevertheless, 10–15% of LHON carriers are thought to be heteroplasmic with the threshold for onset of the disease phenotype being 60% [24]. In MERRF, over 85% mutant loading is typical [25], while in severe multisystem disorder and respiratory chain deficiency syndrome, only 25% mutant loading is required to induce a dramatic phenotype [26]. These contradictory findings may be due to analysis of mutant loading in cybrids using mature differentiated cells, and thus, do not incorporate the period during differentiation, when mtDNA mass accumulates. Whilst studies in mouse models assess events during differentiation and development, they rarely include single cells or specific lineages, thus obscuring significant molecular events. However, one recent study using embryonic stem cell fusion approaches, whereby mutant mtDNA is transferred into mtDNA-depleted embryonic stem cells, has indicated that neuronal differentiation is affected by the mutant mtDNA loading [27], and this may have significant implications for spermatogenesis.

OXPHOS and Sperm Function

In comparison to the mature oocyte and somatic cells, mature mammalian spermatozoa have very few mitochondria, where 22–28 mitochondria are isolated in a helical manner in the midpiece [28]. This is unlike somatic cells where larger numbers of mitochondria are located in the cytoplasm and they have very dynamic roles which are influenced by, amongst other factors, the stage of the cell cycle [29]. Indeed, during spermatogenesis, mitochondria are located within the cytoplasm of these precursor cells [30]. However, a physical relocation takes place during spermiogenesis, just as when the transition between the acrosome and Golgi apparatus takes place [31]. Over the last 30 years, there has been a great deal of debate as to whether these few sperm mitochondria contribute greatly to sperm function, especially as they appear to be isolated in the mature spermatozoa that encapsulates them through rigorous disulphide bonding [32].

The significance of OXPHOS-derived ATP as opposed to anaerobically derived ATP in cells is generally determined using inhibitors that target the specific complexes of the ETC. In a number of classic experiments performed in the 1970s and 1980s, Storey and colleagues overcame the problem of isolating sperm mitochondria by rupturing the cytoplasmic membranes [33–38]. This enabled them to determine the respiration rates and levels of oxygen consumption to predict whether OXPHOS was vital for ATP production and thus motility. Their findings were species-specific, where the requirement for OXPHOS-derived ATP was dependent on the glucose concentration of the female reproductive tract. Nevertheless, in human spermatozoa, we have shown that by using the mitochondrial specific inhibitors, rotenone, potassium cyanide and oligomycin, and culturing spermatozoa in a 2-mM glucose environment, which is indicative of the glucose concentration in the female reproductive tract [39], sperm motility was significantly reduced [40]. However, when spermatozoa were cultured in classic sperm culture media, namely, with 5 mM glucose, it was evident that sperm mitochondrial function was not severely hindered and that spermatozoa could utilise the glucose effectively for motility. Other investigators have also demonstrated an association between the OXPHOS inhibitors for respiratory Complexes I, III and IV and sperm motility [41], and an association with the performance of these complexes and asthenozoospermia [42]. Furthermore, biochemical studies on sperm from a patient harbouring a maternally inherited mtDNA mutation associated with Complex I have shown that the addition of succinate, which enters the ETC at Complex II, increases sperm motility significantly and bypasses the effects of the mutation [43].

The Mitochondrial Nucleoid

In somatic cells, it is thought that the mitochondrial genome persists in multimeric form within the mitochondrion. This would explain the large number of mitochondrial copies that have been observed in somatic tissues, such as cells from skeletal and cardiac muscle which possess 3,650 ± 620 and 6,790 ± 920 mtDNA copies/cell, respectively [44]. These multiple copies of the genome are anchored in the mitochondrial nucleoid, which in turn is likely to be anchored to the inner mitochondrial membrane through ATAD3 proteins [45]. In both spermatozoa [46] and oocytes [47, 48], mtDNA appears to exist in monomeric form in individual mitochondria. The mitochondrial nucleoid consists of not only one or more mitochondrial genomes but also approximately 30 nuclear proteins that are involved in the maintenance and packaging of the genome along with mediating transcription and replication of the genome [49]. In terms of transcription and replication, the key factors are as follows: mitochondrial transcription Factor A (TFAM), the mitochondrial specific Polymerase Gamma (Polg), which has both a catalytic (PolgA) and an accessory subunit (PolgB), the mitochondrial specific RNA Polymerase (mtRNApol), the mitochondrial specific single-stranded binding protein (mtSSB) and the mitochondrial specific helicase, Twinkle.

mtDNA Replication

Currently, two models have been described as mechanistic approaches for the replication of the genome. These are the asymmetric [50] and the coupled leading lagging strand synthesis [51] models. These two quiet distinct mechanisms are controversial with each party disputing each other's approach in the literature [52, 53]. Until recently, the asymmetric model provided the traditional understanding of mtDNA replication (reviewed in [50]). It is initiated from the H strand origin of replication, which is located within the D loop region. In this instance, TFAM interacts with the enhancer of the light strand promoter and this generates a conformational change that exposes the promoter region to mtRNApol. Once the RNA primer has been generated, it is then employed by PolgA to initiate mtDNA replication. Mitochondrial replication then progresses two thirds round the genome to the origin of L-strand replication, which in turn triggers synthesis of the L-strands in the anticlockwise direction. The coupled leading and lagging strands synthesis method proposes that both H- and L-strand synthesis are initiated from the same initiation cluster sites with each strand being replicated in a bidirectional fashion [51]. This model also incorporates the use of replication intermediates to fill gaps within replicating DNA on the lagging strand [54]. Although the proponents of this mechanism do not argue that it is the sole mechanism, they suggest that it operates in addition to the asymmetric model whereby one mechanism would be indicative of accumulation of mtDNA mass as might be the case during the early stages of spermatogenesis, whilst the other may be associated with mtDNA replenishment following mtDNA damage or transcription.

POLG, mtDNA-Type Disease, and Sperm Function

The human chromosomal POLG gene is located at 15q24-15q26 [55] and consists of a 140-KDa catalytic subunit (POLGA) and a 54-KDa accessory subunit (POLGB; [56]). POLGA possesses a 5'-3' exonuclease domain that ensures effective proofreading and DNA repair [57], whilst POLGB is essential for promoting DNA binding and high levels of processivity and fidelity [58, 59]. It also has a putative role in recognising the RNA primers that initiate mtDNA replication [60]. A number of missense mutations have been identified in POLG, and these are associated with large-scale mtDNA deletions and/or mtDNA depletion-type syndromes. These include Progressive External Opthalmoplegia (PEO), mitochondrial neurogastrointestinal encephalomyopathy [61–63], testicular cancer [64], Alper's disease [65–67] and Parkinsonism and premature menopause [68, 69].

POLG activity is severely inhibited by nucleoside analogue reverse transcriptase inhibitors (NRTIs), the compounds that have been used to reduce viral load in HIV-positive patients [70]. For example, the frequently used 2',3'-dideoxycytidine (ddC) can mediate near mtDNA depletion of in vitro cultured cells within a few days [71]. As a result, many HIV-positive patients suffer from mtDNA-depletion type syndromes, such as mitochondrial myopathies and neuropathies [72]. We have also shown that sperm samples from HIV-positive men treated with NRTIs can, after a 12-month period, exhibit large-scale mtDNA deletions and, after a further 6 months, result in complete loss of sperm mtDNA content, rendering the male azoospermic [73].

Characteristic to human POLG, and not to other species, is a series of trinucleotide CAG repeats ($n = 10$), located at the 5′ end, that encode for a polyglutamine tract [74]. The variability of the number of CAG repeats in, for example, the androgen receptor gene, has been proposed as an indicator and putative cause of male infertility (reviewed in [75]). This approach has also been applied to POLG where a series of reports have debated whether it is accountable for some forms of male idiopathic infertility. The initial report suggested an absence of the common allele as the homozygous mutant genotype (not 10/not 10) was observed at an increased frequency in patients presenting with moderate oligozoospermia when compared with fertile men [76]. However, there was no association with extreme oligozoospermia and azoospermia. A subsequent study based on Danish patients identified an association between the loss of the common allele and idiopathic infertility [77]. However, this was not reproducible in two separate cohorts of Italian [78] and French [79] infertile and normozoospermic fertile men. Furthermore, the French study demonstrated that over 50% of the homozygous mutant men were able to produce offspring through intercourse or following assisted reproduction [79]. In addition, as a subsequent Italian study confirmed, there was no association between allelic frequency for oligozoospermia and normozoospermic [80].

As POLG is a mediator of mtDNA replication, it would be anticipated that, as with certain mtDNA depletion syndromes, there would be an increase in either the presence of mtDNA mutations or a decrease in mtDNA copy number in men presenting with the mutant genotype. This is especially in light of studies performed on POLG knockout mice where the homozygous null phenotype is embryonic lethal and the heterozygous knockout suffers from severe mtDNA-depletion type syndrome [81]. Nevertheless, it appears that there are no differences in the numbers of mtDNA nucleotide substitutions for the different POLG CAG genotypes in both normozoospermic and non-normozoospermic men, nor were any mutations identified in the three exonuclease motifs of POLG for such patients [82]. We have, however, taken this a step further by relating gene sequence variation to protein expression and determined that oligoasthenoteratozoospermic men had significantly higher incidences of heterozygosity for CAG repeats, which was coupled to a lower percentage of spermatozoa expressing POLGA [46]. Additionally, these men had higher numbers of mtDNA copy number, which is indicative of poor sperm quality.

TFAM, mtDNA Disease, and Its Role During Spermatogenesis

Human TFAM locates to chromosome 10q21 and its protein is 204 amino acids in size. It is a member of the High Mobility Group (HMG) of proteins and consists of two HMG boxes, a linker and a mitochondrial targeting sequence [83]. Knockout studies in the mouse demonstrate that it has either a direct or indirect role as a regulator of mtDNA copy number. The heterozygous knockout exhibits reduced mtDNA copy number and myocardial OXPHOS deficiency [84]. Homozygous null mice suffer from severe mtDNA depletion and abolished OXPHOS and are embryonic lethal. Depletion of mtDNA in cultured cells also results in decreased expression of TFAM and mtRNApol [9, 85]. TFAM also acts as a regulator of mitochondrial gene expression, [86] but when overexpressed, binds to grooves within the mitochondrial genome, thus inhibiting transcription [87], and as transcription precedes replication, replication will also be inhibited. Other studies have demonstrated that it has a clearly defined role as a packaging protein, characteristic of its HMG family members such as histones [88]. Nevertheless, TFAM is dependent on interaction with other members of the nucleiod for it to be functional. These include mitochondrial transcription factor B1 and B2 [89] and Nuclear Respiratory Factors 1 (NRF-1) and 2 (NRF-2). Indeed, TFAM's promoter possesses recognition sites for NRF-1 and NRF-2 [90, 91], and these sites possess CpG islands, which may control mtDNA transcription and replication through their DNA methylation [92].

Sperm mtDNA Replication

As with oocyte precursor primordial germ cells, male primordial germ cells will have very few copies of mtDNA. However, whilst oocytes accumulate mtDNA mass later during maturation [93], the spermatogonial stem cells maintain higher numbers of mtDNA up to the spermatocyte stage [94]. These are then subsequently reduced once meiosis II has been completed so that, as the round spermatid differentiates into an elongated spermatid, the mature spermatozoa will have tenfold less mtDNA [94]. In the mouse, this loss in mtDNA copy number coincides with the loss of TFAM possessing the mitochondrial targeting sequence that will ensure its translocation to the mitochondria [95]. Instead, its expression is replaced by an isoform that does not possess this targeting sequence, and thus, ensures that TFAM remains located in the head of the spermatozoa and cannot interact with mtDNA. In the human, this is regulated in a somewhat different manner whereby TFAM simply ceases to be expressed [96]. Nevertheless, we have observed that significantly more good-quality spermatozoa express TFAM than poor-quality spermatozoa [46].

Clinically, the regulation of mtDNA copy number during early development may have significant implications for sperm quality. In spermatozoa collected from density gradients that were indicative of progressive motility, the mean mtDNA copy number per spermatozoa was 1.4 for normozoospermic samples, 6.1 when one abnormal sperm parameter was described and 9.1 for samples with two or more abnormal sperm criteria [97]. The spermatozoa present in lower gradient layers possessed higher levels of mtDNA copy number (17.1 copies/spermatozoa). However, another study reported to the contrary, whereby normozoospermics had a mean number of 74.1 DNA copies/spermatozoon, asthenozoospermics possessed a mean of 7.2 molecules [98]. Nevertheless, we have demonstrated that sperm samples from OAT patients exhibited significantly higher mtDNA (>46) content than normozoospermics and conversely they had a lower percentage of spermatozoa expressing POLG and TFAM [46]. On the contrary, good-quality spermatozoa possessed fewer mtDNA copies (<10) but had significantly more spermatozoa that expressed POLG, TFAM and mtDNA-encoded genes. The reduction in mtDNA content in normal samples is most likely indicative of normal spermiogenesis having ensued with the increases in POLG and TFAM expression being a compensatory mechanism for low mtDNA copy number and thus ensuring a form of mitochondrial homeostasis. Similar observations have been made from mtDNA-depletion studies in somatic cells [9].

mtDNA Inheritance

Under normal circumstances, mtDNA is inherited from the population present in the mature metaphase II oocyte just prior to fertilisation.

In mammalian crosses generated from the same strain or breed (intraspecific), sperm mtDNA appears to be eliminated prior to the onset of genome activation in the newly formed embryo, namely, the 2-cell stage in the mouse and [99, 100] and 4–8 cell stages in sheep [101] and non-human primates [102]. This targeted elimination of sperm mtDNA is thought to be through ubiquitination of the spermatozoa's mitochondria [103, 104]. To this extent, it has been proposed that spermatogonial cells maintain a ubiquitin label throughout development, which is recognised by the oocyte's ubiquitination machinery once fertilisation has been initiated [103–105]. This is very similar to an innate immune reaction where foreign particles would be destroyed and, in line with present thinking, indicating a role for mitochondrial or bacterial DNA being initiators of such innate immune responses [106]. Although the ubiquitin label is maintained throughout spermatogenesis, it appears to be suppressed during maturation of spermatozoa in the epididymis, and is then either re-expressed or unmasked in ejaculated spermatozoa [107]. Nevertheless, others have demonstrated in Japanese Medaka embryos the active digestion of sperm mtDNA just after fertilisation, which proceeds destruction of the sperm mitochondria [108]. However, it remains to be determined whether sperm mtDNA elimination is specific or targeted along with oocyte mtDNA elimination during the very early stages of preimplantation development [109, 110].

This process of targeted elimination does not appear to take place in interspecific crosses (i.e. crossings between different strains or breeds) as sperm mtDNA persists, although at low levels, in offspring from a range of mammalian species [99–102]. Interestingly, however, the original sperm mtDNA contribution does not persist in subsequent generations [100], thus indicating that it is not incorporated into the germ line. Interestingly, other species do transmit sperm and oocyte mtDNA in a heteroplasmic manner. *Drosophila* transmit sperm and oocyte mtDNA to their progeny following both intra- and interspecific crossing [111]. Uniquely, mussels transmit both male- and female-specific genomes to male offspring, but female offspring possess female-specific only molecules [112, 113]. Nevertheless, normal and abnormal human embryos can fail to eliminate their sperm mtDNA [114]. When such an outcome occurs, then sperm mtDNA can recombine with oocyte mtDNA resulting in the generation of a new hybrid mtDNA molecule that segregates randomly during development [115]. This has resulted in a male patient suffering from a muscle myopathy [116] and demonstrates the selective replicative advantage that was afforded sperm mtDNA based on its 1:30,000 contribution to the zygote.

Mitochondrial DNA Variants and Their Effect on Sperm Function

Following the initial hypothesis of Cummins et al. [117], it has been demonstrated that mutations associated with a clinical phenotype, such as the A3243G mutation, have effects on sperm quality and their motility [118]. Other studies have analysed large-scale deletions, such as the 4,977 bp common deletion, as an indicator of good- and poor-quality spermatozoa. One group demonstrated a correlation between an increase in the presence of this deletion and poor-quality spermatozoa; however, its incidence was at extremely low levels (0.0032% for the 80% Percoll fractions to 0.0708% for the <50% Percoll fractions; [119]). Two other studies demonstrated that it is not a general predictor for sperm function with the deletion being just as likely to be present at similar levels in semen [120] and sperm [121] samples from normozoospermic men and subfertile patients. Two further deletions, namely, the 7,345 bp and 7,599 bp deletions, were thought to be indicative of poor motility [122], though this was not substantiated in a subsequent study [121].

The long PCR, which allows long regions of the mtDNA genome to be amplified, has been used to identify a range of mtDNA deletions. This technique works on the basis that any deletions present within the region will be amplified and appear as shorter fragments when observed on DNA gels, with the large-scale deletions being represented as the smaller fragments [22]. This approach has been used to analyse sperm samples from a male patient with multiple deletions associated with ptosis, who also exhibited subfertility [123]. This demonstrated a range of multiple deletions, which were symptomatic of poor sperm motility. Equally so, large-scale deletions have been identified in normozoospermic and oligozoospermic men [121, 124]. The presence of large-scale deletions in normozoospermic patients would not preclude the individual from having acceptable levels of motility, as they would still have significant numbers of spermatozoa with wild-type copies present. Nevertheless, it appears that poor-quality sperm samples appear to have a greater number of multiple deletions with oligoasthenoteratozoospermic men having the greatest proportion [121]. Equally so, mutations in the nuclear-encoded mtDNA replication factors, such as Twinkle, can also lead to multiple mtDNA deletions and dysfunctional spermatozoa [125]. However, for a true representation of the number of mtDNA deletions present within a sperm sample, pure populations of sperm mtDNA need to be isolated, as the ejaculate carries a range of somatic cells that would have significantly more copies of the mitochondrial genome, and if mainly WT in composition, it would bias the outcome, thus obscuring the deletions present in spermatozoa [121].

The mechanisms inducing sperm mtDNA deletions still need to be clarified. However, a multitude of studies have indicated a relationship between mtDNA deletions and the levels of mtDNA damage, as characterised by the levels of 8-OH-dG (see, for example, [126]). Many of the large-scale deletions that have been characterised lie between flanking direct repeats, where it has been hypothesised that inefficient proofreading mediates polymerase strand-hopping, i.e., from the heavy to the light strand, resulting in large regions of the genome not being incorporated during replication [127]. 8-OH-dG is a by-product of the hydroxyl (OH) free radical, which arises from H_2O_2 and has been associated with poor sperm quality and function due to increased levels of large-scale mtDNA deletions. In this respect, sperm samples from patients with diabetes mellitus appear to have increased levels of 8-OH-dG and large-scale mtDNA deletions [128]. This outcome is further supported by an increase in the frequency of nucleotide changes in the ATPase 6 and 8, ND 2, 3, 4 and 5 genes of the mtDNA genome in infertile men due to increased levels of free radicals [129]. The sperm mitochondrial genome is likely to be more susceptible to free radical activity, as it is less well-packaged than the chromosomal genome. Consequently, early signs of DNA damage are indicative in the mitochondrial genome rather than the chromosomal genome [130].

Whilst sperm mtDNA appears to be more susceptible to mtDNA deletions, this may not only result from the presence of free radicals present within the ejaculate and the testis, but could result from a decrease in mtDNA copy number during development, where those molecules that are selected for tend to be rearrangements mediated by the nuclear background of the cell [131], such as with other high ATP requiring cells, for example neuronal and muscle cells [12]. Such a mechanism of selection would have a twofold effect: [1] sperm motility becomes dependent on glycolysis, and [2] the mitochondria are rendered dysfunctional, and thus, once they enter the oocyte, are more susceptible to mechanisms such as apoptosis when challenged to generate ATP through the ETC. Consequently, these processes may be a mechanism for ensuring that the paternal genome is not transmitted to the offspring.

Furthermore, it is likely that any mutations and deletions affecting sperm function will arise from the spermatogonial cells, rather than spermatozoa, as they could only be incorporated into the mtDNA genome following mtDNA replication. These molecules would then be randomly selected for during the process of male gamete differentiation and not at later stages when copy number is reduced. We would also hypothesise that, in poor quality spermatogonial cells harbouring rearrangements, failure to regulate mtDNA copy number is indicative of inefficient nucleo-mtDNA interaction or attempts to rescue WT mtDNA at the expense of rearranged mtDNA.

mtDNA Haplotype

It has been argued that specific sequences within mtDNA have evolved and their origins can be traced back to several mitochondrial Eves. This has generated genetic diversity and has potentially provided individual populations with mitochondrial specific genotypes, otherwise known as haplotypes, which afford them specific advantage or disadvantage for survival and function [132]. For example, specific European type haplotypes are associated with tolerance to warmer and colder climates. Other haplotypes have been associated with fertility in a range of species such as pigs [133] and cattle [134], milk quality in cattle [135] and physical performance in mice [136]. A series of studies have indicated that male patients with haplotype H are not associated with asthenozoospermia, whilst individuals with haplotype T have such a predisposition [41]. Furthermore, additional differences in both sperm motility and vitality were identified in a number of sublineages of haplogroup U, perhaps arising from highly conserved missense mutations in the cytochrome C oxidase subunit III and cytochrome B genes [137]. However, similar analysis conducted on a population of Portuguese patients suggested that subpopulation studies can also influence haplogroup association studies, although they reported negative correlations with oligozoospermia when matched with geographic balanced controls [138].

Conclusions

It is evident that OXPHOS has a role to play in mediating sperm function and motility, as demonstrated from biochemical and genetic studies. However, this role needs further defining and characterisation. Specifically, we need to determine how and when rearranged mtDNA is incorporated into the male gamete, and we need to develop elaborate quantification protocols so that we can determine how much rearranged mtDNA is actually present in such samples. We further need to determine whether mtDNA damage is likely to prove a useful clinical diagnostic marker of early-onset DNA damage, which may enable us to warn patients to make lifestyle changes early on if they wish to conceive naturally.

References

1. Anderson S, Bankier AT, Barrell BG, et al. Sequence and organisation of the human mitochondrial genome. Nature. 1981;290:457–65.
2. Pfeiffer T, Schuster S, Bonhoeffer S, et al. Cooperation and competition in the evolution of ATP-producing pathways. Science. 2001;292:504–7.
3. Ojala D, Montoya J, Attardi G. tRNA punctuation model of RNA processing in human mitochondria. Nature. 1981;290:470–4.
4. Clayton DA. Transcription and replication of animal mitochondrial DNAs. Int Rev Cyto. 1992;141:217–32.
5. Clayton DA. Replication of animal mitochondrial DNA. Cell. 1982;28:693–705.
6. Kohnemann S, Pennekamp P, Schmidt PF, et al. qPCR and mtDNA SNP analysis of experimentally degraded hair samples and its application in forensic casework. Int J Legal Med. 2010;124:337–42.
7. Gill P, Ivanov PL, Kimpton C, et al. Identification of the remains of the Romanov family by DNA analysis. Nat Genet. 1994;6:130–5.
8. Brenner CA, Barritt JA, Willadsen S, et al. Mitochondrial DNA heteroplasmy after human ooplasmic transplantation. Fertil Steril. 2000;74: 573–8.
9. Lloyd RE, Lee JH, Alberio R, et al. Aberrant nucleo-cytoplasmic cross-talk results in donor cell mtDNA persistence in cloned embryos. Genetics. 2006; 172:2515–27.
10. Bowles EJ, Lee JH, Alberio R, et al. Contrasting effects of in vitro fertilization and nuclear transfer on the expression of mtDNA replication factors. Genetics. 2007;176:1511–26.
11. Bowles EJ, Tecirlioglu RT, French AJ, et al. Mitochondrial DNA transmission and transcription after somatic cell fusion to one or more cytoplasts. Stem Cells. 2008;26:775–82.
12. Moyes CD, Battersby BJ, Leary SC. Regulation of muscle mitochondrial design. J Exp Biol. 1998;201: 299–307.
13. Lazarou M, Smith SM, Thorburn DR, et al. Assembly of nuclear DNA-encoded subunits into mitochondrial complex IV, and their preferential integration into supercomplex forms in patient mitochondria. FEBS J. 2009;276:6701–13.
14. Birky Jr CW. The inheritance of genes in mitochondria and chloroplasts: laws, mechanisms, and models. Annu Rev Genet. 2001;35:125–48.
15. Birky Jr CW. Uniparental inheritance of mitochondrial and chloroplast genes: mechanisms and evolution. Proc Nat Acad Sci USA. 1992;92:11331–8.

16. He Y, Wu J, Dressman DC, et al. Heteroplasmic mitochondrial DNA mutations in normal and tumour cells. Nature. 2010;464:610–4.
17. Kobayashi Y, Momoi MY, Tominaga K, et al. A point mutation in the mitochondrial tRNA(Leu)(UUR) gene in MELAS (mitochondrial myopathy, encephalopathy, lactic acidosis and stroke-like episodes). Biochem Biophys Res Commun. 1990;173:816–22.
18. Fryer A, Appleton R, Sweeney MG, et al. Mitochondrial DNA 8993 (NARP) mutation presenting with a heterogeneous phenotype including "cerebral palsy". Arch Dis Child. 1994;71:419–22.
19. Wallace DC, Singh G, Lott MT, et al. Mitochondrial DNA mutation associated with Leber's hereditary optic neuropathy. Science. 1988;242:1427–30.
20. Shoffner JM, Lott MT, Lezza AM, et al. Myoclonic epilepsy and ragged-red fiber disease (MERRF) is associated with a mitochondrial DNA tRNA(Lys) mutation. Cell. 1990;61:931–7.
21. Schon EA, Rizzuto R, Moraes CT, et al. A direct repeat is a hotspot for large-scale deletion of human mitochondrial DNA. Science. 1989;244:346–9.
22. Melov S, Shoffner JM, Kaufman A, et al. Marked increase in the number and variety of mitochondrial DNA rearrangements in aging human skeletal muscle. Nucleic Acids Res. 1995;23:4122–6.
23. Yu-Wai-Man P, Griffiths PG, Hudson G, et al. Inherited mitochondrial optic neuropathies. J Med Genet. 2009;46:145–58.
24. Chinnery PF, Andrews RM, Turnbull DM, et al. Leber hereditary optic neuropathy: does heteroplasmy influence the inheritance and expression of the G11778A mitochondrial DNA mutation? Am J Med Genet. 2001;98:235–43.
25. Boulet L, Karpati G, Shoubridge EA. Distribution and threshold expression of the tRNA (Lys) mutation in skeletal muscle of patients with myoclonic epilepsy and ragged red fibres (MERRF). Am J Hum Genet. 1992;51:1187–200.
26. Sacconi S, Salviati L, Nishgaki Y, et al. A functionally dominant mitochondrial DNA mutation. Hum Mol Genet. 2008;17:1814–20.
27. Kirby DM, Rennie KJ, Smulders-Srinivasan TK, et al. Transmitochondrial embryonic stem cells containing pathogenic mtDNA mutations are compromised in neuronal differentiation. Cell Prolif. 2009;42:413–24.
28. Otani H, Tanaka O, Kasai K, et al. Development of mitochondrial helical sheath in the middle piece of the mouse spermatid tail: regular dispositions and synchronized changes. Anat Rec. 1988;222:26–33.
29. Lee S, Kim S, Sun X, et al. Cell cycle-dependent mitochondrial biogenesis and dynamics in mammalian cells. Biochem Biophys Res Commun. 2007;357:111–7.
30. de Sousa Lopes SM, Roelen BA. An overview on the diversity of cellular organelles during the germ cell cycle. Histol Histopathol. 2010;25:267–76.
31. Sutovsky P, Tengowski MW, Navara CS, et al. Mitochondrial sheath movement and detachment in mammalian, but not nonmammalian, sperm induced by disulfide bond reduction. Mol Reprod Dev. 1997;47:79–86.
32. Sinowatz F, Wrobel KH. Development of the bovine acrosome; an ultrastructural and cytochemical study. Cell Tissue Res. 1981;219:511–24.
33. Storey BT. Energy metabolism of spermatozoa. IV. Effect of calcium on respiration of mature epididymal sperm of the rabbit. Biol Reprod. 1975;13:1–9.
34. Storey BT. Effect of ionophores and inhibitors and uncouplers of oxidative phosphorylation on sperm respiration. Arch Androl. 1978;1:169–77.
35. Storey BT. Strategy of oxidative metabolism in bull spermatozoa. J Exp Zool. 1980;212:61–7.
36. Storey BT, Kayne FJ. Energy metabolism of spermatozoa. V. The Embden-Myerhof pathways of glycolysis: activities of pathway enzymes in hypotonically treated rabbit epididymal spermatozoa. Fertil Steril. 1975;26:1257–65.
37. Storey BT, Kayne FJ. Energy metabolism of spermatozoa. VI. Direct intramitochondrial lactate oxidation by rabbit sperm mitochondria. Biol Reprod. 1977;16:549–56.
38. Storey BT, Kayne FT. Properties of pyruvate kinase and flagellar ATPase in rabbit spermatozoa: relation to metabolic strategy of the sperm cell. J Exp Zool. 1980;211:361–7.
39. Quinn P, Kerin JF, Warnes GM. Improved pregnancy rate in human in vitro fertilization with the use of a medium based on the composition of human tubal fluid. Fertil Steril. 1985;44:493–8.
40. St. John JC, Jokhi RP, Barratt CLR. The impact of mitochondrial genetics on male infertility. Int J Androl. 2005;28:65–73.
41. Ruiz-Pesini E, Lapeña AC, Diez-Sanchez C, et al. Human mtDNA haplogroups associated with high or reduced spermatozoa motility. Am J Hum Genet. 2000;67:682–96.
42. Ruiz-Pesini E, Diez C, Lapena AC, et al. Correlation of sperm motility with mitochondrial enzymatic activities. Clin Chem. 1998;44:1616–20.
43. Folgero T, Bertheussen K, Lindal S, et al. Mitochondrial disease and reduced sperm motility. Hum Reprod. 1993;8:1863–8.
44. Miller FJ, Rosenfeldt FL, Zhang C, et al. Precise determination of mitochondrial DNA copy number in human skeletal and cardiac muscle by a PCR-based assay: lack of change of copy number with age. Nucleic Acids Res. 2003;31:e6.
45. He J, Mao CC, Reyes A, et al. The AAA+ protein ATAD3 has displacement loop binding properties and is involved in mitochondrial nucleoid organization. J Cell Biol. 2007;176:141–6.
46. Amaral A, Ramalho-Santos J, St. John JC. The expression of polymerase gamma and mitochondrial transcription factor A and the regulation of mitochondrial DNA content in mature human sperm. Hum Reprod. 2007;22:1585–96.
47. Piko L, Taylor KD. Amounts of mitochondrial DNA and abundance of some mitochondrial gene transcripts in early mouse embryos. Dev Biol. 1987;123:364–74.

48. Piko L, Matsumoto L. Number of mitochondria and some properties of mitochondrial DNA in the mouse egg. Dev Biol. 1976;49:1–10.
49. Kucej M, Butow RA. Evolutionary tinkering with mitochondrial nucleoids. Trends Cell Biol. 2007;17: 586–92.
50. Shadel GS, Clayton DA. Mitochondrial DNA maintenance in vertebrates. Annu Rev Biochem. 1997;66: 409–35.
51. Yasukawa T, Yang MY, Jacobs HT, et al. A bidirectional origin of replication maps to the major noncoding region of human mitochondrial DNA. Mol Cell. 2005;18:651–62.
52. Bogenhagen DF, Clayton DA. The mitochondrial DNA replication bubble has not burst. Trends Biochem Sci. 2003;28:357–60.
53. Holt IJ, Jacobs HT. Response: The mitochondrial DNA replication bubble has not burst. Trends Biochem Sci. 2003;28:355–6.
54. Yasukawa T, Reyes A, Cluett TJ. Replication of vertebrate mitochondrial DNA entails transient ribonucleotide incorporation throughout the lagging strand. EMBO J. 2006;25:5358–71.
55. Walker RL, Anziano P, Meltzer PS. A PAC containing the human mitochondrial DNA polymerase gamma gene (POLG) maps to chromosome 15q25. Genomics. 1997;40:376–8.
56. Gray H, Wong TW. Purification and identification of subunit structure of the human mitochondrial DNA polymerase. J Biol Chem. 1992;267:5835–41.
57. Graves SW, Johnson AA, Johnson KA. Expression, purification, and initial kinetic characterization of the large subunit of the human mitochondrial DNA polymerase. Biochem. 1998;37:6050–8.
58. Lim SE, Longley MJ, Copeland WC. The mitochondrial p55 accessory subunit of human DNA polymerase gamma enhances DNA binding, promotes processive DNA synthesis, and confers N-ethylmaleimide resistance. J Biol Chem. 1999;274: 38197–203.
59. Longley MJ, Nguyen D, Kunkel TA, et al. The fidelity of human DNA polymerase gamma with and without exonucleolytic proofreading and the p55 accessory subunit. J Biol Chem. 2001;276:38555–62.
60. Fan L, Sanschagrin PC, Kaguni LS, et al. The accessory subunit of mtDNA polymerase shares structural homology with aminoacyl-tRNA synthetases: implications for a dual role as a primer recognition factor and processivity clamp. Proc Nat Acad Sci USA. 1999;96:9527–32.
61. Cormier V, Rotig A, Tardieu M, et al. Autosomal dominant deletions of the mitochondrial genome in a case of progressive encephalomyopathy. Am J Hum Genet. 1991;48:643–8.
62. Van Goethem G, Schwartz M, Lofgren A, et al. Novel POLG mutations in progressive external ophthalmoplegia mimicking mitochondrial neurogastrointestinal *encephalomyopathy*. Eur J Hum Genet. 2003;11:547–9.
63. Van Goethem G, Dermaut B, Lofgren A, et al. Mutation of POLG is associated with progressive external ophthalmoplegia characterized by mtDNA deletions. Nat Genet. 2001;28:211–2.
64. Nowak R, Zub R, Skoneczna I, et al. CAG repeat polymorphism in the DNA polymerase γ gene in a Polish population: an association with testicular cancer risk. Ann Oncol. 2005;16:1211–2.
65. Naviaux RK, Nguyen KV. POLG mutations associated with Alpers syndrome and mitochondrial DNA depletion. Ann Neurol. 2005;58:491.
66. Naviaux RK. V. POLG mutations associated with Alpers' syndrome and mitochondrial DNA depletion. Ann Neurol. 2004;55:706–12.
67. Davidzon G, Mancuso M, Ferraris S, et al. POLG mutations and Alpers syndrome. Ann Neurol. 2005; 57:921–3.
68. Luoma P, Melberg A, Rinne JO, et al. Parkinsonism, premature menopause, and mitochondrial DNA polymerase gamma mutations: clinical and molecular genetic study. Lancet. 2004;364:875–82.
69. Pagnamenta AT, Taanman JW, Wilson CJ, et al. Dominant inheritance of premature ovarian failure associated with mutant mitochondrial DNA polymerase gamma. Hum Reprod. 2006;21:2467–73.
70. Chowers M, Gottesman BS, Leibovici L, et al. Nucleoside reverse transcriptase inhibitors in combination therapy for HIV patients: systematic review and meta-analysis. Eur J Clin Microbiol Infect Dis. 2010;29:779–86.
71. Ashley N, Harris D, Poulton J. Detection of mitochondrial DNA depletion in living human cells using PicoGreen staining. Exp Cell Res. 2005;303:432–46.
72. Dalakas MC, Semino-Mora C, Leon-Monzon M. Mitochondrial alterations with mitochondrial DNA depletion in the nerves of AIDS patients with peripheral neuropathy induced by 2'3'-dideoxycytidine (ddC). Lab Invest. 2001;81:1537–44.
73. White DJ, Mital D, Taylor S, et al. Sperm mitochondrial DNA deletions as a consequence of long term highly active antiretroviral therapy. AIDS. 2001;15:1061–2.
74. Ropp PA, Copeland WC. Cloning and characterization of the human mitochondrial DNA polymerase, DNA polymerase gamma. Genomics. 1996;36:449–58.
75. Gottlieb B, Lombroso R, Beitel LK, et al. Molecular pathology of the androgen receptor in male (in)fertility. Reprod Biomed Online. 2005;10:42–8.
76. Rovio AT, Marchington DR, Donat S, et al. Mutations at the mitochondrial DNA polymerase (POLG) locus associated with male infertility. Nat Genet. 2001;29:261–2.
77. Jensen M, Leffers H, Petersen JH, et al. Frequent polymorphism of the mitochondrial DNA polymerase gamma gene (POLG) in patients with normal spermiograms and unexplained subfertility. Hum Reprod. 2004;19:65–70.
78. Krausz C, Guarducci E, Becherini L, et al. The clinical significance of the POLG gene polymorphism in male infertility. J Clin Endocrinol Metab. 2004;89: 4292–7.
79. Aknin-Seifer IE, Touraine RL, Lejeune H, et al. Is the CAG repeat of mitochondrial DNA polymerase gamma (*POLG*) associated with male infertility? A multi-centre French study. Hum Reprod. 2005;20:736–40.

80. Brusco A, Michielotto C, Gatta V, et al. The polymorphic polyglutamine repeat in the mitochondrial DNA polymerase gamma gene is not associated with oligozoospermia. J Endocrinol Invest. 2006;29:1–4.
81. Hance N, Ekstrand MI, Trifunovic A. Mitochondrial DNA polymerase gamma is essential for mammalian embryogenesis. Hum Mol Genet. 2005;14:1775–83.
82. Harris TP, Gomas KP, Weir F. Molecular analysis of polymerase gamma gene and mitochondrial polymorphism in fertile and subfertile men. Int J Androl. 2006;29:421–33.
83. Parisi MA, Clayton DA. Similarity of human mitochondrial transcription factor 1 to high mobility group proteins. Science. 1991;252:965–9.
84. Larsson NG, Wang J, Wilhelmsson H, et al. Mitochondrial transcription factor A is necessary for mtDNA maintenance and embryogenesis in mice. Nat Genet. 1998;18:231–6.
85. Seidel-Rogol BL, Shadel GS. Modulation of mitochondrial transcription in response to mtDNA depletion and repletion in HeLa cells. Nucleic Acids Res. 2002;30:1929–34.
86. Gensler S, Weber K, Schmitt WE, et al. Mechanism of mammalian mitochondrial DNA replication: import of mitochondrial transcription factor A into isolated mitochondria stimulates 7S DNA synthesis. Nucl Acids Res. 2001;29:3657–63.
87. Dairaghi DJ, Shadel GS, Clayton DA. Addition of a 29 residue carboxyl-terminal tail converts a simple HMG box-containing protein into a transcriptional activator. J Mol Biol. 1995;249:11–28.
88. Kaufman BA, Durisic N, Mativetsky JM, et al. The mitochondrial transcription factor TFAM coordinates the assembly of multiple DNA molecules into nucleoid-like structures. Mol Biol Cell. 2007;18:3225–36.
89. Falkenberg M, Gaspari M, Rantanen A, et al. Mitochondrial transcription factors B1 and B2 activate transcription of human mtDNA. Nat Genet. 2002;31:289–94.
90. Evans MJ, Scarpulla RC. Interaction of nuclear factors with multiple sites in the somatic cytochrome c promoter. Characterization of upstream NRF-1, ATF, and intron Sp1 recognition sequences. J Biol Chem. 1989;264:14361–8.
91. Virbasius CA, Virbasius JV, Scarpulla RC. NRF-1, an activator involved in nuclear-mitochondrial interactions, utilizes a new DNA-binding domain conserved in a family of developmental regulators. Genes Dev. 1993;7:2431–45.
92. Choi YS, Lee HK, Pak YK. Characterization of the 5'-flanking region of the rat gene for mitochondrial transcription factor A (Tfam). Biochim et Biophys Acta. 2002;1574:200–4.
93. Smith LC, Alcivar AA. Cytoplasmic inheritance and its effects on development and performance. J Reprod Fertil Suppl. 1993;48:31–43.
94. Hecht NB, Liem H. Mitochondrial DNA is synthesized during meiosis and spermiogenesis in the mouse. Exp Cell Res. 1984;154:293–8.
95. Larsson NG, Garman JD, Oldfors A, et al. A single mouse gene encodes the mitochondrial transcription factor A and a testis-specific nuclear HMG-box protein. Nat Genet. 1996;13:296–302.
96. Larsson NG, Oldfors A, Garman JD, et al. Down-regulation of mitochondrial transcription factor A during spermatogenesis in humans. Hum Mol Genet. 1997;6:185–91.
97. May-Panloup P, Chretien MF, Savagner F, et al. Increased sperm mitochondrial DNA content in male infertility. Hum Reprod. 2003;18:550–6.
98. Kao SH, Chao HT, Liu HW, et al. Sperm mitochondrial DNA depletion in men with asthenospermia. Fertil Steril. 2004;82:66–73.
99. Shitara H, Hayashi JI, Takahama S, et al. Maternal inheritance of mouse mtDNA in interspecific hybrids: segregation of the leaked paternal mtDNA followed by the prevention of subsequent paternal leakage. Genetics. 1998;148:851–7.
100. Gyllensten U, Wharton D, Josefsson A, et al. Paternal inheritance of mitochondrial DNA in mice. Nature. 1991;352:255–67.
101. Zhao X, Li N, Guo W. Further evidence for paternal inheritance of mitochondrial DNA in the sheep (Ovis aries). Heredity. 2004;93:399–403.
102. St. John JC, Schatten G. Paternal mitochondrial DNA transmission during nonhuman primate nuclear transfer. Genetics. 2004;167:897–905.
103. Sutovsky P, Moreno RD, Ramalho-Santos J, et al. Ubiquitin tag for sperm mitochondria. Nature. 1999; 402:371–2.
104. Sutovsky P, Navara CS, Schatten G. Fate of the sperm mitochondria, and the incorporation, conversion, and disassembly of the sperm tail structures during bovine fertilization. Biol Reprod. 1996;55:1195–205.
105. Kaneda H, Hayashi JI, Takahama S, et al. Elimination of paternal mitochondrial DNA in intraspecific crosses during early mouse embryogenesis. Proc Nat Acad Sci USA. 1995;92:4542–6.
106. Zhang Q, Raoof M, Chen Y, et al. Circulating mitochondrial DAMPs cause inflammatory responses to injury. Nature. 2010;464:104–7.
107. Sutovsky P, Moreno R, Ramalho-Santos J, et al. Putative, ubiquitin-dependent mechanism for the recognition and elimination of defective spermatozoa in the mammalian epididymis. J Cell Sci. 2001;114:1665–75.
108. Nishimura Y, Yoshinari T, Naruse K, et al. Active digestion of sperm mitochondrial DNA in single living sperm revealed by optical tweezers. Proc Nat Acad Sci USA. 2006;103:1382–7.
109. May-Panloup P, Vignon X, Chretien MF, et al. Increase of mitochondrial DNA content and transcripts in early bovine embryogenesis associated with upregulation of mtTFA and NRF1 transcription factors. Reprod Biol Endocrinol. 2005;3:65–72.
110. Spikings EC, Alderson J. St John JC. Regulated mitochondrial DNA replication during oocyte maturation is essential for successful porcine embryonic development. Biol Reprod. 2007;76:327–35.
111. Kondo R, Matsuura ET, Chigusa SI. Further observation of paternal transmission of Drosophila

mitochondrial DNA by PCR selective amplification method. Genet Res. 1992;59:81–4.
112. Fisher C, Skibinski DOF. Sex-biased mitochondrial DNA heteroplasmy in the marine mussel Mytilus. Proc Royal Soc Lond B. 1990;242:149–56.
113. Hoeh WM, Blakley KH, Brown WM. Heteroplasmy suggests limited biparental inheritance of Mytilus mitochondrial DNA. Science. 1991;251:1488–90.
114. St. John J, Sakkas D, Dimitriadi K, et al. Abnormal human embryos show a failure to eliminate paternal mitochondrial DNA. Lancet. 2000;355:200.
115. Kraytsberg Y, Schwartz M, Brown TA, et al. Recombination of human mitochondrial DNA. Science. 2004;304:981.
116. Schwartz M, Vissing J. Paternal inheritance of mitochondrial DNA. N Eng J Med. 2002;347:576–80.
117. Cummins JM, Jequier AM, Kan R. Molecular biology of human male infertility: links with aging, mitochondrial genetics, and oxidative stress? Mol Reprod Dev. 1994;37:345–62.
118. Spiropoulos J, Turnbull DM, Chinnery PF. Can mitochondrial DNA mutations cause sperm dysfunction? Mol Hum Reprod. 2002;8:719–21.
119. Kao SH, Chao HT, Wei YH. Mitochondrial deoxyribnucleic acid 4977 bp deletion is associated with diminished fertility and motility of human sperm. Biol Reprod. 1995;52:729–36.
120. Cummins JM, Jequier AM, Martin R, et al. Semen levels of mitochondrial DNA deletions in men attending an infertility clinic do not correlate with phenotype. Int J Androl. 1998;21:47–52.
121. St. John JC, Jokhi RP, Barratt CL. Men with oligoasthenoteratozoospermia harbour higher numbers of multiple mitochondrial DNA deletions in their spermatozoa, but individual deletions are not indicative of overall aetiology. Mol Hum Reprod. 2001;7:103–11.
122. Kao SH, Chao HT, Wei YH. Multiple deletions of mitochondrial DNA are associated with the decline of motility and fertility of human spermatozoa. Mol Hum Reprod. 1998;4:657–66.
123. Lestienne P, Reynier P, Chretien MF, et al. Oligoasthenospermia associated with multiple mitochondrial DNA rearrangements. Mol Hum Reprod. 1997;3:811–4.
124. Reynier P, Chretien MF, Savagner F, et al. Long PCR analysis of human gamete mtDNA suggests defective mitochondrial maintenance in spermatozoa and supports the bottleneck theory for oocytes. Biochem Biophys Res Comm. 1998;252:373–7.
125. Quigley A, Reardon K, Kapsa R, et al. A novel clinical phenotype of myopathy, sensorimotor neuropathy, infertility, and hypogonadism with multiple mitochondrial DNA deletions. J Clin Neuromuscul Dis. 2001;3:77–82.
126. Ozawa T. Mechanism of somatic mitochondrial DNA mutations associated with age and diseases. Biochim Biophys Acta. 1995;1271:177–89.
127. St. John JC, Sakkas D, Barratt CLR. A role for mitochondrial DNA in sperm survival. J Androl. 2000;21: 189–99.
128. Agbaje IM, McVicar CM, Schock BC, et al. Increased concentrations of the oxidative DNA adduct 7,8-dihydro-8-oxo-2-deoxyguanosine in the germ-line of men with type 1 diabetes. Reprod Biomed Online. 2008;16:401–9.
129. Kumar R, Venkatesh S, Kumar M, et al. Oxidative stress and sperm mitochondrial DNA mutation in idiopathic oligoasthenozoospermic men. Indian J Biochem Biophys. 2009;46:172–7.
130. Yakes FM, Van Houten B. Mitochondrial DNA damage is more extensive and persists longer than nuclear DNA damage in human cells following oxidative stress. Proc Natl Acad Sci USA. 1997;94:514–9.
131. Dunbar DR, Moonie PA, Jacobs HT, Holt IJ. Different cellular backgrounds confer a marked advantage to either mutant or wild-type mitochondrial genomes. Proc Natl Acad Sci USA. 1995;92: 6562–6.
132. Wallace DC. Mitochondrial DNA variation in human evolution, degenerative disease, and aging. Am J Hum Genet. 1995;57:201–23.
133. El Shourbagy SH, Spikings EC, Freitas M, et al. Mitochondria directly influence fertilisation outcome in the pig. Reproduction. 2006;131:233–45.
134. Sutarno, Cummins JM, Greeff J, Lymbery AJ. Mitochondrial DNA polymorphisms and fertility in beef cattle. *Theriogenology. 2002;*57:1603–10.
135. Schutz MM, Vanraden PM, Wiggans GR. Genetic variation in lactation means of somatic cell scores for six breeds of dairy cattle. J Dairy Sci. 1994;77: 284–93.
136. Nagao Y, Totsuka Y, Atomi Y, et al. Decreased physical performance of congenic mice with mismatch between the nuclear and the mitochondrial genome. Genes Genet Syst. 1998;73:21–7.
137. Montiel-Sosa F, Ruiz-Pesini E, Enriquez JA, et al. Differences of sperm motility in mitochondrial DNA haplogroup U sublineages. Gene. 2006;368:21–7.
138. Pereira L, Goncalves J, Goios A, et al. Human mtDNA haplogroups and reduced male fertility: real association or hidden population substructuring. Int J Androl. 2005;28:241–7.

The Sperm Epigenome

7

Donovan Chan and Jacquetta Trasler

Abstract

The development of male germ cells from the primordial germ cell stage to that of the mature spermatozoon is a key time of epigenetic reprogramming. Orchestrated by specialized enzymes, DNA methylation and histone modifications undergo dynamic changes throughout gametogenesis. Alterations to any level of the sperm epigenetic coding may affect fertility and the sperm's contribution to normal embryo development. In support of an important role for normal genomic methylation patterns in human sperm, a number of recent studies have reported abnormal DNA methylation in imprinted and other sequences in infertile men. As well, a number of genomic imprinting disorders in offspring, associated with underlying DNA methylation alterations in imprinted genes, have been linked with infertility and the use of assisted reproductive technologies (ARTs). In this chapter, we discuss different aspects of the sperm epigenome, from the timing and mechanisms underlying the acquisition of epigenetic patterns to the consequences of perturbing such patterns. The focus here is on DNA methylation, since it is not only one of the most well-studied epigenetic modifications taking place during male germ cell development but also one that has been clearly linked to infertility in men.

Keywords

Sperm epigenome • Male germ cells, Epigenetics • DNA methylation • Assisted reproductive technologies

J. Trasler (✉)
Departments of Pharmacology and Therapeutics, Pediatrics, and Human Genetics, McGill University and Montreal Children's Hospital of the McGill University Health Centre, 2300 Tupper Street, Montreal, QC, H3H 1P3 Canada
e-mail: jacquetta.trasler@mcgill.ca

The development of male germ cells from the primordial germ cell stage to that of the mature spermatozoon is a key time of epigenetic reprogramming. Orchestrated by specialized enzymes, DNA methylation and histone modifications undergo dynamic changes throughout

A. Zini and A. Agarwal (eds.), *Sperm Chromatin: Biological and Clinical Applications in Male Infertility and Assisted Reproduction*, DOI 10.1007/978-1-4419-6857-9_7,

gametogenesis. Male gamete epigenetic programming plays multiple roles not only in spermatogenesis, including gene expression programs and meiosis, but also in preparing the sperm for its role post fertilization in embryogenesis. Alterations to any level of the sperm epigenetic coding may affect fertility and the sperm's contribution to normal embryo development. In support of an important role for normal genomic methylation patterns in human sperm, a number of recent studies have reported abnormal DNA methylation in imprinted and other sequences in infertile men [1]. As well, a number of genomic imprinting disorders in offspring, associated with underlying DNA methylation alterations in imprinted genes, have been linked with infertility and the use of assisted reproductive technologies (ARTs) [2, 3]. Most of the evidence demonstrating the importance of proper epigenetic marks to reproduction and the general health of the embryo come from the use of animal models. In this chapter, we discuss different aspects of the sperm epigenome, from the timing and mechanisms underlying the acquisition of epigenetic patterns to the consequences of perturbing such patterns. The focus here is on DNA methylation, since it is not only one of the most well-studied epigenetic modifications taking place during male germ cell development but also one that has been clearly linked to infertility in men.

Epigenetics and the Roles of DNA Methylation

The term epigenetics refers to heritable mechanisms that help to control gene expression without an actual change in the underlying DNA sequence. These mechanisms include histone modifications (discussed in Chap. 3), noncoding RNAs (discussed in Chap. 8), and DNA methylation. The different types of epigenetic modifications interact in numerous ways to influence gene expression. The covalent addition of a methyl group to the cytosine residue in DNA is the best studied of the epigenetic modifications. This mark is found at 60–80% of CpG dinucleotides in the genome and plays important roles in many cellular processes. Methylation of the promoter region of genes is invariably associated with gene repression. Deviations from normal epigenetic patterns can result in diseases such as cancer and developmental disorders, fueling the development of a new area of epigenetic therapeutics [4].

The large majority of methylated cytosines is found within transposons and repeat sequences. DNA methylation prevents expression from transposons and their remnants within the genome. These elements have the potential to disrupt gene expression; demethylation of such sequences results in transposon reactivation in animal models [5, 6]. Along with its role in silencing such repeat sequences, DNA methylation may have functions in chromosome organization and structure. Heterochromatin, a densely packed form of DNA, has been associated with mainly gene-free regions and areas of high DNA methylation [7]. By contrast, euchromatin is generally rich in genic sequences showing active transcription, including sequences with low levels of methylation [8].

DNA methylation also contributes to the process of X-inactivation during embryogenesis. The silencing of the second X chromosome is accomplished by repression of genes located on the chromosome, associated with DNA hypermethylation of the underlying sequences [9, 10]. Similarly, genomic imprinting is a phenomenon in which DNA methylation marks at differentially methylated regions (DMRs) allow for the monoallelic expression of genes in a parent-of-origin specific manner [11]. These marks, which are initiated in the germ line, play an important role during embryonic growth and development [12, 13]. In humans, a number of disorders are associated with altered expression of imprinted genes, including the imprinting syndromes Beckwith–Wiedemann, Silver–Russell, Angelman, and Prader–Willi Syndromes, as well as several cancers [14, 15]. Outside of imprinted genes, abnormal methylation is frequently associated with cancers; both genome-wide DNA hypomethylation and site-specific hypermethylation have been reported, associated with the silencing of tumor suppressor genes and the activation of oncogenes [16].

Many mammalian promoter regions contain a high CpG content with approximately 40% containing regions known as CpG islands [17]. Methylation within promoter regions has been shown to affect the transcriptional regulation of genes, mainly through repression. Different mechanisms by which DNA methylation mediates its effect on gene regulation include direct interference with the transcriptional machinery or the recruitment of methyl CpG binding proteins containing transcriptional repression domains [18–22]. DNA methylation may also interact with other epigenetic marks, such as histone modifications, in order to regulate gene expression. Histone 3 lysine 4 (H3K4) methylation and histone acetylation, which are marks of active chromatin structure, are normally associated with a lack of DNA methylation. By contrast, methylation at CpG dinucleotides promotes a closed chromatin structure, blocking H3K4 methyltransferases and thus resulting in transcriptional inhibition [23]. Other histone modifications such as H4K20 and H3K8 methylation are associated with the presence of DNA methylation within the DMRs of imprinted genes [24].

Enzymes Involved in DNA Methylation

The DNA (cytosine-5)-methyltransferases (DNMTs) are the enzymes involved in catalyzing the reaction in which methyl groups from *S*-adenosylmethionine (SAM) are transferred to cytosine residues. Members of this group have been characterized and classified into three groups: DNMT1, DNMT2, and DNMT3 [25]. DNMT1, the first DNA methyltransferase discovered, has a high affinity for hemimethylated sequences and plays a role in maintaining methylation patterns at the time of DNA replication (maintenance methylation) [26–28]; it was also found to be able to *de novo* methylate unmodified DNA residues [29]. DNMT1 is the major form of methyltransferase and is found in all somatic tissues, although the highest levels of mRNA expression are in the testis [30]. DNMT2 has no known role in DNA methylation but has been determined to methylate tRNAs [31]. The DNMT3 family consists of three members: DNMT3A, DNMT3B, and DNMT3L. While DNMT3A and 3B have DNA methyltransferase activity, DNMT3L does not have any catalytic activity [32]. Despite this, DNMT3L improves the *de novo* methylation abilities of the other DNMT3 members [33–36]. Interestingly, DNMT3L has been shown to have higher affinity for the unmethylated lysine 4 of the histone 3 tail (H3K4), helping to direct DNA methylation and providing evidence of interactions between these two epigenetic marks [37].

Germ Cell Expression

From mouse studies, *Dnmt1* expression has been shown to be highly regulated in both male and female gametogenesis. In males, primordial germ cells show high levels of *Dnmt1* during the proliferative phase up to 13.5 days post coitum (dpc). From 14.5 dpc on, levels drop and are undetectable at 18.5 dpc [38, 39]. Postnatally, increased expression is seen when spermatogonia resume mitotic divisions [38, 40]. DNMT1 protein is present during the early stages of meiosis and is depleted in pachytene spermatocytes.

Dnmt3a and *Dnmt3b* show developmental stage-specific differences in expression during gametogenesis. Isoforms of *Dnmt3a* are highly expressed in the prenatal testes at 16.0 dpc, with continued high expression in early postnatal life [41]. *Dnmt3b*, on the other hand, shows minimal expression in prenatal life, but high levels in type A spermatogonia at 6 days postpartum (dpp) [38, 41, 42]. Human *DNMT3A* and *DNMT3B* are highly homologous to their murine counterparts and are expressed in a variety of tissues, including the testes [43].

Expression of *Dnmt3L* in mouse male germ cells is highest before birth. Time course analysis indicated that expression is detected between 13.5 and 18.5 dpc, with a peak at 15.5 dpc [38, 41]. Gene reporter experiments have shown that *Dnmt3L* is also expressed in spermatogonia but that expression is low by 6 dpp [44, 45]. Another study detected *Dnmt3L* expression later in male germ cell development also, in differentiating spermatocytes [46]. *Dnmt3L* expression patterns mimic those of *Dnmt3a*, providing evidence that these two enzymes work together in male germ cells as they do in somatic cells.

DNA Methylation Patterns in Germ Cells

Recent mouse and human studies of numerous types of sequences throughout the genome have shown that a unique pattern of DNA methylation is observed in male germ cells in comparison to that in somatic tissues [47, 48]. For instance, in a study by Weber et al. examining promoter methylation, a unique pattern of DNA methylation was observed in human sperm when compared with that in somatic cells, and a role in gene function was postulated [49]. Indeed, methylation patterns observed at promoters in sperm, such as hypomethylation, would allow for germ cell-specific expression of genes involved with spermatogenesis, whereas hypermethylation would allow the repression of pluripotency and somatic tissue-specific genes [50–52]. Interestingly, many of the sites that were found to be differentially methylated between sperm and somatic tissues were outside genic regions and CpG islands, and therefore, likely to have other roles in addition to those in controlling gene expression. Germ cell-specific DNA methylation patterns at centromeric and intergenic sequences may be necessary for the specialized chromatin structure found in male germ cells as they undergo meiosis and spermiogenesis [48, 53, 54]. Not only are patterns unique in sperm compared to somatic tissues, but spermatozoa from the same individual also exhibit distinctive DNA methylation patterns [55].

Erasure and Acquisition of Germ Cell Patterns

Somatic cell patterns of DNA methylation are established early during embryonic life and are maintained throughout development and into adulthood. Germ cells also follow the early establishment along with the embryo; however, erasure of these patterns subsequently takes place in primordial germ cells to allow the establishment of sex-specific patterns, such as those found on imprinted genes.

Erasure of the inherited somatic cell patterns occurs in mouse primordial germ cells between 10.5 and 13.5dpc [56]. This primordial germ cell hypomethylation was observed in studies using different techniques including Southern blotting, restriction enzyme digests, and PCR approaches, as well as cellular 5-methylcytosine antibody staining [57–60]. Detailed analysis by bisulfite sequencing of several imprinted and nonimprinted genes was also performed indicating a similar time frame for germ cell DNA demethylation [61–63]. This rapid erasure of the methylation patterns over a short period of time suggests an active demethylation process. However, not all epigenetic marks are erased during this time of epigenetic reprogramming of the germ cells. Maatouk et al. demonstrated that methylation at several nonimprinted genes retained relatively high levels of methylation [63]. As well, it was shown that a number of imprinted genes retained low levels of methylation and that several repetitive elements underwent only partial demethylation of their DNA sequences [64–66]. Together, the incomplete reprogramming of the parental DNA methylation patterns in the primordial germ cells allows for the possibility of epigenetic inheritance.

Subsequent to the erasure of epigenetic patterns in primordial germ cells, remethylation of DNA is acquired in a sex-specific manner in germ cells. In females, germ cells begin to acquire their methylation patterns postnatally, following the pachytene phase of meiosis, with imprinted genes acquiring their sex-specific mark during the oocyte growth phase [67–69]. Conversely, male germ cell epigenetic patterns begin to be acquired prenatally. The timing of the initial acquisition follows the expression of both *Dnmt3a* and *DnmtL*, consistent with the role of the DNMT3 class of enzymes as *de novo* DNA methyltransferases. Increases in 5-methyl cytosine immunostaining were observed in gonocytes from 17 to 19 dpc embryos, and bisulfite analysis of the imprinted genes *H19*, *Dlk1-Gtl2*, and *Rasgrf* indicated that acquisition of their paternal methylation imprints occurred between 15.5 and 18.5 dpc [58, 59, 61, 65, 70] The male germ cell methylation patterns are completed after birth by the pachytene phase of meiosis. While most DNA methylation is acquired by the type A spermatogonial phase, several loci still undergo acquisition and loss of methylation marks between this time point and the pachytene spermatocyte phase, at which point similar patterns are observed as those in mature spermatozoa [54].

Compared to studies using animal models, little research has been undertaken on human samples concerning the timing and sequences involved during the erasure, acquisition, and maintenance of DNA methylation marks in male germ cells. However, existing human evidence does support the erasure of methylation patterns in prenatal gonocytes and acquisition and maintenance of such patterns in early and late germ cells. For instance, Kerjean et al. [71] analyzed the DMR of *H19* and found that this sequence was unmethylated in fetal gonocytes and methylated in adult spermatogonia and in later stages of male germ cell development. As discussed in more detail below, imprinted genes that are normally methylated in the female germ line are unmethylated in human sperm as is the case in mouse. Furthermore, DNMT expression shows a similar timing of expression in human fetal gonads as that described in mouse [72].

Histone Modifications and Epigenetic Memory

Several studies have examined the modification of histone marks, in particular histone 3 methylation, during the course of male germ cell development [73–75]. The establishment and the removal of different histone modifications are important for normal spermatogenesis to occur. Transgenic animal models involving the targeting of enzymes involved in histone demethylation have revealed important roles for these enzymes in spermatogenesis and normal fertility [76].

Histone modifications can influence chromatin structure and gene expression in germ cells. In particular for male germ cells, as discussed elsewhere in this volume, extensive chromatin remodeling occurs during spermiogenesis, where histones are replaced by transition proteins, followed by protamines. This replacement allows for the high level of compaction required for packaging the DNA into the sperm head. However, in human sperm, 5–15% of histones remain bound to the genome [77, 78]. Recent studies have suggested that sperm histones and specific methylation modifications of the histones may play important roles post fertilization and "mark" or "poise" genes for expression in the embryo [79, 80]. As well, conservation of these histone modification marks at orthologous genes was seen in mouse spermatozoa [80]. Together, histone modifications in sperm would appear to be important and may contribute to the early stages of embryo development.

Consequences of an Altered Sperm Epigenome for Male Reproductive Function

Animal Models

Gene targeting has been used to examine the function of different DNMT enzymes. Mice with partial ($Dnmt^{n/n}$ and $Dnmt^{s/s}$) and complete ($Dnmt^{c/c}$) loss of function of DNMT1 were developmentally delayed and died at mid-gestation [81], before an effect on germ cells could be examined. DNMT1-deficient embryos also showed abnormal biallelic expression of imprinted genes and expression of normally silent IAP sequences, as well as ectopic X-chromosome inactivation [5, 9, 82]. Embryos obtained from the mating of female mice deficient for the oocyte-specific form of DNMT1, known as DNMT1o, also showed embryonic lethality and abnormal methylation patterns at imprinted loci [83]. Although such studies have not yet been done, with its high and tightly regulated expression in male germ cells, male germ cell-specific targeting of DNMT1 would be likely to help uncover the role of DNMT1 at different times during male germ cell development. Disease-causing mutations in DNMT1 in humans have not been reported yet, with the exception of DNMT1 catalytic domain mutations in certain rare cases of colorectal cancer [84].

DNMT3a-deficient mice do survive to term, although they were underdeveloped and did not survive past the first few weeks of life. While global levels of DNA methylation were normal in these animals, spermatogenesis was impaired [85]. Closer inspection revealed abnormal entry into meiosis as well as decreased methylation at the imprinted *H19* locus, indicating a crucial role of DNMT3a in male germ cell development [86]. Indeed, conditional inactivation of this enzyme in male germ cells resulted in infertility due to spermatogenic failure [87]. While abnormal DNA methylation was observed at the imprinted loci *H19* and *Dlk1-Gtl2*, as well as some repeat regions in spermatogonia, little effect was found at *Rasgfr* and IAP sequences [66, 87].

Consequences of DNMT3b deficiency in mice were dramatic resulting in mid-gestation lethality and demethylation of minor satellite repeats [85]. By contrast, male germ-line conditional elimination of DNMT3b did not appear to have any phenotypic effect, resulting in normal spermatogenesis; overall DNA methylation levels appeared for the most part to be normal, although slight decreases were observed at the *Rasgrf* locus, as well as in minor and major satellite repeats [66, 87]. In humans, mutations in DNMT3B result in an autosomal recessive genetic disorder characterized by immunodeficiency, centromeric instability, and facial anomalies known as ICR syndrome [88]. Pericentric regions, containing normally methylated satellite DNA, and CpG island on the inactive X-chromosome showed aberrant methylation in ICF patients [89, 90]. No studies on fertility have been reported.

Mice with homozygous deficiency for DNMT3L are viable; however, both males and females were infertile [44, 46]. Males had small testes and were azoospermic following the initial wave of spermatogenesis. Early loss of germ cells was observed at 6 dpp and a lack of differentiated spermatocytes was detected in mice at 4-weeks [45, 46, 91]; this loss of spermatocytes occurred after meiotic failure characterized by extensive chromosomal mispairing [45, 92]. Male germ cells of DNMT3L-deficient mice had a lack of methylation of most repetitive elements, leading to their abnormal transcription in early germ cells, as well as hypomethylation of paternally methylated imprinted loci [45, 66, 92]. Loss of methylation at intergenic loci in type A spermatogonia was also observed [91].

One critical factor for all methylation reactions, including the methylation of DNA, is the availability of the methyl donor, SAM. Factors that may influence cellular methyl pools include enzymes within the folic acid pathway. The impact of altered function of some of these enzymes has been studied and shown to be associated with decreased fertility in men [93]. One such enzyme, methylenetetrahydrofolate reductase (MTHFR), is the link between the one-carbon methyl donors of the folate pathway and the formation of SAM from the methionine pathway. Enzymatic activity of MTHFR is highest in testes as compared to other tissues, suggesting a critical role in reproduction. Indeed, homozygosity for one common polymorphism (677C->T), resulting in a thermolabile form of MTHFR, has been shown to be overrepresented in cases of male idiopathic infertility [94, 95]. As well, mice with MTHFR deficiency were created, in which altered SAM levels were observed along with hypomethylation in several tissues including the testes and ovaries [96]. MTHFR-deficient mice show strain-specific pathologies. MTHFR-deficient males of the BALB/c strain had abnormal seminiferous tubules lacking germ cells and were infertile [97]. With the dietary addition of an alternate methyl donor, betaine, some of the spermatogenic defects in the BALB/c strain MTHFR-deficient mice were alleviated, indicating a critical role of methyl donors in male germ cell development. MTHFR mice of the C57BL/6 strain showed normal early germ cell development; however, adverse reproductive outcomes, including decreased testicular weights and sperm counts, were observed starting at about 3.5 months of age [98]. In addition, while normal imprinted gene methylation was found, global methylation analysis revealed both hyper- and hypomethylation at several loci throughout the sperm epigenome.

Drug Targeting

Since abnormal DNA methylation has been associated with a number of disease states, and cancer in particular, interest in epigenetic therapies has emerged. Two inhibitors of DNA methylation, 5-azacytidine and 5-aza-2'deoxycytidine, were first synthesized as potential cancer chemotherapeutic agents [99]. These drugs are cytidine analogs that are incorporated into newly synthesized DNA during replication. When bound with DNA methyltransferases, the drugs inhibit the enzyme activity by forming covalent adducts, thereby depleting cellular pools of available DNMTs [100]. Animal exposures to these chemicals have been shown to cause male reproductive abnormalities and DNA hypomethylation. Treatment of male rats with 5-azacytidine interfered with normal germ cell development; mating with untreated females resulted in decreased fertilization and altered embryo development [101]. An increase in apoptotic germ cells as well as a decrease in global DNA methylation was also observed in mature sperm from treated males [102]. Similar effects were seen in male mice treated with 5-aza-2'deoxycytidine. Kelly et al. observed dose-dependent decreases in testicular weights and abnormal histology in the treated males and reduced pregnancy rates and increased preimplantation loss in females mated with the treated males [103]. A dose-dependent reduction in global sperm DNA methylation was also reported, with the DNA hypomethylation restricted to loci that were shown to acquire methylation marks during spermatogenesis [104]. The results suggested that 5-aza-2'deoxycytidine selectively inhibited *de novo* methylation activity in male germ cells.

Other drugs used for chemotherapy treatment have also been shown to cause epigenetic defects in male germ cells. Cyclophosphamide, an anticancer and immunosuppressive drug, was shown to cause reproductive abnormalities and affect embryo development in a time- and dose-dependent manner [105–107]. Along with increased incidences of chromosomal abnormalities in epididymal rat sperm, epigenetic reprogramming in the early rat embryo was affected [108, 109]. Hyperacetylation of histones and altered DNA methylation were observed in early one- and two-cell rat embryos.

Human Infertility

Idiopathic infertility makes up approximately half of all cases of male infertility. A recent study has looked for genetic causes of infertility examining oligozoospermic, azoospermic, and normospermic men in a genome-wide association study using genotyping microarrays and a gene-centric approach evaluating SNPs associated with male fertility [110]. Results from this and animal models have indicated that although genetics do play a role [111], the causes of male factor infertility are multifactorial and other mechanisms may contribute to the disease. Since epigenetics plays an important role during male germ cell development, and perturbations have been shown to cause abnormal reproductive outcomes, the association of altered epigenetic marks and human infertility has been examined. In particular, the assessment of methylation defects at imprinted gene loci have been the focus of many studies.

One of the first studies analyzed the methylation in sperm at the imprinted locus *H19*, comparing oligozoospermic and normospermic men [112]. Bisulfite sequencing of the *H19* DMR found decreases in methylation at the locus that were associated with decreased sperm numbers; the methylation defects were related to the severity of the oligozoospermia. In a later study, the same researchers analyzed the *H19* locus and a maternally imprinted gene, *PEG1/MEST* [113]. They reported abnormal methylation patterns at both imprinted loci in oligozoospermic men, with a loss and gain of methylation of *H19* and *PEG1/MEST*, respectively, while global methylation (LINE1 transposon) was unaffected. Similarly, a larger study of oligozoospermic men found sperm DNA hypomethylation at *H19* and *GTL2* and hypermethylation of several maternally methylated imprinted loci [114]. In an examination of male idiopathic infertility, Poplinski et al. examined methylation profiles in swim-up purified sperm from 148 idiopathic infertile and 33 normospermic men [115]; again, abnormal methylation at *H19* and *MEST* were associated with low sperm counts. In addition, *MEST* hypermethylation was a marker for decreased motility and abnormal sperm morphology. More widespread changes in DNA methylation were observed in a study of infertile men with abnormal semen parameters, where imprinted loci, gene promoters, and several repetitive elements were shown to be affected [116]. Finally, one recent study reported that altered methylation at different imprinted loci was associated with two different causes of male infertility [117]. Severely oligozoospermic patients had greater alterations at the *MEST* locus, a gene associated with Silver–Russell Syndrome; patients with abnormal chromatin structure were affected at the imprinting sequences of *KCNQ1OT1* (*LIT1*) and *SNRPN*.

From these studies, questions arise as to whether abnormal methylation of the imprinted and nonimprinted loci in sperm may perturb the normal development of the resulting offspring. Changes in sperm methylation profiles may help explain the low birth weight, preterm birth, and other complications reported in babies conceived using ARTs. In an attempt to answer this, Kobayashi et al. examined the methylation of ART-conceived aborted conceptuses as well as the sperm from their fathers [118]. A total of 17 ART-conceived fetal samples were found to have abnormal methylation at imprinting loci; 7 of the 17 (41%) abnormal patterns in ART-conceived fetuses were also found in the sperm DNA profile of fathers. Interestingly, sequence variations in *DNMT3L* were observed in two of these fathers as well. The results suggest that the abnormalities in DNA methylation of the fetus were transmitted from the father. Further evidence comes from a case study in which an infant conceived through in vitro fertilization was born with Silver–Russell Syndrome [119]. It was suggested that abnormal methylation in the *MEST* locus in the father's sperm may have contributed to the imprinting disorder in the child.

Conclusions and Future Directions

Future studies in both human and animal models may help us to better understand the mechanisms underlying the association between altered sperm DNA methylation and infertility. It is currently unclear whether the DNA methylation defects found in the sperm of infertile men are primary or secondary to the cause of the underlying infertility. Understanding the basis of the sperm DNA methylation defects will be important for the development of effective therapies for the associated infertility. Dietary supplementation of the methyl donor folate has been used in the treatment of infertile men [93] and may act by ameliorating abnormal DNA methylation patterns in male germ cells. The high levels of replication that occur during the course of spermatogenesis require an abundant supply of nucleotides that can be produced from the folate pathway. In addition, folic acid supplementation may provide methyl donors for the production of adequate supplies of SAM for germ-cell methylation reactions, including DNA methylation. However, there may be adverse consequences associated with dietary folate supplementation. Data have started to emerge looking at the impact of folate fortification of foods that became mandatory in North America in the late 1990s. While the main reason for fortification was to reduce the incidence of neural tube defects in pregnant women, studies have shown a concurrent increase in the incidence of colorectal cancer with the time just after implementation has begun [120]. Caution may also be warranted before treating infertile men with high doses of folate without appropriate studies showing that such treatments do not lead to abnormal methylation in sperm that might be transmitted to the offspring.

For the future, more studies are required to better understand the role of epigenetic modifications in normal and abnormal male germ cell development. For instance, as next-generation sequencing and bioinformatic resources become more readily available, it will be possible to determine the DNA methylation status at all of the 20–30 million sites in the genome in patients and in germ cells at different stages of development. Such studies may help identify important sites of epigenetic perturbations in the sperm of infertile patients that may be passed on to the offspring. Additionally, genome-wide sequencing studies may help determine which types of methylated sequences are most sensitive to endogenous factors such as age and exogenous factors such as environmental and drug exposures. There is also concern that some epigenetic defects may be passed across generations despite the genome-wide erasure that takes place within the germ line [121].

The mechanisms and potential for transgenerational passage of epigenetic defects will need further study due to the possible adverse consequences for future generations. A better understanding is also needed of the interactions between the different epigenetic modifications and the enzymes involved, in normal male germ cell development, as well as which modifications are important for embryo development.

Acknowledgments Research in the Trasler laboratory is supported by grants from the Canadian Institutes of Health Research (CIHR). D.C. is the recipient of a CIHR Doctoral Research Award. J.M.T. is a James McGill Professor of McGill University. The authors are members of the Research Institute of the McGill University Health Centre, which is supported in part by the Fonds de la Recherche en Santé du Québec (FRSQ).

References

1. Filipponi D, Feil R. Perturbation of genomic imprinting in oligozoospermia. Epigenetics. 2009;4(1):27–30.
2. Gosden R, Trasler J, Lucifero D, Faddy M. Rare congenital disorders, imprinted genes, and assisted reproductive technology. Lancet. 2003;361(9373):1975–7.
3. Lucifero D, Chaillet JR, Trasler JM. Potential significance of genomic imprinting defects for reproduction and assisted reproductive technology. Hum Reprod Update. 2004;10(1):3–18.
4. Egger G, Liang G, Aparicio A, Jones PA. Epigenetics in human disease and prospects for epigenetic therapy. Nature. 2004;429(6990):457–63.
5. Walsh CP, Chaillet JR, Bestor TH. Transcription of IAP endogenous retroviruses is constrained by cytosine methylation. Nat Genet. 1998;20(2):116–7.
6. Walsh CP, Bestor TH. Cytosine methylation and mammalian development. Genes Dev. 1999;13(1):26–34.
7. Sanford J, Forrester L, Chapman V, Chandley A, Hastie N. Methylation patterns of repetitive DNA sequences in germ cells of Mus musculus. Nucleic Acids Res. 1984;12(6):2823–36.
8. Bestor TH, Coxon A. Cytosine methylation: the pros and cons of DNA methylation. Curr Biol. 1993;3(6):384–6.
9. Panning B, Jaenisch R. DNA hypomethylation can activate Xist expression and silence X-linked genes. Genes Dev. 1996;10(16):1991–2002.
10. Sado T, Okano M, Li E, Sasaki H. De novo DNA methylation is dispensable for the initiation and propagation of X chromosome inactivation. Development. 2004;131(5):975–82.
11. Reik W, Dean W, Walter J. Epigenetic reprogramming in mammalian development. Science. 2001;293(5532):1089–93.
12. Smith FM, Garfield AS, Ward A. Regulation of growth and metabolism by imprinted genes. Cytogenet Genome Res. 2006;113(1–4):279–91.
13. Fowden AL, Sibley C, Reik W, Constancia M. Imprinted genes, placental development and fetal growth. Horm Res. 2006;65 Suppl 3:50–8.
14. Paulsen M, Ferguson-Smith AC. DNA methylation in genomic imprinting, development, and disease. J Pathol. 2001;195(1):97–110.
15. Tycko B, Morison IM. Physiological functions of imprinted genes. J Cell Physiol. 2002;192(3):245–58.
16. Feinberg AP, Tycko B. Timeline – The history of cancer epigenetics. Nat Rev Cancer. 2004;4(2):143–53.
17. Fatemi M, Pao MM, Jeong S, et al. Footprinting of mammalian promoters: use of a CpG DNA methyltransferase revealing nucleosome positions at a single molecule level. Nucleic Acids Res. 2005;33(20):e176.
18. Comb M, Goodman HM. CpG methylation inhibits proenkephalin gene expression and binding of the transcription factor AP-2. Nucleic Acids Res. 1990;18(13):3975–82.
19. Nan X, Campoy FJ, Bird A. MeCP2 is a transcriptional repressor with abundant binding sites in genomic chromatin. Cell. 1997;88(4):471–81.
20. Kaffer CR, Srivastava M, Park KY, et al. A transcriptional insulator at the imprinted H19/Igf2 locus. Genes Dev. 2000;14(15):1908–19.
21. Prokhortchouk A, Hendrich B, Jorgensen H, et al. The p120 catenin partner Kaiso is a DNA methylation-dependent transcriptional repressor. Genes Dev. 2001;15(13):1613–8.
22. Filion GJ, Zhenilo S, Salozhin S, Yamada D, Prokhortchouk E, Defossez PA. A family of human zinc finger proteins that bind methylated DNA and repress transcription. Mol Cell Biol. 2006;26(1):169–81.
23. Okitsu CY, Hsieh CL. DNA methylation dictates histone H3K4 methylation. Mol Cell Biol. 2007;27(7):2746–57.
24. Delaval K, Govin J, Cerqueira F, Rousseaux S, Khochbin S, Feil R. Differential histone modifications mark mouse imprinting control regions during spermatogenesis. EMBO J. 2007;26(3):720–9.
25. Goll MG, Bestor TH. Eukaryotic cytosine methyltransferases. Annu Rev Biochem. 2005;74:481–514.
26. Bestor T, Laudano A, Mattaliano R, Ingram V. Cloning and sequencing of a cDNA encoding DNA methyltransferase of mouse cells. The carboxyl-terminal domain of the mammalian enzymes is related to bacterial restriction methyltransferases. J Mol Biol. 1988;203(4):971–83.
27. Bestor TH. Activation of mammalian DNA methyltransferase by cleavage of a Zn binding regulatory domain. EMBO J. 1992;11(7):2611–7.

28. Yoder JA, Soman NS, Verdine GL, Bestor TH. DNA (cytosine-5)-methyltransferases in mouse cells and tissues. Studies with a mechanism-based probe. J Mol Biol. 1997;270(3):385–95.
29. Okano M, Xie S, Li E. Cloning and characterization of a family of novel mammalian DNA (cytosine-5) methyltransferases. Nat Genet. 1998;19(3):219–20.
30. Trasler JM, Alcivar AA, Hake LE, Bestor T, Hecht NB. DNA methyltransferase is developmentally expressed in replicating and non-replicating male germ cells. Nucleic Acids Res. 1992;20(10):2541–5.
31. Goll MG, Kirpekar F, Maggert KA, et al. Methylation of tRNAAsp by the DNA methyltransferase homolog Dnmt2. Science. 2006;311(5759):395–8.
32. Aapola U, Lyle R, Krohn K, Antonarakis SE, Peterson P. Isolation and initial characterization of the mouse Dnmt3l gene. Cytogenet Cell Genet. 2001;92(1–2): 122–6.
33. Chedin F, Lieber MR, Hsieh CL. The DNA methyltransferase-like protein DNMT3L stimulates de novo methylation by Dnmt3a. Proc Natl Acad Sci USA. 2002;99(26):16916–21.
34. Margot JB, Ehrenhofer-Murray AE, Leonhardt H. Interactions within the mammalian DNA methyltransferase family. BMC Mol Biol. 2003;4:7.
35. Suetake I, Shinozaki F, Miyagawa J, Takeshima H, Tajima S. DNMT3L stimulates the DNA methylation activity of Dnmt3a and Dnmt3b through a direct interaction. J Biol Chem. 2004;279(26):27816–23.
36. Chen ZX, Mann JR, Hsieh CL, Riggs AD, Chedin F. Physical and functional interactions between the human DNMT3L protein and members of the de novo methyltransferase family. J Cell Biochem. 2005;95(5): 902–17.
37. Ooi SK, Qiu C, Bernstein E, et al. DNMT3L connects unmethylated lysine 4 of histone H3 to de novo methylation of DNA. Nature. 2007;448(7154):714–7.
38. La Salle S, Mertineit C, Taketo T, Moens PB, Bestor TH, Trasler JM. Windows for sex-specific methylation marked by DNA methyltransferase expression profiles in mouse germ cells. Dev Biol. 2004;268(2):403–15.
39. Sakai Y, Suetake I, Itoh K, Mizugaki M, Tajima S, Yamashina S. Expression of DNA methyltransferase (Dnmt1) in testicular germ cells during development of mouse embryo. Cell Struct Funct. 2001;26(6):685–91.
40. Jue K, Bestor TH, Trasler JM. Regulated synthesis and localization of DNA methyltransferase during spermatogenesis. Biol Reprod. 1995;53(3):561–9.
41. La Salle S, Trasler JM. Dynamic expression of DNMT3a and DNMT3b isoforms during male germ cell development in the mouse. Dev Biol. 2006;296(1):71–82.
42. Watanabe D, Suetake I, Tajima S, Hanaoka K. Expression of Dnmt3b in mouse hematopoietic progenitor cells and spermatogonia at specific stages. Gene Expr Patterns. 2004;5(1):43–9.
43. Xie S, Wang Z, Okano M, et al. Cloning, expression and chromosome locations of the human DNMT3 gene family. Gene. 1999;236(1):87–95.
44. Bourc'his D, Xu GL, Lin CS, Bollman B, Bestor TH. Dnmt3L and the establishment of maternal genomic imprints. Science. 2001;294(5551):2536–9.
45. Webster KE, O'Bryan MK, Fletcher S, et al. Meiotic and epigenetic defects in Dnmt3L-knockout mouse spermatogenesis. Proc Natl Acad Sci USA. 2005; 102(11):4068–73.
46. Hata K, Okano M, Lei H, Li E. Dnmt3L cooperates with the Dnmt3 family of de novo DNA methyltransferases to establish maternal imprints in mice. Development. 2002;129(8):1983–93.
47. Eckhardt F, Lewin J, Cortese R, et al. DNA methylation profiling of human chromosomes 6, 20 and 22. Nat Genet. 2006;38(12):1378–85.
48. Oakes CC, La Salle S, Smiraglia DJ, Robaire B, Trasler JM. A unique configuration of genome-wide DNA methylation patterns in the testis. Proc Natl Acad Sci USA. 2007;104(1):228–33.
49. Weber M, Hellmann I, Stadler MB, et al. Distribution, silencing potential and evolutionary impact of promoter DNA methylation in the human genome. Nat Genet. 2007;39(4):457–66.
50. Maclean 2nd JA, Wilkinson MF. Gene regulation in spermatogenesis. Curr Top Dev Biol. 2005;71: 131–97.
51. Kitamura E, Igarashi J, Morohashi A, et al. Analysis of tissue-specific differentially methylated regions (TDMs) in humans. Genomics. 2007;89(3):326–37.
52. Farthing CR, Ficz G, Ng RK, et al. Global mapping of DNA methylation in mouse promoters reveals epigenetic reprogramming of pluripotency genes. PLoS Genet Jun. 2008;4(6):e1000116.
53. Yamagata K, Yamazaki T, Miki H, et al. Centromeric DNA hypomethylation as an epigenetic signature discriminates between germ and somatic cell lineages. Dev Biol. 2007;312(1):419–26.
54. Oakes CC, La Salle S, Smiraglia DJ, Robaire B, Trasler JM. Developmental acquisition of genome-wide DNA methylation occurs prior to meiosis in male germ cells. Dev Biol. 2007;307(2):368–79.
55. Flanagan JM, Popendikyte V, Pozdniakovaite N, et al. Intra- and interindividual epigenetic variation in human germ cells. Am J Hum Genet. 2006;79(1): 67–84.
56. Hajkova P, Erhardt S, Lane N, et al. Epigenetic reprogramming in mouse primordial germ cells. Mech Dev. 2002;117(1–2):15–23.
57. Monk M, Boubelik M, Lehnert S. Temporal and regional changes in DNA methylation in the embryonic, extraembryonic and germ cell lineages during mouse embryo development. Development. 1987; 99(3):371–82.
58. Kafri T, Ariel M, Brandeis M, et al. Developmental pattern of gene-specific DNA methylation in the mouse embryo and germ line. Genes Dev. 1992;6(5): 705–14.
59. Coffigny H, Bourgeois C, Ricoul M, et al. Alterations of DNA methylation patterns in germ cells and Sertoli cells from developing mouse testis. Cytogenet Cell Genet. 1999;87(3–4):175–81.

60. Seki Y, Hayashi K, Itoh K, Mizugaki M, Saitou M, Matsui Y. Extensive and orderly reprogramming of genome-wide chromatin modifications associated with specification and early development of germ cells in mice. Dev Biol. 2005;278(2):440–58.
61. Davis TL, Yang GJ, McCarrey JR, Bartolomei MS. The H19 methylation imprint is erased and re-established differentially on the parental alleles during male germ cell development. Hum Mol Genet. 2000;9(19):2885–94.
62. Li JY, Lees-Murdock DJ, Xu GL, Walsh CP. Timing of establishment of paternal methylation imprints in the mouse. Genomics. 2004;84(6):952–60.
63. Maatouk DM, Kellam LD, Mann MR, et al. DNA methylation is a primary mechanism for silencing postmigratory primordial germ cell genes in both germ cell and somatic cell lineages. Development. 2006;133(17):3411–8.
64. Lane N, Dean W, Erhardt S, et al. Resistance of IAPs to methylation reprogramming may provide a mechanism for epigenetic inheritance in the mouse. Genesis. 2003;35(2):88–93.
65. Lees-Murdock DJ, De Felici M, Walsh CP. Methylation dynamics of repetitive DNA elements in the mouse germ cell lineage. Genomics. 2003;82(2):230–7.
66. Kato Y, Kaneda M, Hata K, et al. Role of the Dnmt3 family in de novo methylation of imprinted and repetitive sequences during male germ cell development in the mouse. Hum Mol Genet. 2007;16(19):2272–80.
67. Kono T, Obata Y, Yoshimzu T, Nakahara T, Carroll J. Epigenetic modifications during oocyte growth correlates with extended parthenogenetic development in the mouse. Nat Genet. 1996;13(1):91–4.
68. Lucifero D, Mertineit C, Clarke HJ, Bestor TH, Trasler JM. Methylation dynamics of imprinted genes in mouse germ cells. Genomics. 2002;79(4):530–8.
69. Obata Y, Kono T. Maternal primary imprinting is established at a specific time for each gene throughout oocyte growth. J Biol Chem. 2002;277(7):5285–9.
70. Davis TL, Trasler JM, Moss SB, Yang GJ, Bartolomei MS. Acquisition of the H19 methylation imprint occurs differentially on the parental alleles during spermatogenesis. Genomics. 1999;58(1):18–28.
71. Kerjean A, Dupont JM, Vasseur C, et al. Establishment of the paternal methylation imprint of the human H19 and MEST/PEG1 genes during spermatogenesis. Hum Mol Genet. 2000;9:649–59.
72. Galetska D, Weis E, Tralau T, et al. Sex-specific windows for high mRNA expresion of DNA methyltransferases 1 and 3A and methyl-CpG-binding domain proteins 2 and 4 in fetal gonads. Mol Reprod Dev. 2007;74:233–41.
73. Khalil AM, Boyar FZ, Driscoll DJ. Dynamic histone modifications mark sex chromosome inactivation and reactivation during mammalian spermatogenesis. Proc Natl Acad Sci USA. 2004;101(47):16583–7.
74. Payne C, Braun RE. Histone lysine trimethylation exhibits a distinct perinuclear distribution in Plzf-expressing spermatogonia. Dev Biol. 2006;293(2):461–72.
75. Godmann M, Auger V, Ferraroni-Aguiar V, et al. Dynamic regulation of histone H3 methylation at lysine 4 in mammalian spermatogenesis. Biol Reprod. 2007;77(5):754–64.
76. Carrell DT, Hammoud SS. The human sperm epigenome and its potential role in embryonic development. Mol Hum Reprod. 2010;16(1):37–47.
77. Tanphaichitr N, Sobhon P, Taluppeth N, Chalermisarachai P. Basic nuclear proteins in testicular cells and ejaculated spermatozoa in man. Exp Cell Res. 1978;117(2):347–56.
78. Wykes SM, Krawetz SA. The structural organization of sperm chromatin. J Biol Chem. 2003;278(32):29471–7.
79. Hammoud SS, Nix DA, Zhang H, Purwar J, Carrell DT, Cairns BR. Distinctive chromatin in human sperm packages genes for embryo development. Nature. 2009;460(7254):473–8.
80. Brykczynska U, Hisano M, Erkek S, et al. Repressive and active histone methylation mark distinct promoters in human and mouse spermatozoa. Nature Struct Mol Biol. 2010;17(16):679–87.
81. Li E, Bestor TH, Jaenisch R. Targeted mutation of the DNA methyltransferase gene results in embryonic lethality. Cell. 1992;69(6):915–26.
82. Li E, Beard C, Jaenisch R. Role for DNA methylation in genomic imprinting. Nature. 1993;366(6453):362–5.
83. Howell CY, Bestor TH, Ding F, et al. Genomic imprinting disrupted by a maternal effect mutaion in the Dnmt1 gene. Cell. 2001;104:829–38.
84. Kanai Y, Ushijima S, Nakanishi Y, Sakamoto M, Hirohashi S. Mutation of the DNA methyltransferase (DNMT) 1 gene in human colorectal cancers. Cancer Lett. 2003;192(1):75–82.
85. Okano M, Bell DW, Haber DA, Li E. DNA methyltransferases Dnmt3a and Dnmt3b are essential for de novo methylation and mammalian development. Cell. 1999;99(3):247–57.
86. Yaman R, Grandjean V. Timing of entry of meiosis depends on a mark generated by DNA methyltransferase 3a in testis. Mol Reprod Dev. 2006;73(3):390–7.
87. Kaneda M, Okano M, Hata K, et al. Essential role for de novo DNA methyltransferase Dnmt3a in paternal and maternal imprinting. Nature. 2004;429(6994):900–3.
88. Xu GL, Bestor TH, Bourc'his D, et al. Chromosome instability and immunodeficiency syndrome caused by mutations in a DNA methyltransferase gene. Nature. 1999;402(6758):187–91.
89. Jeanpierre M, Turleau C, Aurias A, et al. An embryonic-like methylation pattern of classical satellite DNA is observed in ICF syndrome. Hum Mol Genet. 1993;2(6):731–5.
90. Kondo T, Bobek MP, Kuick R, et al. Whole-genome methylation scan in ICF syndrome: hypomethylation of non-satellite DNA repeats D4Z4 and NBL2. Hum Mol Genet. 2000;9(4):597–604.
91. La Salle S, Oakes CC, Neaga OR, Bourc'his D, Bestor TH, Trasler JM. Loss of spermatogonia and wide-spread DNA methylation defects in newborn male mice deficient in DNMT3L. BMC Dev Biol. 2007;7:104.

92. Bourc'his D, Bestor TH. Meiotic catastrophe and retrotransposon reactivation in male germ cells lacking Dnmt3L. Nature. 2004;431(7004):96–9.
93. Forges T, Monnier-Barbarino P, Alberto JM, Gueant-Rodriguez RM, Daval JL, Gueant JL. Impact of folate and homocysteine metabolism on human reproductive health. Hum Reprod Update. 2007;13(3):225–38.
94. Bezold G, Lange M, Peter RU. Homozygous methylenetetrahydrofolate reductase C677T mutation and male infertility. N Engl J Med. 2001;344(15): 1172–3.
95. Lee HC, Jeong YM, Lee SH, et al. Association study of four polymorphisms in three folate-related enzyme genes with non-obstructive male infertility. Hum Reprod. 2006;21(12):3162–70.
96. Chen Z, Karaplis AC, Ackerman SL, et al. Mice deficient in methylenetetrahydrofolate reductase exhibit hyperhomocysteinemia and decreased methylation capacity, with neuropathology and aortic lipid deposition. Hum Mol Genet. 2001;10(5):433–43.
97. Kelly TL, Neaga OR, Schwahn BC, Rozen R, Trasler JM. Infertility in 5,10-methylenetetrahydrofolate reductase (MTHFR)-deficient male mice is partially alleviated by lifetime dietary betaine supplementation. Biol Reprod. 2005;72(3):667–77.
98. Chan D, Cushnie DW, Neaga OR, Lawrance AK, Rozen R, Trasler JM. Strain-specific defects in testicular development and sperm epigenetic patterns in 5,10-methylenetetrahydrofolate reductase-deficient mice. Endocrinology. 2010;151:3363–73.
99. Cihak A. Biological effects of 5-azacytidine in eukaryotes. Oncology. 1974;30(5):405–22.
100. Gabbara S, Bhagwat AS. The mechanism of inhibition of DNA (cytosine-5-)-methyltransferases by 5-azacytosine is likely to involve methyl transfer to the inhibitor. Biochem J. 1995;307(Pt 1):87–92.
101. Doerksen T, Trasler JM. Developmental exposure of male germ cells to 5-azacytidine results in abnormal preimplantation development in rats. Biol Reprod. 1996;55(5):1155–62.
102. Doerksen T, Benoit G, Trasler JM. Deoxyribonucleic acid hypomethylation of male germ cells by mitotic and meiotic exposure to 5-azacytidine is associated with altered testicular histology. Endocrinology. 2000;141(9):3235–44.
103. Kelly TL, Li E, Trasler JM. 5-aza-2'-deoxycytidine induces alterations in murine spermatogenesis and pregnancy outcome. J Androl. 2003;24(6):822–30.
104. Oakes CC, Kelly TL, Robaire B, Trasler JM. Adverse effects of 5-aza-2'-deoxycytidine on spermatogenesis include reduced sperm function and selective inhibition of de novo DNA methylation. J Pharmacol Exp Ther. 2007;322(3):1171–80.
105. Trasler JM, Hales BF, Robaire B. Paternal cyclophosphamide treatment of rats causes fetal loss and malformations without affecting male fertility. Nature. 1985;316(6024):144–6.
106. Trasler JM, Hales BF, Robaire B. A time-course study of chronic paternal cyclophosphamide treatment in rats: effects on pregnancy outcome and the male reproductive and hematologic systems. Biol Reprod. 1987;37(2):317–26.
107. Trasler JM, Robaire B. Effects of cyclophosphamide on selected cytosolic and mitochondrial enzymes in the epididymis of the rat. J Androl. 1988;9(2):142–52.
108. Barton TS, Wyrobek AJ, Hill FS, Robaire B, Hales BF. Numerical chromosomal abnormalities in rat epididymal spermatozoa following chronic cyclophosphamide exposure. Biol Reprod. 2003;69(4): 1150–7.
109. Barton TS, Robaire B, Hales BF. Epigenetic programming in the preimplantation rat embryo is disrupted by chronic paternal cyclophosphamide exposure. Proc Natl Acad Sci USA. 2005;102(22):7865–70.
110. Aston KI, Carrell DT. Genome-wide study of single-nucleotide polymorphisms associated with azoospermia and severe oligozoospermia. J Androl. 2009;30(6):711–25.
111. O'Flynn O'Brien KL, Varghese AC, Agarwal A. The genetic causes of male factor infertility: a review. Fertil Steril. 2010;93(1):1–12.
112. Marques CJ, Carvalho F, Sousa M, Barros A. Genomic imprinting in disruptive spermatogenesis. Lancet. 2004;363(9422):1700–2.
113. Marques CJ, Costa P, Vaz B, et al. Abnormal methylation of imprinted genes in human sperm is associated with oligozoospermia. Mol Hum Reprod. 2008;14(2):67–74.
114. Kobayashi H, Sato A, Otsu E, et al. Aberrant DNA methylation of imprinted loci in sperm from oligospermic patients. Hum Mol Genet. 2007;16(21): 2542–51.
115. Poplinski A, Tuttelmann F, Kanber D, Horsthemke B, Gromoll J. Idiopathic male infertility is strongly associated with aberrant methylation of MEST and IGF2/H19 ICR1. Int J Androl. 2010;33:642–9.
116. Houshdaran S, Cortessis VK, Siegmund K, Yang A, Laird PW, Sokol RZ. Widespread epigenetic abnormalities suggest a broad DNA methylation erasure defect in abnormal human sperm. PLoS One. 2007;2(12):e1289.
117. Hammoud SS, Purwar J, Pflueger C, Cairns BR, Carrell DT. Alterations in sperm DNA methylation patterns at imprinted loci in two classes of infertility. Fertil Steril. 2010;94:1728–33.
118. Kobayashi H, Hiura H, John RM, et al. DNA methylation errors at imprinted loci after assisted conception originate in the parental sperm. Eur J Hum Genet. 2009;17(12):1582–91.
119. Kagami M, Nagai T, Fukami M, Yamazawa K, Ogata T. Silver-Russell syndrome in a girl born after in vitro fertilization: partial hypermethylation at the differentially methylated region of PEG1/MEST. J Assist Reprod Genet. 2007;24(4):131–6.
120. Mason JB, Dickstein A, Jacques PF, et al. A temporal association between folic acid fortification and an increase in colorectal cancer rates may be illuminating important biological principles: a hypothesis. Cancer Epidemiol Biomarkers Prev. 2007;16(7):1325–9.
121. Richards EJ. Inherited epigenetic variation-revisiting soft inheritance. Nat Rev Genet. 2006;7:395–401.

RNA Expression in Male Germ Cells During Spermatogenesis (Male Germ Cell Transcriptome)

8

Tin-Lap Lee, Albert Hoi-Hung Cheung, Owen M. Rennert, and Wai-Yee Chan

Abstract

Spermatogenesis is a key process in mammalian reproduction. This highly ordered process requires precise and well-controlled programs governed by dynamic patterns of gene expression. Some genes are exclusive to spermatogenic cells, while others are closely related to genes expressed in somatic cells. Although key genes in male germ cell development have been identified, the biological mechanisms and transcripts that govern the programs of spermatogonial stem cell renewal, germ cell differentiation during spermatogenesis, or fertilization remain largely unknown. This is partly due to the lack of information on the identity of genes involved. However, with the advent of various high-throughput genomic assays, it is now possible to obtain the whole-genome RNA expression. This chapter provides a brief account of current knowledge of the male germ cell transcriptome as revealed by studies using expression profiling platforms such as microarray and Serial Analysis of Gene Expression (SAGE). Major findings with regard to transcriptional regulation, transcript diversity, and chromatin-related regulation during male germ cell development are reviewed.

Keywords

RNA expression in germ cells • Spermatogenesis • Germ cell transcriptome • Male germ cell development • Serial analysis of gene expression

Spermatogenesis is a highly regulated developmental process occurring in the seminiferous tubules of the testis. The process begins with the asymmetric division of spermatogonial progenitor cells (spermatogonia), followed by meiosis to form spermatocytes, postmeiotic differentiation to form spermatids, and finally giving rise to mature spermatozoa (Fig. 8.1). Mouse male germ

T.-L. Lee (✉)
Reproduction, Development and Endocrinology Program, School of Biomedical Sciences, The Chinese University of Hong Kong, Shatin, Hong Kong SAR, China

Laboratory of Clinical and Developmental Genomics, Eunice Kennedy Shriver National Institute of Child Health and Human Development, National Institutes of Health, Bethesda, MD, USA
e-mail: leetl@cuhk.edu.hk; leetl@mail.nih.gov

A. Zini and A. Agarwal (eds.), *Sperm Chromatin: Biological and Clinical Applications in Male Infertility and Assisted Reproduction*, DOI 10.1007/978-1-4419-6857-9_8,

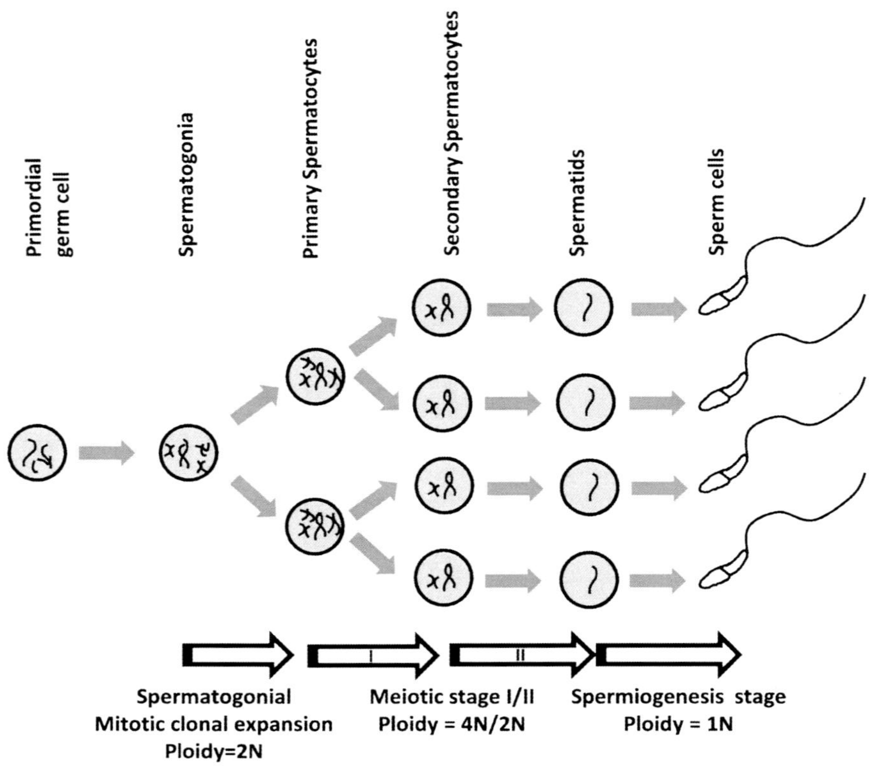

Fig. 8.1 Overview of spermatogenesis. Spermatogenesis is the developmental process by which spermatogonial stem cells differentiate to pachytene spermatocytes, followed by formation of haploid spermatids by meiosis. Male germ cell genome displays several features unique to germ cells only. First, a subset of spermatogonia (type A spermatogonia) undergo mitosis during self-renewal, whereas committed spermatogonia (type B spermatogonia) undergo meiosis to generate haploid spermatids. Primary spermatocytes replicate their genomes during S-phase, followed by meiosis I to form secondary spermatocytes and subsequently meiosis II to form haploid spermatids. Meiosis only occurs in the germ line

cells provide an ideal model for studying the biology of spermatogenesis. This process has been well studied in the mouse with established developmental milestones starting at the derivation of primordial germ cells from the embryonal ectoderm. Embryonic staging of the developing male gonad is accomplished morphologically, or with genetic markers such as the Sex-determining Region Y (*Sry*) [1, 2]. Male germ cells, at different stages of development, have different density, distinct morphology, and stage-specific surface markers. These features serve as the basis of methods for preparation of relatively pure populations of germ cells at different stages of development. Relatively pure preparations of gonocytes can be obtained using laser capture techniques [3, 4]. All germ cells present in the testis of 6-day-old mice are type A spermatogonia (Spga). In adult mice, germ cells at all stages of development are present, and the different cell types can be separated based on their density using the STAPUT procedure [5].

Spermatogenesis consists of a number of hallmark developmental stages: germinal stem cells undergoing self-renewal, Spga progenitor cells at the juncture of renewal and proliferation, pachytene spermatocytes (Spcy) undergoing meiosis, and round spermatids (Sptd) undergoing postmeiotic differentiation. Therefore, studying genetic events occurring during these stages of spermatogenesis will permit a comprehensive look at the genetic events that underlie cellular proliferation and differentiation.

Spermatogenesis is a complicated process. Each step of spermatogenesis is precisely regulated. Studying the genetic programs controlling proliferation and differentiation of male germ cells will provide an insight for understanding infertility and will allow for development of new approaches for male contraception. Demonstration of pathways specific for different stages of spermatogenesis would allow for identification of novel targets and diagnostic markers for intervention or enhancement of the male reproductive process. Knowledge of factors regulating cellular proliferation, as contrasted to differentiation, may be applied to study developmental regulation of other cell types, including stem cells.

The Transcription Landscape in Male Germ Cell Development

Little is known about the underlying mechanisms of stage-specific regulation of gene expression during spermatogenesis despite its biological importance in the genetic regulation of germ cell-specific transcripts during development [6, 7]. Limited knowledge of germ cell gene expression and the lack of a systematic approach for pathway discovery have hampered identification of biological pathways active in these cells. With the completion of various genome projects and the availability of high-throughput expression assays, a better understanding of the male germ cell transcriptome becomes a reality.

Expression Profiling of Male Germ Cell Transcriptome: Past, Present, and Future

From cDNA Library to High-Throughput Sequencing

The first attempt to characterize the male germ cell transcriptome was reported by McCarrey et al. [8]. A total of 23 cDNA libraries representing various developmental stages of mouse and rat testicular cells were constructed. Direct comparison between the cDNA populations in various cells provided the basis for the demonstration of differential gene expression. Though theoretically feasible, this approach is laborious in practice. "Deep sequencing" of cDNA libraries is required before a near-complete picture of the transcriptome can emerge. Because of these considerations, the use of cDNA libraries for transcriptome analysis was not popular.

Transcriptome profiling was a tedious job until the application of microarray platforms. A microarray is a solid support on which DNA probes of known sequence are deposited. The probes may take the form of oligonucleotides, cDNA, or DNA fragments. These probes are hybridized to sequences present in the sample. Depending on its resolution, a whole-genome human microarray chip may contain more than two millions probes. DNA microarrays were originally developed for high-throughput gene expression analysis. But they can also be applied in genetic analysis to detect single-nucleotide polymorphisms or gene copy number variation. Their fast, comprehensive, and flexible nature makes them an indispensable tool in the postgenomic era.

Another popular, widely adopted expression assay is Serial Analysis of Gene Expression (SAGE) [9]. It offers distinct advantages over other expression profiling methods by efficiently detecting polyadenylated transcript populations by sequencing short tags, usually 14–26 bp in length [8, 9]. The tags are first isolated from an anchoring enzyme restriction site (e.g., *NlaIII*) closest to the poly(A) tail of the transcripts. These tags are linked together to form long concatemers that are cloned into vectors generating a SAGE library. A SAGE library is sequenced to the desired depth. Expression of particular transcripts is quantified by the count of the associated SAGE tags in the SAGE library. Once the tags are extracted and counted, the identity of the transcript may be mapped with the SAGEmap database [9]. SAGE provides three important features over microarrays for transcriptome analysis. First, the absolute nature of tag counts allows direct comparison, without normalization and limitation of platform incompatibility, in microarray experiments. Second, since tag-to-transcript mapping in SAGE may be updated with the most current genome information, the transcriptome information provided by SAGE library is eternal. SAGE analysis allows identification of novel transcript species since prior knowledge of transcripts is not required. Finally, microarray analysis provides no orientation information on the transcript, whereas SAGE can differentiate the sense and antisense population in the transcriptome.

While sequence-based transcriptome analysis provides additional advantages, it is cumbersome and slow, with relatively high performance cost ($0.10 per 1,000 bases). However, this is rapidly changing due to the continued improvement of sequencing technologies. The 454 sequencer was introduced in 2005 and was shortly followed by newer and faster sequencers such as Illumina and SOLiD. These technologies are referred to as "next-gen" sequencing [10]. They offer faster (up to 100×) and more cost-effective (up to 1/2,000 of the price) sequencing than conventional methods. Transcriptome analysis at single-base resolution, known as RNA-seq, is now possible [11–13]. Next-gene sequencing will be an important tool for transcriptome analysis in coming years. The huge quantity of generated data by these technologies poses great challenges to experimental biologists [14].

Overview of Germ Cell Transcriptome Studies

A list of male germ cell transcriptome studies is shown in Table 8.1. Most of these were performed on microarray platforms [15–31]. This is because sample preparation and experimental protocols are simpler when using the microarray platform. Additionally, less RNA is required as compared to SAGE [32], cDNA library [8] and differential display [33] methods. Oligonucleotide microarrays are more popular than cDNA microarrays partially because cDNA microarrays are more prone to variation in slide quality and experimental protocol. Another reason is that they are more affordable. Nevertheless, each expression platform has its strength and weakness. Renormalization against known references and correction by using statistical models are required to compare data from different studies.

Key Biological Findings and Implications

Based on the transcriptome data provided by the studies listed (Table 8.1), a number of conclusions about the dynamic changes of the transcriptome of developing male germ cells can be drawn:

Active genome-wide transcription during spermatogenesis. A major observation is that the genome is actively transcribed during germ cell development. It was previously suggested during testis development from birth to adulthood up to 58% of the mouse genome was transcribed [18, 22, 30]. Among the described transcripts, some were either male germ cell-specific or testis-predominant. About 2.3% of the rat testicular transcriptome was testis-specific [30], and ~4% of the mouse genome was only transcribed in male germ cells [18]. Many differentially expressed transcripts were unknown or uncharacterized. Examples include uncharacterized full-length cDNA transcripts, express sequence tags (ESTs), large open reading frames (ORFs), predicted transcripts of hypothetical proteins, and cross-species and predicted transcripts derived from orthologs and homologs. Depending on the cell preparation and experimental platform, the percentage of uncharacterized transcripts ranged from 40 to 60% [18, 32, 34]. Meta-analysis of these transcripts suggested that they demonstrated similar expression trends. These results imply that these transcripts were truly expressed at a particular germ cell stage.

Dynamic expression pattern in conjunction with specific developmental regulation. Transcript overexpression, as revealed by measurement of polyadenylated RNA levels in meiotic and postmeiotic male germ cells, was documented in an earlier study in rats [35]. Such phenomena might be a bystander effect occurring as a consequence of an open chromatin structure, which leads to overall activation of the transcriptional machinery in a specified cell type [7]. Alternatively, it may be a mechanism for maintaining transcript availability in response to cessation of gene transcription due to chromatin condensation during spermiogenesis [36]. Based on global gene expression analyses in various transcriptome studies, germ cell transcriptome exhibited three phases of change. The first phase, peak expression of testicular transcripts, occurred in the mitotic phase, from the day of birth to postnatal day 8, when spermatogonial proliferation predominated. The second phase occurred at the

Table 8.1 Overview of male germ cell transcriptome studies

Samples studied	Expression platform	Reference
Whole mouse adult testes, seminiferous tubule cells from adult testes, combined primary spermatocytes from 18-day-old mouse testes, type A and B spermatogonia, preleptotene, leptotene plus zygotene spermatocytes, juvenile and adult pachytene spermatocytes, round spermatids, Sertoli cells from 6, 8, 17, and 18–20-day-old mice, and peritubular cells from 18- to 20-day-old mice	cDNA library sequencing	McCarrey et al. [8]
Mouse type A spermatogonia, adult mouse wild-type testis, and W/W(v) mutant mouse testis	Differential display	Anway et al. [99]
Mouse and human testes	Microarray	Rockett et al. [15]
Human fetal and adult testes	Microarray (cDNA)	Sha et al. [16]
Mouse type A spermatogonia, pachytene spermatocytes, and round spermatids	Microarray (cDNA)	Pang et al. [17]
Mouse testes from days 1, 4, 8, 11, 14, 18, 21, 26, 29, and 60	Microarray (Oligo)	Schultz et al. [18]
Mouse type A and B spermatogonia, preleptotene and pachytene spermatocytes, round and elongating spermatids	Microarray (cDNA)	Yu et al. [19]
Mouse type A and B spermatogonia, preleptotene and pachytene spermatocytes, round and elongating spermatids	Microarray (cDNA)	Guo et al. [20]
Mouse sertoli cells, spermatogonia, spermatocytes, round spermatids	Microarray (Oligo)	Schlecht et al. [21]
Whole testes from neonates at Days 0, 3, 6, 8, 10, 14, 18, 20, 30, 35, and 56 postpartum	Microarray (Oligo)	Shima et al. [22]
Mouse adult and fetal testes	Microarray (cDNA)	Wang et al. [23]
Mouse type A spermatogonia, pachytene spermatocytes, and round spermatids	SAGE	Wu et al. [32]
Mouse sertoli cells, type A spermatogonia, spermatocytes, round spermatids	Microarray (cDNA)	Clemente et al. [24]
Testes from 17-day-old, 22-day-old, and adult mice	Microarray (Oligo)	Iguchi et al. [25]
Normal testis, patients with maturation arrest or Sertoli-cell-only syndrome	Microarray (cDNA)	Lin et al. [26]
Type A and type B spermatogonia, pachytene spermatocytes, and round spermatids	Microarray (Oligo)	Namekawa et al. [27]
Sertoli cells, spermatogonia, spermatocytes, round spermatids, seminiferous tubules, and total testis from human, rat, and mouse	Microarray (Oligo)	Chalmel et al. [28]
Testicular biopsies obtained from 289 men with azoospermia	Microarray (Oligo)	Feig et al. [29]
Rat seminiferous tubules at various stages, microdissection, sertoli cells, spermatogonia, spermatocytes, pachytene spermatocytes, and round spermatids	Microarray (Oligo)	Johnston et al. [30]
Testis samples of mice aged 4, 9, 18, 35, 54 days and 6 months	Microarray (Oligo)	Xiao et al. [31]

initiation of meiosis, on day 14, during early pachytene spermatocytes development. This was followed by entry into spermiogenesis on day 20 when round spermatids first appeared [37]. Comparison of these three phases showed increased transcript abundance in meiotic and postmeiotic stages. The number of unique genes expressed in these cells was significantly higher than that in spermatogonia [22] when up to 80% of differentially expressed genes, between meiotic and postmeiotic male germ cells, were absent or expressed at relatively low levels in type A spermatogonia [22, 34]. Increased expression of unique genes in meiotic and postmeiotic stages may imply a concomitant increase in the demand of specific gene activities for initiation and maintenance of meiosis-related events, as well as preparation for spermatozoon formation. It is noteworthy that most transcripts first expressed during or after meiosis tended to be testis- or male germ cell-specific [18, 28]. On the contrary, most genes active in spermatogonia (and Sertoli cells) were also expressed in nonreproductive tissues [28].

There was a preferential switch of active genetic loci at different stages of germ cell development. Genes related to meiotic and postmeiotic functions, and displaying higher expression level in testis, are mainly localized to autosomes [38]. By contrast, genes expressed at earlier stages of spermatogenesis are frequently localized to the X chromosome [28, 38, 39]. Similarly, many genes expressed in mitotic and somatic cells were localized on the X chromosome. A similar phenomenon was observed in a particular subset of genes, the X chromosome-derived autosomal retrogenes and their X-linked progenitor genes. Although not all testis-specific autosomal genes were X-derived retrogenes or retrogenes, the absence of X-linkage in general was believed to be a consequence of the selective force imposed by meiotic sex chromosome inactivation (MSCI) [40–42].

The dynamic and specific nature of the germ cell transcriptome was also associated with specific development and regulatory programs. Ontology analysis of the germ cell transcriptome data revealed different categories of biological processes distinctively associated with mitotic, meiotic, and postmeiotic male germ cells [28, 34, 37, 43]. For instance, processes such as integrin signaling, ribosome biogenesis and assembly, carbohydrate metabolism, protein biosynthesis, RNA processing, cell cycle, DNA replication, chromosome organization and biogenesis, and germ cell development were preferentially associated with type A spermatogonia. Surprisingly, genes involved in embryonic development and gastrulation were also found to be prevalent in these cells. On the other hand, biological processes associated with spermatogenesis and reproduction were commonly seen in meiotic and postmeiotic male germ cells. Biological processes such as meiotic cell cycle, chromatin structure and dynamics, chromosome segregation, cytoskeleton and protein degradation (ubiquitin cycle) were overrepresented in pachytene spermatocytes. Genes involved in protein turnover, signal transduction, energy metabolism, intracellular transport, ubiquitin cycle, proteolysis, peptidolysis, and fertilization were more prevalent in round spermatids.

Conserved germ cell transcriptome between human and rodents. The universal features of gametogenesis among mammalian species led to the postulation that a conserved set of genes would be involved in this process. Indeed, recent cross-species whole-genome expression profiling studies of testicular and somatic tissues in human, mouse, and rat revealed hundreds of genes that display concordant meiotic and postmeiotic expression profiles, implying the existence of a "conserved" transcriptome of mammalian spermatogenesis [28, 37]. Conserved genes involved in specific biological transitions during male germ cell development were identified by analysis of gene ontology. For example, doublesex and mab-3 related transcription factor 1 (*Dmrt1*) was found to be essential for testis differentiation; aurora kinase C (*Aurkc*), cyclin A1 (*Ccna1*) and speedy homolog A (*Spdy1*) were associated with meiotic division, whereas genes like *Socs7*, *Ankrd5*, *Fscn3*, and *Spag4l* were involved in postmeiotic regulation. Such findings suggest that rodent models could be used to study aspects of human spermatogenesis. A similar differential expression pattern of testicular genes across species suggests the presence of comparable regulatory mechanism in the control of their transcription.

In addition to changes in expression pattern of protein-encoding genes, emerging evidence identified the prominent presence in testis of non-protein-coding transcripts, including antisense transcript, small and long noncoding RNAs. These novel transcript species have been implicated to play important roles in mammalian testis development [44, 45]. The complexity of the spermatogenic process led to the search for male germ cell-specific transcripts derived from alternative splicing of somatic genes. Additionally, many germ cell genes derived from sex-linked progenitor genes through retroposition to generate testis-specific isoforms of gene products were identified. The limitations of design and probe set information inherent in microarray analysis restrict its capacity to identify non-protein-coding and alternative spliced transcripts. This is a consequence of the need, when using microarray analysis, to have prior sequence knowledge of the transcripts, and whether it is a coding or noncoding sequence to be identified. This problem could be resolved by using the nonstatic and unguided approach of SAGE [46].

Revealing Transcription Complexity of Male Germ Cell Development by Serial Analysis of Gene Expression

Using SAGE, we examined the transcriptomes of mouse Spga, Spcy, and Sptd. SAGE libraries were constructed and sequenced to a comparable depth (~150,000 SAGE tags). A total of 34,619 transcripts were identified among the germ cell libraries. Over 2,700 of them were novel. This represents the most comprehensive male germ cell transcriptome data available. The details and related data of this analysis can be accessed at http://nichddirsage.nichd.nih.gov/publicsage/. The data obtained by the SAGE studies provide a rich resource for germ cell transcriptome discovery. By developing various bioinformatics algorithms, we succeeded in exploiting the SAGE data to decode a number of complex regulatory mechanisms and transcript species that could not be archived by microarray analyses [47–51].

Alternative Splicing

The use of multiple promoters and transcription start sites is one mechanism to create gene diversity in spermatogenesis. Alternative promoter usage allows cells to generate isoforms as well as to establish tissue specificity [52]. A large number of testis-specific splicing variants have been reported. For example, GH-releasing hormone (*GHRH*) is expressed in hypothalamus and placenta of rat. The use of a spermatogenic-specific promoter and alternative transcription initiation allows testicular germ cells to express testis-specific isoforms [53]. Expression of the testis-specific *HEMGN* mRNA (*HEMGN-t*) is developmentally regulated and synchronized with the first wave of meiosis in prepuberal mice. *HEMGN-t* is transcribed by use of alternative promoters and polyadenylation sites, suggesting a role for this testis-specific isoform in spermatogenesis [54]. Calspermin is a Calcium/calmodulin-dependent protein kinase triggering a signaling cascade. A testis-specific isoform is expressed in postmeiotic germ cells, possibly controlled by binding of CREM to the CRE motifs [55].

We reported the global identification and analysis of transcript variants with alternative 3′ end usage based on analysis of SAGE libraries of Spga, Spcy, and Sptd [47]. Unique SAGE tags at each stage of spermatogenesis were mapped to the SAGEmap database to retrieve the unigene cluster. Tags sharing the same unigene cluster within or among the stages were compared against different alternative splicing resources and validated by real-time PCR. The number of genes with 3′ end alternative splicing variants (3′ AS) expressed in Spga, Spcy, and Sptd was 74, 58, and 62, respectively. Two hundred and seven genes with 3′ AS were expressed in both Spga and Spcy. The number of genes expressed in both Spga and Sptd was 249, and the number expressed in both Spcy and Sptd was 158. There were 73 genes with different 3′ AS in all three stages examined. Novel variants involved in developmental and transcriptional control were identified. Examples included heat shock protein 4 (*Hsp4*), H3 histone, family 3B (*H3f3b*), and ubiquitin protein ligase E3A (*Ube3a*). In summary, SAGE not only provides a rapid global survey of the gene expression profile in the germ cell transcriptome but also allows identification of novel alternative splicing variants that may contribute to the unique characteristics of spermatogenesis. Further functional studies of these variants will provide new insight into germ cell development during spermatogenesis.

Antisense Transcription

Though antisense transcription has been recognized in prokaryotes for many years, the widespread occurrence of antisense transcripts in humans and mice has only been recently documented. Most studies on antisense transcription used a computational approach to identify the global presence of antisense transcripts or focused on a single gene. Few reports document the mechanism by which an antisense transcript is generated. A number of processes in spermatogenesis such as genomic imprinting, translation repression, and stage-specific alternative splicing are frequently associated with antisense transcripts [7]. A systematic search for antisense transcripts in spermatogenic cells has not previously been reported.

Utilizing the germ cell SAGE database, our laboratory, employing orientation specific RT-PCR and molecular cloning, demonstrated that a significant percentage (31.1%) of differentially expressed genes in spermatogenic cells are associated with antisense transcripts [48]. Nucleotide sequence analysis of orientation specific RT-PCR products of 19 genes, as well as cloned full-length antisense transcripts, showed that antisense transcripts could potentially arise through a wide spectrum of mechanisms, including reverse transcription of sense mRNA in the cytoplasm, transcription of the opposite strand of the sense gene locus, transcription of a pseudogene, as well as transcription of neighboring genes and the intergenic sequence. Some of the antisense transcripts underwent normal and alternative splicing, 5′ capping, and 3′ polyadenylation like their sense counterparts. There were also antisense transcripts that were not capped and/or polyadenylated in the testis. In all cases, the levels of the sense transcripts were higher than that of the antisense transcripts while the relative expression in nontesticular tissues was variable. Thus antisense transcripts have complex origins and variable structure. Sense and antisense transcripts could be regulated independently.

Noncoding RNA Transcription

Mammalian cells produce thousands of noncoding RNAs (ncRNAs) of unknown function [51, 56–63]. These non-protein-coding portions of the genome often were considered "junk," but present research has highlighted that ncRNAs can have a wide range of regulatory functions. Small ncRNAs such as microRNA (miRNA) [64], Small interfering RNA (siRNAs) [65] and Piwi-interacting RNA (piRNA) [66] have been widely reported to function in various regulatory processes, including male germ cell development [67–74]. Recently, a new class of ncRNAs known as long ncRNAs (>200 bp) has also been demonstrated to function in developmental regulation, such as mouse ESCs pluripotency and differentiation [75–78]. These observations suggest that long ncRNAs may be indispensible in male germ cell development.

To identify potential specific long ncRNA involved in male germ cell development, we searched the SAGE data for the presence of long ncRNA candidates. A computational algorithm was developed to blast, map, and compare the RNA secondary structure of these candidates against various ncRNA databases, including NRED [79], RNAdb [80], fRNAdb [81], and NONCODE [82]. A total of 50, 35, and 24 potential long ncRNA candidates were identified in Spga, Spcy, and Sptd, respectively. These long ncRNA transcripts could be classified based on their association with various genomic features, such as promoter, intronic, intergenic, and antisense. Preliminary functional analysis in a P19 differentiation cell model suggested some long ncRNAs decreased remarkably following induction of differentiation by retinoic acid. The decrease was more obvious in the comparison of testes from vitamin A deficient (VAD) and control animals (Boucheron et al., unpublished). Several ncRNAs exhibited more than a 1000-fold decrease when compared to control testis. These results suggested that long ncRNA might play an active role in male germ cell differentiation and were dependent on retinoic acid-related regulatory pathways.

Germ Cell Transcriptome Informatics

The integration of genome and transcriptome data provides a powerful approach for understanding transcription regulatory networks in germ cell biology. However, the magnitude of this genomic data is a challenge for wet-lab biologists, as they require efficient informatics skills in data handling and processing. Fortunately, an emerging number of online user-friendly tools are available that allow for analysis of transcriptome data from a variety of angles, including static retrieval of data from databases and dynamic analysis at a systems biology level through integration of different biological information.

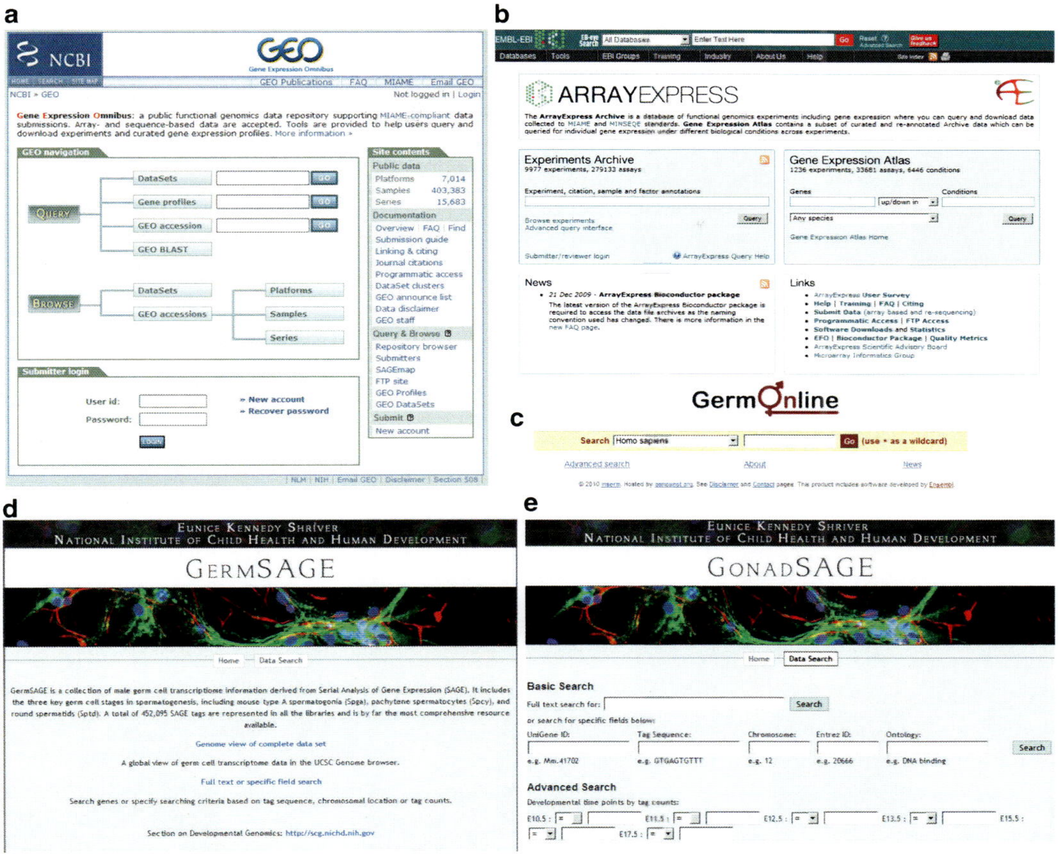

Fig. 8.2 Overview of transcriptome-related databases

Germ Cell Transcriptome Resources

The advent of various high-throughput technologies and completion of various genome projects in recent years have generated a huge amount of information. To allow effective data mining of these data in a standard format and facilitate the sharing of experimental setup and protocols, centralized public database resources were established. Currently, the most popular Web-based public repositories for transcriptome data are Gene Expression Ominbus (GEO, http://www.ncbi.nlm.nih.gov/geo/) at the National Center for Biotechnology Information (NCBI) and ArrayExpress Archive (http://www.ebi.ac.uk/microarray-as/ae/) at the European Bioinformatics Institute (EBI) (Fig. 8.2).

All data in GEO and ArrayExpress are either complied in Minimum Information About a Microarray Experiment [83] (MIAME) or Minimum Information about a high-throughput SeQuencing Experiment [84] (MINSEQE) format to enable the interpretation of experimental results in an unambiguous fashion and to potentially reproduce the experiment. A specific ID, in the form of GSE number (GEO) or E-GEOD-number (ArrayExpress), is assigned to an experiment. At the time of writing this review (Feb, 2010), the GEO and ArrayExpress database contain a total of 15,683 and 9977 experiments, respectively. In addition to GEO and ArrayExpress, online specialized resources on germ cell transcritptome are also available, which can be divided in terms of platform as described in the next section.

Microarray-based transcriptome resource: GermOnline. GermOnline [85, 86] is a microarray expression database that focuses on mitosis-, meiosis-, and gametogenesis-related studies. It adopted a familiar Ensembl-database layout for data presentation. Currently, it covers published transcriptome data from eleven species. The data are presented in Ensembl genome browser format. A microarray information management and annotation system (MIMAS) and a comprehensive system for online editing of database entries (MediaWiki) were applied to describe experimental data from various microarray-based experiments, such as RNA expression levels, transcript start sites and lengths, and exon composition. The database also provides an open environment for scientists to maintain database entries on genes and gene products in a complete and accurate manner by submitting up-to-date curations. The database is accessible at http://www.germonline.org/.

Sequence-based transcriptome resource: GermSAGE and GonadSAGE. GermSAGE [87] (http://germsage.nichd.nih.gov) is a comprehensive Web-based database generated by SAGE. Data deposited represent major stages in mouse male germ cell development, with sequence tag coverage of 150k in each SAGE library. The database covers 452,095 tags derived from type A spermatogonia, pachytene spermatocytes, and round spermatids. It provides an array of easy tools for browsing, comparing, and screening male germ cell transcriptome data. The data can be exported or further analyzed by aligning it with various annotations available in the built-in genome browser of the database. This flexible platform is useful for gaining a better understanding of the genetic networks that regulate spermatogonial cell renewal and differentiation and allows for novel gene discovery.

GonadSAGE [88] (http://germsage.nichd.nih.gov) is another SAGE database on male gonad development. A total of six male mouse embryonic gonad stages were included (E10.5, E11.5, E12.5, E13.5, E15.5, and E17.5). The sequence coverage for each SAGE library is above 150K. A total of 908,453 SAGE tags are represented in all the libraries and is by far the most comprehensive resource available. Altogether, it contains 24,975 known and over 275,583 unannotated transcripts, including an extensive presence of antisense transcripts and splicing variants.

Chromatin Remodeling and Spermatogenesis

Background

Eukaryotic gene regulation is more complicated than prokaryotic gene regulation. Transcriptional regulation is tightly coordinated, determined not only by the genetic information stored in the DNA sequence but also by interactions between a diversity of modifiers on chromatin. The complexity of eukaryotic transcriptional control is reflected by the structure of chromatin, which is composed of small repeating units, the nucleosomes. A nucleosome consists of double-stranded DNA wrapped on histone proteins. Four histone proteins, namely, H3, H4, H2A, and H2B, respectively, form the octamer core of the nucleosome. During DNA packaging, DNA helix is deposited on the H3(2)/H4(2) tetramer, followed by incorporation of two sets of H2A/H2B dimmers. Such packaging allows the huge chromosome to be organized in a compact structure. It is postulated that the linker histone protein H1 further promotes chromatin packaging to a higher-order structure by potentially shielding the negative charge of DNA linking nucleosomes.

Histones are not solely for DNA packaging. The eukaryotic system has evolved another mechanism of gene regulation by changing the chemical nature of histone tails extruding from the nucleosomal core. This is achieved by several posttranslational modifications on the amino residues of the N-terminus of histones. Currently, known covalent modifications on histones include acetylation and ubiquitination of lysine, methylation of arginine or lysine, and phosphorylation of serine or threonine [89]. Together with methylation on cytosine of DNA, these modifications form the epigenetic marks in mammalian genomes. By interacting with different chromatin modifiers, epigenetic marks provide an additional layer of gene regulation through establishing either an active or repressive state of chromatin. Epigenetic control of gene expression permits different cell types to express unique sets of genes despite having the same genome.

A large number of histone variants are found in male germ cells. Many of these variants are testis-specific. Expression of these testis-specific variants suggests the existence of a special nucleosomal architecture during spermatogenesis. There is nuclear reorganization in the chromatin of spermatids, where histone–protamine transition takes place. Postmeiotic haploid spermatids utilize protamine, an arginine-rich H1-like small protein, to replace histone. In addition, mammalian germ cells utilize another testis-specific nuclear protein, the transition proteins (TP1 and TP2), prior to protamine displacement. Transition proteins may not be essential for fertility since knockout of TP1 or TP2 did not result in infertility [90, 91]. It is generally believed that transition proteins replace histones, preparing the chromatin for protamine incorporation. Male germ cells use protamines to create highly compact nuclei, the size of which is about 5% of the somatic nucleus. Unlike transition proteins, protamines are essential for the development of mature sperm. Loss of protamine in mice resulted in male infertility [92]. The creation of a compact nucleus is not favorable for gene transcription. Indeed, HP1 is recruited to the heterochromatic chromocenter of spermatids after the histone–protamine transition, indicative of a silencing mechanism coupled with heterochromatin condensation (HP1 is a transcription repressor binding to methylated H3 in transcriptionally silenced heterochromatin).

Chromatin-Related Transcriptional Regulations in Spermatogenesis

Transcriptional regulation in male germ cells is different from somatic cells as evidenced by the use of histone variants, the expression of testis-specific homologs in the transcriptional machinery, and the use of alternative promoters in spermatogenesis.

For transcription to initiate, nucleosomes must be reorganized to allow access to promoters of transcription factors. Mechanisms include a transient unwrapping of the DNA from histone octamers or shifting nucleosomes along the length of DNA (nucleosome sliding). To accomplish this, chromatin remodeling complexes utilize ATP hydrolysis to disassemble the nucleosomal core, possibly by a mechanism of histone displacement. Chromatin remodeling complexes SWI/SNF, RSC and Pol II are responsible for histone displacement, with histone chapersones as the acceptor. Remarkably, different testis-specific histone variants, such as TH2A, TH2B, H2A.Z, TH3, H1t, H1t2, and HILS1, are incorporated during spermatogenesis and spermiogenesis. Incorporation of variants can change the nucleosomal structure (or the epigenetic modification on the variants' tails), thus influencing gene regulation. For instance, during spermatogenesis canonical H3 is displaced by variants H3.3A and H3.3B. H3.3 variants prominently replacing H3 at active genes [93] probably accounts for active transcription in spermatocytes. Although the role of testis-specific variants on chromatin structure and function of male germ cells is unclear, it is generally believed that the variants result in altered nucleosomal structure, creating a specialized nuclear organization that facilitates binding of chromatin remodeling factors and prepares the sperm genome for subsequent fertilization. Notably, a non-testis-specific H3 variant, CENP-A, localizes to the newly duplicated centromere of germ cells. CENP-A is not displaced during the histone–protamine transition. Its inheritance raises the speculation that it might function in fertilization.

Transcription activation involves three classes of proteins, namely, TATA-binding protein, DNA binding transactivator, and coactivator protein complex. Some transcription factors (TBP, TFIIB, RNA polymerase II) are constitutively expressed but at a much higher level in haploid germ cells [94]. Some are restricted to testis. Some, instead of expressing a regular form in somatic cells, are expressed in testes with a tissue-specific isoform.

A well-studied example is the expression of a testis-specific transcription activator CREM (cyclic AMP-responsive element modulator). CREM is highly expressed in postmeiotic germ cells [8]. It is a homolog of CREB (cAMP response element-binding protein), an activator of cAMP-responsive promoter elements (CREs). In somatic cells, phosphorylation of CREB triggers transcription activation. However, CREB is not expressed in testes. Instead, CREM is actively expressed in haploid germ cells for transcriptional regulation of many genes critical for late spermatogenesis. CREM-mutant mice showed defective spermiogenesis and increased apoptosis of germ cells [95]. Unlike CREB, CREM is phosphorylation-independent, but activated by a coactivator ACT. Notably, ACT is also restricted to male germ cells.

The transcriptional initiation complex in germ cells contains TLF (TBP-like factor), which activates genes with TATA-less promoters. Expression of TLF is developmentally regulated in spermatogenesis. Knockdown of TLF caused complete arrest of late spermiogenesis and fragmentation of the chromocenter in early spermatids [96]. Male germ cells express homologs of other transcriptional factors in the transcriptional machinery. For example, a homolog of TFIIA (Transcription factor II A) is predominantly expressed in testes, the biochemical function of which is indistinguishable from its counterpart [97, 98]. TAF7L (TAF7-like RNA polymerase II), a paralog of TAF7 of the TFIID complex, is X-linked and testis-specific [39]. Since the X chromosome is silenced in spermatocytes and spermatids, an autosomal homolog may exist particularly for spermatogenesis.

Conclusions

Investigation into regulation of gene expression in spermatogenesis is hampered by the lack of a comprehensive understanding of gene expression in germ cells. This is further confounded by limitations of the traditional single gene–single pathway approach. In the past decade, many transcriptome studies have been conducted to examine the biology of germ cell development. With the availability of comprehensive germ cell transcriptome databases, identification and characterization of gene functions in male germ cell development become possible. It is now clear that the germ cell transcriptome is more complex than previously envisioned. It involves not only protein-encoding genes but also non-protein-coding transcripts such as antisense transcripts, small and long noncoding RNAs, etc. Dynamic regulation and usage of the germ cell transcriptome are also obvious. A significant number of male germ cell-specific transcripts undergo alternative splicing or are derived from sex-linked progenitor genes through retroposition to generate testis-specific isoforms, presumably to cope with the specific needs in the spermatogenic process. The application of genome-wide analysis and systems biology approaches should permit elucidation of more novel modes of transcription regulation and identification of biological pathways critical for male germ cell development.

Acknowledgments This work was supported in part by the Intramural Research Program of the National Institutes of Health (NIH), Eunice Kennedy Shriver National Institute of Child Health and Human Development, and the Chinese University of Hong Kong.

References

1. Berta P, Hawkins JR, Sinclair AH, et al. Genetic evidence equating SRY and the testis-determining factor. Nature. 1990;348:448–50.
2. Wallis MC, Waters PD, Graves JA. Sex determination in mammals–before and after the evolution of SRY. Cell Mol Life Sci. 2008;65:3182–95.
3. Gaskell TL, Robinson LL, Groome NP, Anderson RA, Saunders PT. Differential expression of two estrogen receptor beta isoforms in the human fetal testis during the second trimester of pregnancy. J Clin Endocrinol Metab. 2003;88:424–32.
4. Robinson LL, Gaskell TL, Saunders PT, Anderson RA. Germ cell specific expression of c-kit in the human fetal gonad. Mol Hum Reprod. 2001;7:845–52.
5. Bellve AR, Cavicchia JC, Millette CF, O'Brien DA, Bhatnagar YM, Dym M. Spermatogenic cells of the prepuberal mouse. Isolation and morphological characterization. J Cell Biol. 1977;74:68–85.
6. Eddy EM. Male germ cell gene expression. Recent Prog Horm Res. 2002;57:103–28.
7. Kleene KC. A possible meiotic function of the peculiar patterns of gene expression in mammalian spermatogenic cells. Mech Dev. 2001;106:3–23.

8. McCarrey JR, O'Brien DA, Skinner MK. Construction and preliminary characterization of a series of mouse and rat testis cDNA libraries. J Androl. 1999;20: 635–9.
9. Lash AE, Tolstoshev CM, Wagner L, et al. SAGEmap: a public gene expression resource. Genome Res. 2000;10:1051–60.
10. Morozova O, Marra MA. Applications of next-generation sequencing technologies in functional genomics. Genomics. 2008;92:255–64.
11. Nagalakshmi U, Waern K, Snyder M. RNA-Seq: a method for comprehensive transcriptome analysis. Curr Protoc Mol Biol. 2010; Chapter 4:Unit 4 11 11–13.
12. Marguerat S, Bahler J. RNA-seq: from technology to biology. Cell Mol Life Sci. 2010;67:569–79.
13. Wang Z, Gerstein M, Snyder M. RNA-Seq: a revolutionary tool for transcriptomics. Nat Rev Genet. 2009;10:57–63.
14. Lee TL. Big data: open-source format needed to aid wiki collaboration. Nature. 2008;455:461.
15. Rockett JC, Christopher Luft J, Brian Garges J, et al. Development of a 950-gene DNA array for examining gene expression patterns in mouse testis. Genome Biol. 2001;2:RESEARCH0014.
16. Sha J, Zhou Z, Li J, et al. Identification of testis development and spermatogenesis-related genes in human and mouse testes using cDNA arrays. Mol Hum Reprod. 2002;8:511–7.
17. Pang AL, Taylor HC, Johnson W, et al. Identification of differentially expressed genes in mouse spermatogenesis. J Androl. 2003;24:899–911.
18. Schultz N, Hamra FK, Garbers DL. A multitude of genes expressed solely in meiotic or postmeiotic spermatogenic cells offers a myriad of contraceptive targets. Proc Natl Acad Sci USA. 2003;100:12201–6.
19. Yu Z, Guo R, Ge Y, et al. Gene expression profiles in different stages of mouse spermatogenic cells during spermatogenesis. Biol Reprod. 2003;69:37–47.
20. Guo R, Yu Z, Guan J, et al. Stage-specific and tissue-specific expression characteristics of differentially expressed genes during mouse spermatogenesis. Mol Reprod Dev. 2004;67:264–72.
21. Schlecht U, Demougin P, Koch R, et al. Expression profiling of mammalian male meiosis and gametogenesis identifies novel candidate genes for roles in the regulation of fertility. Mol Biol Cell. 2004;15: 1031–43.
22. Shima JE, McLean DJ, McCarrey JR, Griswold MD. The murine testicular transcriptome: characterizing gene expression in the testis during the progression of spermatogenesis. Biol Reprod. 2004;71:319–30.
23. Wang H, Zhou Z, Xu M, et al. A spermatogenesis-related gene expression profile in human spermatozoa and its potential clinical applications. J Mol Med. 2004;82:317–24.
24. Clemente EJ, Furlong RA, Loveland KL, Affara NA. Gene expression study in the juvenile mouse testis: identification of stage-specific molecular pathways during spermatogenesis. Mamm Genome. 2006;17:956–75.
25. Iguchi N, Tobias JW, Hecht NB. Expression profiling reveals meiotic male germ cell mRNAs that are translationally up- and down-regulated. Proc Natl Acad Sci USA. 2006;103:7712–7.
26. Lin YH, Lin YM, Teng YN, Hsieh TY, Lin YS, Kuo PL. Identification of ten novel genes involved in human spermatogenesis by microarray analysis of testicular tissue. Fertil Steril. 2006;86:1650–8.
27. Namekawa SH, Park PJ, Zhang LF, et al. Postmeiotic sex chromatin in the male germline of mice. Curr Biol. 2006;16:660–7.
28. Chalmel F, Rolland AD, Niederhauser-Wiederkehr C, et al. The conserved transcriptome in human and rodent male gametogenesis. Proc Natl Acad Sci USA. 2007;104:8346–51.
29. Feig C, Kirchhoff C, Ivell R, Naether O, Schulze W, Spiess AN. A new paradigm for profiling testicular gene expression during normal and disturbed human spermatogenesis. Mol Hum Reprod. 2007;13:33–43.
30. Johnston DS, Wright WW, Dicandeloro P, Wilson E, Kopf GS, Jelinsky SA. Stage-specific gene expression is a fundamental characteristic of rat spermatogenic cells and Sertoli cells. Proc Natl Acad Sci USA. 2008;105:8315–20.
31. Xiao P, Tang A, Yu Z, Gui Y, Cai Z. Gene expression profile of 2058 spermatogenesis-related genes in mice. Biol Pharm Bull. 2008;31:201–6.
32. Wu SM, Baxendale V, Chen Y, et al. Analysis of mouse germ-cell transcriptome at different stages of spermatogenesis by SAGE: biological significance. Genomics. 2004;84:971–81.
33. Anway MD, Li Y, Ravindranath N, Dym M, Griswold MD. Expression of testicular germ cell genes identified by differential display analysis. J Androl. 2003;24:173–84.
34. Pang AL, Johnson W, Ravindranath N, Dym M, Rennert OM, Chan WY. Expression profiling of purified male germ cells: stage-specific expression patterns related to meiosis and postmeiotic development. Physiol Genomics. 2006;24:75–85.
35. Morales CR, Hecht NB. Poly(A)+ ribonucleic acids are enriched in spermatocyte nuclei but not in chromatoid bodies in the rat testis. Biol Reprod. 1994;50: 309–19.
36. Sassone-Corsi P. Unique chromatin remodeling and transcriptional regulation in spermatogenesis. Science. 2002;296:2176–8.
37. Wrobel G, Primig M. Mammalian male germ cells are fertile ground for expression profiling of sexual reproduction. Reproduction. 2005;129:1–7.
38. Khil PP, Smirnova NA, Romanienko PJ, Camerini-Otero RD. The mouse X chromosome is enriched for sex-biased genes not subject to selection by meiotic sex chromosome inactivation. Nat Genet. 2004;36: 642–6.
39. Wang PJ, McCarrey JR, Yang F, Page DC. An abundance of X-linked genes expressed in spermatogonia. Nat Genet. 2001;27:422–6.
40. Turner JM. Meiotic sex chromosome inactivation. Development. 2007;134:1823–31.

41. Long M, Betran E, Thornton K, Wang W. The origin of new genes: glimpses from the young and old. Nat Rev Genet. 2003;4:865–75.
42. Mueller JL, Mahadevaiah SK, Park PJ, Warburton PE, Page DC, Turner JM. The mouse X chromosome is enriched for multicopy testis genes showing post-meiotic expression. Nat Genet. 2008;40:794–9.
43. Lee TL, Alba D, Baxendale V, Rennert OM, Chan WY. Application of transcriptional and biological network analyses in mouse germ-cell transcriptomes. Genomics. 2006;88:18–33.
44. Amaral PP, Mattick JS. Noncoding RNA in development. Mamm Genome. 2008;19:454–92.
45. Hayashi K, Chuva de Sousa Lopes SM, Kaneda M, et al. MicroRNA biogenesis is required for mouse primordial germ cell development and spermatogenesis. PLoS ONE. 2008;3:e1738.
46. Velculescu VE, Zhang L, Vogelstein B, Kinzler KW. Serial analysis of gene expression. Science. 1995;270:484–7.
47. Chan WY, Lee TL, Wu SM, et al. Transcriptome analyses of male germ cells with serial analysis of gene expression (SAGE). Mol Cell Endocrinol. 2006;250: 8–19.
48. Chan WY, Wu SM, Ruszczyk L, et al. The complexity of antisense transcription revealed by the study of developing male germ cells. Genomics. 2006;87: 681–92.
49. Lee TL, Yeh J, Van Waes C, Chen Z. Epigenetic modification of SOCS-1 differentially regulates STAT3 activation in response to interleukin-6 receptor and epidermal growth factor receptor signaling through JAK and/or MEK in head and neck squamous cell carcinomas. Mol Cancer Ther. 2006;5:8–19.
50. Lee TL, Li Y, Alba D, et al. Developmental staging of male murine embryonic gonad by SAGE analysis. J Genet Genomics. 2009;36:215–27.
51. Lee TL, Pang AL, Rennert OM, Chan WY. Genomic landscape of developing male germ cells. Birth Defects Res C Embryo Today. 2009;87:43–63.
52. Ayoubi TA, Van De Ven WJ. Regulation of gene expression by alternative promoters. FASEB J. 1996;10: 453–60.
53. Srivastava CH, Monts BS, Rothrock JK, Peredo MJ, Pescovitz OH. Presence of a spermatogenic-specific promoter in the rat growth hormone-releasing hormone gene. Endocrinology. 1995;136:1502–8.
54. Yang LV, Heng HH, Wan J, Southwood CM, Gow A, Li L. Alternative promoters and polyadenylation regulate tissue-specific expression of Hemogen isoforms during hematopoiesis and spermatogenesis. Dev Dyn. 2003;228:606–16.
55. Sun Z, Sassone-Corsi P, Means AR. Calspermin gene transcription is regulated by two cyclic AMP response elements contained in an alternative promoter in the calmodulin kinase IV gene. Mol Cell Biol. 1995;15:561–71.
56. Taft RJ, Pang KC, Mercer TR, Dinger M, Mattick JS. Non-coding RNAs: regulators of disease. J Pathol. 2010;220:126–39.
57. Mercer TR, Dinger ME, Mattick JS. Long non-coding RNAs: insights into functions. Nat Rev Genet. 2009;10:155–9.
58. Dinger ME, Amaral PP, Mercer TR, Mattick JS. Pervasive transcription of the eukaryotic genome: functional indices and conceptual implications. Brief Funct Genomic Proteomic. 2009;8:407–23.
59. Meissner A, Mikkelsen TS, Gu H, et al. Genome-scale DNA methylation maps of pluripotent and differentiated cells. Nature. 2008;454:766–70.
60. Carninci P, Non-coding RNA. transcription: turning on neighbours. Nat Cell Biol. 2008;10:1023–4.
61. Mehler MF, Mattick JS. Non-coding RNAs in the nervous system. J Physiol. 2006;575:333–41.
62. Johnson JM, Edwards S, Shoemaker D, Schadt EE. Dark matter in the genome: evidence of widespread transcription detected by microarray tiling experiments. Trends Genet. 2005;21:93–102.
63. Kapranov P, Drenkow J, Cheng J, et al. Examples of the complex architecture of the human transcriptome revealed by RACE and high-density tiling arrays. Genome Res. 2005;15:987–97.
64. Vasudevan S, Tong Y, Steitz JA. Switching from repression to activation: microRNAs can up-regulate translation. Science. 2007;318:1931–4.
65. Hamilton AJ, Baulcombe DC. A species of small antisense RNA in posttranscriptional gene silencing in plants. Science. 1999;286:950–2.
66. Seto AG, Kingston RE, Lau NC. The coming of age for Piwi proteins. Mol Cell. 2007;26:603–9.
67. Wang J, Saxe JP, Tanaka T, Chuma S, Lin H. Mili interacts with tudor domain-containing protein 1 in regulating spermatogenesis. Curr Biol. 2009;19: 640–4.
68. He Z, Kokkinaki M, Pant D, Gallicano GI, Dym M. Small RNA molecules in the regulation of spermatogenesis. Reproduction. 2009;137:901–11.
69. Chi YH, Cheng LI, Myers T, et al. Requirement for Sun1 in the expression of meiotic reproductive genes and piRNA. Development. 2009;136:965–73.
70. Marcon E, Babak T, Chua G, Hughes T, Moens PB. miRNA and piRNA localization in the male mammalian meiotic nucleus. Chromosome Res. 2008;16: 243–60.
71. Iguchi N, Xu M, Hori T, Hecht NB. Noncoding RNAs of the mammalian testis: the meiotic transcripts Nct1 and Nct2 encode piRNAs. Ann N Y Acad Sci. 2007;1120:84–94.
72. Ro S, Park C, Song R, et al. Cloning and expression profiling of testis-expressed piRNA-like RNAs. RNA. 2007;13:1693–702.
73. Carmell MA, Girard A, van de Kant HJ, et al. MIWI2 is essential for spermatogenesis and repression of transposons in the mouse male germline. Dev Cell. 2007;12:503–14.
74. Grivna ST, Pyhtila B, Lin H. MIWI associates with translational machinery and PIWI-interacting RNAs (piRNAs) in regulating spermatogenesis. Proc Natl Acad Sci USA. 2006;103:13415–20.

75. Mercer TR, Dinger ME, Sunkin SM, Mehler MF, Mattick JS. Specific expression of long noncoding RNAs in the mouse brain. Proc Natl Acad Sci USA. 2008;105:716–21.
76. Dinger ME, Amaral PP, Mercer TR, et al. Long noncoding RNAs in mouse embryonic stem cell pluripotency and differentiation. Genome Res. 2008;18:1433–45.
77. Mehler MF, Mattick JS. Noncoding RNAs and RNA editing in brain development, functional diversification, and neurological disease. Physiol Rev. 2007;87:799–823.
78. Mattick JS. A new paradigm for developmental biology. J Exp Biol. 2007;210:1526–47.
79. Dinger ME, Pang KC, Mercer TR, Crowe ML, Grimmond SM, Mattick JS. NRED: a database of long noncoding RNA expression. Nucleic Acids Res. 2009;37:D122–6.
80. Pang KC, Stephen S, Dinger ME, Engstrom PG, Lenhard B, Mattick JS. RNAdb 2.0–an expanded database of mammalian non-coding RNAs. Nucleic Acids Res. 2007;35:D178–82.
81. Mituyama T, Yamada K, Hattori E, et al. The Functional RNA Database 3.0: databases to support mining and annotation of functional RNAs. Nucleic Acids Res. 2009;37:D89–92.
82. He S, Liu C, Skogerbo G, et al. NONCODE v2.0: decoding the non-coding. Nucleic Acids Res. 2008;36: D170–2.
83. Brazma A, Hingamp P, Quackenbush J, et al. Minimum information about a microarray experiment (MIAME)-toward standards for microarray data. Nat Genet. 2001;29:365–71.
84. Brazma A. Minimum information about a microarray experiment (MIAME)–successes, failures, challenges. ScientificWorldJournal. 2009;9:420–3.
85. Gattiker A, Niederhauser-Wiederkehr C, Moore J, Hermida L, Primig M. The GermOnline cross-species systems browser provides comprehensive information on genes and gene products relevant for sexual reproduction. Nucleic Acids Res. 2007;35:D457–62.
86. Wiederkehr C, Basavaraj R, Sarrauste de Menthiere C, et al. GermOnline, a cross-species community knowledgebase on germ cell differentiation. Nucleic Acids Res. 2004;32:D560–7.
87. Lee TL, Cheung HH, Claus J, et al. GermSAGE: a comprehensive SAGE database for transcript discovery on male germ cell development. Nucleic Acids Res. 2009;37:D891–7.
88. Lee TL, Li Y, Cheung HH, et al. GonadSAGE: a comprehensive SAGE database for transcript discovery on male embryonic gonad development. Bioinformatics. 2010;26:585–6.
89. Li B, Carey M, Workman JL. The role of chromatin during transcription. Cell. 2007;128:707–19.
90. Yu YE, Zhang Y, Unni E, et al. Abnormal spermatogenesis and reduced fertility in transition nuclear protein 1-deficient mice. Proc Natl Acad Sci USA. 2000;97:4683–8.
91. Zhao M, Shirley CR, Yu YE, et al. Targeted disruption of the transition protein 2 gene affects sperm chromatin structure and reduces fertility in mice. Mol Cell Biol. 2001;21:7243–55.
92. Cho C, Willis WD, Goulding EH, et al. Haploinsufficiency of protamine-1 or -2 causes infertility in mice. Nat Genet. 2001;28:82–6.
93. Mito Y, Henikoff JG, Henikoff S. Genome-scale profiling of histone H3.3 replacement patterns. Nat Genet. 2005;37:1090–7.
94. Schmidt EE, Schibler U. High accumulation of components of the RNA polymerase II transcription machinery in rodent spermatids. Development. 1995;121:2373–83.
95. Nantel F, Monaco L, Foulkes NS, et al. Spermiogenesis deficiency and germ-cell apoptosis in CREM-mutant mice. Nature. 1996;380:159–62.
96. Martianov I, Brancorsini S, Gansmuller A, Parvinen M, Davidson I, Sassone-Corsi P. Distinct functions of TBP and TLF/TRF2 during spermatogenesis: requirement of TLF for heterochromatic chromocenter formation in haploid round spermatids. Development. 2002;129:945–55.
97. Upadhyaya AB, Lee SH, DeJong J. Identification of a general transcription factor TFIIAalpha/beta homolog selectively expressed in testis. J Biol Chem. 1999;274:18040–8.
98. Ozer J, Moore PA, Lieberman PM. A testis-specific transcription factor IIA (TFIIAtau) stimulates TATA-binding protein-DNA binding and transcription activation. J Biol Chem. 2000;275:122–8.
99. Anway MD, Johnston DS, Crawford D, Griswold MD. Identification of a novel retrovirus expressed in rat Sertoli cells and granulosa cells. Biol Reprod. 2001;65(4):1289–96.

Part II

Laboratory Evaluation of Sperm Chromatin

Sperm Chromatin Structure Assay (SCSA®): 30 Years of Experience with the SCSA®

9

Donald P. Evenson

Abstract
The SCSA® is one of the most widely utilized tests of sperm DNA damage. There are now a number of commercial kits available for testing of sperm DNA fragmentation in which great variations of clinical thresholds exist both within the same test and between tests. This presents a real problem for the clinics in providing a correct diagnosis and prognosis to patients. The greatest utility of the SCSA® has been to suggest when the %DFI is >25% to do changes in lifestyle and/or medical intervention to reduce this value. In addition, such couples should avoid spending time in unsuccessful IUI treatment but instead move on to IVF and preferably ICSI for the greatest success.

Keywords
Sperm chromatin structure assay • Sperm DNA damage

The SCSA® is one of the most widely utilized tests of sperm DNA damage: as recently stated, "the SCSA® remains the most robust test, and the one for which most clinical data are available and, indeed, many of the current indications for sperm DNA fragmentation testing were derived from SCSA® testing – it is the only test of sperm DNA/chromatin for which validated clinical interpretation criteria exist, and these are based on many thousands of tests and hundreds of clinical treatment cycles" [1]. There are now a number of commercial kits available for testing sperm DNA fragmentation, in which great variations of clinical thresholds exist both within the same test and between tests. This presents a real problem for the clinics in providing a correct diagnosis and prognosis to patients.

The SCSA® sperm DNA fragmentation test was invented 30 years ago and has been tested over these years by measuring over 100,000 animal and human sperm samples derived from many etiologies. The SCSA® test was extensively tested for accuracy and precision over decades prior to offering it commercially for human clinical diagnosis and

D.P. Evenson (✉)
SCSA Diagnostics, PO Box 107, 219 Kasan Ave, Volga, SD 57071, USA

Emeritus, South Dakota State University, Brookings, SD, USA

Department of Obstetrics and Gynecology, Sanford Medical School, University of South Dakota, Sioux Falls, SD, USA
e-mail: don@scsatest.com

A. Zini and A. Agarwal (eds.), *Sperm Chromatin: Biological and Clinical Applications in Male Infertility and Assisted Reproduction*, DOI 10.1007/978-1-4419-6857-9_9,

prognosis. In 2005, the SCSA® test was commercialized with a national reference lab, SCSA Diagnostics (http://www.SCSATest.com) and two SCSA licensed European labs: SPZ lab (http://www.spzlab.com) Copenhagen, and Biomnis (http://www.biomnis.com) Lyon, France.

Frozen clinical samples are sent to these centers via overnight courier for processing, and the electronic data are returned to the clinic within a few days following semen collection.

The SCSA® is technically much less demanding than any other DNA fragmentation test and can be conducted within minutes rather than hours. The SCSA® has two straightforward biochemical steps: (1) treat the raw semen dilution with a pH 1.20 buffer for 30 s and then stain with acridine orange (AO). Both the 30-s low-pH-induced opening of the DNA strands at the site of DNA breaks and the AO labeling are highly specific and repeatable in exacting patterns. No other DNA fragmentation test, whether classified artificially as direct or indirect, has this level of biochemical specificity for biochemical probe interaction with damaged chromatin/DNA.

The greatest utility of the SCSA® has been to suggest when the %DFI is >25% to do changes in life style and/or medical intervention to reduce this value. In addition, such couples should avoid spending time in unsuccessful IUI treatment but instead move on to IVF and preferably ICSI for the greatest success.

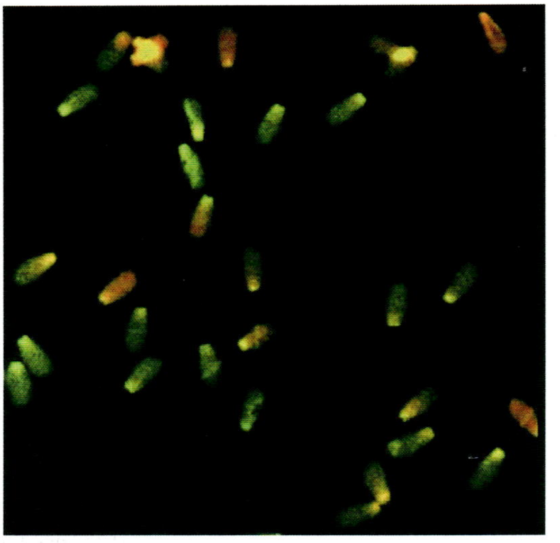

Fig. 9.1 Fluorescence photomicrograph of sperm from a subfertile bull heated at 100°C for 5 min and stained with acridine orange (AO)

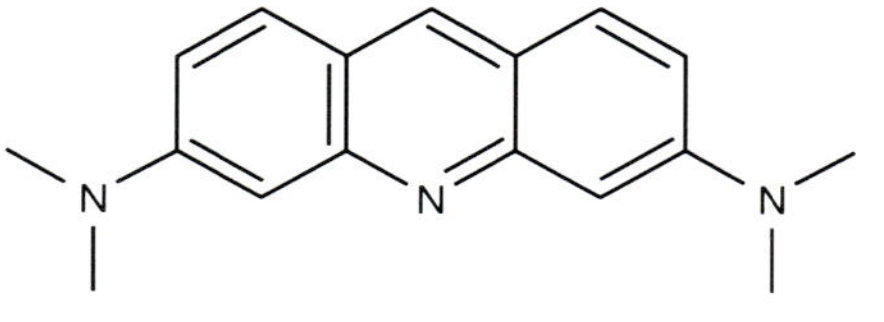

Fig. 9.2 The acridine orange (AO) molecule. Molecular weight (MW) is 265 g/mol

Pioneering the First Sperm DNA Fragmentation Test: SCSA®

Thirty years ago, this author conducted early studies on flow cytometry and acridine orange (AO) biochemistry in collaboration with laboratories that pioneered in the new field of flow cytometry [2, 3]. Following those efforts, we published [4] our pioneering study showing green (intact DNA) and red (damaged DNA) colored sperm in light microscopy, as in Fig. 9.1.

Of much greater significance, we obtained flow cytometry (FCM) data on the susceptibility of sperm obtained from subfertile/infertile men and bulls to heat-induced nuclear DNA denaturation [4]. This DNA denaturation was considered to occur at the sites of sperm double-stranded (ds) and/or single-stranded (ss) DNA breaks. The biochemical probe for detection of DNA strand breaks was AO (Fig. 9.2). AO is a flat planar molecular that intercalates into dsDNA and fluoresces green (F 515–530 nm) when exposed to 488 nm light, while it stacks on single-stranded nucleic acids (DNA and RNA) that then collapses into a crystal that produces a metachromatic shift to red fluorescence emission (F> 630 nm).

Development of SCSA®

After numerous trials with buffers of varying pH, ionic strength, etc., we concluded that a 30-s pretreatment of sperm with pH 1.20 buffer opens up the DNA double strand at the sites of DNA strand breaks followed by staining with AO [5] and that measurement by flow cytometry was the most efficient and effective method to detect DNA strand breaks without known loss of sperm in the heated test tubes.

The last sentence of the Science article (4) stated: "We expect this assay to have application in many research areas, including animal husbandry, human infertility, and environmental and public health." Thirty years later, it is very satisfying to confirm that this prediction has come true and beyond our initial expectations.

In short, with the SCSA®, raw semen aliquots can be flash-frozen, placed in a box with dry ice, and shipped through overnight courier to a SCSA® licensed lab. The samples arrive to the lab by early morning and can be prepared and analyzed in ~10 min each and the results sent back to the doctor via Fax or Web. This method is much more efficient in both time and cost than a clinic sending a few samples to a core FCM facility with no SCSA® experience and poor quality control. All samples analyzed by a SCSA® licensed lab can be precisely referenced to the thousands of other samples sent to SCSA Diagnostics Inc., over the past 6 years.

Before we could claim that the SCSA® was a unique and clinically useful test, we had to show that this new SCSA® test achieved the following:

1. Measured sperm cellular features related to infertility that were not duplicated by existing semen analysis measures.
2. Provided measurements that were practically feasible.
3. Repeat measurements of the same sample had a very low CV (1–2%) between measures.
4. Results provided diagnosis/prognosis for clinic patients.
5. Samples from infertility clinics could easily and quickly be prepared and measured on site, or packaged and sent to a diagnostic lab.

Power of the SCSA® Test: Six Important Parameters

1. An aliquot of fresh, liquefied semen can be measured within a few minutes after collection. Thus, the newly collected sperm sample can be immediately analyzed by a SCSA® trained technician for DNA integrity that may direct a clinical decision regarding treatment.
2. Flow cytometry provides for rapid measures of thousands of single cells resulting in very high statistical robustness, far beyond any light microscope evaluation.
3. In contrast to human eye observations, flow cytometry provides high precision, machine set specifications that gives objective and precise measures (sensitivity = <5/1,024 increments of fluorescence intensity).
4. SCSA® data are dual parameter measures of both green (native DNA) and red (broken DNA) fluorescence – thus providing scattergram patterns that give additional insight into sperm chromatin structure.
5. Uniquely, biochemical interaction between AO and DNA/chromatin is precisely repeatable with any single sample. This is proven by comparing cytograms (X vs. Y scatter plots) of repeat measures of a single semen sample. The dot pattern from replicate measures is virtually identical. Thus, a specific cluster of <1% of the cell population identified in the first measurement will be located on the second measurement at virtually the same X and Y coordinates – this strongly argues against implications from some authors who state that "the acid treatment <u>tends</u> to denature the DNA," as if the DNA denaturation was poorly specific. Both the 30-s low-pH-induced opening of the DNA strands at the site of DNA breaks and the AO labeling are highly specific and repeatable in exacting patterns. No other DNA fragmentation test, whether classified artificially as direct or indirect, has this level of biochemical specificity for biochemical probe interaction with damaged chromatin/DNA.
6. Five populations of sperm are identified as having various classes of DNA integrity and chromatin structure, including the following:
   a. No measurable DNA fragmentation.
   b. Moderate level of DNA fragmentation.
   c. High level of DNA fragmentation.
   d. Total level (% moderate + % high) of DNA fragmentation (the %DFI threshold for reduced natural fecundity is currently set at `~25% DFI).

e. High DNA Staining (HDS) sperm due to abnormally retained histones. This population, identifiable only by the SCSA®, has a threshold of ~15% HDS for increased probability for miscarriage or lack of fertilization; however, HDS data have been equivocal in various studies.

SCSA® Method Overview

1. After arrival of the samples on dry ice, they can be measured that day, or transferred to an ultracold freezer (<−70°C) or preferably a LN2 tank.
2. In a SCSA® licensed flow cytometry laboratory, the samples are individually removed, thawed at 37°C for 30 s, and an aliquot transferred to TNE buffer to a final concentration ~ $1–2 \times 10^6$/ml.
3. 200 μl of this sperm suspension is mixed with 400 μl solution of 0.1% Triton X-100 at pH 1.2.
4. After 30 s 1.20 ml of AO staining solution is added and the sample is placed in the flow cytometer sample chamber and flow is initiated to bring the sheath flow and sample flow to equilibrium.
5. 5,000 sperm are analyzed at an event rate of 100–200 cells/sec. If the event rate is above 250, a new sample must be prepared to ensure precise equilibrium between the AO dye and the sperm.
6. The data are analyzed for the % of cells with (%DFI) measurable increased red fluorescence (sperm with fragmented DNA).

SCSA® Data

SCSA® Raw and Computer Reoriented Data

When the sperm are passing through the flow cytometer, small variations in the green and red emission light will occur due to the flattened shape of the sperm head [5]. This problem is overcome by use of the SCSAsoft® software where the DFI (red/red+green) signal is analyzed against the total fluorescence from the sperm (red plus green).

Typical examples of good sperm DNA integrity and poor DNA integrity are shown in Figs. 9.3 and 9.4, respectively. Two analyses are performed from each patient sample to ensure that no instrument or biochemical problems exist. Note the extremely high repeatability between the two replicates for each patient. This level of precision is not accomplished by any other measure in the andrology lab.

It is very important for the SCSA® that the flow cytometer is set up according to a reference sperm sample each day and that repeated analyses of the reference sample is performed after measuring every 6–10 patient samples. When the software analysis is performed with SCSAsoft®, the gates are set according to the reference samples. Subsequently, all analyses of patient samples are done in a batch without changing the gates. This procedure ensures that no bias is introduced during the software analysis. Figure 9.5 shows an example of an analysis where 78% of the sperm displayed moderate DNA fragmentation. In this case, it was virtually impossible to correctly gate between the populations without fragmentation and the ones with moderate levels of DNA fragmentation in the dot plot from the FCM. However, the SCSAsoft® gating between these two populations was unproblematic [34].

Characterization of Sperm Populations Identified in a SCSA® Analysis

We conducted an experiment [6] with sorted sperm to characterize more precisely the different sperm populations indentified in the SCSA® analysis. A SCSA® analysis was performed on a FACSort flow cytometer (BD Biosciences, San Jose, CA, USA), and each sperm population was sorted into a test tube and aliquots were cytocentrifuged onto glass microscope slides. One aliquot was Feulgen-stained for computer image analysis, while the second aliquot was prepared for comet assay. The image analysis photos are shown for each population in Fig. 9.6.

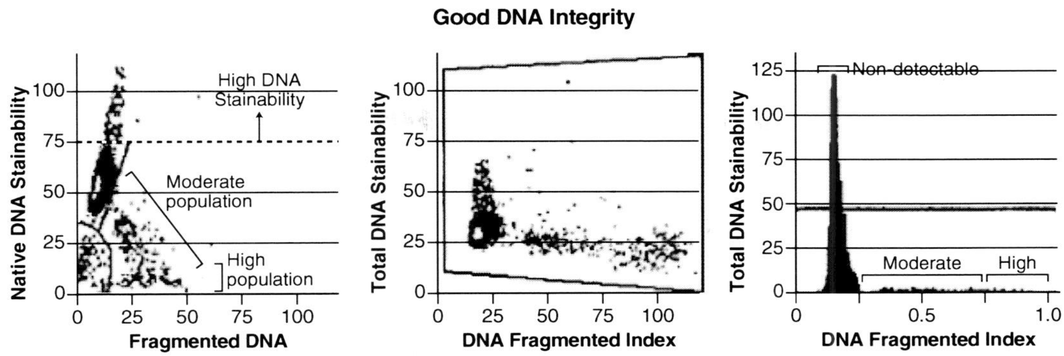

Patient	Date	Measurement	DFI	SD DFI	% DFI	% HDS
7272-87	## ##	1	213.5	111.6	5.8	5
		2	221.2	118.1	8.3	5.4
		mean	217.4	114.8	7.5	5.2
		sd	5.4	4.6	1.1	0.2

Fig. 9.3 *Left panel*: Green vs. red scattergram (cytogram) showing 5,000 dots, each representing a single event with specific green (native DNA) and red (fragmented DNA) coordinates on a scale from 0 to 1024. The *horizontal dashed line* lays at the top of the highest green fluorescence values for normal sperm. Sperm above this line have "High DNA Stainabilty" (HDS) and are characterized by immature sperm lacking full protamination. *Center panel*: SCSAsoft® software (SCSA Diagnostics., Brookings, SD) converts the data in the *left panel* to total DNA stainability vs. the DNA Fragmentation Index (DFI). This reorients the data into a vertical/horizontal pattern of dots. *Right panel*: The data in the middle panel is converted to a frequency histogram of DFI which is divided into (**a**) nondetectable DNA fragmentation, (**b**) moderate level of DNA fragmentation, and (**c**) high level of DNA fragmentation. Total %DFI is Moderate + High level of DNA fragmentation, a parameter that is most frequently used in expressing the extent of sperm DNA fragmentation in a sample. This method, derived from SCSAsoft®, provides a much more accurate calculation of total %DFI due to the difficulties for a significant proportion of the samples to gate between the populations with no or moderate fragmentation in the *left hand panel*

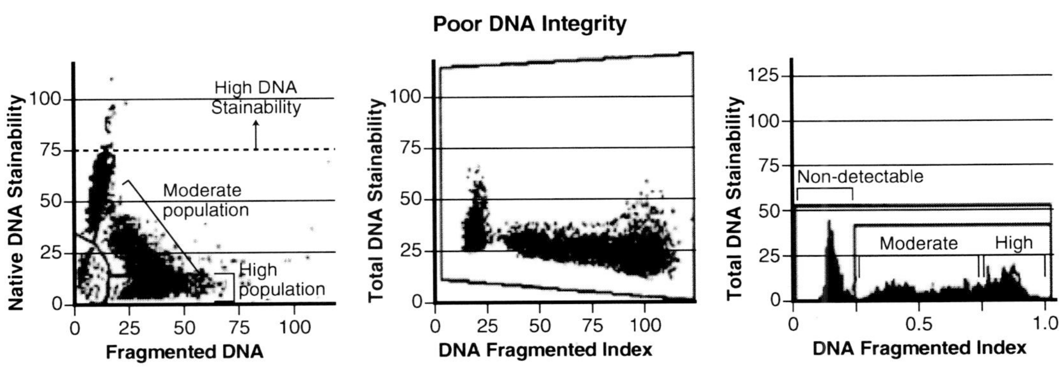

Patient	Date	Measurement	DFI	SD DFI	% DFI	% HDS
7272-113	## ##	1	563.7	307	64.9	6.4
		2	561.4	304.8	64.9	7.2
		mean	562.6	605.9	64.9	6.8
		sd	1.2	1.1	0	0.4

Fig. 9.4 SCSA® data from a sample with very poor DNA integrity, in this case, 64.9% of sperm demonstrate sperm DNA fragmentation. In this case, the two replicates provided exactly the same %DFI, resulting in a SD of 0%

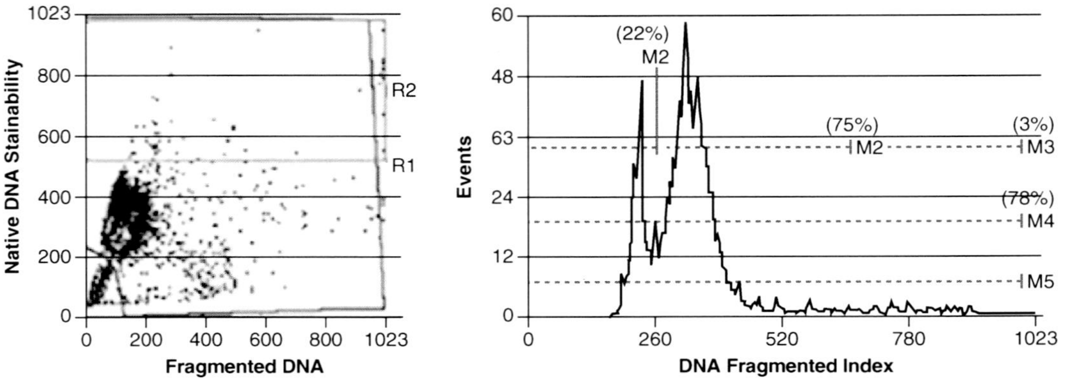

Fig. 9.5 SCSA® data from a sample with a high frequency of sperm with moderate DNA fragmentation. In this case, it is nearly impossible to gate between sperm with no or moderate DNA fragmentation in the FCM dot plot (*left panel*). With the SCSAsoft®, gating between the two populations is unproblematic (*right panel*, [34])

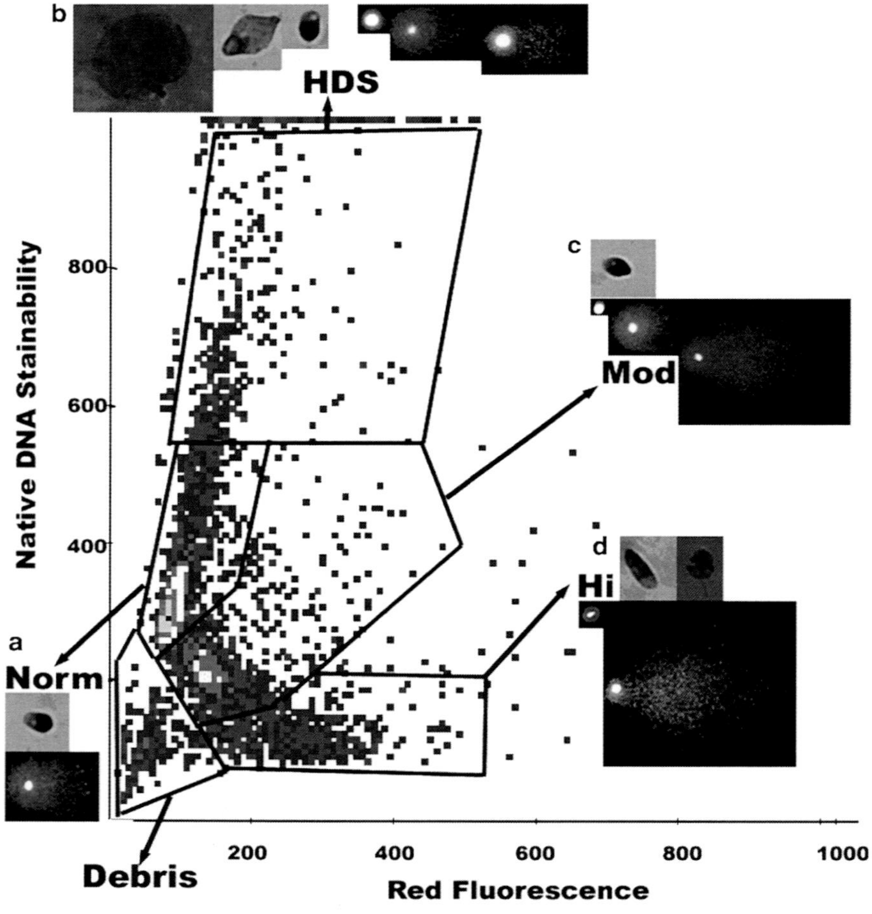

Fig. 9.6 The figure shows computer gating around each population of SCSA® measured sperm including: (**a**) normal population (Norm), (**b**) HDS population (HDS) (**c**) sperm with moderate DNA fragmentation (Mod) (**d**) sperm with high DNA fragmentation (Hi). Examples of sperm morphology and Feulgen-stained sperm and comets are shown for each population

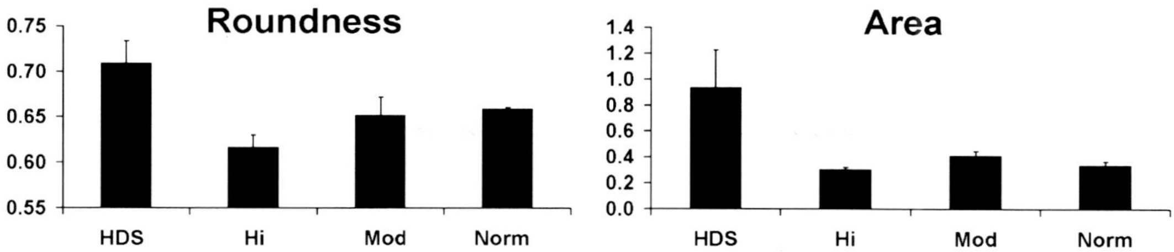

Fig. 9.7 Bar graphs for 500 sperm per category stained with Feulgen and analyzed for various nuclear parameters with a Nikon E800 fluorescence microscope fitted with a digital camera and image analysis software. Roundness and area are shown for sperm populations without DNA fragmentation (Norm), with moderate (Mod) or high DNA fragmentation (Hi), as well as sperm with high DNA stainability (HDS)

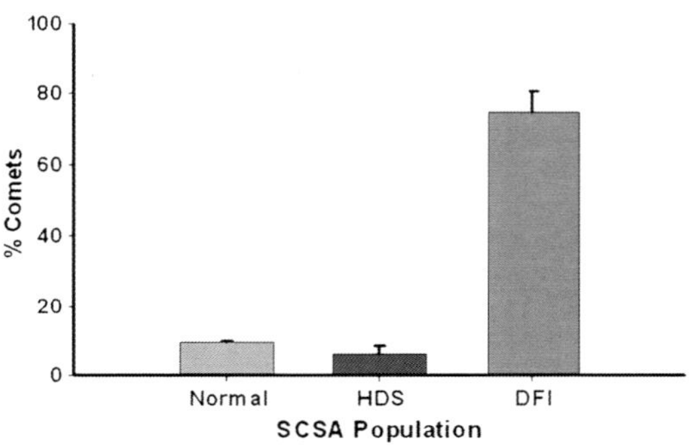

Fig. 9.8 Percent positive comets in the SCSA® populations of sperm without DNA fragmentation (Normal), with moderate or high DNA fragmentation (DFI) or sperm with HDS

The Feulgen-stained slides were examined with a Nikon E800 fluorescence microscope fitted with a digital camera and computer image analysis system. Various sperm nuclear parameters were analyzed for 500 sperm, and the data for nuclear roundness and area are shown in Fig. 9.7. It was observed that the populations of normal and moderate DNA fragmentation essentially had the same morphology. However, the population of sperm with high DNA fragmentation had a smaller area. The HDS fraction, known to be immature sperm, had significantly more area and roundness as would be expected from immature sperm.

The comet analysis showed that approximately 75% of the sperm with moderate and high DNA fragmentation also had positive comets (Fig. 9.8). The population without sperm DNA fragmentation (Norm) and the population of sperm with high DNA stainability (HDS) only showed a minor degree of background noise level of comets. The "noise" in the mechanical FACSort FCM system probably caused a less than unity between % comets and %DFI. Several conclusions can be drawn from this: (1) Sperm with fragmented DNA in a SCSA® analysis demonstrate true DNA strand breaks, (2) HDS sperm, lacking full protamination and having increased ratio of histones to protamines, do not have any significant amount of DNA strand breaks.

Other Probes that Shed Light on SCSA® Data

Disulfide Bonding of Chromatin

Mammalian sperm are unique cells that have highly condensed chromatin and other unique structures. Transmission electron microscope images of human sperm show a great variation of chromatin condensation [7]. Flow cytometry of sperm treated with dithiothreitol (DTT) and/or proteases shows great variation of decondensation [8]. It may be questioned whether such variations of chromatin packaging allow biochemical probes of chromatin structure to interact with the chromatin as equally as a nucleus with highly compacted chromatin. The highly condensed chromatin and/or intertwined mesh of fibers may inhibit access for large DNA probes (enzymes and tagged antibodies); furthermore, this same meshwork may inhibit the complete washing out of nonreacted labeled probes from this meshwork. A great advantage of the SCSA® test is that it requires no washing, fixing, and centrifugation or digestion steps. Following the highly repeatable opening of the DNA strands at the sites of damage, the very small AO molecules are kept in equilibrium (~2 AO molecules DNA base pair) during the measuring time, making the entire procedure highly exacting and independent of agents such as enzymes and tagged antibodies.

A unique feature of mammalian sperm nuclei is the high level of disulfide bonding (S=S) between the cysteine residues of nuclear protamines, which provides high structural strength and protection to paternal genome DNA. A study was done [9] on stallion sperm to determine the relationship between the extent of free nuclear –SH groups and SCSA® data. Semen samples from 30 stallions were sonicated to liberate sperm nuclei, purified through a 60% sucrose gradient, stained with an –SH-specific fluorochrome (CPM (7-diethylamino-3-(4'-maleimidylphenyl)-4-methyl-coumarin)) and the blue fluorescence of 5,000 sperm per sample was measured by flow cytometry. If S=S bonds stabilized chromatin, and thus, inhibited the low-pH-induced DNA strand separation, low blue intensity would correlate with low DFI values. However, this study showed no significant correlation (Fig. 9.9, $r=-0.199$, $P=0.31$). Another study [10] claimed a correlation between these two parameters; however, this study, was done on whole sperm which included the measurement of a high level of –SH groups on the sperm tails.

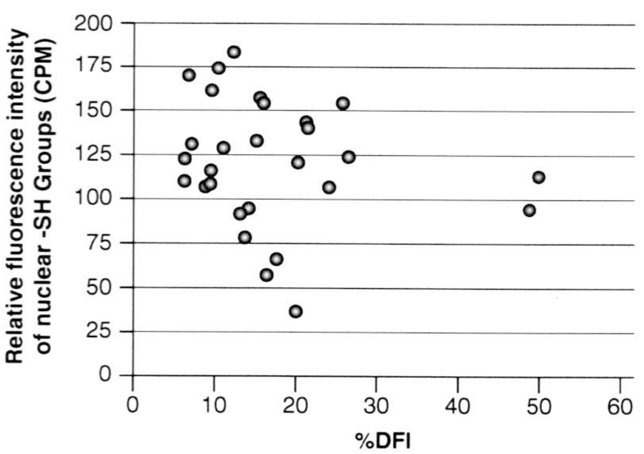

Fig. 9.9 A plot of the coordinates for each stallion semen sample related to %DFI vs. relative fluorescence intensity of CPM labeled nuclear –SH groups [8]

Chromomycin A3 (CMA3) Staining of HDS Sperm

The HDS sperm have an increased histone to protamine ratio [11]. Chromomycin A_3 (CMA) staining is thought to reflect underprotamination of sperm DNA, a phenomenon that could result in incomplete condensation. To further examine this relationship, semen samples from 182 men (aged 18–40) were analyzed by SCSA® and CMA staining [12]. The %DFI and %HDS were not significantly correlated ($r=0.038$, $P=0.61$), showing that they measure independent features of sperm nuclei. %HDS, on the other hand, was significantly correlated with %CMA+ sperm ($r=0.610$, $P<0.0001$ [9]. This correlation suggests that these two assays measure a common feature of sperm nuclei. As has been reported previously, %DFI correlated with neither sperm morphology nor sperm concentration. By contrast, both %HDS and %CMA+ were significantly correlated with both of these routine measures. Together, these observations provide insights into the interpretation of sperm nuclear integrity assays. As has been shown in infertility patients, DNA fragmentation may be present in the absence of other semen abnormalities; therefore, %DFI can be considered a relatively independent predictor of infertility or abnormal pregnancy outcomes. On the other hand, %HDS and %CMA appear to be less independent of routine semen measures such as sperm concentration and morphology.

Comparison Between SCSA® and TUNEL

In some TUNEL assays [13–15], sperm are first washed with phosphate-buffered saline (PBS), resuspended in paraformaldehyde, and fixed for approximately one hour. The sperm are then washed again to remove the paraformaldehyde, resuspended in ETOH, and stored. The sperm are then washed twice to remove the ETOH and the final sperm pellet is resuspended in a staining solution containing TdT enzyme/reaction buffer and FITC-tagged dUTP for an hour. The resulting batch of specimens are then washed again in rinse buffer, resuspended in a propidium iodide/RNAse solution, incubated for 30 min, and then measured by flow cytometry. A potential concern for worldwide utility of the TUNEL assay is that TdT enzyme kits may vary in activity not only between batches from commercial firms but also between products. Data in Fig. 9.10 show a comparison of TUNEL and SCSA® data. Although there was a statistical significant relationship between the two tests, correlations were relatively moderate and varied from 0.56 to 0.78, with the lowest correlation observed for human semen and the highest for bull semen. In conclusion, the measures obtained by TUNEL and SCSA® should not be regarded as identical.

Validation of Flow Cytometry and AO Biochemistry on Sperm DNA Integrity

Requirements for Validating a New DNA Fragmentation Test

The requirements are as follows:

1. Precision of interaction between the detector probe and the damaged DNA.
2. Repeatability of different sources and lots of kits used for DNA damage detection (SCSA® is the only assay not susceptible to commercial kit variation).
3. High repeatability (low CV) between repeat measures, both within a diagnostic lab, and importantly, between labs.

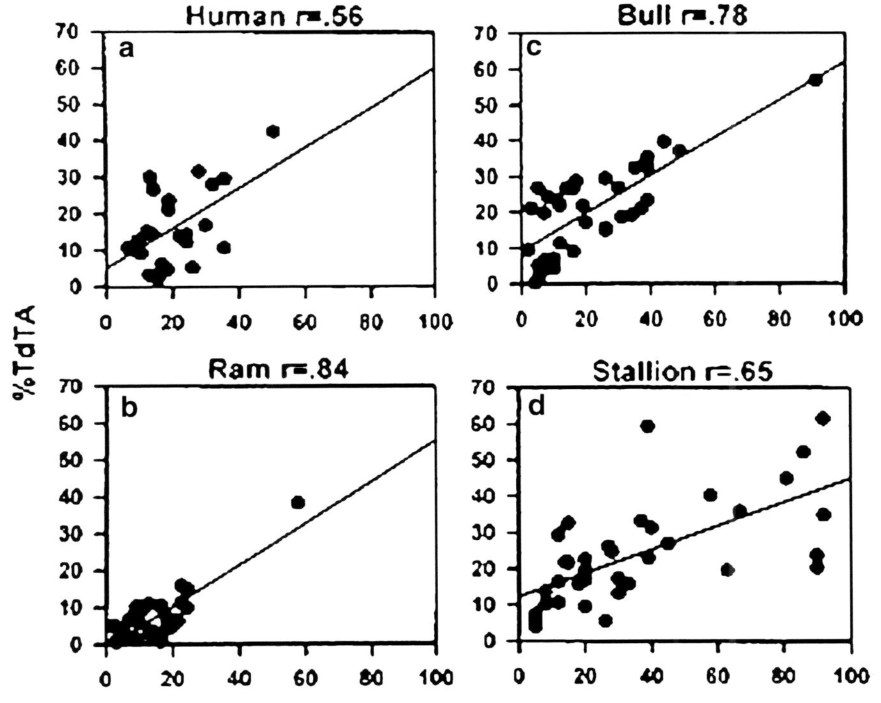

Fig. 9.10 Regression analysis showing the relationship between percentage of sperm with fragmented DNA after SCSA® analysis (x-axis) and TUNEL (y-axis, 12). (**a**) Human sperm ($n=25$, $r=0.56$, $P=0.004$). (**b**) Ram sperm ($n=29$, $r=0.84$, $P=0.002$). (**c**) Bull sperm ($n=36$, $r=0.78$, $P<0.001$). (**d**) Stallion sperm ($n=36$, $r=0.65$, $P<0.001$)

4. Meaningful detection of DNA damage with a variety of etiologies including toxicology, disease, and environmental-induced damage.

The development and validation of the SCSA® has extensively gone through all the above required steps over the past 20 years with well over a hundred thousand SCSA® measures of sperm obtained from animals and humans of known fertility and those being exposed to a variety of reproductive toxicants.

Examples of Repeatable High-Quality SCSA® Data

Genotoxicant Exposure

An excellent means to determine the precision and utility of the SCSA® was to conduct studies on sperm from animals exposed in a time–dosage fashion to genotoxicants. In addition, repeatability studies were done between samples measured as freshly collected sperm and frozen aliquots accumulated over time and then measured at a single time period. What makes the male particularly susceptible to toxin-induced damage is that the testis is characterized by a very high rate of cell proliferation with millions of sperm produced daily. Furthermore, the precursor stem cells undergo highly complex cell differentiation with specific steps known to be highly susceptible to certain types of chemical exposures.

Mouse

1. Genotoxic actions of triethylenemelamine (TEM) on mouse sperm DNA integrity was studied [16] by examining effects of TEM for 44 weeks after exposure. Fresh epididymal sperm were assayed by SCSA® each week over 44 weeks and the data were compared to samples frozen each week and then measured at one time period. As shown in Fig. 9.11, freezing had little to no effect on SCSA® data. Correlation of %DFI between fresh and frozen sperm for 1.0 mg/kg treated mice ($n=55$) collected over 44 weeks (no controls included) was 0.93 ($P<0.001$). This evidence also shows that instrument settings over the 11 months study period can be adjusted to provide highly repeatable measurements.
2. X-radiation. The scrotal region of male mice was exposed to X-rays ranging from 0 to 400 rads [17]. Forty days after exposure, the mice were killed and the caudal epididymal sperm were removed. The SCSA® detected increased DNA fragmentation after 12.5 rads of X-ray exposure, with significant increases following 25 rads. These data not only show that the SCSA® is a very sensitive method of detecting X-ray damage to sperm DNA but also show the very high repeatability of the measurements (Fig. 9.12.).
3. Dominant lethal agents. The effects of 150 mg/ml methyl methane sulfate on mouse epididymal

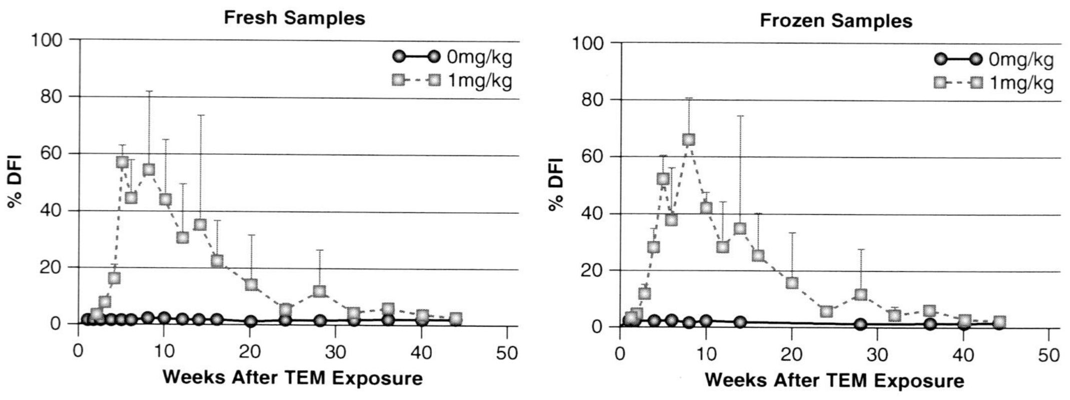

Fig. 9.11 Effects of 1.0 mg/kg (daily ×5) TEM on %DFI in epididymal sperm during a 44-week period. *Left*: %DFI on fresh samples. *Right*: Aliquots of the same samples frozen and measured later at a single time period

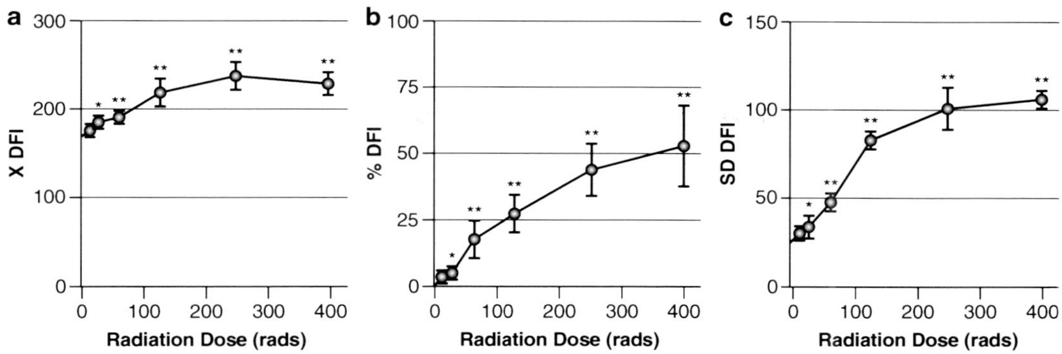

Fig. 9.12 SCSA® data on epididymal sperm from scrotum-exposed mice to 0–400 rads X-ray. Epididymal sperm were surgically removed from mice at 40 days post exposure. $n=3$ for each point

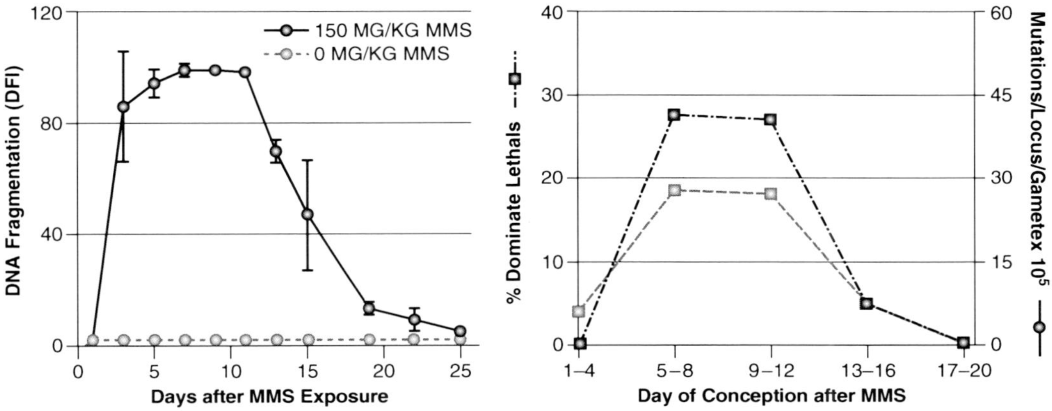

Fig. 9.13 Effect of methyl methanesulfonate on mouse sperm chromatin structure and subsequent embryo death

sperm DNA fragmentation can be seen in Fig. 9.13 [18]. By day 3 post exposure, about 85% of the sperm have extensive DNA damage; however, mating of these exposed mice to nonexposed females did not result in embryo death until 5 days post conception [18a]. Thus, the molecular events leading to embryo death can be derived from SCSA® data. Of interest, glutathione depletion potentiates ethyl methanesulfonate induced susceptibility of rat sperm DNA fragmentation [19].

Human

1. Pesticides: Men exposed to various insecticides and pesticides showed significantly increased levels of sperm DNA fragmentation. A dramatic effect of exposure to organophosphorous pesticides showed that 3/4 pesticide operators, not using protective gear, had DFI values above 30%, whereas those not exposed showed an average of 9.9% DFI [20] Fig. 9.14 shows SCSA® cytograms from a nonexposed and an exposed worker.
2. Air Pollution: For the first time, SCSA® data showed a dose–response relationship for men exposed to winter time air pollution [21]. Residents of Teplice, Czech Republic, a town with heavy winter-time air pollution, generated by burning soft brown coal, experienced a higher than normal rate of infertility and spontaneous miscarriages. Czech army conscripts, 18–20 years of age, provided semen samples in a 2-year longitudinal study that went through periods of clean summer air and polluted winter air. Sperm DNA fragmentation measured by the SCSA® was the only semen quality

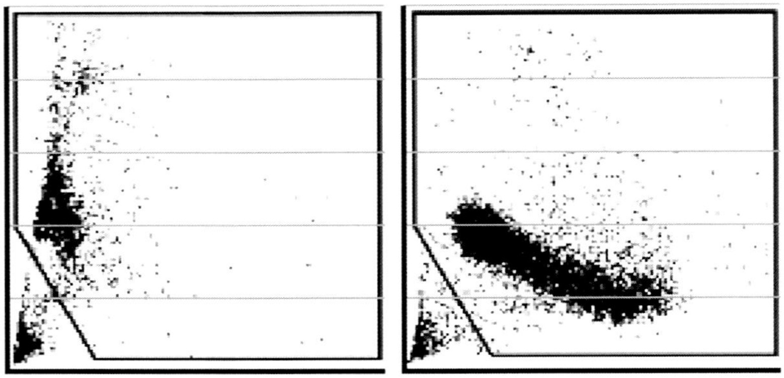

Fig. 9.14 SCSA® cytograms from a nonexposed (*left*) and an exposed (*right*) worker

measure to detect a statistically significant correlation between air pollution levels and semen quality in these young men. One fourth of these young men had %DFI above 30, placing them in a statistical group known to be at an increased risk for infertility.

Potential RNA Staining Artifacts for SCSA®

Since AO stains both single-stranded DNA and RNA in the fluorescent color red, it was very important to know if cytoplasmic or nuclear RNA contributed to the red fluorescence that might be erroneously attributed to denatured DNA.

First, any small amount of nuclear mRNA should be of small consequence to the total red fluorescence and, furthermore, should be a constant amount making only a constant background. RNAse treatment of mouse sperm did not reduce the red fluorescence caused by genotoxicant treatment [5]. Also of question was whether any residual cytoplasmic RNA contributed to ssDNA values. We addressed this question [22] by sonicating whole bull, mouse, stallion, and human sperm, purifying each sample of nuclei through a sucrose gradient and measuring both the sonicated and nonsonicated sperm by SCSA®. Somatic cells are fully destroyed and removed from the purified nuclei fraction. As illustrated in Fig. 9.15, the unsonicated and sonicated sperm produced cytograms that were practically identical. In the upper panel, note that the small percent of sperm with increased red fluorescence is in the same location (just to right of lower edge of main sperm population) after as before sonication. Of importance, note in the bottom panel that sperm with a high level of red fluorescence also produced essentially the same cytogram pattern. Given the rigor of sonication that destroys somatic cells and frees sperm nuclei, these data may suggest that sperm with a high %DFI are not fragmented nuclei that would be ripped apart by sonication. Since histone complexed chromatin is likely broken apart by sonication, it is hypothesized that the 15% histone complex in human sperm nuclei is not at the nuclear periphery where it might be highly susceptible to being removed by the sonication process.

Repeatability of SCSA® Data Over Time for Men

Forty-five men provided a semen sample once per month for nine months [22]. While the CV for the classic sperm parameters varied considerably, the SCSA® data showed that, as discussed above, the AO/DNA biochemistry as well as the flow cytometry measurements were highly repeatable with great precision. The sample to sample variation was only 3.4%, indicating that %DFI is

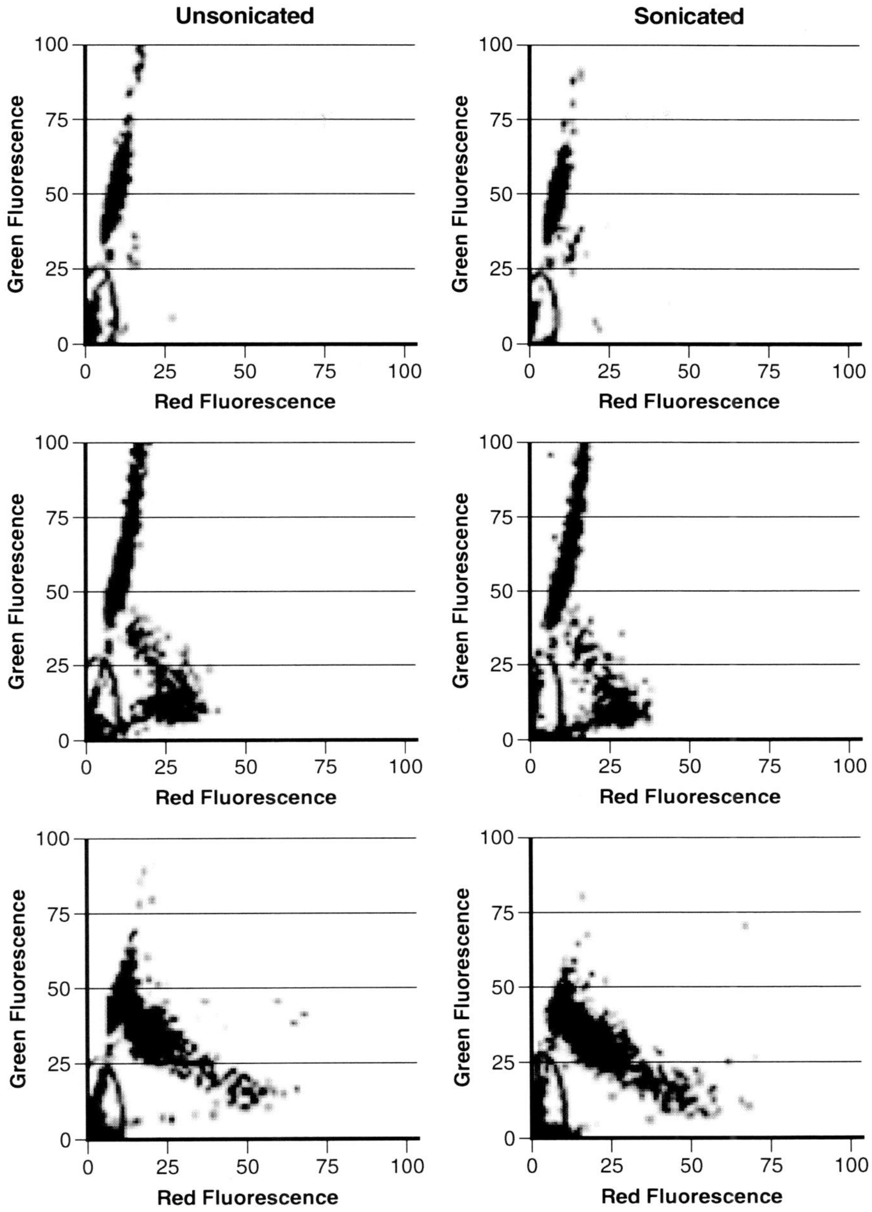

Fig. 9.15 Each human semen sample was diluted to a final volume of 0.5 ml with TNE buffer to obtain a count of approximately 2×10^6 sperm/ml. The samples in the right column were sonicated for 30 s with a Branson 450 Sonifier operating at a power setting of 3 and utilizing 70% of 1-s pulses

a much more stable parameter than the classic sperm parameters. Of significant interest in the cytograms seen in Fig. 9.16 is the repeatability of the pattern of the scattergrams within a man, and also, the repeatability of a very small percent of the populations appearing exactly with the same green and red values – it is speculated from these data that a fraction of a percent of germ cells have a mutation such that the altered chromatin has a highly distinct pattern of DNA damage.

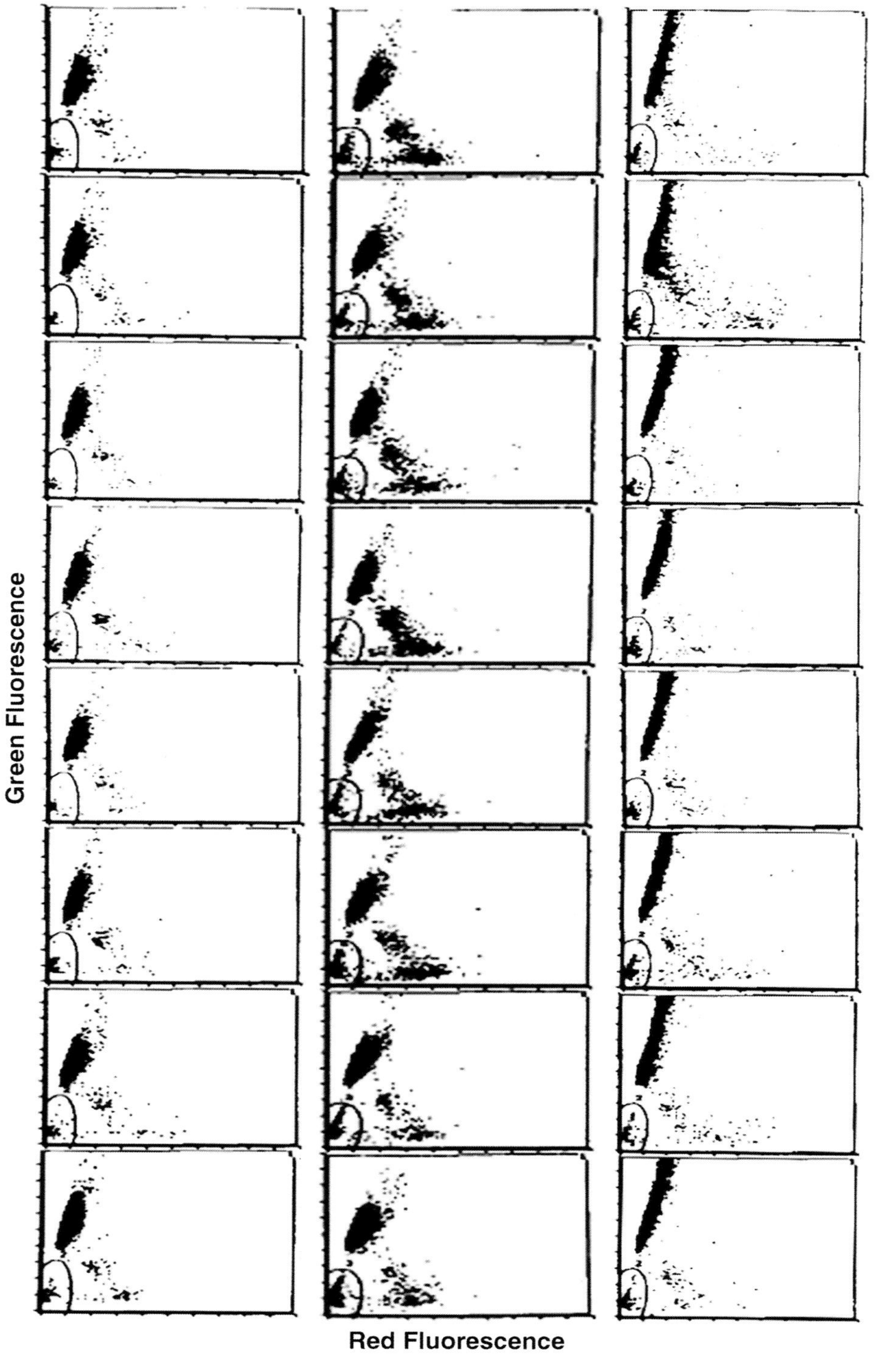

Fig. 9.16 Green vs. Red fluorescence cytograms from monthly semen samples provided by three donors. Examples are selected from the 45-men illustration of different types of cytogram patterns by Evenson et al. [22]

Repeatability of %DFI Values of Human Sperm Samples from Two Commercial SCSA® Laboratories

While repeatability of SCSA® values, as well as other sperm DNA fragmentation tests, may be repeatable within a laboratory, it is important for any used test to be highly repeatable between laboratories that may have different types of flow cytometers and different technicians. SCSA Diagnostics has a licensed agreement with SPZ Lab in Copenhagen, Denmark). As part of this agreement, all SCSA® samples must be done according to strict protocols to ensure that any patient gets a result that is repeatable. Figure 9.17 shows the correlation of %DFI obtained from aliquots of the same sample in the SCSA Diagnostics Inc. lab in Brookings, SD and SPZ Lab (Copenhagen, Denmark). The data show a $R^2=0.98$ solidifying the high repeatability of the SCSA® between two SCSA® certified laboratories.

Animal Fertility

Given the great complexity of human fertility, we considered it important to conduct mammalian animal fertility trials for validation of the SCSA® prior to doing human clinical studies.

Bulls

Semen from individual bulls is often used for hundreds to thousands of cow inseminations. Thus, fertility rankings can be made between bulls in a stud service. Following the preliminary study of bull fertility as reported in the Science paper [4], the relationship between nuclear chromatin structure and fertility was evaluated in two groups of Holstein bulls: Group 1, 49 mature bulls, and Group 2, 18 young bulls [23]. Fertility ratings had been estimated for Group 1 and nonreturn rates were known for group 2. Intraclass correlations of the SCSA® values were high (>0.70), based on four collections obtained over several years from Group 1 bulls. Negative correlations were seen between fertility ratings and both SD DFI (−0.58, $P<0.01$) and %DFI (−0.40, $P<0.01$) in Group 1, and between nonreturn rates and both SD DFI (0.65, $P<0.01$) and %DFI in Group 2 (−0.53, $P<0.05$). These data showed that the SCSA® is a useful tool for identification of low fertility bulls and poor quality semen samples (Fig. 9.18).

Inherent in studies mentioned above, and much more so with human studies, are the variables in the females and a host of other factors such as experience of the artificial insemination team. To get around this problem, animal studies can use what is known as heterospermic insemination protocols in which equal numbers of motile sperm from two or more phenotypically different bulls are mixed prior to insemination. The parentage of calves resulting from these matings is determined, and based on the number of calves sired with each phenotype, a competitive fertility index is derived for each bull [24]. Correlations of SD DFI and %DFI with competitive index were −0.94 ($P<0.01$) and −0.74 ($P<0.05$), respectively.

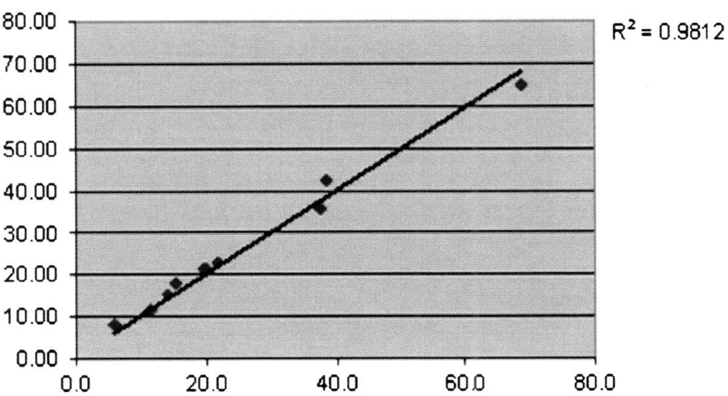

Fig. 9.17 SCSA® %DFI values obtained from SCSA Diagnostics Inc. (Brookings, SD, *x*-axis) and SPZ Lab (Copenhagen, Denmark, *y*-axis) Two aliquots were made for each human semen sample, which were frozen in LN2. One aliquot was measured in each laboratory and the results are mean values of two replicates per aliquot

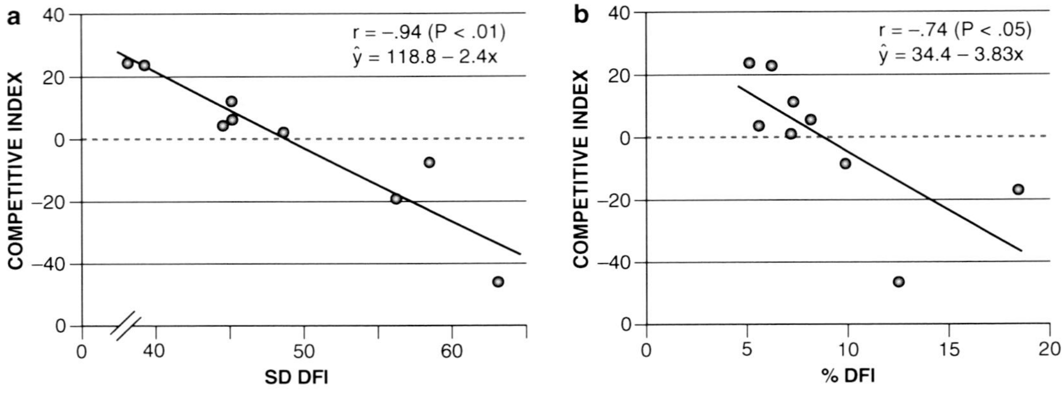

Fig. 9.18 Relationship of the competitive fertility index for bulls with (**a**) Standard deviation of DFI (SD DFI) and (**b**) %DFI

Boars

The advantage of investigating the relationship between SCSA® data and boar fertility is that pigs are multiparous, thus allowing a determination of both fertility rate and number of piglets per litter. The SCSA® was used [25] retrospectively to characterize sperm from 18 sexually mature boars having fertility information. Boar fertility was defined by farrow rate (FR) and average total number of pigs born (ANB) per litter of gilts and sows mated to individual boars. Fertility data were compiled for 1,867 matings across the 18 boars. In contrast to humans and other mammals studied, where the threshold for reduced fertility is an approximate 25–30% DFI, the threshold for boars is about 6% DFI. The %DFI and SD DFI showed the following significant negative correlations with FR and ANB; %DFI vs. FR, $r=-0.55$, $P<0.01$; SD DFI vs FR, $r=-0.67$; %DFI vs. ANB, $r=-0.54$, $P<0.01$ and SD DFI vs. ANB, $r=-0.54$, $P<0.02$. The present data suggest that boar sperm possessing fragmented DNA can affect embryonic development corroborating earlier studies in mice showing that fertilization occurs whether the sperm has damaged DNA or not [26] but may cause embryonic death. In a recent study by Boe-Hansen et al. [27], fertility has been studied for 155 boars with 2,593 experimental litters. Using a threshold of 3% for DFI, it was found that the number of piglets born decreased from 14.94 piglets per litter (below threshold to 13.90 piglets per litter ($P<0.01$).

Human Fertility

As stated above, the SCSA® or any other sperm DNA fragmentation test cannot predict fertility for a couple. Good fertility for the couple also depends on many female factors, and a low DFI value for a couple attending a fertility clinic may, therefore, imply that another cause of the infertility exists. However, the SCSA® can be predictive of male subfertility or infertility. Other chapters in this book provide more details than that outlined here.

Natural Conception

The SCSA® was the first flow-cytometric test to suggest that abnormal sperm chromatin structure was predictive of failed natural conception [4]. Following the pioneering study described in Science, the Georgetown fertility study [28] suggested an odds ratio of approximately 8 if the %DFI was above 30%. In this study [28], 200 couples with no known infertility factors were enrolled in a natural conception male factor infertility study. Monthly semen samples were obtained for the first 3 months or up to the time of biochemical or clinical pregnancy. Pregnancies were recorded over the first 12 months. The results showed that the men who had a <15% DFI had the shortest time to establish a pregnancy. Men with DFI between 15 and 30% had the next longest time period, while men with DFI above 30% had the longest time to pregnancy or no pregnancy. This latter group also had the highest level of miscarriages.

The "first pregnancy planner" study by Spano et al [29] also suggested for natural conception an odds ratio of 8–10 when the DFI was between 30 and 40%. A lower level of %DFI (20%) as a significant clinical threshold has been very recently reported by Giwercman et al. [30]. A value of 20–25% DFI appears to be a clinically significant threshold for natural conception.

SCSA® Test and ART Clinics

The first studies relating %DFI with IVF pregnancies consisted of 26 patients [31], IUI and IVF patients [28], and 89 IVF patients [31, 33] for a total of 148 patients with no pregnancies when DFI was above 27%. This led to the early concept that pregnancies were difficult to obtain when %DFI was above 27–30%. Boe-Hansen et al. [33] used SCSA in a clinical study for IUI, IVF, and ICSI treatments with reproductive outcomes of biochemical pregnancy (BP), clinical pregnancy (CP) and implantation ratio (IR). 385 semen samples from 234 couples were frozen for SCSA, and smears were prepared for morphology: 48 IUI, 139 IVF, and 47 ICSI. The results showed no significant difference in the fertility variables BP, CP, and IR when <27% DFI was used between the IVF and ICSI groups. A low number of patients received IUI with low success rate, and statistical analysis was therefore not performed. Ongoing pregnancy was achieved for both IVF and ICSI couples with DFI levels >27%, and six couples in ICSI treatment achieved CP full-term. DFI >27% had a high prognostic power for predicting no CP for IVF patients, with a specificity of 97%. Similar results were obtained from a study of 249 couples undergoing their first IVF and/or ICSI cycle conducted in the Markham clinic [35]. However, later studies showed that SCSA® values above 30% DFI could result in pregnancies after ART treatment.

While the TUNEL test has shown a wide variation of thresholds for clinical pregnancy outcomes ranging from about 4 to 36%. By contrast, the threshold for human semen with the SCSA® appears to be close to 30% and has changed only slightly downward (25%) since it was estimated many years ago. The SCSA® is now implemented routinely for all couples considered for IUI in the Southern Sweden hospital region, and a threshold of 25% was selected as a compromise. Bungum et al. [36] observed that the success for IUI started to decrease at a DFI value of 20% and approached zero when the DFI was 30%. A recent study by Giwercman et al. [31] also included information regarding sperm morphology in the assessments and suggested that the SCSA® %DFI threshold for reduced fecundity appears to be at 20%.

The greatest utility of the SCSA®, as shown by Bungum et al. [36] is that couples with a DFI above 25% should move on to IVF and preferably ICSI for the greatest success. IUI for these couples may not be cost-effective.

One hypothesis as to why ICSI can achieve a pregnancy when the %DFI >25%, is that the ICSI technician will pick up sperm with the best morphology and the greatest motility. Also, ICSI fertilization avoids potential additional DNA damage from oxidative stress either in the female reproductive tract or during in IVF. Finally, one to several of the best-grade embryos will be transferred to the female.

TESA for Failed ICSI Cycles with High %DFI

The %DFI thresholds for ICSI are likely to be higher than for IUI or natural conception since ICSI is the best method for avoiding potential additional DNA damage to the sperm prior to fertilization. However, a precise threshold for ICSI is difficult to establish, since only 3–5% of fertility patients have a %DFI above 50.

Previous and new data show that the use of testicular sperm in combination with ICSI provides an efficient treatment option for couples who fail multiple IVF cycles due to high levels of sperm DNA fragmentation. Initially, Greco et al. [37] found that for couples with failed ICSI cycles and the man had a high TUNEL defined %DFI, pregnancy success was dramatically increased with the use of testicular sperm (TESA). The overall incidence of DNA fragmentation in the testicular sperm samples was 4.8+3.6%, which was significantly lower ($P<0.001$) compared with the ejaculated sperm samples from the same individuals (23.6+5.1%). (Note: DFI levels reported here cannot be compared directly DFI levels reported for the SCSA®). Greco et al. [37] did not observe differences in fertilization and cleavage rates and in embryo morphological grade found between the ICSI attempts performed with ejaculated and with testicular spermatozoa. However, eight ongoing clinical pregnancies (four singletons and four twins) were achieved by ICSI with testicular spermatozoa (44.4% pregnancy rate; 20.7% implantation rate), whereas ICSI with ejaculated spermatozoa led to only one pregnancy that was spontaneously aborted.

A recent study [38] has included couples who had undergone between one and seven prior ICSI attempts with a mean of three failed cycles. A pregnancy rate of 62.5% was achieved when testicular sperm were used. An 83% pregnancy rate was achieved when the SCSA® defined DFI was >65%. A 75% pregnancy rate was achieved in couples who underwent four or more prior failed IVF cycles. Likewise, among the thousands of measurements done at our SCSA Diagnostics lab, we have numerous ad hoc cases where several to a dozen unsuccessful ICSI cases have failed when the %DFI is above 50–60%. Thus, there is utility for the SCSA® for those patients that have had several ICSI failures. As noted by Carrell et al. [39], those patients that had two or more failed ICSI cycles, the %DFI by TUNEL was about fourfold higher than that found in sperm donors.

SCSA® Defined Etiologies of Increased DNA Fragmentation

The most likely common factor in causing sperm DNA fragmentation is oxidative stress [40] in response to reactive oxygen species (ROS). Simply stated, we need oxygen to live, but excess ROS activity is a negative consequence of this fact. Many of the environmental factors discussed here are related to increased oxidative stress. Thus, many physicians and patients are well aware of the need to have a diet rich in antioxidants.

Age

While it has become socially acceptable to father children at an older age, this increased age of fatherhood has been correlated with an increased time to establish a pregnancy or no pregnancy. Since 1980, US birth rates have increased up to 40% for men aged 35–49 years and have decreased up to 20% for men under 30 years of age.

The first study on the relationship between age of nonsmoking, healthy men and sperm DNA integrity [41] showed that among all the sperm genomic end points measured, age had the strongest effects on sperm DNA integrity. A healthy 20 year old man typically has about 5% DFI. A gradual upward trend in the average frequency of sperm with increased %DFI was observed, beginning in the early reproductive years as seen in Fig. 9.19.

In this age study, men in their 50s ranged from excellent %DFIs (5%) to very poor levels (73%). Even men in their 20s and 30s had abnormal DFI values, suggesting they too might experience diminished fertility and/or abnormal pregnancy outcomes. This factor is likely related to the other factors as discussed below.

The statistical odds in this study to reach the 30% DFI threshold for negative natural pregnancy outcome was age 48 as seen in Fig. 9.20, even though these men may have fathered children in their 20s. Thus, the reproductive biological clock also ticks for men, but the time window is not as narrow as for women.

Genetics

Although the evidence is very limited, it would be fully expected that genetics plays an important role in susceptibility to sperm DNA fragmentation. One example is from a study [42] on a group of men who were participants in the Teplice, Czech Republic study described above [21]. The hypothesis was as follows: men who are homozygous null for glutathione-S-transferase M1 (GSTM1-) are less able to detoxify reactive metabolites of carcinogenic polycyclic aromatic

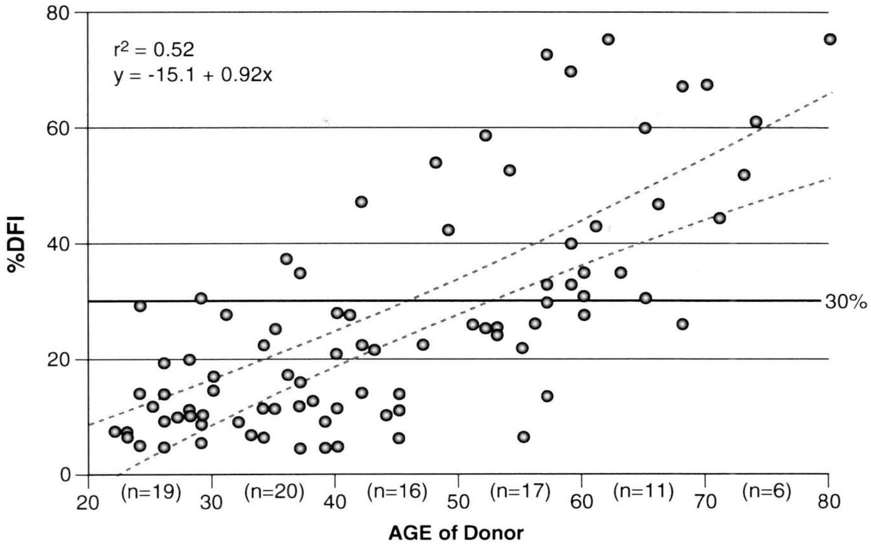

Fig. 9.19 Age of men vs. %DFI. The horizontal line is placed at 30% DFI, the approximate clinical threshold for risk of reduced natural fertility potential

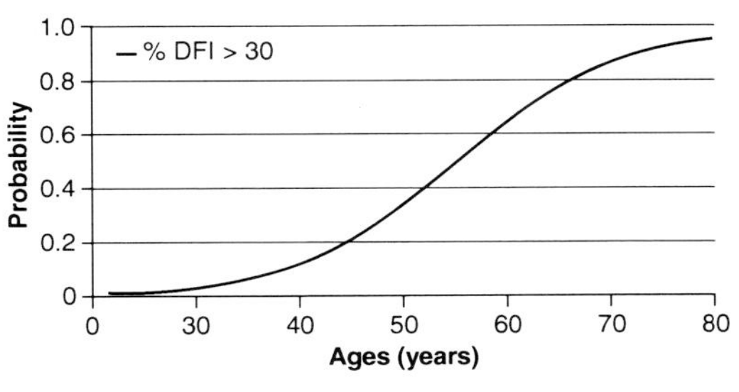

Fig. 9.20 Statistical probability of a man reaching a 30% DFI by age alone

hydrocarbons (c-PAHs) found in air pollution. Consequently, they are more susceptible to the effects of air pollution on sperm chromatin. Using a longitudinal study design in which men provided semen samples during periods of both low (baseline) and episodically high air pollution, this study revealed a statistically significant association between GSTM1 null genotype and increased SCSA®-defined %DFI (beta=0.309; 95% CI: 0.129, 0.489). Furthermore, GSTM1 null men also showed higher %DFI in response to exposure to intermittent air pollution (beta=0.487; 95% CI: 0.243, 0.731). This study, thus, provides novel evidence for a gene–environment interaction between GSTM1 and air pollution (presumably c-PAHs).

Varicocele

Varicoceles are found in approximately 15% and 19–41% of the general and infertile populations, respectively, and have long been recognized as a common cause of infertility.

The exact pathways of damage by varicocele are difficult to explain and may be due to apoptotic events, oxidative stress, or heat [40, 43]. Zini et al. found that sperm DNA fragmentation was significantly increased in infertility patients with varicocele in comparison with patients with normal results on genital examination [44]. Furthermore, it has been shown that sperm DNA fragmentation decreases after varicocele repair [45]. Recently, Werthman et al. [46] have found a 31% increase in pregnancy rate after varicocelectomy, whereas no pregnancy occurred before surgery. In this study, %DFI values were assessed by SCSA® before and after varicocelectomy (Fig. 9.21). Although this study was small, 10 of the 11 patients with varicocele showed a significant decrease in sperm DNA fragmentation after varicocele repair.

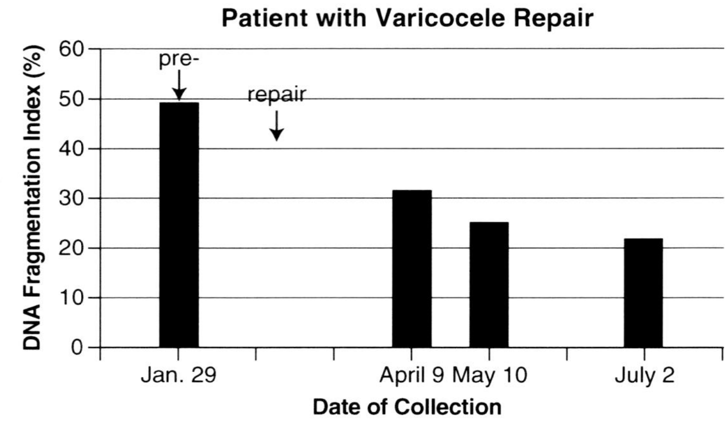

Fig. 9.21 %DFI values obtained from a man with a varicocele prior to surgical repair and at later time points. Note that a return to the lowest level %DFI shown occurred at 5 months post surgery

Cancer

Not unexpectedly, the majority of young patients with newly diagnosed testicular cancer is concerned about future fertility and wants to be informed about the different treatment modalities' influence on spermatogenesis. In the first study of effects of cancer on sperm DNA fragmentation, 14 patients with testicular cancer, assessed after orchiectomy but before further treatment [47], displayed considerable variability in the SCSA® results, most often revealing an increased percentage of sperm cells with abnormal chromatin structure.

As a follow-up to this initial study [48], semen samples from 39 patients with testicular cancer were analyzed by the SCSA® after orchiectomy but before further treatment, and in 28 patients the SCSA® was repeated 12–26 months after orchiectomy. Figure 9.22 shows the pretreatment %DFI for the patients compared to %DFI for 18 healthy semen donors.

The results from 19 patients undergoing cytotoxic treatment (radiotherapy, 13 chemotherapy, 6) indicate that posttreatment recovery of spermatogenesis (recovery in 4 of 5 patients) is observed more often in patients with a normal pretreatment chromatin structure than in those with abnormal SCSA® values before treatment. This study suggested that pretreatment SCSA® results may help clinicians to identify those testicular cancer patients with a high risk of long-lasting posttreatment disturbance of spermatogenesis.

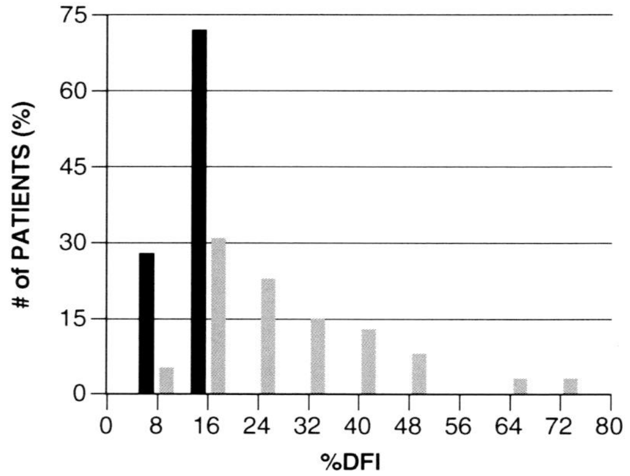

Fig. 9.22 Distribution of individual %DFI values. *Black bars*, 18 samples from healthy semen donors; *gray bars*, 39 samples from patients with testicular cancer after unilateral orchiectomy and before further treatment

It is not known whether childhood cancer and its treatment are associated with sperm DNA damage, which subsequently affects fertility and might be transmitted to the offspring. In 99 children cancer survivors (CCS) and 193 age-matched healthy controls, %DFI was assessed using the SCSA® [49]. In the whole group of CCS, %DFI was increased compared with the controls, with borderline statistical significance. Those treated with radiotherapy only or surgery only had statistically significantly higher %DFI than the controls. The odds ratio (OR) for having DFI >20%, which is associated with reduced fertility, was significantly increased in CCS compared with the control group. (OR, 2.2) For the radiotherapy-only group, the OR was even higher (OR, 4.9). %DFI was not associated with dose of scattered testicular irradiation or type of chemotherapy given. It was concluded that %DFI was increased in CCS, with those treated with chemotherapy being the only exception. This sperm DNA impairment may be associated with the disease per se, rather than due to the treatment, and may have negative consequences in terms of fertility and risk of transmission to the offspring.

Environmental Heat

The purpose of a scrotum is to keep sperm function and maturation at an approximate 2°C lower than body temperature. Mammalian sperm, including maturing epididymal sperm, are very sensitive to excess heat. Studies on bulls that had a wool sock placed over the scrotum for 48 h showed significant damage to sperm DNA [50]. Three samples were collected for 3 time periods and the %DFI measured. For day 0=4%, days 3–9=11%, and days 12–21=22% DFI. These data clearly show environmentally induced sperm DNA damage.

In another experiment [51], mice were anesthetized and the scrota exposed on the underside of a Styrofoam raft floating in a high precision water bath at 2° and 4° degrees above body temperature for 60 min. The higher temperature caused a significant amount of SCSA® defined sperm DNA damage. Figure 9.23 shows significantly increased epididymal sperm DNA damage after 3 days post exposure. Caudal epididymal sperm at this time point would have been traversing the caput and corpus epididymides during exposure to the elevated temperatures. Sperm at this stage of maturation would be undergoing further condensation including intra- and intermolecular S–S bonding between protamine cysteine–SH residues. The 38°C mice exhibited SCSA® values close to controls for most days. The SD DFI values showed the largest difference between controls and 40°C treated mice with a significant increase in value by day 11 ($P<0.001$) and a return to control values by day 35, or about one spermatogenic cycle.

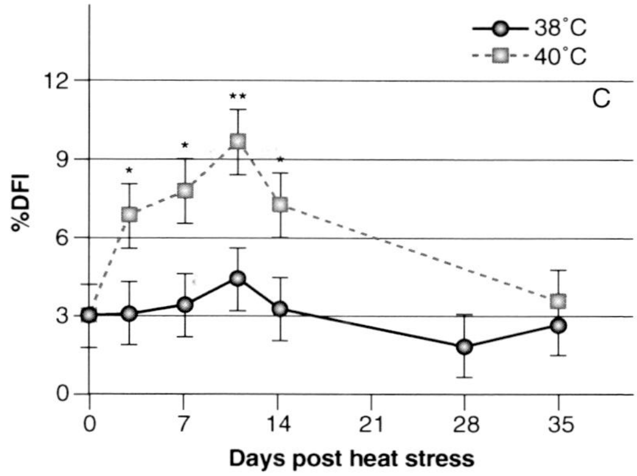

Fig. 9.23 SCSA® data on epididymal sperm obtained from scrotal heated mice. The scrotal regions of anesthetized mice were placed on the underside of a Styrofoam raft floating on a water bath (38° or 40°C) for 60 min. Three mice were used for each time point studied for each temperature

Fever

High fever has long been known to be a negative factor for pregnancy. A man who had a 104°F fever for 1 day showed [52] a dramatic increase to 36% DFI 18d post fever (dpf). The %DFI then decreased, while the %HDS increased to 49% at 33 dpf (Fig. 9.24).

Sperm nuclear proteins were isolated from this 33 dpf sample; amino acid sequencing of the first 8 N-terminal residues identified this unique protein as the precursor to protamine 2. Flowcytometric measurements of nuclear –SH groups revealed the greatest reduction in free nuclear thiols at 33 dpf, and then returned to normal by 45 dpf. Increased DNA staining is likely due to the increased histone/protamine ratio. By 60 days the sperm chromatin structure was back to normal – an approximate waiting time that physicians should suggest to such patients until trying to achieve conception.

Medications

Given the myriad of prescription and over-the-counter medications, it would not be surprising that some single agents or unstudied combination of agents will cause sperm DNA damage. Publications are sparse in this area.

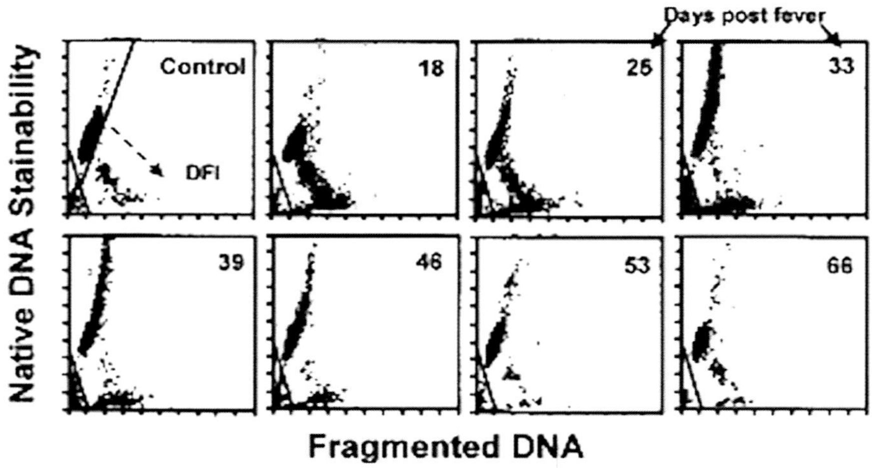

Fig. 9.24 Native DNA stainability vs. fragmented DNA for 66 days post fever

Recently, several manuscripts have been published on the effects of SSRI's on sperm DNA fragmentation. Tanrikut et al. [53] showed that the mean sperm DNA fragmentation index (TUNEL assay) was significantly higher for men while on paroxetine (30.3%) vs. baseline (13.8%). Before paroxetine, 9.7% of patients had a TUNEL score ≥ 30% compared with 50% at week 4 of treatment. The odds ratio (OR) of having abnormal DNA fragmentation while taking paroxetine was 9.33 (95% confidence interval, 2.3–37.9). Multivariate logistic regression correcting for age and body mass index confirmed this correlation (OR, 11.12). Of interest, standard semen parameters were not significantly altered during paroxetine treatment; however, the fertility potential of a substantial number of men on paroxetine may be adversely affected by these changes in sperm DNA integrity.

Diabetes and Insulin Resistance

Agbaje et al. [54] studied a cohort of 27 diabetic and 29 nondiabetic men. The level of sperm DNA fragmentation was significantly different between the two groups. Pittleloud et al. [55] reported that insulin resistance leads to a decrease in testosterone secretion at the testicular level (Leydig cell). Stigsby (personal communication, 2010) have also observed a link between insulin resistance (as measured by blood c-peptide level) and DFI value in a group of 10 men. When these men were consuming a diet with low glycemic index (GI) for a period of 4 months, both the c-peptide as well as the DFI values deceased. Although this study was very small, it appears that reduction of dietary intake of carbohydrates with a high GI may be advisable. It is recommended that identification of such individuals is based on blood levels of c-peptide (normal reference 200–700 pm/l). C-peptide is the "connecting peptide" that is cleaved from proinsulin when this is activated to insulin. This is a more stable parameter than traditional measurements of blood sugar or insulin. High insulin seems to increase the level of tumor necrosis factor alfa (TNF-alfa). TNF-alfa has a negative effect on sperm motility [56] and induces DNA fragmentation [57].

Conclusions

Thirty years ago, human infertility was considered to be a female problem if the man's semen analysis was within a reasonable range of normal. Today, couple infertility is almost equally shared between the man and woman. The routine semen analysis may in some cases identify subfertility or infertility when sperm motility is very poor or sperm concentration is very low. However, in many cases the cause of the decreased or absent fertility remains undetected unless sperm DNA fragmentation is considered. According to our experience and the data from Bungum et al. [36], sperm DNA fragmentation is the cause for every fourth couple attending the infertility clinic. In many cases, this problem is overlooked because other problems coexist, e.g., PCOS. In such cases, detection of sperm DNA fragmentation is essential for successful treatment of the couple.

The SCSA® is technically the easiest sperm DNA fragmentation test, and the repeatability of the assay is high within and between certified SCSA® laboratories. This is in contrast to many laboratories performing TUNEL where threshold ranging from 4 to 36% in DFI has been reported.

Clinical Utility of the SCSA®

The SCSA® has currently established a 20–30% DFI threshold for reduced pregnancy via natural or IUI. When %DFI reaches 20%, fertility starts to decline, and at 30% it reaches a very low level.

It appears that most unsuccessful IUI treatments can be avoided if couples with a DFI above 25% go on to IVF, or even better, ICSI treatment. However, if DFI is below 20–25% and no other causes of subfertility or infertility are detected for the couple, IUI treatment is likely to be successful.

A %DFI close to or above 50 is found in 3–5% of the couples with failed ART cycles. Currently, no threshold for %DFI is detected for ICSI treatment, but when %DFI is above 50, standard IVF is likely to be unsuccessful. Couples with >50% DFI might consider combination of TESA with ICSI, although there are few data to support this practice.

It is recommended that possible causes and lifestyle factors producing a high DFI be ruled out early in the treatment process. Repair of varicocele and corrections of other factors are likely to reduce the DFI level and will maximize the chances of a successful fertility treatment.

References

1. Bjorndahl L, Mortimer D, Barratt C, et al. A practical guide to basic laboratory andrology. Cambridge: Cambridge University Press; 2010.
2. Kametsky LA, Melamed MR. Spectrophotometer: spectrophotometer cell sorter. Science. 1967;156:1364–5.
3. Darzynkiewicz Z, Traganos F, Sharpless T, et al. Thermal denaturation of DNA in situ as studied by acridine orange staining and automated cytofluorometry. Exp Cell Res. 1975;90:411.
4. Evenson DP, Darzynkiewicz Z, Melamed MR. Relation of mammalian sperm chromatin heterogeneity to fertility. Science. 1980;240:131–1133.
5. Evenson DP, Higgins PH, Grueneberg D, et al. Flow cytometric analysis of mouse spermatogenic function following exposure to ethylnitrosourea. Cytometry. 1985;6:238–53.
6. Evenson DP, Tritle D. Platform Presentation Abstract: "Characterization of SCSA Resolved Sperm Populations by Comet Assay and Image Analysis". IFFS 8th World Congress on Fertility and Sterility, Palais des congres de Montreal, Montreal, Quebec Canada. 2004; May 23/28.
7. Evenson D, Witkin S, de Harven E, et al. Ultrastructure of partially decondensed human spermatozoal chromatin. Ultrastructure.1978;63:178–87.
8. Evenson D, Darzynkiewicz Z, Melamed M. Comparison of human and mouse chromatin structure by flow cytometry. Chromosoma. 1980;78:225–38.
9. Evenson DP, Jost LK, Varner DD. Stallion sperm nuclear protamine -SH status and susceptibility to DNA denaturation are not strongly correlated. J Reprod Fertility Suppl. 2000;56:401–6.
10. Love CC, Kenny RM. Scrotal Heat stress induces altered sperm chromatin structure associated with a decrease in protamine disulfide bonding in the stallion. Biol Reprod. 1999;60:615–20.
11. Evenson DP, Jost LK, Corzett M, et al. Characteristics of human sperm chromatin structure following an episode of influenza and high fever: a case study. J Androl. 2001;21:739–46.
12. Jeffay SC, Strader LF, Buus RM, et al. Relationships among semen endpoints used as indicators of sperm nuclear integrity. Am Soc Androl. Abstract. 2006.
13. Gorczyca W, Gong J, Darzynkiewicz Z. Detection of DNA strand breaks in individual apoptotic cells by the in situ terminal deoxynucleotidyl transferase and nick translational assays. Cancer Res. 1993;53:1945–51.
14. Sharma RK, Sabenegh E, Mahfouz R, et al. TUNEL as a test for sperm DNA damage in the evaluation of male infertility. Urology. 2010;76:1380–86.
15. Sailer BL, Jost LK, Evenson DP. Mammalian sperm DNA susceptibility to in situ denaturation associated with the presence of DNA strand breaks as measured by the terminal deoxynucleotidyl transferase assay. J Andrology. 1995;16:80–7.
16. Evenson DP, Baer RK, Jost LK. Long term effects of triethylenemelamine exposure on mouse testis cells

and sperm chromatin structure assayed by flow cytometry. Environ Mol Mutagen. 1989;14:79–89.
17. Sailer BL, Jost LK, Erickson KR, et al. Effects of X-ray irradiation on mouse testicular cells and sperm chromatin structure. Environ Mol Mutagen. 1995;25:23–30.
18. Evenson DP, Jost L, Baer R. Effect of methyl methanesulfonate on mouse sperm chromatin structure and testicular cell kinetics. Environ Mol Mutagen. 1993;21:144–53.
18a. Sega GA, Owens JG. Methylation of DNA and protamine by methyl methane sulfonate in the germ cells of male mice. Mutat Res. 1983;111:227–44.
19. Evenson DP, Jost LK, Gandy JG. Glutathione depletion potentiates ethyl methanesulfonate-induced susceptibility of rat sperm DNA denaturation in situ. Reprod Toxicol. 1993;7:297–304.
20. Sanchez-Pena LC, Reyes BE, Lopez-Carrillo L, et al. Organophosphorous pesticide exposure alters sperm chromatin structure in Mexican agricultural workers. Toxicol Appl Pharmacol. 2004;196:108–13.
21. Rubes J, Selevan SG, Evenson DP, et al. Episodic air pollution is associated with increased DNA fragmentation in human sperm without other changes in semen quality. Hum Reprod. 2005;20:2776–83.
22. Evenson DP, Jost L, Baer R, et al. Individuality of DNA denaturation patterns in human sperm as measured by the sperm chromatin structure assay. Reprod Toxicol. 1991;5:115–25.
23. Ballachey BE, Hohenboken WD, Evenson DP. Heterogeneity of sperm nuclear chromatin structure and its relationship to fertility of bulls. Biol Reprod. 1987;36:915–25.
24. Ballachey BE, Saacke RG, Evenson DP. The sperm chromatin structure assay: relationship with alternate tests of sperm quality and heterospermic performance of bulls. J Androl. 1988;9:109–115.
25. Didion B, Kasperson K, Wixon R, et al. Boar fertility and sperm chromatin structure status: a retrospective report. J Androl. 2009;30:655–60.
26. Ahmadi A. Ng S-C Fertilizing ability of DNA-damaged spermatozoa. J Exp Zool. 1999;284:696–704.
27. Boe-Hansen GB, Christensen P, Vibjerg D, et al. Sperm chromatin structure integrity in liquid stored boar semen and its relationships with field fertility. Theriogenology. 2008;69:728–36.
28. Evenson DP, Jost LK, Zinaman MJ, et al. Utility of the sperm chromatin structure assay (SCSA) as a diagnostic and prognostic tool in the human fertility clinic. Hum Reprod. 1999;14(4):1039–49.
29. Spano M, Bonde J, Hjollund HI, et al. Sperm chromatin damage impairs human fertility. Fertil Steril. 2000;73:43–50.
30. Giwercman A, Lindstedt L, Larsson M, et al. Sperm chromatin structure assay as an independent predictor of fertility in vivo: a case-control study. Int J Androl. 2010;33:221–7.
31. Larson KL, DeJonge CJ, Barnes AM, et al. Sperm chromatin structure assay parameters as predictors of failed pregnancy following assisted reproductive techniques. Hum Reprod. 2000;15(8):1717–22.
32. Larson-Cook K, Brannian JD, Hansen KA, et al. Relationship between assisted reproductive techniques (ART) outcomes and DNA fragmentation (DFI) as measured by the sperm chromatin structure assay (SCSA). Fertil Steril. 2003;80:895–902.
33. Boe-Hansen GB, Ersboll AK, Greve T, Christensen P. Increasing storage time of extended boar semen reduces sperm DNA integrity. Theriogenology. 2005;26(3):360–8.
34. Boe-Hansen GB, Fedder J, Ersboll AK, et al. The sperm chromatin structure assay as a diagnostic tool in the human fertility clinic. Hum Reprod. 2006; 21(6):1576–82.
35. Virro MR, Larson-Cook KL, Evenson DP. Sperm chromatin structure assay (SCSA®) related to blastocyst rate, pregnancy rate and spontaneous abortion in IVF and ICSI cycles. Fertil Steril. 2004;81:1289–95.
36. Bungum M, Humaidan P, Axmon A, et al. Sperm DNA integrity assessment in prediction of assisted reproduction technology outcome. Hum Reprod. 2007;22:174–9.
37. Greco E, Scarselli F, Iacobelli M, et al. Efficient treatment of infertility due to sperm DNA damage by ICSI with testicular spermatozoa. Hum Reprod. 2005;20:226–30.
38. Werthman P, Boostanfar R, Chang W. Use of testicular sperm/intracytoplasmic sperm injection yields high pregnancy rates in couples who failed multiple in vitro fertilization cycles owing to high levels of sperm DNA Fragmentation. 2010 Pacific Coast Reproductive Society Abstract.
39. Carrell DT, Liu L, Peterson CM, et al. Sperm DNA fragmentation is increased in couples with unexplained recurrent pregnancy loss. Arch Androl. 2003;49:49–55.
40. Saleh RA, Agarwal A, Nada EA, et al. Negative effects of increased sperm DNA damage in relation to seminal oxidative stress in men with idiopathic and male-factor infertility. Fertil Steril. 2003;79: 1597–605.
41. Wyrobek AJ, Eskenazi B, Young S, et al. Advancing age has differential effects on DNA damage, chromatin integrity, gene mutations, and aneuploidies in sperm. Proc Natl Acad Sci USA. 2006;103:9601–6.
42. Rubes J, Selevan SG, Sram RJ, et al. GSTM1 genotype influences the susceptibility of men to sperm DNA damage associated with exposure to air pollution. Mutat Res. 2007;625:20–8.
43. Chen SS, Huang WJ, Chang LS, et al. 8-hydroxy-20-deoxyguanosine in leukocyte DNA of spermatic vein as a biomarker of oxidative stress in patients with varicocele. J Urol. 2004;172:1239–40.
44. Zini A, Blumenfeld A, Libman J. et al; Beneficial effect of microsurgical varicocelectomy on human sperm DNA integrity. Hum Reprod. 2005;20:1018–21.
45. Yamamoto M, Hibi H, Tsuji Y, et al. The effect of varicocele ligation on oocyte fertilization and pregnancy after failure of fertilization in in vitro fertilization–embryo transfer. 1994;40:683–7.

46. Werthman P, Wixon R, Kasperson K, et al. Significant decreases in sperm deoxyribonucleic acid fragmentation after varicocelectomy. Fertil Steril. 2008;90: 1880–4.
47. Evenson DP, Klein FA, Whitmore WF, et al. Flow cytometric evaluation of sperm from patients with testicular carcinoma. J Urol. 1984;132:1220–25.
48. Fossa SD, De Angelis P, Kraggerud SM. Predication of post treatment spermatogenesis in patients with testicular cancer by flow cytometric sperm chromatin structure assay. Cytometry (Communications in Clinical Cytometry). 1997;30:192–6.
49. Romerius P, Stahl O, Moell C, et al. Sperm DNA integrity in men treated for childhood cancer. Clin Cancer Res. 2010;16:3843–7.
50. Karabinus DS, Vogler CJ, Saacke RG, et al. Chromatin structural changes in sperm after scrotal insulation of holstein bulls. J Androl. 1997;18: 549–55.
51. Sailer B, Sarkar LJ, Bjordahl JA, et al. Effects of heat stress on mouse testicular cells and sperm chromatin structure. J Androl. 1997;18:294–301.
52. Evenson DP, Jost LK, Corzett M, Balhorn R. Characteristics of human sperm chromatin structure following an episode of influenza and high fever: a case study. J Androl. 2000;21:739–46.
53. Tanrikut C, Feldman AS, Altemus M, et al. Adverse effect of paroxetine on sperm. Fertil Steril. 2010; 94:1021–6.
54. Agbaje IM, Rogers DA, McVicar CM, McClure N, Atkinson AB, Mallidis C, et al. Insulin dependant diabetes mellitus: implications for male reproductive function. Hum Reprod. 2007;22:1–7.
55. Pitteloud N, Hardin M, Dwyer AA, et al. Increasing insulin resistance is associated with decrease in Leydig cell testosterone secretion in men. J Clin Endrocrinol Metab. 2005;90:2636–41.
56. Koçak I et al. Relationship between seminal plasma interleukin-6 and tumor necrosis factor alpha levels with semen parameters in fertile and infertile men. Urol Res. 2002;30:263–7.
57. Perdichizzi A et al. Effects of tumour necrosis factor-alpha on human sperm motility and apoptosis. J Clin Immunol. 2007;27(2):152–62.

Sperm Chromatin Dispersion Test: Technical Aspects and Clinical Applications

10

Jaime Gosálvez, Carmen López-Fernández, and José Luís Fernández

Abstract

Sperm DNA damage has been connected, among other things, with an increased incidence of miscarriage and enhanced risk of disease in the offspring. However, its occurrence is multifaceted and many of the variable consequences it has for fertility are as yet not fully understood. Tests that assess sperm quality should identify not only the ability of spermatozoa to reach the oocyte with an intact DNA molecule but also their ability to fertilize the oocyte and activate embryo growth. Sperm DNA fragmentation should be considered a parameter of sperm quality. Compared to other methods of assessing DNA fragmentation, the sperm chromatin dispersion (SCD) test can be conducted promptly and without the need for complex and expensive laboratory equipment. The SCD test is a powerful and versatile approach for investigating DNA fragmentation, allowing for the assessment of damaged DNA over a diverse range of clinical situations. The technique can be easily adapted to incorporate new research directions, and the analysis of sperm DNA can be performed on a wide range of species.

Keywords

Sperm DNA damage • Sperm DNA fragmention • DNA fragmentation in sperm • Sperm chromatin dispersion test

Sperm DNA Fragmentation: Now and Then

After more than 30 years using different approaches to assess sperm DNA fragmentation (SDF), the scientific community still has serious doubts about which technique produces the most

J. Gosálvez (✉)
School of Biological Sciences,
Madrid Autonoma University,
St Darwin 2, Cantoblanco, Madrid 28049, Spain
e-mail: jaime.gosalvez@uam.es

A. Zini and A. Agarwal (eds.), *Sperm Chromatin: Biological and Clinical Applications in Male Infertility and Assisted Reproduction*, DOI 10.1007/978-1-4419-6857-9_10,

reliable results, and most importantly, what value these results have in a clinical context [1–4]. Several techniques have been used effectively to detect SDF in humans and several animal species: (1) The sperm chromatin structure assay (SCSA; [5–7]) was one of the first experimental approaches performed to assess SDF. The underlying principle for this method involves subjecting the DNA to mild acid in order to denature double-stranded or single-stranded breaks. Subsequent staining with acridine orange, which fluoresces green with double-stranded non-denatured DNA or red with single-stranded denatured DNA, allows for the quantification of sperm cells with fragmented DNA using a flow cytometer. (2) Another approach that has been successfully implemented to assess sperm DNA breakage is based upon the enzymatic addition of labelled nucleotides to the end of a DNA break. This includes techniques such as terminal deoxynucleotidyl transferase (TdT)-mediated nick-end labelling (TUNEL) or in situ nick translation (ISNT) using *E. coli* DNA polymerase [8, 9]. (3) The comet assay consists in performing single-cell gel electrophoresis (SCGE). Because of the differential resistance encountered by DNA molecules of different sizes when moving through the gel, a characteristic "comet" distribution is formed after fluorescent staining, with a dense head containing long molecules of DNA and a tail of varying length with shorter fragments of DNA. Thus, DNA breakage can be evaluated by measuring the number of cells with migration tails, as well as the length of the tail and/or percentage of DNA contained in the tail [10, 11]. A modification of this technique based on a two-dimensional displacement of the DNA fragments offers the possibility of differentiating single- and double-strand breaks on the DNA molecule [12, 13]. (4) Lastly, the sperm chromatin dispersion (SCD) test [14–16] and the improved commercially available version of this test, Halosperm® (Halotech, Madrid, Spain), constitute a fast method based on a controlled DNA denaturation and protein depletion to determine SDF. As detailed in the following section, this procedure gives rise to halos of chromatin dispersion due to the spreading of nuclear DNA loops and/or fragments of DNA when the spermatozoa contain fragmented DNA. The size of the halo is related to the amount of sperm DNA damage. Other approaches to measure sperm DNA damage and chromatin alterations have also been described but warrant no further mention due to their restricted use.

As researchers, we are aware that there exists a tendency in the laboratory to use those methods or techniques with which we feel most confident, even though these may present certain constraints. This is the reason why, in our opinion, a sterile debate has evolved over the capacity of the different technologies to measure "real" vs. "potential" sperm DNA damage [17]. It has been claimed that tests that measure "real" DNA damage, such as TUNEL, ISNT or the comet assay (neutral conditions), have a higher predictive value than tests that measure "potential" DNA damage, such as the SCSA, SCD, DBD-FISH, Chromomycin A3 staining or the comet assay (alkaline conditions). It is important to clarify whether DNA breakage is simply present or not; it can exist as a single-strand or double-strand DNA break. In either case, this damage is "real". A similar debate has arisen over how the different techniques measure this damage – whether by a "direct" or "indirect" method. We believe that all existing techniques to assess SDF are "indirect", and that each one has its own particular set of limitations. The TUNEL assay, for instance, is not "direct", as it requires an enzymatic mediator to incorporate labelled nucleotides into DNA breaks. The substrate for the terminal transferase must be a clean hydroxyl 3′ end that has not been chemically modified, and so, the TUNEL assay may underestimate the amount of DNA damage. In addition, the TUNEL protocol used in most laboratories has been designed for use with DNA from somatic cells where the chromatin is arranged with histones, but this protocol may not be as effective when used on highly protected protaminated sperm DNA, given that the enzymes used in this assay are large molecules that may not reach all DNA targets equally [18]. For example, in Fig. 10.1g, a TUNEL labelled sperm cell is shown after partial protein removal. The efficiency of DNA labelling is notably improved

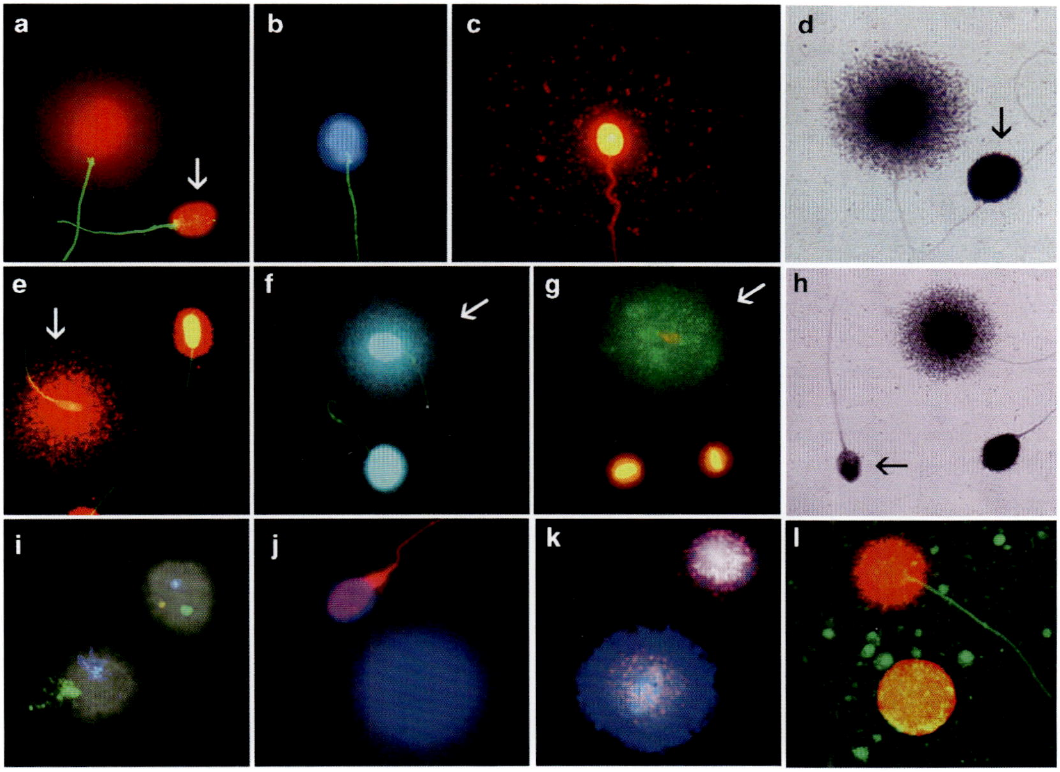

Fig. 10.1 Versatility of the SCD test. (**a**) Classical SCD in human sperm showing normal sperm (halo of dispersed chromatin) or a fragmented sperm (absence of halo; *arrow*) under fluorescence microscopy. (**b**, **c**) A comparison of two fragmented human sperm stained with DAPI (**b**) and GelRed (**c**) to highlight the presence of a massive halo of dispersed and atomized chromatin remnants in a fragmented sperm cell. (**d**) Classical SCD under bright-field microscopy (*arrow* denotes fragmented sperm cell). (**e**) Modified SCD for animal species (boar). The presence of a halo is correlated with sperm DNA fragmentation (SDF) (*arrow*). (**e–g**) Direct correlation between the presence of haloes of chromatin dispersion and in situ DNA labelling. In animal species, the expanded halo of dispersed chromatin (**f**; koala) could be highly labelled by in situ extension of the DNA breaks using polymerase (**g**; deer). (**h**) Classical SCD in human sperm under bright-field microscopy showing the presence of a degraded spermatozoon (*arrow*). (**i–k**) Direct correlation of SDF and specific DNA targets in human sperm. (**i**) SCD combined with FISH for aneuploidy detection. (**j**) SCD combined with McAbs for detection of 8-oxoguanosine. (**k**) SCD combined with McAbs for detection of 5-methyl citosine. (**l**) Dual staining (DNA *red*, proteins *red*) to differentiate histonized (*yellow*) and protaminized (*red*) cells. In this case, yellow fluorescence corresponds to a leukocyte

with respect to that obtained using paraformaldehyde-fixed samples. In fact, a recent report has demonstrated this very point by showing that there is increased TUNEL labelling when sperm samples are treated with the disulphide bond reducing agent DTT [19]. The only explanation for this is that the terminal transferase is not reaching all the available DNA breaks in the damaged sperm. Lastly, differences in SDF levels have been reported when the results of the TUNEL assay are assessed by flow cytometry or optical microscopy [20]. On the other hand, the so-called "indirect" methodologies, based on the susceptibility of DNA to denaturation, have been extensively used in mutagenesis [21]. The established dogma is that acid denaturation does not create any "potential" DNA break, but rather DNA breakage makes DNA more susceptible to DNA denaturation, DNA mobilization or enzyme-mediated incorporation of nucleotides.

Despite their different approaches and their limitations, the techniques have been shown to produce results that are highly correlated [8, 22, 23]. Thus, the main advantages or disadvantages of each procedure will largely depend on the time to obtain results, cost and the requirement for technical equipment or qualified personnel. The SCSA is not easily implemented in every laboratory, since it is a complex procedure that requires an expensive flow cytometer and highly specialized personnel. Alternatively, the samples may be shipped and analyzed in reference laboratories, but this prolongs considerably the time to obtain results. The comet assay requires trained personnel to perform the methodology with a certain level of reproducibility. The requirement of an electrophoresis unit and specific software for image analysis also limit the quick production of results. The methods based on in situ hybridization or enzyme-mediated extension of the DNA molecule also have the limitation of being complex, time-consuming and requiring specialized personnel. As a result, these procedures are best suited for research purposes and are therefore considered unsuitable for routine use in the andrology laboratory.

Technical Basis of the SCD Test

The technical basis of the SCD test rests on two observations: the first is that DNA strands that contain breaks or nicks are more easily denatured, since the ends of the breaks behave as origins of denaturation. This is the rationale for the classical unwinding assays that have been employed for many years for the quantification of DNA breaks in radiobiology and mutagenesis [24]; the second is that partial protein depletion from chromatin results in a characteristic pattern of DNA loops, spreading around a nucleoid of DNA that remains attached to protein residues, as described by Cook and Brazell [25].

The SCD test has been adapted for the nuclei of human spermatozoa and the methodology comprises three main steps: (1) inclusion of sperm cells in an inert semi-solid medium spread over a glass slide, (2) sperm sample incubation in HCl for acid denaturation, (3) treatment in a lysing solution for controlled nuclear protein removal and a final staining step [14]. The acid solution produces a controlled DNA denaturation only when this DNA contains extensive breakage. The subsequent incubation in the lysis solution removes protamines. If the sperm DNA is intact, a characteristic halo of DNA loops is formed around a dense central core (Fig. 10.1a). On the other hand, if the sperm nucleus contains fragmented DNA, the halos are absent or they are very small (arrow in Fig. 10.1a). This differential chromatin behaviour is the base of the SCD test. In actual fact, halos are also produced when the DNA is fragmented and susceptible to denaturation by acid (Fig. 10.1c). In this case, however, the DNA fragments diffuse further from the central core and because they are smaller, they are faintly stained to the point that they remain invisible using standard fluorochromes such as propidium iodide, diamidino phenyl indole (DAPI) or Diff-Quick under bright-field microscopy (Fig. 10.1b). Nevertheless, this pattern can be revealed using more efficient fluorescent DNA binding molecules such as GelRed (Biotium, Hayward, CA, USA) or Synergy Brand derived molecules (Invitrogen, Carlsbad, CA, USA) and captured with high-performance CCD (cooled charge-coupled-device; Fig. 10.1c).

This methodology has also been used with sperm from other mammalian species including Eutheria [26–29], Metatheria [30, 31] and Prototheria [32] to produce similar halos of chromatin dispersion. The methodology needs to be adapted for each species, although commercial procedures have been developed for each mammalian species (Halomax®, Halotech, Madrid, Spain). For mammalian species, the SCD test was simplified so that only a species-specific modified lysing solution is used for protein depletion. This is because each species contains different protamine residues that require a different strength of lysis solution to produce efficient protein removal, and this is enough to produce a differential chromatin dispersion pattern without the need to subject the DNA to acid denaturation. The result is that, unlike the SCD test adapted for use with human sperm, large halos of spotty dispersed chromatin are associated to fragmented DNA (arrow in Fig. 10.1e, f) and small, compact halos of chromatin loops correspond to sperm cells with intact DNA (Fig. 10.1e, f; [33]). Therefore, the expanded halos of dispersed chromatin are positive for TUNEL labelling (arrow in Fig. 10.1g). This serves a direct control to demonstrate that the presence of halos is associated to DNA damage. Similarly, the presence of halos of chromatin dispersion in this test is correlated with the characteristic migration tails denoting DNA fragmentation in the comet assay [31, 34, 35].

Validation of the SCD Test

The SCD test has the unique advantage that it can be directly validated by other techniques applied on the same sperm cell. Such experiments have been conducted using DNA breakage detection-fluorescence in situ hybridization (DBD-FISH). In this procedure, breaks in the DNA molecule are transformed into restricted single-stranded DNA areas by a denaturing acid or alkaline solution. These areas are targets for hybridization with a fluorescent-labelled whole genome probe or even using DNA probes for specific genome domains [15, 36]. The intensity of fluorescence after hybridization is related to the amount of DNA damage [15]. Incubation with a whole-genome probe following the SCD test – the acid used in the SCD test is sufficient to reveal the single-stranded targets for the probe – results in strong hybridization only in those nucleoids with a small or absent halo, demonstrating in situ that these sperm cells contain fragmented DNA. Validation was also obtained using enzymatic labelling of DNA breaks on SCD-processed nucleoids. The sequential incubation with the TdT, DNA polymerase I or the Klenow fragment, following the TUNEL, ISNT or Klenow-end labelling procedures, respectively, also resulted in intense labelling of those nucleoids that presented a small or no halo [36–38].

The SCD test was also validated using agents that are known to induce DNA breakage. When sperm samples were exposed to hydrogen peroxide, sodium nitroprusside (SNP) or DNaseI, a concomitant dose-dependent increase was observed in the frequency of sperm cells with no halo or small halos [15, 39]. Lastly, the SCD test was validated indirectly by comparing the results with those obtained using other techniques with the aliquots from the same semen sample. The percentage of sperm cells with fragmented DNA as measured with TUNEL and SCSA correlated highly with the number obtained using the SCD test adapted for human sperm samples [23, 40] and for other animal species [13, 27, 32]. Results obtained with ISNT and the comet assay also correlate with those obtained with the SCD test adapted for stallion [27], ram [29], marsupials [30, 32] rhinoceros [41] or fish [35].

Methodological Versatility

Assessing DNA Damage Intensity

The amount of DNA damage differs from one sperm cell to another in any given semen sample. Such variation accounts for the dispersion in colour ratio values obtained with SCSA and the different amount of DNA labelling obtained with the TUNEL assay. Similarly, the different halo sizes produced by the SCD test are indicative of the level of DNA damage [15]. In addition to the differences in halo size, the SCD test also reveals a distinct class of sperm cells referred to in the literature as "degraded sperm", which are characterized by a residual nuclear core after protein depletion (arrow in Fig. 10.1h; [42]). This extreme level of nuclear damage may involve damage of the nuclear matrix. Such degraded sperm cells have been observed in both fertile and infertile patients but are especially prevalent in cases of varicocele [15, 42].

Assessing Chromosomal Abnormalities

Conventional FISH may be performed on sperm cells that have been previously processed by the

SCD test because the protein-depleted sperm chromatin exposes the DNA in such a way as to allow efficient hybridization of fluorescent DNA probes. Thus, it is possible to simultaneously determine the level of fragmentation and the presence of aneuploidies (Fig. 10.1i) or structural chromosome rearrangements [43] in the same sperm cell. In patients presenting genomic unbalances in their sperm, SCD-processed slides were subjected to FISH against chromosomes X, Y and 18. The authors describe a 4.4 ± 1.9-fold increase in diploidy rate, and a 5.9 ± 3.5-fold increase in disomy rate in sperm containing fragmented DNA, with the overall aneuploidy rate being 4.6 ± 2.0-fold higher in sperm with fragmented DNA (Wilcoxon rank test: $p < 0.001$ in the three comparisons; Muriel et al. [43]). A similar correlation between SDF and the incidence of aneuploidies has been shown using FISH and SCSA, although this study did not measure both parameters simultaneously in the same cell and so the correlation is only indirect [44]. These results suggest that the occurrence of numerical chromosome abnormalities during meiosis may lead to SDF as part of a genomic screening mechanism conducted to genetically inactivate sperm with a defective genomic background.

Assessing Oxidative DNA Base Damage

Intense oxidative stress may give rise to DNA modifications such that the guanine residues at C-8 are hydroxylated to form 8-oxo-7,8-dihydro-2′-deoxyguanosine (8-oxoG) [45]. Thus, the presence of 8-oxoG is considered an indirect marker of oxidative stress [46], and monoclonal antibodies have been developed against these modified residues [47]. The anti-8-oxoG antibodies have been effectively used to show the presence of 8-oxoG in somatic tissue samples using liver sections [48]. The SCD test may be used together with specific antibodies against 8-oxoG to investigate the link between oxidative stress and DNA damage (Fig. 10.1j). A recent study has shown that increased levels of 8-oxoG were mostly present in those spermatozoa that had fragmented DNA, suggesting a close relationship between both DNA lesion types [39]. The presence of 8-oxoG was also associated with decreased sperm motility and lower embryo quality after in vitro fertilization (IVF) or intracytoplasmic sperm injection (ICSI) [49]. As a positive control, sperm cells were subjected to H_2O_2, to produce DNA fragmentation and a concomitant 8-oxoG base modification. As a negative control, SNP produced similar DNA damage, but an 8-nitroguanine base modification rather than 8-oxoG, and DNAase I produced only DNA breakage.

Assessing DNA Methylation

DNA methylation is an important base modification closely related to gene regulation during mammalian development, and its presence is related with diverse processes such as gene expression and genomic imprinting [50, 51]. Abnormal DNA methylation levels in sperm have been associated with decreased pregnancy rates in IVF [52]. The SCD method can be combined with the use of antibodies directed against 5-methylcytosine for the sequential assessment of DNA methylation and DNA fragmentation. The intensity of the signal can be quantified to provide a semi-quantitative estimate of DNA methylation levels in each sperm cell (Fig. 10.1h; Kumar, personal communication).

Assessing Sperm Protein Matrix

The classical SCD protocol can be modified to omit the acid denaturation step resulting in an extensive spreading of DNA loops [53]. With this protocol, the use of a fluorochrome specific for proteins enriched in disulphide bonds (2,7-dibrom-4-hydroxy-mercury-fluorescein) reveals that remnants of other nuclear proteins tend to remain within the core of the nucleoid only in those spermatozoa with fragmented DNA [53]. This suggests that the nuclear matrix of sperm containing fragmented DNA is more resistant to protein removal by the lysis solution. Spermatozoa with fragmented DNA may thus have a modified nuclear protein matrix, suggesting that the processes that initiate DNA fragmentation are also

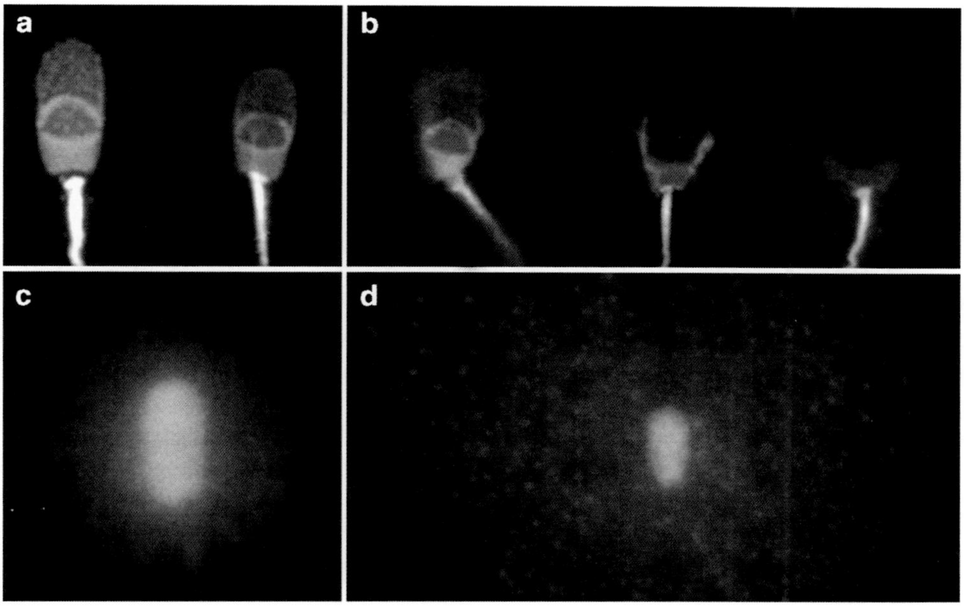

Fig. 10.2 Residual protein matrix (**a**, **b**) and SDF (**c**, **d**) in boar spermatozoa after the SCD test. Fragmented sperm (**d**) show an altered residual protein scaffold (**b**) when compared with unfragmented spermatozoa (**c**)

expressed at the nuclear matrix level. In other mammalian species such as the boar, this effect is also present. Thus, the residual protein matrix is more intensely damaged when the sperm DNA is more fragmented ([54]; Fig. 10.2).

In leukocytes or other somatic cells, the DNA is coiled around histones rather than the protamines of sperm cells. The DNA denaturing and protein lysis treatments of the SCD do not remove the nuclear histone proteins in these cell types. The leukocyte nucleoids, therefore, show no halos of chromatin dispersion. Double fluorescent staining can thus be used on SCD-processed slides to discriminate, for example, leukocytes from sperm cells in patients with leukocytospermia. If SCD processed slides are stained with a mixture of fluorochromes directed against proteins (green emission) and DNA (red emission), cells that contain histones will have overlapping protein and DNA labelling and exhibit yellow fluorescence, while sperm cell heads will exhibit red fluorescence. This methodological approach was used to analyze a Kartagener syndrome patient. In this case, a baseline SDF of 76.4% and a proportion of 1:4 germ cells to somatic cells were observed [55]. This methodological variant may be used to study those patients with high leukocyte counts, since these cells may release reactive oxygen species (ROS) or stimulate their production by spermatozoa, thus producing DNA fragmentation [56]. The scenario could be of particular interest, since Henkel et al. [57] have suggested that the threshold value of leukocytospermia of 1×10^6/mL should be re-evaluated because lower leukocyte counts can compromise DNA integrity.

The SCD and Low Sperm Counts

The SCD can easily be applied to assess SDF in sperm samples obtained from critical clinical situations where the number of spermatozoa is very low. Thus, this should be the procedure of choice in severe oligozoospermia, immotile sperm samples, TESA/TESE samples [55], sorted spermatozoa for sexed semen production [58], samples to be selected using intracytoplasmic morphologically selected sperm injection (IMSI) or even postmortem epididymal samples.

In the case of IMSI or high magnification sperm selection, a direct correlation can be established between the selected sperm and SCD results. In collaboration with Dr Monica Antinori and the Ginemed Clinic (Sevilla, Spain), we are investigating the correlation between SDF and the presence of sperm vacuolization in the same sperm cell. The preliminary results suggest that high sperm vacuolization and abnormal sperm morphology may be associated with increased SDF (Fig. 10.3).

The SCD test, due to its technical simplicity, reliability and lack of requirement of technical equipment, is quite adequate to accomplish large epidemiological studies or screening of specific male populations exposed to presumed toxic agents or environmental contaminants. This is true not only for humans but also for different domestic, farming or endangered animal species. The SCD methodology has been used outdoors in the field, where electric-powered facilities such as freezers, microscopes or heaters are not available. With only minor modifications to the standard protocol, the SCD test can be performed readily in the field, offering reliable information on SDF. An LED-equipped microscope attached to a laptop, a gas heater and a CO_2 spray for cooling are sufficient to assess the quality of sperm DNA. The results obtained after assessing ram semen samples under different conditions (30°C in the laboratory and at 17 and 4°C in the field) showed that, except when processing at 4°C, the technique was highly reproducible [59]. This opens up the possibility to study the fertility potential of sperm samples post-mortem since mature spermatozoa collected from the *caudal epididymis* have been used successfully for artificial insemination [60]. A decision can be made on site based on DNA quality to inseminate, cryopreserve or reject the sample. This decision can be made within 30 min of sperm recovery.

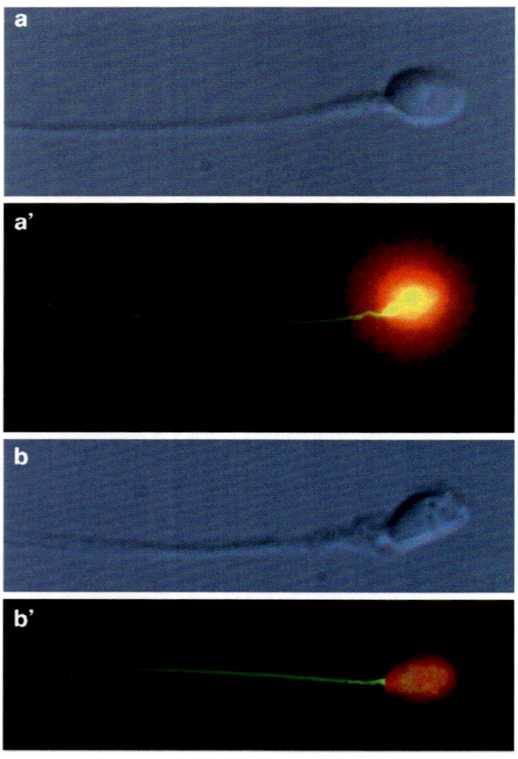

Fig. 10.3 High magnification selected sperm (**a**, **b**) and the characterization of SDF in the same sperm (**a´**, **b´**). The SCD test allows the direct assessment of the DNA status and the sperm morphology

The SCD Test in the Assisted Reproductive Technology (ART) Laboratory

The SCD test produces results that correlate highly with those obtained with other methodologies [23, 40]; however, relatively few studies have been published with this technique. Therefore, when discussing the clinical applications of SDF, we have considered studies performed using other techniques as well, making particular mention of those that use the SCD test.

Fertility Assessment

Infertile men possess significantly more SDF than their fertile counterparts [61]. It therefore follows that DNA damage may adversely affect reproductive outcomes. Numerous groups have suggested that there may be a threshold level of DNA damage above which pregnancy is impaired [4]. Indeed, the percentage of sperm cells with fragmented DNA has been suggested as a complementary parameter to the standard semen quality parameters (sperm concentration, motility and morphology) in predicting the success of natural conception. In a recent study [61] using 127 men from infertile couples with no known female factor, it was demonstrated that in men with normal standard semen parameters the odds ratio (OR) for infertility was significantly higher than in control patients when the percentage of sperm cells with DNA fragmentation was above 20% (OR 5.1, 95% confidence interval [CI] 1.2–23). Moreover, if one of the standard semen parameters was abnormal, the OR for infertility was significant above 10% (OR 16, 95% CI: 4.2–60). Such findings have been corroborated by similar studies that suggest that SDF above approximately 30% is associated with low success for natural conception and prolonged time to pregnancy [62].

Elevated values of SDF have also been associated with decreased success rates in intrauterine insemination (IUI). Sperm samples with SDF values over approximately 30% have been shown to reduced the efficacy of IUI from 16 to 4% [63] or lower [64]. Probably, one of the most robust studies investigating the influence of SDF on IUI outcome was conducted by Bungum and colleagues [65]. Using data collected from 387 cycles of insemination, the authors demonstrated that there was a significant decrease in the percentage of biochemical pregnancies, clinical pregnancies and deliveries (19.0–1.5%, OR 9.9, 95% CI 2.37–41.51, $p<0.001$) when SDF was above the 30% threshold.

The influence of SDF on the outcome of IVF and ICSI may perhaps have received the most attention [65–72]. A detailed study performed using the SCD test on 85 couples subjected to IVF and ICSI demonstrated that the percentage of sperm cells with fragmented DNA was inversely correlated with the fertilization rate of the oocyte (r −0.245, $p<0.05$). Higher DNA fragmentation was associated to type IV zygotes with asynchronous nucleolar precursor bodies (73.8 vs. 28.8%, $p<0.001$). Moreover, high SDF was correlated with slower embryo development and day-6 embryos classified as lower quality by morphological assessment (47.7 vs. 29.4%, $p<0.05$). Lastly, high DNA fragmentation was negatively correlated with implantation rate (r −0.250, $p<0.05$) [70]. This study was later expanded to 622 couples, collected from five clinics in France [72], and the results obtained were in line with those from the previous report.

Interestingly, despite the clear impact of SDF on fertilization and the development of the embryo, neither study found a significant correlation with pregnancy outcome in IVF or ICSI. Along the same lines, a systematic review and meta-analysis of nine IVF studies suggests that sperm DNA damage is only weakly associated with lower IVF pregnancy rates (combined OR 1.57, 95% CI 1.18–2.07, $p<0.05$ [4]). The same meta-analysis reviewing 11 ICSI studies revealed that sperm DNA damage is not associated with ICSI pregnancy rates (combined OR 1.14, 95% CI 0.86–1.54, $p=0.65$). The explanation for this apparent contradiction is that there exist several processes in these techniques that mitigate the effect of SDF: (1) Sperm selection by swim-up before IVF or ICSI reduces the percentage of sperm cells with DNA damage [73]; (2) The selection of sperm cells for ICSI based on morphology is likely to result in the selection of a sperm cell with minimal DNA fragmentation, as abnormal morphology has been shown to correlate with DNA damage and the presence of aneuploidies [74, 75]; (3) Since embryos with poor morphology and slower development are associated with SDF, it is likely that the embryos selected for transfer have resulted from fertilization by sperm cells with less DNA damage [70, 72]; (4) As we shall discuss below, SDF is a dynamic process that increases over time such that a semen sample assessed for SDF immediately after ejaculation will have a lower percentage of damaged sperm cells than when assessed following a few hours. In this way, the effect of SDF is much more pronounced in IUI where the time to fertilize the oocyte is much longer than IVF or ICSI [76, 77].

Thus, assessment of SDF may serve to evaluate the most appropriate assisted reproduction technique given that SDF is highly correlated with pregnancy outcome in IUI but not in IVF

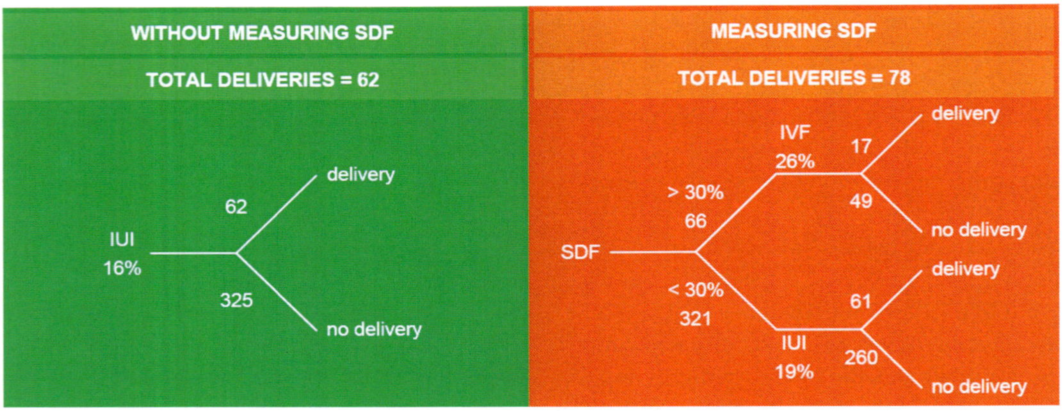

Fig. 10.4 Number of deliveries after the first cycle, taking into account SDF in the choice of assisted reproduction technique. Selecting IUI for all patients (*left panel*) yields a total of 62 live births, whereas only subjecting couples with high SDF to IVF and couples with low SDF to IUI (*right panel*) yields a total of 78 live births (adapted from Bungum et al. [65], with permission)

or ICSI. Couples presenting values of SDF above the 30% threshold should undergo IVF or ICSI in their first cycle, avoiding unnecessary IUI cycles. If one considers the results obtained by Bungum et al. [65] by selecting IVF rather than IUI in the first cycle for couples presenting SDF values over 30%, there is a significant increase from 62 to 78 deliveries, that is, a 25.8% increase in the efficacy of the first cycle of ART (Fig. 10.4).

Much debate remains and it is clear that the clinical applications of measuring SDF require more study. The long-term effects of SDF on techniques that bypass the natural selection barriers to fertilization such as ICSI are as yet unknown. A recent experimental study in mice has shown that ICSI performed using semen with a high percentage of cells with fragmented DNA resulted in reduced pre-implantation embryo development and less offspring [78]. Most interestingly, this study demonstrated that offspring from animals produced from semen with high SDF performed less well in a battery of behavioural tests than control animals. These animals also presented tumours and aged prematurely, suggesting that despite the ability of the oocyte to repair sperm DNA damage [79], incomplete repair may lead to long-term pathologies. In line with this, a recent study has demonstrated that a 10% increase in SDF increased the probabilities of not achieving pregnancy by an order of 1.31 times, but this effect was absent when using donor oocytes (Meseguer personal communication and submitted). This points to the fact that oocyte quality is a conditioning factor to be taken into account, as the capacity of oocytes to repair DNA lesions in both quantity and fidelity, may be compromised, especially in oocytes from older women or with certain fertility problems. The concurrence of undetected female factor may influence the results from the different reports measuring SDF and pregnancy outcome.

Lastly, as alluded previously, the majority of studies fail to take into account the progressive increase in sperm cells with fragmented DNA over time after ejaculation or thawing. The rate of SDF and shape of the curve of dynamic progression of SDF over time has a unique pattern, but remarkable differences may exist among individuals [80–82] and species [83]. Thus, the sperm DNA longevity may be quite different when different individuals are compared, and individuals with a similar baseline level of SDF may exhibit large differences when SDF is assessed some hours after ejaculation. A differential amount of iatrogenic SDF may therefore be embedded into the results cited in these studies depending on the time taken handling the sperm sample in the laboratory. This factor may partially explain the controversial correlations obtained in different reports when trying to establish correlations between sperm DNA damage and fertility or pregnancy outcome. Owing to its outstanding implications, the dynamic approach of SDF is further developed in a subsequent section.

The SCD Test in the Andrology Laboratory

Varicocele

Varicocele is the dilation of the pampiniform venous plexus above and around the testicle. It occurs in approximately 15–20% of the general male population, mainly in adolescents. Moreover, 19–41% of men seeking infertility treatment and around 80% of men with secondary infertility experience this pathology. Thus, this anatomical abnormality is perhaps one of the most common causes of poor sperm production and decreased semen quality. When the SCD test was applied to sperm samples collected from a group of infertile males with varicocele, it was found that 32.4 ± 2.3% of the spermatozoa had fragmented DNA [42]. These values are more than double those measured in control fertile subjects. Such values are similar to those obtained from infertile men with other pathologies. However, varicocele patients exhibit a higher proportion of degraded sperm cells (1 in every 4.2 cells) compared to fertile (1 in 8.2) or infertile patients with other pathologies [42]. The effect of increased SDF has been claimed to be a consequence of an increase in ROS production and a decrease in the antioxidant capacity [84–86]. Moreover, the dilated veins may produce high levels of nitric oxide and peroxinitrite, which also attack sperm DNA [87, 88].

Thus, varicocele promotes SDF in such a manner that nuclear injury tends to be very intense. Given that in certain cases varicocelectomy decreases the frequency of sperm cells with fragmented DNA and increases pregnancy rate [71, 89], while in other cases the difference between preoperative and postoperative values is not so evident [90], it should be of great interest to evaluate the presence of this degraded sperm class after surgery.

Genitourinary Infections

Chlamydia trachomatis is the most prevalent sexually transmitted bacterium with nearly 90 million cases detected worldwide annually. This infection is the main cause of subfertility in both males and females [91] and is frequently associated with other pathogens such as *Mycoplasma*. In males, *Chlamydia* is responsible for 50% of non-gonococcal urethritis and the majority of post-gonococcal urethritis. Furthermore, it may be associated with epididymitis, prostatitis and orchitis, as well as stenosis of the ducts. The standard semen parameters are only very subtly altered, so this cannot account for subfertility in infected males. In vitro studies of co-incubation of *Chlamydia* or its lipopolysaccharide with sperm cells demonstrated an induction of phosphatidylserine membrane translocation and DNA fragmentation [92, 93]. To gain information about the situation in vivo, 143 patients infected with *Chlamydia trachomatis* and *Mycoplasma* were evaluated for standard semen parameters and SDF using the SCD test [94]. While the traditional semen parameters were only slightly affected, infected males displayed a percentage of sperm cells with DNA fragmentation of 35.2 ± 13.5%; that is, 3.2 times higher than in the control fertile group (10.8 ± 5.6%). A group of 95 patients was then further evaluated after antibiotic therapy, and the mean frequency of sperm cells with fragmented DNA significantly decreased from 37.7 ± 13.6 to 24.2 ± 11.2% [94]. This improvement was most pronounced after the first 3 months of treatment. These results suggest that the improvement in the DNA integrity of sperm cells after therapy could underlie an improvement in pregnancy rates. The mechanism of DNA fragmentation in vivo following infection may be complex. The bacterium's own components or toxins may induce the DNA fragmentation. Moreover, the accompanying acute or chronic inflammatory reaction in the genital tract may result in oxidative stress by overproduction of ROS by the epithelium or activated leukocytes. Local heat and systemic fever may also have an influence. If this is true, other genitourinary infections originated by different bacteria [95], viruses, fungi such as *Candida albicans* [96] or protozoa could also affect sperm DNA integrity. As demonstrated in the *Chlamydia* infection, the SCD test may be useful to evaluate the possible affectation of sperm DNA integrity and its recovery after therapy.

Sperm DNA Damage and Cancer

Induction of DNA damage is the main mechanism of cell death produced by most drugs or local radiotherapy used for cancer treatment. It is known that cancer itself is linked to disruption of spermatogenesis [97] and that chemotherapy usually results in temporary or permanent azoospermia. The determination of SDF may be useful to monitor the toxicogenetic effect of cancer therapy on sperm cells and to evaluate their recovery in terms of DNA integrity [98]. Sperm cryopreservation before radio/chemo-treatment remains the best option for cancer patients to preserve their fertility. With the introduction of IVF and ICSI, even the poorer sperm samples might be frozen with good expectations of success [99]. In spite of this, the quality of sperm DNA may be affected in tumorous cancers (non-seminoma type), seminoma and others. The mean SDF in these patients was 35.8%, which is comparable to what has been reported in infertile patients, and higher than that of fertile donors. The percentage of SDF was 46.2% in leukaemia and 48.8% for other types, but was lowest in Hodgkin lymphoma (28.08%). A recent study with the SCSA has also reported similar results [100]. In conclusion, the presence of cancer, regardless of its origin, affects sperm DNA quality and could perhaps be an underlying cause of temporary infertility. SDF should therefore be evaluated in the sperm samples to be frozen before therapy, in order to choose those samples with the best DNA quality.

Azoospermia

Azoospermia may be due to testicular failure or due to duct obstruction. In any case, foci of spermatogenesis may still exist within the testicle, and so, sperm cells may be obtained from testicular biopsies. The SCD technique is especially adequate to analyze samples with low amounts of spermatozoa and much debris. Testicular sperm samples from 62 patients were analyzed with the SCD test. The patients with obstructive azoospermia ($n = 40$) showed 35.9 ± 2.6% of sperm cells with fragmented DNA, whereas those with non-obstructive azoospermia ($n = 22$) revealed 46.9 ± 4.5% of cells with SDF [101]. Thus, the incidence of DNA damage in testicular sperm populations from infertile men with azoospermia is much lower in normal and active spermatogenic testis than in testis with incomplete sperm production. A recent study by Smit et al. [102] has also confirmed that SDF is higher in patients with poor spermatogenesis than in those with normal spermatogenesis. It is possible that defective spermatozoa are sensed by a genomic screening mechanism that triggers DNA fragmentation to genetically inactivate sperm cells with a defective genomic makeup. In fact, sperm cells containing aneuploidies are more prone to contain fragmented DNA [43]. A study by Greco et al. [103] showed that the incidence of DNA fragmentation was lower in testicular spermatozoa compared with ejaculated spermatozoa, proposing its use in ICSI for patients with high levels of SDF in the ejaculate. Both studies clearly show that sperm DNA damage may be detected just after finishing telophase II at the onset of spermiogenesis or can occur during the epididymal sperm passage.

Toxicogenetics

Reproductive toxicology is a discipline of remarkable interest, with strong implications on the potential adverse reproductive health effects of exposure to internal or environmental toxic agents. SDF is an ideal parameter to monitor, as it is a very sensitive marker of reproductive toxicants. Many agents that affect germs cells at different stages of meiosis or spermiogenesis induce genome modifications that will later be translated as DNA fragmentation in the sperm cell [104]. For example, exposure to anticancer chemotherapy [98], air pollution [105], pesticides such as

DDT [106], mobile phone radiation [107], and treatment with the serotonine reuptake inhibitor paroxetine [108], have all been shown to induce SDF. Interestingly, in many cases, DNA fragmentation is observed without any significant effect on standard seminal parameters. In a study by Viloria et al. [109], 99 males provided semen samples that were analyzed by the SCD test before and after swim-up treatment. The results were correlated with the patient's cigarette smoking habits. Although no differences were detected before swim-up, in the capacitated samples, smokers and especially heavy smokers (≥20 cigarettes per day) showed significantly impaired DNA quality compared to non-smokers. The fact that differences are observed after swim-up but not in the ejaculate may be due to the fact that the incubation time necessary for the swim-up technique allows cryptic DNA damage to be expressed. This highlights the potential interest in a dynamic evaluation of DNA fragmentation as a more sensitive assay for reproductive toxicology.

The effect of vaccination on SDF was assessed in rams vaccinated with Miloxan (*Clostridium perfringens* type C, D and *C. oedematiens* type B), using the SCD test [110]. Miloxan increased the percentage of sperm cells with fragmented DNA by tenfold on average (from 6.5±7.9 to 63.4±24.2%). However, the negative impact of vaccination on SDF was reversible, decreasing to 21.7±10.6% 40 days after vaccination. The effects of vaccination on sperm quality and particularly on sperm DNA integrity probably consist of many factors and effectors, such as the genetic background, and the capacity to respond to oxidative stress or temperature variations. This result has important implications in the use of semen samples from vaccinated animals and the same implications for post-vaccination in humans.

Sperm DNA Fragmentation Dynamics

Semen parameters such as motility, viability, etc., are usually evaluated once at different periods in time after sperm collection. However, these values may change during the useful lifespan of a sperm sample. Measurements are therefore of value when performed (1) at the time of ejaculation and (2) at the time of insemination, IVF or intracytoplasmic injection. Usually, ART logistics generate a time lapse between both periods and a clear reference to the time of assessment is generally not precisely stated in the literature. Similarly, when values for SDF are quoted, clear references about the time of assessment following ejaculation are seldom included.

When SDF is assessed immediately after ejaculation, using the SCD test, donors with proven fertility show significantly less fragmentation than infertile patients (Fig. 10.5a, b). One would assume that sperm DNA is unstable when maintained in a para-biological environment such as those used to store a semen sample after ejaculation. The conditions of sperm storage influence the sperm DNA longevity and a certain amount of iatrogenic DNA damage is to be expected. Some reports indicate that when the kinetics of sperm DNA damage are analyzed, DNA degrades progressively when incubated in identical conditions to those used for IVF. The use of semen samples incubated at 37°C during a period of 24 h produce a cumulative increase in the level of the DNA in the order of 2–8% during the first 4 h of incubation [77] depending on the individual analyzed. In donors of proven fertility, the rate of SDF behaves independently of other sperm parameters such as the dynamic loss of sperm viability, although a certain degree of negative correlation exists [80].

This dynamic loss of sperm DNA quality has also been observed in other animal species such as stallion [27], ram [29], boar [34, 111], donkey [28], rhinoceros [41], koala [30, 31], echidnas [32] and fish [35]. In all species analyzed to date, two important factors must be taken into account (1) the existence of a large variation in the species-specific rate for SDF and (2) the variability in the inter-individual rate for SDF. Thus, while in fish the increase in SDF is triggered after a few minutes of sperm activation, in boar, the increase SDF is triggered after days incubated at 37°C in the appropriate semen extender. In humans, there exists large

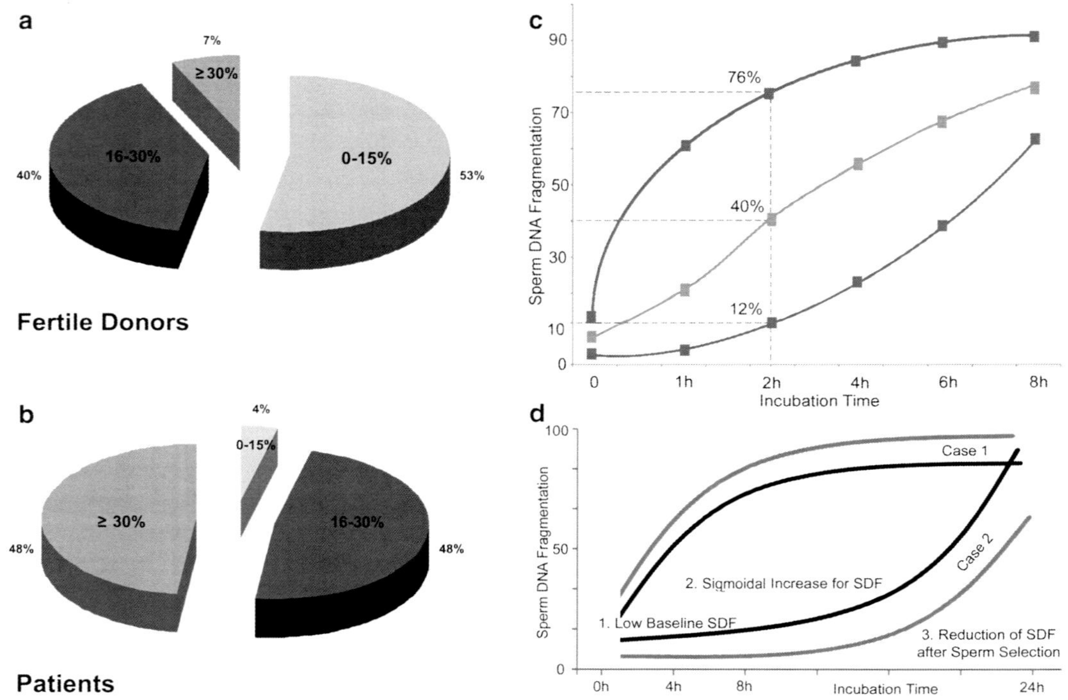

Fig. 10.5 Distribution of SDF a time 0 (baseline SDF) in fertile donors (**a**; $n=55$) and patients (**b**; $n=75$). Distribution ranges were fixed to <15, 16–30 and >31% of SDF (**c**) Different values for SDF obtained at different incubation times in three individuals showing different dynamics for SDF. Note that large differences are obtained at different incubation times (values at 2 h are represented). (**d**) Recommended criteria to discriminate between a "good" and a "bad" sperm sample considering the dynamic behaviour of the SDF after sperm selection

variability in the rate of SDF from individual to individual. Thus, sperm samples with a similar level of SDF as measured immediately after ejaculation will behave differently when incubated at 37°C. As shown in Fig. 10.5c, the SDF level obtained after 4 h of incubation is 15% in one individual and 70% in the other. The general figure depicts three main patterns for SDF increase that can be adjusted to a logarithmic, linear or sigmoidal curve (Fig. 10.5d). Individuals presenting a sigmoidal tendency for the increase in SDF would have a lower percentage of sperm cells with damaged DNA at any given incubation time. Thus, as depicted in Fig. 10.5d, the best donor would be one that (1) presented the lowest level of baseline sperm DNA damage, (2) exhibited a sigmoidal tendency for increase in SDF and (3) showed a decrease in the level of SDF but maintained the sigmoidal tendency after sperm selection (swim-up or gradient). The analysis of the rate of SDF increase may provide useful information when used for IVF or IUI. Although this requires further inspection, there exists the possibility that the dynamic increase of SDF serves as a possible explanation to some of the discrepancies observed in the literature about the role of SDF and ART outcome.

The first clue about the impact of the dynamics of sperm DNA damage was offered by Young et al. [112]. The authors of this study demonstrated that semen collection away from the laboratory with overnight mail delivery could lead to sperm DNA damage and this had subsequent implications on fertilization. In particular, the longevity of the DNA molecule could be highly compromised in cases such as the use of samples from testicular sperm extraction or aspiration. In testicular sperm from men with obstructive azoospermia, DNA fragmentation after cryopreservation is increased by 4 and 24-h incubations, and this effect is intensified by post-thaw incubation. In such circumstances, it is recommended that testicular sperm samples for ICSI should be used with the minimum delay in sperm capacitation [113, 114]. Bungum et al. [115] found that co-incubation of 777 sibling oocytes from 81 women undergoing IVF produced good fertilization rates using co-incubation for either 30 s or for 90 min and significantly lower rates of polyspermy. All these inputs indicate that there may be beneficial effects for short sperm/oocyte co-incubation in IVF. Although more relevant studies are needed, taking into account the dynamic increase of SDF, the probability of fertilization with a damaged sperm would diminish using short incubation periods.

The debate over whether cryopreservation induces direct damage on the DNA molecule is still open. A comparison of the dynamics of SDF in fresh and cryopreserved semen samples from the same donor showed that sperm DNA tends to degrade very quickly after thawing. In practice, sperm DNA degradation could be detected at the onset of thawing and temperature recovery to 37°C. However, large differences in the level of SDF were not observed when the semen sample was assessed for SDF just after thawing [82]. This indicates that cryopreservation does not change the baseline rate of SDF when analyzed just after thawing but may change the dynamics of SDF [29, 80].

In conclusion, the dynamic behaviour of SDF indicates that when the semen sample is used for IUI or IVF, the level of SDF may be higher at the time of fertilization than when assessed in the clinical practice. In natural reproduction and IUI, only a small fraction of the sperm cells will enter the cervix, pass into the uterus, and progress to the uterotubal junctions to reach the Fallopian tubes. In this environment, the selected sperm fraction is maintained in a fully functional state by connecting with endosalpingeal epithelium [116, 117]. To reduce the delay in fertilization and mitigate the effect of a rapid rate of SDF, full synchronization of the oestrus and time of insemination is required, reducing the handling of semen ex vivo. The role of semen plasma in connection with the female tract and its implications in sperm protection for SDF is largely unknown. There are indications that semen plasma proteins are absent in the oviduct. This indicates that their presence is probably restricted to uterine environments and not to other female reproductive regions closer to the oocyte [118]. These considerations should be taken into account when making extrapolations about the stability of sperm DNA ex vivo and in vivo.

Finally, we want to draw attention to the fact that the comparison of results for SDF from different laboratories or even those obtained within the same laboratory may be biased if clear references to the time of measurement are not precisely given. This could be aggravated if details of the storage or thawing conditions are not clearly communicated.

Conclusion: Value of the SCD Test

Sperm DNA damage has been connected, among other things, with an increased incidence of miscarriage and enhanced risk of disease in the offspring. However, its occurrence is multifaceted, and many of the variable consequences it has for fertility are as yet not fully understood [119–121]. Fertility is a multifactorial phenomenon that usually involves both members of the couple, and assessment of sperm DNA integrity is only one piece of a complex puzzle. Tests that assess sperm quality should identify not only the ability of spermatozoa to reach the oocyte with an intact DNA molecule but also their ability to fertilize the oocyte and activate embryo growth. To paraphrase Makhlouf and Niederberger [122] when referring to the sperm as a whole functional cell, it is not just the carrier but also the content that is important. With the appearance of ICSI, however, the content seems to have taken a preponderant role. SDF should therefore be considered a parameter of sperm quality. Its determination may provide beneficial information in andrological pathology, complementary to that obtained from standard semen parameters. SDF must be evaluated concurrently and examined within the clinical context of each patient or couple.

Compared to other methods of assessing DNA fragmentation, the SCD test can be conducted promptly and without the need for complex and expensive laboratory equipment. The SCD test is a powerful and versatile approach for investigating DNA fragmentation, allowing the assessment of damaged DNA over a diverse range of clinical situations. The technique can be easily adapted to incorporate new research directions, and the analysis of sperm DNA can be performed on wide range of species. The SCD test has the unique ability to allow direct observations to be made of the spermatozoa and the corresponding DNA damage; this technical advantage allows direct correlations between DNA fragmentation and DNA sequence variations, nucleotide modification and/or protein status. The SCD is a procedure that allows researchers the flexibility to use their creative imagination when designing and conducting experiments to disentangle the obscure topic of sperm DNA damage.

References

1. Agarwal A, Allamaneni SS. The effect of sperm DNA damage on assisted reproduction outcomes. A review. Minerva Ginecol. 2004;56:235–45.
2. Zini A, Libman J. Sperm DNA damage: clinical significance in the era of assisted reproduction. CMAJ. 2006;175:495–500.
3. Shafik A, Shafik AA, El Sibai O, et al. Sperm DNA fragmentation. Arch Androl. 2006;52:197–208.
4. Zini A, Sigman M. Are tests of sperm DNA damage clinically useful? Pros and Cons. J Androl. 2009;30: 219–29.
5. Evenson DP, Jost L, Baer RK, et al. Individuality of DNA denaturation patterns in human sperm as measured by the sperm chromatin structure assay. Reprod Toxicol. 1991;5:115–25.
6. Evenson DP, Jost LK, Marshal D, et al. Utility of the sperm chromatin structure assay as a diagnostic and prognostic tool in the human fertility clinic. Hum Reprod. 1999;14:1039–49.
7. Evenson DP, Larson KL, Jost LK. Sperm chromatin structure assay: its clinical use for detecting sperm DNA fragmentation in male infertility and comparisons with other techniques. J Androl. 2002;23:25–43.
8. Gorczyca W, Gong J, Darzynkiewicz Z. Detection of DNA strand breaks in individual apoptotic cells by the in situ terminal deoxynucleotidyl transferase and nick translation assays. Cancer Res. 1993;53:945–51.
9. Manicardi GC, Bianchi PG, Pantano S, et al. Underprotamination and nicking of DNA in ejaculated human spermatozoa are highly related phenomena. Biol Reprod. 1995;52:864–7.
10. Singh NP, Danner DB, Tice RR, et al. Abundant alkali-sensitive sites in DNA of human and mouse sperm. Exp Cell Res. 1989;184:461–70.
11. Tice RR, Agurell E, Anderson D, et al. Single cell gel/comet assay: guidelines for in vitro and in vivo genetic toxicology testing. Environ Mol Mutagen. 2000;35:206–21.
12. Rivero MT, Vázquez-Gundín F, Muriel L, et al. Patterns of DNA migration in two-dimensional single-cell gel electrophoresis analyzed by DNA breakage detection-fluorescence in situ hybridization. Environ Mol Mutagen. 2003;42:223–7.
13. Enciso M, Sarasa J, Agarwal A, et al. A two-tailed Comet assay for assessing DNA damage in spermatozoa. Reprod Biomed Online. 2009;18:609–16.
14. Fernández JL, Muriel L, Rivero MT, et al. The sperm chromatin dispersion test: a simple method for the determination of sperm DNA fragmentation. J Androl. 2003;24:59–66.
15. Fernández JL, Muriel L, Goyanes V, et al. Simple determination of human sperm DNA fragmentation with an improved sperm chromatin dispersion test. Fertil Steril. 2005;84:833–42.
16. Fernández JL, Muriel L, Goyanes V, et al. Halosperm® is an easy, available and cost-effective alternative to determine sperm DNA fragmentation. Fertil Steril. 2005;884:883.
17. Álvarez JG, Lewis S. Sperm chromatin structure assay parameters measured after density gradient centrifugation are not predictive of the outcome of ART. Hum Reprod. 2008;23:1235–6.
18. Gosálvez J, López-Fernández C, Ferrucci L, et al. DNA base sequence is not the only factor for restriction endonuclease activity on metaphase chromosomes: evidence using isoschizomeres. Cytogenet Cell Genet. 1989;50:142–4.
19. Mitchell LA, De Iuliis GN, Aitken RJ. The TUNEL assay consistently underestimates DNA damage in human spermatozoa and is influenced by DNA compaction and cell vitality: development of an improved methodology. Int J Androl. 2011;34:2–3. doi:10.1111/j.1365-2605.2009.01042.
20. Domínguez-Fandos D, Camejo MI, Ballesca JL, et al. Human sperm DNA fragmentation: correlation of TUNEL results as assessed by flow cytometry and optical microscopy. Cytom A. 2007;71:1011–8.
21. Ahnström G. Techniques to measure DNA strand breaks in cells: a review. Int J Radiat Biol. 1988;54: 695–707.
22. Aravindan GR, Bjordahl J, Jost LK, et al. Susceptibility of human sperm to in situ DNA denaturation is strongly correlated with DNA strand breaks identified by single-cell electrophoresis. Exp Cell Res. 1997;236:231–7.
23. Chohan KR, Griffin JT, Lafromboise M, et al. Comparison of chromatin assays for DNA fragmentation evaluation in human sperm. J Androl. 2006; 27:53–9.

24. Ahnström G, Erixon K. Radiation-induced single-strand breaks in DNA determined by rate of alkaline strand separation and hydroxylapatite chromatography: an alternative to velocity sedimentation. Int J Radiat Biol. 1973;36:197–9.
25. Cook PR, Brazell IA. Supercoils in human DNA. J Cell Sci. 1975;19:261–79.
26. Enciso M, López-Fernández C, Fernández JL, et al. A new method to analyze boar sperm DNA fragmentation under bright-field or fluorescence microscopy. Theriogenology. 2006;65:308–16.
27. López-Fernández C, Crespo F, Arroyo F, et al. Dynamics of sperm DNA fragmentation in domestic animals II: the stallion. Theryogenology. 2007;68:1240–50.
28. Cortés-Gutiérrez E, Crespo F, Gosálvez J, et al. DNA fragmentation in frozen sperm of *Equus asinus*: Zamorano-Leonés, a breed at risk of extinction. Theryogenology. 2008;8:1022–32.
29. López-Fernández C, Fernández JL, Gosálbez A, et al. Dynamics of sperm DNA fragmentation in domestic animals III. Ram. Therigenology. 2008;70:898–908.
30. Johnston SD, López-Fernández C, Gosálbez A, et al. The relationship between sperm morphology and chromatin integrity in the koala (*Phascolarctos cinereus*) as assessed by the sperm chromatin dispersion (SCD) test. J Androl. 2007;28:891–9.
31. Zee YP, López-Fernández C, Arroyo F, et al. Evidence that single-stranded DNA breaks are a normal feature of koala sperm chromatin, while double-stranded DNA breaks are indicative of DNA damage. Reproduction. 2009;138:267–78.
32. Johnston SD, López-Fernández C, Gosálbez A, et al. Directional mapping of DNA nicking in ejaculated and cauda epididymidal spermatozoa of the short-beaked echidna (*Tachyglossus aculeatus*: Monotremata). Reprod Fertil Dev. 2009;21:1–7.
33. Fernández JL, Gosálvez J, Santiso R, et al. Adaptation of the sperm chromatin dispersion (SCD) test to determine DNA fragmentation from of bovine sperm. In: Valentino RG, editor. New cell apoptosis research. New York: Nova Biochemical Books; 2007. p. 69–84.
34. Pérez-Llano B, López Fernández C, García-Casado P, et al. Dynamics of sperm DNA fragmentation in the boar: ejaculate and temperature effects. Anim Reprod Sci. 2010;119:235–43. doi:10.1016/j.anireprosci. 2010.01.002.
35. López-Fernández C, Gage MJ, Arroyo F, et al. Rapid rates of sperm DNA damage after activation in tench (*Tinca tinca*: Teleostei, Cyprinidae) measured using a sperm chromatin dispersion test. Reproduction. 2009;138:257–66.
36. Fernández JL, Gosálvez J. Application of FISH to detect DNA damage: DNA breakage detection-FISH (DBD-FISH). In: Didenko V, editor. In situ detection of DNA damage: methods and protocols. Methods in Molecular Biology, vol. 203. Totowa: Humana; 2002. p. 203–16.
37. Fernández JL, Goyanes V, Segrelles E, et al. DNA breakage detection-fluorescence in situ hybridization (DBD-FISH): an in situ procedure to explore genome organization and DNA damage within specific DNA sequence areas. Curr Top Genet. 2005;1:139–43.
38. Fernández JL, Goyanes V, Gosálvez J. Determination of sperm DNA fragmentation into the IVF and andrology laboratory, with the halosperm® kit. Clin Embriologist. 2006;9:23–7.
39. Santiso R, Tamayo M, Gosálvez J, et al. Simultaneous determination in situ of DNA fragmentation and 8-oxoguanine in human sperm. Fertil Steril. 2010;93:314–8.
40. Zhang LH, Qiu Y, Wang KH, et al. Measurement of sperm DNA fragmentation using bright-field microscopy: comparison between sperm chromatin dispersion test and terminal uridine nick-end labeling assay. Fertil Steril. 2010;94:1027–32. doi:10.1016/j.fertnstert.2009.04.034 DOI:dx.doi.org.
41. Portas T, Johnston S, Hermes R, et al. Frozen-thawed rhinoceros sperm exhibit DNA damage shortly after thawing when assessed by the sperm chromatin dispersion assay. Theriogenology. 2009;72:711–20.
42. Enciso M, Muriel L, Fernández JL, et al. Infertile men with varicocele show a high relative proportion of sperm cells with more intense nuclear damage level in the population with fragmented DNA, evidenced by the sperm chromatin dispersion (SCD) test. J Androl. 2006;27:106–11.
43. Muriel L, Goyanes V, Segrelle E, et al. Increased aneuploidy rate in sperm with fragmented DNA as determined by the sperm chromatin dispersion (SCD) test and fish analysis. J Androl. 2007;28:38–49.
44. Wyrobek A, Eskenazi B, Young S, et al. Advancing age has differential effects on DNA damage, chromatin integrity, gene mutations, and aneuploidies in sperm. PNAS. 2006;103:9601–6.
45. Breen AP, Murphy JA. Reactions of oxyl radicals with DNA. Free Radic Biol Med. 1995;18:1033–77.
46. Kasai H. Analysis of a form of oxidative DNA damage, 8-hydroxy-2′-deoxyguanosine, as a marker of cellular oxidative stress during carcinogenesis. Mutat Res. 1997;387:147–63.
47. Bespalov IA, Bond JP, Purmal AA, et al. Fabs specific for 8-oxoguanine: control of DNA binding. J Mol Biol. 1999;293:1085–95.
48. Gao D, Wei C, Chen L, et al. Oxidative DNA damage and DNA repair enzyme expression are inversely related in murine models of fatty liver disease. Am J Physiol Gastrointest Liver Physiol. 2004;287:1070–7.
49. Meseguer M, Martínez-Conejero JA, O'Connor JE, et al. The significance of sperm DNA oxidation in embryo development and reproductiveoutcome in an oocyte donation program: a new model to study a male infertilityprognostic factor. Fertil Steril. 2008;89:1191–9.
50. Li E. Chromatin modification and epigenetic reprogramming in mammalian development. Nat Rev Genet. 2002;3:662–73.
51. Klose RJ, Bird AP. Genomic DNA methylation: the mark and its mediators. Trends Biochem Sci. 2006;31:89–97.

52. Benchaib M, Braun V, Ressnikof D, et al. Influence of global sperm DNA methylation on IVF results. Hum Reprod. 2005;20:768–73.
53. Santiso R, Muriel L, Goyanes V, et al. Evidence of modified nuclear protein matrix in human spermatozoa with fragmented deoxiribonucleic acid. Fertil Steril. 2006;87:191–4.
54. De La Torre J, López Fernández C, Pita M, et al. Simultaneous observation of DNA fragmentation and protein loss in the boar spermatozoon following application of the sperm chromatin dispersion (SCD) test. J Androl. 2007;28:533–40.
55. Nuñez R, López-Fernández C, Arroyo F, et al. Characterization of sperm DNA damage in Kartagener´s syndrome with recurrent fertilization failure: case revisited. Sex Reprod Health. 2010;1:73–5.
56. Alvarez JG, Sharma RK, Ollero M, et al. Increased DNA damage in sperm from leukocytospermic semen samples as determined by the sperm chromatin structure assay. Fertil Steril. 2002;78:319–29.
57. Henkel R, Kierspel E, Stalf T, et al. Effect of reactive oxygen species produced by spermatozoa and leukocytes on sperm functions in non-leukocytospermic patients. Fertil Steril. 2005;83:635–42.
58. De Ambrogi M, Spinaci M, Galeati G, et al. Viability and DNA fragmentation in differently sorted boar spermatozoa. Theriogenology. 2006;66:1994–2000.
59. Gosálvez J, Gosálbez A, Arroyo F, et al. Assessing sperm DNA fragmentation in the field: an adaptation of sperm chromatin dispersion technology. Biotech Histochem. 2008;83:247–52.
60. Santiago-Moreno J, Toledano-Díaz T, Pulido-Pastor A, et al. Birth of live Spanish ibex (*Capra pyrenaica* hispanica) derived from artificial insemination with epididymal spermatozoa retrieved after death. Theriogenology. 2006;66:283–91.
61. Giwercman A, Lindstedt L, Larsson M, et al. Sperm chromatin structure assay as an independent predictor of fertility in vivo: a case-control study. Int J Androl. 2009;33:e221–7. doi:10.1111/j.1365-2605.2009.00995.x.
62. Evenson DP. Data analysis of two in vivo fertility studies using sperm chromatin structure assay-derived DNA fragmentation index vs. pregnancy outcome. Fertil Steril. 2008;90:1229–31.
63. Bungum M, Humaidan P, Spano M, et al. The predictive value of sperm chromatin structure assay (SCSA) parameters for the outcome of intrauterine insemination, IVF and ICSI. Hum Reprod. 2004;19: 1401–8.
64. Duran EH, Morshedi M, Taylor S, et al. Sperm DNA quality predicts intrauterine insemination outcome: a prospective cohort study. Hum Reprod. 2002;17: 3122–8.
65. Bungum M, Humaidan P, Axmon A, et al. Sperm DNA integrity assessment in prediction of assisted reproduction technology outcome. Hum Reprod. 2007;22:174–9.
66. Benchaib M, Lornage J, Mazoyer C, et al. Sperm deoxyribonucleic acid fragmentation as a prognostic indicator of assisted reproductive technology outcome. Fertil Steril. 2007;87:93–100.
67. Frydman N, Prisant N, Hesters L, et al. Adequate ovarian follicular status does not prevent the decrease in pregnancy rates associated with high sperm DNA fragmentation. Fertil Steril. 2008;89:92–7.
68. Gandini L, Lombardo F, Paoli D, et al. Full-term pregnancies achieved with ICSI despite high levels of sperm chromatin damage. Hum Reprod. 2004;19: 1409–17.
69. Check JH, Graziano V, Cohen R, et al. Effect of an abnormal sperm chromatin structural assay (SCSA) on pregnancy outcome following (IVF) with ICSI in previous IVF failures. Arch Androl. 2005;51:121–4.
70. Muriel L, Garrido N, Fernández JL, et al. Value of the sperm deoxyribonucleic acid fragmentation level, as measured by the sperm chromatin dispersion test, in the outcome of in vitro fertilization and intracytoplasmic sperm injection. Fertil Steril. 2006;85:371–83.
71. Zini A, Meriano J, Kader K, et al. Potential adverse effect of sperm DNA damage on embryo quality after ICSI. Hum Reprod. 2005;20:3476–80.
72. Velez de la Calle JF, Muller A, Walschaerts M, et al. Sperm deoxyribonucleic acid fragmentation as assessed by the sperm chromatin dispersion test in assisted reproductive technology programs: results of a large prospective multicenter study. Fertil Steril. 2008;90:1792–9.
73. Spano M, Bonde JP, Hjollund HI, et al. Sperm chromatin damage impairs human fertility. The Danish First Pregnancy Planner Study Team. Fertil Steril. 2000;73:43–50.
74. Varghese AC, Bragais FM, Mukhopadhyay D, et al. Human sperm DNA integrity in normal and abnormal semen samples and its correlation with sperm characteristics. Andrologia. 2009;41:207–15.
75. Tang SS, Gao H, Zhao Y, et al. Aneuploidy and DNA fragmentation in morphologically abnormal sperm. Int J Androl. 2010;33:163–79.
76. Gosálvez J, Fernández JL, Gosálbez A, et al. Dynamics of sperm DNA fragmentation in mammalian species as assessed by the SCD methodology. Fertil Steril. 2007;88:S365–1365.
77. Gosálvez J, Cortés-Gutierrez E, López-Fernádez C, et al. Sperm DNA fragmentation dynamics in fertile donors. Fertil Steril. 2009;92:170–3.
78. Fernández-González R, Moreira PN, Pérez-Crespo M, et al. Long-term effects of mouse intracytoplasmic sperm injection with DNA-fragmented sperm on health and behaviour of adult offspring. Biol Reprod. 2008;78:761–72.
79. Ahmadi A, Ng SC. Fertilizing ability of DNA-damaged spermatozoa. J Exp Zool. 1999;284:696–704.
80. Gosálvez J, Cortés-Gutiérrez E, Rocio Nuñez R, et al. A dynamic assessment of sperm DNA fragmentation versus sperm viability in proven fertile human donors. Fertil Steril. 2009;92:1915–9.
81. Gosálvez J, Núñez R, Fernández JL, et al. Dynamics of sperm DNA damage in fresh versus frozen-thawed

and gradient processed ejaculates in human donors. Andrologia. 2010 (in press).

82. Gosálvez J, de la Torre J, López-Fernández C, et al. DNA fragmentation dynamics in fresh versus frozen thawed 1 plus gradient-isolated human spermatozoa. Syst Biol Reprod Med. 2010;56:27–36.
83. Fernández JL, Vélez de la Calle JF, Tamayo M, et al. Sperm DNA integrity and male infertility: current perspectives. Arch Med Sci. 2009;1A:S55–62.
84. Barbieri ER, Hidalgo ME, Venegas A, et al. Varicocele associated decrease in antioxidant defenses. J Androl. 1999;20:713–7.
85. Hendin BN, Kolettis PN, Sharma RK, et al. Varicocele is associated with elevated spermatozoal reactive oxygen species production and diminished seminal plasma antioxidant capacity. J Urol. 1999;161:1831–4.
86. Allamaneni SS, Naughton CK, Sharma RK, et al. Increased seminal reactive oxygen species levels in patients with varicoceles correlate with varicocele grade but not with testis size. Fertil Steril. 2004;82: 1684–6.
87. Mitropoulos D, Deliconstantinos G, Zervas A, et al. Nitric oxide synthase and xanthine oxidase activities in the spermatic vein of patients with varicocele: a potential role for nitric oxide and peroxynitrite in sperm dysfunction. J Urol. 1996;156:1952–8.
88. Romeo C, Ientile R, Santoro GI, et al. Nitric oxide production is increased in the spermatic veins of adolescents with left idiopathic varicocele. J Pediatr Surg. 2001;36:389–93.
89. Smit M, Romjin JC, Wildhagen MF, et al. Decreased sperm DNA fragmentation after surgical varicocelectomy is associated with increased pregnancy rate. J Urol. 2010;183:270–4.
90. Grober E, O'brien J, Jarvi KA, et al. Preservation of testicular arteries during subinguinal microsurgical varicocelectomy: clinical considerations. J Androl. 2004;25:740–3.
91. World Health Organization. Global prevalence and incidence of selected curable sexually transmitted infections: overview and estimates. Geneva: WHO; 2001. p. 1–43.
92. Eley A, Hosseinzadeh S, Hakimi H, et al. Apoptosis of ejaculated human sperm is induced by co-incubation with *Chlamydia trachomatis* lipopolysaccharide. Hum Reprod. 2005;20:2601–7.
93. Satta A, Stivala A, Garozzo A, et al. Experimental *Chlamydia trachomatis* infection causes apoptosis in human sperm. Hum Reprod. 2006;21:134–7.
94. Gallegos G, Ramos B, Santiso R, et al. Sperm DNA fragmentation in infertile men with genitourinary infection by *Chlamydia trachomatis* and *Mycoplasma* sp. Fertil Steril. 2008;90:328–34.
95. Moretti E, Capitani S, Figura N, et al. The presence of bacteria species in semen and sperm quality. J Assist Reprod Genet. 2009;26:47–56.
96. Burrello N, Calogero AE, Perdichizzi A, et al. Inhibition of oocyte fertilization by assisted reproductive techniques and increased sperm DNA fragmentation in the presence of *Candida albicans*: a case report. Reprod Biomed Online. 2004;8:569–73.
97. Agarwal A, Allamaneni SS. Disruption of spermatogenesis by the cancer disease process. J Natl Cancer Inst Monogr. 2005;34:9–12.
98. Delbes G, Hales BF, Robaire B. Effects of the chemotherapy cocktail used to treat testicular cancer on sperm chromatin integrity. J Androl. 2007;28: 241–9.
99. Meseguer M, Molina N, Garcia-Velasco JA, et al. Sperm cryopreservation in oncological patients: a 14-year follow-up study. Fertil Steril. 2006;85: 640–5.
100. Said TM, Tellez S, Evenson DP, et al. Assessment of sperm quality, DNA integrity and cryopreservation protocols in men diagnosed with testicular and systemic malignancies. Andrologia. 2009;41:377–82.
101. Meseguer M, Santiso R, Garrido N, et al. Sperm DNA fragmentation levels in testicular sperm samples from azoospermic males, assessed by the sperm chromatin dispersion (SCD) test. Fertil Steril. 2009; 92:1638–45.
102. Smit M, Romijn JC, Wildhagen MF, et al. Sperm chromatin structure is associated with the quality of spermatogenesis in infertile patients. Fertil Steril. doi:10.1016/j.fertnstert.2009.10.030.
103. Greco E, Scarselli F, Iacobelli M, et al. Efficient treatment of infertility due to sperm DNA damage by ICSI with testicular spermatozoa. Hum Reprod. 2005;20:226–30.
104. Evenson DP, Wixon R. Environmental toxicants cause sperm DNA fragmentation as detected by the sperm chromatin structure assay (SCSA). Toxicol Appl Pharmacol. 2005;207:S532–7.
105. Rubes J, Selevan SG, Evenson DP, et al. Episodic air pollution is associated with increased DNA fragmentation in human sperm without other changes in semen quality. Hum Reprod. 2005;20:2776–83.
106. De Jager C, Aneck-Hahn NH, Bornman MS, et al. Sperm chromatin integrity in DDT-exposed young men living in a malaria area in the Limpopo Province, South Africa. Hum Reprod. 2009;24:2429–38.
107. De Iuliis GN, Newey RJ, King BV, et al. Mobile phone radiation induces reactive oxygen species production and DNA damage in human spermatozoa in vitro. PLoS One. 2009;4:6446.
108. Tanrikut C, Feldman AS, Altemus M, et al. Adverse effect of paroxetine on sperm. Fertil Steril. 2010; 94:1021–6. doi:10.1016/j.fertnstert.2009.04.039 DOI: dx.doi.org.
109. Viloria T, Garrido N, Fernández JL, et al. Sperm selection by swim-up in terms of DNA fragmentation as measured by the sperm chromatin dispersion (SCD) test is altered in heavy smokers. Fertil Steril. 2007;88:523–5.
110. Gosálvez J, Vázquez JM, Enciso M, Fernández JL, et al. Sperm DNA fragmentation in rams vaccinated with Miloxan. Open Vet Sci J. 2008;2:7–10.
111. Pérez-Llano B, Enciso M, García-Casado P, et al. Sperm DNA fragmentation in boars is delayed or

abolished by using sperm extenders. Theriogenology. 2006;66:2137–43.
112. Young KE, Robbins WA, Xun L, et al. Evaluation of chromosome breakage and DNA integrity in sperm: an investigation of remote semen collection conditions. J Androl. 2003;24:853–61.
113. Dalzell LH, Thompson-Cree M, McClure N, et al. Effects of 24-hour incubation after freezethawing on DNA fragmentation of testicular sperm from infertile and fertile men. Fertil Steril. 2003;79: 1670–2.
114. Dalzell LH, McVicar CM, McClure N, et al. Effects of short and long incubations on DNA fragmentation of testicular sperm. Fertil Steril. 2004;82:1443–5.
115. Bungum M, Bungum L, Humaidan P. A prospective study, using sibling oocytes, examining the effect of 30 seconds versus 90 minutes gamete co-incubation in IVF. Hum Reprod. 2006;21:518–23.
116. Hawk HW. Sperm survival and transport in the female reproductive tract. J Dairy Sci. 1987;66:2645–60.
117. Suarez SS, Pacey AA. Sperm transport in the female reproductive tract. Hum Reprod Update. 2006;12: 23–37.
118. Carballada R, Esponda P. Fate and distribution of seminal plasma proteins in the genital tract of the female rat after natural mating. J Reprod Fertil. 1997;109:325–35.
119. O'Flynn O'Brien KL, Varghese AC, Agarwal A, et al. The genetic causes of male factor infertility: a review. Fertil Steril. 2010;93:1–12.
120. Aitken RJ, De Iuliis GN. On the possible origins of DNA damage in human spermatozoa. Mol Hum Reprod. 2010;16:3–13.
121. Schulte RT, Ohl DA, Sigman M, et al. Sperm DNA damage in male infertility: etiologies, assays, and outcomes. J Assist Reprod Genet. 2010;27:3–12.
122. Makhlouf AA, Niederberger C. DNA integrity tests in clinical practice: it is not a simple matter of black and white (or red and green). J Androl. 2006;27: 316–23.

Basic and Clinical Aspects of Sperm Chromomycin A3 Assay

11

Gian Carlo Manicardi, Davide Bizzaro, and Denny Sakkas

Abstract

Semen quality is conventionally determined according to the number, motility, and morphology of spermatozoa in an ejaculate. In turn, it is generally accepted that an association exists between these semen parameters and fertilizing ability. With the advent of in vitro fertilization (IVF) and related techniques such as intracytoplasmic sperm injection (ICSI), it has become increasingly apparent that the number, motility, and morphology of spermatozoa are not always indicative of a male's fertility status. Methods exploring sperm DNA stability and integrity have been applied during the last decade to evaluate fertility disorders and to increase the predictive value of sperm analysis for procreation in vivo and in vitro. It has been shown that infertile men have an increased sperm histone–protamine ratio compared to fertile counterparts. This alteration of histone–protamine ratio, also called abnormal packing, increases susceptibility of sperm DNA to external stresses due to poorer chromatin compaction. Recent studies have also underlined the link between protamine deficiency and sperm DNA damage that resulted in poor fertilizing capacity.

Keywords

Sperm chromomycin A3 assay • A3 assay • Semen quality • Fluorochromes • Histone–protamine ratio in sperm

Semen quality is conventionally determined according to the number, motility, and morphology of spermatozoa in an ejaculate [1]. In turn, it is generally accepted that an association exists between these semen parameters and fertilizing ability [1]. With the advent of in vitro fertilization

G.C. Manicardi (✉)
Department of Agricultural and Food Science, University of Modena and Reggio Emilia, Via Amendola, 2, 42100 Reggio Emilia, Italy
e-mail: giancarlo.manicardi@unimore.it

A. Zini and A. Agarwal (eds.), *Sperm Chromatin: Biological and Clinical Applications in Male Infertility and Assisted Reproduction*, DOI 10.1007/978-1-4419-6857-9_11,

(IVF) and related techniques such as intracytoplasmic sperm injection (ICSI), it has become increasingly apparent that the number, motility, and morphology of spermatozoa are not always indicative of a male's fertility status. Significantly different fertilization rates have been reported for patients with similar semen parameters, suggesting that a more sensitive test is needed to identify the inherent defects that render certain spermatozoa unable to fertilize [2, 3]. A failure of the conventional semen parameters to predict fertilization indicates that hidden anomalies, lying at the sperm membrane level or at the chromatin level, should also be evaluated. Methods exploring sperm DNA stability and integrity have been applied during the last decade to evaluate fertility disorders and to increase the predictive value of sperm analysis for procreation in vivo and in vitro [4]. With these new techniques, it was shown that normozoospermic infertile men, in addition to those having poor semen parameters, have higher percentages of spermatozoa with DNA fragmentation compared to the individuals presenting with normal semen quality [5–9]. Moreover, a number of studies have shown that spermatozoa with abnormal nuclear chromatin organization are more frequent in infertile men than in fertile men [10–13] (and enclosed references). Sperm chromatin is a highly organized, compact structure, consisting of DNA and heterogeneous proteins. In somatic cells, DNA is normally wrapped around an octamer of histones to form nucleosomes that eventually give rise to a solenoid DNA structure. During spermatogenesis, sperm nuclei undergo drastic modifications as histones are replaced by protamines, leading to a highly packaged chromatin in mature spermatozoa [14–16]. Chromatin compaction in mammalian spermatozoa is acquired by replacement of histones by protamines leading to a DNA–protamine complex that is highly compact, inert, and transcriptionally inactive [17]. Further stabilization is obtained by the oxidation of the protamine cysteine residues to disulfides [18–21]. Protamines are small arginine-rich nuclear proteins that replace histones in developing spermatozoa to achieve a high level of chromatin compaction [22, 23]; this is made possible by DNA charge neutralization when protamines complex with DNA. Mature human and mouse sperm nuclei contain 0.85% and 0.95% of protamines in their nucleoprotein component respectively [24–26]. In mouse, they allow the mature sperm nuclei to adopt a volume 40-fold smaller than that of a normal somatic nucleus [23]. It has been shown that infertile men have an increased sperm histone–protamine ratio compared to fertile counterparts [27]. This alteration of histone–protamine ratio, also called abnormal packing, increases susceptibility of sperm DNA to external stresses due to poorer chromatin compaction. Recent studies have also underlined the link between protamine deficiency and sperm DNA damage that resulted in poor fertilizing capacity [4, 28].

Fluorochromes as Indicators of Sperm Chromatin Compaction

The accessibility of different fluorochromes has been used widely to establish the relative packaging quality of sperm nuclei in mammals. Different dye and fluorochrome patterns have been established during spermiogenesis in mammals [20, 29–34]. One successfully used fluorochrome is Acridine Orange, which displays an increase in the number of red-staining spermatozoa in infertile males [10]. Acridine Orange fluorescence is related to the thiol-disulfide status of sperm nuclei [35] and has been shown to change from red to green during sperm maturation [36]. Other widely used fluorochromes and dyes are Aniline Blue, which stains histones [11, 37, 38], and MBB [39–41] and Toluidine Blue [32, 42, 43], which are specific for examining the status of the disulfide bridges. In studies examining different fluorochromes a direct correlation between increased fluorochrome accessibility and protamine loss has not been established. It has also been well documented that fluorochrome accessibility differs in cases of subfertility in mammals, particularly in man [10, 44–46]. Of the numerous fluorochromes used, we have developed an interest in CMA3 [12, 47]. The interactions between this polymerase inhibitor and herring sperm DNA were analyzed for the first time by Hayasaka and Inoue in 1969 [48].

Successively, Evenson [47, 49], while examining changes in accessibility to various fluorochromes during spermiogenesis, found that the guanine–cytosine (GC)-specific externally binding dyes [mithramycin and chromomycin A3 (CMA3)] were better able to distinguish round and elongating spermatids and vas deferens spermatozoa when compared to certain intercalating dyes. Monaco and Rasch [50] also suggested that the decline in mithramycin and CMA3 staining intensity observed in maturing spermatozoa of fish, frogs, and rabbits reflected changes in protein composition and in DNA packaging ratios. Our research group has repeatedly shown that CMA3 is a useful tool for the rapid screening of subfertility in man, as it seems to allow an indirect visualization of protamine-deficient, nicked, and partially denatured DNA [5, 12, 47, 51]. In addition, CMA3 accessibility differs during spermiogenesis in the mouse suggesting that it varies according to the level of protamination. In mouse, testicular spermatids are highly CMA3 positive, while mature spermatozoa are completely negative and fertilizing spermatozoa stain with fluorescence only when decondensation begins in the oocyte [12, 34]. Interestingly, in both human spermatozoa and testicular mouse spermatids, in situ protamination of fixed spermatozoa can inhibit the access of CMA3 to the sperm chromatin. Displacement of sperm nucleoproteins, including protamines, can be achieved in vitro by treating mouse sperm preparations with NaCl under reducing conditions [52]. This simple technique leads to the swelling of the sperm head and displacement of the nuclear basic proteins. A modification of the above-mentioned technique has been used to investigate the relationship between the amount of bound protamine on mouse and human sperm DNA and the level of CMA3 fluorescence [53]. This was accomplished by performing a competition assay between salmon protamine and the fluorochromes CMA3 and DAPI on decondensed spermatozoa that had had their nuclear proteins extracted and were fixed on slides. In this study, we had shown that the extraction of nucleoproteins from both mature mouse and human spermatozoa led to an expanded flattened appearance, not unlike that of macrocephalic spermatozoa observed in human semen samples [5, 11]. Remarkably, even though mammals use two classes of protamines (protamine 1 and 2) to compact their sperm DNA [54–56], in situ protamination of the decondensed spermatozoa with salmon protamines, which only represent an example of the protamine 1 class, led to a partial recoiling of the DNA sperm head, resulting in a condensed, rounded morphology. Furthermore, this coincided with a sharp decrease in accessibility of the CMA3 fluorochrome until it was unable to stain the sperm chromatin, as is routinely observed in normal fully mature mouse and normal human spermatozoa [5, 12]. When using CMA3, an all-or-none fluorescence can be distinguished readily when performing in situ protamination of deprotaminated spermatozoa. On the contrary, the DAPI fluorochrome and ethidium bromide fail to provide this distinction [12, 53]. This would suggest that CMA3 can be used as a feasible indicator of protamine-depleted sperm chromatin even in laboratories that do not have microfluorometric or flow-cytometric equipment, as a standard fluorescent microscope would suffice. Other fluorochromes such as mithramycin, which binds to DNA in a similar manner to CMA3, and 7-amino-actinomycin D may also show the same pattern of competition with protamines. In fact, actinomycin-D binding has been shown to be restricted in spermiogenesis during protamine deposition in mouse [57], as has also been shown with CMA3 [34]. In this connection, it must be noted that tritiated-labeled actinomycin D (H-3-AMD) incorporation into the sperm nuclei was used to assess the chromatin status of frozen-thawed boar spermatozoa [58, 59].

Interaction Between CMA3 and Sperm DNA

How CMA3 acts in distinguishing areas in the chromatin that lack protamine is not completely clear and its mode of action can only be postulated on the basis of the literature data available. This molecule has been shown to bind as a

Mg_2-coordinated dimer at the minor groove of GC-rich DNA and induces a conformational perturbation in the DNA helix resulting in a wider and shallower minor groove at its binding site [60–63]. It was formerly proposed that protamines bind through the minor groove [57]; hence, as CMA3 has also been shown to bind through the minor groove, it was supposed that both molecules compete for the same site [12]. However, Fita et al. [64] and Hud et al. [65] have proposed a new model stating that protamine binds within the major groove, producing conformational changes in the B-form DNA, which lead to a certain degree of base unstacking. They proposed that the arginine residues within the DNA binding domain of each protamine molecule interact with phosphate groups of both DNA strands, locking the two phosphodiester strands in a rigid form with respect to each other. When the CMA3 dimer binds it needs to induce a conformational perturbation in the DNA helix resulting in a wider and shallower minor groove at its binding site [61, 63, 66]. The conformational arrangement adopted by the DNA–protamine complex could limit the access of CMA3 to the minor groove, as it would impede the conformational change required for it to bind effectively. In addition, a study employing the oligonucleotide decamer d(CATGGCCATG) has shown that when CMA3 binds it also compresses the wide major groove of the double helix [67]. In contrast to CMA3, DAPI did not show an all-or-none response even though a decrease was observed when spermatozoa were treated with high concentrations of protamine. DAPI binds with high affinity to the minor groove in AT-rich sequences and at a lower affinity by a GC-specific intercalation [68–71]. The minor groove associated with A–T regions is narrower than G–C regions of B-DNA, leading to a snug fit of the flat aromatic rings of DAPI between the walls of the groove [69]. We could, therefore, postulate that the smaller size of DAPI is only minimally impeded at the higher concentrations of protamine when sufficient conformational changes occur in the chromatin to limit access to the minor groove. In conclusion, in situ protamination of deprotaminated spermatozoa could be used as an effective tool for studying the interactions of certain fluorochromes with sperm DNA. This experimental evidence supports our previous hypothesis that CMA3 can be effectively used as an indicator of underprotaminated spermatozoa [12, 34, 47, 72]. In the context of human infertility, this may be an important form of assessing spermatozoa from male-factor patients because the current use of ICSI means that some of the previous methods of assessing spermatozoa are not useful. ICSI overrides deficiencies in sperm motility, zona and oolemma binding and leaves the onus on the sperm nucleus to complete fertilization. Hence, accurate tests that measure the quality of the sperm nucleus take on greater importance.

CMA3 and DNA Damage

The molecular basis of the DNA fragmentation observed in the ejaculated spermatozoa is largely unresolved. This is an issue of some importance because knowledge of the mechanisms responsible for inducing DNA strand breakage in the male germ line would inform our attempts to understand the etiology of this damage and develop therapeutic approaches for its amelioration. A large body of experimental evidence supports the hypothesis that the presence of DNA damage in mature spermatozoa is correlated to poor chromatin packaging (see Sakkas and Alvarez [73] and references herein). Previous studies have indicated that one of the major problems of sperm displaying abnormal morphology is their protamine depleted state [5, 12, 28, 47, 51, 72, 74]. CMA3 would, therefore, be a useful tool as an adjunct to sperm morphology assessment to help characterize further a patient's sperm sample, particularly for male-factor patients [12, 28, 75]. In addition to the protamine deficiency in abnormal sperm is the higher incidence of damaged DNA (see Sakkas and Alvarez [73] and references herein). The presence of nicks in sperm DNA has also been shown in numerous animal studies [34, 76–78] and their appearance is believed to facilitate the packaging of the DNA into a very small volume during spermiogenesis [79, 80]. Nicks are present during the elongating spermatid stage in mouse and rat, and they disappear by the late spermatid stage [34, 76]. On the basis of these studies, it can be postulated that the sperm possessing damaged DNA may represent a population of sperm that have failed to complete maturation. Moreover, from the results presented in these studies, it appears that abnormal sperm morphology is an overall indicator of spermatozoa that have failed to progress through a complete spermiogenesis. As a consequence, they display many properties present in immature sperm. Correlations between CMA3 staining, sperm morphology, fertilization, and assisted reproduction outcome have been found in patients undergoing routine IVF, subzonal insemination (SUZI), or ICSI [5, 46, 81, 82]. Thus, CMA3 has been generally considered as a useful tool for evaluating infertile patients ([83] and references herein). In this context, a hypothesis to explain the relationship between CMA3 positivity and DNA damage in human spermatozoa has recently been proposed [84]. According to this model, the first stage in the cascade of events leading to DNA damage involves an error in chromatin remodeling during spermiogenesis leading to the generation of spermatozoa with poorly protaminated nuclear DNA. This creates a state of vulnerability in affected cells such that they are then susceptible to oxidative attack. The oxidative stress associated with the latter could originate in a number of different ways including the following: (1) the generation of reactive oxygen species (ROS) by leukocytes as a consequence of male genital tract infections, (2) electromagnetic radiation, including heat or radio-frequency radiation in the mobile phone range, (3) redox cycling metabolites or xenobiotics such as catechol estrogens or quinones, (4) ROS generated as a consequence of electron leakage from the sperm mitochondria, and (5) a deficiency in the antioxidant protection afforded to these vulnerable cells during their transit through the male reproductive tract [85–89]. This hypothesis predicts that there should be close relationships between the efficiency of chromatin remodeling, oxidative base damage to sperm DNA, and DNA fragmentation in human spermatozoa. Accordingly, staining with CMA3 has been shown to be positively correlated with the presence of nuclear histones [90] and ultrastructural evidence of poor chromatin compaction [91] but negatively correlated with the presence of protamines [53]. Our observation that the binding of CMA3 correlates with the presence of DNA strand breaks [12, 34, 47] is in keeping with previous studies in indicating that impaired chromatin remodeling during spermiogenesis is a consistent feature of defective human spermatozoa possessing fragmented DNA [12, 27, 47, 92, 93]. The dependence of sperm DNA damage on fundamental errors that occur during spermatogenesis would also explain why this pathology is correlated with elements of the conventional semen profile, particularly sperm count [7]. Considering that one of the potential consequences of underprotamination is an increased susceptibility to sperm DNA damage [56], a direct relationship between protamine deficiency and DNA damage is not surprising. However, the contrary is not always true: spermatozoa with DNA fragmentation, which may derive from a number of causes (reviewed in Aitken and De Iuliis, [89]), are not necessarily cells with abnormal protamination [94]. In fact, sperm DNA damage is multifactorial and may be due to many conditions: in addition to poor chromatin packaging, sperm DNA fragmentation may be a consequence of high levels of free radicals, produced by both spermatozoa and leukocytes, or aberrant endonuclease activity, associated with abortive apoptosis [94, 95]. On the contrary, experimental evidence [96–98] stated that despite abnormal sperm protamination and sperm DNA fragmentation being positively correlated, they affect the reproductive outcome in different ways: while sperm DNA fragmentation seems to affect ICSI outcome, sperm chromatin underprotamination affects fertilization and pregnancy in IVF. This result may be explained considering the different nature of sperm DNA damage and sperm protamine deficiency: these two conditions are distinct aspects of chromatin alteration, so they probably have a different impact on biological quality of spermatozoa; additionally, the different technical features of the laboratory procedures used to assist fertilization (IVF and ICSI) and the contribution of the operator performing assisted reproduction procedures have to be taken into account.

References

1. WHO. World Health Organization Laboratory manual for examination of human semen. 5th ed. Cambridge: Cambridge University Press; 2010.
2. Wolf JP, Ducot B, Kunstmann JM, Frydman R, Jouannet P. Influence of sperm parameters on outcome of subzonal insemination in the case of previous IVF failure. off. Hum Reprod. 1992;7(10): 1407–13.
3. Wolf JP, Bulwa S, Ducot B, Rodrigues D, Jouannet P. Fertilizing ability of sperm with unexplained in vitro fertilization failures, as assessed by the zona-free hamster egg penetration assay: its prognostic value for sperm-oolemma interaction. Fertil Steril. 1996;65(6):1196–201.
4. Zini A, Libman J. Sperm DNA damage: clinical significance in the era of assisted reproduction. CMAJ. 2006;175(5):495–500.
5. Bianchi PG, Manicardi GC, Urner F, Campana A, Sakkas D. Chromatin packaging and morphology in ejaculated human spermatozoa: evidence of hidden anomalies in normal spermatozoa. Mol Hum Reprod. 1996;2(3):139–44.
6. Sun JG, Jurisicova A, Casper RF. Detection of deoxyribonucleic acid fragmentation in human sperm: correlation with fertilization in vitro. Biol Reprod. 1997;56(3):602–7.
7. Irvine DS, Twigg JP, Gordon EL, Fulton N, Milne PA, Aitken RJ. DNA integrity in human spermatozoa: relationships with semen quality. J Androl. 2000;21(1): 33–44.
8. Zini A, Bielecki R, Phang D, Zenzes MT. Correlations between two markers of sperm DNA integrity, DNA denaturation and DNA fragmentation, in fertile and infertile men. Fertil Steril. 2001;75(4):674–7.
9. Varghese AC, Bragais FM, Mukhopadhyay D, et al. Human sperm DNA integrity in normal and abnormal semen samples and its correlation with sperm characteristics. Andrologia. 2009;41(4):207–15.
10. Evenson DP, Darzynkiewicz Z, Melamed MR. Relation of mammalian sperm chromatin heterogeneity to fertility. Science. 1980;210(4474):1131–3.
11. Foresta C, Zorzi M, Rossato M, Varotto A. Sperm nuclear instability and staining with aniline blue: abnormal persistence of histones in spermatozoa in infertile men. Int J Androl. 1992;15(4):330–7.
12. Bianchi PG, Manicardi GC, Bizzaro D, Bianchi U, Sakkas D. Effect of deoxyribonucleic acid protamination on fluorochrome staining and in situ nick-translation of murine and human mature spermatozoa. Biol Reprod. 1993;49(5):1083–8.
13. Barratt CL, Aitken RJ, Bjorndahl L, et al. Sperm DNA: organization, protection and vulnerability: from basic science to clinical applications–a position report. Hum Reprod. 2010;25(4):824–38.
14. Kumaroo KK, Jahnke G, Irvin JL. Changes in basic chromosomal proteins during spermatogenesis in the mature rat. Arch Biochem Biophys. 1975;168(2): 413–24.
15. Goldberg RB, Geremia R, Bruce WR. Histone synthesis and replacement during spermatogenesis in the mouse. Differentiation. 1977;7(3):167–80.
16. Poccia D. Remodeling of nucleoproteins during gametogenesis, fertilization, and early development. Int Rev Cytol. 1986;105:1–65.
17. Ward WS, Zalensky AO. The unique, complex organization of the transcriptionally silent sperm chromatin. Crit Rev Eukaryot Gene Expr. 1996;6(2–3): 139–47.
18. Calvin HI, Bedford JM. Formation of disulphide bonds in the nucleus and accessory structures of mammalian spermatozoa during maturation in the epididymis. J Reprod Fertil Suppl. 1971;13:Suppl-75.
19. Marushige Y, Marushige K. Transformation of sperm histone during formation and maturation of rat spermatozoa. J Biol Chem. 1975;250(1):39–45.
20. Pellicciari C, Hosokawa Y, Fukuda M, Manfredi Romanini MG. Cytofluorometric study of nuclear sulphydryl and disulphide groups during sperm maturation in the mouse. J Reprod Fertil. 1983;68(2): 371–6.
21. Bertelsmann H, Kuehbacher M, Weseloh G, Kyriakopoulos A, Behne D. Sperm nuclei glutathione peroxidases and their occurrence in animal species with cysteine-containing protamines. Biochim Biophys Acta. 2007;1770(10):1459–67.
22. Pogany GC, Corzett M, Weston S, Balhorn R. DNA and protein content of mouse sperm. Implications regarding sperm chromatin structure. Exp Cell Res. 1981;136(1):127–36.
23. Ward WS, Coffey DS. DNA packaging and organization in mammalian spermatozoa: comparison with somatic cells. Biol Reprod. 1991;44(4):569–74.
24. Gatewood JM, Cook GR, Balhorn R, Bradbury EM, Schmid CW. Sequence-specific packaging of DNA in human sperm chromatin. Science. 1987;236(4804): 962–4.
25. Bellve AR, McKay DJ, Renaux BS, Dixon GH. Purification and characterization of mouse protamines P1 and P2. Amino acid sequence of P2. Biochemistry. 1988;27(8):2890–7.
26. Debarle M, Martinage A, Sautiere P, Chevaillier P. Persistence of protamine precursors in mature sperm nuclei of the mouse. Mol Reprod Dev. 1995;40(1): 84–90.
27. Zini A, Gabriel MS, Zhang X. The histone to protamine ratio in human spermatozoa: comparative study of whole and processed semen. Fertil Steril. 2007;87(1):217–9.
28. Tavalaee M, Razavi S, Nasr-Esfahani MH. Influence of sperm chromatin anomalies on assisted reproductive

technology outcome. Fertil Steril. 2009;91(4): 1119–26.
29. Krzanowska H. Toluidine blue staining reveals changes in chromatin stabilization of mouse spermatozoa during epididymal maturation and penetration of ova. J Reprod Fertil. 1982;64(1):97–101.
30. Balhorn R, Weston S, Thomas C, Wyrobek AJ. DNA packaging in mouse spermatids. Synthesis of protamine variants and four transition proteins. Exp Cell Res. 1984;150(2):298–308.
31. Evenson D, Darzynkiewicz Z, Jost L, Janca F, Ballachey B. Changes in accessibility of DNA to various fluorochromes during spermatogenesis. Cytometry. 1986;7(1):45–53.
32. Barrera C, Mazzolli AB, Pelling C, Stockert JC. Metachromatic staining of human sperm nuclei after reduction of disulphide bonds. Acta Histochem. 1993;94(2):141–9.
33. Yossefi S, Oschry Y, Lewin LM. Chromatin condensation in hamster sperm: a flow cytometric investigation. Mol Reprod Dev. 1994;37(1):93–8.
34. Sakkas D, Manicardi G, Bianchi PG, Bizzaro D, Bianchi U. Relationship between the presence of endogenous nicks and sperm chromatin packaging in maturing and fertilizing mouse spermatozoa. Biol Reprod. 1995;52(5):1149–55.
35. Kosower NS, Katayose H, Yanagimachi R. Thiol-disulfide status and acridine orange fluorescence of mammalian sperm nuclei. J Androl. 1992;13(4): 342–8.
36. Evenson DP. Flow cytometry of acridine orange stained sperm is a rapid and practical method for monitoring occupational exposure to genotoxicants. Prog Clin Biol Res. 1986;207(3):121–32.
37. Auger J, Mesbah M, Huber C, Dadoune JP. Aniline blue staining as a marker of sperm chromatin defects associated with different semen characteristics discriminates between proven fertile and suspected infertile men. Int J Androl. 1990;13(6):452–62.
38. Wong A, Chuan SS, Patton WC, Jacobson JD, Corselli J, Chan PJ. Addition of eosin to the aniline blue assay to enhance detection of immature sperm histones. Fertil Steril. 2008;90(5):1999–2002.
39. Shalgi R, Seligman J, Kosower NS. Dynamics of the thiol status of rat spermatozoa during maturation: analysis with the fluorescent labeling agent monobromobimane. Biol Reprod. 1989;40(5):1037–45.
40. Seligman J, Kosower NS, Weissenberg R, Shalgi R. Thiol-disulfide status of human sperm proteins. J Reprod Fertil. 1994;101(2):435–43.
41. Zubkova EV, Wade M, Robaire B. Changes in spermatozoal chromatin packaging and susceptibility to oxidative challenge during aging. Fertil Steril. 2005;84 suppl 2:1191–8.
42. Erenpreiss J, Jepson K, Giwercman A, Tsarev I, Erenpreisa J, Spano M. Toluidine blue cytometry test for sperm DNA conformation: comparison with the flow cytometric sperm chromatin structure and TUNEL assays. Hum Reprod. 2004;19(10):2277–82.
43. Tsarev I, Bungum M, Giwercman A, et al. Evaluation of male fertility potential by Toluidine Blue test for sperm chromatin structure assessment. Hum Reprod. 2009;24(7):1569–74.
44. Engh E, Clausen OP, Scholberg A, Tollefsrud A, Purvis K. Relationship between sperm quality and chromatin condensation measured by sperm DNA fluorescence using flow cytometry. Int J Androl. 1992;15(5):407–15.
45. Spano M, Evenson DP. Flow cytometric studies in reproductive toxicology. Prog Clin Biol Res. 1991;372:497–511.
46. Sakkas D, Urner F, Bizzaro D, et al. Sperm nuclear DNA damage and altered chromatin structure: effect on fertilization and embryo development. Hum Reprod. 1998;13 Suppl 4:11–9.
47. Manicardi GC, Bianchi PG, Pantano S, et al. Presence of endogenous nicks in DNA of ejaculated human spermatozoa and its relationship to chromomycin A3 accessibility. Biol Reprod. 1995;52(4):864–7.
48. Hayasaka T, Inoue Y. Chromomycin A3 studies in aqueous solutions. Spectrophotometric evidence for aggregation and interaction with herring sperm deoxyribonucleic acid. Biochemistry. 1969;8(6):2342–7.
49. Evenson DP. Male germ cell analysis by flow cytometry: effects of cancer, chemotherapy, and other factors on testicular function and sperm chromatin structure. Ann N Y Acad Sci. 1986;468:350–67.
50. Monaco PJ, Rasch EM. Differences in staining with DNA-specific fluorochromes during spermiogenesis. J Histochem Cytochem. 1982;30:585.
51. Sakkas D, Urner F, Bianchi PG, et al. Sperm chromatin anomalies can influence decondensation after intracytoplasmic sperm injection. Hum Reprod. 1996;11(4):837–43.
52. Rodman TC, Pruslin FH, Allfrey VG. Mechanisms of displacement of sperm basic nuclear proteins in mammals. An in vitro simulation of post-fertilization results. J Cell Sci. 1982;53:227–44.
53. Bizzaro D, Manicardi GC, Bianchi PG, Bianchi U, Mariethoz E, Sakkas D. In-situ competition between protamine and fluorochromes for sperm DNA. Mol Hum Reprod. 1998;4(2):127–32.
54. Kolk AH, Samuel T. Isolation, chemical and immunological characterization of two strongly basic nuclear proteins from human spermatozoa. Biochim Biophys Acta. 1975;393(2):307–19.
55. Oliva R. Protamines and male infertility. Hum Reprod Update. 2006;12(4):417–35.
56. Carrell DT, Emery BR, Hammoud S. Altered protamine expression and diminished spermatogenesis: what is the link? Hum Reprod Update. 2007;13(3): 313–27.
57. Balhorn R. A model for the structure of chromatin in mammalian sperm. J Cell Biol. 1982;93(2):298–305.
58. Fraser L, Strzezek J. Is there a relationship between the chromatin status and DNA fragmentation of boar spermatozoa following freezing-thawing? Theriogenology. 2007;68(2):248–57.

59. Glogowski J, Strzezek J, Jazdzewski J. Intensity of 3 H-actinomycin D (3 H-AMD) binding to chromatin of bull spermatozoa. Reprod Domest Anim. 1994;29: 396–403.
60. Gao XL, Patel DJ. Chromomycin dimer-DNA oligomer complexes. Sequence selectivity and divalent cation specificity. Biochemistry. 1990;29(49):10940–56.
61. Gao XL, Mirau P, Patel DJ. Structure refinement of the chromomycin dimer-DNA oligomer complex in solution. J Mol Biol. 1992;223(1):259–79.
62. Chakrabarti S, Bhattacharyya D, Dasgupta D. Structural basis of DNA recognition by anticancer antibiotics, chromomycin A(3), and mithramycin: roles of minor groove width and ligand flexibility. Biopolymers. 2000;56(2):85–95.
63. Hou MH, Robinson H, Gao YG, Wang AH. Crystal structure of the [Mg2+–(chromomycin A3)2]-d(TTGGCCAA)2 complex reveals GGCC binding specificity of the drug dimer chelated by a metal ion. Nucleic Acids Res. 2004;32(7):2214–22.
64. Fita I, Campos JL, Puigjaner LC, Subirana JA. X-ray diffraction study of DNA complexes with arginine peptides and their relation to nucleoprotamine structure. J Mol Biol. 1983;167(1):157–77.
65. Hud NV, Milanovich FP, Balhorn R. Evidence of novel secondary structure in DNA-bound protamine is revealed by Raman spectroscopy. Biochemistry. 1994;33(24):7528–35.
66. Gao XL, Patel DJ. Solution structure of the chromomycin-DNA complex. Biochemistry. 1989;28(2): 751–62.
67. Goodsell DS, Kopka ML, Cascio D, Dickerson RE. Crystal structure of CATGGCCATG and its implications for A-tract bending models. Proc Natl Acad Sci USA. 1993;90(7):2930–4.
68. Wilson WD, Tanious FA, Barton HJ, et al. DNA sequence dependent binding modes of 4',6-diamidino-2-phenylindole (DAPI). Biochemistry. 1990;29(36): 8452–61.
69. Kapuscinski J. DAPI: a DNA-specific fluorescent probe. Biotech Histochem. 1995;70(5):220–33.
70. Trotta E, D'Ambrosio E, Ravagnan G, Paci M. Evidence for DAPI intercalation in CG sites of DNA oligomer [d(CGACGTCG)]2: a 1H NMR study. Nucleic Acids Res. 1995;23(8):1333–40.
71. De Castro LF, Zacharias M. DAPI binding to the DNA minor groove: a continuum solvent analysis. J Mol Recognit. 2002;15(4):209–20.
72. Bianchi PG, Manicardi G, Bizzaro D, Campana A, Bianchi U, Sakkas D. Use of the guanine-cytosine (GC) specific fluorochrome, chromomycin A3, as an indicator of poor sperm morphology. J Assist Reprod Genet. 1996;13(3):246–50.
73. Sakkas D, Alvarez JG. Sperm DNA fragmentation: mechanisms of origin, impact on reproductive outcome, and analysis. Fertil Steril. 2010;93(4):1027–36.
74. Franken DR, Franken CJ, de la Guerre H, de Villiers A. Normal sperm morphology and chromatin packaging: comparison between aniline blue and chromomycin A3 staining. Andrologia. 1999;31(6):361–6.
75. Esterhuizen AD, Franken DR, Lourens JG, Van Zyl C, Muller I, van Rooyen LH. Chromatin packaging as an indicator of human sperm dysfunction. J Assist Reprod Genet. 2000;17(9):508–14.
76. McPherson SM, Longo FJ. Nicking of rat spermatid and spermatozoa DNA: possible involvement of DNA topoisomerase II. Dev Biol. 1993;158(1):122–30.
77. Lopez-Fernandez C, Crespo F, Arroyo F, et al. Dynamics of sperm DNA fragmentation in domestic animals II. The stallion. Theriogenology. 2007;68(9):1240–50.
78. Lopez-Fernandez C, Fernandez JL, Gosalbez A, et al. Dynamics of sperm DNA fragmentation in domestic animals III. Ram. Theriogenology. 2008;70(6):898–908.
79. Iseki S. DNA strand breaks in rat tissues as detected by in situ nick translation. Exp Cell Res. 1986;167(2): 311–26.
80. McPherson S, Longo FJ. Chromatin structure-function alterations during mammalian spermatogenesis: DNA nicking and repair in elongating spermatids. Eur J Histochem. 1993;37(2):109–28.
81. Nasr-Esfahani MH, Razavi S, Mozdarani H, Mardani M, Azvagi H. Relationship between protamine deficiency with fertilization rate and incidence of sperm premature chromosomal condensation post-ICSI. Andrologia. 2004;36(3):95–100.
82. Nasr-Esfahani MH, Salehi M, Razavi S, et al. Effect of sperm DNA damage and sperm protamine deficiency on fertilization and embryo development post-ICSI. Reprod Biomed Online. 2005;11(2):198–205.
83. Nasr-Esfahani MH, Aboutorabi R, Razavi S. Credibility of chromomycin A3 staining in prediction of fertility. Int J Fertil Steril. 2010;3:5–10.
84. De Iuliis GN, Thomson LK, Mitchell LA, et al. DNA damage in human spermatozoa is highly correlated with the efficiency of chromatin remodeling and the formation of 8-hydroxy-2'-deoxyguanosine, a marker of oxidative stress. Biol Reprod. 2009;81(3):517–24.
85. Banks S, King SA, Irvine DS, Saunders PT. Impact of a mild scrotal heat stress on DNA integrity in murine spermatozoa. Reproduction. 2005;129(4):505–14.
86. Aitken RJ, Bennetts LE, Sawyer D, Wiklendt AM, King BV. Impact of radio frequency electromagnetic radiation on DNA integrity in the male germline. Int J Androl. 2005;28(3):171–9.
87. Aitken RJ, Wingate JK, De Iuliis GN, Koppers AJ, McLaughlin EA. Cis-unsaturated fatty acids stimulate reactive oxygen species generation and lipid peroxidation in human spermatozoa. J Clin Endocrinol Metab. 2006;91(10):4154–63.
88. Aitken RJ, De Iuliis GN, McLachlan RI. Biological and clinical significance of DNA damage in the male germ line. Int J Androl. 2009;32(1):46–56.
89. Aitken RJ, De Iuliis GN. On the possible origins of DNA damage in human spermatozoa. Mol Hum Reprod. 2010;16(1):3–13.
90. Singleton S, Zalensky A, Doncel GF, Morshedi M, Zalenskaya IA. Testis/sperm-specific histone 2B in the sperm of donors and subfertile patients: variability and relation to chromatin packaging. Hum Reprod. 2007;22(3):743–50.

91. Iranpour FG, Nasr-Esfahani MH, Valojerdi MR, al-Taraihi TM. Chromomycin A3 staining as a useful tool for evaluation of male fertility. J Assist Reprod Genet. 2000;17(1):60–6.
92. Aoki VW, Emery BR, Liu L, Carrell DT. Protamine levels vary between individual sperm cells of infertile human males and correlate with viability and DNA integrity. J Androl. 2006;27(6):890–8.
93. Carrell DT, De JC, Lamb DJ. The genetics of male infertility: a field of study whose time is now. Arch Androl. 2006;52(4):269–74.
94. Henkel R, Bastiaan HS, Schuller S, Hoppe I, Starker W, Menkveld R. Leucocytes and intrinsic ROS production may be factors compromising sperm chromatin condensation status. Andrologia. 2010;42(2):69–75.
95. Aitken RJ, De Iuliis GN. Value of DNA integrity assays for fertility evaluation. Soc Reprod Fertil Suppl. 2007;65:81–92.
96. Borini A, Tarozzi N, Bizzaro D, et al. Sperm DNA fragmentation: paternal effect on early post-implantation embryo development in ART. Hum Reprod. 2006;21(11):2876–81.
97. Tarozzi N, Bizzaro D, Flamigni C, Borini A. Clinical relevance of sperm DNA damage in assisted reproduction. Reprod Biomed Online. 2007;14(6):746–57.
98. Tarozzi N, Nadalini M, Stronati A, et al. Anomalies in sperm chromatin packaging: implications for assisted reproduction techniques. Reprod Biomed Online. 2009;18(4):486–95.

Cytochemical Tests for Sperm Chromatin Maturity

12

Igor Tsarev and Juris Erenpreiss

Abstract

Normal structure and maturity of sperm chromatin is essential for the fertilizing ability of spermatozoa in vivo. It is a relatively independent measure of semen quality that yields additional prognostic information complementary to standard sperm parameters – concentration, motility, and morphology. Several methods are used to assess sperm chromatin status. At present, indirect methods for sperm DNA fragmentation assessment are routinely used in andrological workup. However, several simple and efficient tests for chromatin maturation status are also available. The normality ranges and predictive thresholds for male fertility potential for these assays still need to be established or clarified

Keywords

Chytochemical tests for sperm chromatin maturity • Sperm chromatin maturity • Sperm chromatin, structural probes • Planar ionic dyes in sperm chromatin

Infertility is a major medical problem that affects approximately 15% of couples trying to conceive, and a male cause is believed to be a contributing factor in approximately half of these cases [1]. In andrological practice, visual light microscopic examination of semen quality plays principal role in male fertility potential evaluation. This consists of measuring seminal volume, pH, sperm concentration, motility, morphology, and vitality. However, often a diagnosis of male fertility cannot be made as a result of basic semen analysis. This is caused by a significant overlap in the values of sperm concentration, motility, and morphology between fertile and infertile men, as it has been demonstrated by several studies [2]. In addition, quality control introduction within and between laboratories has highlighted the subjectivity and variability of traditional semen parameters.

It has been demonstrated that abnormalities in the male genome, characterized by disturbed chromatin packaging and damaged sperm

I. Tsarev (✉)
Andrology Laboratory, Riga Stradins University, Riga, Latvia 1089, Latvia
e-mail: Igor.tsarev@gmail.com

A. Zini and A. Agarwal (eds.), *Sperm Chromatin: Biological and Clinical Applications in Male Infertility and Assisted Reproduction*, DOI 10.1007/978-1-4419-6857-9_12,

deoxyribonucleic acid (DNA), may be a cause for male infertility regardless of routine semen parameters [3, 4]. Sperm chromatin abnormalities have been studied extensively in the past several years as a cause of male infertility [5]. Focus on the chromatin integrity and maturity of the male gamete has been especially intensified by the growing concern about transmission of damaged DNA through assisted reproductive techniques (ARTs) such as intracytoplasmic sperm injection (ICSI). Accumulating evidence suggests a negative relationship between disorganization of the chromatin material in sperm nuclei and the fertility potential of spermatozoa both in vivo and in vitro [4–12].

Abnormalities in the sperm chromatin organization, characterized both by damaged DNA and incompletely remodeled chromatin in mature sperm cells, may be indicative of male infertility regardless of normal semen parameters [3, 13]. Evaluation of sperm chromatin structure is an independent measure of sperm quality that provides good diagnostic and prognostic capabilities. Therefore, it may be considered a reliable predictor of a couple's inability to conceive [14, 15]. Sperm chromatin quality correlates with pregnancy outcome in in vitro fertilization (IVF) [14–18].

Many techniques have been described for evaluation of the chromatin status and maturity. In andrological practice, the most popular are indirect methods for estimation of DNA integrity in sperm chromatin. These methods are based on the ability of some stains to test the conformation of sperm chromatin, which in turn depends on DNA strand breaks and DNA interaction with proteins [19–22]. However, since some studies had demonstrated that spermatozoa with abnormal nuclear chromatin packaging are more frequent in infertile men than in fertile men, a number of techniques have been developed to test sperm chromatin maturation status. These techniques help to evaluate male reproductive status and might be also useful for ART outcome prediction [23, 24]. These assays, often referred as "cytochemical," include acidic aniline blue (AAB), Chromomycin A3, and Toluidine Blue (TB) tests.

Cytochemical Properties of Human Sperm Chromatin and Basis of its Testing by Planar Ionic Dyes

In many mammals, spermatozoa nuclei are highly homogenous and compact. This allows mature sperm nuclei to adopt a volume 40 times less than that of normal somatic nuclei [25]. This highly compact packaging of the primary sperm DNA filament is produced by DNA–protamine complexes [26]. Human sperm nuclei, on the contrary, contain considerably fewer protamines (around 85%) than sperm nuclei of several other mammals (such as bull, stallion, hamster, and mouse) [27, 28], and therefore, they are less regularly compacted and frequently contains DNA strand breaks [29, 30]. Sperm DNA is packed in specific toroids, each containing 50–60 kilobases of DNA. Individual toroids represent the DNA loop-domains, highly condensed by protamines and fixed at the nuclear matrix. Toroids are cross-linked by disulfide bonds formed by oxidation of sulfhydryl groups of cysteine present in the protamines [25, 26, 31]. Such condensed, insoluble, and highly organized structure of sperm chromatin is necessary to protect the genetic integrity during transport of the paternal genome through the male and female reproductive tracts [32–34].

However, in comparison to other species [35], human sperm chromatin packaging is exceptionally variable. This variability has been mostly attributed to its basic protein component. The retention of 15% histones, which are less basic than protamines, leads to the formation of a less-compact chromatin structure [28]. Moreover, human spermatozoa contain two types of protamines, P1 and P2, with a second type deficient in cysteine residues [36]. This results in diminished disulfide cross-linking if compared with species in which sperm contain only P1 group of proteins [37].

Chromatin structural probes using planar ionic dyes allow to analyze chromatin structure in terms of protein packaging correctness and disulfide cross-linking density. Their cytochemical background, however, is quite complex. Several factors influence the staining of chromatin by planar ionic dyes: (1) secondary structure of DNA, (2) regularity and density of chromatin packaging, and (3) binding of DNA to chromatin proteins, which influences its charge.

DNA Secondary Structure and Conformation – Fragmented DNA is easily denaturable [38]. However, even a single DNA strand break causes conformational transition of the DNA loop-domain from a supercoiled state to a relaxed state. Supercoiled DNA avidly takes up intercalating dyes (such as acridine orange [AO]) because this reduces the free energy of torsion stress. By contrast, the affinity for intercalation is low in relaxed DNA and is lost in fragmented DNA. In this case, an external mechanism of dye binding to DNA phosphate residues and dye polymerization (metachromasy) is favored [39, 40]. Nevertheless, fragmentation of DNA is not the only factor affecting the choice between metachromatic vs. orthochromatic staining. Chromatin packaging density also influences this balance.

Chromatin Packaging Density – in the regularly arranged and sufficiently densely packed sperm chromatin, coplanar dye polymerization providing metachromatic shift (change of color) is favored [41, 42]. However, in even more densely (as in normal sperm) packaged chromatin, the polymerization of the dye is hindered [43] and may even impair dye binding and coplanar polymerization. The latter is seen with aniline blue (AB) at low pH where it stains basic proteins loosely associated with DNA and is unable to bind to the chromatin of normal sperm, which is very densely packaged and uncharged. Substitution of histones to more cationic protamines occurring during spermiogenesis neutralizes DNA charge and decreases the accessibility of DNA-specific dyes. However, after removal of nuclear proteins, increase in sperm DNA stainability can vary depending on the chemical structure of the dye and the binding type which the dye forms with the DNA substrate [19, 44–46].

Chromatin Proteins affect the binding of DNA dyes in the way that they themselves bind differently to relaxed, fragmented, or supercoiled DNA. DNA supercoiling requires covalent binding of some nuclear matrix proteins and tighter ionic interactions between DNA and chromatin proteins to support negative supercoils [47]. Relaxed and fragmented DNA has looser ionic interactions with chromatin proteins, which can be easily displaced from the DNA, favoring external metachromatic binding of the dye to DNA phosphate groups. Both mechanisms of dye binding, external and intercalating, compete within each other within constraint loop-domain (toroid) depending on its conformational state.

Sperm Chromatin Structural Probes

Chromatin proteins in sperm nuclei with the impaired DNA appear to be more accessible to binding with the acidic dye, as found by the AB test [48]. An increase in the ability to stain sperm by acid AB indicates a looser chromatin packaging and increased accessibility of the basic groups of the nucleoproteins. This is due to the presence of residual histones [49], and correlates well with the AO test [50]. Chromomycin A3 (CMA3) is another staining technique that has been used as a measure of sperm chromatin condensation anomalies. CMA3 is a fluorochrome specific for GC-rich sequences and is believed to compete with protamines for binding to the minor groove of DNA. The extent of staining is, therefore, related to the degree of protamination of mature spermatozoa [51, 52]. In turn, phosphate residues of sperm DNA in nuclei with loosely packed chromatin and/or impaired DNA will be more liable to binding with basic dyes. Such conclusions were also deduced from the results of staining with basic dyes, such as TB, methyl green, and Giemsa stain [52, 53].

Acidic Aniline Blue

The AAB stain discriminates between lysine-rich histones and arginine/cysteine-rich protamines. This technique provides a specific positive reaction for lysine and reveals differences in the basic nuclear protein composition of human spermatozoa. Histone-rich nuclei of immature spermatozoa are rich in lysine and will consequently take up the blue stain. On the contrary, protamine-rich nuclei of mature spermatozoa are rich in arginine and cysteine and contain relatively low levels of lysine, which means they will not take up the stain [54].

Technique: slides are prepared by smearing 5 μL of either raw or washed semen sample. The slides are air-dried and fixed for 30 min in 3% glutaraldehyde in phosphate-buffered saline (PBS). The smear is dried and stained for 5 min in 5% aqueous AB solution (pH 3.5). Sperm heads containing immature nuclear chromatin stain blue and those with mature nuclei do not. The percentage of spermatozoa stained with AB is determined by counting 200 spermatozoa per slide under bright-field microscopy [55].

Results of AAB staining have shown a clear association between abnormal sperm chromatin and male infertility [56]. However, the correlation between the percentage of AB-stained spermatozoa and other sperm parameters remains controversial. Immature sperm chromatin may or may not correlate with asthenozoospermic samples and abnormal morphology patterns [55, 56]. Most important is the finding that chromatin condensation as visualized by AB staining is a good predictor for IVF outcome, although it cannot determine the fertilization potential and the cleavage and pregnancy rates following ICSI [57].

Toluidine Blue Stain Assay

TB is a basic planar nuclear dye used for metachromatic and orthochromatic staining of the chromatin. The phosphate residues of sperm DNA in nuclei with loosely packed chromatin and/or impaired DNA become more liable to binding with basic TB, providing a metachromatic shift due to dimerization of the dye molecules from light blue to purple–violet color [58]. This stain is a sensitive structural probe for DNA structure and packaging.

Technique: thin sperm smears are prepared on precleaned defatted slides and then air-dried for 30–60 min. Dried smears are fixed with freshly made 96% ethanol–acetone (1:1) at 4°C for 30 min to 12 h and air-dried. Hydrolysis is performed with 0.1 mol/L HCl at 4°C for 5 min followed by three changes of distilled water, 2 min each. TB (0.05% in 50% McIlvain's citrate phosphate buffer at pH3.5, is applied for 5 min. The slides are rinsed briefly in distilled water, lightly blotted with filter paper, dehydrated in tertiary butanol at 37°C (2 and 3 min) and xylene at room temperature (2 and 3 min), and mounted with DPX.

The results of the TB test are estimated using oil-immersion (10 and 100) light microscope. Sperm heads with good chromatin integrity stain light blue and those with diminished integrity stain violet (purple) [59]. The proportion of cells with violet heads (high optical density) are calculated based on 200 sperm cells examined per sample. Based on the different optical densities of cells stained by the TB, the image analysis cytometry test has been elaborated [60] (Figs. 12.1 and 12.2).

TB staining may be considered a fairly reliable method for assessing sperm chromatin. Abnormal nuclei (purple–violet sperm heads) have been shown to be correlated with counts of red–orange sperm heads as revealed by the AO method [58]. Also, correlations between the results of the TB, sperm chromatin structure assay (SCSA), and terminal deoxynucleotidyl transferase dUTP nick-end labeling (TUNEL) tests have been demonstrated. The proportion of

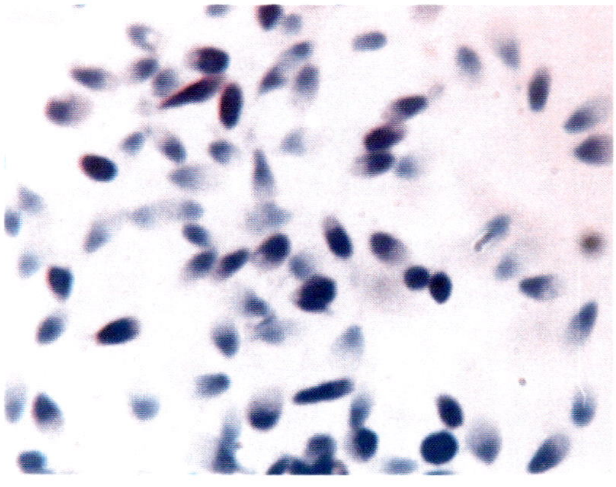

Fig. 12.1 Toluidine blue staining example

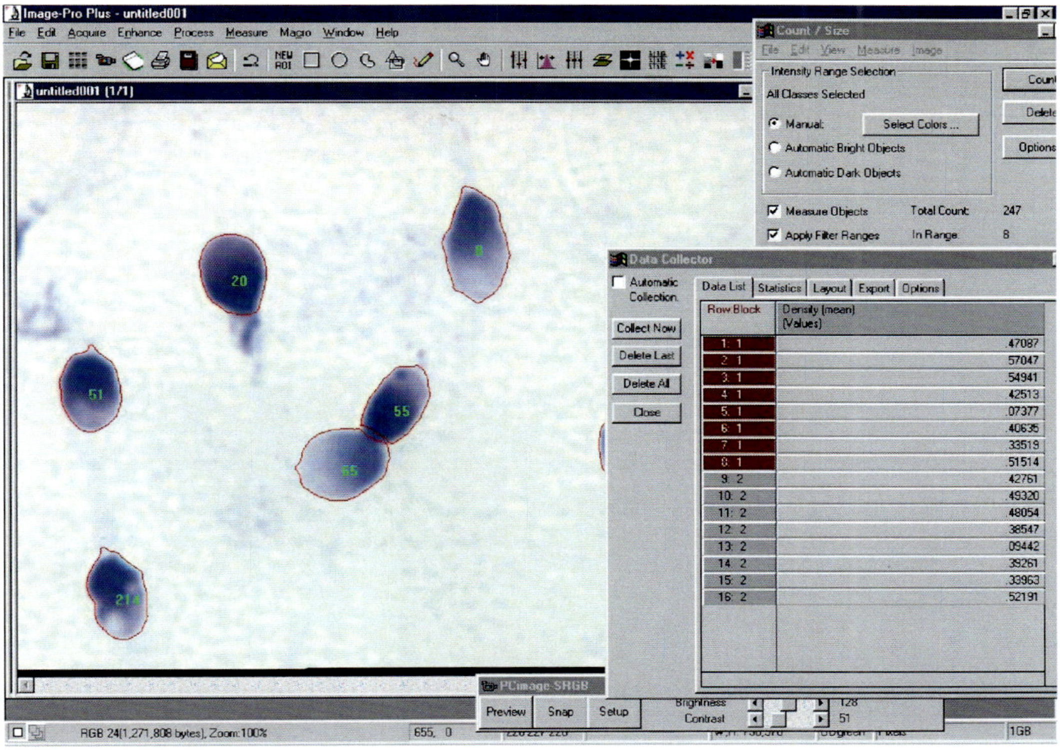

Fig. 12.2 Image cytometry for toluidine blue staining

sperm cells with abnormal DNA conformation, detected by the TB test (violet heads), correlated significantly with the proportion of spermatozoa containing denaturable DNA detected as SCSA percentage DFI ($r=0.84$, $P<0.001$) and with the fraction of spermatozoa with fragmented DNA in the FCM TUNEL test ($r=0.80$, $P<0.001$) [59]. Thresholds for the TB test between fertile and infertile men also were set. A threshold for proportion of cells with violet heads was set at 45%; it provides 92% specificity and 42% sensitivity for infertility detection [61].

TB staining is simple and inexpensive and has the advantage of providing permanent preparations for use with an ordinary microscope. The smears stained with the TB method can also be used for morphological assessment of the cells. However, these methods may have the inherent limits of repeatability dictated by a limited number of cells, which can be reasonably scored.

Chromomycin A3 Assay

Chromomycin A3 is a fluorochrome that specifically binds to guanine–cytosine DNA sequences. It reveals chromatin that is poorly packaged in human spermatozoa by visualization of protamine-deficient DNA. Chromomycin A3 and protamines compete for the same binding sites in the DNA. Therefore, high CMA3 fluorescence is a strong indicator of the low protamination state of spermatozoa chromatin [62].

Technique: for CMA3 staining, semen smears are first fixed in methanol–glacial acetic acid (3:1) at 4°C for 20 min and are then allowed to air-dry at room temperature for 20 min. The slides are treated for 20 min with 100 µL of CMA3 solution. The CMA3 solution consists of 0.25 mg/mL CMA3 in McIlvain's buffer (pH 7.0) supplemented with 10 mmol/L $MgCl_2$. The slides are rinsed in buffer and mounted with 1:1 v/v PBS-glycerol. The slides are then kept at 4°C overnight.

Fluorescence is evaluated using a fluorescent microscope. A total of 200 spermatozoa are randomly evaluated on each slide. CMA3 staining is evaluated by distinguishing spermatozoa that stain bright yellow (CMA3 positive) from those that stain dull yellow (CMA3 negative) [62].

As a discriminator of IVF success (>50% oocytes fertilized), CMA3 staining has a sensitivity of 73% and specificity of 75%. Therefore, it can distinguish between IVF success and failure [63]. In cases of ICSI, percentage of CMA3 positivity does not indicate failure of fertilization entirely and suggested that poor chromatin packaging contributes to a failure in the decondensation process and probably reduced fertility [64]. It appears that semen samples with high CMA3 positivity (>30%) may have significantly lower fertilization rates if used for ICSI [65].

The CMA3 assay yields reliable results as it is strongly correlated with other assays used in the evaluation of sperm chromatin. In addition, the sensitivity and specificity of the CMA3 stain are comparable with those of the AAB stain (75 and 82%, 60 and 91%, respectively) if used to evaluate the chromatin status in infertile men [66]. However, the CMA3 assay is limited by observer subjectivity.

Conclusion

Normal structure and maturity of sperm chromatin is essential for the fertilizing ability of spermatozoa in vivo. It is a relatively independent measure of semen quality that yields additional prognostic information complementary to standard sperm parameters – concentration, motility, and morphology. Several methods are used to assess sperm chromatin status. At present, indirect methods for sperm DNA fragmentation assessment are routinely used in andrological workup. However, several simple and efficient tests for chromatin maturation status are also available. The normality ranges and predictive thresholds for male fertility potential for these assays still need to be established or clarified.

References

1. Oehninger S. Strategies for the infertile man. Semin Reprod Med. 2001;19:231–7.
2. Guzick DS, Overstreet JW, Factor-Litvak P, et al. National Cooperative Reproductive Medicine Network. Sperm morphology, motility, and concentration in fertile and infertile men. N Engl J Med. 2001;345:1388–93.
3. Sakkas D, Tomlinson M. Assessment of sperm competence. Semin Reprod Med. 2000;18:133–9.
4. Agarwal A, Allamaneni SSR. Sperm DNA damage assessment: a test whose time has come. Fertil Steril. 2005;84:850–3.
5. Agarwal A, Said Tamer M. Role of sperm chromatin abnormalities and DNA damage in male infertility. Hum Reprod Update. 2003;9:331–45.
6. Erenpreiss J, Spano M, Erenpreisa J, Bungum M, Giwercman A. Sperm chromatin structure and male fertility: biological and clinical aspects. Asian J Androl. 2006;8:11–29.
7. Evenson D, Regina W. Meta-analysis of sperm DNA fragmentation using the sperm chromatin structure assay. Reprod Biomed Online. 2006;12:466–72.
8. Borini A, Tarozzi N, Bizzaro D, Bonu MA, Fava L, Flamigni C, et al. Sperm DNA fragmentation: paternal effect on early post-implantation embryo development in ART. Hum Reprod. 2006;21:2876–81.
9. Aitken RJ, De Iuliis GN. Origins and consequences of DNA damage in male germ cells. Reprod Biomed Online. 2007;14:727–33.
10. Bungum M, Humaidan P, Axmon A, Spano M, Bungum L, Erenpreiss J, et al. Sperm DNA integrity assessment in prediction of assisted reproduction technology outcome. Hum Reprod. 2007;22:174–9.
11. Ozmen B, Koutlaki N, Youssry M, Diedrich K, Al-Hasani S. DNA damage of human spermatozoa in assisted reproduction: origins, diagnosis, impacts and safety. Reprod Biomed Online. 2007;14:384–95.
12. Tarozzi N, Bizzaro D, Flamigni C, Borini A. Clinical relevance of sperm DNA damage in assisted reproduction. Reprod Biomed Online. 2007;14:746–57.
13. Lopes S, Jurisicova A, Sun JG, Casper RF. Reactive oxygen species: potential cause for DNA fragmentation in human spermatozoa. Hum Reprod. 1998;13:896–900.
14. Evenson DP, Jost LK, Marshall D, Zinaman MJ, Clegg E, Purvis K, et al. Utility of the sperm chromatin structure assay as a diagnostic and prognostic tool in the human fertility clinic. Hum Reprod. 1999;14:1039–49.
15. Sun JG, Jurisicova A, Casper RF. Detection of deoxyribonucleic acid fragmentation in human sperm: correlation with fertilization in vitro. Biol Reprod. 1997;56:602–7.
16. Duran EH, Morshedi M, Taylor S, Oehninger S. Sperm DNA quality predicts intrauterine insemination outcome: a prospective cohort study. Hum Reprod. 2002;17:3122–8.

17. Larson KL, DeJonge CJ, Barnes AM, Jost LK, Evenson DP. Sperm chromatin structure assay parameters as predictors of failed pregnancy following assisted reproductive techniques. Hum Reprod. 2000;15:1717–22.
18. De Jonge C. The clinical value of sperm nuclear DNA assessment. Hum Fert. 2002;5:51–3.
19. Evenson D, Darzynkiewicz Z, Jost L, Janca F, Ballachey B. Changes in accessibility of DNA to various fluorochromes during spermatogenesis. Cytometry. 1986;7:45–53.
20. Erenpreiss J, Bars J, Lipatnikova V, Erenpreisa J, Zalkalns J. Comparative study of cytochemical tests for sperm chromatin integrity. J Androl. 2001;22:45–53.
21. Gledhill BL, Gledhill MP, Rigler R, Ringertz NR. Atypical changes of deoxyribonucleo-protein during spermiogenesis associated with a case of infertility in the bull. J Reprod Fertil. 1966;12:575–8.
22. Gledhill BL, Gledhill MP, Rigler R, Ringertz NR. Changes in deoxyribonucleoprotein during spermiogenesis in the bull. Exp Cell Res. 1966;41:652–65.
23. Nijs M, Creemers E, Cox A, Franssen K, Janssen M, Vanheusden E, et al. Chromomycin A3 staining, sperm chromatin structure assay and hyaluronic acid binding assay as predictors for assisted reproductive outcome. Reprod Biomed Online. 2009;19(5):671–84.
24. Iranpour FG, Nasr-Esfahani MH, Valojerdi MR, al-Taraihi TM. Chromomycin A3 staining as a useful tool for evaluation of male fertility. J Assist Reprod Genet. 2000;17(1):60–6.
25. Ward WS, Coffey DS. DNA packaging and organization in mammalian spermatozoa: comparison with somatic cells. Biol Reprod. 1991;44:569–74.
26. Fuentes-Mascorro G, Serrano H, Rosado A. Sperm chromatin. Arch Androl. 2000;45:215–25.
27. Gatewood JM, Cook GR, Balhorn R, et al. Sequence-specific packaging of DNA in human sperm chromatin. Science. 1987;236:962–4.
28. Bench GS, Friz AM, Corzett MH, et al. DNA and total protamine masses in individual sperm from fertile mammalian subjects. Cytometry. 1996;23:263–71.
29. Sakkas D, Mariethoz E, Manicardi G, et al. Origin of DNA damage in ejaculated human spermatozoa. Rev Reprod. 1999;4:31–7.
30. Irvine DS, Twigg JP, Gordon EL, et al. DNA integrity in human spermatozoa: relationships with semen quality. J Androl. 2000;21:33–44.
31. Ward WS. Deoxyribonucleic acid loop domain tertiary structure in mammalian spermatozoa. Biol Reprod. 1993;48:1193–201.
32. Solov'eva L, Svetlova M, Bodinski D, et al. Nature of telomere dimers and chromosome looping in human spermatozoa. Chromosome Res. 2004;12:817–23.
33. Ward WS, Zalensky AO. The unique, complex organization of the transcriptionally silent sperm chromatin. Crit Rev Eukaryot Gene Expr. 1996;6:139–47.
34. De Jonge CJ. Paternal contributions to embryogenesis. Reprod Med Rev. 2000;8:203–14.
35. Lewis JD, Song Y, de Jong ME, et al. A walk through vertebrate and invertebrate protamines. Chromosoma. 1999;111:473–82.
36. Corzett M, Mazrimas J, Balhorn R. Protamine 1: protamine 2 stoichiometry in the sperm of eutherian mammals. Mol Reprod Dev. 2002;61:519–27.
37. Jager S. Sperm nuclear stability and male infertility. Arch Androl. 1990;25:253–9.
38. Darzynkiewicz Z. Acid-induced denaturation of DNA in situ as a probe of chromatin structure. Methods Cell Biol. 1994;41:527–41.
39. Erenpreisa EA, Zirne RA, Zaleskaia ND, S'iakste TG. Effect of single-stranded breaks on the ultrastructural organization and cytochemistry of the chromatin in tumor cells. Biull Eksp Biol Med. 1988;106:591–3. [Article in Russian].
40. Erenpreisa EA, Sondore O, Zirne RA. Conformational changes in the chromatin of tumor cells and the phenomenon of nuclear achromasia. Eksp Onkol. 1988;10:54–7. [Article in Russian].
41. Sculthore HH. Metachromasia. Med Lab Sci. 1978;35:365–70.
42. Erenpreisa J, Zaleskaya N. Effect of triton X-100 on cytochemical and ultrastructural pattern of chromatin. Acta Morphol Hung. 1983;31:387–93.
43. Erenpreisa J, Freivalds T, Selivanova G. Influence of chromatin condensation on the absorption spectra of nuclei stained with toluidine blue. Acta Morphol Hung. 1992;40:3–10.
44. Brewer LR, Corzett M, Balhorn R. Protamine-induced condensation and decondensation of the same DNA molecule. Science. 1999;286:120–3.
45. Brewer L, Corzett M, Balhorn R. Condensation of DNA by spermatid basic nuclear proteins. J Biol Chem. 2002;277:38895–900.
46. Brewer L, Corzett M, Lau EY, Balhorn R. Dynamics of protamine 1 binding to single DNA molecules. J Biol Chem. 2003;278:42403–8.
47. Benyajati C, Worcel A. Isolation, characterization, and structure of the folded interphase genome of Drosophila melanogaster. Cell. 1976;9:393–407.
48. Terquem T, Dadoune JP. Aniline blue staining of human spermatozoa chromatin. Evaluation of nuclear maturation. In: Adre J, editor. The sperm cell. The Hague: Martinus Nijhoff Publishers; 1983. p. 249–52.
49. Liu DY, Baker HW. Sperm nuclear chromatin normality: relationship with sperm morphology, sperm-zona pellucida binding, and fertilization rates in vitro. Fertil Steril. 1992;58:1178–84.
50. Manicardi GC, Bianchi PG, Pantano S, Azzoni P, Bizzaro D, Bianchi U, et al. Presence of endogenous nicks in DNA of ejaculated human spermatozoa and its relationship to chromomycin A3 accessibility. Biol Reprod. 1995;52:864–7.
51. Bianchi PG, Manicardi GC, Bizzaro D, Bianchi U, Sakkas D. Effect of deoxyribonucleic acid protamination on fluorochrome staining and in situ nick-translation of murine and human mature spermatozoa. Biol Reprod. 1993;49:1083–8.

52. Mello ML. Induced metachromasia in bull spermatozoa. Histochemistry. 1982;74:387–92.
53. Andreetta AM, Stockert JC, Barrera C. A simple method to detect sperm chromatin abnormalities: cytochemical mechanism and possible value in predicting semen quality in assisted reproductive procedures. Int J Androl. 1995;18 Suppl 1:23–8.
54. Hammadeh ME, Zeginiadov T, Rosenbaum P, et al. Predictive value of sperm chromatin condensation (aniline blue staining) in the assessment of male fertility. Arch Androl. 2001;46:99–104.
55. Baker H, Liu D. Assessment of nuclear maturity. In: Acosta A, Kruger T, editors. Human spermatozoa in assissted reproduction. London: CRC Press; 1996. p. 193–203.
56. Foresta C, Zorzi M, Rossato M, et al. Sperm nuclear instability and staining with aniline blue: abnormal persistence of histones in spermatozoa in infertile men. Int J Androl. 1992;15:330–7.
57. Hammadeh ME, Stieber M, Haidl G, et al. Association between sperm cell chromatin condensation, morphology based on strict criteria, and fertilization, cleavage and pregnancy rates in an IVF program. Andrologia. 1998;30:29–35.
58. Erenpreiss J, Bars J, Lipatnikova V, et al. Comparative study of cytochemical tests for sperm chromatin integrity. J Androl. 2001;22:45–53.
59. Erenpreiss J, Jepson K, Giwercman A, et al. Toluidine blue cytometry test for sperm DNA conformation: comparison with the flow cytometric sperm chromatin structure and TUNEL assays. Hum Reprod. 2004;19:2277–82.
60. Erenpreisa J, Erenpreiss J, Freivalds T, et al. Toluidine blue test for sperm DNA integrity and elaboration of image cytometry algorithm. Cytometry. 2003;52:19–27.
61. Tsarev I, Bungum M, Giwercman A, Erenpreisa J, Ebessen T, Ernst E, et al. Evaluation of male fertility potential by Toluidine Blue test for sperm chromatin structure assessment. Hum Reprod. 2009;24:1569–74.
62. Manicardi GC, Bianchi PG, Pantano S, et al. Presence of endogenous nicks in DNA of ejaculated human spermatozoa and its relationship to chromomycin A3 accessibility. Biol Reprod. 1995;52:864–7.
63. Esterhuizen AD, Franken DR, Lourens JG, et al. Sperm chromatin packaging as an indicator of in-vitro fertilization rates. Hum Reprod. 2000;15:657–61.
64. Sakkas D, Urner F, Bianchi PG, et al. Sperm chromatin anomalies can influence decondensation after intracytoplasmic sperm injection. Hum Reprod. 1996;11:837–43.
65. Sakkas D, Urner F, Bizzaro D, et al. Sperm nuclear DNA damage and altered chromatin structure: effect on fertilization and embryo development. Hum Reprod. 1998;13:11–9.
66. Fernandez JL, Vazquez-Gundin F, Delgado A, et al. DNA breakage detection-FISH (DBD-FISH) in human spermatozoa: technical variants evidence different structural features. Mutat Res. 2000;453:77–82.

Acridine Orange Test for Assessment of Human Sperm DNA Integrity

13

Alex C. Varghese, C. Fischer-Hammadeh, and M.E. Hammadeh

Abstract

Acridine orange test (AOT) is a simple microscopic procedure based on acid conditions to denaturant DNA followed by staining with acridine orange. The AOT measures the metachromatic shift of AO fluorescence from green (native DNA) to red (denatured DNA). Acridine Orange fluoresce green when it binds to native DNA and red when it binds to the fragmented DNA. Many authors observed that ≥50% green fluorescence in sample is a normal cut-off value for AOT in sample from fertile donors. AOT using fluorescence microscopy provides a general picture of the status of DNA denaturation.

Similarly, the sperm chromatin structure assay (SCSA) using cytometry and SCSA Software measures the intensity of Acridine Orange (AO) fluorescence. The ratio of red/red + green yields the percentage of DNA fragmentation, referred to as a DNA fragmentation index (DFI%).

The semen samples with SCSA value of less than or equal 15% DFI represent low level, greater than 15% to less than or equal 30% DFI values represent moderate, and more than or equal 30% DFI values represent high levels of DNA fragmentation. In the present article the history, principle, mechanism, technique and troubleshooting points and the clinical significance of Acridine Orange Test are thoroughly discussed.

Keywords

Acridine orange test • Sperm DNA integrity • Acridine orange staining technique • Sperm chromatin denaturation

Introduction

Acridine orange (AO) is a fluorescent cationic cytochemical stain that is specific for cell nuclei, and specifically, DNA. It is used as a supravital stain and in fluorescence cytochemistry. The compound binds to genetic material and can

A.C. Varghese (✉)
Fertility Clinic and IVF Division, AMRI Medical Centre (A Unit of AMRI Hospitals), Kolkata, India
e-mail: alexcv2008@gmail.com

A. Zini and A. Agarwal (eds.), *Sperm Chromatin: Biological and Clinical Applications in Male Infertility and Assisted Reproduction*, DOI 10.1007/978-1-4419-6857-9_13,

differentiate between deoxyribonucleic acid (DNA) and ribonucleic acid (RNA) and reflects sperm chromatin denaturation. AO staining fluoresces green when it intercalates into native double-stranded and normal DNA as a monomer, and red when it binds to denatured single-stranded DNA as an aggregate [1]. Thus, the maturity of mammalian sperm nuclei can be assessed by the AO nuclear fluorescence of sperm. AO staining is a simplified microscopic and cytochemical method for determining sperm DNA integrity, which allows the differentiation between normal, double-stranded and abnormal, and single-stranded sperm DNA, using the metachromatic properties of the dye [2].

History

Over the past 25 years, various methods have been developed to measure sperm DNA strand breaks in situ. Currently, there are several tests of sperm DNA fragmentation, including the Comet, TUNEL, sperm chromatin structure assay (SCSA), and the acridine orange test (AOT).

AO has been used for many years to label nucleic acids of somatic cells [3, 4]. Evenson et al. [5] first reported deferential staining of human semen sample with AO, based on the amount of denatured DNA in spermatozoa; with this technique, spermatozoa from infertile men displayed increased red fluorescence when compared to those from fertile men [5]. Tejada et al. [6] have proposed a new simple test for the study of sperm head chromatin heterogeneity by evaluating the resistance of the chromatin to denaturing agents.

Later, it was introduced as an indicator of the DNA status of human spermatozoa [7, 8]. This dye produces green fluorescence when AO monomers intercalated between parallel bases in an expanded double-strand DNA helix. Orange or red fluorescence indicates ionic blondes between AO polymers and single–stranded DNA [9]. This reflects the process of Protamine binding to the external groove of DNA [10], which in turn replaced the histones in somatic cells in the spermatids stage during spermatogenesis [11]. AO has been used to determine nuclear maturity and DNA condensation of sperm; red (AO) staining increase in sperm when the sperm's nuclear is immature and contains more single–stranded, thiol-containing protamine nucleoprotein Kosower et al. [12].

Therefore, higher level of red staining sperm in the ejaculate would suggest higher levels of immature sperm and would also suggest that fewer functionally mature sperm would be present in the ejaculate [13]. Using the AO metachromatic properties, some investigators have applied this test for visualizing the spermatic fragmented DNA on fluorescence microscope [6, 14, 15].

However, as stated by Evenson et al. [16], disadvantages of the microscopy technique are the high intraobserver variations and low numbers of spermatozoa analyzed, resulting in low statistical value [16, 17]. The major problem of this technique is the interobserver variability because there are several intermediate colors associated with different levels of sperm denaturation. Moreover, the results are not highly reproducible, as they can change with time, and do not allow one to distinguish between infertile patients and donors [18].

A flow cytometrics method for evaluation of the degree of sperm chromatin condensation by AO was developed, which also identified some specific chromatin abnormalities that may be related to some specific clinical entities [19].

The SCSA is also more likely to identify frailties in what appear to be normal spermatozoa as it challenges them with exposure to heat- or acid-induced denaturation in situ. Although the assessment of sperm chromatin appears to correlate strongly with a number of fertility parameters, it also suffers, like many other available tests, from the drawback that it may be technically difficult to perform.

Also, computer- interfaced flow cytometry (FCM) has entered the andrology laboratory and several studies have used this technique for evaluation of chromatin structure [20].

Although some DNA-chromatin assessment techniques can be performed by simple staining of a smear of sperm on a slide, other techniques such as the SCSA need FCM equipment and the necessary expertise linked with this equipment. Unless these techniques can be better automated and made less expensive, they may be difficult to utilize for routine assessment of sperm. However, until now, no sole laboratory test on its own can assess fertility potential. Therefore, multiple assays have been developed to measure sperm chromosomal aberrations, abnormal chromatin packaging, and chromatin structural integrity by using FCM [21].

AO fluorescence has been suggested as a screening test to predict human fertilization. Several studies have shown differential AO staining in human semen samples, with the sperm of subfertile men showing an increase in red fluorescence [5, 6]. However, these results were not confirmed by others [22].

Principle

AO is a nucleic acid selective metachromatic stain useful for cell cycle determination. AO interacts with DNA and RNA by intercalation and electrostatic attraction, respectively. DNA intercalated AO fluoresces green (525 nm); RNA electrostatically bound AO fluoresces red (>630 nm). It may distinguish between quiescent and activated, proliferating cells, and may also allow differential detection of multiple G_1 compartments [23].

AO staining is an established cytochemical method for determining sperm DNA integrity, allowing the differentiation between normal, double-stranded and abnormal, and single-stranded sperm DNA, using the metachromatic properties of the dye [6]. AO fluoresces green when it intercalates into native DNA (double-stranded and normal) as a monomer and red when it binds to denatured (single-stranded DNA) as an aggregate. In spermatozoa the thiol-disulfide status of the nuclear protamines determines the AO fluorescence pattern [12]. This procedure optimally stains cells for analysis by FCM. Besides, AO may also be useful as a method for measuring apoptosis, and for detecting intracellular pH gradients and the measurement of proton-pump activity [24].

There are multiple assays that may be used for the evaluation of the sperm chromatin status. The choice of which assay to be performed depends on many factors such as the expense, the available laboratory facilities, and the presence of experienced technicians. The establishment of a cut-off point between normal levels in the average fertile population and the minimal levels of sperm DNA integrity required for achieving pregnancy still remains to be investigated. Such an average range or value is still lacking for most of these assays except for the SCSA [25].

Mechanism

A fluorescent dye such as AO absorbs the energy of incoming light. The energy of the light passes into the dye molecules. This energy cannot be accommodated by the dye forever, and so is released. The released energy is at a different wavelength than was the incoming light, and so is detected as a different color (Fig. 13.1).

AO absorbs the incoming radiation because of its ring structure. The excess energy effectively

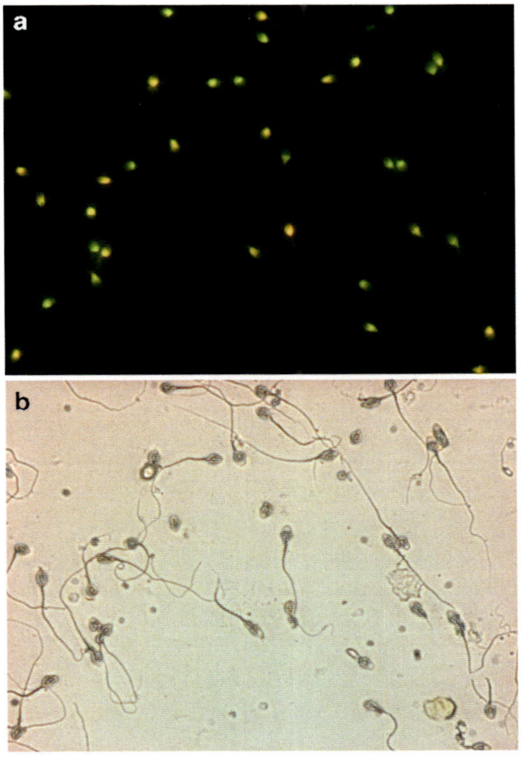

Fig. 13.1 (**a**, **b**) Human spermatozoa stained by acridine orange–fluorescent microscope and bright field microscope view of similar field of observation

passes around the ring, being distributed between the various bonds that exist within the ring. However, the energy must be dissipated to preserve the stability of the dye structure.

Binding of AO to the nucleic acid occurs in living and dead bacteria and other microorganisms. Thus, the dye is not a means of distinguishing living from dead microbes. Nor does AO discriminate between one species of microbe vs. a different species. The tightness of fit between protein and DNA can be assessed by the degree of exclusion of the dye AO, which bind to DNA [8, 9].

With this technique, it has been demonstrated that a significant portion of chromatin condensation in hamster spermatozoa occurred during passage of the spermatozoa through the epididymal lumen [26].

Mature sperms contain predominantly protamine nucleoproteins, as compared to somatic cells, which contain histones. During sperm transport through the epididymus, thiols associated with protamine nucleoproteins gradually shift to disulfides [27]. Thus, in the mature nucleus, disulfide-rich protamines dominate; in contrast, thiol-rich protamines are more prevalent in the immature sperm nucleus. The presence of disulfide–rich protamines in the mature sperm decreases the DNA's susceptibility to denaturation in the presence of acid or heat.

This denaturation can be detected by the color of AO fluorescence; AO intercalates into double-stranded DNA as a monomer and fluoresces green, whereas AO intercalates into single–stranded DNA as an aggregate and fluoresces red. Thus the presence of red or green fluoresces in a sperm population reflects nuclear maturity and the presence of single or double–stranded DNA. High levels of red staining sperm in the ejaculate would suggest higher levels of immature sperm, and would also suggest that fewer functionally mature sperm would be present in these ejaculate [13].

Acridine Orange Staining Technique

The AO assay measures the susceptibility of sperm nuclear DNA to acid-induced denaturation *in situ* by quantifying the metachromatic shift of AO fluorescence from green (native DNA) to red (denatured DNA). The fluorochrome AO intercalates into double-stranded DNA as a monomer and binds to single-stranded DNA as an aggregate. The monomeric AO bound to native DNA fluoresces green, whereas the aggregated AO on denatured DNA fluoresces red.

Procedure

The AO assay may be used for either by fluorescence microscope or FCM.

Acridine Orange Test by Fluorescence Microscope

Reagent Preparation

Add 1% AO stock solution in distilled water to a mixture of 40 mL of 0.1 M citric acid and 2.5 mL of 0.3 M $Na_2HPO_47H_2O$, pH 2.5. Store the 1% AO stock solution in the dark at 4–8°C for 4 weeks.

Sperm Preparation and AO Staining

1. Allow the semen to liquefy for 20–30 min. Semen could be selected with swim-up or discontinuous gradient centrifugation concentration gradient (80/40%) technique and then washed in 5 mL of culture media. After centrifugation, the sperm pellet should be resuspended in 0.5 mL of culture media. A small aliquot (10 μL) of sperm suspension can be spread on a glass slide.
2. Prepare a smear from each sample on a clean, glass slide and allow to air dry for 20 min.
3. Fix the slides in Carnoy's solution for at least 2 h, preferably overnight.

   Carnoy's solution constitutes three parts of methanol and one part of glacial acetic acid.
4. Wash the slides in distilled water and stained with AO solution for 5 min. The AO staining solution should be prepared daily.
5. Gently rinse the slides in a stream of deionized water. After washing and drying, the slides can be examined using a fluorescent microscope.

6. Place a coverslip before the slide dries. Place a paper towel over the mounted slide and firmly squeeze the excess water using a rubber roller.
7. Seal the coverslip with nail polish.
8. Read the slides on the same day on a fluorescent microscope using a 490-nm excitation filter and a 530-nm barrier filter. Observation time per field should be no longer than 40 s.
9. At least 200 cells should be counted so that the estimate of the number of sperm with green and red fluorescence is accurate.
10. Calculate the percentage of spermatozoa with normal DNA at ×400 magnification. Spermatozoa with normal, intact, double-stranded DNA stain green and those with denatured DNA show red or orange fluorescence. Three types of staining patterns have been identified; green sperm (double-stranded DNA), yellow and red sperm (single-stranded DNA) (Tejada et al. 1984).

Critical and Troubleshooting Points

Since several steps in AO staining method can affect the results, critical care is taken during the whole procedure. It is important to fix sperm smears on the same day of analysis and stain on the very next day. Storage of either fixed or nonfixed smears for later staining could affect the results. It is also important to use clean, grease-free and high quality microscopic slides for making the sperm smears, and ideally, stained smears should be evaluated immediately, and in a dark room. Storage of slides can cause fading of fluorescence. If unavoidable, slides could be stored at dark in 4°C but not more than 24 h. Since high background staining is a major hurdle in AOT, use of freshly prepared AO solution and removal of seminal plasma by slow speed centrifugation may be useful. Also, observer subjectivity may hinder the results if fluorescent microscopy is used.

Acridine Orange Test by Flow Cytometry (Sperm Chromatin Structure Assay)

The AO assay, also named as SCSA, is a functional assay that measures sperm quality. The SCSA measures the susceptibility of sperm nuclear DNA to heat- or acid-induced denaturation in situ, followed by staining with AO.

Although SCSA and AOT both use acid conditions to denature DNA followed by staining with AO, the reason they have no correlation for results might be the different evaluation procedure. Evenson et al. [16] suggested that fluorescence microscopy under AOT provides a general picture of the status of DNA denaturation. AOT is limited to only two to three classifications (green, red, yellow) compared with SCSA, which evaluates 1,024 discrete channels of red and green fluorescence using a flow cytometer.

As stated by Cordelli et al. "FCM is an automated approach able to measure the amount of one or more fluorescent stains associated with cells in an unbiased manner, offering unmatched properties of precision, sensitivity, accuracy, rapidity, and multiparametric analysis on a statistically relevant number of cells" [28].

Chromatin Anomalies and Clinical Significance of Acridine Orange Test

Sperm chromatin is a highly organized, condensed, and compact structure, which is considered to be an important factor for the normal fertilization and pregnancy outcome [29].

Sperm chromatin structure and DNA integrity are known to have a crucial influence on the fertilizing process [30–32] and on individual fertility capabilities [16, 33]. Infertile men are reported to have a higher fraction of sperm with chromatin defects and DNA breaks than fertile controls [34–36]. Sperm donors have also been found to exhibit lower levels of nuclear DNA damage when specifically compared to infertility patients [37]. The incidence of DNA fragmented sperm in human ejaculate is documented, in particular in men with poor semen quality [37–40]. Poor chromatin packaging has been shown to correlate with numerous reproductive outcomes: the fertility of couples after intercourse [16, 33], poor fertilization after IVF and intracytoplasmic sperm injection (ICSI) [39, 41], and a higher incidence of pregnancy loss [16].

Early onset paternal effects on zygote development [42] and early cleavage [43] have also been described. An increased number of embryos arrested at the 2–6 cell stage in the increased sperm single-stranded DNA group is likely to be related to the switch from maternal to embryonic genome at the 4–8 cell stage [44].

Moreover, in human reproduction, poor sperm quality as judged by the conventional DNA integrity assays is often found to be linked to reduced cleavage/blastocyst rates [36, 38, 45–47], reduced in vivo fertility and ART outcome [16, 36, 48–51] (for reviews see [43] and [52]). Also, a link between a paternal factor and poor embryo quality [53] resulting in reduced pregnancy rates, has been observed [54]. However, others found weak or no significant relationship between sperm DNA test results and outcome of either IVF or ICSI [55–57].

Sperm quality assessments based on the basic WHO sperm parameters are often supported by DNA integrity measurements [18]. Variations in the degree of nuclear condensation can be evaluated by several sperm nuclear maturity assessments, including AO fluorescence staining, aniline blue staining [58], Chromomycin A3 (CMA3) [59], and sodium dodecyl sulfate (SDS) analysis [60]. For the past decades, the DNA integrity of the sperm nucleus has been measured by numerous techniques, i.e., in situ nick translation, terminal deoxynucleotidyl transferase dUTP end labeling (TUNEL), single-cell electrophoresis (SCE, or comet assay in alkaline and "neutral" variants), sperm chromatin dispersion test and SCSA [61, 62]. Sharma et al. [63] summarized various assay for assessment of DNA integrity (for more details see [63]) (Table 13.1).

However, assessment of sperm chromatin integrity using the metachromatic dye AO has also been examined using both manual and automated techniques, and is claimed to be an independent measure of semen quality [64–66].

However, the clinical significance of the AO test as a sperm quality test has been controversial. Following the first report of Tejada et al. [6], who demonstrated that the AO testing of semen samples is one of the practical and clinically significant procedures to determine sperm quality, other studies have also reported its usefulness.

The literature shows several studies on sperm DNA integrity using AOT. The recent study by

Table 13.1 Illustrate a different technique for evaluate sperm chromatin maturity/DNA integrity

Technique	Assay principle	Detection method
In situ nick translation	Single-strand DNA breaks	Fluorescence microscopy
Acridine orange staining	Differentiates between single and double stranded DNA	Fluorescence microscopy
TUNEL assay	DNA fragmentation, single- and double-strand DNA breaks	Flowcytometry/ fluorescence microscopy
Alkaline single-cell gel electrophoresis (Comet assay)	Evaluates DNA integrity, single- and double-strand DNA breaks	Fluorescence microscopy
8-oxo-7,8 dihydro-2 deoxyguanosine (8-OH-dG)	HPLC with electrochemical detection	HPLC with electrochemical detection
Sperm chromatin structure assay (SCSA)	Acid DNA denaturation	Flowcytometry
DNA breakage detection-fluorescence in situ hybridization	DNA breaks	Fluorescence microscopy and image analyzer
Sperm chromatin decondensation	Intact spermatozoa with nonfragmented DNA produce characteristic DNA decondensation halo	Fluorescence microscopy
Chromamycin A3	Indirect visualization of nicked, denatured DNA	Fluorescence microscopy
Toluidine blue stain	The stain, which is a sensitive structural probe for DNA structure and packaging, becomes incorporated in the damaged dense chromatin	Optical microscopy

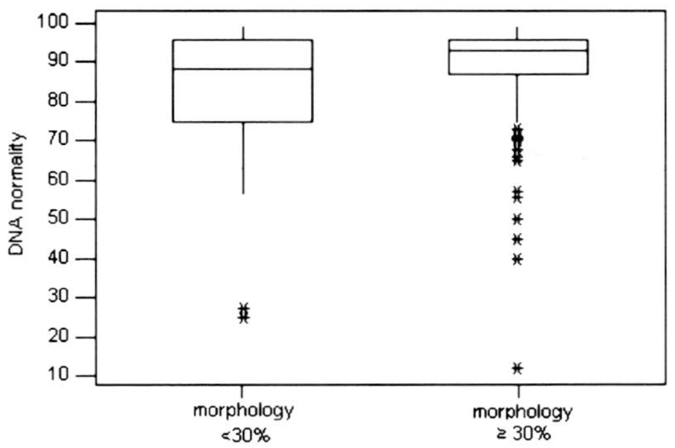

Fig. 13.2 Showing the characteristics of data on DNA normality for patients with normal morphology <30% and with normal morphology ≥30%. (P-value =0.020 for parametric test, and =0.011 for nonparametric test)

Varghese et al. [67] reported a significant correlation between DNA normality and sperm concentration (r=0.18, P=0.000), motility (r=0.21, P=0.0001), rapid motility (0.19, P=0.000), normal morphology by World Health Organization (r=0.15, P=0.019) and head defects (r=0.15, P=0.023). A significant difference was noted in AO levels between donors and patients with asthenozoospermia (P=0.002) and oligoasthenozoospermia (P=0.001). Besides, significant difference in DNA integrity was noted in samples having <30 and >30% normal morphology. A wide range of % DNA normality was observed in the patient group [67] (Fig. 13.2).

A negative correlation between semen quality and abnormal DNA integrity (ADI) assessed by AO test has been reported recently in 187 men (mostly infertile) by Erenpreiss et al. [29]. They also found a negative effect of leukocyte concentration on sperm DNA integrity (ADI: 50±10.7), especially in samples with abnormal sperm quality.

Some investigators suggested that sperm from subfertile men showed an increase in red fluorescence [5, 6, 65]. Sperm single-stranded DNA, detected by AO staining, affects the fertilization process in a classical IVF program negatively [65, 66, 68]. However, the ability of the AO test to predict fertilization and pregnancy outcome after in vitro fertilization (IVF) is controversial [22, 65, 66]. Previous studies have also shown that AOT cannot be recommended as a screening test for sperm quality and functional capacity and that AOT has a very low clinical significance for infertility testing [22, 68]. Angelopoulos et al. [69] believed that AO staining does not predict fertilization efficiency or pregnancy outcome in IVF cycles. In contrast, some studies show that sperm single-stranded DNA, detected by AO staining, affects the fertilization process in a classical IVF program negatively [2].

Hoshi et al. [66] also reported that in vitro fertilization (IVF) was successful when sperm exhibited more than or equal 50% green AO fluorescence and no pregnancies were obtained when green-fluorescing sperm were less than 50% even though an average 26% of oocytes were able to be fertilized using ICSI. Gopalkrishnan et al. [15] observed greater than 50% green fluorescence in samples from fertile donors and used this as a normal cut-off value for AOT.

Katayose et al. [70] used diamide-AO staining to detect DNA abnormalities in human sperm. A positive correlation was observed between the fertilization rate after conventional IVF and the green-type increase ratio (percentage of green-pattern sperm after diamide-AO staining/percentage of green-pattern sperm after AO staining). In addition, when the level of spermatozoa with single-stranded DNA was increased, there was a significantly lower fertilization rate and a higher percentage of fragmentation of ICSI-derived embryos.

However, no correlation was found between the level of spermatozoa with single-stranded DNA, pregnancy rate, and live-birth rate achieved by ICSI [2]. Among sperm chromatin related tests, much emphasis has been given to the SCSA test that has been widely used to assess fertility potential of both infertile and fertile individuals and is based on the ability of sperm to undergo DNA deterioration upon heat or acid treatment [6, 62]. SCSA results are reported to be closely related to fertility in both animals and humans [71–74].

Spano et al. [33] and Evenson et al. [16, 71], using the SCSA, which assesses the integrity of the nuclear DNA, showed that patients with high proportions of sperm with abnormal DNA in their ejaculate are less likely to father a child. There is a significant body of data indicating that nuclear integrity measurement, using diagnostic tools such as SCSA, TUNEL, or nick translation, adds significantly to the diagnostic power of the semen analysis. Possibly the best reported test is that used for many years by Evenson's group [16, 71]. Evenson et al. and Spano et al. have provided strong evidence of a relationship between sperm nuclear DNA integrity, as assessed by using the SCSA and fertility after both normal intercourse [16, 33] and ART [75].

However, using SCSA, Evenson et al. [16, 75] found no strong relationships between DNA damage and WHO semen parameters. Saleh et al. [76] used SCSA to assess DNA damage in 92 men seeking infertility treatment, of whom 21 had normal semen parameters and 71 had abnormal semen parameters, and in 16 fertile volunteers.

A threshold value of COMα_t >30% is reported to identify subfertile men and predict poor results with IVF [16, 33, 50, 75].

The SCSA was done at the Reproductive Medicine Center, University of Minnesota, following exactly the established protocol of Evenson and Jost [77]. More than 5,000 sperm were evaluated for each semen sample and the results were expressed as percent DNA fragmentation index (%DFI) using SCSA Software (SCSA Diagnostics Inc, Brookings, SD). The semen samples with SCSA value of less than or equal 15% DFI represent low levels, greater than 15% to less than or equal 30% DFI values represent moderate, and more than or equal 30% DFI values represent high levels of DNA fragmentation. However, the DNA damage was measured as DNA fragmentation index (% DFI), which is the percentage of cells outside the main population of αt, which represents the population of cells with DNA damage. It is interesting that % DFI in sperm was statistically significantly higher in infertile men with normal semen parameters as compared with the fertile volunteers, but it was not statistically significantly different from infertile men with abnormal semen parameters. Hence, information on sperm DNA quality might provide a good explanation for idiopathic infertility in men with normal conventional semen parameters [40].

Although, SCSA is widely used in clinical studies and increased DFI levels are reported to be associated with decreased in vivo fertilizing potential [16, 33] and intrauterine insemination (IUI) results [78]. Whether or not DFI is prognostic for the outcome of assisted reproduction remains controversial [79]. DFI showed no significant relationship to fertilization [36, 80], implantation rates [81], and embryo quality [82]. A recent study found no association between DFI levels and IVF/ICSI outcome [57]. Interestingly, Evenson et al. [16, 33] found cases in which the classical criteria (concentration, motility, and morphology) were within the normal ranges but the SCSA values were poor and not compatible with good fertility after intercourse. The microscopic and the FCM-based AO tests can uncover sperm chromatin defects in men with a normal standard semen analysis.

References

1. Ichimura S, Zama M, Fujita H. Quantitative determination of single-stranded sections in DNA using the fluorescent probe acridine orange. Biochem Biophysiol Acta. 1971;240:485–95.
2. Virant-Klun I, Tomazevic T, Meden-Vrtovec H. Sperm single- stranded DNA, detected by acridine orange staining, reduces fertilization and quality of ICSI-derived embryos. J Assist Reprod Genet. 2002; 19:319–28.
3. Rigler R, Killander D, Bolund S, Ringertz NR. Cytochemical characterization of desoxyribonuclein in individual cell nuclei. Exp Cell Res. 1969;55: 215–24.
4. Ichimura S. Differences in the red fluorescence of acridine orange bound to single stranded RNA and DNA. Biopolymers. 1975;14:1033–47.
5. Evenson DP, Darzynkiewicz Z, Melamed MR. Relation of mammalian sperm chromatin heterogeneity to fertility. Science. 1980;210:1131–1133.
6. Tejada RI, Mitchell JC, Normal A, et al. A test for the practical evaluation of male fertility by acridine orange (AO) fluorescence. Fertil Steril. 1984;42: 87–91.
7. Royere D, Hamamah S, Nicolle JC, et al. Freezing and thawing alter chromatin stability of ejaculated human spermatozoa: fluorescence acridine orange staining and Feulgen-DNA cytophotometric studies. Gamete Res. 1988;21:51–7.
8. Evenson DP. Flow cytometric analysis of male germ cell quality. Methods Cell Biol. 1990;33:40–410.
9. Evensen DP, Darzynkiewicz Z, Jost L, et al. Changes of accessibilitxy of DNA to various flourochromes during spermatogenesis. Cytometry. 1986;7:45–53.

10. Ward WS, Coffey DS. DNA packaging and organization in mammalian spermatozoa: comparison with somatic cells. Biol Reprod. 1991;44:569–74.
11. Green CM, Cockle SM, Watson PF, Fraser LR. Stimulating effect of pyroglutamylglutamylprolineamide, a prostatic, TRH-related tripeptide, on mouse sperm capacitation and fertilizing ability in vitro. Mol Reprod Dev. 1994;38(2):215–21.
12. Kosower NS, Katayose H, Yanagimachi R. Thiol-sulfide status and acridine orange fluorescence of mammalian sperm nuclei. J Androl. 1992;13:342–8.
13. Golan R, Cooper TG, Oschry Y, Oberpenning F, Schulze H, Shochat L, et al. Changes in chromatin condensation of human spermatozoa during epididymal transit as determined by flow cytometry. Hum Reprod. 1996;11:1457–62.
14. Solis EA, Bouvet BR, Brufman AS, Feldman R, Gatti VN. The possible macrophage role in seminal fluid. Actas Urol Esp. 2003;27(3):185–9.
15. Gopalkrishnan K, Hurkadli K, Padwal V, Balaiah D. Use of acridine orange to evaluate chromatin integrity of human spermatozoa in different groups in infertile men. Andrologia. 1999;31:277–82.
16. Evenson DP, Jost LK, Marshall D, Zinaman MJ, Clegg E, Purvis K, et al. Utility of the sperm chromatin structure assay as a diagnostic and prognostic tool in the human fertility clinic. Hum Reprod. 1999;14: 1039–49.
17. Amann RP. Can the fertility potential of a seminal sample be predicted accurately? J Androl. 1989;10: 89–98.
18. Chohan KR, Griffin JT, Lafromboise M, De Jonge CJ, Carrell DT. Comparison of chromatin assays for DNA fragmentation evaluation in human sperm. J Androl. 2006;27:53–9.
19. Golan R, Shochat L, Weissenberg R, Soffer Y, Marcus Z, Oschry Y, et al. Evaluation of chromatin condensation in human spermatozoa: a flow cytometric assay using acridine orange staining. Mol Hum Reprod. 1997;3:47–54.
20. Evenson DP, Emerick RJ, Jost LK, Kayongo-Male H, Stewart SR. Zinc- Silicon interactions influencing sperm chromatin integrity and testicularcell development in the rat as measured by flow cytometry. J Anim Sci. 1993;71:955–62.
21. Perreault SD, Aitken RJ, Baker HW, Evenson DP, Huszar G, Irvine DS, et al. Integrating new tests of sperm genetic integrity into semen analysis: breakout group discussion. Adv Exp Med Biol. 2003;518: 253–68.
22. Eggert-Kruse W, Rohr G, Kerber H, Schwalbach B, Demirakca T, Klinga K, et al. The acridine orange test: a clinically relevant screening method for sperm quality during infertility investigation? Hum Reprod. 1996;11:784–9.
23. Darzynkiewicz Z. Differential staining of DNA and RNA in intact cells and isolated cell nuclei with acridine orange. Meth Cell Biol. 1990;33:285–98.
24. Darzynkiewicz Z, Bruno S, Del Bino G, Gorczyca W, Hotz MA, Lassota P, et al. Features of apoptotic cells measured by flow cytometry. Cytometry. 1992;13(8): 795–808.
25. Agarwal A, Said T. Sperm chromatin assessment. In: Gardner D, Weissman A, Howles C, Shoham Z, editors. Textbook of assisted reproductive techniques. London: Martin Dunitz; 2004. p. 93–106.
26. Yossefi S, Oschry Y, Lewin LM. Chromatin condensation in hamster sperm: a flow cytometric investigation. Mol Reprod Dev. 1994;37:93–8.
27. Calvin HL, Bedford JM. Formation of disulfide bonds in the nucleus and accessory structures of mammalian spermatozoa during maturation in the epididymis. J Reprod Fert. 1971;Suppl 13:65–75.
28. Cordelli E, Eleuteri P, Leter G, Rescia M, Spanò M. Flow cytometry applications in the evaluation of sperm quality: semen analysis, sperm function and DNA integrity. Contraception. 2005;72:273–9.
29. Erenpreiss J, Spano M, Erenpreisa J, Bungum M, Giwercman A. Sperm chromatin structure and male fertility: biological and clinical aspects. Asian J Androl. 2006;8:11–29.
30. Twigg J, Irvine D, Aitken J. Oxidative damage to DNA in human spermatozoa does not preclude pronucleus formation at intracytoplasmic sperm injection. Hum Reprod. 1998;13:1864–71.
31. Agarwal A, Said T. Role of sperm chromatin abnormalities and DNA damage in male infertility. Hum Reprod Update. 2003;9:331–45.
32. Sakkas D, Seli E, Bizzaro D, Tarozzi N, Manicardi GC. Abnormal spermatozoa in the ejaculate: abortive apoptosis and faulty nuclear remodeling during spermatogenesis. Reprod Biomed Online. 2003;7:428–32.
33. Spano M, Bonde JP, Hjollund HI, et al. Sperm chromatin damage impairs human fertility. The Danish First Pregnancy Planner Study Team. Fertil Steril. 2000;73:43–50.
34. Saleh R, Agarwal A. Oxidative stress and male infertility: from research bench to clinical practice. J Androl. 2002;23:737–52.
35. Zini A, Fischer M, Sharir S, et al. Prevalence of abnormal sperm DNA denaturation in fertile and infertile men. Urology. 2002;60:1069–72.
36. Virro MR, Larson-Cook KL, Evenson DP. Sperm chromatin structure assay (SCSA) parameters are related to fertilization, blastocyst development, and ongoing pregnancy in in vitro fertilization and intracytoplasmic sperm injection cycles. Fertil Steril. 2004;81:1289–95.
37. Irvine DS, Twigg JP, Gordon EL, Fulton N, Milne PA, Aitken RJ. Integrity in human spermatozoa: relationships with semen quality. J Androl. 2000;21:33–44.
38. Sun JG, Jurisicova A, Casper RF. Detection of deoxyribonucleic acid fragmentation in human sperm: correlation with fertilization in vitro. Biol Reprod. 1997;56:602–7.
39. Lopes S, Sun J, Jurisicova A, et al. Sperm deoxyribonucleic acid fragmentation is increased in poor quality semen samples and correlates with failed fertilization in intracytoplasmic sperm injection. Fertil Steril. 1998;69:528–32.

40. Host E, Lindenberg S, Kahn J, et al. DNA strand breaks in human sperm cells: a comparison between men with normal and oligozoospermic sperm samples. Acta Obstet Gynecol Scand. 1999;78:336–9.
41. Esterhuizen AD, Franken DR, Lourens JG, et al. Sperm chromatin packaging as an indicator of in vitro fertilization rates. Hum Reprod. 2000;15:657–61.
42. Tesarik J, Mendoza C, Greco E. Paternal effects acting during the first cell cycle of human preimplantation development after ICSI. Hum Reprod. 2002;17:184–9.
43. Lewis SE, Aitken RJ. DNA damage to spermatozoa has impacts on fertilization and pregnancy. Cell Tissue Res. 2005;322:33–41.
44. Bolton VN, Hawes SM, Taylor CT, Parsons JH. Development of spare human preimplantation embryos in vitro: an analysis of the correlations among gross morphology, cleavage rates, and development to the blastocyst. J In Vitro Fert Embryo Transf. 1989;6:30–5.
45. Morris ID, Ilott S, Dixon L, Brison DR. The spectrum of DNA damage in human sperm assessed by single cell gel electrophoresis (Comet assay) and its relationship to fertilization and embryo development. Hum Reprod. 2002;17:990–8.
46. Seli E, Gardner DK, Schoolcraft WB, Moffatt O, Sakkas D. Extent of nuclear DNA damage in ejaculated spermatozoa impacts on blastocyst development after in vitro fertilization. Fertil Steril. 2004;82: 378–83.
47. Zini A, Meriano J, Kader K, Jarvi K, Laskin CA, Cadesky K. Potential adverse effect of sperm DNA damage on embryo quality after ICSI. Hum Reprod. 2005;20:3476–80.
48. Tomlinson MJ, Moffatt O, Manicardi GC, Bizzaro D, Afnan M, Sakkas D. Interrelationships between seminal parameters and sperm nuclear DNA damage before and after density gradient centrifugation: implications for assisted conception. Hum Reprod. 2001;16:2160–5.
49. Duran EH, Morshedi M, Taylor S, Oehninger S. Sperm DNA quality predicts intrauterine insemination outcome: a prospective cohort study. Hum Reprod. 2002;17:3122–8.
50. Larson-Cook KL, Brannian JD, Hansen KA, Kasperson KM, Aamold ET, Evenson DP. Relationship between the outcomes of assisted reproductive techniques and sperm DNA fragmentation as measured by the sperm chromatin structure assay. Fertil Steril. 2003;80:895–902.
51. Tesarik J, Greco E, Mendoza C. Late, but not early, paternal effect on human embryo development is related to sperm DNA fragmentation. Hum Reprod. 2004;19:611–5.
52. Spano M, Seli E, Bizzaro D, Manicardi GC, Sakkas D. The significance of sperm nuclear DNA strand breaks on reproductive outcome. Curr Opin Obstet Gynecol. 2005;17:255–60.
53. Shoukir Y, Chardonnens D, Campana A, Sakkas D. Blastocyst development from supernumerary embryos after intracytoplasmic sperm injection: a paternal influence? Hum Reprod. 1998;13:1632–7.
54. Loutradi KE, Tarlatzis BC, Goulis DG, Zepiridis L, Pagou T, Chatziioannou E, et al. The effects of sperm quality on embryo development after intracytoplasmic sperm injection. J Assist Reprod Genet. 2006;23:69–74.
55. Chohan KR, Griffin JT, Lafromboise M, De Jonge CJ, Carrell DT. Sperm DNA damage relationship with embryo quality and pregnancy outcome in IVF patients. Fertil Steril. 2004;82 Suppl 2:S55.
56. Gandini L, Lombardo F, Paoli D, Caruso F, Eleuteri P, Leter G, et al. Full-term pregnancies achieved with ICSI despite high levels of sperm chromatin damage. Hum Reprod. 2004;19:1409–17.
57. Payne JF, Raburn DJ, Couchman GM, Price TM, Jamison MG, Walmer DK. Redefining the relationship between sperm deoxyribonucleic acid fragmentation as measured by the sperm chromatin structure assay and outcomes of assisted reproductive techniques. Fertil Steril. 2005;84:356–64.
58. Terquem A, Dadoune JP. Aniline blue staining of human spermatozoa chromatin: evaluation of nuclear maturation. In: Andr J, editor. The sperm cell. London: Martinus-Nijhoff Publishers; 1983. p. 696–701.
59. Manicardi GC, Bianchi PG, Pantano S, Azzoni P, Bizzaro D, Bianchi U, et al. Presence of endogenous nicks in DNA of ejaculated human spermatozoa and its relationship to chromomycin A3 accessibility. Biol Reprod. 1995;52:864–7.
60. Gonzales GF, Salirrosas A, Dicina-Torres LN, Sanchez A, Villena A. Use of clomiphene citrate in the treatment of men with high sperm chromatin stability. Fertil Steril. 1998;69:1109–14.
61. Fernandez JL, Muriel L, Goyanes V, Segrelles E, Gosalvez J, Enciso M, et al. Simple determination of human sperm DNA fragmentation with an improved sperm chromatin dispersion test. Fertil Steril. 2005;84:833–42.
62. Evenson DP, Wixon R. Clinical aspects of sperm DNA fragmentation detection and male infertility. Theriogenology. 2006;65:979–91.
63. Sharma RK, Said T, Agarwal A. Sperm DNA damage and its clinical relevance in assessing reproductive outcome. Asian J Androl. 2004;6:139–48.
64. Evensen DP, Jost LK, Baer PK, et al. Individuality of DNA denaturation patterns in human sperm as measured by the chromatin structure assay. Reprod Toxicol. 1991;5:115–25.
65. Liu DY, Baker HWG. Sperm nuclear chromatin normality: relationship with sperm morphology, sperm – zona pellucida binding, and fertilization rates in vitro. Fertil Steril. 1992;58:1178–84.
66. Hoshi K, Katayose H, Yanagida K, et al. The relationship between acridine orange fluorescence of sperm nuclei and the fertilizing ability of human sperm. Fertil Steril. 1996;66:634–9.
67. Varghese AC, Bragais FM, Mukhopadhyay D, Kundu S, Pal M, Bhattacharyya AK, et al. Human sperm DNA integrity in normal and abnormal semen samples and its correlation with sperm characteristics. Andrologia. 2009;41(4):207–15.
68. Claassens OE, Menkveld R, Franken DR, Pretorius E, Swart Y, Lombard CJ, et al. The acridine orange test:

determining the relationship between sperm morphology and fertilization in vitro. Hum Reprod. 1992;7: 242–7.

69. Angelopoulos T, Moshel YA, Lu L, Macanas E, Grifo JA, Krey LC. Simultaneous assessment of sperm chromatin condensation and morphology before and after separation procedures: effect on the clinical outcome after in vitro fertilization. Fertil Steril. 1998;69: 740–7.
70. Katayose H, Yanagida K, Hashimoto S, et al. Use of diamide–acridine orange fluorescence staining to detect aberrant protamination of human ejaculated sperm nuclei. Fertil Steril. 2003;79:670–6.
71. Evenson DP, Larson K, Jost LK. The sperm chromatin structure assay (SCSATM): clinical use for detecting sperm DNA fragmentation related to male infertility and comparisons with other techniques. Androl Lab Corner J Androl. 2002;23:25–43.
72. Ballachey BE, Hohenboken WD, Evensen DP. Heterogeneity of sperm nuclear chromatin structure and its relationship to bull fertility. Biol Reprod. 1987;36:915–25.
73. Dobriniski I, Hughes HP, Barth AD. Flow cytometric and microscopic evaluation and effect on fertility of abnormal chromatin condensation in bovine sperm nuclei. J Reprod Fertil. 1994;101:531–8.
74. Sailer BL, Jost LK, Evenson DP. Bull sperm head morphometry related to abnormal chromatin structure and fertility. Cytometry. 1996;24:167–73.
75. Larson KL, DeJonge CJ, Barnes AM, Jost LK, Evenson DP. Sperm chromatin structure assay parameters as predictors of failed pregnancy following assisted reproductive techniques. Hum Reprod. 2000;15:1717–22.
76. Saleh RA, Agarwal A, Nelson DR, Nada EA, El-Tonsy MH, Alvarez JG, et al. Increased sperm nuclear DNA damage in normozoospermic infertile men: a prospective study. Fertil Steril. 2002;78: 313–8.
77. Evenson D, Jost L. Sperm chromatin structure assay: DNA denaturability. Methods Cell Biol. 1994;42 (Pt B):159–76.
78. Bungum M, Humaidan P, Axmon A, Spano M, Bungum L, Erenpreiss J, et al. Sperm DNA integrity assessment in prediction of assisted reproduction technology outcome. Hum Reprod. 2007;22: 174–9.
79. Collins JA, Barnhart KT, Schlegel PN. Do sperm DNA integrity tests predict pregnancy with in vitro fertilization? Fertil Steril. 2008;89:823–31.
80. Henkel R, Hajimohammad M, Stalf T, Hoogendijk C, Mehnert C, Menkveld R, et al. Influence of deoxyribonucleic acid damage on fertilization and pregnancy. Fertil Steril. 2004;81:965–72.
81. Bungum M, Humaidan P, Spano M, Jepson K, Bungum L, Giwercman A. The predictive value of sperm chromatin structure assay (SCSA) parameters for the outcome of intrauterine insemination, IVF and ICSI. Hum Reprod. 2004;19:1401–8.
82. Benchaib M, Braun V, Lornage J, Hadj S, Salle B, Lejeune H, et al. Sperm DNA fragmentation decreases the pregnancy rate in an assisted reproductive technique. Hum Reprod. 2003;18(5):1023–8.

Laboratory Evaluation of Sperm Chromatin: TUNEL Assay

14

Rakesh Sharma and Ashok Agarwal

Abstract

Routine semen analysis is unable to assess alterations in sperm chromatin organization such as DNA damage. Because fertility is based not only on the absolute number of spermatozoa but also on their functional capability, methods for exploring sperm DNA stability and integrity are being used to evaluate fertility disorders. A large number of direct and indirect tests that measure sperm DNA damage have been developed. This chapter focuses on one of those tests – the terminal deoxytransferase mediated deoxyuridine triphosphate (dUTP) nick end-labeling or TUNEL assay – which is increasingly being used in many laboratories.

Keywords

Sperm chromatin • TUNEL assay • Semen analysis • Laboratory evaluation of sperm chromatin • Terminal deoxytransferase mediated deoxyuridine triphosphate assay

One of the possible causes of infertility in men with normal semen parameters is abnormal sperm DNA. Fortunately, a number of sperm function tests are available to assess sperm DNA integrity. One of the most commonly used tests is the terminal deoxytransferase mediated deoxyuridine triphosphate (dUTP) nick end-labeling assay, which is otherwise called TUNEL. This test identifies in situ DNA strand breaks resulting from apoptotic signaling cascades by labeling the 3′-hydroxyl (3′-OH) free ends with a fluorescent label. The fluorescence, which is proportional to the number of strand breaks, can be measured either with microscopy or with flow cytometry. This chapter discusses the TUNEL assay in detail, including clinical protocols, clinical outcomes, and future strategies aimed at optimizing this test and increasing its application as the test of choice in clinical andrology.

R. Sharma (✉)
Andrology Laboratory and Center for Reproductive Medicine, Glickman Urological and Kidney Institute, OB-GYN and Women's Health Institute, Cleveland Clinic, Cleveland, OH, USA
e-mail: Sharmar@ccf.org

A. Zini and A. Agarwal (eds.), *Sperm Chromatin: Biological and Clinical Applications in Male Infertility and Assisted Reproduction*, DOI 10.1007/978-1-4419-6857-9_14,

Mechanisms of Sperm DNA Damage

When sperm DNA is damaged, infertility, miscarriage, and birth defects in offspring can occur [1]. The main cause of sperm DNA damage is oxidative stress [2–5]. Oxidative stress occurs when levels of reactive oxygen species (ROS) increase, when levels of antioxidant decrease, or both. A number of factors can lead to oxidative stress, including infection (viral or bacterial), exposure to xenobiotics, and tobacco and alcohol consumption.

DNA fragmentation may also occur during spermiogenesis. During this process, torsional stress increases, as DNA is condensed and packaged into the differentiating sperm head. Endogenous endonucleases (topoisomerases) may induce DNA fragmentation as a way of relieving this stress [6, 7].

Spermatozoa are transcriptionally and translationally inactive and cannot undergo conventional programmed cell death or "regulated cell death" called "apoptosis" but are capable of exhibiting some of the hallmarks of apoptosis including caspase activation and phosphatidylserine exposure on the surface of the sperm. This form of apoptosis is termed as "abortive apoptosis" [8, 9]. Sperm cells are able to repair some DNA damage during spermatogenesis, but once they mature, they lose this innate ability [10, 11]. Therefore, posttesticular sperm are more vulnerable to DNA damage. Studies show that DNA damage is lowest in testicular sperm and that it increases in epididymal and ejaculated sperm [12–15].

Measuring Sperm DNA Damage with TUNEL

Sperm DNA damage can be assessed with a number of techniques that measure different aspects of DNA damage (Table 14.1). Each assay has its own advantages and disadvantages (Table 14.2). One of the most commonly used assays is the TUNEL assay. The quantity of DNA 3′-OH free ends can be assessed in spermatozoa using this assay in which the terminal deoxytransferease (TdT) enzyme incorporates a fluorescent UTP at the 3′-OH end, and the fluorescence is proportional to the number of DNA strand breaks (Fig. 14.1). This assay can be run either as a slide-based (fluorescent microscopy) (Fig. 14.2) or flow-cytometry assay [16] (Fig. 14.3, Table 14.3). TUNEL identifies what is termed as "real" or actual DNA damage – that is, damage that has already occurred – as opposed to "potential" damage caused by exposing sperm to denaturing conditions (Table 14.4).

All of the assays shown in Table 14.1 have a strong correlation with one another. Unfortunately, none of them are able to selectively differentiate clinically important DNA fragmentation from clinically insignificant fragmentation. The assays also cannot differentiate the DNA nicks that occur normally (physiological) from pathological nicking, nor can they evaluate the genes that may be affected by DNA fragmentation. These assays, including TUNEL, can only determine the amount of DNA fragmentation that occurs with the assumption that higher levels of DNA fragmentation are pathological.

Measurement of DNA Damage in Spermatozoa by TUNEL Assay

DNA damage can be measured using the TUNEL assay by various protocols such as the following:

1. Biotin-d(UTP)/avidin system.
2. BrdUTP/anti-Br-dUTP-FITC system.
3. Fluorescein isothiocynate labeled (FITC) dUTP system (In Situ Cell Detection kit, Catalog No. 11 684 795 910, Roche Diagnostics GmBH, Mannheim, Germany or Roche Diagnostics, Indianapolis, IN).
4. Apoptosis detection kit (Apo-Direct kit; Catalog No. 556381; BD Pharmingen, San Diego, CA).

We describe protocol #3 and #4 because they are commonly used tests for measuring sperm DNA damage in sperm. The detection of sperm DNA fragmentation by flow cytometry and epifluorescence microscopy methods will also be described.

Table 14.1 Basics of common sperm DNA integrity assays

	Basis of assay	Measured parameter
Direct assays		
TUNEL	Adds labeled nucleotides to free DNA ends Template independent Labels SS and DS breaks	% Cells with labeled DNA
Comet	Electrophoresis of single sperm cells DNA fragments form tail Intact DNA stays in head Alkaline Comet Alkaline conditions, denatures all DNA Identifies both DS and SS breaks Neutral Comet Does not denature DNA Identifies DS breaks, maybe some SS breaks	% Sperm with long tails (tail length, % of DNA in tail)
In situ nick translation	Incorporates biotinylated dUTP at SS DNA breaks with DNA polymerase I Template-dependent Labels SS breaks, not DS breaks	% Cells with incorporated dUTP (fluorescent cells)
Indirect assays		
DNA break detection FISH	Denatures nicked DNA Whole genome probes bind to SS DNA	Amount of fluorescence proportional to number of DNA breaks
SCD	Individual cells immersed in garose Denatured with acid then lysed Normal sperm produce halo	% Sperm with small or absent halos
Acridine orange flow cytometric assays (e.g., SCSA, SDFA)	Mild acid treatment denatures DNA with SS or DS breaks Acridine orange binds to DNA DS DNA (nondenatured) fluoresces green SS DNA (denatured) fluoresces red Flow cytometry counts thousands of cells	DFI – the percentage of sperm with a ratio of red to (red + green) fluorescence greater than the main cell population
Acridine orange test	Same as above, hand-counting of green and red cells	% Cells with red fluorescence

DFI DNA fragmentation index; *DS* double-stranded; *FISH* fluorescence in situ hybridization; *SCD* sperm chromatin dispersion test; *SCSA* sperm chromatin structure assay; *SDFA* sperm DNA fragmentation assay; *SS* single-stranded; *TUNEL* terminal deoxynucleotidyl transferase-mediated dUTP nick end-labeling
From Zini and Sigman [17], with permission

First, the semen specimen is collected:

(a) Ideally, the sample should be collected after a minimum of 48 h and not more than 72 h of sexual abstinence. The name of the patient, period of abstinence, date, and time and place of collection should be recorded on the form accompanying each semen analysis.

(b) The sample should be collected in private in a room near the laboratory. If not, it should be delivered to the laboratory within 1 h of collection.

(c) The sample should be obtained by masturbation and ejaculated into a clean, wide-mouth plastic specimen cup. Lubricants should not be used to facilitate semen collection.

Table 14.2 Advantages and disadvantages of various DNA integrity assays

Direct assays	Pros	Cons
TUNEL	Can perform on few sperms Expensive equipment not required Simple and fast High sensitivity Indicative of apoptosis Correlated with semen parameters Associated with fertility Available in commercial kits	Thresholds not standardized Variable assay protocols Not specific to oxidative damage Special equipment required (flow cytometer)
COMET	High Sensitivity Simple and inexpensive Correlates with seminal parameters Small number of cells required Can perform on few sperm Alkaline: identifies all breaks Neutral: may identify more clinically relevant breaks	Labor intensive Not specific to oxidative damage Subjectiveness in data acquired No evident correlation in fertility Lack of standard protocols Requires imaging software Variable assay protocols Alkaline: may identify clinically unimportant fragmentation May induce breaks at "alkaline-labile" sites
In situ nick translation	Simple	Unclear thresholds
Indirect assays		Less sensitive
DNA break detection FISH	Can perform on few sperm	Limited clinical data
SCD	Easy, can use bright-field microscopy	Limited clinical data
Acridine orange flow cytometric assays	Many cells rapidly examined Most published studies reproducible	Expensive equipment required Small variations in lab conditions affect results Calculations involve qualitative decisions
Manual acridine orange test	Simple	Difficulty with indistinct colors, rapid fading, heterogeneous staining
8-OHdg analysis	High specificity Quantitative High sensitivity Correlated with sperm function Associated with fertility	Large amount of sample required Introduction of artifacts Special equipment required Lack of standard protocols

FISH fluorescence in situ hybridization; *SCD* sperm chromatin dispersion test; *TUNEL* terminal deoxynucleotidyl transferase-mediated dUTP nick end-labeling
From Zini and Sigman [17], with permission

(d) Coitus interruptus is not acceptable as a means of collection because it is possible that the first portion of the ejaculate, which usually contains the highest concentration of spermatozoa, will be lost. Moreover, cellular and bacteriological contamination of the sample and the acid pH of the vaginal fluid adversely affect sperm quality.

(e) The sample should be protected from extreme temperatures (not less than 20°C and not more than 40°C) during transport to the laboratory.

(f) Any unusual collection or condition of the specimen should be noted on the report form.

Protocol #3: In Situ Death Detection kit (Roche Diagnostics, Indianapolis, IN)

Reagents and Equipment

(a) Flow cytometry tubes (12 × 75 mm)

(b) Pipettes and pipette tips (1,000, 100, and 50 μL)

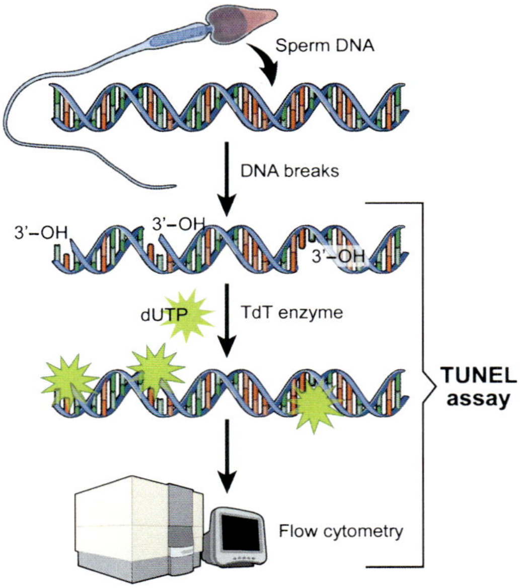

Fig. 14.1 Schematic of the TUNEL assay

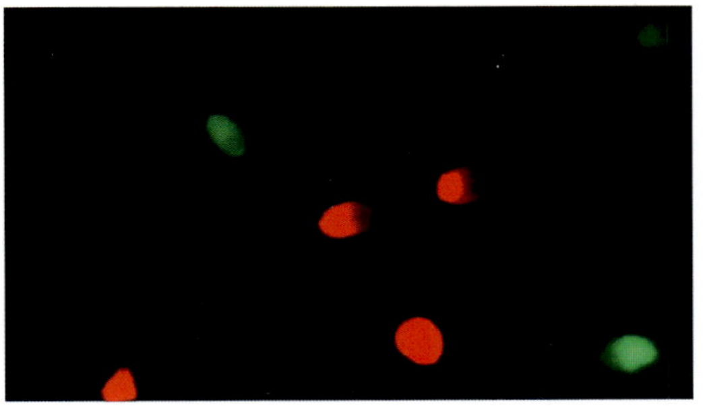

Fig. 14.2 Fluorescence microscopic staining with TUNEL and propidium iodide. TUNEL-positive sperm stain green and TUNEL-negative samples stain red

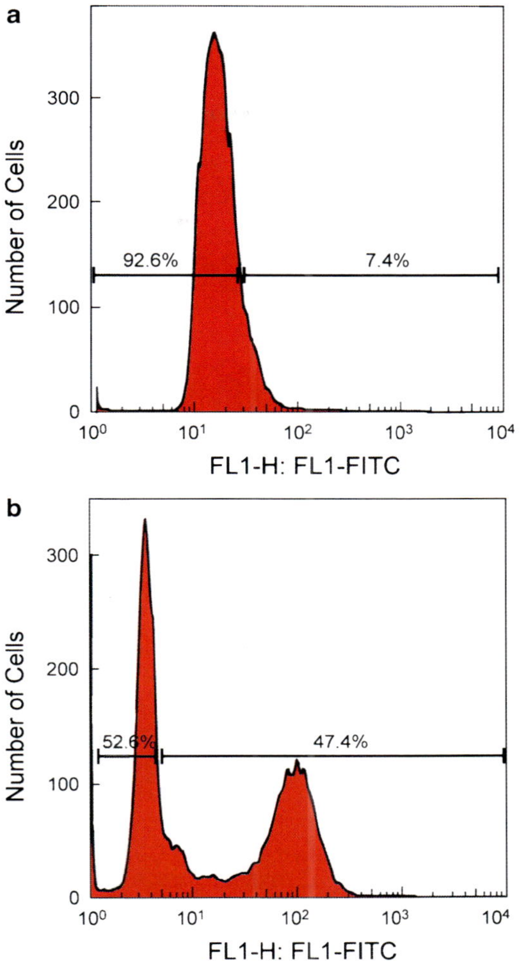

Fig. 14.3 Representative histogram showing (**a**) TUNEL negative and (**b**) TUNEL positive sample

(c) Serological pipettes (2 and 5 mL)
(d) Sperm counting chamber (MicroCell; Conception Technologies, San Diego, CA)
(e) Paraformaldehyde in phosphate-buffered saline, pH 7.4
(f) Ethanol (70%)

In Situ Death Detection Kit

1. Blue vial/cap (Enzyme solution): It contains the terminal deoxynucleotidyl transferase (TdT) enzyme solution. It is ×10 concentration and contains 5×50 μL aliquots.
2. Violet vial (Label solution): It consists of a nucleotide mixture in a reaction buffer of ×1 concentration and has 5×550 μL aliquots.
3. Benchtop centrifuge.
4. Flow cytometer.
5. Phase and epifluorescence microscope.

Assay Principle

The cleavage of genomic DNA during apoptosis leads to both single-strand breaks (nicks) and double-stranded, low-molecular-weight DNA fragments. These DNA strand breaks can be

Table 14.3 TUNEL positivity using flow cytometry or fluorescence microscopy technique

Reference	Sample size	TUNEL positive (%)	
		Flow cytometry	Microscopy
Muratori et al. [18]	140	11.07 ± 8.00 (0.79–42.64)	–
Muratori et al. [19]	43	15	
Lopes et al. [20] (swim up)	150	–	14.5 ± 1.5 (0.5–75)
Barroso et al. [21]	10	–	11.7 ± 7 (low motility sample)
Donnelly et al. [22]	25	–	35 (15–60); 18 (7–45) (gradient)
Gandini et al. [23]	52	–	11 (infertile) and ~2.5 (fertile)
Oosterhuis et al. [24]	34	20 ± 15 (1.3–64)	–
Sergerie et al. [25]	97	15% controls	–
Ramos and Wetzels [26]	11	–	10.0 (controls)
Zini et al. [27]	40	–	25.4 (infertile) and 10.2 (fertile)
Duran et al. [28]	119	–	7.3 ± 3.5 (pregnancy) vs. 13.9 ± 10.8 (no pregnancy) (gradient separation)
Sakkas et al. [29]	68	20.7(1.0–71.7)	–
Shen et al. [30]	60	~15	–
Weng et al. [9]	34	–	10 (patient) and 7 (control) with high motility vs. 33 (patient) and ~25 (control) in low motility samples
Benchaib et al. [31]	108	–	12–15 (abnormal samples) and 6–7 (normal after gradient separation)
Carrell et al. [32]	21	–	~38.4 (miscarriages)
Erenpreisa et al. [33]	6	–	10–40 (range) (methanol:ethanol fixed)
Erenpreiss et al. [34]	18	10.5 (4–27) (frozen)	
Lachaud et al. [35]	7	–	12.5 ± 2.2 (0 h; washed); 7.6 ± 1.1 (0 h; gradient); 1.7 ± 0.8 (0 h; swim up)
Tesarik et al. [36]	18	–	8.9 ± 3.7 (patient) and 8.7 ± 3.6 (control)
Greco et al. [37]	18	–	23.6 ± 5.1 (in ejaculates) vs. 4.8 ± 3.6 (testicular sperm)
Sergerie et al. [38]	73	40.6 (patients) vs. 13.0 (controls)	–
Sergerie et al. [38]	113	40.9 ± 14.3 (patients) and 13.1 ± 7.3 (controls)	–
Sergerie et al. [38]	15	22.44 ± 29.48 (patients) and 13.1 ± 17.56 (controls)	–
Stahl et al. [39]	24	11 (2.5–31) (control)	–
Sepaniak et al. [40]	108	–	25.9 (nonsmokers) and 32 (smokers)
Chohan et al. [41]	67	–	19.5 ± 1.3 (infertile) and 11.1 ± 0.9 (fertile)
De Paula et al. [42]	77	–	8.6 ± 3.6 (patient) and 5.4 ± 2.7 (control)
Aoki et al. [43]	79		40.8 ± 4.9 (low P1/P2) and 21.6 ± 1.7 (normal P1/P2) and 28.3 ± 3.1 (high P1/P2)
Spermon et al. [44]	22	–	21.0 (8.0–66.0) (pretreatment) and 25.0 (10–47) (posttreatment)
Dominguez et al. [45]	66	39.7 ± 23.1	15.3 ± 10.3
Sakamoto et al. [46]	15	–	79.6 ± 13.6 prevaricocele and 27.5 ± 19.4 (postvaricocele)

P1 protamine-1; *P2* protamine-2

Table 14.4 Why TUNEL methods for DNA damage is the method of choice?

	TUNEL	SCSA	Comet
Principal	1. Adds labeled nucleotides to free DNA ends 2. Template independent 3. Labels SS and DS breaks	1. Mild acid treatment denatures DNA with SS or DS breaks 2. Acridine orange binds to DNA 3. Double-stranded DNA (nondenatured) fluoresces green Single-stranded DNA (denatured) fluoresces red 4. Flow cytometry counts thousands of cells	1. Electrophoresis of single sperm cells 2. DNA fragments form tail 3. Intact DNA stays in head Alkaline Comet 1. Alkaline conditions, denatures all DNA 2. Identifies both DS and SS breaks Neutral Comet 1. Does not denature DNA 2. Identifies DS breaks, maybe some SS breaks
What is measured	% Cells with labeled DNA	DFI – the percentage of sperm with a ratio of red to (red + green) fluorescence greater than the main cell population	% sperm with long tails (tail length, % of DNA in tail)
Type of assay	Direct Objective	Indirect Objective	Direct Subjective
Ease of assay	Many labs run this assay	Samples have to be shipped to reference Lab	Very few labs perform this assay
Instrumentation	Flow cytometry	Flow cytometry	Microscopy
Nature of assay	TUNEL kit available	Only in reference or designated labs	Manual, no assay kits available
Reference alues	Ranges from 10 to 30%	Robust, >30% DFI indicative of decreased pregnancies	Clinically useful reference values not established
Type of samples	Fresh or frozen	Fresh or frozen	Fresh
Repeatability of assay	Good	Good	Poor
Cost	Inexpensive	Expensive	Inexpensive

identified by labeling the free 3′-OH termini with modified nucleotides in an enzymatic reaction (Fig. 14.1).

This occurs in two stages: (1) Labeling of DNA strand breaks with TdT, which catalyzes the polymerization of labeled nucleotides to free 3′-OH DNA ends in a template-independent manner (TUNEL reaction) and (2) Fluorescein isothiocynate (FITC)-dUTP is incorporated into nucleotide polymers, and it can be directly detected and quantified by fluorescence microscopy or flow cytometry.

This kit is designed to be a precise, fast, and simple nonradioactive technique to detect and quantify the number apoptotic cells. It is specific as it labels DNA strand breaks generated during apoptosis, which enables the test to discriminate between apoptotic and necrotic cells.

Sample Preparation

(a) Wash the semen aliquot containing 2×10^6 spermatozoa by centrifuging at 800 g at room temperature for 5 min with phosphate-buffered saline.

(b) After removing the seminal plasma, wash the pellet twice in PBS with 1% bovine serum albumin (BSA).

(c) Suspend the pellet in 100 μL of PBS/BSA (1%) and fix in 100 μL of 4% paraformaldehyde in PBS (pH 7.4) for 1 h at room temperature by vortexing.

(d) Resuspend the pellet in 100 μL of PBS and permeabilize with 100 μL of 0.1% Triton–X 100 in 0.1% sodium citrate in PBS for 2 min in ice. Repeat two washes in PBS/BSA (1%).

(e) Preparation of the staining solution
One pair of tubes (vial 1: Enzyme solution, (50 μL) and vial 2: Label solution (550 μL)) is sufficient for staining ten samples. The TUNEL reaction mixture is prepared by adding 50 μL of enzyme solution to 450 μL of label solution to give a total volume of 500 μL.

(f) Preparation of negative and positive controls
Negative control: Incubate fixed and permeabilized cells with 50 μL of label solution (without TdT).
Positive control: Incubate fixed and permeabilized cells with DNase I (3–3,000 U/mL in 50 mM Tris-HCl, pH 7.5, 1 mg/mL BSA) for 10 min at 25°C to induce DNA damage.

(g) Resuspend the pellet in 50 μL of the staining solution for 1 h at 37°C in the dark and mix them.

(h) After staining, rinse twice in PBS/BSA (1%) and resuspend in 200–500 μL PBS/BSA (1%).

(i) The samples can be directly analyzed under a fluorescence microscope or by flow cytometry.

Note: The kit is stable at −15 to −25°C.

Note: The enzyme solution (TdT) must be kept on ice and should be discarded after use.

Note: The samples can be counterstained with 0.5 μg/mL of propidium iodide to provide background DNA staining.

Protocol #4: APO-DIRECT™ kit (BD Pharmingen, Catalog # 556381)

Principal

Fragmented DNA can be detected with a reaction catalyzed by exogenous TdT and refereed as end labeling. The assay kit consists of two parts: Part A (Component No. 6536AK) that must be stored at 4°C and part B (Component No. 6536BK) that must be stored at −20°C (Table 14.5).

1. Sample preparation
   (a) Following liquefaction, load a 5-μL aliquot of the sample on a Microcell slide chamber for manual evaluation of concentration and motility. Check the concentration of sperm in the sample. Adjust it to $2\text{–}3 \times 10^6$/mL.
   (b) Suspend the cells in 3.7% (w/v) paraformaldehyde prepared in PBS (pH 7.4).
   (c) Place the cell suspension on ice for 30–60 min.
   (d) Centrifuge to pellet the cells at 300 g for 7 min. Discard the supernatant and suspend the pellet in 1 mL of ice-cold 70% (v/v) ethanol at −20°C until use. Cells can be stored at −20°C several days before use.

Table 14.5 Components of the Apo-direct kit

Component No.	Size (mL)	Description	Color code
51-6551AZ[a]	25	PI/RNase staining buffer (5 μg/mL PI, 200 μg/mL RNase)	Amber bottle
51-6549AZ[a]	0.50	Reaction buffer (contains cacodylate acid) (dimethylarsenic)	Green cap
51-6550AZ[a]	100	Rinsing buffer (contains 0.05% sodium azide)	Red cap
51-6548AZ[a]	100	Wash buffer (contains 0.05% sodium azide)	Blue cap
51-6555EZ[b]	0.40	FITC-dUTP (0.25 nmol/reaction; contains 0.05% sodium azide)	Orange cap
51-6553LZ[b]	5	Negative control cells (contains 70% vol./vol. ethanol)	Clear cap
51-6552LZ[b]	5	Positive control (contains 70% vol./vol. ethanol)	Brown cap
51-6554EZ[b]	0.038	TdT enzyme (10,000 U/mg) (20 μg/mL in 50% vol./vol. glycerol solution)	Yellow cap

[a]Component No. 6536AK to be stored at 4°C
[b]Component No. 6536BK to be stored at −20°C

Table 14.6 Preparation of staining solution for the TUNEL test

Staining solution	1 assay (μL)	6 assays (μL)	12 assays (μL)
Reaction buffer (green cap)	10.00	60.00	120.00
TdT enzyme (yellow cap)	0.75	4.50	9.00
FITC-dUTP (orange cap)	8.00	48.00	96.00
Distilled H_2O	32.25	193.00	387.00
Total volume	51.00	306.00	612.00

2. Staining Protocol
   (a) Resuspend the positive (6552LZ) and negative (6553LZ) control cells supplied in the kit by swirling the vials. Remove 2-mL aliquots of the control cell suspensions (approximately 1×10^6 cells/mL) and place in 12×75 mm centrifuge tubes. Centrifuge the control cell suspensions for 5 min at $300 \times g$ and remove the 70% (v/v) ethanol by aspiration, being careful to not disturb the cell pellet.
   (b) Resuspend each tube of control and sample tubes with 1.0 mL of Wash Buffer (6548AZ) (Blue cap) for each tube. Centrifuge as before and remove the supernatant by aspiration.
   (c) Repeat the Wash Buffer treatment.
   (d) Resuspend each tube of the control cell pellets in 50 μL of the Staining Solution (prepared as described below).
3. Staining Solution (single assay)
   (a) Prepare the staining solution by mixing the appropriate amounts of the staining reagents (Table 14.6).
   (b) Incubate the sperm in the Staining Solution for 60 min at 37°C.
   (c) At the end of the incubation time, add 1.0 mL of Rinse Buffer (6550AZ) (Red cap) to each tube and centrifuge each tube at $300 \times g$ for 5 min. Remove the supernatant by aspiration.
   (d) Repeat rinsing with 1.0 mL of Rinse Buffer, centrifuge, and then remove the supernatant by aspiration.
   (e) Resuspend the cell pellet in 0.5 mL of the PI/RNase Staining Buffer (6551AZ).
   (f) Incubate the cells in the dark for 30 min at room temperature.
   (g) Analyze the cells in PI/RNase solution by flow cytometry.

In addition to the negative and positive controls provided with the kit, it is also important to include the negative and positive sperm control samples.

- *Negative control*: In this the TdT enzyme is omitted from the reaction mixture.
- *Positive control*: DNA damage is induced by adding 100 μL of DNase I (1 mg/mL) for 1 h at 37°C.

Note: The volume of staining solution needed can be adjusted based on the number of tubes prepared and multiplying with the component volumes needed for one assay.

Note: Mix only enough staining solution necessary to complete the number of assays prepared.

Note: The staining solution is active for approximately 24 h at 4°C.

Note: If the sperm density is low, decrease the amount of PI/RNase Staining Buffer to 0.3 mL.

Note: The cells must be analyzed within 3 h of staining. The cells may begin to deteriorate if left overnight before the analysis.

Measurement of Sperm DNA Damage

Flow Cytometry

A minimum of 10,000 events are examined for each measurement at a flow rate of about 100 events/s on a flow cytometer (fluorescence activated cell sorting [FACS]) (Becton and Dickinson, San Jose, CA). The excitation wavelength is 488 nm supplied by an argon laser at 15 mW. Green fluorescence (480–530 nm) is measured in the FL-1 channel and red fluorescence (580–630 nm) in the FL-2 channel. Spermatozoa obtained in the plots are gated using a forward-angle light scatter (FSC) and a side-angle light scatter (SSC) dot plot to gate out debris, aggregates, and other cells different from spermatozoa. TUNEL-positive spermatozoa in the population are measured after converting the data into a histogram (Fig. 14.3). The percentage of positive cells (TUNEL-positive) are calculated on a 1,023-channel scale using the appropriate flow cytometer software (FlowJo Mac version 8.2.4) (FlowJo, LLC, Ashland, OR) as described by us earlier [47] (Fig. 14.3).

Fluorescence Microscopy

The sperm suspension is counterstained with 4,6 diamidoino-2-phenylindole (DAPI), 2 μg/mL in vecta shield (Vector, Burlingame, CA) or propidium iodide (5 μL). A minimum of 500 spermatozoa per sample are scored under 40x objective of the epifluorescence microscope. For the green signal (FITC), an excitation wavelength in the range of 450–500 nm (e.g., 488 nm) and detection in the range of 525–565 nm are adequate (green). The number of spermatozoa per field stained with DAPI (blue) or PI (red) is first counted and then the number of cells emitting green fluorescence (TUNEL-positive) is counted; and the numbers are expressed as percentage of total count of the sample (Fig. 14.2).

Protocol for Shipping Semen Samples for TUNEL Test

Semen samples can be shipped to labs that perform the TUNEL assay. Following liquefaction, the sperm count should be checked using these steps:

1. Fixation protocol:
   (a) Suspend the sperm cells ($2–3\times10^6$ cells/mL) in 3.7% (weight/vol.) paraformaldehyde prepared in PBS (pH 7.4).
   (b) Place the cell suspension on ice for 30–60 min.
   (c) Centrifuge the cells for 5 min at $300\times g$ and discard the supernatant.
   (d) Adjust the cell concentration to $2–3\times10^6$ cells/mL in 70% (vol./vol.) ice cold ethanol.
   (e) Store the cells in 70% (v/v) ethanol at −20°C until use. The cells can be stored at −20°C several days before use.
   (f) Label the cryovials with the sample information (i.e., date, name, type of sample, volume, etc.).
   (g) At the time of shipping, place cryovials in the cryoboxes, place these in adequate amount of dry ice and ship it by overnight courier.
   (h) Enclose the list of the samples being shipped.

(i) Ensure that the quantity of ice is sufficient to last 2–3 days in case of an unexpected delay in delivery.

Reference Ranges of Sperm Damage

We have established sperm DNA damage reference ranges using the protocol described for the apoptosis detection kit (protocol #4). Unprocessed or "raw" liquefied seminal ejaculates were used, and healthy donors of proven and unproven fertility were included. A receiver operating characteristic (ROS) curve was used to establish the cutoff values (Fig. 14.4).

Normal sperm DNA damage: <20%

Abnormal DNA damage: >20%

Sensitivity and Specificity

The sensitivity of the TUNEL test was 64% with specificity of almost 100%. This cutoff value is specific to our program; other centers should establish their own lab cutoffs, as this will vary with the methodology, assay reagents, staining steps, and patient population (Fig. 14.4).

Factors Affecting TUNEL Assay Results

The methods used for DNA damage assessment were originally developed and validated for the investigation of DNA in somatic cells. The TUNEL assay includes specific detection of free DNA ends ("nicks") by enzymatic incorporation of marked nucleotides. Therefore, an important question is whether these adaptations are adequate to allow reagents to access the highly compacted sperm DNA without inducing damage. Owing to a lack of standardization in methods, it is difficult to determine if the variations in the findings are real (related to biology) or due to differences in methods. Therefore, an important question is whether the treatments used to prepare the sperm may themselves induce DNA damage.

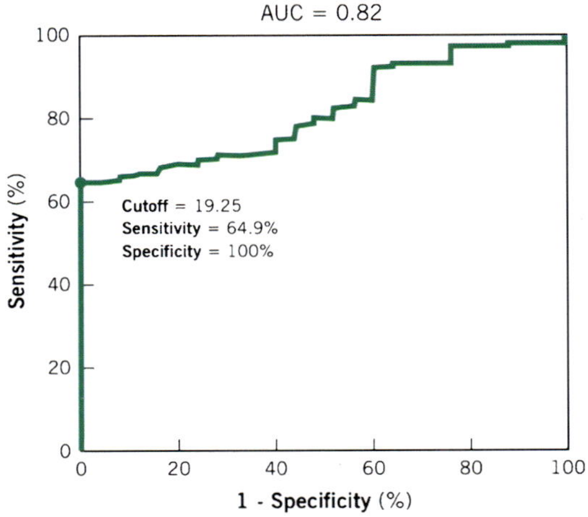

Fig. 14.4 Receiver operating characteristic (ROC) curve showing cutoff of 19.26%, sensitivity of 64%, and specificity of 100%

1. Accessibility of the DNA.

   Reagents used in the TUNEL test such as terminal deoxynucleotidyl transferase (TdT) can be a limiting factor. If the protamine bound chromatin is resistant to nucleases, it would be resistant to enzymes such as TdT. However, improperly stabilized protamine and histone bound DNA can be expected to be accessible to TdT as TdT can access DNA breaks in nucleosomal DNA. Therefore, the TUNEL assay can be expected to reveal this type of chromatin structure.
2. Sperm preparation.

   It is important to understand that the results will be different if the test is performed before or after sperm preparation. This will also influence the predictive potential of assisted reproductive technology (ART) success. The TUNEL assay has been shown to be discriminative for clinical pregnancy using either raw semen or cohorts of spermatozoa prepared by density-gradient centrifugation for clinical use [28, 48].
3. Presence of dead cells.

   Interpretation of the results can be confounded by the presence of dead cells as in the case of tests performed on unprocessed semen. Dead cells contain fragmented DNA, and this may bias the overall results.
4. Number of cells examined.

   A large number of spermatozoa (approximately 400) must be counted for accuracy. If a lower number is counted, the confidence limits will increase. Counting by flow cytometry is faster and more accurate and robust than counting with optical microscopy. In fact, flow cytometry results were shown to be 2.6 fold higher than the results from fluorescence microscopy [45].

5. Inter-and intraobserver as well as the inter- and intraassay variations.
   Establishing inter- and intraobserver as well as inter- and intraassay variations is extremely important [47]. For example, the total variation among a set of healthy normal men (control) would tend to be much less than the total variability among a set of infertile patients, so even if separate observers provided a similar degree of interobserver variability in the two populations with respect to their absolute difference in assigned TUNEL values, the results would look more impressive in the more variable patient sample, since the variance components are compared to total variability.
6. Other factors.
   Similarly, in establishing the cutoff or normal threshold values as well as the sensitivity, specificity, NPV and PPV, it is important to determine whether a test will be utilized as a screening/diagnostic test or to predict an established end point. Sensitivity is important and must be high for a test to be used for screening or diagnostic purposes so that it can be offered to a large population. However, specificity becomes critical for a test to be offered as a predictive marker of a defined end point. Positive and negative predictive values are dependent on the prevalence of infertility in the tested population, so they will be different in populations where the percentage of fertile subjects may be higher [47].

Sergerie et al. [38] examined the threshold values of TUNEL in 47 men with proven fertility and in 66 infertile men. The infertile men had higher mean level of DNA damage than the proven fertile men (40.9 ± 14.3% vs. 13.1 ± 7.3%) ($P<0.001$). The area under curve was 0.93 for a 20% cutoff; the specificity was 89.4% and the sensitivity was 96.9%. The positive and negative predictive values were 92.8 and 95.5%, respectively. Our study [47] with 194 infertile patients and a control population consisting of men with proven and unproven fertility showed a very similar cutoff value (19.2 vs. 20%) to those reported by Sergerie et al. [38].

These values are similar to those reported earlier (20%; [31] and 24.3%; [49]; and 24%; [38]). These values are much lower than the threshold established for SCSA (≥30%). Both sensitivity and specificity are associated with intrinsic performance of the TUNEL assay. However, the PPV and NPV are strongly associated with the prevalence of the sample [38].

Standardized methods that allow researchers to compare results from different laboratories are needed. It is important to understand and control for changes in sperm chromatin that occur after ejaculation and to distinguish between genuine and artifactual variations caused by a lack of reagent access to DNA. Standardized protocols and appropriate external quality controls are necessary to implement findings worldwide. For useful clinical cutoff limits, it is also necessary that the test can distinguish between affected and unaffected individuals. In this context, correlations by themselves are not adequate and other means of interpreting results such as predictive value, likelihood ratios, odds ratios, ROS cutoff value are more valuable [50].

Future of TUNEL Assay

The literature suggests that this test is a safe and effective means of measuring sperm DNA damage. Additional research needs to be generated to further fine tune the lower thresholds and minimize the variations in the methodology. The highly specialized and compacted nature of sperm chromatin makes it less permeable and less sensitive to allow the terminal enzyme (TdT) to access the DNA stand breaks deep within the sperm nucleus. This particular issue may be one factor that contributes to variation in results. To overcome this challenge, a recent study [51] has used Dithiothreitol (DTT) to expose the chromatin for 45 min prior to the fixation step. This simple additional step significantly improved the signals generated by the spermatozoa. Furthermore, the TUNEL methodology was refined to include a vitality stain (Live/Dead Fixable Dead cell stain) that remained associated with the spermatozoa during fixation and processing of the TUNEL assay, thereby allowing both DNA integrity and vitality to be simultaneously detected in the same flow-cytometry assay. This modification allows the assay to be more sensitive and robust. Measuring viability in the spermatozoa tested for DNA damage by TUNEL is critical, as this may help further improve the predictive value of this test.

Conclusions

Sperm DNA integrity is essential for the accurate transmission of genetic information. Sperm chromatin is a highly specialized and compact structure that is essential for protection and transmission of the human genome. A large number of tests are available to assess different aspects of sperm DNA integrity, but there is no consensus on the optimal technique or appropriate clinical cutoff levels. We review the use of TUNEL test by flow cytometry and fluorescence microscopy as used by different laboratories, its advantages and challenges and highlight further improvements to make it more robust. This test has the potential of being offered to a select group of infertile patients presenting with idiopathic infertility or in cases where oxidative stress may be an underlying issue. The use of this test can be cost-effective in establishing the DNA integrity of the sperm in selected cases of male infertility by any fertility testing facility with access to flow cytometry before considering other more expensive ART procedures. Further research is needed to create a platform for andrology labs and other testing centers to use with this test in measuring sperm DNA damage.

References

1. Aitken RJ, De Iuliis GN. On the possible origins of DNA damage in human spermatozoa. Mol Hum Reprod. 2010;16:3–13.
2. Ollero M, Gil-Guzman E, Lopez MC, et al. Characterization of subsets of human spermatozoa at different stages of maturation: implications in the diagnosis and treatment of male infertility. Hum Reprod. 2001;16:1912–21.
3. Saleh RA, Agarwal A, Nada EA, El-Tonsy MH, Sharma RK, Meyer A, Nelson DR, Thomas AJ, Jr. Negative effects of increased sperm DNA damage in relation to seminal oxidative stress in men with idiopathic and male factor infertility. Fertil Steril. 2003;79:1597–1605.
4. Aitken RJ, Sawyer D. The human spermatozoon-not waving but drowning. Adv Exp Med Biol. 2003; 518:85–98.
5. De Iuliis GN, Thomson LK, Mitchell LA, Finnie JM, Koppers AJ, Hedges A, Nixon B, Aitken RJ. DNA damage in human spermatozoa is highly correlated with the efficiency of chromatin remodeling and the formation of 8-hydroxy-2'-deoxyguanosine, a marker of oxidative stress. Biol Reprod. 2009;81:517–24.
6. Sakkas D, Mariethoz E, Manicardi G, et al. U. Origin of DNA damage in ejaculated human spermatozoa. Rev Reprod. 1999;4:31–7.
7. Marcon L, Boissonneault G. Transient DNA strand breaks during mouse and human spermiogenesis new insights in stage specificity and link to chromatin remodeling. Biol Reprod. 2004;70:910–8.
8. Sakkas D, Seli E, Bizzaro D Tarozzi N, Manicardi GC.. Abnormal spermatozoa in the ejaculate: abortive apoptosis and faulty nuclear remodelling during spermatogenesis. Reprod Biomed Online 2003;7:428–32.
9. Weng SL, Taylor SL, Morshedi M, Schuffner A, Duran EH, Beebe S, Oehninger S. Caspase activity and apoptotic markers in ejaculated human sperm. Mol Hum Reprod. 2002;8:984–91.
10. Matsuda Y, Tobari I, Maemori M, Seki N. Mechanism of chromosome aberration induction in the mouse egg fertilized with sperm recovered from postmeiotic germ cells treated with methyl methanesulfonate. Mutat Res. 1989;214:165–80.
11. Aitken RJ. Founders' Lecture. Human spermatozoa: fruits of creation, seeds of doubt. Reprod Fertil Dev. 2004;16:655–64.
12. Steele EK, Lewis SE, McClure N. Science versus clinical adventurism in treatment of azoospermia. Lancet. 1999;13;353:516–7.
13. O'Connell M, McClure N, Powell LA, Steele EK, Lewis SE. Differences in mitochondrial and nuclear DNA status of high-density and low-density sperm fractions after density centrifugation preparation. Fertil Steril. 2003;79:754–62.
14. Lewis SE, O'Connell M, Stevenson M, Thompson-Cree L, McClure N. An algorithm to predict pregnancy in assisted reproduction. Hum Reprod. 2004;19:1385–94.
15. Suganuma R, Yanagimachi R, Meistrich ML. Decline in fertility of mouse sperm with abnormal chromatin during epididymal passage as revealed by ICSI. Hum Reprod. 2005;20:3101–8.
16. Sailer BL, Jost LK, Evenson DP. Mammalian sperm DNA susceptibility to in situ denaturation associated with the presence of DNA strand breaks as measured

by the terminal deoxynucleotidyl transferase assay. J Androl. 1995;16:80–7.

17. Zini A, Sigman M. Are tests of sperm DNA damage clinically useful? Pros and cons. J Androl. 2009;30: 219–29.
18. Muratori M, Piomboni P, Baldi E, Filimberti E, Pecchioli P, Moretti E, Gambera L, Baccetti B, Biagiotti R, Forti G, Maggi M. Functional and ultrastructural features of DNA-fragmented human sperm. J Androl. 2000;21:903–12.
19. Muratori M, Maggi M, Spinelli S, Filimberti E, Forti G, Baldi E. Spontaneous DNA fragmentation in swim-up selected human spermatozoa during long term incubation. J Androl. 2003;24:253–262.
20. Lopes S, Sun JG, Jurisicova A, Meriano J, Casper RF. Sperm deoxyribonucleic acid fragmentation is increased in poor-quality semen samples and correlates with failed fertilization in intracytoplasmic sperm injection. Fertil Steril. 1998;69:528–532.
21. Barroso G, Morshedi M, Oehninger S. Analysis of DNA fragmentation, plasma membrane translocation of phosphatidylserine and oxidative stress in human spermatozoa. Hum Reprod. 2000;15:1338–44.
22. Donnelly ET, O'Connell M, McClure N, Lewis SE. Differences in nuclear DNA fragmentation and mitochondrial integrity of semen and prepared human spermatozoa. Hum Reprod. 2000;15:1552–1561.
23. Gandini L, Lombardo F, Paoli D, Caponecchia L, Familiari G, Verlengia C, Dondero F, Lenzi A. Study of apoptotic DNA fragmentation in human spermatozoa. Hum Reprod. 2000;15:830–839.
24. Oosterhuis GJ, Mulder AB, Kalsbeek-Batenburg E, Lambalk CB, Schoemaker J, Vermes I. Measuring apoptosis in human spermatozoa: a biological assay for semen quality? Fertil Steril. 2000;74:245–50.
25. Sergerie M, Ouhilal S, Bissonnette F, Brodeur J, Bleau G. Lack of association between smoking and DNA fragmentation in the spermatozoa of normal men. Hum Reprod. 2000;15:1314–21.
26. Ramos L, Wetzels AM. Low rates of DNA fragmentation in selected motile human spermatozoa assessed by the TUNEL assay. Hum Reprod. 2001;16:1703–1707.
27. Zini A, Bielecki R, Phang D, Zenzes MT. Correlations between two markers of sperm DNA integrity, DNA denaturation and DNA fragmentation, in fertile and infertile men. Fertil Steril. 2001a;75:674–677.
28. Duran EH, Morshedi M, Taylor S, et al. Sperm DNA quality predicts intrauterine insemination outcome: a prospective cohort study. Hum Reprod. 2002;17: 3122–8.
29. Sakkas D, Moffatt O, Manicardi GC, Mariethoz E, Tarozzi N, Bizzaro D. Nature of DNA damage in ejaculated human spermatozoa and the possible involvement of apoptosis. Biol Reprod. 2002;66:1061–1067.
30. Shen HM, Dai J, Chia SE, Lim A, Ong CN. Detection of apoptotic alterations in sperm in subfertile patients and their correlations with sperm quality. Hum Reprod. 2002;17:1266–1273.
31. Benchaib M, Braun V, Lornage J, et al. Sperm DNA fragmentation decreases the pregnancy rate in an assisted reproductive technique. Hum Reprod. 2003;18:1023–8.
32. Carrell DT, Liu L, Peterson CM, Jones KP, Hatasaka HH, Erickson L, Campbell B. Sperm DNA fragmentation is increased in couples with unexplained recurrent pregnancy loss. Arch Androl. 2003;49:49–55.
33. Erenpreisa J, Erenpreiss J, Freivalds T, Slaidina M, Krampe R, Butikova J, Ivanov A, Pjanova D. Toluidine blue test for sperm DNA integrity and elaboration of image cytometry algorithm. Cytometry. 2003;52:19–27.
34. Erenpreiss J, Jepson K, Giwercman A, Tsarev I, Erenpreisa J, Spano M. Toluidine blue cytometry test for sperm DNA conformation: comparison with the flow cytometric sperm chromatin structure and TUNEL assays. Hum Reprod. 2004; 19:2277–82.
35. Lachaud C, Tesarik J, Cañadas ML, Mendoza C. Apoptosis and necrosis in human ejaculated spermatozoa. Hum Reprod. 2004;19:607–10.
36. Tesarik J, Greco E, Mendoza C. Late, but not early, paternal effect on human embryo development is related to sperm DNA fragmentation. Hum Reprod. 2004;19:611–5.
37. Greco E, Iacobelli M, Rienzi L, Ubaldi F, Ferrero S, Tesarik J. Reduction of the incidence of sperm DNA fragmentation by oral antioxidant treatment. J Androl. 2005;26:349–53.
38. Sergerie M, Laforest G, Bujan L, et al. Sperm DNA fragmentation: threshold value in male fertility. Hum Reprod. 2005;20:3446–51.
39. Stahl O, Eberhard J, Jepson K, Spano M, Cwikiel M, Cavallin Stahl E, Giwercman A. The impact of testicular carcinoma and its treatment on sperm DNA integrity. Cancer 2004;100:1137–1144.
40. Sépaniak S, Forges T, Monnier-Barbarino P. Cigarette smoking and fertility in women and men. Gynecol Obstet Fertil. 2006;34:945–9.
41. Chohan KR, Griffin JT, Lafromboise M, De Jonge CJ, Carrell DT. Comparison of chromatin assays for DNA fragmentation evaluation in human sperm. J Androl. 2006;27:53–9.
42. de Paula TS, Bertolla RP, Spaine DM, Cunha MA, Schor N, Cedenho AP. Effect of cryopreservation on sperm apoptotic deoxyribonucleic acid fragmentation in patients with oligozoospermia. Fertil Steril. 2006; 86:597–600.
43. Aoki VW, Emery BR, Liu L, Carrell DT. Protamine levels vary between individual sperm cells of infertile human males and correlate with viability and DNA integrity. J Androl. 2006;27:890–8.
44. Spermon JR, Ramos L, Wetzels AM, Sweep CG, Braat DD, Kiemeney LA, Witjes JA. Sperm integrity pre- and post-chemotherapy in men with testicular germ cell cancer. Hum Reprod. 2006;21:1781–6.
45. Domínguez-Fandos D, Camejo MI, Ballescà JL, Oliva R. Human sperm DNA fragmentation: correlation of TUNEL results as assessed by flow cytometry and optical microscopy. Cytometry. 2007;71:1011–8.
46. Sakamoto Y, Ishikawa T, Kondo Y, Yamaguchi K, Fujisawa M. The assessment of oxidative stress in

infertile patients with varicocele. BJU Int. 2008;101: 1547–52.

47. Sharma RK, Sabanegh E, Mahfouz R, Gupta S, Thiyagarajan A, Agarwal A. TUNEL as a test for sperm DNA damage in the evaluation of male infertility. Urology 2010;76:1380–6.
48. Borini A, Tarozzi N, Bizzaro D, Bonu MA, Fava L, Flamigni C, Coticchio G. Sperm DNA fragmentation: paternal effect on early post-implantation embryo development in ART. Hum Reprod. 2006;21:2876 –2881.
49. Henkel R, Kierspel E, Hajimohammad M, Stalf T, Hoogendijk C, Mehnert C, Menkveld R, Schill WB, Kruger TF. DNA fragmentation of spermatozoa and assisted reproduction technology. Reprod Biomed Online. 2003;7:477–484.
50. Barratt CL, Aitken RJ, Björndahl L, Carrell DT, de Boer P, Kvist U, Lewis SE, Perreault SD, Perry MJ, Ramos L, Robaire B, Ward S, Zini A. Sperm DNA: organization, protection and vulnerability: from basic science to clinical applications–a position report. Hum Reprod. 2010;25:824–38.
51. Mitchell LA, De Iuliis GN, Aitken RJ. The TUNEL assay consistently underestimates DNA damage in human spermatozoa and is influenced by DNA compaction and cell vitality: development of an improved methodology. Int J Androl. 2011;34:2–13.

Basic and Clinical Aspects of Sperm Comet Assay

15

Luke Simon and Sheena E.M. Lewis

Abstract

Sperm DNA damage is associated with poorer assisted reproductive treatment (ART) outcomes including reduced fertilization rates, embryo quality, and pregnancy rates and higher rates of spontaneous miscarriage and childhood diseases. It shows promise as a more robust biomarker of infertility than conventional semen parameters. Among the sperm DNA testing methods, the alkaline comet assay is a sensitive, reliable, and powerful tool to detect even low levels of DNA damage within individual sperm. The present chapter provides an overview of the use of the alkaline comet assay in sperm. This includes the need for standardization of the alkaline comet assay protocol and its present strengths and weaknesses. Since sperm DNA damage is often the result of increased oxidative stress in the male reproductive tract, primarily formed due to an imbalance between reactive oxygen species generation and antioxidant depletion, a novel addition to the comet assay to measure oxidized bases is explored. The potential use of antioxidant therapy to protect against such damage is also described. Finally, the diagnostic and prognostic values of sperm DNA damage measures in determining the assisted reproductive technology (ART) success are discussed.

Keywords

Sperm comet assay • Sperm DNA • Assisted reproductive treatment • Comet assay • Oxidative stress in male reproduction

The Need for Novel Diagnostic and Prognostic Tests

Male infertility is implicated in more than 40% of couples presenting for treatment with assisted reproductive technology (ART). Conventional

S.E.M. Lewis (✉)
Centre for Public Health, Queen's University of Belfast, Room 208, Institute of Clinical Science, Belfast, Northern Ireland, UK
e-mail: s.e.lewis@qub.ac.uk

A. Zini and A. Agarwal (eds.), *Sperm Chromatin: Biological and Clinical Applications in Male Infertility and Assisted Reproduction*, DOI 10.1007/978-1-4419-6857-9_15,

semen analysis continues to be the only routine test to diagnose male infertility. However, semen analysis cannot discriminate between the sperm of fertile and infertile men [1]. Recent evidence has suggested that instability in the genomic material of the sperm nuclei is a more robust parameter in measuring the fertility potential of sperm, either in vivo or in vitro. For a test to be useful diagnostically or prognostically, it must have a threshold value that provides a discriminatory power above or below the threshold value with little overlap between groups of fertile and infertile men and couples with ART success and failure. However, neither the routine semen analysis nor the available sperm DNA tests yet meet these standards (reviewed in references [2, 3]).

The primary function of the sperm is to deliver the paternal genome to the oocyte. Recent studies have shown a number of sperm nuclear abnormalities such as DNA strand breaks, Y chromosome microdeletions, alterations in chromosome number, distorted epigenetic regulation and sperm's environmental milieu during epididymal transport and ejaculation. Factors such as increased oxidative stress or low levels of antioxidants may have implications on male reproductive health [4]. As the structural organization of the sperm chromatin is also essential for the normal function of the sperm [5], characterization of sperm DNA quality has gained importance. In recent years, comet assay, TUNEL, SCSA, and SCDA or Halo assay, in situ nick end labeling have been studied extensively to analyze sperm chromatin integrity. Each of these tests determines different aspects of DNA integrity, but to date, combining all the studies available in meta-analysis shows that these tests lack the statistical power and diagnostic potential necessary to incorporate them into routine clinical use.

Causes of Sperm DNA Damage

In recent years, the generation of reactive oxygen species (ROS) has been widely studied in the male reproductive tract and reported to be a concern because of their toxic effects on sperm quality and function (reviewed by Saleh and Agarwal [6]). They have been shown to cause DNA fragmentation in the reproductive tract as well as damage in ejaculated sperm [7]. High levels of ROS have also been reported in the seminal plasma of infertile men [8]. Sperm are vulnerable to the oxidative-stress-mediated damage, due to their structure with a high proportion of polyunsaturated fatty acids in their plasma membranes [9]. As sperm cannot repair such damage, sperm DNA has evolved to protect itself by compact packaging of the sperm DNA by protamines [10, 11].

The exact mechanisms by which ROS induces DNA damage are poorly understood, However, ROS-induced sperm DNA damage is exemplified by DNA cross-links, frameshifts, production of base free sites, chromosomal rearrangements and DNA base-pair oxidation [12–14]. It is also well known to cause strand breaks, with the levels of ROS correlated with increased percentage of single and double-strand damage in sperm [15–17]. ROS-mediated DNA damage is also seen in the formation of modified bases, which are often converted into strand breaks and considered to be important biomarkers for oxidative DNA damage [18]. Finally, ROS cause gene mutations such as point mutations and polymorphism [19, 20].

Seminal plasma is contaminated with ROS [21, 22] primarily produced by leukocytes and defective sperm [23]. The presence of elevated levels ($>1\times10^6$/mL) of leukocytes in the semen is defined as leukocytospermia [24] and is associated with increased levels of ROS, leading to sperm DNA damage [25]. Cytoplasmic droplets are also associated increased ROS generation and poor sperm quality [26, 27].

Environmental and Lifestyle Hazards

It has recently been reported that male fertility declines with age, even though spermatogenesis continues [28]. An increase in male age has been associated with increased genetic and chromosomal defects [29, 30]. Men over 37 years have been shown to three times more sperm DNA damage then men aged <37 years [31, 32]. Male germ cells are particularly vulnerable to environmental chemicals and xenobiotics that cause DNA damage [33]. Studies also show the adverse impact of some occupations to increase the sperm DNA damage, for example among coke oven workers [34]. Oh et al. [35] concluded that there are elevated levels of DNA damage among waste incineration workers, when compared with men from similar origin. Further, men working in the factories with organic molecules such as styrene show a significant amount of increase in sperm DNA damage [36]. Similarly, men working in the insecticide and pesticide industries have higher levels of sperm DNA damage [37, 38].

A further hazard for sperm DNA is by pharmacological exposure to drugs. This has become very common as molecular medicine advances, especially in the field of cancer. Chemotherapeutic drugs are genotoxic to the male germ cells. A well-known example for such an intervention is the cyclophosphamide [39, 40] in animal model. Hellman et al.'s [41] cyclophosphamide treatment resulted in a five fold increase in DNA damage. Environmental exposure of xenobiotics cannot be avoided in our contemporary lifestyle because these pollutants are present in our food, water, and air. Studies have shown the association between environmental estrogens and derived compounds and male infertility through elevated sperm DNA damage [42]. Environmental pollutants such as organochlorides [43] and smog [44] also have the ability to induce DNA damage. Bennetts et al. [45] showed that estrogenic compounds such as 2-hydroxyestradiol induce redox cycling activities and concomitant sperm DNA damage. These examples support the belief that exposure to xenobiotics has powerful impacts on sperm DNA and sperm functions, leading to male infertility.

Lifestyle choices also play an important role in male infertility. For example, smoking and consumption of alcohol and caffeine have been associated to the increase in nuclear DNA damage of the white blood cells [46, 47]; on the contrary, very little is known about their effect on sperm DNA [48]. There is a very strong and significant correlation between smoking and genetic defects in the sperm [49, 50]. Smoking increases oxidative stress, which results in depletion of antioxidants in the seminal plasma, thereby inducing oxidative DNA damage to the sperm [15] and mutagenic adducts [51]. Recent studies have also suggested a possible link between cell-phone use associated with electromagnetic radiations and sperm DNA damage [52–56]. Finally, physical factors such as mild scrotal heating [57] and radio frequencies [55] have also been proven to diminish sperm DNA integrity.

The Comet Assay: What Does It Measure?

For a sperm DNA test to be clinically useful, (a) it should measure both single- and double-strand breaks, as both may be important and the oocyte has limited ability to repair fragmented paternal DNA, (b) it should measure the level of DNA fragmentation in each sperm, as an ejaculate is known to show a high degree of variation, (c) the methodology should be appropriate for cell lysis and DNA decondensation for full extent of damage to be determined, (d) the test must have strong predictive capacity for pregnancy outcome and little overlap between fertile and infertile samples. Among the tests currently available, the alkaline comet assay addresses the first three above-mentioned issues but useful thresholds have not been established yet to validate the assay.

Initially, the comet assay [58] was designed to characterize the structure of the nucleus. However, when electrophoresis of DNA strands after alkaline denaturation came into existence in 1988 by Singh et al., the detection of DNA damage within the nucleus became a possibility. Collins et al. [59] suggested that the migrated comet tail after electrophoresis consists of fragments originated from relaxation of supercoiled loops and single-stranded DNA formed under alkaline conditions. Some studies suggest that double-strand DNA breaks alone may be detected under neutral conditions (pH 8–9) [60, 61], and in these studies the level of measurable DNA damage is low compared to the alkaline comet assay. This is due to either the measurement of additional DNA damage by the alkaline condition or the relatively higher migration of DNA strands under alkaline conditions [62].

The extent of DNA damage in individual cells could be monitored by the use of image analyzing system. Presently, different commercial software packages are available to measure the comet parameters. A fully automated comet analyzing system has also been developed [63]. In the past, different methodologies were used to measure the extent of DNA damage such as the proportion of cells with altered tail DNA migration, approaches classifying comets into several categories based on the tail migration [64, 65]. However, these approaches are generally limited to electrophoretic conditions. Hughes et al. [66] reported that the evidence for intact DNA is considered more important in relation to fertility status than measurement of other comet parameters that could be altered by the experiment conditions.

The commonly used comet parameters are percentage head DNA, percentage tail DNA, tail length, and olive tail moment. The software system analyzes the light intensities (fluorescence) in the head relative to the tail to determine the percentage of DNA present in the head and tail. The background light intensity is subtracted from head and tail intensities to get the actual value. Also, the sperm populations are known to be more heterogeneous, and the baseline values of DNA damage of sperm population in an ejaculate are substantially higher than those in somatic cells [67]. Although, few number of sperm could be analyzed in the comet assay, Hughes et al. [66] demonstrated that the analysis of 50 sperm is sufficient to provide a measurement of DNA damage of the total sperm population with a coefficients of variation lower than 4%.

The comet assay is highly sensitive to detect extensive fragmented cell in the form of nonexistent heads or a large diffused tail termed as "ghost" or "clouds" or "hedgehogs" [68]. In such cases, the comet image system cannot interpret the full extent of DNA damage [69]; therefore, it is advisable to consider the ghosts as completely damaged cells. In sperm, such highly damaged cells should not be excluded during analysis [70]. The DNA-specific fluorescent dyes are used for comet visualization. The most frequently used fluorescent dyes are ethidium bromide, propidium iodide, DAPI, SYBR Green I, and benzoxazolium-4-quinolinum oxazole yellow homodimer [71]. Addition of an antifade reagent along with fluorescent dyes could significantly reduce fluorescence quenching [72]. No fluorescent dyes such as silver nitrate are also reported for comet assay; however, the efficiency of the assay is reduced [73]. Excess of fluorescence dye could increase the background intensity of the slides thereby very low-molecular-weight DNA fragments could not be measured. Hence, standardization of the comet assay is required for accurate performance.

Strengths of Comet Assay

The comet assay is one of the most sensitive techniques available to measure DNA damage, and according to Aravindan et al. [74], the results of comet assay are also related to the results obtained from the TUNEL assay. The alkaline comet assay could be used in all the cell types and also in the sperm [75]. The assay requires only a few numbers of cells; hence, the assay is possible in cases of oligospermia and testicular biopsy. The DNA damage data can be collected at the level of individual cells, making the analysis efficient. The removal of protamines and histones during the assay reveals the total DNA damage in the cell. The range of DNA damage measured in sperm using the alkaline comet assay varies from 0–100% showing its capacity to identify sperm with much or little damage. A further advantage is that, unlike the TUNEL and SCSA, which detect primarily breaks in histone-associated chromatin, the comet assay has a broader use in detecting breaks in both protamine- and histone-bound chromatin equally.

Weaknesses of Comet Assay

One disadvantage of the comet assay is that it lacks standardized protocols, which makes it difficult to combine the results from different laboratories [76]. This should be resolved by agreement on an optimal protocol (see next section). The assay is criticized for the use of high pH conditions, which is known to denature the alkaline-labile sites measurable after electrophoresis [77], making it difficult to discriminate between endogenous and induced DNA breaks. However, labile sites may be considered as another form of potential damage, and some consider this as a strength, in that an indication of existing and potential damage may be more important clinically. The assay is also criticized for an underestimation of DNA damage that may occur through entangling of DNA strands or the presence of proteins and cross-linked DNA strands, which could restrict the movement of DNA fragments during electrophoresis. In some protocols, incomplete chromatin decondensation will not allow all strand breaks to be revealed. Overlapping comet tails decrease the accuracy of the assay, and few small tail fragments are lost or too small fragments are difficult to be visualized. As in other DNA tests, strong reducing agents are sometimes used to remove protamines, and they may increase what is perceived to be baseline damage. Also, the assay requires a laborious process of analysis and shows high interlaboratory variation and, hence, is not used clinically [78]. Owing to a labor-intensive and sensitive protocol, the assay requires skilled technicians for accuracy. Finally, the available software to measure DNA damage cannot recognize "Ghost cells" without head DNA and overlapping comet tails, making the scoring difficult. However, most of these weaknesses can be corrected with appropriate protocols and training.

Need for Standardized Methodology for the Comet

The comet assay is currently used primarily for genotoxic studies, although it is a test with great potential for ART [79]. For use with sperm, a number of academic and methodological issues need to be addressed, as there is no generally accepted protocol for the assay, even though international groups of scientists [53, 80–82] have used it extensively.

The first variation relates to lysis conditions. Absence of cytoplasm in sperm makes it difficult to optimize lysis conditions compared to the somatic cells. For example, in some labs, lysis of plasma membranes is performed by incubating cells with a buffer (usually containing Proteinase K, Triton X-100, and high concentrations of NaCl) for a short time (3 h), in others a long, even overnight period (18 h) [83–85].

As discussed previously, the sperm genomic DNA is more highly condensed than somatic cells preventing the migration of the comet tail, so for use with sperm it requires the use of additional steps to decondense the tightly packed DNA. A wide range of strong reagents (Proteinase K, Triton X-100, Dimethyl Sulfoxide, DTT, and LIS) have been used to remove protamines and histones [67, 83, 84, 86–88], but these agents may also induce damage. The presence of these different approaches prevents interlaboratory comparisons.

To reduce the level of laboratory-induced damage and make the assay more reproducible, our group has replaced Proteinase K with DTT and LIS and for a shorter duration of 3 h [85].

Another difference between labs is the pH at which the assay is performed. Currently, electrophoresis is carried out with wide range of buffers with pH ranging from pH 8.0 to 13.5 [66, 67, 84–86, 89]. Such a wide range of pH conditions again makes results difficult to compare, as the extent of DNA migration is highly influenced by the degree of alkali denaturation and the pH value.

A further confusion from "comet" studies comes from the lack of standardization of comet parameters described in different studies. There are several parameters used in comet studies. McKelvey et al. [90] described it as "DNA migration can be determined visually by the categorization of comets into different "classes" of migration. The percentage of DNA in the tail (percent migrated DNA), tail length and tail moment (fraction of migrated DNA multiplied by some measure of tail length). Of these, tail moment and/or tail length measurements are the most commonly reported, but there is much to recommend the use of per cent DNA in tail, as this gives a clear indication of the appearance of the comets and, in addition, is linearly related to the DNA break frequency over a wide range of levels of damage. The approach or parameter used must be clearly defined and, if not typical, be justified."

Hughes et al. [67] recommended the use of percent tail DNA, as its coefficients of variation was less than 4%. Measurement of fifty comets from a single slide is reported to have a coefficient of variation of less than 6% within a sperm population [67]. They also reported the reproducibility of the image analysis software with repeated analysis of individual sample showed a coefficient of variation of less than 5.4%.

Tice et al. [71] recommended the measurement of tail length, percent tail DNA and tail moment, finding different results between tail DNA and tail moment. However, Kumaravel and Jha [91] did not find any statistical difference with olive tail moment and percentage tail DNA to analyze the extent of DNA damage. The percentage of tail DNA is reported to be directly proportional to the dosage of radiation and concentration of hydrogen peroxide. By contrast, the olive tail moment is highly influenced by the study conditions, so it is not consistent between labs and, thus, not advisable for use.

In summary, agreement on a standardized protocol for the comet to necessary to compare results between groups. To reduce the additional DNA damage caused during the assay procedure, the duration of lysis, the composition of the lysis buffer, the method of decondensation, the pH for unwinding, and electrophoresis condition and parameters to be reported should be standardized.

Clinical Significance of DNA Fragmentation Measured by the Comet Assay

The alkaline comet assay is proving to be a useful diagnostic tool for male infertility. The clinical importance of the comet assay in assessing male infertility has been demonstrated by a number of authors [79, 92–94]. However, until recently, its predictive value in assisted reproduction outcome has been assessed by few [86, 95].

In a recent study from our group [82] of 360 couples having IVF or ICSI we reported that sperm DNA damage is associated with poorer ART outcomes and promises to be a more robust biomarker of infertility than conventional semen parameters. We found significant inverse correlations between DNA fragmentation, fertilization rate, and embryo quality assessed by the alkaline comet assay (to detect both double and single strand breaks) following IVF treatment. A decrease in fertilization rates were observed as DNA damage of native semen and DGC sperm increased. Low DNA damage (0–20%) showed a significantly higher fertilization rate compared with DNA damage >60%. Our work supports that of Morris et al. [88] who also reported a significant correlation between fertilization and DNA damage when measured by the neutral comet assay (measuring double-strand breaks only). However, by contrast, no correlations were observed between fertilization rates and DNA fragmentation measured in alkaline comet assay by Tomsu et al. [95].

Our study [82] also showed a decrease in embryo quality following IVF treatment, as DNA fragmentation increased both in native semen and DGC sperm. The embryo quality showed a significant decrease, when DNA damage was greater than 60% in the native semen. The embryo cumulative score calculated according to Steer et al. [96] was 15.5 in the group where sperm DNA fragmentation was <20% and was only 7.3 where sperm DNA fragmentation was >60% in DGC sperm. Similarly, Tomsu et al. [95] showed a negative correlation between embryo quality and DNA fragmentation in both the native semen and the DGC sperm. However, Morris et al. [88] did not find any association in embryo quality and DNA damage. In contrast to associations following IVF, we did not find any correlation between sperm DNA damage and fertilization rate or embryo quality when ICSI was used as a treatment of choice [82].

Using pregnancy as the outcome measure, Morris et al. [88] did not find an association between clinical pregnancy and sperm DNA fragmentation measured by the neutral comet assay. Similarly, Tomsu et al. [95] in a small study ($n=40$) no associations were found. However, we found a significant difference in DNA fragmentation of clinically pregnant and nonpregnant couples following IVF [82]. By contrast, although couples undergoing ICSI who failed to achieve a clinical pregnancy tended to have more DNA fragmentation but it was not statistically significant.

Further Uses of the Comet to Measure DNA Adducts

A major cause of sperm DNA damage is oxidative stress due to the generation of the ROS from contaminating leukocytes, defective sperm, and antioxidant depletion [23, 97]. FPG is the commonly used bacterial repair enzyme that could recognize and excise 8-OHdG and other modified bases generated by ROS. This FPG enzyme has been shown to possess affinities toward the various modified DNA bases [98, 99]. The catalytic activity of FPG involves a three-step process: (a) hydrolysis of the glycosidic bond between the damaged base and the deoxyribose, (b) incision of DNA at abasic sites, leaving a gap at the 3′ and 5′ ends by phosphoryl groups, and (c) removal of terminal deoxyribose 5′-phosphate from 5′ terminal site to excise the damaged base showed by Kuznetsov [100].

When a eukaryotic or prokaryotic base repair enzyme or glycosylase is introduced as an intermediate step during the alkaline comet assay, the modified bases can be converted into single-strand breaks [101, 102]. Addition of base repair enzymes can increase the sensitivity of the assay by including the modified bases, resulting in total DNA damage measured after the alkaline comet assay [103]. Among the modified bases, 8-OHdG is the most commonly studied biomarker and is often selected as a representative of oxidative DNA damage due to its high specificity, potent mutagenicity, and relative abundance in DNA [33, 104]

Clinical Significance of Existing Strand Breaks Plus Adducts Measured by the Comet Assay

To analyze modified bases in the sperm DNA, we have used the prokaryotic repair enzyme (FPG) as an intermediate step during the alkaline comet assay, to introduce breaks at sites of modified bases [82]. We found inverse relationships between total DNA damage (existing strand breaks plus modified bases) and IVF and ICSI outcomes after conversion of modified bases to DNA strand breaks by FPG. There was a significant increase in DNA damage after treatment with the DNA glycosylase FPG in both native and DGC samples. In the IVF patients, addition of the FPG enzyme showed a significant increase in mean percentage of sperm DNA fragmentation in nonpregnant compared with that from pregnant couples (55 vs. 72) in the native semen and (42 vs. 56) in DGC sperm. Similarly, in ICSI couples, when modified bases were included, the percent DNA damage between pregnant and nonpregnant couples was markedly different (63 vs. 80 in native semen, and 50 vs. 65 in DGC sperm), in contrast to comet assay without FPG where it was not significant.

The Risks of Using Sperm with Damaged DNA

Sperm DNA damage measured by SCSA, TUNEL, and alkaline and neutral comet assays has been closely associated with all the stages of ART outcome such as fertilization, embryo quality, implantation pregnancy, and spontaneous abortion [105, 106]. A limited amount of sperm DNA damage can be repaired by the oocyte post fertilization, but above a threshold limit this process is either incomplete or inappropriate, resulting in genetic mutations and may impact the viability of the embryo and the health of the offspring [107]. Men suffering from male infertility have high levels of sperm with DNA damage, which result in an negative impact on their ART outcome [25, 108–112].

In recent years, sperm DNA damage has gained interest to understand the fertilization process to improve fertility diagnostics. The influence of DNA damage on fertilization rates in assisted reproduction is still controversial. A number of papers have analyzed the possible association between sperm DNA damage and fertilization rates in vitro [16, 106, 111, 113–127]. But, many of these papers suggest that sperm DNA damage does not affect fertilization rates [106, 111, 115, 117–120, 126, 127]. Sperm with damaged DNA are still capable of fertilization [93] but its effect is prominent in the later stages [128]. Sperm with abnormal chromatin packing and DNA damage is showed to result in decondensation failure, which results in fertilization failure [25]. It is also showed that that a significantly proportion of nondecondensed sperm in human oocytes has a higher DNA damage, compared to decondensed sperm and higher degree of chromatin damage, this may prevent the initiation or completion of decondensation, and may be an important factor leading to a failure in fertilization [129]. A negative correlation between the proportion of sperm having DNA strand breaks and the proportion of oocytes fertilized after IVF is established [114].

Measurement of sperm DNA damage has been shown to have a significant negative effect on the developing embryo [130]. Poor sperm DNA quality is associated with poor blastocyst development and the failure to achieve a clinical pregnancy. Sperm DNA damage has a significant impact on embryo development [16, 95, 105, 114, 126, 129, 131–133]. However, a number of studies have contradicted the influence of DNA damage on embryo development [106, 108, 109, 112, 115–119, 122, 123, 126, 127]. Abnormalities in the embryo seen in vitro can be more directly related to male factors because the results can be assessed without the interference of female factors such as uterine and endocrine abnormalities that may lead to miscarriage after embryo transfer [134]. The embryonic genome is activated on day three, and its transcriptional products take over from the regulatory control provided by maternal messages stored in the oocyte [132]. The effect of sperm DNA damage has been attributed to embryo development, particularly between four and eight cell stage of preimplantation development until which the embryonic genome is transcriptionally inactivated and the paternal genome plays a significant contributory role in embryo function during the transcriptional activity [133]. Therefore, the effect of sperm DNA damage impacts more on pregnancy rates than embryo quality [115].

Couples who failed to achieve a pregnancy are known to have a higher mean level of DNA fragmentation than pregnant couple after IVF treatment [105, 112, 115, 118–120, 122, 133, 135]. This implies that sperm with DNA fragmentation can still fertilize an oocyte but that when paternal genes are "switched on," further embryonic development stops, resulting in failed pregnancy [121]. In contrast to these reports, no significant association between sperm DNA damage and clinical pregnancies has been reported [88, 95, 114, 116, 123, 124, 126, 127, 136, 137]. Studies using animal models show that oocytes and developing embryos can repair sperm DNA damage; however, there is a threshold beyond which sperm DNA cannot be repaired [138]. They also reported that sperm with defective DNA can fertilize an oocyte and produce high-quality early-stage embryos, but then, as the extent of the DNA damage increases, the likelihood of a successful pregnancy decreases. Virro et al. [132] have shown that high levels of sperm DNA damage significantly decrease the pregnancy rates and results in higher rate of spontaneous abortions. An increase in sperm DNA damage is associated with decreased implantation, thereby a decrease in pregnancy rates [118]. By contrast, Bungum et al. [136] and Boe-Hansen et al. [137] showed a decrease in implantation rates with increase in DNA damage but no effect is seen on clinical pregnancies. Frydman et al. [106] showed increase in DNA damage not only decrease implantation and pregnancy rates but also increase spontaneous miscarriage rates. Lin et al. [127] also observed an increase in miscarriage rates with an increase in DNA damage.

It is also shown that damage in the paternal genome could result in abnormalities occur during postimplantation development [139]. Genetic abnormalities in the paternal genome in the form of strand breaks are a significant cause of miscarriages [134]. Sperm DNA damage could likely be the cause of infertility in a large percentage of patients [140]. However, these studies may not causal, but simply associations between DNA damage and reduced ART outcomes. Are the tests clinically useful?

The Clinical Usefulness of the Comet Test

Two recent systematic reviews have shown that the impact of sperm DNA damage on ART outcomes decreases from IUI to IVF and is least useful in ICSI [3, 141]. In IVF, using TUNEL and SCSA assays, the odds ratios is 1.57 (95% CI 1.18–2.07; $p<0.05$). However, in our study using the alkaline comet [82] we obtained an odds ratio of 4.52 (1.79–11.92) in native semen and 6.20 (1.74–26.30) in DGC sperm for clinical pregnancy following IVF, indicating its promise as a prognostic test. Owing to the high sensitivity of the test and level of damage observed when both strand breaks and modified bases were measured it was not possible to establish thresholds for our novel combined test. Following ICSI, the odds ratio for clinical pregnancy was 1.97 (0.81–4.77) using native semen and 2.08 (0.93-4.68) in DGC sperm showing less strength and supporting the combined odds ratio of 1.14 from the meta-analyses by Collins et al. [141] and Zini and Sigman [3]. This supports the belief that ICSI bypasses genetic, as well as functional defects, but the results are counterintuitive. Given the many animal studies showing adverse effects of DNA damage on the long-term health of offspring (reviewed by Aitken et al. [142]; Fernadez-Gonzalez et al. [143]), we need to follow-up the children born by ISCI to make sure that this genetic heritage does not have long-term adverse effects of these children's health even if short-term success in terms of pregnancies is achieved.

Two People but Just One Prognostic Test

The quest for one perfect test to predict a outcome with multifactorial input is particularly unachievable when this outcome involves not just one individual but, in the case of ART, two partners. Since female factors such as age, occyte and embryo quality, and uterine competence all impact significantly on pregnancy, it is not surprising that if one test on the male partner is not acceptably strong. The current literature exemplifies how the controversies as to the usefulness of sperm DNA testing are exacerbated by flawed experimental design. Couples undergoing IVF treatment can be divided into those with female, male, and unexplained infertility. A large proportion of couples undergoing IVF treatment are due to female causes.

In many studies, couples with male, female, and idiopathic infertility have been grouped together. In order to assess the clinical usefulness of a test for one partner of the infertile couple, the appropriate patient population should be identified. Future studies should be designed to minimize the variation in these female factors. Only then can we accurately determine the effects of sperm DNA and thereby maximize the usefulness of the test.

Protection of DNA from Damage

In the male reproductive tract, oxidative stress is due to the increase in the production of ROS, rather than the decrease in the seminal antioxidants. Owing to the lack of cytoplasm excluded during spermatogenesis, there is no self DNA repair mechanism in the sperm; therefore, antioxidants in the seminal plasma are essential to reduce the oxidative stress, and it is the only available mechanism for the sperm to protest against oxidative-stress-mediated DNA damage. Naturally, the concentration of antioxidants in seminal plasma is 10 times greater than in blood plasma [144], and the presence of antioxidants in the seminal plasma protects the functional integrity of the sperm against the oxidative stress [145]. Several other studies showed the role of antioxidants against ROS [21, 146–148]. However, some studies show limited protection of antioxidants against induced ROS [149].

Low levels of antioxidants in semen are associated with suboptimal semen parameters (Kao et al. [189]) and increased sperm DNA damage [150]. Oral administration of the antioxidants has been shown to significantly increase antioxidant levels in the seminal plasma and an improvement in the semen quality [151–155]. Specifically, antioxidant treatment to infertile patients by oral administration of vitamins significantly improved their sperm motility [152, 154, 156–158], sperm concentration [12, 159, 160], and normal morphology of the sperm [152, 159]. Improvement in semen parameters by administration of oral antioxidants were seen in volunteers as well as patients [154, 161]. Studies by Lenzi et al. [162–164] reported a protective function of antioxidants on semen quality due to a reduction of ROS and a reduction in the lipid peroxidation of the membrane. By contrast, other studies have shown no significant effects of oral antioxidant treatment on semen parameters [165–167]. The absence of effects in these studies may be due to shorter duration of treatment [167, 168] and/or very low dosage of antioxidants used [169].

Administration of oral antioxidants had been shown to significantly decrease sperm DNA damage [12, 170–173] and to reduce sperm DNA adducts [174] and the incidence of aneuploidy in sperm [175], thereby increasing the assisted reproductive success [176, 177].

Protection of sperm from DNA damage should also be monitored during sperm processing and cryopreservation when they are especially vulnerable. The absence of antioxidant protection in these procedures has been shown to increase sperm DNA damage [15]. Zalata et al. [178] showed that high-speed centrifugation and removal of sperm from the protective seminal plasma resulted in ROS-mediated DNA damage. Addition of antioxidants in the sperm medium could decrease oxidative stress [179] and damage to sperm [180]. Donnelly et al. [181] showed that addition of vitamins in the sperm suspension media could protect the sperm from DNA damage. This in turn would have a positive effect on male infertility [174]. Cryopreservation of sperm is known to increase the level of sperm DNA damage [93, 182–184].

Oxidative stress occurs when the level of ROS exceeds the antioxidant protection resulting in sperm DNA damage. Approximately, half of infertile men exhibit oxidative stress [185]. In light of these considerations, future research to determine the best regime of antioxidant therapy so be pursued to find an effective treatment [186–188].

Conclusions and Future Recommendations

Clinical evidence shows the negative impact of sperm DNA fragmentation on reproductive outcomes, and sperm from infertile men show higher levels of DNA fragmentation than the sperm of fertile or donor men. Recent studies have shown that the use of alkaline comet assay to test sperm DNA fragmentation is a useful tool for male infertility diagnosis and early predictor of ART outcomes. Below novel "comet" threshold values of sperm DNA fragmentation in both native semen and DGC sperm obtained from the alkaline comet assay, there is evidence of infertility in vivo and in vitro. Therefore, it is beneficial to assess sperm DNA fragmentation in couples presenting with infertility problems and also in patients undergoing ART. We encourage studies to analyze the impact of sperm DNA fragmentation and to validate the current protocol of the alkaline comet assay through large multicenter trials, using good quality control, with standardized protocols.

References

1. Guzick DS, Verstreet JWO, Factor-Litvak P, et al. Sperm morphology, motility, and concentration in fertile and infertile men. N Engl J Med. 2001; 345(19):1388–93.
2. Lewis SEM. Is sperm evaluation useful in predicting human fertility? Reproduction. 2007;134:31–40.
3. Zini A, Sigman M. Are tests of sperm DNA damage clinically useful? Pros and cons. J Androl. 2009; 30(3):219–29.

4. Agarwal A, Prabakaran SA. Mechanism, measurement and prevention of oxidative stress in male reproductive physiology. Ind J Exp Biol. 2005; 43:963–74.
5. Ward WS, Coffey DS. DNA packaging and organization in mammalian spermatozoa: comparison with somatic cell. Biol Reprod. 1991;44:569–74.
6. Saleh R, Agarwal A. Oxidative stress and male infertility: from research bench to clinical practice. J Androl. 2002;23:737–52.
7. Barroso G, Morshedi M, Oehringer S. Analysis of DNA fragmentation, plasma membrane translocation of phosphatidylserine and oxidative stress in human spermatozoa. Hum Reprod. 2000;15: 1338–44.
8. Padron OF, Brackett NL, Sharma RK, et al. Seminal reactive oxygen species, sperm motility and morphology in men with spinal cord injury. Fertil Steril. 1997;67:115–1120.
9. Sharma RK, Agarwal A. Role of reactive oxygen species in male infertility. Urology. 1996;48: 835–50.
10. Balhorn R. A model for the structure of chromatin in mammalian sperm. J Cell Biol. 1982;93: 298–305.
11. Brewer LR, Corzett M, Balhorn R. Protamine-induced condensation and decondensation of the same DNA molecule. Science. 1999;286:120–3.
12. Kodama H, Yamaguchi R, Fukuda J, et al. Increased oxidative deoxyribonucleic acid damage in the spermatozoa of infertile male patients. Fertil Steril. 1997;68:519–24.
13. Twigg J, Fulton N, Gomez E, et al. Analysis of the impact of intracellular reactive oxygen species generation on the structural and functional integrity of the human spermatozoa: lipid peroxidation, DNA fragmentation and effectiveness of antioxidants. Hum Reprod. 1998;13:1429–36.
14. Duru NK, Morshedi M, Oehninger S. Effects of hydrogen peroxide on DNA and plasma membrane integrity of human spermatozoa. Fertil Steril. 2000;74:1200–7.
15. Fraga CG, Motchnik PA, Wyrobek AJ, et al. Smoking and low antioxidant levels increase oxidative damage to sperm DNA. Mutat Res. 1996;351:199–203.
16. Sun JG, Jurisicova A, Casper RF. Detection of deoxyribonucleic acid fragmentation in human sperm: correlation with fertilization In Vitro. Biol Reprod. 1997;56:602–7.
17. Aitken RJ, Krausz C. Oxidative stress, DNA damage and the Y chromosome. Reproduction. 2001;122: 497–506.
18. Ames BN, Shigenaga MK, Hagen TM. Oxidants, antioxidants and the degenerative disease and aging. Proc Natl Acad Sci USA. 1993;90:7915–22.
19. Spiropoulos J, Turnbull DM, Chinnery PF. Can mitochondrial DNA mutations cause sperm dysfunction? Mol Hum Reprod. 2002;8:719–21.
20. Sharma RK, Said T, Agarwal A. Sperm DNA damage and its clinical relevance in assessing reproductive outcome. Asian J Androl. 2004;6:139–48.
21. Aitken RJ. Free radicals, lipid peroxidation and sperm function. Reprod Fertil Dev. 1995;7:659–68.
22. Lopes S, Jurisicova A, Sun J, et al. Reactive oxygen species: a potential cause for DNA fragmentation in human spermatozoa. Hum Reprod. 1998;13:896–900.
23. Garrido N, Meseguer M, Simon C, et al. Pro-oxidative and anti-oxidative imbalance in human semen and its relation with male fertility. Asian J Androl. 2004;6:59–65.
24. Thomas J, Fishel SB, Hall JA, et al. Increased polymorphonuclear granulocytes in seminal plasma in relation to sperm morphology. Hum Reprod. 1997; 12:2418–21.
25. Lopes S, Sun JG, Jurisicova A, et al. Sperm deoxyribonucleic acid fragmentation is increased in poorquality semen samples and correlates with failed fertilization in intracytoplasmic sperm injection. Fertil Steril. 1998;69(3):528–32.
26. Gomez E, Buckingham DW, Brindle J, et al. Development of an image analysis system to monitor the retention of residual cytoplasm by human spermatozoa: correlation with biochemical markers of the cytoplasmic space, oxidative stress, and sperm function. J Androl. 1996;17:276–87.
27. Agarwal A, Makker K, Sharma R. Clinical relevance of oxidative stress in male factor infertility: an update. Am J Reprod Immunol. 2008;59:2–11.
28. Sloter E, Nath J, Eskenazi B, et al. Effects of male age on the frequencies of germinal and heritable chromosomal abnormalities in humans and rodents. Fertil Steril. 2004;81(4):925–43.
29. Crow JF. The origins, patterns and implications of human spontaneous mutation. Nat Rev Genet. 2000; 1(1):40–7.
30. Bosch M, Rajmil O, Egozcue J, et al. Linear increase of structural and numerical chromosome 9 abnormalities in human sperm regarding age. Eur J Hum Genet. 2003;11(10):754–9.
31. Singh NP, Muller CH, Berger RE. Effects of age on DNA doublestrand breaks and apoptosis in human sperm. Fertil Steril. 2003;80(6):1420–31.
32. Wyrobek AJ, Evenson D, Arnheim N, et al. Advancing male age increase the frequencies of sperm with DNA fragmentation and certain gene mutations, but not aneuploidies or diploidies. Proc Natl Acad Sci USA. 2006;103(25):9601–6.
33. Aitken R, de Iuliis GN. Origin and consequences of DNA damage in male germ cells. Reprod Biomed Online. 2007;14(6):727–33.
34. Hsu PC, Chen IY, Pan CH, et al. Sperm DNA damage correlates with polycyclic aromatic hydrocarbons biomarker in coke oven workers. Int Arch Occup Environ Health. 2006;79:349–56.
35. Oh E, Lee E, Im H, et al. Evaluation of immuno and reproductive toxicities and association between immunotoxicological and genotoxicological parameters in waste incineration workers. Toxicology. 2005;210:65–80.
36. Migliore L, Naccarali A, Zanello A, et al. Assessment of sperm DNA integrity in workers exposed to styrene. Hum Reprod. 2002;17:2912–8.

37. Bian Q, Xu LC, Wang SL. Study on the relation between occupational fenvalerate exposure and spermatozoa DNA damage of pesticide factory workers. Occup Environ Med. 2004;61:999–1005.
38. Xia Y, Cheng S, Bian Q, et al. Genotoxic effects on spermatozoa of carbaryl-exposed workers. Toxicol Sci. 2005;85:615–23.
39. Harrouk W, Codrington A, Vinson R, et al. Paternal exposure to cyclophosphamide induces DNA damage and alters the expression of DNA repair genes in the rat preimplantation embryo. Mutat Res. 2000; 461(3):229–41.
40. Hales BF, Barlon TS, Robaire B. Impact of paternal exposure to chemotherapy on offspring in the rat. J Natl Cancer Inst Monogr. 2005;34:28–31.
41. Hellman B, Vaghef H, Bostrom B. The concepts of tail moment and tail inertia in the single electrophoresis assay. Mutat Res. 1995;336:123–31.
42. Anderson D. Overview of male-mediated developmental toxicity. Adv Exp Med Biol. 2003;518: 11–24.
43. Spano M, Toft G, Hagmar L, et al. Exposure to PCB and p.p′-DDE in European and Inuit populations: impact on human sperm chromatin integrity. Hum Reprod. 2005;20:3488–99.
44. Evenson DP, Wixon R. Environmental toxicants cause sperm DNA fragmentation as detected by the sperm chromatin structure assay (SCSA(R)). Toxicol Appl Pharmacol. 2005;207:532–7.
45. Bennetts LE, de Iuliis GN, Nixon B, et al. Impact of estrogenic compounds on DNA integrity in human spermatozoa: evidence for cross-linking and redox cycling activities. Mutat Res. 2008;641:1–11.
46. Park E, Kang MH. Smoking and high plasma triglyceride levels as risk factors for oxidative DNA damage in the Korean population. Ann Nutr Metab. 2004;48(1):36–42.
47. Wyrobek AJ, Schmid TE, Marchetti F. Cross-species sperm-FISH assays for chemical testing and assessing paternal risk for chromosomally abnormal pregnancies. Environ Mol Mutagen. 2005;45:271–83.
48. Wyrobek AJ, Schmid TE, Marchetti F. Relative susceptibilities of male germ cells to genetic defects induced by cancer chemotherapies. J Natl Cancer Inst Monogr. 2005;34:31–5.
49. Rubes J, Lowe X, Moore DII, et al. Smoking cigarettes is associated with increased sperm disomy in teenage men. Fertil Steril. 1998;70(4):715–23.
50. Shi Q, Ko E, Barclay L, et al. Cigarette smoking and aneuploidy in human sperm. Mol Reprod Dev. 2001; 59(4):417–21.
51. Zenzes MT. Smoking and reproduction: gene damage to human gametes and embryos. Hum Reprod Update. 2000;6:122–31.
52. Sykes PJ, McCallum BD, Bangay MJ, et al. Effect of exposure to 900 MHz radiofrequency radiation on intrachromosomal recombination in pKZ1 mice. Radiat Res. 2001;156:495–502.
53. Tice RR, Hook GG, Donner M, et al. Genotoxicity of radiofrequency signals. I. Investigation of DNA damage and micronuclei induction in cultured human blood cells. Bioelectromagnetics. 2002;23:113–26.
54. Mashevich M, Folkman D, Kesar A, et al. Exposure of human peripheral blood lymphocytes to electromagnetic fields associated with cellular phones leads to chromosomal instability. Bioelectromagnetics. 2003;24:82–90.
55. Aitken RJ, Bennetts LE, Sawyer D, et al. Impact of radio frequency electromagnetic radiation on DNA integrity in the male germline. Int J Androl. 2005;28: 171–9.
56. Agarwal A, Desai N, Makker K, et al. Effects of radiofrequency electromagnetic waves (RF-EMW) from cellular phones on human ejaculated semen: an in vitro pilot study. Fertil Steril. 2008;92(4): 1318–25.
57. Banks S, King SA, Irvine DS, et al. Impact of a mild scrotal heat stress on DNA integrity in murine spermatozoa. Reproduction. 2005;129:505–14.
58. Cook PR, Brazell IA, Jost E. Characterization of nuclear structures containing superhelical DNA. J Cell Sci. 1976;22:303–24.
59. Collins AR, Dobson VL, Dusinska M, et al. The comet assay: what can it really tell us? Mutat Res. 1997;375:183–93.
60. Singh NP, Stephens RE. X-ray induced DNA double-strand breaks in human sperm. Mutagenesis. 1998;13:75–9.
61. McVicar CM, McClure N, Williamson K, et al. Incidence of Fas positivity and deoxyribonucleic acid double-stranded breaks in human ejaculated sperm. Fertil Steril. 2004;81(1):767–74.
62. Angelis KJ, Dusinska M, Collins AR. Single cell gel electrophoresis: detection of DNA damage at different levels of sensitivity. Electrophoresis. 1999;20:2133–8.
63. Bocker W, Rolf W, Bauch T, et al. Automated comet assay analysis. Cytometry. 1999;35:134–44.
64. Gedik CM, Ewen SWB, Collins AR. Single-cell gel electrophoresis applied to the analysis of UV-C damage and its repair in human cells. Int J Radiat Biol. 1992;62:313–20.
65. Kobayashi H, Sugiyama C, Morikawa Y, et al. A comparison between manual microscopic analysis and computerized image analysis in the single cell gel electrophoresis assay. MMS Commun. 1995;3: 103–15.
66. Hughes CM, Lewis SEM, McKelvey-Martin VJ, et al. Reproducibility of human sperm DNA measurements using the alkaline single cell gel electrophoresis assay. Mutat Res. 1997;374:261–8.
67. Hughes CM, Lewis SEM, McKelvey-Martin VJ, et al. A comparison of baseline and induced DNA damage in human spermatozoa from fertile and infertile men, using a modified comet assay. Mol Hum Reprod. 1996;2:613–9.
68. Olive PL, Banath JP. Sizing highly fragmented DNA in individual apoptotic cells using the comet assay and a DNA crosslinking agent. Exp Cell Res. 1995;221:19–26.

69. Choucroun P, Gillet D, Dorange G, et al. Comet assay and early apoptosis. Mutat Res. 2001;478:89–96.
70. Sakkas D, Seli E, Bizzaro D, et al. Abnormal spermatozoa in the ejaculate: abortive apoptosis and faulty nuclear remodelling during spermatogenesis. Reprod Biomed Online. 2003;7:428–32.
71. Tice RR, Agurell E, Anderson D, et al. Single cell gel/comet assay: guidelines for in vitro and in vivo genetic toxicology testing. Environ Mol Mutagen. 2000;35:206–21.
72. Tebbs RS, Pederson RA, Cleaver JE, et al. Modification of the comet assay for the detection of DNA strand breaks in extremely small samples. Mutagenesis. 1999;14:437–8.
73. Kizilian N, Wilkins RC, Reinhardt P, et al. Silver-stained comet assay for detection of apoptosis. Biotechniques. 1999;27:926–30.
74. Aravindan GR, Bjordhal J, Jost LK, et al. Susceptibility of human sperm to in situ DNA denaturation is strongly correlated with DNA strand breaks identified by single-cell electrophoresis. Exp Cell Res. 1997;10:231–7.
75. Singh NP, Danner DB, Tice RR, et al. Abundant alkali-sensitive sites in DNA of human and mouse sperm. Exp Cell Res. 1989;184:461–70.
76. Tarozzi N, Bizzaro D, Flamigni C, et al. Clinical relevance of sperm DNA damage in assisted reproduction. Reprod Biomed Online. 2007;14(6):746–57.
77. Fairbairn DW, Olive PL, O'Neill KL. The comet assay: a comprehensive review. Mutat Res. 1995;339: 37–59.
78. Olive PL, Durand RE, Banath JP, et al. Analysis of DNA damage in individual cells. Methods Cell Biol. 2001;64:235–49.
79. Irvine DS, Twigg JP, Gordon EL, et al. DNA integrity in human spermatozoa: relationships with semen quality. J Androl. 2000;21:33–44.
80. Hartmann A, Agurell E, Beevers C, et al. Recommendations for conducting the in vivo alkaline Comet assay. 4th International Comet Assay Workshop. Mutagenesis. 2003;18:45–51.
81. Burlinson B, Tice RR, Speit G, et al. Fourth International Workgroup on Genotoxicity testing: results of the in vivo Comet assay workgroup. Mutat Res. 2007;627:31–5.
82. Simon L, Lutton D, McManus J. Sheena EM. Lewis Clinical significance of sperm DNA damage in reproductive outcome. Hum Reprod. 2010;25(7):1594–608.
83. Haines GB, Daniel MP, Morris I. B, Daniel MP, Morris IDNA damage in human and mouse spermatozoa after in vitro-irradiation assessed by the comet assay. Adv Exp Med Biol. 1998;444:79–91.
84. Duty SM, Singh NP, Ryan L, et al. Reliability of the comet assay in cryopreserved human sperm. Hum Reprod. 2002;17:1274–80.
85. Baumgartner A, Schmid TE, Cemeli E, et al. Parallel evaluation of doxorubicin-induced genetic damage in human lymphocytes and sperm using the comet assay and spectral karyotyping. Mutagenesis. 2004; 19:313–8.
86. Donnelly ET, McClure N, Lewis SE. Antioxidant supplementation in vitro does not improve human sperm motility. Fertil Steril. 1999;72:484–95.
87. Hughes CM, McKelvey-Martin VJ, Lewis SEM. Human sperm DNA integrity assessed by the Comet and ELISA assays. Mutagenesis. 1999;14: 71–5.
88. Anderson D, Dobrzynska MM, Basaran N, et al. Flavonoids modulate comet assay responses to food mutagens in human lymphocytes and sperm. Mutat Res. 1998;402:269–77.
89. Morris ID, Ilott S, Dixon L, et al. The spectrum of DNA damage in human sperm assessed by single cell gel electrophoresis (Comet assay) and its relationship to fertilization and embryo development. Hum Reprod. 2002;17:990–8.
90. Agbaje IM, Rogers DA, McVicar CM, et al. Insulin dependant diabetes mellitus: implications for male reproductive function. Hum Reprod. 2007;22: 1871–7.
91. McKelvey-Martin VJ, Green MHL, Schmezer P, et al. The single cell gel electrophoresis assay (COMET) assay: a European review. Mutagenesis. 1993;288:47–63.
92. Kumaravel TS, Jha AN. Reliable comet assay measurements for detecting DNA damage induced by ionising radiation and chemicals. Mutat Res. 2006; 605:7–16.
93. Chan PJ, Corselli JU, Patton WC, et al. A simple Comet assay for archived sperm correlates DNA fragmentation to reduced hyperactivation' and penetration of zona-free hamster oocytes. Fertil Steril. 2001;75:186–92.
94. Donnelly ET, Steele EK, McClure N, et al. Assessment of DNA integrity. and morphology of ejaculated spermatozoa from fertile and infertile men before and after cryopreservation. Hum Reprod. 2001;16:1191–9.
95. Lewis SEM, Agbaje IM. Using the alkaline comet assay in prognostic test for male infertility and assisted reproductive technology outcome. Mutagenesis. 2008;23(3):163–70.
96. Tomsu M, Sharma V, Miller D. Embryo quality and IVF treatment outcomes may correlate with different sperm comet assay parameters. Hum Reprod. 2002; 17:1856–62.
97. Steer CV, Mills CL, Tan SL, et al. The cumulative embryo score: a predictive embryo scoring technique to select the optimal number of embryos to transfer in an in-vitro fertilization and embryo transfer programme. Hum Reprod. 1992;7(l):117–9.
98. Lewis SEM, Boyle PM, McKinney KA, et al. Total antioxidant capacity of seminal plasma is different in fertile and infertile men. Fertil Steril. 1995;64: 868–70.
99. Boiteux S, O'Connor TR, Lederer F. Homogeneous Escherichia coli FPG protein. A DNA glycosylase which excises imidazole ring-opened purines and nicks DNA at apurinic/apyrimidinic sites. J Biol Chem. 1990;265:3916–22.

100. Olsen AK, Duale N, Bjoras M, et al. Limited repair of 8-hydroxy-7,8-dihydroguanine residues in human testicular cells. Nucleic Acids Res. 2003;31: 1351–63.
101. Kuznetsov SV, Sidorkina OM, Jurado J, et al. Effect of single mutations on the structural dynamics of a DNA repair enzyme, the *Escherichia coli* formamidopyrimidine-DNA glycosylase: A fluorescence study using tryptophan residues as reporter groups. Eur J Biochem. 1998;253:413–20.
102. Collins AR, Duthie SJ, Dobson VL. Direct enzymic detection of endogenous oxidative base damage in human lymphocyte DNA. Carcinogenesis. 1993;14: 1733–5.
103. Aukrust P, Luna L, Ueland T, et al. Impaired base excision repair and accumulation of oxidative base lesions in CD4+ T cells of HIV-infected patients. Blood. 2005;105:4730–5.
104. Collins AR, Ma AG, Duthie SJ. The kinetics of repair of oxidative DNA damage (strand breaks and oxidised pyrimidines) in human cells. Mutat Res. 1995;336:69–77.
105. Floyd RA. Tine role off 8-hydroxyguanine to carcinogenesis. Carcinogenesis. 1990;11(9):1447–50.
106. Saleh RA, Agarwal A, Nada EA, et al. Negative effects of increased sperm DNA damage in relation to seminal oxidative stress in men with idiopathic and male factor infertility. Fertil Steril. 2003;79(3): 1597–606.
107. Frydman N, Prisant N, Hesters L, et al. Adequate ovarian follicular status does not prevent the decrease in pregnancy rates associated with high sperm DNA fragmentation. Fertil Steril. 2008;89(1):93–8.
108. Aitken RJ, Koopman P, Lewis SEM. Seeds of concern. Nature. 2004;432:48–52.
109. Larson KL, de Jonge CJ, Barnes AM, et al. Sperm chromatin structure assay parameters as predicted of failed pregnancy following assisted reproductive techniques. Hum Reprod. 2000;15(8):1717–22.
110. Larson-Cook KL, Brannian JD, Hansen KA, et al. Relationship between the outcomes of assisted reproductive techniques and sperm DNA fragmentation as measured by the sperm chromatin structure assay. Fertil Steril. 2003;80(4):895–902.
111. Muratori M, Maggi M, Spinelli S, et al. Spontaneous DNA fragmentation in swim-up selected human spermatozoa during long term incubation. J Androl. 2003;24:253–62.
112. Bungum M, Humaidan P, Axmon A, et al. Sperm DNA integrity assessment in prediction of assisted reproduction technology outcome. Hum Reprod. 2007;22:174–9.
113. Bakos HW, Thompson JG, Feil D, et al. Sperm DNA damage is associated with assisted reproductive technology pregnancy. Int J Androl. 2008;31(5): 518–26.
114. Esterhuizen AD, Franken DR, Lourens JGH, et al. Sperm chromatin packing as an indicator of *in-vitro* fertilization rate. Hum Reprod. 2000;15(3):657–61.
115. Host E, Lindenberg S, Smidt-Jensen S. The role of DNA strand breaks in human spermatozoa used for IVF and ICSI. Acta Obstet Gynecol Scand. 2000;79: 559–63.
116. Tomlinson MJ, Moffatt O, Manicardi GC, et al. Interrelationships between seminal parameters and sperm nuclear DNA damage before and after density gradient centrifugation: implications for assisted conception. Hum Reprod. 2001;16(10):2160–5.
117. Benchaib M, Braun V, Lornage J, et al. Sperm DNA fragmentation decreases the pregnancy rate in an assisted reproductive technique. Hum Reprod. 2003;18(5):1023–8.
118. Henkel R, Kierspel E, Hajimohammad M, et al. DNA fragmentation of spermatozoa and assisted reproduction technology. Reprod Biomed Online. 2003;7:477–84.
119. Adams C, Anderson L, Wood S. High, but not moderate, levels of sperm DNA fragmentation are predictive of poor outcome in egg donation cycles. Fertil Steril. 2004;82(2):S44.
120. Chohan KR, Griffin JT, Lafromboise M, et al. Sperm DNA damage relationship with embryo quality and pregnancy outcome in IVF patients. Fertil Steril. 2004;82(2):S55–6.
121. Gandini L, Lombardo F, Paoli D, et al. Full-term pregnancies achieved with ICSI despite high levels of sperm chromatin damage. Hum Reprod. 2004; 19(6):1409–17.
122. Henkel R, Hajimohammad M, Stalf T, et al. Influence of deoxyribonucleic acid damage on fertilization and pregnancy. Fertil Steril. 2004;81(4):965–72.
123. Huang CC, Lin DPC, Tsao HM, et al. Sperm DNA fragmentation negatively correlates with velocity and fertilization rates but might not affect pregnancy rates. Fertil Steril. 2005;84(1):130–40.
124. Payne JF, Raburn DJ, Couchman GM, et al. Redefining the relationship between sperm deoxyribonucleic acid fragmentation as measured by the sperm chromatin structure assay and outcomes of assisted reproductive techniques. Fertil Steril. 2005;84:356–64.
125. Borini A, Tarozzi N, Bizzaro D, et al. Sperm DNA fragmentation: paternal effect on early post-implantation embryo development in ART. Hum Reprod. 2006;21(11):2876–81.
126. Bakos HW, Thompson JG, Lane M. Sperm DNA damage: is altered carbohydrate metabolism to blame? Fertil Steril. 2007;88(1):S48–9.
127. Benchaib M, Lornage J, Mazoyer C, et al. Sperm deoxyribonucleic acid fragmentation as a prognostic indicator of assisted reproductive technology outcome. Fertil Steril. 2007;87(1):93–101.
128. Lin HH, Lee RK, Li SH, et al. Sperm chromatin structure assay parameters are not related to fertilization rates, embryo quality, and pregnancy rates in in vitro fertilization and intracytoplasmic sperm injection, but might be related to spontaneous abortion rates. Fertil Steril. 2008;90(2):352–9.

129. Cebesoy FB, Aydos K, Unlu C. Effect of sperm chromatin damage on fertilization ratio and embryo quality post-ICSI. Arch Androl. 2006;52:397–402.
130. Sakkas D, Urner F, Bianchi PG, et al. Sperm chromatin anomalies can influence decondensation after intracytoplasmic sperm injection. Hum Reprod. 1996;11:837–43.
131. Jones GM, Trounson AO, Lolatgis N, et al. Factors affecting the success of human blastocyst development and pregnancy following in vitro fertilization and embryo transfer. Fertil Steril. 1998;70:1022–9.
132. Seli E, Gardner DK, Schoolcraft WB, et al. Extent of nuclear DNA damage in ejaculated spermatozoa impacts on blastocyst development after *in vitro* fertilization. Fertil Steril. 2004;82(2):378–83.
133. Virro MR, Larson-Cook KL, Evenson DP. Sperm chromatin structure assay (SCSA) parameters are related to fertilization, blastocyst development, and ongoing pregnancy in *in vitro* fertilization and intracytoplasmic sperm injection cycles. Fertil Steril. 2004;81(5):1289–95.
134. Braude P, Bolton V, Moore S. Human gene expression first occurs between the four and eight-cell stages of preimplantation development. Nature. 1988;332:459–61.
135. Muriel L, Garrido N, Fernández JL, et al. Value of the sperm DNA fragmentation level, measured by the sperm chromatin dispersion (SCD) test, in the IVF and ICSI outcome. Fertil Steril. 2006;85:371–83.
136. Li TC. Guides for practitioners. Recurrent miscarriage: principles of management. Hum Reprod. 1998;13:478–82.
137. Duran EH, Morshedi M, Taylor S, et al. Sperm DNA quality predicts intrauterine insemination outcome: a prospective cohort study. Hum Reprod. 2002;17:3122–8.
138. Bungum M, Humaidan P, Spano M, et al. The predictive value of sperm chromatin structure assay (SCSA) parameters for the outcome of intrauterine insemination, IVF and ICSI. Hum Reprod. 2004;19(6):1401–8.
139. Boe-Hansen GB, Fedder J, Ersboll AK. The sperm chromatin structure assay as a diagnostic tool in the human fertility clinic. Hum Reprod. 2006;21(6):1576–82.
140. Ahmadi A, Ng SC. Fertilizing ability of DNA-damaged spermatozoa. J Exp Zool. 1999;284:696–704.
141. Banerjee S, Lamond S, McMahon A, et al. Does blastocyst culture eliminate paternal chromosomal defects and select good embryos? Inheritance of an abnormal paternal genome following ICSI. Hum Reprod. 2000;15:2455–9.
142. Agarwal A, Said TM. Role of sperm chromatin abnormalities and DNA damage in male infertility. Hum Reprod Update. 2003;9:331–45.
143. Collins JA, Barnhart KT, Schlegel PN. Do sperm DNA integrity tests predict pregnancy with in vitro fertilization? Fertil Steril. 2008;89(4):823–31.
144. Aitken RJ, de Iuliis GN, McLachlan RI. Biological and clinical significance of DNA damage in the male germ line. Int J Androl. 2008;32:46–56.
145. Fernadez-Gonzalez R, Moreira PN, Perez-Crespo M, et al. Long-term effects of mouse intracytoplasmic sperm injection with DNA-fragmented sperm on health and behavior of adult off-spring. Biol Reprod. 2008;78:761–72.
146. Lewis SE, Sterling ES, Young IS, et al. Comparison of individual antioxidants of sperm and seminal plasma in fertile and infertile men. Fertil Steril. 1997;67:142–7.
147. Wolf AM, Asoh S, Hiranuma H. Astaxanthin protects mitochondrial redox state and functional integrity against oxidative stress. J Nutr Biochem. 2010;21(5):381–9.
148. Kovalski NN, de Lamirande E, Gagnon C. Reactive oxygen species generated by human neutrophils inhibit sperm motility: protective effect of seminal plasma and scavengers. Fertil Steril. 1992;5S:809–16.
149. Aitken RJ, Buckingham D, Harkiss D. Use of a xanthine oxidase free radical generating system to investigate the cytotoxic effects of reactive oxygen species on human spermatozoa. J Reprod Fertil. 1993;97:441–50.
150. Griveau JF, Le Lannou D. Effects of antioxidants on human sperm preparation techniques. Int J Androl. 1994;17:225–31.
151. Cao G, Alessio HM, Cutler RG. Oxygen-radical absorbance capacity assay for antioxidants. Free Radic Biol Med. 1993;14:303–11.
152. Song GJ, Norkus EP, Lewis V. Relationship between seminal ascorbic acid and sperm DNA integrity in infertile men. Int J Androl. 2006;29:569–75.
153. Dawson EB, Harris WA, Teter MC, et al. Effect of ascorbic acid supplementation on the sperm quality of smokers. Fertil Steril. 1992;58:1034–9.
154. Vezina D, Mauffette F, Roberts KD, et al. Selenium-vitamin E supplementation in infertile men. Effects on semen parameters and micronutrient levels and distribution. Biol Trace Elem Res. 1996;53:65–83.
155. Lenzi A, Lombardo F, Sgro P, Salacone P, et al. Use of carnitine therapy in selected cases of male factor infertility: a double-blind crossover trial. Fertil Steril. 2003;79:292–300.
156. Eskenazi B, Kidd SA, Marks AR, et al. Antioxidant intake is associated with semen quality in healthy men. Hum Reprod. 2005;20:1006–12.
157. Safarinejad MR, Safarinejad S. Efficacy of selenium and/or Nacetyl-cysteine for improving semen parameters in infertile men: a double-blind, placebo controlled, randomized study. J Urol. 2009;181:741–51.
158. Scott R, MacPherson A, Yates RW, et al. The effect of oral selenium supplementation on human sperm motility. Br J Urol. 1998;82:76–80.
159. Comhaire FH, Christophe AB, Zaiata AA, et al. The effects of combined conventional treatment, oral antioxidants and essential fatty acids on sperm biology

in subfertile men. Prostaglandins Leukot Essent Fatty Acids. 2000;63:159–65.

160. Lenzi A, Sgro P, Salacone P, et al. A placebo-controlled double-blind randomized trial in the use of combined L-carnitineand l-acetyl-carnitine treatment in men with asthenozoospermia. Fertil Steril. 2004;81:1578–84.

161. Wong WY, Merkus HMWM, Thomas CMG, et al. Effects of folic acid and zinc sulphate on male factor subfertility: a double-blind, randomized, placebo-controlled trial. Fertil Steril. 2002;77:491–8.

162. Ebisch IMW, Pierik FH, de Jong FH, et al. Does folic acid and zinc sulphate intervention affect endocrine parameters and sperm characteristics in men? Int J Androl. 2006;29:339–45.

163. Keskes-Ammar L, Feki-Chakroun N, Rebai T, et al. Sperm oxidative stress and the effect of an oral vitamin E and selenium supplement on semen quality in infertile men. Arch Androl. 2003;49:83–94.

164. Lenzi A, Culasso F, Gandini L, et al. Placebo-controlled, double-blind, crossover trial of glutathione therapy in male infertility. Hum Reprod. 1993; 8:1657–62.

165. Lenzi A, Lombardo F, Gandini L, et al. Glutathione therapy for male infertility. Arch Androl. 1992;29: 65–8.

166. Lenzi A, Picardo M, Gandini L, et al. Glutathione treatment of dyspermia: effect on the lipoperoxidation process. Hum Reprod. 1994;9:2044–50.

167. Abel BJ, Carswell G, Elton R, et al. Randomised trial of clomiphene citrate treatment and vitamin C for male infertility. Br J Urol. 1982;54:780–4.

168. Kessopoulou E, Powers HJ, Sharma KK, et al. A double-blind randomized placebo crossover controlled trial using the antioxidant vitamin E to treat reactive oxygen species associated male infertility. Fertil Steril. 1995;64:825–31.

169. Moilanen J, Hovatta O. Excretion of alpha-tocopherol into human seminal plasma after oral administration. Andrologia. 1995;27:133–6.

170. Rolf C, Cooper TG, Yeung CH, et al. Antioxidant treatment of patients with asthenozoospermia or moderate oligoasthenozoospermia with high-dose vitamin C and vitamin E: a randomized, placebo-controlled, double-blind study. Hum Reprod. 1999;14:1028–33.

171. Geva E, Bartoov B, Zabludovsky N, et al. The effect of antioxidant treatment on human spermatozoa and fertilization rate in an in vitro fertilization program. Fertil Steril. 1996;66:430–4.

172. Greco E, Iacobelli M, Rienzi L, et al. Reduction of the incidence of sperm DNA fragmentation by oral antioxidant treatment. J Androl. 2005;26(3):349–53.

173. Greco E, Romano S, Iacobelli M. ICSI in cases of sperm DNA damage: beneficial effect of oral antioxidant treatment. Hum Reprod. 2005;20(9):2590–4.

174. Menezo YJR, Hazout A, Panteix G, et al. Antioxidants to reduce sperm DNA fragmentation: an unexpected adverse effect. Reprod Biomed Online. 2007;14: 418–21.

175. Piomboni P, Gambera L, Serafini F, et al. Sperm quality improvement after natural anti-oxidant treatment of asthenoteratospermic men with leukocytospermia. Asian J Androl. 2008;10:201–6.

176. Comhaire FH, El Garem Y, Mahmoud A, et al. Combined conventional/antioxidant "Astaxanthin" treatment for male infertility: a double blind, randomized trial. Asian J Androl. 2005;7:257–62.

177. Young SS, Eskenazi B, Marchetti FM, et al. The association of foliate, zinc, and antioxidant intake with sperm aneuploidy in healthy non-smoking men. Hum Reprod. 2008;23:1014–22.

178. Suleiman SA, Ali ME, Zaki ZM, et al. Lipid peroxidation and human sperm motility: protective role of vitamin E. J Androl. 1996;17:530–7.

179. Tremellen K, Miari G, Froiland D, et al. A randomized control trial of an antioxidant (Menevit) on pregnancy outcome during IVF-ICSI treatment. Aust N Z J Obstet Gynaecol. 2007;47:216–21.

180. Zalata A, Hafez T, Comhaire F. Evaluation of the role of reactive oxygen species in male infertility. Hum Reprod. 1995;10:1444–51.

181. Chi HJ, Kim JH, Ryu CS, et al. Protective effect of antioxidant supplementation in sperm-preparation medium against oxidative stress in human spermatozoa. Hum Reprod. 2008;23(5):1023–8.

182. Lamond S, Watkinson M, Rutherford T, et al. Gene-specific chromatin damage in human spermatozoa can be blocked by antioxidants that target mitochondria. Reprod Biomed Online. 2003;7: 407–18.

183. Donnelly ET, McClure N, Lewis SE. The effect of ascorbate and alpha-tocopherol supplementation in vitro on DNA integrity and hydrogen peroxide-induced DNA damage in human spermatozoa. Mutagenesis. 1999;14:505–12.

184. Hammadeh ME, Askari AS, Georg T, et al. Effect of freeze-thawing procedure on chromatin stability, morphological alteration and membrane integrity of human spermatozoa in fertile and subfertile men. Int J Androl. 1999;22:155–62.

185. de Paula TS, Bertolla RP, Spaine DM. Effect of cryopreservation on sperm apoptotic deoxyribonucleic acid fragmentation in patients with oligozoospermia. Fertil Steril. 2006;86:597–600.

186. Thomson LK, Fleming SD, Aitken RJ. Cryopreservation-induced human sperm DNA damage is predominantly mediated by oxidative stress rather than apoptosis. Hum Reprod. 2009;24(9): 2061–70.

187. Tremellen K. Oxidative stress and male infertility – a clinical perspective. Hum Reprod Update. 2008; 14(3):243–58.

188. Kefer JC, Agarwal A, Sabanegh E. Role of antioxidants in the treatment of male infertility. Int J Urol. 2009;16:449–57.

189. Kao SH, Chao HT, Chen HW, Hwang TIS, Liao TL, Wei YH. Increase of oxidative stress in human sperm with lower motility. Fertil Steril. 2008;89(5):1183–90.

Assays Used in the Study of Sperm Nuclear Proteins

Timothy G. Jenkins, Benjamin R. Emery, and Douglas T. Carrell

16

Abstract

Male factor infertility is a complex, multifactorial disease with over 2/3 of the cases being classified idiopathic. The idiopathic category of infertile males includes men who have compromised testicular function resulting in mature sperm with decreased functional parameters. One well established correlate to decreased sperm function is altered protamination in the mature, ejaculated sperm. The process of protamination involves an elegant interplay of several proteins: histones (both canonical and testis-specific), transition proteins, and protamines. Each of these proteins work in concert to ensure that chromatin is packaged efficiently and stably to facilitate normal sperm motility and fertilization, and ultimately, to be able to contribute the paternal genome to the embryo. A developing area of interest in the field of sperm chromatin compaction is elucidating how protamination and retained histones affect the epigenetic status of the mature sperm. Nucleoprotein assays can be broken down into two main categories: assays that involve protein isolation and quantification techniques, and assays that involve in situ staining of nuclear proteins, which are discussed in this chapter.

Keywords

Sperm nuclear proteins • Male infertility • Assays used in sperm nuclear proteins • Protamination in sperm • Protamines

Male factor infertility is a complex, multifactorial disease with over 2/3 of the cases being classified idiopathic [1–3]. The idiopathic category of infertile males includes men who have compromised testicular function resulting in mature sperm with decreased functional parameters. One well-established correlate to decreased sperm

D.T. Carrell (✉)
Andrology and IVF Laboratories, Department of Surgery, University of Utah, Salt Lake City, UT 84108, USA
e-mail: douglas.carrell@hsc.utah.edu

A. Zini and A. Agarwal (eds.), *Sperm Chromatin: Biological and Clinical Applications in Male Infertility and Assisted Reproduction*, DOI 10.1007/978-1-4419-6857-9_16,

function is altered protamination in the mature, ejaculated sperm [4]. The process of protamination involves an elegant interplay of several proteins: histones (both canonical and testis-specific), transition proteins, and protamines. Each of these proteins work in concert to ensure that chromatin is packaged efficiently and stably to facilitate normal sperm motility and fertilization, and ultimately, to be able to contribute the paternal genome to the embryo.

A developing area of interest in the field of sperm chromatin compaction is elucidating how protamination and retained histones affect the epigenetic status of the mature sperm [5]. Epigenetic changes, including histone modifications and DNA methylation, are intimately related to chromatin compaction in the mature sperm and appear to "poise" paternal genes expressed during embryogenesis for expression [6]. Hence, evaluation of the chromatin packaging in sperm not only is of interest as a measure of normal spermiogenesis but may prove to be important in evaluating the potential normality of the epigenetic contribution of the paternal genome.

Human sperm go through dramatic chromatin reorganization during spermiogenesis. Chromatin is taken from a relatively decondensed state (histone bound) and then packaged in an extremely condensed conformation by the incorporation of protamine 1 (P1) and protamine 2 (P2). The unique architecture of the spermatocyte chromatin begins early in spermatogenesis with the incorporation of testis-specific histones replacing the majority of canonical histones. Transition proteins 1 and 2 are detectable for a short time during late spermatogenesis [7–10]. The transition proteins are fully removed by the end of spermatogenesis when the total genome is compacted with 90–95% of the chromatin condensed with P1 and P2. The remaining chromatin is composed of linker, canonical, and testis-specific histones. The result is a transcriptionally silent chromatin structure that is at least 6 times more compact then its nonprotaminated counterpart [11, 12]. This tight structure is thought to serve several functions in the mature sperm: to protect the DNA from damage, to silence transcription, and to facilitate efficient movement of the cell, allowing safe delivery to the oocyte [13]. As one might expect, aberrant protamination has been linked in multiple studies to male factor infertility, demonstrating the important role of these proteins and their influence on chromatin structure [4, 14–16].

Histones

Histones play an important role in both somatic cells and gametes. There are however, distinct differences in the way these proteins are utilized in sperm compared with other cell types. These differences involve unique testis-specific histones as well canonical histones. In the mature sperm, protamines are far more prevalent than histones; this has led many researchers to focus on the role of protamine in sperm, but there is growing interest in the important role that sperm histones may play in the maturing sperm and possibly in fertilization and early embryo development.

In recent studies that have evaluated patterns of histone retention in the sperm of fertile men, it has been found that histones are retained in a nonrandom way and that their retention is not just a result of inefficient machinery. In fertile men, histones are retained at the promoters of microRNAs, embryonic developmental genes, and imprinted loci [6, 17]. The fact that these regions retain histones is intriguing, since their lack of protamine results in a chromatin structure that is far less compact in relation to the rest of the DNA. This less dense structure may allow for increased accessibility of transcriptional machinery so that the genes most readily transcribed are those that retain histone.

Since abnormal histone retention leaves the chromatin far less dense and more accessible to DNA damage, it is thought that increased histone retention could be linked to increased DNA damage in some patients. In fact, histone staining techniques have been utilized to observe testis-specific histone variants in fertile and infertile patients (asthenospermic or asthenoteratozoospermic). These studies have revealed a more diffuse, but intense staining pattern in infertile patients indicating higher levels of histone, and a more random distribution of those histones in the infertile population [15, 18]. This diffuse staining pattern was also correlated with increased DNA damage, which suggests that the DNA damage found in these patients, at the very least, is associated with abnormal histone levels.

Histones play an essential role in the formation of the unique chromatin structure in sperm. Further study of these unique proteins and their possible involvement in early embryo development is needed to effectively understand the role of epigentics in male factor infertility.

Protamines

Protamines are arginine- and cysteine-rich proteins that form tight disulfide bonds [19, 20]. This tight positively charged structure allows for an extremely tight chromatin conformation that is transcriptionally silent. The incorporation of P1 and P2 into the sperm genome during spermatogenesis is strictly regulated. The quantity of human P1 and P2 is approximately equal resulting in a 1:1 ratio [4].

There have been studies designed to look at both the quantity of protamine in a given patient's sample, as well as the ratio of expression for P1 and P2. Through these studies, it appears that the most significant factor to overall male factor fertility is the ratio of these proteins. Abnormal P1–P2 ratios are tightly associated with male factor infertility. Studies have shown that in humans the incidence of low motility, low concentration, poor morphology, and reduced fertilization capacity is increased in patients with aberrant P1–P2 ratios [4, 16, 21–23].

The protamination process is essential in ensuring proper chromatin compaction, which in turn facilitates normal sperm function. If protamines are aberrantly expressed, then the desired level of chromatin compaction could not be reached, and thus, normal function would be inhibited.

Although much is known regarding the function of protamine in the mature sperm and how it relates to fertility, there remain unanswered questions that require additional study. A creative utilization of techniques that are currently available is required to effectively investigate these questions. The following information outlines these techniques and how they are being used in the study of nuclear proteins in mature sperm.

Assays

Nucleoprotein assays can be broken down into two main categories: assays that involve protein isolation and quantification techniques and assays that involve in situ staining of nuclear proteins (Fig. 16.1).

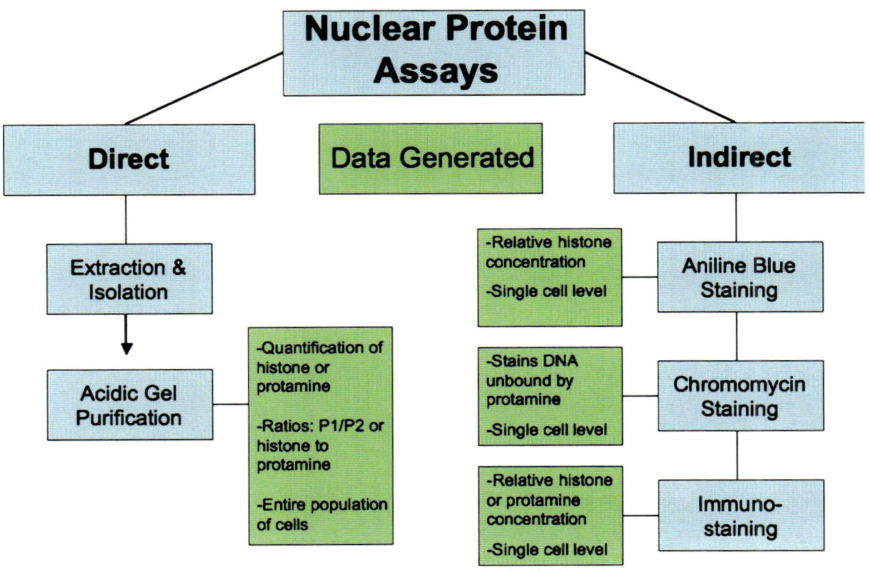

Fig. 16.1 Nucleoprotein assays can be broken down into two main categories: assays that involve protein isolation and quantification techniques and assays that involve in situ staining of nuclear proteins

Isolation Techniques

Isolation techniques have yielded important information in the field male factor infertility. These techniques have facilitated the quantification of histones and protamines in the mature sperm and have demonstrated the importance of the P1–P2 ratio and the histone to protamine ratio in proper sperm function. The main benefit of these assays is that they can accurately determine the nuclear protein makeup of an entire population of cells in any given sample. One of the main drawbacks of the test is the inability to generate the same information for individual cells.

Utilizing these techniques to isolate and quantitate protamines from sperm of fertile and infertile men, Aoki et al. [16] demonstrated a link between P1–P2 ratios and fertilization ability. Another study from 2006 used similar techniques to measure the histone to protamine ratio in mature sperm. This study found that a high histone to protamine ratio was seen more frequently in an infertile population than was seen in a healthy control group [15]. These studies and others have generated promising results that are proving informative to our understanding of sperm chromatin composition and its effects in fertile and infertile men.

Isolation and Purification of Protamines

The following protocol is adapted by Carrell and Liu [23] from the original work done by de Yebra and Oliva [24].

- Semen sample is diluted 3:1 with washing medium and centrifuged at 500 × *g* for 10 min at 4°C.
  - This wash removes the seminal plasma.
- 40 million sperm are removed from the resuspended pellet and are again washed in distilled water containing 1 mM phenylmethylsulfonyl fluoride (PMSF), which acts as a protease inhibitor.
  - The wash in distilled water causes osmotic shock to the cells and disrupts the cell membrane. The sperm suspension is then centrifuged 500 × *g* for 10 min and the supernatant is discarded.
- The pellet is resuspended in 100 μL of 100 mM Tris buffer with 20 mM Ethylenediamine tetraacetic acid (EDTA) and 1 mM PMSF.
  - Tris–EDTA buffers the conditions of the upcoming treatments.
- 100 mL of 6 M Guanidine and 575 mM dithiothreitol (DTT) is added to the suspension and mixed. 200 μL of 552 mM sodium iodoacetate is then added to the suspension and mixed. The mixture is incubated at room temperature for 30 min, protected from the light.
  - DTT is used to disrupt disulfide bonds, and once these bonds are broken sodium iodoacetate caps the now-free sulfur residues, ensuring that the bond cannot be re-formed.
- After incubation, the suspension is mixed with cold 100% ethanol and centrifuged at maximum speed for 10 min.
  - Ethanol removes salt from the solution.
- The pellet is resuspended in 0.8 mL of 0.5 M HCl and incubated for 15 min at 37°C, then centrifuged at a maximum speed for 10 min.
  - The acid treatment removes the nuclear proteins from the rest of the mix due to the increased positively charged residues found in nuclear proteins.
- The supernatant is removed and added to 200 μL of 100% trichloroacetic acid (TCA) to a final concentration of 20% TCA. The solution is incubated at 4°C for 5 min and then centrifuged at maximum speed at 4°C for 15 min.
  - TCA is used to precipitate the nuclear proteins.
  - The pellet is then washed twice in 1 μL of acetone with 1% beta-mercaptoethanol and centrifuged at a maximum speed at 4°C for 15 min.
  - Final washes remove any additional acids and reagents.
- The final pellet is air-dried and stored at −20°C until running it on an acid gel for analysis.
- When ready to run the sample on the acid gel, resuspend the pellet in 20 μL of loading buffer made up of 0.375 M potassium acetate, 15% sucrose, and 0.05% pyronin Y.

Acid Gel Preparation

These gel conditions are optimized for the Multiphor II horizontal gel apparatus:

- Separating gel: 15–20% acrylamide (with the addition of 0.625 M acetic acid), stacking gel: 7.5% acrylamide
- Loading buffer: 0.375 M potassium acetate, 15% sucrose, 0.05% pyronin Y
- Running buffer: 0.9 M Acetic acid
- Electrophoresis: prerun – 300 V for 30 min, stacking run – 100 V for 30 min, separation – 200 V for 4.5 h
- Stained with Coomassie Blue

Isolation and Purification of Histones

The same protocol as used for protamines can be applied to isolate histones from mature sperm. Owing to the acidic conditions of the nuclear protein extraction mentioned above, all positively charged nuclear proteins will be extracted, and as a result both histones and protamines will be removed. The isolation of P1, P2, and histone occurs via gel purification in the acid gel.

Protamine Quantification and Ratio Generation

Once the sample has been gel-purified, both P1 and P2 are quantified and a P1–P2 ratio is determined.

Quantification

A set of P1 and P2 standards are run on the same gel as the sample. The gel is stained with Coomassie Blue and then scanned. The intensity of the bands is measured with an imaging software; this can be easily done using National Institute of Health Image-J software [25]. A standard curve is generated and the samples are quantified based on the standard curve [23, 26].

Generation of P1–P2 Ratio

Using the image analysis software, the relative amounts of P1 and P2 can be assessed. This is done by measuring the bands in the following manner: (P1 band – background)/(P2 band – background).

Generation of the Histone to Protamine Ratio

A purified sperm nuclear fraction is run on a gel as described above for protamines, but the gel is prepared with each sample being run in duplicate. Once the separation has occurred, the gel is cut into two halves. One half of the gel is stained with Coomassie Blue as described in the protamine assays. The other half of the gel is transferred to a polyvinylidene fluoride membrane in 0.7% acetic acid, 30% ethanol at 200 mA for 60 min. The membrane is blocked and then incubated with primary antibodies for P1, P2, and H2B for 1 h. The membrane is washed and a horseradish peroxidase conjugated secondary antibody is applied. Positive bands are detected by chemiluminescence. Once positive bands have been identified, they are used to identify the P1, P2, and H2B bands on the Coomassie Blue stained gel. The bands are quantitated using imaging software, and the histone to protamine ratio is determined from these relative quantities as follows: H2B/(P1 + P2 + H2B) [14, 15].

Staining Techniques

There are three main classes of staining techniques that are commonly utilized to evaluate sperm nuclear proteins: Chromomycin staining, Aniline Blue staining, and imunocytochemistry. Each has its own benefits and drawbacks that are described in this section.

Chromomycin

Chromomycin is a fluorochrome that binds specifically to Guanine–Cysteine dinucleotides [27]. Protamine proteins bind competitively to this same dinucleotide region of the DNA, so sperm that have high concentrations of protamine will theoretically have only small amounts of chromomycin fluorescence. Because of its unique properties this fluorochrome can be used to detect cells that have poor chromatin compaction due to aberrant protamination. Studies have demonstrated that sperm displaying increased chromomycin fluorescence are associated with decreased fertilization rates, increased protamine abnormalities, and an increased frequency of recurrent pregnancy [28, 29].

The major benefit of the assay is its relative simplicity in looking at possible protamine abnormalities. One of the main concerns with the use of chromomycin staining is its target, any available G–C dinucleotides in the genome. Increased chromomycin accessibility to G–Cs may be the result of a number of factors in addition to altered protamination such as DNA fragmentation. The argument can be made that the DNA damage resulting in increased fluorescence could have originated from a protamine deficiency. Though this may be true, it still calls into question the power of the staining technique to deliver informative data.

Chromomycin Protocol. This staining protocol is described by Sakkas et al. [29], previously reported by Bianchi et al. [27]

- The sample is smeared on a slide and allowed to air-dry.
- Once dry each slide is treated with 100 mL of chromomycin A3 solution (0.25 mg/mL in Mcilvaine buffer, pH 7.0, containing 10 mM $MgCl_2$) for 20 min.
- The slides are then rinsed and mounted with buffered glycerol.
- Fluorescence is analyzed via microscopy with a filter set appropriate for visualizing chromomycin.
- A minimum of 100 cells is counted as either positive (fluorescence observed) or negative (no fluorescence observed). The resulting percentage of positive cells is reported.

Aniline Blue

Aniline Blue (AB) selectively stains histone proteins due to their high lysine content. Since most histones are replaced with protamines during spermiogenesis, staining for histones can be a good method to detect possible problems in histone replacement in the mature sperm. Increased histone retention in individual cells would be expected to increase AB staining. Cells with increased AB should likely be more susceptible to DNA damage. With the use of AB staining, recent studies have shown that patients with recurrent pregnancy loss have an increased percentage of cells that stain positively for AB [28]. Hammadeh et al. [30] showed that positive AB staining occurred significantly more frequently in a patient population than it did in a healthy control population. Like other staining techniques, AB staining is useful because of its simplicity. The main drawback of the technique is that the threshold of histone retention resulting in positive AB staining has not been established. The literature indicates that there are some cells that show no staining at all even though there is always some degree of histone retention in mature sperm. This raises some concern about the sensitivity of the method and its clinical relevance. More research will be necessary to determine how the technique can best be used.

Aniline Blue Staining Technique

- Semen sample is smeared onto a slide and fixed with 4% formalin
- The slide is washed and stained with 5% AB in 4% acetic acid (pH 3.5) for 5 min
- At this point the slide can be dried and viewed, but Wong et al. [31] suggests an additional staining step with eosin to help enrich the signal
  - Stain in 0.5% eosin for 1 min
- Once the slide is dried, it can be viewed with bright-field microscopy under oil immersion
- Cells are counted as positive (nuclear staining) or negative (no nuclear staining) [31]

Immunocytochemistry

Immunocytochemistry can be used to detect any of the nuclear proteins. Both histones and protamines have been stained by immunocytochemistry in previous studies [18, 32]. Immunostaining allows the researcher to evaluate the protein content of individual cells using fluorescently labeled antibodies. Immunocytochemistry allows the observation of nuclear protein makeup of single cells and also facilitates the utilization of other assays on those same cells at the same time. DNA damage (with the use of the Terminal deoxynucleotidyl transferase dUTP nick end labeling or TUNEL assay), viability, and sperm chromatin structure can all be analyzed alongside the immunostaining. This allows the researcher to correlate abnormal protamine or histone levels with a number of other abnormalities. The main limitation of this technique is the inability of precise quantification, although the results still yield meaningful data.

Much have already been learned with the use of these techniques. Aoki et al. reported dramatic variations in the protamine state between individual cells, within a single ejaculate. Additionally, individual cells that were found to have the lowest protamine levels were also those that demonstrated decreased viability and showed the highest levels of DNA damage [32]. In 2008 Zini et al. [18] demonstrated a significant relationship between cells with diffuse, but intense, histone H2B (sperm nuclear histone) staining and an increased DNA fragmentation index as well as increased DNA stainability. This increased stainability and DNA fragmentation index is likely the result of increased DNA damage. The ability to look at multiple factors in a single cell is a powerful tool and will allow for future creative approaches to better understand the role of nuclear proteins.

Immunocytochemistry Preparation

While protocols will vary widely based on the proteins being targeted or the antibodies being used, there are a few simple steps that will likely be part of any preparation that falls under this category. Those steps are as follows:

- The sperm sample will generally be washed and centrifuged in PBS and will then be smeared onto a slide where it is fixed and allowed to air-dry.
- To gain proper access to the DNA, the sperm will be decondensed with incubation in DTT.
- The slides can then be incubated in a mix containing the desired primary antibody along with other essential compounds required for the researchers' specific tests.
- The slides would then be washed and incubated in the secondary antibody mix.
- The use of fluorescence microscopy can then be utilized to observe the results [18, 32].

Conclusions

The study of sperm nuclear proteins in the field of male factor infertility is exciting and is yielding important and interesting results. To fully understand what is occurring in these cells, we must understand what their chromatin structure is to ensure accuracy in describing what influences the structure has on sperm function.

The various assays described in this chapter all provide insight into our current understanding of sperm chromatin structure (Table 16.1). Each of the assays is uniquely informative, but they all have limitations. The nucleoprotein isolation techniques can be laborious but can provide important and accurate diagnostic information about the population of sperm in an ejaculate. They allow us to generate ratios of protamine proteins and histones in addition to quantifying the different nuclear protein species. The assay is limited, however, in that it cannot be used to examine single cells, but can only provide a general average of the total sperm in a given patient's sample.

Staining with the use of chromomycin and AB are useful techniques that are simple to use and provide quick results that help describe general deficits in the sperm chromatin. Chromomycin is an indirect staining method that allows us to see free G–C nucleotides that may be a result of incomplete protamination. This indirect approach can raise some questions as to what the real cause of increased staining may be (since the protamines are not being examined directly). However, the increased stainability still is descriptive of overall chromatin abnormalities, which could lead to DNA damage or a host of

Table 16.1 Test results and sperm assays

Test results	Effects on individual/diagnosis	Assay used to discover abnormality	References
High P1–P2 ratio	Low fertilization rate, decreased general semen parameters	Nuclear protein extraction and isolation	Aoki et al. [16, 26, 32]
Low P1–P2 ratio	Low fertilization rate, low pregnancy rate, increased DNA fragmentation, decreased general semen parameters, poor pregnancy outcome	Nuclear protein extraction and isolation	Aoki et al. [16, 26, 32], de Mateo et al. [33]
High histone to protamine ratio	General infertility	Nuclear protein extraction and isolation	Zhang et al. [15], Zini et al. [14]
Increased chromomycin staining	Low fertility rate, high protamine abnormalities, high recurrent pregnancy loss	Chromomycin staining	Sakkas et al. [29], Kazerooni et al. [28]
Increased AB staining	High recurrent pregnancy loss, general infertility	AB staining	Hammadeh et al. [30], Kazerooni et al. [28]
Decreased immunostaining for protamine	Low viability, high DNA damage	Immunostaining	Aoki et al. [16, 32]
Increased immunostaining for histone	High DNA stainability, high DNA fragmentation	Immunostaining	Zini et al. [18]

other problems in the sperm. As a result, this staining technique still provides relevant, easy-to-generate data. AB stains histones directly and is used generally to analyze cells for increased histone concentrations. This data can be of use in diagnostics, but there are a few limitations. The main drawback is the overall sensitivity and selectivity of the assay. There is no real threshold of histone retention that results in positive AB staining that has been established. Despite this, the assay offers rapid analysis of histone retention that is informative of relative abnormalities in chromatin composition.

Immunocytochemistry has yielded many interesting results in the past. Like the other staining techniques already discussed, it allows the researcher to analyze nuclear proteins at the single cell level, but unlike other methods immunocytochemistry allows the simultaneous observation of other abnormalities in the same cell. This facilitates a study of nuclear protein changes in a single cell that can be correlated directly to other chromatin or DNA changes. The one main drawback of this technique is the inability of precise quantification.

Much progress has been made in this field with the use of the assays described in this chapter, but there is still much to learn. A creative utilization of these techniques and others will allow us to gain more insight into the dynamics of nuclear proteins in the mature sperm; both in fertile men and in various classes of infertility. A more detailed understanding of the complex sperm chromatin structure is essential in generating new ideas for clinically relevant diagnostic and treatment tools.

References

1. Carrell DT, Emery BR, Hammoud S. The aetiology of sperm protamine abnormalities and their potential impact on the sperm epigenome. Int J Androl. 2008; 31(6):537–45.
2. Dohle GR, Halley DJ, Van Hemel JO, et al. Genetic risk factors in infertile men with severe oligozoospermia and azoospermia. Hum Reprod. 2002;17(1):13–6.
3. Erenpreiss J, Spano M, Erenpreisa J, Bungum M, Giwercman A. Sperm chromatin structure and male fertility: biological and clinical aspects. Asian J Androl. 2006;8(1):11–29.
4. Balhorn R, Reed S, Tanphaichitr N. Aberrant protamine 1/protamine 2 ratios in sperm of infertile human males. Experientia. 1988;44(1):52–5.

5. Carrell DT, Hammoud SS. The human sperm epigenome and its potential role in embryonic development. Mol Hum Reprod. 2010;16(1):37–47.
6. Hammoud SS, Nix DA, Zhang H, Purwar J, Carrell DT, Cairns BR. Distinctive chromatin in human sperm packages genes for embryo development. Nature. 2009;460(7254):473–8.
7. Steger K. Transcriptional and translational regulation of gene expression in haploid spermatids. Anat Embryol (Berl). 1999;199(6):471–87.
8. Dadoune JP. The nuclear status of human sperm cells. Micron. 1995;26(4):323–45.
9. Oliva R, Dixon GH. Vertebrate protamine gene evolution I. Sequence alignments and gene structure. J Mol Evol. 1990;30(4):333–46.
10. Hecht NB. Regulation of 'haploid expressed genes' in male germ cells. J Reprod Fertil. 1990;88(2):679–93.
11. Powell D, Cran DG, Jennings C, Jones R. Spatial organization of repetitive DNA sequences in the bovine sperm nucleus. J Cell Sci. 1990;97(Pt 1):185–91.
12. Ward WS, Coffey DS. DNA packaging and organization in mammalian spermatozoa: comparison with somatic cells. Biol Reprod. 1991;44(4):569–74.
13. Carrell DT, Emery BR, Hammoud S. Altered protamine expression and diminished spermatogenesis: what is the link? Hum Reprod Update. 2007;13(3):313–27.
14. Zini A, Gabriel MS, Zhang X. The histone to protamine ratio in human spermatozoa: comparative study of whole and processed semen. Fertil Steril. 2007;87(1):217–9.
15. Zhang X, San Gabriel M, Zini A. Sperm nuclear histone to protamine ratio in fertile and infertile men: evidence of heterogeneous subpopulations of spermatozoa in the ejaculate. J Androl. 2006;27(3):414–20.
16. Aoki VW, Liu L, Jones KP, et al. Sperm protamine 1/protamine 2 ratios are related to in vitro fertilization pregnancy rates and predictive of fertilization ability. Fertil Steril. 2006;86(5):1408–15.
17. Miller D, Brinkworth M, Iles D. Paternal DNA packaging in spermatozoa: more than the sum of its parts? DNA, histones, protamines and epigenetics. Reproduction. 2010;139(2):287–301.
18. Zini A, Zhang X, San Gabriel M. Sperm nuclear histone H2B: correlation with sperm DNA denaturation and DNA stainability. Asian J Androl. 2008;10(6):865–71.
19. Balhorn R, Brewer L, Corzett M. DNA condensation by protamine and arginine-rich peptides: analysis of toroid stability using single DNA molecules. Mol Reprod Dev. 2000;56(2 Suppl):230–4.
20. Vilfan ID, Conwell CC, Hud NV. Formation of native-like mammalian sperm cell chromatin with folded bull protamine. J Biol Chem. 2004;279(19):20088–95.
21. Belokopytova IA, Kostyleva EI, Tomilin AN, Vorob'ev VI. Human male infertility may be due to a decrease of the protamine P2 content in sperm chromatin. Mol Reprod Dev. 1993;34(1):53–7.
22. Carrell DT, Emery BR, Liu L. Characterization of aneuploidy rates, protamine levels, ultrastructure, and functional ability of round-headed sperm from two siblings and implications for intracytoplasmic sperm injection. Fertil Steril. 1999;71(3):511–6.
23. Carrell DT, Liu L. Altered protamine 2 expression is uncommon in donors of known fertility, but common among men with poor fertilizing capacity, and may reflect other abnormalities of spermiogenesis. J Androl. 2001;22(4):604–10.
24. de Yebra L, Oliva R. Rapid analysis of mammalian sperm nuclear proteins. Anal Biochem. 1993;209(1):201–3.
25. Abramoff MD, Magelhaes PJ, Ram SJ. Image processing with ImageJ. Biophotonics Int. 2004;11(7):36–42.
26. Aoki VW, Liu L, Carrell DT. Identification and evaluation of a novel sperm protamine abnormality in a population of infertile males. Hum Reprod. 2005;20(5):1298–306.
27. Bianchi PG, Manicardi GC, Bizzaro D, Bianchi U, Sakkas D. Effect of deoxyribonucleic acid protamination on fluorochrome staining and in situ nick-translation of murine and human mature spermatozoa. Biol Reprod. 1993;49(5):1083–8.
28. Kazerooni T, Asadi N, Jadid L, et al. Evaluation of sperm's chromatin quality with acridine orange test, chromomycin A3 and aniline blue staining in couples with unexplained recurrent abortion. J Assist Reprod Genet. 2009;26(11–12):591–6.
29. Sakkas D, Urner F, Bianchi PG, et al. Sperm chromatin anomalies can influence decondensation after intracytoplasmic sperm injection. Hum Reprod. 1996;11(4):837–43.
30. Hammadeh ME, Zeginiadov T, Rosenbaum P, Georg T, Schmidt W, Strehler E. Predictive value of sperm chromatin condensation (aniline blue staining) in the assessment of male fertility. Arch Androl. 2001;46(2):99–104.
31. Wong A, Chuan SS, Patton WC, Jacobson JD, Corselli J, Chan PJ. Addition of eosin to the aniline blue assay to enhance detection of immature sperm histones. Fertil Steril. 2008;90(5):1999–2002.
32. Aoki VW, Emery BR, Liu L, Carrell DT. Protamine levels vary between individual sperm cells of infertile human males and correlate with viability and DNA integrity. J Androl. 2006;27(6):890–8.
33. de Mateo S, Gázquez C, Guimerà M, Balasch J, Meistrich ML, Ballescà JL, Oliva R. Protamine 2 precursors (Pre-P2), protamine 1 to protamine 2 ratio (P1/P2), and assisted reproduction outcome. Fertil Steril. 2009;91:715–22.

Sperm Epigenetic Profile

17

Cristina Joana Marques, Alberto Barros, and Mário Sousa

Abstract

Spermatogenesis is a unique process by which diploid mitotically dividing spermatogonia give rise to mature haploid male gametes by a process involving several complex events such as proliferation (mitotic divisions of spermatogonia), meiotic divisions, and differentiation steps. During this process, several epigenetic modifications of the paternal genome occur, which lead to a compacted nuclear structure and transcriptionally inactive genome. It involves histone variants and histone to protamine exchanges. In addition, sperm carry important epigenetic information such as paternal imprinting marks that are crucial for the normal development of the future embryo.

Keywords

Sperm epigenetic profile • Spermatogonia • Histone in sperm • Gene expression in epigenetic regulation

Epigenetic Regulation of Gene Expression

Epigenetic modifications of the genome control gene expression by governing the chromatin structure and transcription factors' accessibility to promoters and key regulatory regions of genes. Two widely studied epigenetic modifications are DNA methylation and histone modifications.

C.J. Marques (✉)
Genetics Department, Faculty of Medicine, University of Porto, Portugal, UK
e-mail: cmarques@med.up.pt

DNA Methylation

DNA methylation is a heritable epigenetic modification that controls gene expression and genome stability. Methylation of DNA occurs through the addition of a methyl group, provenient from the *S*-adenosyl-methionine (SAM) donor, to the carbon 5 position of cytosines [1]. In mammals, this modification is thought to occur mainly on cytosines located in CpG dinucleotides, although a recent study suggests the existence of methylation in non-CG contexts in embryonic stem (ES) cells [2]. A large majority of CpG sequences are methylated and are located in transposons and repetitive sequences, leading to transposon

A. Zini and A. Agarwal (eds.), *Sperm Chromatin: Biological and Clinical Applications in Male Infertility and Assisted Reproduction*, DOI 10.1007/978-1-4419-6857-9_17,

repression and genomic stability to the genome. On the contrary, unmethylated CpGs are usually found in CpG islands (regions with a greater CpG content) and associate with gene promoters. DNA methylation regulates the expression of genes in two different ways: (1) by physically impeding the binding of transcription factors to the promoter regions and (2) by the recruitment of methyl-CpG-binding domain (MBD) proteins that recruit other chromatin remodeling factors such as histone deacetylases (HDACs) to create a compact silent structure.

DNA methyltransferases (DNMTs) are the enzymes responsible for adding methyl groups to the cytosines located on CpG dinucleotides. Two main classes of DNMTs have been identified: maintenance (DNMT1) and de novo (DNMT3A, DNMT3B, and DNMT3L) DNMTs.

DNMT1 is known as the maintenance methyltransferase. Dnmt1 is assumed to have the main role of maintaining methylation due to its catalytic preference for hemimethylated DNA [3, 4] and association with replication foci during S-phase, having a diffuse nucleoplasmic distribution in non-S-phase cells [5]. However, it was shown that Dnmt1 also has substantial de novo methylation activity, about 5–20% of the activity on hemimethylated DNA [3]. Three splicing variants were identified in mice: Dnmt1s (somatic), Dnmt1p (pachytene), and Dnmt1o (oocyte). Dnmt1p and Dnmt1o are two germ-line-specific isoforms of Dnmt1 and result from alternative splicing of sex-specific 5′ exons [6]. Dnmt1p transcription is restricted to pachytene spermatocytes but does not result in detectable levels of protein despite the high levels of mRNA [6, 7]. Dnmt1o is transcribed from an oocyte specific promoter and encodes a truncated at the N-terminus but enzymatically active version of Dnmt1 that accumulates in the cytoplasm during the later stages of oocyte growth [6].

Dnmt3a and Dnmt3b encode essential de novo methyltransferases, since inactivation of these enzymes in mice was shown to cause embryonic lethality [8]. Dnmt3a was shown to be essential for the establishment of both maternal and paternal imprints [9]. Furthermore, Dnmt3a was shown to be essential for normal spermatogenesis since seminiferous tubules from conditional mutant mice presented only spermatogonia and absence of spermatocytes, spermatids, and spermatozoa [9]. Dnmt3b is specifically required for methylation of pericentromeric repetitive regions and was also suggested to be involved in the methylation of Rasgrf1 imprinted gene (paternally methylated), together with Dnmt3a [8, 10]. In humans, mutations in DNMT3B cause ICF (immunodeficiency, centromeric instability, facial anomalies) syndrome, a rare autosomal recessive disorder characterized by hypomethylation at the pericentromeric satellite regions of chromosomes 1, 9, and 16 [11, 12].

Dnmt3L encodes a protein with regions of homology to Dnmt3a and Dnmt3b, but lacking enzymatic activity itself due to the absence of conserved catalytic motifs [13]. DNMT3L mRNA is present, albeit at low levels, in the testis, ovary, and thymus. However, expression in the testis is increased about 100-fold when compared with the other tissues [13, 14]. In fact, Dnmt3L was shown to be essential for normal spermatogenesis and for the establishment of maternal imprints during oogenesis [15]. Male germ cells from Dnmt3l-deficient mice show reactivation of retrotransposon expression and several problems in synapsis at the meiotic prophase leading to meiotic arrest [16]. Dnmt3l-deficient prospermatogonia have been shown to lose methylation at paternally imprinted genes and repetitive sequences [10].

Histone Modifications

In addition to the importance of DNA methylation in epigenetic control of gene expression, differential histone tail modifications such as methylation, phosphorylation, acetylation, and ubiquitinylation are also key regulators of chromatin states and are referred to as the histone code [17]. DNA methylation and histone modifications are likely to interplay to establish a repressive/active chromatin state, with hypermethylated regions being rich in histone marks associated with silencing, such as H3 di- or trimethylation at lysine 9 (H3K9me2/3). Histone acetylation and other marks such as H3K4me2 and H3K4me3 are normally associated with transcriptionally active genes [18]. The histone methyltransferases (HMTs), such as SUV39H1 and SUV39H2 (suppressor of variegation 3–9 homolog) [19], and histone demethylases, such as JHDM1 (Jumonji C domain containing histone demethylase 1) [20], mediate the establishment and removal of methylation at arginine and lysine residues in histones N-terminal regions. Histone acetyltransferases (HATs) establish the acetylation active mark, and histone deacetyltransferases (HDACs) reverse this active status to a repressive one [21].

Genomic Imprinting Mechanism in Mammals

Genomic imprinting is a mechanism of gene regulation that causes a subset of mammalian genes to be expressed from only one of the two parental chromosomes. Some imprinted genes are expressed from the maternal copy, while others are expressed from the paternally inherited chromosome. This mechanism was first described in 1984, after the findings that androgenotes (zygotes with two male pronuclei) and gynogenotes (zygotes with two female pronuclei) could not develop to term, suggesting that parental genomes were not functionally equivalent and both were required for normal embryogenesis to occur [22, 23]. Strikingly, it was also noticed that gynogenotes and androgenotes presented a nearly opposite phenotype: while gynogenotes gave rise to a normal but small embryo with extraembryonic tissues severely deficient, androgenotes presented better development of extraembryonic tissues but poor embryo development. These observations led to the speculation that the paternal genome is essential for normal development of extraembryonic tissues, while the maternal genome plays a more important role in embryo development [24]. Additional work generating embryos with uniparental disomies for individual chromosomes or chromosome regions demonstrated that parental imprinting is restricted to some parts of the genome and leads to differential functioning of genes within those regions [25]. These functional differences between the parental genomes were later shown to be heritable and retained following the activation of the embryonic genome at the two-cell stage [26].

Monoallelic expression of imprinted genes depends on an epigenetic mark that allows distinguishing both parental alleles. This imprinting mark must be heritable and reversible and must be interpreted by the transcription machinery to lead to expression/repression. Although the whole complex nature of the imprint itself remains elusive, the involvement of DNA methylation and allele-specific differential chromatin structure has been described [27]. Many imprinted genes contain one or more differentially methylated regions (DMRs) displaying an allele-specific DNA methylation pattern that determines the expression status of the genes [28].

Parental specific imprints are combined at fertilization in the zygote and propagated thereafter during embryogenesis in somatic tissues. In both germ lines, imprinting marks must be erased and reestablished according to parental sex (Fig. 17.1). The process of erasure occurs in primordial germ cells (PGCs) of developing male and female mice embryos, between 11.5 and 12.5 days postcoitum, following their migration into the genital ridge [29, 30] in a process that seems to be caused by active DNA demethylation. The mechanism behind active DNA demethylation remains largely unknown despite extensive efforts to unravel this process. One of the pathways that might be involved is the deamination of 5-methylcytosine to thymine followed by DNA-repair events such as base excision repair (BER) or nucleotide excision repair (NER) [31]. The DNA deaminases AID (activation-induced cytidine deaminase) and APOBEC1 (apolipoprotein B mRNA-editing enzyme, catalytic polypeptide 1) were shown to be able to deaminate in vitro 5-meC into thymine, thus leading to T:G mismatches [32], which could be later repaired by the thymine DNA glycosylase (TDG) or other repair enzymes that can specifically remove T from a T:G mismatch within a CpG context [33].

Supporting this notion, a recent report has shown that *Aid* deficiency interferes with

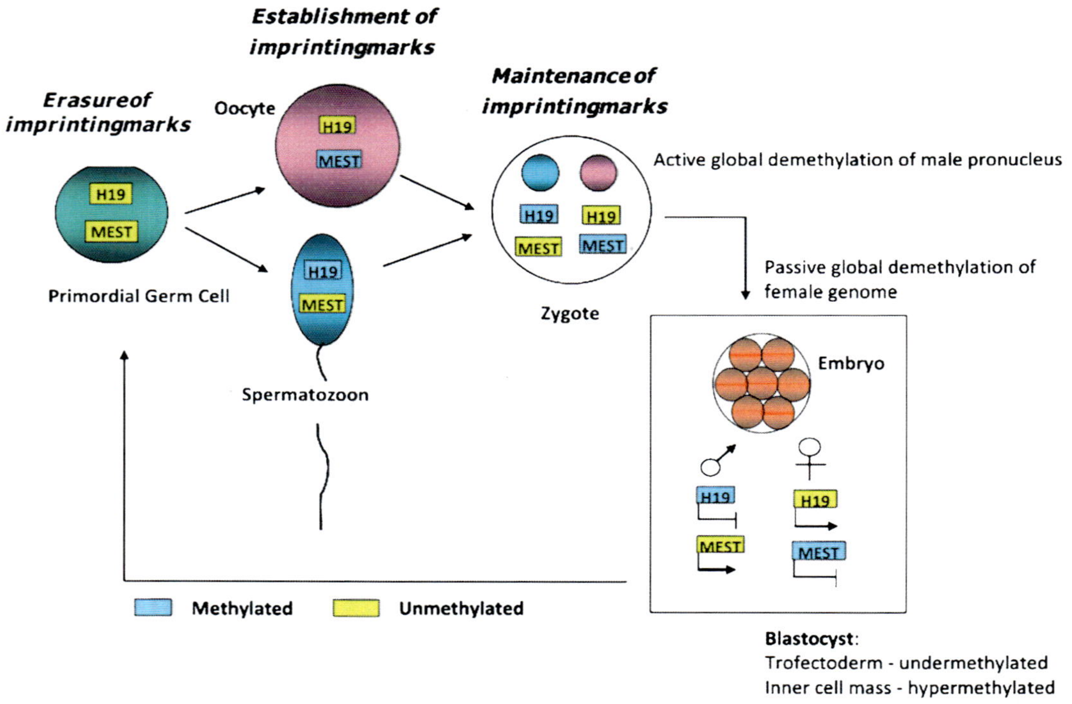

Fig. 17.1 Schematic representation of the life cycle of imprinting marks. Imprints are erased in primordial germ cells (PGCs) and later reestablished during gametogenesis according to the sex of the germ line. Parental-specific marks are combined at fertilization and resist the DNA-demethylation events that occur in early embryogenesis, later leading to monoallelic expression of imprinted genes (e.g., *H19* – paternally methylated and maternally expressed; *MEST* – maternally methylated and paternally expressed)

genome-wide erasure of DNA methylation, with *Aid–/–* male and female PGCs showing up to three times more methylation than wild-type cells [34]. In addition, the process of DNA demethylation in PGCs was shown to be accompanied by an extensive erasure of several histone modifications and exchange of histone variants [35]. Reestablishment of imprints occurs later during gametogenesis, in a strictly sex-specific manner. After erasure, de novo methylation begins in both germ lines at late fetal stages and continues after birth. In oocytes, *de novo* methylation of imprinted genes occurs in the postnatal ovary during oocyte growth phase, corresponding to meiotic prophase I [36–38], whereas in the male germ line this process seems to occur earlier, beginning in fetal spermatogonia and being completed before meiosis occur [39, 40]. In the human, methylation of imprinted genes seems to be completely reestablished at the oocyte GV (germinal vesicle) stage [41, 42] and at the primary spermatocyte stage [39].

In the mouse zygote, there is a drastic decrease of DNA methylation in the paternal genome within few hours after fertilization [43, 44], while the maternal genome undergoes progressive demethylation during segmentation stages [45, 46]. Despite genome-wide demethylation, imprinted genes are exempt from this process and remain methylated, as do certain retrotransposons such as IAPs (intracisternal particle A) [47–49]. De novo methylation completes the cycle of methylation reprogramming during preimplantation development, occurring in cells from the inner cell mass but not from the trophectoderm of the blastocysts [46]. The passive demethylation of the maternal genome might be linked to the absence of the maintenance DNMT,

Dnmt1, in the nucleus, since it is retained in the cytoplasm from the oocyte to the blastocyst stage, with the exception of the eight-cell stage [50]. This transient nuclear localization of Dnmt1o (oocyte-specific isoform) has been suggested to provide maintenance methyltransferase activity specifically at imprinted loci, specifically during the fourth embryonic S-phase [51]. However, recent studies have suggested that the somatic isoform of Dnmt1 (Dnmt1s) is present in the embryo from the one-cell stage and has nuclear localization from the two-cell stage onward, which might protect methylation imprints from being erased [52–54]. Active demethylation of the paternal genome is linked to the formation of the paternal pronucleus [46] and occurs after removal of protamines and acquisition of female histones by the paternal genome, during G1 phase, before DNA replication [55].

It is also intriguing as to how the maternal genome and the paternal methylation of imprinted genes resist this wave of active DNA demethylation. One possible explanation is the chromatin conformation being different between the maternal and paternal genomes, since the first contains histones bound to the DNA, while the paternal genome contains mostly protamines. In this regard, it would be interesting to study whether the paternally methylated imprinted genes retain their histones during sperm differentiation and are, thus, resistant to DNA demethylation. Additionally, a recent paper [56] has provided evidence of a protein named Stella that might protect imprinted genes and other genome sequences from demethylation in the zygote. Stella is present in large amounts in oocytes and, after fertilization, translocates to both pronuclei.

Epigenetic Modifications in Sperm

Establishment of Paternal Imprints in the Male Germ Line

DNA Methylation Imprinting Marks in the Male Germ Line

Up until now, only the DMRs of three imprinted regions were reported to acquire methylation in the male gametes in mice – *H19-Igf2*, *Dlk1-Gtl2*, and *Rasgrf1* [57]. The acquisition of *H19* methylation imprinting marks has been extensively characterized.

Although acquiring methylation, *H19* was shown to be biallelically expressed in spermatogonia isolated from 7-day-old mice testes, suggesting that the imprint may not be recognized by germ cells. However, pachytene spermatocytes and spermatids isolated from adult mice testes did not present *H19* RNA, supporting the idea that the imprint is reprogrammed in the male germ line prior to the production of spermatozoa [58].

Davis et al. have described the acquisition of *H19* methylation in mice spermatogenesis [59]. They have reported that *H19* methylation imprint is acquired differentially on the parental alleles, with the paternal allele being remethylated earlier than the maternal allele. It was demonstrated that the paternal allele acquires methylation in primitive type A spermatogonia, obtained from the testes of 6-day-old mice, whereas the maternal allele is not hypermethylated until the completion of meiosis I. These results indicated that, albeit devoid of methylation, the parental alleles can retain their identity and be recognized by the methylation machinery.

Additionally, full methylation of *H19*, *Rasgrf1* and *Gtl2* was reported in mature spermatozoa [57]. Methylation on these imprinted genes is acquired progressively and is not completed at embryonic day 17.5 germ cells. Methylation at the *Dlk1-Gtl2* DMR in mice was shown to be established in prospermatogonia of embryonic day 19.5, and the two parental alleles were reported to acquire methylation in an identical way [60]. Oakes et al. have shown that these three paternally methylated imprinted genes have already acquired the imprinting mark by the pachytene spermatocyte stage of mice spermatogenesis [61].

Analysis of *H19* methylation in human spermatogenesis has shown that fetal spermatogonia are completely unmethylated, whereas adult primary spermatocytes are already fully methylated [39]. On the contrary, it was demonstrated that *MEST/PEG1* gene (which is maternally methylated and paternally expressed) is already completely unmethylated in fetal spermatogonia.

Several studies have also addressed genome-wide methylation patterns of sperm DNA in comparison to somatic cell DNA [62–64]. Remarkably, the sperm genome seems to be hypomethylated in comparison to somatic cells and to resemble ES and EG (embryonic germ) methylation patterns at promoter regions. This suggests that, although sperm cells are differentiated into a highly specialized function, their epigenome resembles the pluripotent states.

It was also shown that DNA hypomethylation in male germ cells, induced by treatment with 5-aza-2′-deoxycytidine, leads to infertility and/or a decreased ability to support preimplantation embryonic development [65]. It was later shown that although the treatment adversely affected sperm motility and the survival of embryos to the blastocyst stage, the major contributor to infertility was a marked decrease in the sperm fertilization ability [66].

Additionally to DNA methylation marks, histone modifications were also shown to mark imprinted genes in murine spermatogenesis [67]. Specifically, the authors observed that, in stages preceding the global histone-to-protamine exchange (spermatocytes, round and elongating spermatids), H3 lysine 4 methylation and H3 acetylation are enriched at maternally methylated ICRs (namely *Igf2r* and KvDMR1) but are absent at paternally methylated ICRs.

DNA Methyltransferases Expression in the Male Germ Line

The 5.2-kb DNA MTase (currently known as Dnmt1) mRNA, characteristic of somatic cells, is present in type A and B spermatogonia, in meiotic preleptotene and leptotene/zygotene spermatocytes and in haploid round spermatids. In adult spermatogenesis, the 5.2-kb form is more abundant in preleptotene and leptotene/zygotene spermatocytes [68]. A specific testicular form of DNA MTase, 6.2 kb long, has been observed in prepubertal mouse testis and is restricted to pachytene spermatocytes (prepubertal and adult) [7, 68]. The presence of this testis-specific DNA MTase mRNA coincides with active *de novo* methylation of testis-specific genes (namely, transition protein, protamine 1, and protamine 2) [69]. In adult spermatogenesis, the testis-specific 6.2 kb form is more abundant in pachytene spermatocytes [68]. DNA MTase protein is present in spermatogonia A and B, in preleptotene and leptotene/zygotene spermatocytes and is absent in pachytene spermatocytes. It is later detected in round spermatids, albeit in a lower level than in the previous spermatogenic cell stages [68, 70].

Concerning the prenatal period, Dnmt1 is present in prenatal gonocytes but is downregulated between 14.5 and 18.5 days of gestation, being absent at the time of acquisition of methylation in the male germ line, implicating other enzymes in the de novo methylation of DNA that is initiated in the prenatal period [71]. Expression profiles showing concomitant peaks of *Dnmt3a* and *Dnmt3l* expression in the prenatal testis, at E15.5 (embryonic day 15.5), suggests that these two enzymes may interact to establish paternal DNA methylation patterns. *Dnmt1* and *Dnmt3b* expression levels peak in the early postnatal period in the male suggesting a role for these enzymes in the maintenance of methylation patterns in rapidly proliferating spermatogonia. It is possible that Dnmt3b plays a role at these early times during spermatogenesis in actively methylating centromeric regions to ensure proper pairing and recombination between homologous chromosomes [71].

A recent study has described, by quantitative RT-PCR, the expression of *Dnmt1*, *Dnmt3a* and *Dnmt3b* during postnatal male germ cell development in the mouse [72]. *Dnmt1*, *Dnmt3a*, and *Dnmt3b* have their peak of expression in type A spermatogonia, decrease in type B spermatogonia and preleptotene spermatocytes, and increase in leptotene/zygotene spermatocytes; levels of transcripts decreased as pachynema progressed and increased again in round spermatids, being almost undetected in elongated spermatids. For the *Dnmt3b* transcripts, the increases and decreases in expression were more pronounced than for *Dnmt1* and *Dnmt3a*. When specific primers were used to discriminate between the two *Dnmt3a* transcripts, *Dnmt3a* was found to be expressed relatively constantly, whereas

Dnmt3a2 showed the variations described before. When the authors compared the expression of the three enzymes, it was clear that *Dnmt3a* and/or *Dnmt3b* were more expressed than *Dnmt1* in all cell stages. Analysis at the protein level revealed that *Dnmt3a2* is present in all stages except in elongating spermatids. Dnmt3a is expressed from type A spermatogonia until prepubertal pachytene and then is absent, and Dnmt3b is present throughout spermatogenesis except in elongated spermatids.

It was recently shown that Dnmt3L also uses three sex-specific promoters [73]. A promoter active in prospermatogonia drives transcription of an mRNA encoding the full-length protein in perinatal testis, where de novo methylation occurs. Late pachytene spermatocytes activate a second promoter in intron 9 of the *Dnmt3L* gene. After this stage, the predominant transcripts are three truncated mRNAs, which appear to be noncoding.

In humans, it has been shown that in normal spermatogenesis, *DNMT1* mRNA is present in spermatogonia, pachytene spermatocytes, and round spermatids, while DNMT1 protein is present only in the nuclei of spermatogonia and in the cytoplasm of round spermatids [74].

Chromatin Organization of the Sperm Nucleus

One important process occurring during spermiogenesis is the compaction of the sperm genome into the sperm head, achieved through the replacement of histones by protamines [75]. During spermatogenesis, there is a replacement of somatic histones by testis-specific variants, followed by the replacement of most histones (85% in human sperm) by transition proteins and then with protamines [76]. Some histone variants were found to be crucial for normal spermatogenesis, mainly phosphorylated H2AX (γH2AX) and H3.3, which are involved in the mechanism of MSCI (meiotic sex chromosome inactivation), which occurs to silence transcription in the XY body during the pachytene stage of meiotic prophase [77]. H2AX is an important part of the nucleosome of meiotic cells, and it gets phosphorylated in response to double-strand breaks (DSB) in DNA [78]. During spermatogenesis, it accumulates in the sex (XY) body in leptotene/diplotene spermatocytes and allows efficient accumulation of DNA repair proteins [79]. H3.3 incorporation into the XY body promotes extensive chromatin remodeling and is essential for gene silencing on the XY body during the later stages of MSCI and the postmeiotic stages of spermatogenesis [80]. Histone-to-protamine exchange is associated with core histone acetylation, as acetyl groups turn the basic state of histones into a neutral one that, as a consequence, decreases the affinity of histones for DNA and allows protamines to interact with DNA [81]. After meiosis, the beginning of spermiogenesis is characterized by a massive wave of transcriptional activity, which results in the activation of a number of essential postmeiotic genes in early haploid cells [75].

Epigenetic Defects in Assisted Reproduction Techniques (ART)

Imprinting Syndromes in ART Children

Deregulation of imprinted genes in two imprinting domains, located on chromosomes 11p15.5 and 15q11-q13 is the cause of Beckwith–Wiedemann syndrome (BWS) and Prader–Willi/Angelman syndromes (PWS/AS), respectively. A higher incidence of BWS and AS cases has been recently reported in children born after Assisted Reproduction Techniques (ART) than in the normal population (reviewed in [82]). The major defect found was hypomethylation on the maternal allele of KCNQ1OT1 gene (BWS) and on small nuclear ribonucleoprotein polypeptide N (SNRPN) gene (AS), leading to the hypothesis that ART procedures, such as hormonal stimulation of the ovaries, could affect the establishment of imprints in the oocytes or in vitro culture of the embryos could lead to loss of maternal methylation on imprinted genes [83]. However, a large-scale population study reporting a threefold increase in the incidence of imprinting syndromes

(PWS, AS, and BWS) in ART children in the Dutch population has suggested that this increase is particularly associated with fertility problems and not with the use of ART treatments [84].

More recently, another syndrome – Silver–Russell syndrome (SRS, OMIM 180860) – has been linked to epigenetic alterations in the *H19-IGF2* domain, particularly to hypomethylation of the paternal allele [85–89]. This syndrome is characterized by intrauterine and postnatal growth retardation with reduced cranial growth, dysmorphic features, and frequent body asymmetry [90]. Some SRS cases have been described in children born after ART that presented not only *H19* hypomethylation [85, 91, 92] but also one case showing hypermethylation of the *MEST* gene [93]. Previously, it has also been observed maternal uniparental disomy (mUPD) of chromosome 7 in about 10% of the SRS, suggesting the involvement of imprinted genes located in this region, such as *MEST*, *COPG2IT1*, and *GRB10* [94]. This suggests that multiple genetic causes might be involved in the etiology of this syndrome.

Imprinting Errors in Male infertility

There has been increasing evidence that abnormal spermatogenesis leading to oligozoospermia (low sperm counts) is associated with sperm carrying methylation defects at imprinted genes. Several groups have described both hypomethylation at paternally methylated imprinted genes and hypermethylation at maternally methylated imprinted genes in sperm cells from patients presenting a myriad of spermiogram defects such as decreased number or absence of spermatozoa, abnormal morphology and/or motility, and abnormal protamine replacement (Table 17.1).

Although the first report analyzing methylation at the imprinted gene *SNRPN* (maternally methylated and paternally expressed) in sperm from oligozoospermic patients did not describe any alterations, this was possibly due to the restraints of the technique applied (MSP – methylation-specific PCR) [95]. Even so, the authors did observe some abnormal methylation using a more sensitive approach (heminested PCR), but since this was also present in normal sperm samples, it was regarded as possibly being caused by somatic cell contamination [95]. The first description of methylation imprinting defects in sperm from infertile men has reported *H19* hypomethylation in 24% (23/96) of oligozoospermic patients, while normozoospermic individuals showed complete methylation at this locus [96]. Moreover, this hypomethylation affected one of the CTCF binding sites in 11% (11/96) of the patients. CTCF (CCCTC-binding factor) is an insulator protein that binds to the unmethylated maternal *H19* DMR and prevents another imprinted gene, *IGF2* (insulin-like growth factor 2) from accessing common enhancers, hence repressing its expression [97]. Hypomethylation at the CTCF binding sites might lead to the inactivation of the *IGF2* paternal copy, causing biallelic repression in the embryo. A second study has corroborated these findings and extended the number of imprinted genes showing methylation errors [98]. These authors observed an increased incidence of imprinting errors in sperm from oligozoospermic patients, in two paternally methylated (*H19* and *GTL2*) and three maternally methylated (*PEG1/MEST*, *ZAC*, and *SNRPN*) imprinted genes. Hypomethylation errors were found in 14% (14/97) and hypermethylation in 21% (20/97) of patients. Moreover, global sperm DNA methylation, evaluated by LINE1 and Alu regions, was within normal levels suggesting that these defects were restricted to imprinted genes. Additionally, five of the six patients presenting severe oligozoospermia had methylation errors at both paternally and maternally methylated imprinted genes. A subsequent study by our group [99] showed that imprinting errors, consisting of complete lack of methylation at the *H19* gene and complete methylation at *MEST* gene, were restricted to sperm from patients presenting less than ten million sperm per ml of semen. We have also shown that hypomethylation is restricted to imprinted genes through the analysis of methylation levels at the LINE1 transposon element.

Several other studies were subsequently undertaken that showed methylation errors at imprinted genes in sperm from patients with oligozoospermia

Table 17.1 Summary of imprinting errors described in sperm from infertile patients

References	Genes/regions	Patients	Alterations in DNA methylation
Marques et al. [96]	H19, MEST	27 NZ 96 OZ	Hypomethylation of H19 (and CTCF-6) in 24% OZ
Kobayashi et al. [98]	H19, GTL2, PEG1, LIT1, ZAC, PEG3, SNRPN	79 NZ 18 OZ	Hypomethylation in 14% and hypermethylation in 21% patients
Houshdaran et al. [103]	repetitive elements, promoter CpG islands, and DMRs of imprinted genes	65[a]	Hypermethylation in nine regions associated with decreased sperm concentration and motility
Marques et al. [99]	H19, MEST, LINE1	5 NZ 20 OZ	Hypomethylation of H19 (and CTCF-6) and/or hypermethylation of MEST in 47% of patients with sperm count below 10×10^6 Sz/ml
Marques et al. [104]	H19, MEST	24 AZO	Hypomethylation of H19 (and CTCF-6) in SAZ and hypermethylation of MEST in OAZ
Boissonnas et al. [102]	IGF2 (DMR0 and DMR2), H19 (CTCF-3 and CTCF-6), PEG3, LINE1	17 NZ 19 T 22 OAT	Loss of methylation of IGF2-DMR2 and/or CTCF-6 in 58% T patients and loss of methylation of CTCF-6 in 73% of OAT patients
Poplinski et al. [101]	IGF2/H19 (ICR1), MEST	33 NZ 148 idiopathic infertile	Hypomethylation at ICR1 and hypermethylation of MEST associated with low sperm counts. MEST hypermethylation associated with low sperm motility and abnormal morphology
Hammoud et al. [100]	LIT1, MEST, SNRPN, PLAGL1, PEG3, H19, IGF2	Five fertile 10 OZ 10 APR	Alterations in OZ and APR in all imprinted genes except IGF2

NZ normozoospermia; *OZ* oligozoospermia; *AZO* azoospermia; *SAZ* secretory azoospermia (germinal hypoplasia); *OAZ* obstructive azoospermia; *T* teratozoospermia; *OAT* oligozoospermia and/or asthenozoospermia and/or teratozoospermia; *APR* abnormal protamine replacement

[a]Patients were not classified in terms of spermiogram results, but genes were correlated with all types of spermiogram abnormalities

but also with abnormal protamine replacement (altered P1:P2 ratio) (Table 17.1) [100–103]. Interestingly, one of the studies also found a strong association between loss of methylation at the sixth CTCF binding site and decreased sperm counts, as we described before [99, 102]. In addition, we have also analyzed sperm retrieved from testicular biopsies of azoospermic patients for their imprinting status [104]. Imprinting errors such as complete lack of methylation at *H19* and the CTCF binding site-6 were found in sperm from a patient presenting secretory (nonobstructive) azoospermia due to germinal hypoplasia.

Another interesting recent study analyzed 78 paired sperm and abortion (6–9 weeks of gestation) DNA samples, from ART treatments [105]. Importantly, the authors report that 17 fetal samples (22%) presented imprinting methylation errors at one or more imprinted loci and that in seven of these (41%) the same alteration was also found in the sperm. This important observation supports our hypothesis that the increased incidence of imprinting syndromes observed in ART children might be related to inherent gametic defects and not only to the specific techniques involved, such as ovarian hormonal stimulation and in vitro embryo culture. However, most of the imprinting syndromes described in ART children present alterations at the maternal allele; although technically challenging, it would be interesting to analyze if poor quality oocytes also present imprinting errors. So far, there has been some evidence that in vitro maturation of oocytes interferes with correct establishment of maternal methylation marks, both by hypermethylation of the *H19* gene and by hypomethylation at the KvDMR1 of *KCNQ1OT1* [106, 107].

Methods for Assessing Epigenetic Modifications in Sperm

DNA methylation can be assessed by several methodologies, the most common one being bisulfite genomic sequencing. Sodium bisulfite converts unmethylated cytosines into thymines while methylated cytosines remain unchanged. Site-specific analysis can be performed by PCR (polymerase chain reaction) amplification of the modified DNA, either using primers that bind specifically if the region is methylated or unmethylated (MSP) or designing primers for regions without CpGs, therefore amplifying both previously methylated and unmethylated molecules. The first approach is more limiting, since it gives information on only the (usually) few CpGs that are located in the primer binding sequence. The second approach is more informative as is generally followed by cloning of the PCR products and sequencing, providing information on the methylation status of all CpGs located in the amplified region and giving a theoretical estimation of the percentage of methylation at each site in the original DNA sample. This latter methodology has been routinely used in the studies analyzing methylation at imprinted genes in sperm from infertile men. However, new techniques are now emerging that take advantage of the next-generation sequencing (NGS) methods providing a genome-wide coverage of the epigenome, such as MeDIP-seq (methyl-DNA immunoprecipitation-sequencing) and Bis-Seq (Bisulfite-sequencing) [108]. These approaches will provide a broader spectrum of methylated regions in the sperm genome and hopefully contribute to a greater understanding on the methylation errors associated with abnormal spermatogenesis.

On the contrary, histone modifications can be analyzed using ChIP (chromatin immunoprecipitation), a method for assaying DNA-protein binding in vitro. However, this technique has been proved difficult to apply in sperm cells, possibly due to the dense compacted structure of the sperm DNA [67].

Clinical Importance of Sperm Epigenetic Profiling in ART

As described before, several types of imprinting errors have been found in sperm from infertile patients presenting spermiogram abnormal parameters, such as oligozoospermia. The appropriate establishment of imprinting marks during male and female gametogenesis is essential for correct expression of these genes in the embryo and

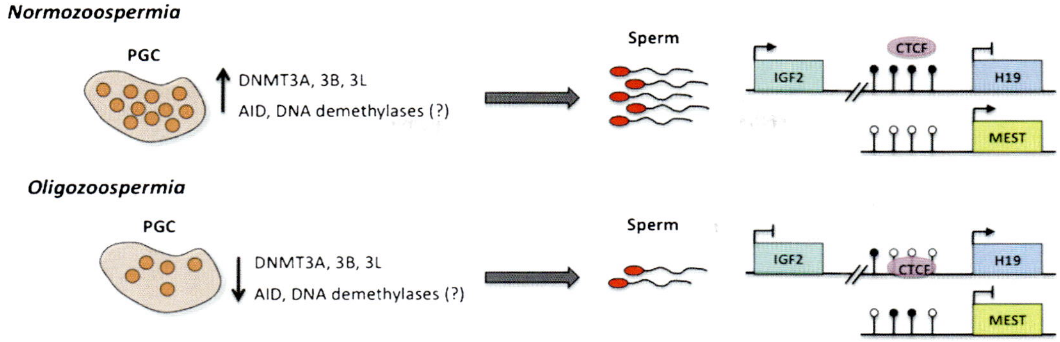

Fig. 17.2 Model for the occurrence of imprinting errors in sperm from patients with oligozoospermia. Compromised development of PGCs in the embryonic gonad, leading to a decreased number of sperm cells in the adult, might be associated with abnormal expression of epigenetic reprogramming enzymes such as de novo DNA methyltransferases (DNMT3A, 3B and 3L) or enzymes potentially involved in DNA demethylation (such as Aid deaminase). Consequently, hypomethylation (in *H19* and CTCF-binding site) and hypermethylation (in *MEST*) errors in imprinted genes occur and might lead to their abnormal expression or repression, respectively, from the paternal allele

later on. Imprinted genes play important roles in the regulation of growth and development, particularly in regulating embryonic growth and placental function [83]. To our knowledge, there is no evidence of a repair mechanism that could correct imprinting errors transmitted by the gametes and therefore, it is likely that embryos produced with sperm that carry abnormal imprints will be developmentally affected. Indeed, one of the studies describing imprinting errors in sperm from oligozoospermic patients reports that the outcome of ART was poor in these cases [98]. We have also observed that one azoospermic patient that presented *H19* complete unmethylation did not produce viable embryos, since these were arrested in development [104]. Additionally, *H19* hypomethylation has been linked to SRS [90]. In light of all these findings, it is advisable that methylation analysis of imprinted genes be added to spermiogram analysis of the infertile men, especially in cases where severe oligozoospermia is detected.

Conclusion

Despite all the advances in characterizing methylation imprinting errors associated with abnormalities in spermatogenesis, it is still unclear why these occur. Since PGCs are a crucial stage for the erasure and reestablishment of methylation imprints, it is expected that developmentally compromised PGCs, leading to low sperm counts, might have lower levels of de novo DNMTs expression and enzymes involved in DNA demethylation, leading to the incorrect erasure and reestablishment of imprinting marks that are later detected in sperm cells (Fig. 17.2). Indeed, Kobayashi et al. have found alterations in *DNMT3L* gene in patients presenting imprinting errors in their sperm [105]. In this regard, analysis of DNMTs expression in germ cells from infertile patients could provide further evidence on the mechanism behind the occurrence of imprinting errors in association with abnormal spermatogenesis.

Acknowledgments The authors wish to thank Filipa Carvalho, Susana Fernandes, Joaquina Silva, Luís Ferraz, Mariana Cunha, Paulo Viana and Ana Gonçalves for their valuable help. The authors have been partially supported by the Portuguese Foundation for Science and Technology (FCT) with a PhD fellowship to CJM (SFRH/BD/19967/2004) and research grants to MS (POCI/SAU-MMO/60555/04, 60709/04, 59997/04; UMIB).

References

1. Bird A. The essentials of DNA methylation. Cell. 1992;70:5–8.
2. Lister R, Pelizzola M, Dowen RH, et al. Human DNA methylomes at base resolution show widespread epigenomic differences. Nature. 2009;462:315–22.

3. Yoder JA, Soman NS, Verdine GL. DNA (cytosine-5)-methyltransferases in mouse cells and tissues. Studies with a mechanism-based probe. J Mol Biol. 1997;270:385–95.
4. Pradhan S, Bacolla A, Wells RD. Recombinant human DNA (cytosine-5) methyltransferase. I. Expression, purification, and comparison of de novo and maintenance methylation. J Biol Chem. 1999; 274:33002–10.
5. Leonhardt H, Page AW, Weier HU, Bestor TH. A targeting sequence directs DNA methyltransferase to sites of DNA replication in mammalian nuclei. Cell. 1992;71:865–73.
6. Mertineit C, Yoder JA, Taketo T, Laird DW, Trasler JM, Bestor TH. Sex-specific exons control DNA methyltransferase in mammalian germ cells. Development. 1998;125:889–97.
7. Trasler JM, Alcivar AA, Hake LE, Bestor T, Hecht NB. DNA methyltransferase is developmentally expressed in replicating and non-replicating male germ cells. Nucleic Acids Res. 1992;20:2541–5.
8. Okano M, Bell DW, Haber DA, Li E. DNA methyltransferases Dnmt3a and Dnmt3b are essential for de novo methylation and mammalian development. Cell. 1999;99:247–57.
9. Kaneda M, Okano M, Hata K, et al. Essential role for de novo DNA methyltransferase Dnmt3a in paternal and maternal imprinting. Nature. 2004;429:900–3.
10. Kato Y, Kaneda M, Hata K, et al. Role of the Dnmt3 family in de novo methylation of imprinted and repetitive sequences during male germ cell development in the mouse. Hum Mol Genet. 2007;16: 2272–80.
11. Xu GL, Bestor TH, Bourc'his D, et al. Chromosome instability and immunodeficiency syndrome caused by mutations in a DNA methyltransferase gene. Nature. 1999;402:187–91.
12. Hansen RS, Wijmenga C, Luo P, et al. The DNMT3B DNA methyltransferase gene is mutated in the ICF immunodeficiency syndrome. Proc Natl Acad Sci USA. 1999;96:14412–7.
13. Aapola U, Kawasaki K, Scott HS, et al. Isolation and initial characterization of a novel zinc finger gene, DNMT3L, on 21q22.3, related to the cytosine-5-methyltransferase 3 gene family. Genomics. 2000;65: 293–8.
14. Aapola U, Lyle R, Krohn K, Antonarakis SE, Peterson P. Isolation and initial characterization of the mouse Dnmt3l gene. Cytogenet Cell Genet. 2001;92:122–6.
15. Bourc'his D, Xu GL, Lin CS, Bollman B, Bestor TH. Dnmt3L and the establishment of maternal genomic imprints. Science. 2001;294:2536–9.
16. Bourc'his D, Bestor TH. Meiotic catastrophe and retrotransposon reactivation in male germ cells lacking Dnmt3L. Nature. 2004;431:96–9.
17. Jenuwein T, Allis CD. Translating the histone code. Science. 2001;293:1074–80.
18. Nightingale KP, O'Neill LP, Turnaer BM. Histone modifications: signalling receptors and potential elements of a heritable epigenetic code. Curr Opin Genet Dev. 2006;16:125–36.
19. Peters AH, O'Carroll D, Scherthan H, et al. Loss of the Suv39h histone methyltransferases impairs mammalian heterochromatin and genome stability. Cell. 2001;107:323–37.
20. Tsukada Y, Fang J, Erdjument-Bromage H, et al. Histone demethylation by a family of JmjC domain-containing proteins. Nature. 2006;439:811–6.
21. Kuo MH, Allis CD. Roles of histone acetyltransferases and deacetylases in gene regulation. Bioessays. 1998;20:615–26.
22. McGrath J, Solter D. Completion of mouse embryogenesis requires both the maternal and paternal genomes. Cell. 1984;37:179–83.
23. Surani MA, Barton SC, Norris ML. Development of reconstituted mouse eggs suggests imprinting of the genome during gametogenesis. Nature. 1984;308: 548–50.
24. Barton SC, Surani MA, Norris ML. Role of paternal and maternal genomes in mouse development. Nature. 1984;311:374–6.
25. Cattanach BM, Kirk M. Differential activity of maternally and paternally derived chromosome regions in mice. Nature. 1985;315:496–8.
26. Surani MA, Barton SC, Norris ML. Nuclear transplantation in the mouse: heritable differences between parental genomes after activation of the embryonic genome. Cell. 1986;45:127–36.
27. Delaval K, Feil R. Epigenetic regulation of mammalian genomic imprinting. Curr Opin Genet Dev. 2004;14:188–95.
28. Edwards CA, Ferguson-Smith AC. Mechanisms regulating imprinted genes in clusters. Curr Opin Cell Biol. 2007;19:281–9.
29. Hajkova P, Erhardt S, Lane N, et al. Epigenetic reprogramming in mouse primordial germ cells. Mech Dev. 2002;117:15–23.
30. Lee J, Inoue K, Ono R, et al. Erasing genomic imprinting memory in mouse clone embryos produced from day 11.5 primordial germ cells. Development. 2002;129:1807–17.
31. Gehring M, Reik W, Henikoff S. DNA demethylation by DNA repair. Trends Genet. 2009;25:82–90.
32. Morgan HD, Dean W, Coker HA, Reik W, Petersen-Mahrt SK. Activation-induced cytidine deaminase deaminates 5-methylcytosine in DNA and is expressed in pluripotent tissues: implications for epigenetic reprogramming. J Biol Chem. 2004;279: 52353–60.
33. Millar CB, Guy J, Sansom OJ, et al. Enhanced CpG mutability and tumorigenesis in MBD4-deficient mice. Science. 2002;297:403–5.
34. Popp C, Dean W, Feng S, et al. Genome-wide erasure of DNA methylation in mouse primordial germ cells is affected by AID deficiency. Nature. 2010;463:1101–5.
35. Hajkova P, Ancelin K, Waldmann T, et al. Chromatin dynamics during epigenetic reprogramming in the mouse germ line. Nature. 2008;452:877–81.

36. Hiura H, Obata Y, Komiyama J, Shirai M, Kono T. Oocyte growth-dependent progression of maternal imprinting in mice. Genes Cells. 2006;11:353–61.
37. Lucifero D, Mann MR, Bartolomei MS, Trasler JM. Gene-specific timing and epigenetic memory in oocyte imprinting. Hum Mol Genet. 2004;13: 839–49.
38. Lucifero D, Mertineit C, Clarke HJ, Bestor TH, Trasler JM. Methylation dynamics of imprinted genes in mouse germ cells. Genomics. 2002;79: 530–8.
39. Kerjean A, Dupont JM, Vasseur C, et al. Establishment of the paternal methylation imprint of the human H19 and MEST/PEG1 genes during spermatogenesis. Hum Mol Genet. 2000;9:2183–7.
40. Ueda T, Abe K, Miura A, et al. The paternal methylation imprint of the mouse H19 locus is acquired in the gonocyte stage during foetal testis development. Genes Cells. 2000;5:649–59.
41. Geuns E, De Rycke M, Van Steirteghem A, Liebaers I. Methylation imprints of the imprint control region of the SNRPN-gene in human gametes and preimplantation embryos. Hum Mol Genet. 2003;12: 2873–9.
42. Geuns E, Hilven P, Van Steirteghem A, Liebaers I, De Rycke M. Methylation analysis of KvDMR1 in human oocytes. J Med Genet. 2007;44:144–7.
43. Mayer W, Niveleau A, Walter J, Fundele R, Haaf T. Demethylation of the zygotic paternal genome. Nature. 2000;403:501–2.
44. Oswald J, Engemann S, Lane N, et al. Active demethylation of the paternal genome in the mouse zygote. Curr Biol. 2000;10:475–8.
45. Rougier N, Bourc'his D, Gomes DM, et al. Chromosome methylation patterns during mammalian preimplantation development. Genes Dev. 1998;12:2108–13.
46. Santos F, Hendrich B, Reik W, Dean W. Dynamic reprogramming of DNA methylation in the early mouse embryo. Dev Biol. 2002;241:172–82.
47. Olek A, Walter J. The pre-implantation ontogeny of the H19 methylation imprint. Nat Genet. 1997;17: 275–6.
48. Warnecke PM, Biniszkiewicz D, Jaenisch R, Frommer M, Clark SJ. Sequence-specific methylation of the mouse H19 gene in embryonic cells deficient in the Dnmt-1 gene. Dev Genet. 1998;22:111–21.
49. Lane N, Dean W, Erhardt S, et al. Resistance of IAPs to methylation reprogramming may provide a mechanism for epigenetic inheritance in the mouse. Genesis. 2003;35:88–93.
50. Cardoso MC, Leonhardt H. DNA methyltransferase is actively retained in the cytoplasm during early development. J Cell Biol. 1999;147:25–32.
51. Howell CY, Bestor TH, Ding F, et al. Genomic imprinting disrupted by a maternal effect mutation in the Dnmt1 gene. Cell. 2001;104:829–38.
52. Cirio MC, Ratnam S, Ding F, Reinhart B, Navara C, Chaillet JR. Preimplantation expression of the somatic form of Dnmt1 suggests a role in the inheritance of genomic imprints. BMC Dev Biol. 2008;8:9.
53. Hirasawa R, Chiba H, Kaneda M, et al. Maternal and zygotic Dnmt1 are necessary and sufficient for the maintenance of DNA methylation imprints during preimplantation development. Genes Dev. 2008;22: 1607–16.
54. Kurihara Y, Kawamura Y, Uchijima Y, et al. Maintenance of genomic methylation patterns during preimplantation development requires the somatic form of DNA methyltransferase 1. Dev Biol. 2008;313:335–46.
55. Reik W. Stability and flexibility of epigenetic gene regulation in mammalian development. Nature. 2007;447:425–32.
56. Nakamura T, Arai Y, Umehara H, et al. PGC7/Stella protects against DNA demethylation in early embryogenesis. Nat Cell Biol. 2007;9:64–71.
57. Li JY, Lees-Murdock DJ, Xu GL, Walsh CP. Timing of establishment of paternal methylation imprints in the mouse. Genomics. 2004;84:952–60.
58. Szabo PE, Mann JR. Biallelic expression of imprinted genes in the mouse germ line: implications for erasure, establishment, and mechanisms of genomic imprinting. Genes Dev. 1995;9:1857–68.
59. Davis TL, Trasler JM, Moss SB, Yang GJ, Bartolomei MS. Acquisition of the H19 methylation imprint occurs differentially on the parental alleles during spermatogenesis. Genomics. 1999;58:18–28.
60. Hiura H, Komiyama J, Shirai M, Obata Y, Ogawa H, Kono T. DNA methylation imprints on the IG-DMR of the Dlk1-Gtl2 domain in mouse male germline. FEBS Lett. 2007;581:1255–60.
61. Oakes CC, La Salle S, Smiraglia DJ, Robaire B, Trasler M. Developmental acquisition of genome-wide DNA methylation occurs prior to meiosis in male germ cells. Dev Biol. 2007;307:368–79.
62. Oakes CC, La Salle S, Smiraglia DJ, Robaire B, Trasler JM. A unique configuration of genome-wide DNA methylation patterns in the testis. Proc Natl Acad Sci USA. 2007;104:228–33.
63. Farthing CR, Ficz G, Ng RK, et al. Global mapping of DNA methylation in mouse promoters reveals epigenetic reprogramming of pluripotency genes. PLoS Genet. 2008;4:e1000116.
64. Hammoud SS, Nix DA, Zhang H, Purwar J, Carrell DT, Cairns BR. Distinctive chromatin in human sperm packages genes for embryo development. Nature. 2009;460:473–8.
65. Kelly TL, Li E, Trasler JM. 5-aza-2′-deoxycytidine induces alterations in murine spermatogenesis and pregnancy outcome. J Androl. 2003;24:822–30.
66. Oakes CC, Kelly TL, Robaire B, Trasler JM. Adverse effects of 5-aza-2′-deoxycytidine on spermatogenesis include reduced sperm function and selective inhibition of de novo DNA methylation. J Pharmacol Exp Ther. 2007;322:1171–80.
67. Delaval K, Govin J, Cerqueira F, Rousseaux S, Khochbin S, Feil R. Differential histone modifications

mark mouse imprinting control regions during spermatogenesis. EMBO J. 2007;26:720–9.

68. Jue K, Bestor TH, Trasler JM. Regulated synthesis and localization of DNA methyltransferase during spermatogenesis. Biol Reprod. 1995;53:561–9.
69. Trasler JM, Hake LE, Johnson PA, Alcivar AA, Millette CF, Hecht NB. DNA methylation and demethylation events during meiotic prophase in the mouse testis. Mol Cell Biol. 1990;10:1828–34.
70. Benoit G, Trasler JM. Developmental expression of DNA methyltransferase messenger ribonucleic acid, protein, and enzyme activity in the mouse testis. Biol Reprod. 1994;50:1312–9.
71. La Salle S, Mertineit C, Taketo T, Moens PB, Bestor TH, Trasler JM. Windows for sex-specific methylation marked by DNA methyltransferase expression profiles in mouse germ cells. Dev Biol. 2004;268: 403–15.
72. La Salle S, Trasler JM. Dynamic expression of DNMT3a and DNMT3b isoforms during male germ cell development in the mouse. Dev Biol. 2006;296: 71–82.
73. Shovlin TC, Bourc'his D, La Salle S, et al. Sex-specific promoters regulate Dnmt3L expression in mouse germ cells. Hum Reprod. 2007;22:457–67.
74. Omisanjo OA, Biermann K, Hartmann S, et al. DNMT1 and HDAC1 gene expression in impaired spermatogenesis and testicular cancer. Histochem Cell Biol. 2007;127:175–81.
75. Sassone-Corsi P. Unique chromatin remodeling and transcriptional regulation in spermatogenesis. Science. 2002;296:2176–8.
76. Kimmins S, Sassone-Corsi P. Chromatin remodelling and epigenetic features of germ cells. Nature. 2005;434:583–9.
77. Zamudio NM, Chong S, O'Bryan MK. Epigenetic regulation in male germ cells. Reproduction. 2008; 136:131–46.
78. Celeste A, Petersen S, Romanienko PJ, et al. Genomic instability in mice lacking histone H2AX. Science. 2002;296:922–7.
79. Mahadevaiah SK, Turner JM, Baudat F, et al. Recombinational DNA double-strand breaks in mice precede synapsis. Nat Genet. 2001;27:271–6.
80. van der Heijden GW, Derijck AA, Posfai E, et al. Chromosome-wide nucleosome replacement and H3.3 incorporation during mammalian meiotic sex chromosome inactivation. Nat Genet. 2007;39: 251–8.
81. Biermann K, Steger K. Epigenetics in male germ cells. J Androl. 2007;28:466–80.
82. Manipalviratn S, DeCherney A, Segars J. Imprinting disorders and assisted reproductive technology. Fertil Steril. 2009;91:305–15.
83. Arnaud P, Feil R. Epigenetic deregulation of genomic imprinting in human disorders and following assisted reproduction. Birth Defects Res C Embryo Today. 2005;75:81–97.
84. Doornbos ME, Maas SM, McDonnell J, Vermeiden JP, Hennekam RC. Infertility, assisted reproduction technologies and imprinting disturbances: a Dutch study. Hum Reprod. 2007;22:2476–80.
85. Bliek J, Terhal P, van den Bogaard MJ, et al. Hypomethylation of the H19 gene causes not only Silver-Russell syndrome (SRS) but also isolated asymmetry or an SRS-like phenotype. Am J Hum Genet. 2006;78:604–14.
86. Eggermann T, Schonherr N, Meyer E, et al. Epigenetic mutations in 11p15 in Silver-Russell syndrome are restricted to the telomeric imprinting domain. J Med Genet. 2006;43:615–6.
87. Gicquel C, Rossignol S, Cabrol S, et al. Epimutation of the telomeric imprinting center region on chromosome 11p15 in Silver-Russell syndrome. Nat Genet. 2005;37:1003–7.
88. Zeschnigk M, Albrecht B, Buiting K, et al. IGF2/H19 hypomethylation in Silver-Russell syndrome and isolated hemihypoplasia. Eur J Hum Genet. 2008;16:328–34.
89. Penaherrera MS, Weindler S, Van Allen MI, et al. Methylation profiling in individuals with Russell-Silver syndrome. Am J Med Genet A. 2010;152A: 347–55.
90. Rossignol S, Netchine I, Le Bouc Y, Gicquel C. Epigenetics in Silver-Russell syndrome. Best Pract Res Clin Endocrinol Metab. 2008;22:403–14.
91. Kallen B, Finnstrom O, Nygren KG, Olausson PO. In vitro fertilization (IVF) in Sweden: risk for congenital malformations after different IVF methods. Birth Defects Res A Clin Mol Teratol. 2005;73:162–9.
92. Svensson J, Bjornstahl A, Ivarsson SA. Increased risk of Silver-Russell syndrome after in vitro fertilization? Acta Paediatr. 2005;94:1163–5.
93. Kagami M, Nagai T, Fukami M, Yamazawa K, Ogata T. Silver-Russell syndrome in a girl born after in vitro fertilization: partial hypermethylation at the differentially methylated region of PEG1/MEST. J Assist Reprod Genet. 2007;24:131–6.
94. Smith FM, Garfield AS, Ward A. Regulation of growth and metabolism by imprinted genes. Cytogenet Genome Res. 2006;113:279–91.
95. Manning M, Lissens W, Liebaers I, Van Steirteghem A, Weidner W. Imprinting analysis in spermatozoa prepared for intracytoplasmic sperm injection (ICSI). Int J Androl. 2001;24:87–94.
96. Marques CJ, Carvalho F, Sousa M, Barros A. Genomic imprinting in disruptive spermatogenesis. Lancet. 2004;363:1700–2.
97. Bell AC, Felsenfeld G. Methylation of a CTCF-dependent boundary controls imprinted expression of the Igf2 gene. Nature. 2000;405:482–5.
98. Kobayashi H, Sato A, Otsu E, et al. Aberrant DNA methylation of imprinted loci in sperm from oligospermic patients. Hum Mol Genet. 2007;16:2542–51.
99. Marques CJ, Costa P, Vaz B, et al. Abnormal methylation of imprinted genes in human sperm is associated

with oligozoospermia. Mol Hum Reprod. 2008; 14:67–74.
100. Hammoud SS, Purwar J, Pflueger C, Cairns BR, Carrell DT. Alterations in sperm DNA methylation patterns at imprinted loci in two classes of infertility. Fertil Steril. 2010;94:1728–33.
101. Poplinski A, Tuttelmann F, Kanber D, Horsthemke B, Gromoll J. Idiopathic male infertility is strongly associated with aberrant methylation of MEST and IGF2/H19 ICR1. Int J Androl. 2010;33:642–9.
102. Boissonnas CC, Abdalaoui HE, Haelewyn V, et al. Specific epigenetic alterations of IGF2-H19 locus in spermatozoa from infertile men. Eur J Hum Genet. 2010;18:73–80.
103. Houshdaran S, Cortessis VK, Siegmund K, Yang A, Laird PW, Sokol RZ. Widespread epigenetic abnormalities suggest a broad DNA methylation erasure defect in abnormal human sperm. PLoS ONE. 2007;2:e1289.
104. Marques CJ, Francisco T, Sousa S, Carvalho F, Barros A, Sousa M. Methylation defects of imprinted genes in human testicular spermatozoa. Fertil Steril. 2010;94:585–94.
105. Kobayashi H, Hiura H, John RM, et al. DNA methylation errors at imprinted loci after assisted conception originate in the parental sperm. Eur J Hum Genet. 2009;17:1582–91.
106. Borghol N, Lornage J, Blachere T, Sophie Garret A, Lefevre A. Epigenetic status of the H19 locus in human oocytes following in vitro maturation. Genomics. 2006;87:417–26.
107. Khoueiry R, Ibala-Rhomdane S, Mery L, et al. Dynamic CpG methylation of the KCNQ1OT1 gene during maturation of human oocytes. J Med Genet. 2008;45:583–8.
108. Laird PW. Principles and challenges of genome-wide DNA methylation analysis. Nat Rev Genet. 2010;11:191–203.

Clinical Significance of Sperm RNA

18

Jean-Pierre Dadoune, Isabelle Galeraud-Denis,
and Serge Carreau

Abstract

Controversies about the potential involvement of mRNAs in functional spermatozoa are numerous, but recent findings suggest that both transcriptional and translational activities could occur at least in mitochondria. In spite of a high degree of chromatin compaction in spermatozoa, the existence of isolated domains in more DNAse-I-sensitive open conformations suggests a potential transcriptional state for specific genes involved in early embryogenesis. All the data provided here reflect the complexity and heterogeneity of the RNA transcripts present in spermatozoa. Further investigations are necessary to understand the significance and the differential role of these mRNA present in ejaculated and uncapacitated spermatozoa. Analysis of mRNA profiles by a genome-wide approach using microarrays technique and/or evaluation of individual transcripts using real-time RT-PCR in infertile patients could be helpful as a diagnostic tool to evaluate male infertility and/or as a tool of prognostic value for fertilization and embryo development, since mRNAs could be delivered to the oocyte.

Keywords

Sperm RNA • Spermatogenesis • mRNA in spermatozoa • RT-PCR in infertile patients • Infertility • Sperm

Mammalian young spermatids contain high levels of extremely various transcripts that are produced either throughout early spermatogenesis [1] or during spermiogenesis from the haploid genome [2–4]. The arrest of transcription that is concomitant with major changes in the chromatin organization occurs during mid-spermiogenesis [5, 6]. However, over the past 15 years, in line with earlier observations [7–9], an increasing number of works have reported the presence of various RNA populations, namely, mRNA, antisense, and microRNA, in the sperm nucleus [10–21].

It is now commonly accepted that RNA profiles obtained from mature ejaculated spermatozoa

J.-P. Dadoune (✉)
Department of Histology, Biology of Reproduction, Hôpital Tenon 4, rue de la Chine, 75970 Paris Cedex 20, France
e-mail: jean-pierre.dadoune@upmc.fr

A. Zini and A. Agarwal (eds.), *Sperm Chromatin: Biological and Clinical Applications in Male Infertility and Assisted Reproduction*, DOI 10.1007/978-1-4419-6857-9_18,

reflect spermatogenic gene expression [16, 18, 22]. Besides genomic imprinting errors and abnormal sperm nuclear packaging, altered mRNA profiles represent another type of epigenetic abnormality that may contribute to idiopathic male infertility and eventually affect in vitro fertilization outcome [23]. The application of microarray technology to spermatozoal RNA has provided a unique opportunity to assess alterations in male fertility [22, 24, 25].

The first part of this chapter presents the diverse RNA populations identified within the sperm nucleus and discusses the functional significance of these RNAs in the spermatozoon itself and in the early embryo following fertilization. The second part deals with the clinical significance of the sperm transcriptome.

Presence of Various RNAs in Spermatozoa

The presence of transcripts in human spermatozoa has been initially shown using reverse transcription PCR [26, 27] and in situ hybridization (ISH) [27, 28]. More recently, it has been demonstrated that diverse interfering RNAs (iRNAs) are also accumulated in spermatozoa of humans [19] and mice [20].

Multiple Origins of the Sperm Transcripts

Most of the spermatozoal RNAs that are seemingly remnants of previous transcription events occurring throughout spermatogenesis may be sequentially expressed in premeiotic, meiotic, or postmeiotic male germ cells [1, 29, 30]. It is well known that production of RNAs during meiosis and sperm differentiation may take place in different manners. In addition to a continuous mRNA synthesis during spermatogenesis, certain mRNAs may be generated in meiotic cells and then translated or stored before delayed translation in early and mid-spermatids, while others are transcribed exclusively from the haploid genome [31].

Transcripts from Genes Coexpressed in Somatic and Male Germ Cells

Some transcripts that are synthesized in variable amounts in spermatocytes and spermatids are the products of genes expressed ubiquitously. Transcription factors belonging to the STAT (signal transducer and activator of transduction) family exemplify this kind of expression pattern [32]. However, *Stat4* is abundantly and exclusively expressed in round and elongating mouse spermatids [33]. The transcript coding for STAT4 [34] as well as the related protein [35] has been identified in human spermatozoa.

Transcripts from Male Germ-Cell-Specific Homologous Genes

Certain genes expressed only in spermatogenic cells are homologues of genes transcribed in somatic cells and are members of gene families. Examples are genes encoding members of the 70-kDa heat-shock protein family (HSP70-2 and HSC70T) and enzymes in the glycolytic pathway (lactate dehydrogenase-C [LDH-C], phosphoglycerate kinase-2 [PGK-2], and glyceraldehyde 3-phosphate dehydrogenase-S [GAPD-S]) [1, 36]. Unlike the genes encoding enzymes in the glycolytic pathway, the genes encoding members of the HSP70 family may be expressed without inactivation of the homologue. HSP70-2, encoded by the *hsp70-2* gene, is specifically expressed at high levels in spermatogenic cells during meiosis [37], and the *Hcs70t* transcript, another kind of *Hsp70*-related transcript, first appears in haploid male germ cells without changes in the expression of other *Hsp70* family genes [38]. Transcripts coding for HSP70 have been also found in human spermatozoa [16].

Transcript Variants

Some genes expressed in somatic cells also produce transcripts encoding protein isoforms found only in spermatogenic cells including enzymes (for example, angiotensin-converting enzyme [ACE] and hexokinase-1 [HK-1s]) and transcription factors (for example, cyclic-AMP response element binding protein [CREB] and cyclic-AMP response element modulator protein [CREMt]) [29, 39]. Testis ACE mRNA is first detected in late-pachytene spermatocytes, whereas testis ACE protein appears only in elongating spermatids [40]. However, RT-PCR and ISH analyses have revealed the persistence of ACE transcripts in human sperm nuclei [34].

Transcripts from Testis-Specific Genes

Of the genes expressed only in spermatogenic cells, some are unique and have nucleotide sequences that show little similarity to genes expressed elsewhere. Several testis-specific genes are specifically activated during spermiogenesis. Indeed, the haploid genome of the spermatid not only continues transcribing genes activated during meiosis but also initiates transcription of genes that are related to edification of differentiating spermatozoa. Examples of unique genes are those encoding synaptonemal complex proteins (e.g., synaptonemal complex protein 1 [SCP1] and chromosomal core protein 1 [COR1]) present during the meiotic phase [41], and basic nuclear proteins such as transition proteins 1 and 2 (TP1 and TP2), and protamines 1 and 2 (P1 and P2) or sperm tail-specific cytoskeletal proteins present only during the postmeiotic phase [6].

Numerous investigations using ISH and immunochemistry have clearly shown that the mRNAs of both TPs and protamines are transcribed and stored in the cytoplasm of round spermatids and that the related proteins are expressed with a temporal delay in the nucleus of elongating spermatids. In man, the mRNAs for both TPs and protamines can be found from step-3 to early step-4 spermatids [42–45]. Protein TP1 is expressed in spermatids at steps 3 and 4 and protein TP2 from step 1 to 5 [43]. Protamines P1 and P2 are deposited within the nucleus from step-4 elongating spermatids to step-8 mature spermatids [46–49]. Nevertheless, the transcripts encoding P1, P2, and TP2 have been detected in human ejaculated spermatozoa [16, 28, 34].

Spermiogenic RNA Retention

Assuming that the diversity of spermatozoal transcripts reflect the multiple origin of the transcripts found at high levels in spermatids, accumulation of mRNAs within mature sperm cells may be the result of a global overtranscription process in the spermatogenic genome [50] and/or of the presence of intercellular cytoplasmic bridges among the germ cells. These bridges allow transcript sharing among genetically different spermatids and provide a mechanism by which these cells develop synchronously into biochemically and functionally equivalent sperm. Although sharing may not be a global phenomenon for all spermatid-expressed genes, as illustrated by the transmission ratio distortion, TRD (a deviation from Mendelian ratio) for the *Spam1* RNA [51], studies of spermatid-expressed genes for protamines [52, 53], and several X-linked sperm-specific proteins [54–57] provide strong evidence for transcript sharing.

Classes of Spermatozoal Transcripts

There are now many reports listing the different mRNA species identified in human ejaculated spermatozoa by using RT-PCR and/or ISH (Table 18.1). The presence of RNA in high-quality preparations of human sperm has been reaffirmed in a single study combining ISH, RT-PCR, and macroarrays [58].

The most comprehensive description of the sperm transcripts present in normal fertile men has been provided by microarray analyses [18, 22, 23, 59, 60]. Sperm RNAs that are required for growth, signal transduction, cell proliferation, oncogenesis, and transcriptional regulation are highly represented, and a great part of them appears to be similar to mRNAs found in spermatids. Identical results have been obtained from the molecular analysis of the population of mRNAs in bovine spermatozoa [61].

Serial analysis of gene expression (SAGE), an alternative approach, has been successfully employed to characterize and quantify the mRNA transcripts in ejaculated spermatozoa of healthy fertile men. After data processing by DAVID software, SAGE data has evidenced a lot of transcription-regulation-related DNA-binding protein genes and protein-synthesis-related ribosomal subunit genes. Transcripts coding for catalytic activity proteins (e.g., COX5B, a subunit of the

Table 18.1 Identification of the mRNAs present in human ejaculated spermatozoa following RT-PCR and/or ISH

Transcripts coding for	References
c-Myc	Kumar et al. [27]
Human leukocyte-associated antigen (HLA-A)	Chiang et al. [26]
b1-Integrins	Rohwedder et al. [159]
Human protamine 1 (HP1)	Wykes et al. [28] and Siffroi and Dadoune [34]
Human protamine 2 (HP2)	Wykes et al. [28], Miller et al. [16], and Siffroi and Dadoune [34]
Transition protein 2 (TP2)	Wykes et al. [28] and Siffroi and Dadoune [34]
β-Actin	Miller et al. [16]
Heat-shock proteins (HSP70, HSP90)	Richter et al. [160]
Estrogen receptor (ERa)	Richter et al. [160]
Cyclic nucleotide phosphodiesterase	Richter et al. [160]
N-cadherin	Goodwin et al. [161]
L-type voltage-dependent Ca2+channel alpha-1C subunit	Goodwin et al. [162]
Progesterone receptor	Sachdeva et al. [163] and Luconi et al. [143]
Transcription factor Stat 4	Siffroi and Dadoune [34]
Cyclin B1	Siffroi and Dadoune [34]
Angiotensin-converting enzyme	Siffroi and Dadoune [34]
Transition protein 1 (TP1)	Siffroi and Dadoune [34]
Estrogen receptor (ERs)	Hirata et al. [164], Aquila et al. [141], and Solakidi et al. [142]
Cytochrome P450 aromatase	Carreau et al. [132], Aquila et al. [137], and Jedrzejczak et al. [139]
Deleted in azoospermia-like protein (DAZL protein)	Lin et al. [165]
Variable charge Y chromosome (VCY)	Wong et al. [166]
Ubiquitine protein ligase (UBE3A)	Park et al. [167]
Voltage-activated Ca2+ channel	Park et al. [167]
Antimicrobial defensins (HNP1-3, HD-5, HBD-1)	Com et al. [168]
CC chemokine receptor 5 (CCR5)	Januchowski et al. [169]
Endothelial and neuronal nitric oxide synthases	Lambard et al. [133, 152] and Carreau et al. [170]
Transcription factors NFκB, HOX2A, ICSBP, protein kinase JNK2, growth factor HBEGF, receptors RXRβ, ErbB3	Dadoune et al. [58]
SRY	Modi et al. [70]
Progesterone receptor B isoform	Shah et al. [71]
Mineralocorticoid receptor	Fiore et al. [171]
Fertilin β, spermatid-specific linker histone H1-like protein (HILS1)	Depa-Martynow et al. [172]
Leptin receptor	De Ambrogi et al. [173]
CDC25B isoforms	Teng et al. [174]
Potassium channels	Yeung and Cooper [175]
Human Y chromosome gene mRNAs (DBY, SRY, RPS4Y)	Yao et al. [176]

terminal mitochondrial respiratory transport enzyme) and transcription factors (e.g., TFAM; the mitochondrial transcription factor A) have been found in high quantities among the 30 most abundant unique transcripts detected in sperm cells [21].

Localization of the RNA Within Sperm Cells

RNA has been visualized within the sperm nuclei by ISH [9, 28, 34, 58]. In somatic cells, RNA is closely associated with a proteinaceous structure interior to the nuclear envelope, termed the nuclear matrix. This dynamic nuclear compartment that organizes the chromatin into functional loops of DNA [62, 63] is believed to be involved in many nuclear functions including DNA replication, transcription, repair, and pre-mRNa processing/transport [64–66]. This loop domain structure is present throughout the entire sperm chromatin, even though the tertiary structure of most of the DNA is very different in spermatozoa [67]. The sperm nuclear matrix plays an important structural and functional role in fertilization and development [68]. The assumption that sperm RNA is a part of the nuclear matrix [69] has been recently confirmed. After extraction of histones and protamines by treatment with high salt and reducing reagent followed by ISH using an RNA-specific dye, RNA is clearly detected as an integral component of the nuclear matrix and is degraded after prior treatment with RNAse [15].

However, certain ISH observations also reveal that the spermatozoon midpiece is another site of RNA accumulation [27, 70, 71]. In this segment, the mitochondria appear to be a preferential cell compartment of RNA storage, as shown by immunoelectron microscopical studies [72].

RNA Involvement in Paternal Genome Packaging

Nuclear condensation during spermiogenesis is accomplished by replacing most somatic and testis-specific histones with transition proteins, and subsequently, protamines [5, 73]. Protamines facilitate the packaging of the male haploid genome within the sperm nucleus. They contain several cysteines that are though to confer an increased stability on sperm chromatin by intermolecular disulfide cross-links [74]. The vast majority of sperm DNA is coiled into toroids by protamines [75]. Each toroidal subunit represents one DNA-loop domain that is attached to the sperm nuclear matrix at MARs (matrix attachment regions) through a DNase-sensitive linker of chromatin [76–79].

However, the mature sperm nucleus retains some chromatin domains containing histones that are assembled with the DNA in a typical nucleosomal organization [80–83]. Depending on the species, between 2 and 15% of mammalian sperm chromatin is bound to histones [81, 82, 84–87]. These include H2A and its variants, H2B and a TH2B variant, as well as highly acetylated forms of H3 and H4 [81, 83, 88]. In both mouse and human sperm, histones have been localized to the nuclear periphery in association with LINE/L1 elements [82] and telomeric sequences [83, 89], respectively. Consistent with data from comparative genomic hybridization (CGH) studies of histone and protamine-bound sperm DNA [15], very recent works have indicated that histones are nonrandomly distributed in the sperm genome and are associated with specific genes [87, 90].

A relationship between spermatozoal RNA, gene potentiation, and differential chromatin packaging has been suggested, which explains the peripheral location of both spermatozoal RNA and histone-bound DNA in close association with the nuclear envelope. Spermiogenic RNA, just after transcription shutdown during mid-spermiogenesis, might have essentially a structural role aimed at saving nascent potentiated histone-bound sequences from repackaging by protamines [69].

Interfering RNA in Mature Spermatozoa

RNA interference (RNAi), also called posttranscriptional gene silencing, is a process within living cells that takes part in controlling gene expression. Two types of small RNA molecules, microRNAs (miRNAs) and small interfering RNAs (siRNAs), are central to RNAi [91]. MicroRNAs are noncoding single-stranded RNAs (ssRNAs) of ~22 nt in length that are generated from endogenous hairpin-shaped transcripts. These small RNAs function as guide molecules in posttranscriptional gene regulation by base-pairing with the target mRNAs, usually in the 3′ untranslated region (uTR). Binding of a miRNA to the target mRNA typically leads to translational repression and exonucleolytic mRNA decay, although highly complementary targets can be cleaved endonucleolytically [92]. There is increasing evidence indicating that proper small RNA processing is essential for normal spermatogenesis and male fertility [93, 94].

Microarray analysis of spermatozoal RNAs from six normal fertile men has evidenced 68 shared RNAs, some of which are similar to those previously defined as microRNAs in human and mouse testis [95], whereas others are the antisense of previously *in silico*-predicted transcripts. The identification of spermatozoal miRNAs, such as an antisense *IGF-2 receptor* (*IGF-2R*) RNA and an antisense sequence for the *Dickkopf-2* (*DKK2*) gene, has led to the speculation that the delivery of this class of RNAs to the ovocyte enables their participation in early postfertilization processes and/or establishment of imprints in early embryos [96]. Subsequent works have confirmed the presence of miRNAs in spermatozoa [10, 20] as well as mouse testes [97–99].

Apart from miRNAs, another type of small RNAs has also been isolated from mouse testis [100–103]. These are piwi-interacting RNA (piRNA), of approximately 26–31 bp, which are specifically expressed in the testis. PiRNAs interact with piwi-family proteins such as Miwi, Miwi2, and Mili. These piwi-family proteins play an essential role in spermatogenesis [104–106]. Piwi protein and piRNA synthesis are directly implicated in maintaining transposon silencing in the germ-line genome [104, 107, 108]. However, piRNAs have not been found in murine epididymal spermatozoa, suggesting an absence from maturing and mature sperm [101]. Another hypothesis is that the sensitivity of the detection method was not sufficient to detect this type of small RNAs [109].

Functional Significance of the RNA During Embryo Development

The functional role of the spermatozoal RNAs in fertilization and early development remains a subject of discussion. It has been generally assumed that, compared to the large stores of ovocyte mRNAs prior to zygotic genome activation [110], these RNAs are too few in number to be functional in embryo development. However, the data now available are consistent with the assumption that the RNA performs functions for the zygote following fertilization.

Delivery of Sperm RNAs to the Ovocyte

It is now commonly accepted that, in addition to essential genomic and some sperm components required for further development such as the centriole (in humans and primates) [111], the perinuclear theca [33, 112, 113], and the phospholipase Cζ (PLC-ζ) protein [114], male gametes can transmit some RNAs to the ovocyte during fertilization, as shown in both mice [115–117] and humans [118, 119].

The demonstration of the delivery of spermatozoal RNAs to the ovocyte at fertilization has been essential to support the hypothesis that they could be important in early zygotic and embryonic development [119]. Using the hamster sperm penetration assay, the authors have shown that the clusterin and protamine-2 transcripts, present in sperm cells but not in hamster ovocytes, are consistently detected in zygotes at 30 min and 3 h post fertilization [119]. With the same experimental procedure, a more recent investigation has confirmed that some human sperm transcripts coding for molecules known to be involved in implantation and early embryogenesis (pregnancy-specific β-1-glycoprotein and human leukocyte antigen-E) are selectively retained in the newly formed zygote for at least 24 h [118]. But it has been also proven that various paternal transcripts including those encoding P1, P2, TP2, ropporin, and glyceraldehyde 3-phosphate dehydrogenase are removed from the embryo at the four-cell stage [115]. In this respect, many other sperm RNAs may await the same fate.

Nevertheless, considering the high number and diversity of sperm transcripts detected by large-scale analyses, it cannot be totally excluded that a few of them play a functional role in the zygote. An example of this group of RNAs is the mRNA encoding PLC-ζ. Injection of this RNA into mouse eggs causes Ca2+ oscillations and egg activation [120], and the PLC-ζ transcript has been detected in human spermatozoa [60]. Other examples of this group of detected sperm

RNAs include STAT4, which could modulate transcription from the male pronucleus, and cyclin B1, which ensures progression through the G2/M phase of the cell cycle [34].

RNA-Mediated Epigenetic Effects on the Embryo

Evidence for RNA-mediated inheritance of an epigenetic change in the mouse [116, 117] strengthens the hypothesis that RNAs of paternal origin, including microRNAs, can play a role in modulating gene expression in the embryo [69]. The results from these studies have been interpreted as a paramutation phenomenon that has been demonstrated in mice for the first time. Paramutation is a stable and heritable epigenetic change of the phenotype initiated by an interaction between alleles in a heterozygous parent [121]. Rassoulzadegan et al. have examined alterations in the expression profile of the *Kit* gene in the progeny of heterozygotes carrying the *tm1Alf* mutation, which abolishes the synthesis of the Kit tyrosine kinase receptor involved in melanogenesis. In spite of a homozygous wild-type genotype, most of their offsprings have maintained the white-spotted phenotype characteristic of the mutant heterozygote. The modified phenotype has resulted from the accumulation of nonpolyadenylated RNA molecules of abnormal size in brain and testis, as well as from unusual amounts of *Kit* RNA in sperm cells. Microinjection into fertilized ovocytes either of brain and sperm RNA from heterozygous mutants or of *Kit*-specific microRNAs has induced a heritable mutant phenotype [116, 117]. However, contrary to this important finding, another investigation has failed to find any effect of sperm-borne miRNAs on pronuclear activation or preimplantation development, suggesting that if there is any miRNA contribution from spermatozoa during fertilization, it is limited [10].

On the other hand, the spontaneous reverse transcription-mediated process, recently named SMRGT (sperm-mediated reverse gene transfer) [122], in which the reverse transcriptase (RT) originally identified in human sperm [123] plays a central role, provides a novel route for the introduction of non-Mendelian traits in subsequent offspring. Immunogold electron microscopy using anti-RT antibody has shown that RT molecules are stably associated with the sperm nuclear scaffold [124]. After incubation of epididymal spermatozoa with exogenous RNA molecules, the sperm endogenous RT can retrotranscribe cDNA copies that can be transferred into eggs during in vitro fertilization [124]. When sperm cells are directly incubated with RNA molecules harboring β-gal sequences and used in IVF assays to produce embryos, and adult animals, nonintegrated β-gal cDNAs are generated in spermatozoa, transferred to embryos, and propagated in tissues of both F0 and F1 animal populations [125]. Surprisingly, new evidence has appeared, indicating that an RT-dependent process is also triggered when sperm cells are exposed to exogenous DNA. Following incubation with a plasmid harboring a green fluorescent protein (EGFP) retrotransposition cassette interrupted by an intron in the opposite orientation to the EGFP gene, reverse-transcribed spliced EGFP DNA sequences are generated in sperm cells and transmitted to embryos in IVF assays. Thus, it has been proven that efficient machinery is present in spermatozoa, which can transcribe, splice, and reverse-transcribe exogenous DNA molecules [126]. Together, all these results support the view that the sperm endogenous RT is implied in the genesis and non-Mendelian propagation of new genetic information.

Given that RT activities operate throughout embryogenesis [126–128], as well as in adult tissues [129, 130], Spadafora [122] has recently suggested the possibility that the RT-dependent mechanism that underlines the SMRGT process could be involved in an RNA-mediated inheritance phenomenon by ensuring the replication of RNA molecules through DNA intermediates generated during a reverse transcription step.

Clinical Significance of the Sperm Transcriptome

As early as 1994, it was demonstrated that it became possible to investigate gene expression in human spermatogenesis by differential RNA fingerprinting of ejaculated spermatozoa [131]. Following a number of works using RT-PCR and/or ISH (see above), subsequent microarray studies have established the existence of a stable subset of spermatozoal full-length transcripts that could be useful for prognostic male factor infertility assessments [18, 19, 96, 132]. Large-scale microarray analysis in sperm from fertile and infertile men has confirmed that this diagnostic strategy would prove valuable for understanding failure in human spermatogenesis [22, 24, 25, 60]. Compared with the microarray technology approach, the evaluation of specific sperm transcripts such as aromatase and estrogen receptors (ERs) mRNAs in relation with the classical semen parameters could provide valuable information for a rational initial diagnosis, and thus, for clinical management of infertility.

Transcripts of Aromatase and Estrogen Receptors

The difficulty in analyzing the spermatozoal mRNAs concerns the preparation, which should be devoid of any other somatic cells or immature germ cells since individually they contain a greater amount of RNA than a single human spermatozoon [14]. Thus, spermatozoa from native semen could be purified on density gradient centrifugation followed by the identification of specific somatic cell markers (CD 45 and E-cadherin) as reported [133].

In monkey [134] and human testis, aromatase has been described not only in Leydig cells [135] but also in Sertoli cells [136], as well as in immature germ cells [133] and ejaculated spermatozoa [132, 133, 137]. The aromatase enzyme complex, which transforms irreversibly androgens into estrogens, comprises two proteins: a specific cytochrome P450 (P450arom) encoded by the *CYP19* gene and a ubiquitous NADPH cytochrome P450 reductase. In humans, *CYP19* gene is located in the 21.2 region of the long arm of the chromosome 15 and is approximately 123 kb length [138]. Spermatozoa functions such as motility could also be related to the mRNA profile. Thus, the presence of aromatase and ERs both in human immature germ cells and ejaculated spermatozoa has been described [132]. A 30% decrease of aromatase mRNAs was observed in immotile sperm fraction recorded in all samples studied; moreover, the aromatase activity determined in vitro was also diminished, of 34%. Using real-time quantitative PCR, we have recently analyzed 57 samples (18 normospermia N, 12 teratospermia T, 16 asthenospermia A, and 11 asthenoteratospermia AT). A significant decrease of the aromatase/GAPDH (A/G) ratio was recorded in the group T (52%) and AT (67%). In the latter group, most of the samples are devoid of detectable aromatase transcripts (Galeraud-Denis, unpublished data). Moreover, a negative correlation (−0.56) has been observed between the levels of aromatase transcript and the spermatozoal morphology (microcephaly or acrosome malformations). It is noteworthy that a twofold decrease of the amount of aromatase transcripts has been also observed in a group of infertile men from Poland [139].

As can be seen in Fig. 18.1, there is a dual immunohistolocalization of aromatase in ejaculated spermatozoa with strong staining in the midpiece and an annular presence of aromatase

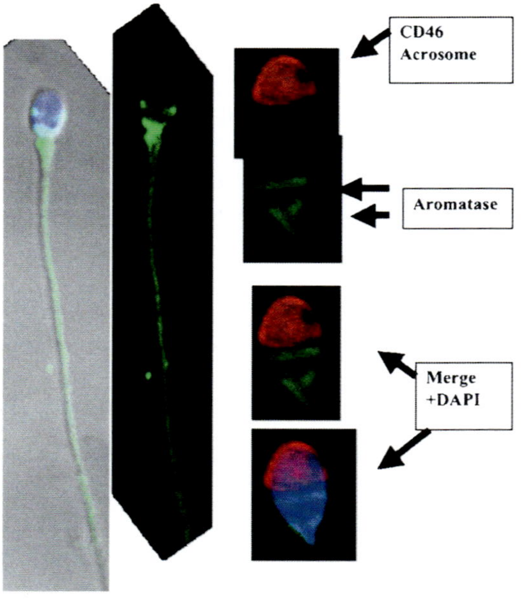

Fig. 18.1 Confocal localization of aromatase in ejaculated spermatozoa. Chromatin is localized using DAOI (*blue*), inner acrosomal membrane is depicted with CD 46 (*red*), and aromatase is revealed using a polyclonal antibody (*green*)

at the acrosomal membrane–nucleus interface [140]. Future studies should be realized to correlate the amount of aromatase transcripts in relation with the evaluation of the nucleus quality.

In contrast with rodent spermatozoa, the human spermatozoa express a functional aromatase, which is still active after ejaculation, and together with the presence of ERs [141, 142], these data open new considerations about the role of estrogens all along the male genital tract and likely also in the sperm mobility and the fertilizing ability.

Therefore, the effects of estrogens in human ejaculated spermatozoa are more and more obvious: besides the classical genomic effects, membrane ERs are connected with numerous signal transduction pathways involving quick responses [143, 144], and among them, the MEK pathway, calcium channel and a calcium/calmodulin complex, known to be concerned for instance in sperm mobility and capacitation [145]. Aquila et al. [141] have also shown a rapid membrane effect of estrogens which in turn activate the PI3K/AKT pathway in human ejaculated spermatozoa.

In this respect, Fraser et al. [146] have demonstrated that genistein improves the capacitation and acrosome loss of human spermatozoa. In addition, the existence of ERs in mitochondria [147] could be relevant to significance for an estrogen role in male gamete motility. The observations of decreased sperm motility in men with aromatase deficiency [148], which is a feature in common with the knockout models of mice [149], together with our data showing a significant decrease of aromatase in immotile spermatozoa could suggest that aromatase is involved in the acquisition of sperm motility [132]. All these reports are in fitting with old works demonstrating the involvement of estrogens in man spermatozoa motility [150, 151].

Significance of Other Transcripts

We have compared the levels of different transcripts coding for molecules involved in nuclear condensation (Prm-1 and Prm-2), capacitation (eNOS, nNOS, and c-myc), motility (estrogens) in high and low motile fractions from normospermic patients [152]. C-myc, was one of the first transcripts [27], as well as its protein [153], described in spermatozoa. We have found a partial or complete disappearance of c-myc transcripts after 4 h of capacitation, whereas the amount of Prm-2 transcripts was unchanged [152]. Moreover, the levels of c-myc transcripts were roughly identical to those measured before capacitation when spermatozoa were incubated with cycloheximide (protein synthesis inhibitor), therefore suggesting that this "marker" is likely used during capacitation. No significant change in the c-myc/Prm-2 ratio between the two populations of spermatozoa was observed. In sperm samples from healthy men, an increase of Prm-1 mRNA in low motile population compared to high motile fraction is recorded, whereas Prm-2 remains identical. An important decrease of Prm-1 gene expression has been observed in testicular biopsies from nonobstructive azoospermia compared to obstructive azoospermia associated with a normal spermatogenesis [154]. Thus, these data confirmed the absence of modification of Prm-2 transcripts, suggesting that Prm-1 is one of the main factors that could be studied in male infertility.

Recently, Aoki et al. [155] have analyzed the protamine levels in sperm cells from fertile and infertile patients and showed a relation between the quality of the sperm (viability and DNA damage), the presence of protamines, and the fertility status as recently reviewed by Oliva [156].

In most of high motile sperm samples analyzed, eNOS and nNOS transcripts were undetectable whereas they were observed in low motile sperm. Nitric oxide synthesized by NOS is a potential modulator of spermatozoa function mainly in the acquisition of motility and capacitation. The high levels of eNOS and nNOS transcripts in low motile spermatozoa could be related to the excessive production of NO responsible for an inhibition of the sperm motility [157]. The accumulation of high amounts of transcripts such as eNOS or Prm1 in low motile spermatozoa could be the consequence of an altered translation during spermiogenesis consecutive to either a defective histone/protamine exchange and/or an impaired chromatin condensation.

Concluding Remarks

The semen analysis is the initial routine male investigation in couples with a history of infertility. Sperm functions are related to the compartmentalized structure of the spermatozoa: head implicated in fertilization steps (capacitation, acrosome reaction, and/or fusion), tail whose motility is responsible for the transport of chromosomal material and midpiece involved in energy metabolism (Fig. 18.2). Controversies about the potential involvement of mRNAs in functional spermatozoa are numerous (see above), but recent findings suggest that both transcriptional and translational activities could occur at least in mitochondria (see for review [156, 158]). In spite of a high degree of chromatin compaction in spermatozoa, the existence of isolated domains in more DNAse-I sensitive open conformations suggests a potential transcriptional state for specific genes involved in early embryogenesis.

All the data provided here reflect the complexity and heterogeneity of the RNA transcripts present in spermatozoa. Further investigations are necessary to understand the significance and the differential role of these mRNAs present in ejaculated and uncapacitated spermatozoa. Some of them could be considered only as the fingerprint of spermatogenesis and/or spermiogenesis events, while others could be important for the final events just before and after fertilization.

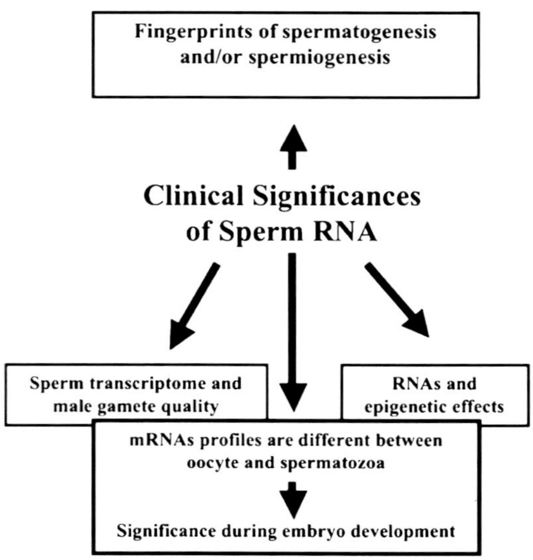

Fig. 18.2 Clinical significance of sperm RNAs

To conclude, male infertility is a today's world problem. Consequently, analysis of mRNA profiles by a genome-wide approach using microarrays technique and/or evaluation of individual transcripts using real-time RT-PCR in infertile patients could be helpful as a diagnostic tool to evaluate male infertility and/or as a tool of prognostic value for fertilization and embryo development, since mRNAs could be delivered to the oocyte.

Acknowledgments Special thanks to our collaborators (S. Lambard, L. Said, A. Saad, A. Chocat, and C. de Vienne) and the financial support from French Ministry of Education and Research.

References

1. Eddy EM. Male germ cell gene expression. Recent Prog Horm Res. 2002;57:103–28.
2. Eddy EM, Toshimori K, O'Brien DA. Fibrous sheath of mammalian spermatozoa. Microsc Res Tech. 2003;61:103–15.
3. Petersen C, Fuzesi L, Hoyer-Fender S. Outer dense fibre proteins from human sperm tail: molecular cloning and expression analyses of two cDNA transcripts encoding proteins of approximately 70 kDa. Mol Hum Reprod. 1999;5:627–35.
4. Steger K. Haploid spermatids exhibit translationally repressed mRNAs. Anat Embryol (Berl). 2001;203:323–34.
5. Dadoune JP. Expression of mammalian spermatozoal nucleoproteins. Microsc Res Tech. 2003;61:56–75.
6. Dadoune JP, Siffroi JP, Alfonsi MF. Transcription in haploid male germ cells. Int Rev Cytol. 2004;237:1–56.
7. Concha II, Urzua U, Yanez A, Schroeder R, Pessot C, Burzio LO. U1 and U2 snRNA are localized in the sperm nucleus. Exp Cell Res. 1993;204:378–81.
8. Dadoune JP, Alfonsi MF, Fain-Maurel MA. Cytochemical variations in the nucleolus during spermiogenesis in man and monkey. Cell Tissue Res. 1991;264:167–73.
9. Pessot CA, Brito M, Figueroa J, Concha II, Yanez A, Burzio LO. Presence of RNA in the sperm nucleus. Biochem Biophys Res Commun. 1989;158:272–8.

10. Amanai M, Brahmajosyula M, Perry AC. A restricted role for sperm-borne microRNAs in mammalian fertilization. Biol Reprod. 2006;75:877–84.
11. Boerke A, Dieleman SJ, Gadella BM. A possible role for sperm RNA in early embryo development. Theriogenology. 2007;68 Suppl 1:S147–55.
12. Dadoune JP. Spermatozoal RNAs: what about their functions. Microsc Res Tech. 2009;8:536–51.
13. Galeraud-Denis I, Lambard S, Carreau S. Relationship between chromatin organization, mRNAs profile and human male gamete quality. Asian J Androl. 2007;9:587–92.
14. Krawetz SA. Paternal contribution: new insights and future challenges. Nat Rev Genet. 2005;6:633–42.
15. Lalancette C, Miller D, Li Y, Krawetz SA. Paternal contributions: new functional insights for spermatozoal RNA. J Cell Biochem. 2008;104:1570–9.
16. Miller D, Briggs D, Snowden H, Hamlington J, Rollinson S, Lilford R, et al. A complex population of RNAs exists in human ejaculate spermatozoa: implications for understanding molecular aspects of spermiogenesis. Gene. 1999;237:385–92.
17. Miller D, Ostermeier GC. Towards a better understanding of RNA carriage by ejaculate spermatozoa. Hum Reprod Update. 2006;12:757–67.
18. Ostermeier GC, Dix DJ, Miller D, Khatri P, Krawetz SA. Spermatozoal RNA profiles of normal fertile men. Lancet. 2002;360:772–7.
19. Ostermeier GC, Goodrich RJ, Moldenhauer JS, Diamond MP, Krawetz SA. A suite of novel human spermatozoal RNAs. J Androl. 2005;26:70–4.
20. Yan W, Morozumi K, Zhang J, Ro S, Park C, Yanagimachi R. Birth of mice after intracytoplasmic injection of single purified sperm nuclei and detection of messenger RNAs and MicroRNAs in the sperm nuclei. Biol Reprod. 2008;78:896–902.
21. Zhao Y, Li Q, Yao C, Wang Z, Zhou Y, Wang Y, et al. Characterization and quantification of mRNA transcripts in ejaculated spermatozoa of fertile men by serial analysis of gene expression. Hum Reprod. 2006;21:1583–90.
22. Wang H, Zhou Z, Xu M, Li J, Xiao J, Xu ZY, et al. A spermatogenesis-related gene expression profile in human spermatozoa and its potential clinical applications. J Mol Med. 2004;82:317–24.
23. Emery BR, Carrell DT. The effect of epigenetic sperm abnormalities on early embryogenesis. Asian J Androl. 2006;8:131–42.
24. Garrido N, Martinez-Conejero JA, Jauregui J, Horcajadas JA, Simon C, Remohi J, et al. Microarray analysis in sperm from fertile and infertile men without basic sperm analysis abnormalities reveals a significantly different transcriptome. Fertil Steril. 2009;91:1307–10.
25. Moldenhauer JS, Ostermeier GC, Johnson A, Diamond MP, Krawetz SA. Diagnosing male factor infertility using microarrays. J Androl. 2003;24:783–9.
26. Chiang MH, Steuerwald N, Lambert H, Main EK, Steinleitner A. Detection of human leukocyte antigen class I messenger ribonucleic acid transcripts in human spermatozoa via reverse transcription-polymerase chain reaction. Fertil Steril. 1994;61:276–80.
27. Kumar G, Patel D, Naz RK. c-MYC mRNA is present in human sperm cells. Cell Mol Biol Res. 1993;39:111–7.
28. Wykes SM, Visscher DW, Krawetz SA. Haploid transcripts persist in mature human spermatozoa. Mol Hum Reprod. 1997;3:15–9.
29. Eddy EM, O'Brien DA. Gene expression during mammalian meiosis. Curr Top Dev Biol. 1998;37:141–200.
30. Erickson RP. Post-meiotic gene expression. Trends Genet. 1990;6:264–9.
31. O'Brien DA. Stage-specific protein synthesis by isolated spermatogenic cells throughout meiosis and early spermiogenesis in the mouse. Biol Reprod. 1987;37:147–57.
32. Schindler C, Darnell Jr JE. Transcriptional responses to polypeptide ligands: the JAK-STAT pathway. Annu Rev Biochem. 1995;64:621–51.
33. Herrada G, Wolgemuth DJ. The mouse transcription factor Stat4 is expressed in haploid male germ cells and is present in the perinuclear theca of spermatozoa. J Cell Sci. 1997;110:1543–53.
34. Siffroi JP, Dadoune JP. Accumulation of transcripts in the mature human sperm nucleus: implication of the haploid genome in a functional role. Ital J Anat Embryol. 2001;106:189–97.
35. D'Cruz OJ, Vassilev AO, Uckun FM. Members of the Janus kinase/signal transducers and activators of transcription (JAK/STAT) pathway are present and active in human sperm. Fertil Steril. 2001;76:258–66.
36. Eddy EM. Regulation of gene expression during spermatogenesis. Semin Cell Dev Biol. 1998;9:451–7.
37. Dix DJ, Rosario-Herrle M, Gotoh H, Mori C, Goulding EH, Barrett CV, et al. Developmentally regulated expression of Hsp70-2 and a Hsp70-2/lacZ transgene during spermatogenesis. Dev Biol. 1996;174:310–21.
38. Matsumoto M, Kurata S, Fujimoto H, Hoshi M. Haploid specific activations of protamine 1 and hsc70t genes in mouse spermatogenesis. Biochim Biophys Acta. 1993;1174:274–8.
39. Eddy EM. "Chauvinist genes" of male germ cells: gene expression during mouse spermatogenesis. Reprod Fertil Dev. 1995;7:695–704.
40. Langford KG, Zhou Y, Russell LD, Wilcox JN, Bernstein KE. Regulated expression of testis angiotensin-converting enzyme during spermatogenesis in mice. Biol Reprod. 1993;48:1210–8.
41. Moens PB, Pearlman RE, Heng HH, Traut W. Chromosome cores and chromatin at meiotic prophase. Curr Top Dev Biol. 1998;37:241–62.
42. Siffroi JP, Alfonsi MF, Dadoune JP. Co-localization of HP1 and TP1 transcripts in human spermatids by double electron microscopy in situ hybridization. Int J Androl. 1999;22:83–90.
43. Steger K, Klonisch T, Gavenis K, Drabent B, Doenecke D, Bergmann M. Expression of mRNA and protein of nucleoproteins during human spermiogenesis. Mol Hum Reprod. 1998;4:939–45.

44. Steger K, Pauls K, Klonisch T, Franke FE, Bergmann M. Expression of protamine-1 and -2 mRNA during human spermiogenesis. Mol Hum Reprod. 2000;6:219–25.
45. Wykes SM, Nelson JE, Visscher DW, Djakiew D, Krawetz SA. Coordinate expression of the PRM1, PRM2, and TNP2 multigene locus in human testis. DNA Cell Biol. 1995;14:155–61.
46. Le Lannic G, Arkhis A, Vendrely E, Chevaillier P, Dadoune JP. Production, characterization, and immunocytochemical applications of monoclonal antibodies to human sperm protamines. Mol Reprod Dev. 1993;36:106–12.
47. Lescoat D, Blanchard Y, Lavault MT, Quernee D, Le Lannou D. Ultrastructural and immunocytochemical study of P1 protamine localization in human testis. Andrologia. 1993;25:93–9.
48. Prigent Y, Muller S, Dadoune JP. Immunoelectron microscopical distribution of histones H2B and H3 and protamines during human spermiogenesis. Mol Hum Reprod. 1996;2:929–35.
49. Roux C, Gusse M, Chevaillier P, Dadoune JP. An antiserum against protamines for immunohistochemical studies of histone to protamine transition during human spermiogenesis. J Reprod Fertil. 1988;82:35–42.
50. Dass B, Attaya EN, Michelle Wallace A, MacDonald CC. Overexpression of the CstF-64 and CPSF-160 polyadenylation protein messenger RNAs in mouse male germ cells. Biol Reprod. 2001;64:1722–9.
51. Martin-DeLeon PA, Zhang H, Morales CR, Zhao Y, Rulon M, Barnoski BL, et al. Spam1-associated transmission ratio distortion in mice: elucidating the mechanism. Reprod Biol Endocrinol. 2005;3:32.
52. Braun RE, Behringer RR, Peschon JJ, Brinster RL, Palmiter RD. Genetically haploid spermatids are phenotypically diploid. Nature. 1989;337:373–6.
53. Caldwell KA, Handel MA. Protamine transcript sharing among postmeiotic spermatids. Proc Natl Acad Sci USA. 1991;88:2407–11.
54. Hendriksen PJ, Hoogerbrugge JW, Themmen AP, Koken MH, Hoeijmakers JH, Oostra BA, et al. Postmeiotic transcription of X and Y chromosomal genes during spermatogenesis in the mouse. Dev Biol. 1995;170:730–3.
55. Moss SB, VanScoy H, Gerton GL. Mapping of a haploid transcribed and translated sperm-specific gene to the mouse X chromosome. Mamm Genome. 1997;8:37–8.
56. Turner RM, Johnson LR, Haig-Ladewig L, Gerton GL, Moss SB. An X-linked gene encodes a major human sperm fibrous sheath protein, hAKAP82. Genomic organization, protein kinase A-RII binding, and distribution of the precursor in the sperm tail. J Biol Chem. 1998;273:32135–41.
57. Westbrook VA, Diekman AB, Klotz KL, Khole VV, von Kap-Herr C, Golden WL, et al. Spermatid-specific expression of the novel X-linked gene product SPAN-X localized to the nucleus of human spermatozoa. Biol Reprod. 2000;63:469–81.
58. Dadoune JP, Pawlak A, Alfonsi MF, Siffroi JP. Identification of transcripts by macroarrays, RT-PCR and in situ hybridization in human ejaculate spermatozoa. Mol Hum Reprod. 2005;11:133–40.
59. Mao XM, Ma WL, Feng CQ, Zou YG, Zheng WL. An initial examination of the spermatozoal gene expression profile. Di Yi Jun Yi Da Xue Xue Bao. 2004;24:1033–6.
60. Platts AE, Dix DJ, Chemes HE, Thompson KE, Goodrich R, Rockett JC, et al. Success and failure in human spermatogenesis as revealed by teratozoospermic RNAs. Hum Mol Genet. 2007;16:763–73.
61. Gilbert I, Bissonnette N, Boissonneault G, Vallee M, Robert C. A molecular analysis of the population of mRNA in bovine spermatozoa. Reproduction. 2007;133:1073–86.
62. Dijkwel PA, Hamlin JL. Origins of replication and the nuclear matrix: the DHFR domain as a paradigm. Int Rev Cytol. 1995;162A:455–84.
63. Gerdes MG, Carter KC, Moen Jr PT, Lawrence JB. Dynamic changes in the higher-level chromatin organization of specific sequences revealed by in situ hybridization to nuclear halos. J Cell Biol. 1994;126:289–304.
64. Anachkova B, Djeliova V, Russev G. Nuclear matrix support of DNA replication. J Cell Biochem. 2005;96:951–61.
65. Stief A, Winter DM, Stratling WH, Sippel AE. A nuclear DNA attachment element mediates elevated and position-independent gene activity. Nature. 1989;341:343–5.
66. Tsutsui KM, Sano K, Tsutsui K. Dynamic view of the nuclear matrix. Acta Med Okayama. 2005;59:113–20.
67. Ward WS, Partin AW, Coffey DS. DNA loop domains in mammalian spermatozoa. Chromosoma. 1989;98:153–9.
68. Ward WS. Function of sperm chromatin structural elements in fertilization and development. Mol Hum Reprod. 2010;16:30–6.
69. Miller D, Ostermeier GC, Krawetz SA. The controversy, potential and roles of spermatozoal RNA. Trends Mol Med. 2005;11:156–63.
70. Modi D, Shah C, Sachdeva G, Gadkar S, Bhartiya D, Puri C. Ontogeny and cellular localization of SRY transcripts in the human testes and its detection in spermatozoa. Reproduction. 2005;130:603–13.
71. Shah C, Modi D, Sachdeva G, Gadkar S, D'Souza S, Puri C. N-terminal region of progesterone receptor B isoform in human spermatozoa. Int J Androl. 2005;28:360–71.
72. Gur Y, Breitbart H. Mammalian sperm translate nuclear-encoded proteins by mitochondrial-type ribosomes. Genes Dev. 2006;20:411–6.
73. Wouters-Tyrou D, Martinage A, Chevaillier P, Sautiere P. Nuclear basic proteins in spermiogenesis. Biochimie. 1998;80:117–28.
74. Balhorn R, Corzett M, Mazrimas J, Watkins B. Identification of bull protamine disulfides. Biochemistry. 1991;30:175–81.

75. Hud NV, Downing KH, Balhorn R. A constant radius of curvature model for the organization of DNA in toroidal condensates. Proc Natl Acad Sci USA. 1995;92:3581–5.
76. Martins RP, Ostermeier GC, Krawetz SA. Nuclear matrix interactions at the human protamine domain: a working model of potentiation. J Biol Chem. 2004;279:51862–8.
77. McCarthy S, Ward WS. Functional aspects of mammalian sperm chromatin. Hum Fertil (Camb). 1999;2: 56–60.
78. Nadel B, de Lara J, Finkernagel SW, Ward WS. Cell-specific organization of the 5S ribosomal RNA gene cluster DNA loop domains in spermatozoa and somatic cells. Biol Reprod. 1995;53:1222–8.
79. Sotolongo B, Lino E, Ward WS. Ability of hamster spermatozoa to digest their own DNA. Biol Reprod. 2003;69:2029–35.
80. Allen MJ, Bradbury EM, Balhorn R. The chromatin structure of well-spread demembranated human sperm nuclei revealed by atomic force microscopy. Scanning Microsc. 1996;10:989–94.
81. Gineitis AA, Zalenskaya IA, Yau PM, Bradbury EM, Zalensky AO. Human sperm telomere-binding complex involves histone H2B and secures telomere membrane attachment. J Cell Biol. 2000;151:1591–8.
82. Pittoggi C, Renzi L, Zaccagnini G, Cimini D, Degrassi F, Giordano R, et al. A fraction of mouse sperm chromatin is organized in nucleosomal hypersensitive domains enriched in retroposon DNA. J Cell Sci. 1999;112(Pt 20):3537–48.
83. Zalenskaya IA, Bradbury EM, Zalensky AO. Chromatin structure of telomere domain in human sperm. Biochem Biophys Res Commun. 2000;279: 213–8.
84. Adenot PG, Mercier Y, Renard JP, Thompson EM. Differential H4 acetylation of paternal and maternal chromatin precedes DNA replication and differential transcriptional activity in pronuclei of 1-cell mouse embryos. Development. 1997;124:4615–25.
85. Bench GS, Friz AM, Corzett MH, Morse DH, Balhorn R. DNA and total protamine masses in individual sperm from fertile mammalian subjects. Cytometry. 1996;23:263–71.
86. Churikov D, Siino J, Svetlova M, Zhang K, Gineitis A, Morton Bradbury E, et al. Novel human testis-specific histone H2B encoded by the interrupted gene on the X chromosome. Genomics. 2004;84:745–56.
87. Hammoud SS, Nix DA, Zhang H, Purwar J, Carrell DT, Cairns BR. Distinctive chromatin in human sperm packages genes for embryo development. Nature. 2009;460:473–8.
88. Gatewood JM, Cook GR, Balhorn R, Schmid CW, Bradbury EM. Isolation of four core histones from human sperm chromatin representing a minor subset of somatic histones. J Biol Chem. 1990;265:20662–6.
89. Li Y, Lalancette C, Miller D, Krawetz SA. Characterization of nucleohistone and nucleoprotamine components in the mature human sperm nucleus. Asian J Androl. 2008;10:535–41.
90. Arpanahi A, Brinkworth M, Iles D, Krawetz SA, Paradowska A, Platts AE, et al. Endonuclease-sensitive regions of human spermatozoal chromatin are highly enriched in promoter and CTCF binding sequences. Genome Res. 2009;19:1338–49.
91. Carthew RW, Sontheimer EJ. Origins and mechanisms of miRNAs and siRNAs. Cell. 2009;136: 642–55.
92. Kim VN, Han J, Siomi MC. Biogenesis of small RNAs in animals. Nat Rev Mol Cell Biol. 2009;10: 126–39.
93. He Z, Kokkinaki M, Pant D, Gallicano GI, Dym M. Small RNA molecules in the regulation of spermatogenesis. Reproduction. 2009;137:901–11.
94. Maatouk DM, Loveland KL, McManus MT, Moore K, Harfe BD. Dicer1 is required for differentiation of the mouse male germline. Biol Reprod. 2008;79:696–703.
95. Liu CG, Calin GA, Meloon B, Gamliel N, Sevignani C, Ferracin M, et al. An oligonucleotide microchip for genome-wide microRNA profiling in human and mouse tissues. Proc Natl Acad Sci USA. 2004;101:9740–4.
96. Ostermeier GC, Goodrich RJ, Diamond MP, Dix DJ, Krawetz SA. Toward using stable RNAs for prognostic assessment of male factor fertility. Fertil Steril. 2005;83:1686–94.
97. Mishima T, Takizawa T, Luo SS, Ishibashi O, Kawahigashi Y, Mizuguchi Y, et al. MicroRNA cloning analysis reveals sex differences in microRNA expression profiles between adult mouse testis and ovary. Reproduction. 2008;136:811–22.
98. Ro S, Park C, Sanders KM, McCarrey JR, Yan W. Cloning and expression profiling of testis-expressed microRNAs. Dev Biol. 2007;311:592–602.
99. Yan N, Lu Y, Sun H, Tao D, Zhang S, Liu W, et al. A microarray for microRNA profiling in mouse testis tissues. Reproduction. 2007;134:73–9.
100. Aravin A, Gaidatzis D, Pfeffer S, Lagos-Quintana M, Landgraf P, Iovino N, et al. A novel class of small RNAs bind to MILI protein in mouse testes. Nature. 2006;442:203–7.
101. Girard A, Sachidanandam R, Hannon GJ, Carmell MA. A germline-specific class of small RNAs binds mammalian Piwi proteins. Nature. 2006;442: 199–202.
102. Grivna ST, Beyret E, Wang Z, Lin A. A novel class of small RNAs in mouse spermatogenic cells. Genes Dev. 2006;20:1709–14.
103. Watanabe T, Takeda A, Tsukiyama T, Mise K, Okuno T, Sasaki H, et al. Identification and characterization of two novel classes of small RNAs in the mouse germline: retrotransposon-derived siRNAs in oocytes and germline small RNAs in testes. Genes Dev. 2006;20:1732–43.
104. Carmell MA, Girard A, van de Kant HJ, Bourc'his D, Bestor TH, de Rooij DG, et al. MIWI2 is essential for spermatogenesis and repression of transposons in the mouse male germline. Dev Cell. 2007;12: 503–14.

105. Deng W, Lin H. Miwi, a murine homolog of piwi, encodes a cytoplasmic protein essential for spermatogenesis. Dev Cell. 2002;2:819–30.
106. Kuramochi-Miyagawa S, Kimura T, Ijiri TW, Isobe T, Asada N, Fujita Y, et al. Mili, a mammalian member of piwi family gene, is essential for spermatogenesis. Development. 2004;131:839–49.
107. Aravin AA, Sachidanandam R, Girard A, Fejes-Toth K, Hannon GJ. Developmentally regulated piRNA clusters implicate MILI in transposon control. Science. 2007;316:744–7.
108. O'Donnell KA, Boeke JD. Mighty Piwis defend the germline against genome intruders. Cell. 2007;129: 37–44.
109. Miller D. Ensuring continuity of the paternal genome: potential roles for spermatozoal RNA in mammalian embryogenesis. Soc Reprod Fertil Suppl. 2007;65:373–89.
110. Stitzel ML, Seydoux G. Regulation of the oocyte-to-zygote transition. Science. 2007;316:407–8.
111. Schatten G. The centrosome and its mode of inheritance: the reduction of the centrosome during gametogenesis and its restoration during fertilization. Dev Biol. 1994;165:299–335.
112. Mujica A, Navarro-Garcia F, Hernandez-Gonzalez EO, De Lourdes Juarez-Mosqueda M. Perinuclear theca during spermatozoa maturation leading to fertilization. Microsc Res Tech. 2003;61:76–87.
113. Sutovsky P, Manandhar G, Wu A, Oko R. Interactions of sperm perinuclear theca with the oocyte: implications for oocyte activation, anti-polyspermy defense, and assisted reproduction. Microsc Res Tech. 2003;61:362–78.
114. Saunders CM, Swann K, Lai FA. PLCzeta, a sperm-specific PLC and its potential role in fertilization. Biochem Soc Symp. 2007;74:23–36.
115. Hayashi S, Yang J, Christenson L, Yanagimachi R, Hecht NB. Mouse preimplantation embryos developed from oocytes injected with round spermatids or spermatozoa have similar but distinct patterns of early messenger RNA expression. Biol Reprod. 2003;69:1170–6.
116. Rassoulzadegan M, Grandjean V, Gounon P, Cuzin F. Inheritance of an epigenetic change in the mouse: a new role for RNA. Biochem Soc Trans. 2007;35: 623–5.
117. Rassoulzadegan M, Grandjean V, Gounon P, Vincent S, Gillot I, Cuzin F. RNA-mediated non-Mendelian inheritance of an epigenetic change in the mouse. Nature. 2006;441:469–74.
118. Avendano C, Franchi A, Jones E, Oehninger S. Pregnancy-specific {beta}-1-glycoprotein 1 and human leukocyte antigen-E mRNA in human sperm: differential expression in fertile and infertile men and evidence of a possible functional role during early development. Hum Reprod. 2009;24:270–7.
119. Ostermeier GC, Miller D, Huntriss JD, Diamond MP, Krawetz SA. Reproductive biology: delivering spermatozoan RNA to the oocyte. Nature. 2004; 429:154.
120. Sone Y, Ito M, Shirakawa H, Shikano T, Takeuchi H, Kinoshita K, et al. Nuclear translocation of phospholipase C-zeta, an egg-activating factor, during early embryonic development. Biochem Biophys Res Commun. 2005;330:690–4.
121. Chandler VL. Paramutation: from maize to mice. Cell. 2007;128:641–5.
122. Spadafora C. Sperm-mediated "reverse" gene transfer: a role of reverse transcriptase in the generation of new genetic information. Hum Reprod. 2008;23:735–40.
123. Witkin SS, Traganos F, Bendich A. Isolation of a nuclear DNA synthesizing complex from human sperm. Biochem Biophys Res Commun. 1977;77:1404–10.
124. Giordano R, Magnano AR, Zaccagnini G, Pittoggi C, Moscufo N, Lorenzini R, et al. Reverse transcriptase activity in mature spermatozoa of mouse. J Cell Biol. 2000;148:1107–13.
125. Sciamanna I, Barberi L, Martire A, Pittoggi C, Beraldi R, Giordano R, et al. Sperm endogenous reverse transcriptase as mediator of new genetic information. Biochem Biophys Res Commun. 2003;312:1039–46.
126. Pittoggi C, Beraldi R, Sciamanna I, Barberi L, Giordano R, Magnano AR, et al. Generation of biologically active retro-genes upon interaction of mouse spermatozoa with exogenous DNA. Mol Reprod Dev. 2006;73:1239–46.
127. Beraldi R, Pittoggi C, Sciamanna I, Mattei E, Spadafora C. Expression of LINE-1 retroposons is essential for murine preimplantation development. Mol Reprod Dev. 2006;73:279–87.
128. Pittoggi C, Sciamanna I, Mattei E, Beraldi R, Lobascio AM, Mai A, et al. Role of endogenous reverse transcriptase in murine early embryo development. Mol Reprod Dev. 2003;66:225–36.
129. Banerjee S, Thampan RV. Reverse transcriptase activity in bovine bone marrow: purification of a 66-kDa enzyme. Biochim Biophys Acta. 2000;1480: 1–5.
130. Medstrand P, Blomberg J. Characterization of novel reverse transcriptase encoding human endogenous retroviral sequences similar to type A and type B retroviruses: differential transcription in normal human tissues. J Virol. 1993;67:6778–87.
131. Miller D, Tang PZ, Skinner C, Lilford R. Differential RNA fingerprinting as a tool in the analysis of spermatozoal gene expression. Hum Reprod. 1994; 9:864–9.
132. Carreau S, Delalande C, Galeraud-Denis I. Mammalian sperm quality and aromatase expression. Microsc Res Tech. 2009;72:552–7.
133. Lambard S, Galeraud-Denis I, Saunders PTK, Carreau S. Human immature germ cells and ejaculated spermatozoa contain aromatase and oestrogen receptors. J Mol Endocrinol. 2004;32:279–89.
134. Pereyra-Martinez AC, Roselli CE, Stadelman HL, Resko JA. Cytochrome P450 aromatase in testis and epididymis of male rhesus monkeys. Endocrine. 2001;16(1):15–9.

135. Payne PI, Dyer TA. Evidence for the nucleotide sequence of 5-S rRNA from the flowering plant Secale cereale (Rye). Eur J Biochem. 1976;71(1): 33–8.
136. Foucault P, Carreau S, Kuczynski W, Guillaumin JM, Bardos P, Drosdowsky MA. Human sertoli cells in vitro. Lactate, estradiol-17 beta and transferrin production. J Androl. 1992;13(5):361–7.
137. Aquila S, Sisci D, Gentile M, Middea E, Siciliano L, Ando S. Human ejaculated spermatozoa contain active P450 aromatase. J Clin Endocrinol Metab. 2002;87:3385–90.
138. Sebastian S, Bulun SE. A highly complex organization of the regulatory region of the human CYP19 (aromatase) gene revealed by the Human Genome Project. J Clin Endocrinol Metab. 2001;86:4600–2.
139. Jedrzejczak P, Januchowski R, Taszarek-Hauke G, Laddach R, Pawelczyk L, Jagodzinski PP. Quantitative analysis of CCR5 chemokine receptor and cytochrome P450 aromatase transcripts in swim-up spermatozoa isolated from fertile and infertile men. Arch Androl. 2006;52:335–41.
140. Galeraud-Denis I, de Vienne CM, Said L, Chocat A, Carreau S. Differential expression of aromatase in human spermatozoa from normozoospermic and teratozoospermic patients. Int J Androl. 2008;31(S1):1.
141. Aquila S, Sisci D, Gentile M, Middea E, Catalano S, Carpino A, et al. Estrogen receptor (ER)α and ERβ are both expressed in human ejaculated spermatozoa: evidence for their direct interaction with phosphatidylinositol-3-OH Kinase/Akt pathway. J Clin Endocrinol Metab. 2004;89:1443–51.
142. Solakidi S, Psarra AMG, Nikolaropoulos S, Sekeris CE. Estrogen receptors alpha and beta (ERalpha and ERbeta) and androgen receptor (AR) in human sperm: localization of ERbeta and AR in mitochondria of the midpiece. Hum Reprod. 2005;20:3481–5.
143. Luconi M, Bonaccorsi L, Bini L, Liberatori S, Pallini V, Forti G, et al. Characterization of membrane nongenomic receptors for progesterone in human spermatozoa. Steroids. 2002;67:505–9.
144. Luconi M, Forti G, Baldi E. Pathophysiology of sperm mobility. Front Biosci. 2006;11:1433–47.
145. Revelli A, Massobrio M, Tesarik J. Nongenomic actions of steroid hormones in reproductive tissues. Endocr Rev. 1998;19:3–17.
146. Fraser LR, Beyret E, Milligan SR, Adeoya-Osiguwa SA. Effects of estrogenic xenobiotics on human and mouse spermatozoa. Hum Reprod. 2006;21: 1184–93.
147. Chen JQ, Yager JD, Russo J. Regulation of mitochondrial respiratory chain structure and function by estrogens/estrogen receptors and potential physiological/pathophysiological implications. Biochim Biophys Acta. 2005;1746:1–17.
148. Rochira V, Granata ARM, Madeo B, Zirilli L, Rossi G, Carani C. Estrogens in males: what we have learned in last 10 years? Asian J Androl. 2005;7:3–20.
149. O'Donnell L, Robertson KM, Jones ME, Simpson ER. Estrogen and spermatogenesis. Endocr Rev. 2001;22:289–318.
150. Beck KJ, Herscel S, Hungershofer R, Schwinger E. The effect of steroid hormones on motility and selective migration of X-and Y-bearing human spermatozoa. Fertil Steril. 1976;27:407–12.
151. Idaomar M, Guerin JF, Lornage J, Czyba JC. Stimulation of motility and energy metabolism of spermatozoa from asthenozoospermic patients by 17 beta-estradiol. Arch Androl. 1989;22(3):197–202.
152. Lambard S, Galeraud-Denis I, Martin G, Levy R, Chocat A, Carreau S. Analysis and significance of mRNA in human ejaculated sperm from normozoospermic donors: relationship to sperm motility and capacitation. Mol Hum Reprod. 2004;10:535–41.
153. Naz RK, Ahmad K, Kumar G. Presence and role of *c-myc* proto-oncogene product in mammalian sperm cell function. Biol Reprod. 1991;44:842–50.
154. Steger K, Fink L, Failing K, Bohle RM, Kliesch S, Weidner W, et al. Decreased protamine-1 transcript levels in testes from infertile men. Mol Hum Reprod. 2003;9:331–6.
155. Aoki VW, Emery BR, Liu L, Carrell DT. rotamine levels vary between individual sperm cells of infertile human males and correlate with viability and DNA integrity. J Androl. 2006;27:890–8.
156. Oliva R. Protamines and male infertility. Hum Repod Update. 2006;12:417–35.
157. Rosselli M, Dubey RK, Imthurn B, Macas E, Keller PJ. Effects of nitric oxide on human spermatozoa: evidence that nitric oxide decreases sperm motility and induces sperm toxicity. Hum Reprod. 1995;10:1786–90.
158. Gur Y, Breitbart H. Protein synthesis in sperm: dialog between mitochondria and cytoplasm. Mol Cell Endocrinol. 2008;282:45–55.
159. Rohwedder A, Liedigk O, Schaller J, Glander HJ, Werchau H. Detection of mRNA transcripts of beta 1 integrins in ejaculated human spermatozoa by nested reverse transcription-polymerase chain reaction. Mol Hum Reprod. 1996;2:499–05.
160. Richter W, Dettmer D, Glander H. Detection of mRNA transcripts of cyclic nucleotide phosphodiesterase subtypes in ejaculated human spermatozoa. Mol Hum Reprod. 1999;5:732–6.
161. Goodwin LO, Karabinus DS, Pergolizzi RG. Presence of N-cadherin transcripts in mature spermatozoa. Mol Hum Reprod. 2000a;6:487–97.
162. Goodwin LO, Karabinus DS, Pergolizzi RG, Benoff S. L-type voltage-dependent calcium channel alpha-1C subunit mRNA is present in ejaculated human spermatozoa. Mol Hum Reprod. 2000b;6: 127–36.
163. Sachdeva G, Shah CA, Kholkute SD, Puri CP. Detection of progesterone receptor transcript in human spermatozoa. Biol Reprod. 2000;62:1610–4.
164. Hirata S, Shoda T, Kato J, Hoshi K. The multiple untranslated first exons system of the human estrogen

receptor beta (ER beta) gene. J Steroid Biochem Mol Biol. 2001;78:33–40.

165. Lin YM, Chen CW, Sun HS, Tsai SJ, Lin JS, Kuo PL. Presence of DAZL transcript and protein in mature human spermatozoa. Fertil Steril. 2002;77:626–9.
166. Wong EY, Tse JY, Yao KM, Tam PC, Yeung WS. VCY2 protein interacts with the HECT domain of ubiquitin-protein ligase E3A. Biochem Biophys Res Commun. 2002;296:1104–11.
167. Park JY, Ahn HJ, Gu JG, Lee KH, Kim JS, Kang HW, Lee JH. Molecular identification of Ca2+ channels in human sperm. Exp Mol Med. 2003;35:285–92.
168. Com E, Bourgeon F, Evrard B, Ganz T, Colleu D, Jegou B, Pineau C. Expression of antimicrobial defensins in the male reproductive tract of rats, mice, and humans. Biol Reprod. 2003;68:95–104.
169. Januchowski R, Breborowicz AK, Ofori H, Jedrzejczak P, Pawelczyk L, Jagodzinski PP. Detection of a short CCR5 messenger RNA isoform in human spermatozoa. J Androl. 2004;25:757–60.
170. Carreau S, Lambard S, Said L, Saad A, Galeraud-Denis I. RNA dynamics of fertile and infertile spermatozoa. Bioch Soc Trans; 2007;35:634–36
171. Fiore C, Sticchi D, Pellati D, Forzan S, Bonanni G, Bertoldo A, Massironi M, Calo L, Fassina A, Rossi GP, Armanini D. Identification of the mineralocorticoid receptor in human spermatozoa. Int J Mol Med. 2006;18:649–52.
172. Depa-Martynow M, Kempisty B, Lianeri M, Jagodzinski PP, Jedrzejczak P. Association between fertilin beta, protamines 1 and 2 and spermatid-specific linker histone H1-like protein mRNA levels, fertilization ability of human spermatozoa, and quality of preimplantation embryos. Folia Histochem Cytobiol. 2007;45 Suppl 1:S79–85.
173. De Ambrogi M, Spinaci M, Galeati G, Tamanini C. Leptin receptor in boar spermatozoa. Int J Androl. 2007;30:458–61.
174. Teng YN, Chung CL, Lin YM, Pan HA, Liao RW, Kuo PL. Expression of various CDC25B isoforms in human spermatozoa. Fertil Steril. 2007; 88:379–82.
175. Yeung CH, Cooper TG. Potassium channels involved in human sperm volume regulation--quantitative studies at the protein and mRNA levels. Mol Reprod Dev. 2008;75:659–68.
176. Yao C, Wang Z, Zhou Y, Xu W, Li Q, Ma D, Wang L, Qiao Z. A study of Y chromosome gene mRNA in human ejaculated spermatozoa. Mol Reprod Dev. 2010;77:158–66.

Part III

Biological and Clinical Determinants of Sperm Chromatin Damage

Role of Oxidative Stress in the Etiology of Sperm DNA Damage

19

R. John Aitken and Geoffry N. De Iuliis

Abstract

DNA damage in human spermatozoa is a source of some concern because of its association with impaired conception, disrupted embryonic development, increased rates of miscarriage, and morbidity in the offspring. Oxidative stress appears to be the single most important cause of sperm DNA damage, although the factors responsible for the creation of this stress are currently unresolved. What appears certain is that there is an extremely close link between DNA damage in spermatozoa and impaired spermiogenesis, leading to poor chromatin remodeling. We propose that the latter is associated with the creation of vulnerable spermatozoa that readily default to an apoptotic pathway characterized by the generation of reactive oxygen species by the mitochondria and the creation of oxidative DNA adducts that subsequently result in strand breakage. The impairment of spermiogenesis may itself be the result of oxidative stress in the testes created by a wide range of clinical and environmental factors including testicular torsion, varicocele, diabetes, lifestyle factors such as smoking, and excess exposure to a range of environmental toxicants. If oxidative stress is at the heart of DNA damage in the germ line, then there should be a place for antioxidant therapy in the amelioration of this condition. Carefully controlled trials designed to examine the efficacy of different antioxidant formulations are now urgently needed.

Keywords

Human spermatozoa • Oxidative stress • DNA damage • Male infertility

Male infertility is the single largest defined cause of human infertility and, along with maternal age, is the major reason why patients are referred for assisted conception therapy. Maternal age is a significant factor in the etiology of human infertility because it affects the quality of the oocytes and their capacity to support normal embryonic development. Importantly, the fertilizability of such

R.J. Aitken (✉)
Discipline of Biological Sciences, School of Environmental and Life Sciences, University of Newcastle, University Drive, Callaghan, NSW 2308, Australia

ARC Centre of Excellence in Biotechnology and Development, Priority Research Centre in Reproductive Science, University of Newcastle, Callaghan, NSW 2308, Australia
e-mail: john.aitken@newcastle.edu.au

A. Zini and A. Agarwal (eds.), *Sperm Chromatin: Biological and Clinical Applications in Male Infertility and Assisted Reproduction*, DOI 10.1007/978-1-4419-6857-9_19,

oocytes is not impaired by advances in maternal age. As a consequence, even when conception is facilitated in such patients using assisted reproductive technologies (ARTs) such as in vitro fertilization (IVF) or intracytoplasmic sperm injection (ICSI), the live birth rate declines with maternal age much as it does in the natural population [1]. The fact is that an old oocyte cannot be rescued by facilitating contact with a spermatozoon because achieving fertilization is not the limiting issue with such patients; it is the establishment of normal embryonic development. As a result, the use of ART to treat age-related infecundity is of questionable utility. On the other hand, ART is a perfectly rational treatment for male infertility, which generally involves defects in the fertilizing potential of the spermatozoa that can be effectively remedied by facilitating contact with an egg, even if that treatment involves bypassing the entire physiology of fertilization by physically injecting a spermatozoon into the ooplasm.

Even though defective sperm function is recognized as the largest single defined cause of human infertility [2], relatively little is known about the etiology of this condition. A majority of infertile men produce spermatozoa in sufficient numbers to fertilize the egg; however, in this subpopulation of individuals, the fertilizing potential of these cells has been compromised for reasons that are still not fully elucidated. The only major breakthrough we have seen in the past half-century is the awareness that one of the major causes of defective sperm function is oxidative stress [3, 4]. Analysis of the impact of oxidative stress on the male gamete initially focused on the impaired fertilizing potential of these cells as a consequence of lipid peroxidation in the plasma membrane [5–7]. Spermatozoa are particularly vulnerable to lipid peroxidation because they possess a high cellular content of unsaturated fatty acids, particularly arachidonic and decosahexaenoic acids [5, 8]. As a consequence of free radical attack and the initiation of a lipid peroxidation cascade, the sperm plasma membrane loses its fluidity and hence its capacity for engaging in the membrane fusion events associated with fertilization including acrosomal exocytosis and the act of sperm–oocyte fusion itself [9]. This association between oxidative stress and male infertility has been established in a large number of independent studies [10–12], and as a result, we can now safely conclude that the fertilizing potential of human spermatozoa is frequently impaired by the excessive generation of reactive oxygen species (ROS) and peroxidative damage. However, this is not the whole story.

The initial emphasis on lipid peroxidation and lost fertilizing potential has recently given way to the realization that polyunsaturated fatty acids are not the only target for free radical attack. A second vulnerable substrate for free radical attack in spermatozoa is the DNA in the sperm nucleus and mitochondria [13–15]. Sperm DNA damage is now recognized as a major attribute of the human condition, which is significantly elevated in the spermatozoa of subfertile males and highly correlated with a number of adverse clinical outcomes including poor fertilization rates, poor development of the preimplantation embryo, high rates of miscarriage, and an increased incidence of disease in offspring [12, 16–19]. The consequences of DNA damage in the paternal genome for the F1 generation are many and varied but include cancer and complex neurological conditions such as autism, spontaneous schizophrenia, bipolar disease, and epilepsy [17]. The existence of these correlations has served to broaden our concept of what constitutes a normal fertile male. Normal reproductive function is not just about producing spermatozoa that will fertilize the egg. It is also about producing spermatozoa that will support normal embryonic development and the birth of normal, healthy children.

Since sperm DNA damage is highly represented in the subfertile population and since DNA integrity cannot be determined in the spermatozoon that achieves fertilization in vitro, there is a high probability that DNA-damaged spermatozoa are being used in ART. Such involvement of DNA-damaged spermatozoa in assisted conception may explain the increased risk of abnormalities in the offspring conceived by such methods. Thus, we already know that the incidence of birth defects following assisted conception is double that seen in the naturally conceived population [20] and that imprinting disorders, notably the Beckwith–Wiedemann and Angelman syndromes, appear to be increased in such children [21]. Infants produced by ART are also significantly more likely to be admitted to a neonatal intensive care unit, to be hospitalized, and to stay in hospital longer than their naturally conceived counterparts [22]. Recent studies using record linkage have also shown an increase in the hospitalization of ART offspring in infancy and early childhood compared with spontaneously conceived children [23–25]. Additional independent investigations have also revealed abnormal retinal vascularization in such children, while another study has uncovered an eightfold increase in the incidence of undescended testicles in boys conceived by ICSI [26, 27].

In light of this information, it is clearly important that we understand the etiology of DNA damage in spermatozoa and take steps to reduce its incidence. At present the factors contributing to this damage are poorly understood, although paternal age certainly plays a major role, as does infection, lifestyle (e.g., smoking), and exposure to environmental pollutants. A common denominator that cuts across all of the factors thought to contribute to DNA damage in the male germ line is that they are all capable of generating a state of oxidative stress. In keeping with this assertion is the recent observation that DNA fragmentation in human spermatozoa is highly correlated with oxidative DNA damage as reflected by the presence of 8-hydroxy 2′ deoxyguanosine (8OHdG), a marker of oxidative stress. Indeed, this correlation is so high that we have been forced to conclude that oxidative stress is the major cause of DNA damage in the male germ line [28, 29]. This finding raises a number of questions about the detection, cause, prevention, and treatment of DNA damage in the germ line that are addressed in this review. Before these biological issues are discussed, we first examine the fundamental chemistry of free radicals and consider how they precipitate a state of oxidative stress.

The Chemistry of Oxidative Stress

Reactive Oxygen Species

The term reactive oxygen species (ROS) covers a wide range of metabolites derived from the reduction of molecular oxygen, including free radicals, such as the superoxide anion ($O_2^{-\bullet}$) and powerful oxidants such as hydrogen peroxide (H_2O_2). The term also covers molecules derived from the reaction of carbon-centered radicals with oxygen including peroxyl radicals (ROO$^\bullet$), alkoxyl radicals (RO$^\bullet$), and organic hydroperoxides (ROOH). It may also refer to other powerful oxidants such as peroxynitrite ($ONOO^-$) or hypochlorous acid (HOCl), as well as the highly biologically active free radical, nitric oxide (NO$^\bullet$).

The specific term "free radicals" refers to any atom or molecule containing one or more unpaired electrons. As unpaired electrons are highly energetic and seek out other electrons with which to pair, they confer upon free radicals considerable reactivity. Thus, free radicals and related "reactive species" have the ability to react with, and modify the structure of, many different kinds of biomolecules including proteins, lipids, and nucleic acids. The wide range of targets that can be attacked by ROS is a critical aspect of their chemistry that contributes significantly to the pathological significance of these metabolites. In this context, it is important to emphasize that ROS are not discrete single entities but, by virtue of their very reactivity, react with one another to generate complex mixtures of reactive metabolites, classic examples being the dismutation (reaction with itself) of $O_2^{-\bullet}$ to generate H_2O_2 or the reaction of NO$^\bullet$ and $O_2^{-\bullet}$ to generate $ONOO^-$. One of the most important such processes is the reaction of $O_2^{-\bullet}$ with H_2O_2 in the presence of transition metals to generate the hydroxyl radical (OH$^\bullet$). The latter is extremely reactive and a major factor in the initiation of oxidative damage to vulnerable substrates including polyunsaturated fatty acids and DNA.

Lipid Peroxidation

Since most biological molecules only have paired electrons, free radicals are also likely to be involved in chain reactions that can propagate the damage induced by ROS. A classic example of such a chain reaction is the peroxidation of lipids in biological membranes. In this process, a ROS-mediated attack on unsaturated fatty acids generates peroxyl (ROO$^\bullet$) and alkoxyl (RO$^\bullet$) radicals that, in order to stabilize, abstract a hydrogen atom from an adjacent carbon, generating the corresponding acid (ROOH) or alcohol (ROH). The abstraction of a hydrogen atom from an adjacent lipid creates a carbon-centered radical that combines with molecular oxygen to re-create another lipid peroxide. In order to stabilize, the latter must again abstract a hydrogen atom from a nearby lipid, creating another carbon radical that combines with molecular oxygen to create yet another lipid peroxide. In this manner, a chain reaction is created that, if unchecked, would propagate the peroxidative damage throughout the plasma membrane, leading to a rapid loss of membrane-dependent functions.

Such chain reactions are promoted by the presence of transition metals such as iron and copper that can vary their valency states by gaining or losing electrons. Significantly, there is sufficient free iron and copper in human seminal plasma to promote lipid peroxidation once this process has been initiated [30]. When iron sulfate and ascorbate (added as a reductant to maintain the iron in a reduced state) are added to suspensions of human spermatozoa, large amounts of lipid peroxide are generated. A majority of these peroxides arise from the iron-catalyzed propagation, rather than de novo initiation, of lipid peroxidation cascades [31], according to the following equations:

$$\underset{\text{lipid hydroperoxide}}{\mathrm{ROOH}} + \mathrm{Fe}^{2+} \rightarrow \underset{\text{alkoxyl radical}}{\mathrm{RO}^{\bullet}} + \mathrm{OH}^{-} + \mathrm{Fe}^{3+}$$

$$\underset{\text{lipid hydroperoxide}}{\mathrm{ROOH}} + \mathrm{Fe}^{3+} \rightarrow \underset{\text{peroxyl radical}}{\mathrm{ROO}^{\bullet}} + \mathrm{H}^{+} + \mathrm{Fe}^{2+}$$

Thus, the amounts of lipid peroxide generated on addition of transition metals, such as iron, to human sperm suspensions will reflect the amount of lipid peroxide present in these cells at the moment the catalyst was added. The lipid peroxide content of these cells will, in turn, reflect differences in the amount of oxidative stress the spermatozoa have suffered during their life history. As a result, transition metals such as iron have been used to promote lipid peroxidation cascades in human spermatozoa in order to generate sufficient reaction product (e.g., malondialdehyde or 4-hydroxyalkenals) to monitor for diagnostic purposes. Such measurements of the "lipoperoxidative potential" of human spermatozoa have been shown to have clear diagnostic value [32, 33].

Oxidative DNA Damage

DNA fragmentation can be induced enzymatically, as that occurs during apoptosis, or be initiated by free radical attack. Like lipid peroxidation, the latter can also be catalyzed by transition metals, which serve to localize these reactions at the DNA molecule, vastly increasing the efficiency of the generated $OH^{\bullet}$ to attack DNA. As in the case of lipid peroxidation, such attacks create carbon radicals that, in the presence of oxygen, form peroxyl radicals. The initiating radical, $OH^{\bullet}$, can attack sugars, purines, and pyrimidines, generating a wide variety of oxidatively damaged DNA metabolites. One of the most important metabolites from a diagnostic perspective is 8OHdG, formed by the ability of $OH^{\bullet}$ to add to the C-8 carbon in the purine ring of guanine. One of the eventual consequences of free radical attack on bases such as guanine is to labilize the glycosyl bond that attaches the base to the ribose unit with the resultant generation of an abasic site. Abasic sites have a strong destabilizing effect on the DNA backbone and can subsequently result in strand breaks. Strand breaks can also occur through free-radical-mediated attacks of the DNA sugar moiety.

Antioxidant Protection

Protection against oxidative stress includes membrane-associated antioxidants epitomized by α-tocopherol, a hydrophobic vitamin that is capable of intercepting alkoxyl and peroxyl radicals and terminating the peroxidation chain reaction. Significantly, this vitamin has been shown to significantly improve the fertility of males selected on the basis of high levels of lipid peroxidation in their spermatozoa [34]. Moreover, this vitamin has been known since the 1940s to be essential for male reproduction. Of the small molecular mass scavengers involved in the protection of human spermatozoa while they are suspended in seminal plasma, the most important are vitamin C, uric acid, tryptophan, and taurine [35, 36]. In terms of antioxidant enzymes, spermatozoa possess both the mitochondrial and cytosolic forms of superoxide dismutase (SOD) and the enzymes of the glutathione cycle, but little catalase.

SOD catalyzes the dismutation of $O_2^{-\bullet}$ to generate H_2O_2. Such dismutation can occur spontaneously without SOD; however, the reaction proceeds much more slowly in the absence of this enzyme. There is sufficient SOD activity in the mitochondria and cytosol of human spermatozoa to account for most, if not all, of the H_2O_2 produced by these cells [2]. Although SOD is usually thought of in antioxidant terms, this is only true if this enzyme is tightly coupled with additional enzymes that can metabolize the H_2O_2 generated as a consequence of $O_2^{-\bullet}$ dismutation. In isolation, SOD converts a short-lived, rather inert, membrane-impermeable free radical ($O_2^{-\bullet}$) into a powerful, membrane-permeable oxidant, H_2O_2. Although the latter is not a free radical, it is, nevertheless, a potentially pernicious molecule. If not rapidly metabolized, it has the potential to both initiate lipid peroxidation in the sperm plasma membrane and, in the presence of transition metals, trigger DNA damage to both the nuclear and mitochondrial genomes of these cells.

Some insight into the relative importance of $O_2^{-\bullet}$ and H_2O_2 in the initiation of peroxidative damage in human spermatozoa has come from studies employing xanthine oxidase to generate an extracellular mixture of ROS in vitro [37]. In the presence of this ROS-generating system, human spermatozoa rapidly lose their motility as a consequence of the initiation and propagation of peroxidative damage. If SOD is added to the medium to remove $O_2^{-\bullet}$, motility loss still occurs. However, if catalase is added to the incubation mixture to remove the H_2O_2, then lipid peroxidation is suppressed and sperm motility is fully maintained. The implication of such experiments is that H_2O_2 is the major cytotoxic species of ROS as far as spermatozoa are concerned. This conclusion has been confirmed by experiments in which the direct addition of this oxidant has been shown to disrupt the movement of human spermatozoa, their competence for oocyte fusion, and the integrity of their DNA [38, 39].

Given the damaging nature of H_2O_2, it is obviously important that this oxidant is rapidly removed from spermatozoa before it can initiate lipid peroxidation or DNA damage. The enzymes of the glutathione cycle (glutathione peroxidase and reductase) are responsible for peroxide metabolism in these cells. Under normal circumstances, sufficient nicotinamide adenine dinucleotide phosphate (NADPH) is generated by the oxidation of glucose through the hexose monophosphate shunt to fuel glutathione reductase and maintain an adequate pool of reduced glutathione (GSH) to counteract the H_2O_2 and lipid peroxides generated as a consequence of sperm metabolism [40]. It should also be noted that the detoxification of lipid peroxides by glutathione peroxidase requires the concerted action of an additional enzyme in the form of phospholipase A2. This enzyme is required to cleave the lipid peroxide away from the parent phospholipid so that it becomes available for the detoxifying action of glutathione peroxidase.

In addition to these intracellular antioxidants, spermatozoa are also protected by highly specialized extracellular antioxidant enzymes secreted by the male reproductive tract. These enzymes include glutathione peroxidase 5 (GPX5) as well as the extremely large amounts of extracellular SOD present in epididymal and seminal plasma [41, 42]. Indeed, seminal plasma contains more SOD than any other fluid in biology.

Measurement of Oxidative Stress in Spermatozoa

Assessment of Reactive Oxygen Species Generation

Confounding Effect of Leukocyte Contamination

If oxidative stress is such a major factor in the etiology of human infertility, the measurement of free radical generation by human spermatozoa should feature in the routine diagnostic workup of male infertility patients. Unfortunately, this is much more difficult than it sounds. One of the major reasons for this is that most human sperm populations are contaminated by leukocytes, particularly neutrophils and macrophages. These phagocytes are much more powerful generators of ROS than spermatozoa, so only a small level of white cell contamination can overwhelm the signal generated by the spermatozoa and obfuscate the analysis. Although seminal leukocytes are clearly capable of generating ROS [43], the presence of these cells in subclinical concentrations ($<1 \times 10^6$/mL) does not appear to have any impact on sperm quality [44]. The reason for this is that under normal circumstances a majority of seminal phagocytes originate from the secondary sexual glands and only enter the seminal fluid and make contact with the spermatozoa at the moment of ejaculation. At this juncture, spermatozoa are protected from leukocyte-derived ROS by the powerful antioxidants present in seminal plasma. Once the seminal plasma has been removed, however, as occurs when spermatozoa are being prepared for assisted conception therapy, then the free radicals generated by the leukocyte population have unfettered access to the spermatozoa and are capable of inducing significant damage to these cells [45]. Thus, the use of a formyl peptide provocation test to examine the presence of leukocytes in sperm preparations used for assisted conception purposes has confirmed not only that such cells are present in these suspensions but also that their presence significantly disrupts fertilization [46]. Experimentally, the addition of activated leukocytes to human sperm suspensions has been found to suppress sperm function [47], while the physical removal of these cellular contaminants using magnetic beads or ferrofluids coated with a monoclonal antibody against the common leukocyte antigen significantly increases fertilization rates [48]. In addition, it has also been shown that the disruptive effect of leukocytes in vitro can be reversed by the addition of antioxidants to the medium including GSH, N-acetylcysteine, hypotaurine, and catalase [47].

There are important implications in these findings for the methods used to prepare spermatozoa for ART. In order to avoid a leukocyte-mediated free radical attack on spermatozoa, it is essential that the spermatozoa are separated from these cells while still protected by the antioxidants present in seminal plasma. Thus, separation of spermatozoa by discontinuous gradient centrifugation or swim-up from semen, are superior to swim-up from a washed pellet, where the spermatozoa would have no protection against attack by free-radical-generating leukocytes [45]. Importantly, preparation of human spermatozoa in the absence of seminal plasma has been found to significantly increase the levels of DNA damage sustained by the spermatozoa as well as their potential for fertilization [49]. Given the importance of sperm DNA damage to the ultimate health and well-being of the embryo, every precaution should be taken during assisted conception therapy to prevent such iatrogenically generated DNA damage from occurring.

Chemiluminescence

One of the earliest techniques used to detect ROS generation by human sperm suspensions was chemiluminescence [3]. This technique involves the use of probes such as lucigenin or luminol, which ostensibly generate light in the presence of ROS. Luminol is often used in conjunction with horseradish peroxidase, in order to sensitize the assay for H_2O_2 [50], although lucigenin appears to be the more capable of identifying populations of defective spermatozoa [51]. Such assays are simple, convenient, sensitive, and cheap; however, there are major problems associated with their clinical application. To begin with, the precise redox activity measured by these probes is open to question. In the case of lucigenin, for example, we have demonstrated that the chemiluminescent signals generated in the presence of this probe do not reflect the generation of ROS. Rather, this probe detects the presence of oxidoreductases including cytochrome b5 reductase [52] and cytochrome P450 reductase [53] that are capable of effecting the one-electron reduction of lucigenin to generate the corresponding lucigenin radical ($LucH^{\bullet}+$). The latter will readily give up its electron to ground-state oxygen to generate

$O_2^{-\bullet}$ and regenerate the parent lucigenin molecule (Luc^{2+}). $O_2^{-\bullet}$ will then react with another lucigenin radical ($LucH^{\bullet}$+) to create dioxetane that, in turn, decomposes with the generation of light (chemiluminescence). Similar issues apply to luminol when used in isolation as a probe for ROS. Thus, luminol chemiluminescence can also be activated by any one of a number of factors capable of inducing univalent oxidation of the probe, including ferricyanide, persulfate, hypochlorite, $ONOO^-$, and xanthine oxidase, as well as H_2O_2. It is therefore impossible to determine whether the intense chemiluminescence signals generated by populations of defective human spermatozoa represent the excessive generation of ROS or redox cycling of the probes [54].

A second problem with chemiluminescence is that it is impossible to accurately calibrate the output from conventional luminometers because the readout from the photomultipliers used in these machines is in relative units. Thus, the results generated by individual luminometers will differ in terms of sensitivity and number of counts recorded in accordance with the properties of the individual photomultiplier used in their construction. While brave attempts have been made to provide diagnostic thresholds for chemiluminescent assays, the numbers described in such publications are only relevant for the luminometer used in their calculation and do not have wider application.

Finally, because luminescence gives an integrated picture of redox activity in the entire sperm suspension, the results will be profoundly influenced by the presence of any leukocytes that are present in the same sperm suspension. Any chemiluminescent studies of ROS production that have not rigorously removed all contaminating leukocytes beforehand cannot generate meaningful data on ROS generation by the spermatozoa. If it is the latter we are interested in, then techniques need to be used that focus on these cells to the exclusion of all others. In this context, flow cytometry is the technique of choice.

DHE and Mitosox Red

Flow cytometers can be set up in such a way that only spermatozoa are analyzed by virtue of their unique size and light scattering characteristics. In this context, we have recently described and validated an improved assay for the generation of ROS by spermatozoa [55], which utilizes the fluorogenic probe, dihydroethidium (DHE). In the presence of ROS, DHE generates DNA-sensitive fluorochromes that stain the nuclei of free-radical-generating cells red (Fig. 19.1). Molecular analysis of the fluorescent products of DHE oxidation in the presence of spermatozoa revealed the generation of ethidium (the 2 electron oxidation product of DHE) and 2-hydroxyethidium. The latter is significant because it is a unique reaction product created by the interaction between DHE and $O_2^{-\bullet}$. Its presence is a conclusive proof that spermatozoa can generate ROS and, specifically, the $O_2^{-\bullet}$ [55].

A further refinement of the DHE method is to use a charged variant of this molecule, MitoSox red, to monitor free radical generation by the sperm mitochondria. We had originally thought that because $O_2^{-\bullet}$ production by human spermatozoa was insensitive to rotenone and the inhibition

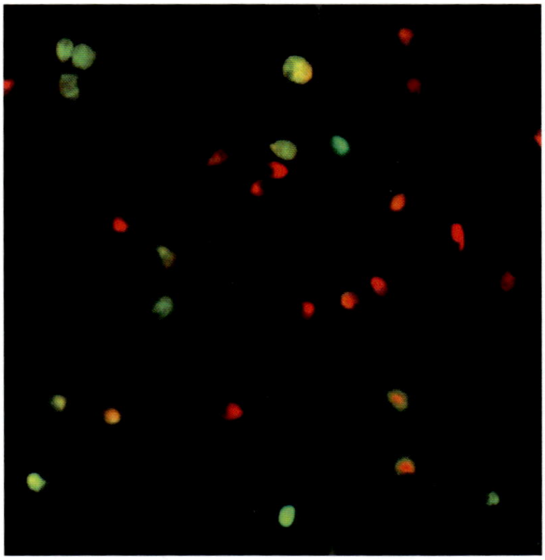

Fig. 19.1 Superoxide anion generation by human spermatozoa. This sperm suspension was stained with dihydroethidium (DHE) and, as a vitality stain, Sytox green. In the presence of superoxide anion, DHE generates DNA-sensitive fluorochromes (ethidium and 2-hydroxyethidium) that stain the sperm nuclei red. The cells in this micrograph that have *red* nuclei, and no trace of *green* staining, are therefore, viable and generating superoxide anion. *Green* cells are nonviable. Magnification ×1,000

of mitochondrial membrane potential (MMP), the source must be nonmitochondrial [55]. However, subsequent studies demonstrated that the source is indeed largely mitochondrial but is unexpectedly impervious to changes in MMP and is actually stimulated by rotenone [56]. Our current hypothesis is that the mitochondria *are* the major source of free radicals in human spermatozoa and that mitochondrial ROS are involved in both the etiology of defective sperm function [56] and the induction of DNA damage [29].

Measurement of DNA Damage in Spermatozoa

Analysis of sperm DNA damage in a majority of laboratories focuses on the measurement of DNA strand breaks. For this purpose, a wide variety of assays have been developed including sperm chromatin dispersion assays [57], sperm chromatin structure assays (SCSA) [58], comet [15, 59] and TUNEL (terminal deoxynucleotidyl transferase dUTP nick-end labeling) assays [13, 60]. The SCSA assay measures the existence of single-stranded DNA following denaturation of the chromatin under pH stress (around pH 1.2). Importantly, preexisting, acid-labile DNA modifications, which are not represented as strand breaks in the original sperm sample, will contribute to the DNA fragmentation index readout with this method. The comet assay exists in two forms, the neutral and the alkaline. The alkaline version, like the SCSA assay, yields information on strand breaks but also encompasses the presence of DNA adducts or abasic sites that transform into strand breaks at high pH and contribute to the overall DNA fragmentation readout. The TUNEL assay measures the existence of preexisting 3′-OH ends but cannot discriminate whether these are double- or single-strand breaks or provide information on the origins of the DNA damage. This assay is performed by adding to the spermatozoa a terminal nucleotidyl transferase and a fluorescently labeled UTP substrate. The transferase attaches the fluorescently tagged UTP to any accessible 3′-OH phosphate group and the resulting fluorescent signal intensity is monitored by microscopy or flow cytometry. The conventional version of this assay underestimates DNA damage because the terminal transferase cannot adequately penetrate the condensed chromatin in the sperm nucleus. However, a modified version of this assay, involving relaxation of the chromatin with a reducing agent (dithiothreitol) prior to performing the TUNEL assay, is able to detect DNA damage induced by clastogens such as H_2O_2 [60]. Furthermore, this version of the assay is readily able to distinguish between semen samples produced by donors or ART patients, detecting significantly higher levels of DNA damage in the latter [61]. The DNA fragmentation detected with this assay is also highly correlated with levels oxidative DNA damage in the form of 8OHdG expression [61]. Oxidative DNA adducts of this type are not only potentially mutagenic but also destabilize the nucleic acid structure, resulting in fragmentation of the DNA and leaving it more vulnerable to further attack. This type of DNA damage has been identified as being central to the initiation of cancer in other cell types [62].

Criteria for Diagnosing Oxidative DNA Damage in the Germ Line

Given that oxidative stress appears to be a major cause of DNA damage in human spermatozoa, it is now important that we development robust criteria for assessing the incidence of this damage in the spermatozoa of male infertility patients, including the establishment of thresholds of normality for diagnostic purposes. This is more problematical than it seems because the distribution of DNA damage among human sperm donors is not bimodal, i.e., there is no easily identifiable subpopulation of males suffering from oxidative damage to their sperm DNA. In every ejaculate some spermatozoa are 8OHdG positive and in the population at large these data are normally distributed (Fig. 19.2). This raises the obvious question as to how much DNA damage is too much, and therefore, requiring some form of therapeutic intervention. In order to address this question, we have

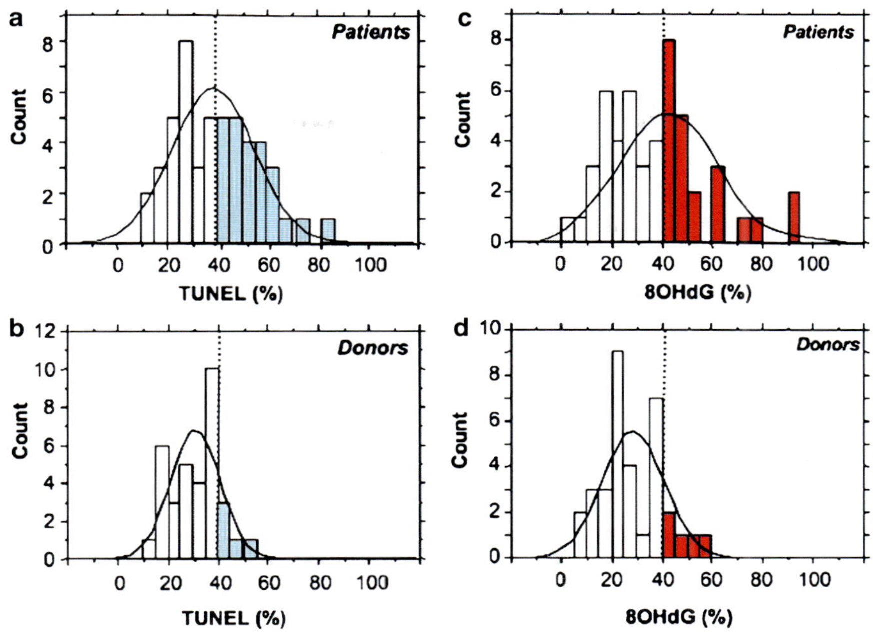

Fig. 19.2 Frequency distribution data for 8-hydroxy, 2′-deoxyguanosine (8OHdG) expression and TUNEL positivity in the spermatozoa of assisted conception patients and semen donors. Panels (**a**, **b**), present the TUNEL data, while panels (**c**, **d**), focus on the 8OHdG results. *Dotted line* represents the diagnostic threshold of around 40% positivity for the optimal discrimination of patient and donors samples, as determined by Youden's J statistic following Receiver Operating Characteristic curve analysis. *Colored bars* represent those samples that would have been identified as abnormal. *Solid line* represents normal distribution. For both TUNEL and 8OHdG frequency distributions, the difference between patients and donors was highly significant ($P<0.001$)

compared the frequency distribution of 8OHdG positivity in human spermatozoa recovered from normal donors and a random cross section of patients attending an assisted conception clinic. The assumption unpinning this analysis was that although the frequency distribution 8OHdG data in each of these populations would be normally distributed and overlapping (since fertile men would be present in both groups of subjects), the patient samples would be shifted to the right because this population would be enriched with samples exhibiting excessively high levels of oxidative DNA damage and DNA fragmentation. Using Receiver Operating Curve analyses, the frequency distribution of data for both the TUNEL and 8OHdG assays was indeed found to be extended to the right in the patient population, as anticipated (Fig. 19.2). Applying Youden's J statistic, we were able to determine those threshold values for the TUNEL and 8OHdG assays that optimally separated the patient and donor populations. The results of this analysis were consistent in recommending a diagnostic threshold of around 40% positive cells for both the TUNEL and 8OHdG assays in sperm suspensions prepared by repeated centrifugation in medium BWW. Using this threshold, populations of spermatozoa suffering from oxidative DNA damage could be readily identified, proving a rational means of selecting patients for whom antioxidant therapy would be rational and appropriate (Fig. 19.2).

Origins of DNA Damage in the Germ Line

While the development of robust protocols for a diagnosing oxidative DNA damage and DNA fragmentation in spermatozoa is important, the development of preventative measures can only be achieved through an understanding of the cause of this damage. Some of the major theories that have been constructed to explain the etiology of DNA damage in human spermatozoa are presented below.

Physiological DNA Strand Breaks

DNA fragmentation in spermatozoa may be the result of unresolved strand breaks created during the normal process of spermiogenesis in order to relieve the torsional stresses involved in packaging a large amount of DNA into the head of the smallest cell in the body. Normally, these "physiological" strand breaks are corrected by a complex process involving H2Ax phosphorylation and the subsequent activation of nuclear poly (ADP-ribose) polymerase and topoisomerase [63]. However, if spermiogenesis should be disrupted for some reason, then the restoration of these cleavage sites might be impaired, and the spermatozoa, lacking any capacity for DNA repair in their own right, would be released from the germinal epithelium still carrying their unresolved strand breaks.

Antioxidant Depletion

A second possible cause of DNA damage is the creation of oxidative stress due to the poor availability of antioxidant protection. The spermatozoon is very vulnerable to a lack of antioxidants because, while it might possess some SOD and glutathione peroxidase activities, these enzymes are in short supply given the limited volume and restricted distribution of cytoplasm in these highly specialized cells. As a result, spermatozoa are very dependent on extracellular antioxidant protection, particularly while they are being matured and stored in the epididymis. Any disruption in the availability of these extrinsic antioxidants leads to a state of oxidative stress within the male reproductive tract and oxidative DNA damage to the spermatozoa. This chain of cause and effect has recently been demonstrated in the GPx5 knockout mouse. GPx5 is one of the major antioxidant enzymes present in the mammalian epididymis. Its functional deletion results in an age-related phenotype associated with a significant increase in the incidences of miscarriage and birth defects in the offspring as a consequence of high levels of oxidative DNA damage in the spermatozoa [64]. Clinically, systemic antioxidant depletion is observed in men who smoke heavily [65] and is correlated with high levels of oxidative DNA damage in their spermatozoa and the appearance of severe pathology in their offspring, including cancer [66]. Although there are many other examples in the literature supporting the notion that a loss of antioxidant protection leads to oxidative stress and male infertility, as in the GPx4 knockout mouse or the aging brown Norway rat [67, 68], very few clinical analyses have been performed on patients where idiopathic infertility is involved. The limited data available to date suggest that GPx4 deficiency in the spermatozoa of infertile patients could be involved in the etiology of their oxidative stress [69]. Whether oxidative DNA damage can result from such a deficiency has not yet been examined in clinical material. However, it has been shown experimentally that removal of seminal antioxidant protection through surgical ablation of the secondary sexual glands in an animal model leads to a state of oxidative stress characterized by high rates of DNA damage in the spermatozoa [70]. Some data are also available to suggest that the antioxidant status of human seminal plasma is inversely correlated with DNA damage in the spermatozoa [71]. More specifically, men with insufficient seminal ascorbic acid frequently possess high levels of sperm DNA damage [72]. Furthermore, the presence of varicocele has been linked with a loss of antioxidant protection from seminal plasma and the induction DNA damage in the spermatozoa, via mechanisms that can be reversed by varicocele ligation [73, 74].

Overall, the current literature suggests that DNA damage in the male germ line can, and occasionally is, induced as a consequence of systemic antioxidant depletion. Whether this is a major factor in the idiopathic DNA damage we encounter regularly in the patient population is still an open question. It is also debatable whether a patient's antioxidant status can be gleaned from an analysis of their seminal plasma for two major reasons. First, spermatozoa, especially those destined for fertilization, spend very little time in seminal plasma before colonizing the female reproductive tract. Second, although many authors have argued that oxidative stress in the ejaculate is generated by a decline in antioxidant protection, it is just as likely that the antioxidant status of human seminal plasma is a consequence of oxidative stress, not its cause. In other words, ROS production in the ejaculate rapidly consumes antioxidant equivalents from seminal plasma lowering the level of protection that can be afforded to the spermatozoa. In this context, the major culprits responsible for lowering the antioxidant capacity of human semen are not the spermatozoa, but infiltrating leucocytes.

Leukocytic Infiltration

Since every human semen sample is contaminated with leukocytes and these cells are actively generating ROS, a relationship between DNA damage and leukocytic infiltration would seem rational. For reasons given above, subclinical seminal leukocyte contamination ($<1 \times 10^6$/mL) does not seem to have a profound effect on DNA damage in spermatozoa [75, 76], although some sperm samples may be more vulnerable to free radical attack than others [77]. However, when levels of leukocyte infiltration are high, as in cases of leukocytospermia, then the presence of these cellular contaminants appears to overwhelm the male tract's antioxidant defenses and induce significant levels of DNA damage in the spermatozoa [78]. This relationship could reflect a direct effect of leukocyte-derived ROS on sperm DNA integrity and/or the indirect creation of oxidative stress through the consumption of seminal antioxidants. However, we should also recognize the possibility that there may be no direct causal relationship between DNA damage and leukocytic infiltration. Rather, the leukocytes could be attracted into the seminal fluid by the presence of DNA damaged spermatozoa that are prematurely undergoing a program of regulated senescence, similar to apoptosis.

Apoptosis

The role that apoptosis plays in the etiology of DNA damage in the germ line has been a subject of some confusion and controversy. It has been postulated that as spermatozoa enter the postmeiotic stages of differentiation, they lose the capacity to complete the process of apoptosis [79]. As a result, differentiating germ cells may enter the apoptotic pathway in response to stress within the germinal epithelium of the testes, and this process may then proceed to the point where endonucleases have been activated and the DNA has become cleaved. However, because the germ cell has lost some of the cellular machinery needed to effect cell death, it is proposed that spermiogenesis and spermiation continue normally with the result that viable spermatozoa are released from the germinal epithelium still carrying the DNA strand breaks left over from their abortive attempt at apoptosis-mediated suicide.

There can be no doubt that spermatozoa can exhibit many of the characteristics of apoptosis including activation of caspases 1, 3, 8, and 9, annexin-V binding, mitochondrial generation of ROS, and DNA fragmentation [56, 80–83]. Although many of the reagents that have been shown to induce apoptosis in somatic cells (staurosporine, lipopolysaccharide, 3-deoxy-D-manno-octulosonic acid, and genistein) are ineffective with human spermatozoa, these cells will default to the intrinsic apoptotic pathway in response to oxidative stress. Thus, exposure of human spermatozoa to H_2O_2 can readily trigger an apoptotic cascade characterized by the activation of caspase 3 and the appearance of annexin-V binding positivity [84]. Furthermore preexposure of human spermatozoa to antioxidants, such as melatonin or catalase, will prevent this apoptotic response to oxidative stress [85, 86]. Such an apoptotic cascade can also be precipitated by a variety of factors that induce oxidative stress in spermatozoa by triggering free radical generation, including exposure to radio-frequency electromagnetic radiation [87], unsaturated fatty acids [88] and exposure to the PI3 kinase inhibitor, wortmannin (A. Koppers and R.J. Aitken, unpublished observations).

Whether the activation of apoptosis is a cause or consequence of DNA cleavage in the germ line is a matter of debate. If it is a potential cause, then we might anticipate that apoptosis would have to be activated in the testes before chromatin remodeling and sperm morphogenesis has reached completion. In the mature gamete, it is physically unlikely that endonucleases activated in the cytosol or released from the mitochondria as a consequence of apoptosis could damage the DNA for two reasons. First, the spermatozoon is unique in that the mitochondria and surrounding cytoplasm are located in a different compartment of the cell, the midpiece, from the nucleus in the sperm head. As illustrated in Fig. 19.3, it is extremely difficult to imagine how endonucleases could move out of the midpiece and penetrate the sperm head to induce DNA cleavage. Second, the chromatin present in mature spermatozoa is so densely compacted that it would be difficult to imagine how an enzyme might penetrate into the heart of this structure and induce DNA fragmentation (Fig. 19.3). This problem would be solved if spermatozoa possessed a nuclease that was already integrated into the structure of the chromatin as described by Sotolongo et al. [89]. Such an enzyme could be activated when the spermatozoa are losing vitality in order to ensure the complete destruction of the DNA, as an aid to cell disposal.

The only other way in which apoptosis could induce DNA damage would be through an oxidative attack mediated by mitochondrial ROS generation. When apoptosis is induced in human spermatozoa, the mitochondria generate $O_2^{-\bullet}$, which then rapidly dismutates to H_2O_2. Such a mechanism fits comfortably with the fact that most DNA damage in human spermatozoa is oxidatively induced [29] and supports the apparent ameliorating effect of antioxidant treatment on DNA damage in the germ line [90].

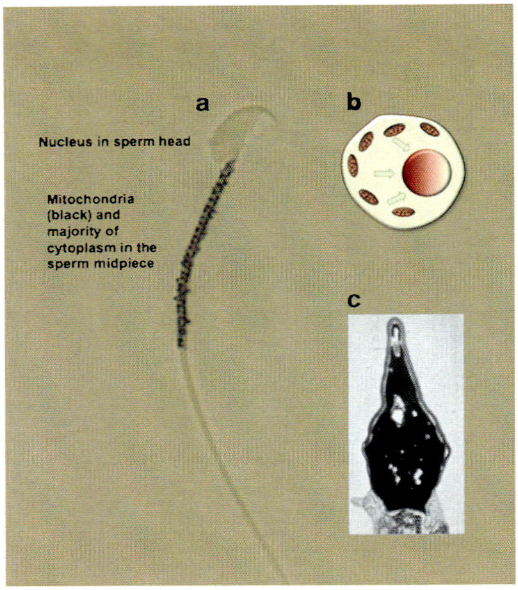

Fig. 19.3 Apoptosis and DNA cleavage in spermatozoa. (**a**) High-power image of a mouse spermatozoon stained to reveal the location of the mitochondria (stained *black*) in the sperm midpiece, emphasizing the separation of these organelles from the nucleus in the sperm head (magnification ×4,000). It is difficult to envisage how nucleases released from the mitochondria or activated in the cytoplasm could make their way to the nucleus to induce DNA cleavage. (**b**) This situation contrasts with most somatic cells in which the nucleus is typically surrounded by cytoplasm and mitochondria, and nuclease migration to the nucleus is a characteristic feature of apoptosis. (**c**) The sperm chromatin is also so densely packed that nucleases would find it difficult to penetrate this structure to induce DNA fragmentation (magnification ×12,000)

Impaired Spermiogenesis

A final piece of the DNA damage puzzle is the tight correlation that has been observed by several authors concerning the relationship between DNA damage in the male germ line and impaired chromatin remodeling during spermiogenesis, as measured with the chromomycin A3 (CMA) assay [29, 91]. The latter is a fluorescent probe that competes with protamines for binding sites on the minor groove of DNA so that cells with inadequately protaminated chromatin fluoresce brightly and can be readily identified by flow cytometry. Such signals correlate extremely well with measures of DNA damage [29]. This association between defective spermiogenesis and DNA damage is further supported by the fact that several independent studies have recorded correlations between DNA damage in human spermatozoa and elements of the conventional semen profile (specifically sperm count and morphology) that, in turn, reflect the efficiency of the spermatogenic process [15, 92, 93].

That defective chromatin remodeling should be associated with DNA damage is not surprising because the efficient protamination and compaction of DNA is known to protect this material from oxidative attack [94]. DNA that is poorly protaminated will possess domains that are relatively open and relaxed as a consequence of the presence of residual histones, and are therefore vulnerable to free radical attack – but why would such an attack occur? One possibility is that poorly differentiated spermatozoa have a tendency to default to an apoptotic pathway that features the generation of mitochondrial ROS as discussed above (Fig. 19.4). A second possibility is that impaired spermatogenesis and DNA fragmentation share a common cause in the presence of oxidative stress within the testes. Spermiogenesis is highly susceptible to

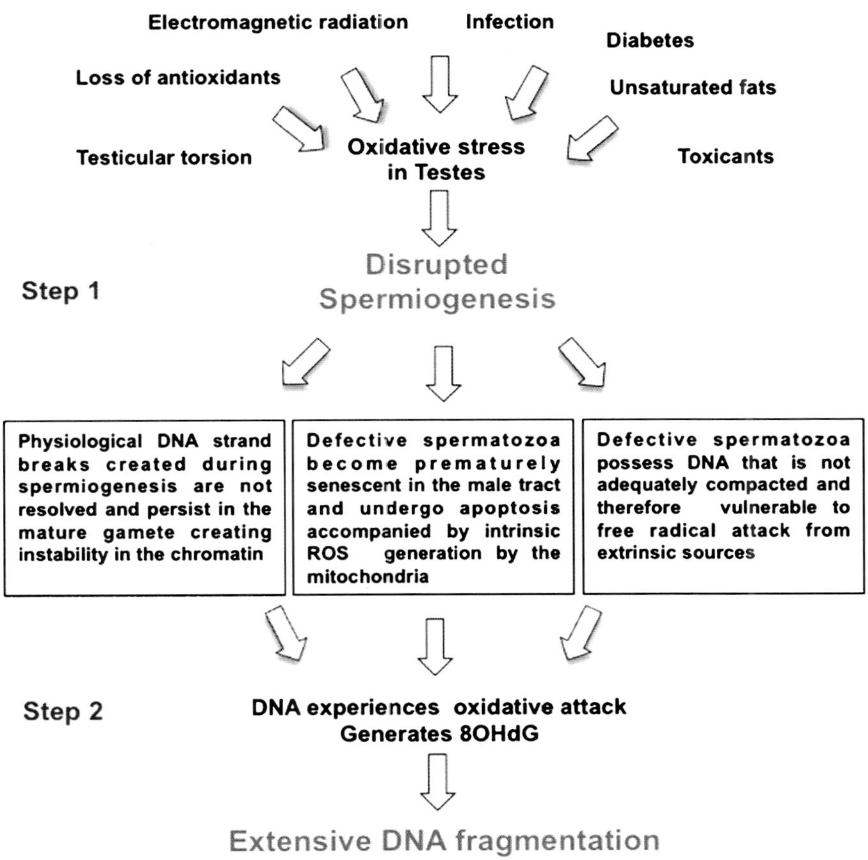

Fig. 19.4 Hypothesis to explain the etiology of DNA fragmentation in the male germ line. The core of this concept is a two-step hypothesis: Step 1, the disruption of spermiogenesis as a consequence of oxidative stress within the testes created via a large array of lifestyle and environmental factors, as well as pathological conditions such as diabetes or testicular torsion. The result of this disrupted spermiogenic process is the production of spermatozoa with poorly remodeled chromatin that are themselves vulnerable to oxidative attack; Step 2 refers to this oxidative attack. It may involve the intrinsic generation of reactive oxygen species by the sperm mitochondria, as these defective cells default to the apoptotic pathway. Alternatively, the free radicals may come from extrinsic sources such as infiltrating leukocytes or redox-cycling xenobiotics

oxidative stress because isolated spermatids have a limited capacity for both DNA repair and glutathione replenishment [95]. It may also be significant that spermiogenesis is entirely dependent on the regulated translation of preexisting mRNA species. Recent studies have indicated that severe oxidative stress can induce protein mistranslation through impairment of an aminoacyl-tRNA synthetase editing site [96]. If protein translation should be disrupted in this way when differentiating spermatids are placed under oxidative stress, it would explain the close relationship between such stress and disrupted spermiogenesis. Situations where oxidative stress in the testes might both disrupt spermiogenesis, creating vulnerability in the gametes, and then trigger DNA fragmentation in the spermatozoa, include varicocele, testicular torsion, cryptorchidism, hyperthyroidism, diabetes, infection, inflammation, physical exertion, impaired gonadotrophic support, reduced testosterone production, lifestyle factors such as smoking, chemotherapeutic agents, heavy metals, and the presence of xenobiotics that either redox-cycle and generate ROS directly or trigger aberrant metabolism that results in the generation of ROS [97]. This two-step hypothesis for the etiology of DNA damage in the germ line is set out in Fig. 19.4.

Conclusions and Future Recommendations

In conclusion, there is a great deal of evidence indicating that most of the idiopathic DNA damage we see in the spermatozoa of male patients is oxidatively induced. We have proposed a two-step hypothesis to explain the etiology of this DNA damage. Step 1 features the generation of a state of oxidative stress in the testes that impairs spermiogenesis, leading to the generation of vulnerable spermatozoa with poorly remodeled chromatin. In Step 2, this vulnerable DNA is oxidatively attacked, possibly as a result of the generation of mitochondrial ROS as these vulnerable cells succumb to apoptosis (Fig. 19.4).

If oxidative stress is a major cause of DNA damage in the male germ line, then antioxidants should be part of the cure. It is remarkable that despite the current awareness of the importance of oxidative stress in the etiology of male infertility, there have still been no definitive assessments of the therapeutic value of antioxidant therapy using a double-blind, placebo-controlled, crossover design. The field urgently needs such studies to be conducted.

Acknowledgments We gratefully acknowledge the support of the NHMRC and ARC Centre of Excellence in Biotechnology and Development.

References

1. Jansen RP. The effect of female age on the likelihood of a live birth from one in-vitro fertilisation treatment. Med J Aust. 2003;178:258–61.
2. Hull MGR, Glazener CMA, Kelly NJ, Conway DI, Foster PA, Hunton RA, et al. Population study of causes, treatment and outcome of infertility. BMJ. 1985;291:1693–7.
3. Alvarez JG, Touchstone JC, Blasco L, Storey BT. Spontaneous lipid peroxidation and production of hydrogen peroxide and superoxide in human spermatozoa. J Androl. 1987;8:338–48.
4. Aitken RJ, Clarkson JS. Cellular basis of defective sperm function and its association with the genesis of ROS by human spermatozoa. J Reprod Fertil. 1987; 81:459–69.
5. Jones R, Mann T, Sherins RJ. Peroxidative breakdown of phospholipids in human spermatozoa: spermicidal effects of fatty acid peroxides and protective action of seminal plasma. Fertil Steril. 1979;31:531–7.
6. Aitken RJ, Harkiss D, Buckingham D. Relationship between iron-catalysed lipid peroxidation potential and human sperm function. J Reprod Fertil. 1993;98: 257–65.
7. de Lamirande E, Gagnon C. Impact of reactive oxygen species on spermatozoa: a balancing between beneficial and detrimental effects. Hum Reprod. 1995;10:15–21.
8. Koppers AJ, Garg ML, Aitken RJ. Stimulation of mitochondrial reactive oxygen species production by unesterified, unsaturated fatty acids in defective human spermatozoa. Free Radic Biol Med. 2010;48:112–9.
9. Aitken RJ, Clarkson JS, Fishel S. Generation of reactive oxygen species, lipid peroxidation and human sperm function. Biol Reprod. 1989;41:183–7.
10. Tremellen K. Oxidative stress and male infertility – a clinical perspective. Hum Reprod Update. 2008; 14:243–58.
11. Agarwal A, Sharma RK, Desai NR, Prabakaran S, Tavares A, Sabanegh E. Role of oxidative stress in

pathogenesis of varicocele and infertility. Urology. 2009;73:461–9.
12. Aitken RJ, De Iuliis GN. On the possible origins of DNA damage in human spermatozoa. Mol Hum Reprod. 2010;16:3–13.
13. Sun JG, Jurisicova A, Casper RF. Detection of deoxyribonucleic acid fragmentation in human sperm: correlation with fertilization in vitro. Biol Reprod. 1997;56:602–7.
14. Sawyer DE, Roman SD, Aitken RJ. Relative susceptibilities of mitochondrial and nuclear DNA to damage induced by hydrogen peroxide in two mouse germ cell lines. Redox Rep. 2001;6:182–4.
15. Irvine DS, Twigg JP, Gordon EL, Fulton N, Milne PA, Aitken RJ. DNA integrity in human spermatozoa: relationships with semen quality. J Androl. 2000;21:33–44.
16. Aitken RJ, De Iuliis GN. Origins and consequences of DNA damage in male germ cells. Reprod Biomed Online. 2007;14:727–33.
17. Aitken RJ, De Iuliis GN, McLachlan RI. Biological and clinical significance of DNA damage in the male germ line. Int J Androl. 2009;32:46–56.
18. Zini A, Boman JM, Belzile E, Ciampi A. Sperm DNA damage is associated with an increased risk of pregnancy loss after IVF and ICSI: systematic review and meta-analysis. Hum Reprod. 2008;23:2663–8.
19. Zini A, Sigman M. Are tests of sperm DNA damage clinically useful? Pros and cons. J Androl. 2009; 30:219–29.
20. Hansen M, Kurinczuk JJ, Bower C, Webb S. The risk of major birth defects after intracytoplasmic sperm injection and in vitro fertilization. N Engl J Med. 2002;346:725–30.
21. Shiota K, Yamada S. Intrauterine environment-genome interaction and children's development (3): assisted reproductive technologies and developmental disorders. J Toxicol Sci. 2009;34 Suppl 2: SP287–91.
22. Hansen M, Colvin L, Petterson B, Kurinczuk JJ, de Klerk N, Bower C. Admission to hospital of singleton children born following assisted reproductive technology (ART). Hum Reprod. 2008;23:297–1305.
23. Ericson A, Nygren KG, Olausson PO, Kallen B. Hospital care utilization of infants born after IVF. Hum Reprod. 2002;7:929–32.
24. Kallen B, Finnstrom O, Nygren KG, Olausson PO. In vitro fertilization in Sweden: child morbidity including cancer risk. Fertil Steril. 2005;84:605–10.
25. Klemetti R, Sevon T, Gissler M, Hemminki E. Health of children born as a result of in vitro fertilization. Pediatrics. 2006;118:1819–27.
26. Ludwig AK, Katalinic A, Thyen U, Sutcliffe AG, Diedrich K, Ludwig M. Physical health at 5.5 years of age of term-born singletons after intracytoplasmic sperm injection: results of a prospective, controlled, single-blinded study. Fertil Steril. 2009;91:115–24.
27. Wikstrand MH, Niklasson A, Strömland K, Hellström A. Abnormal vessel morphology in boys born after intracytoplasmic sperm injection. Acta Paediatr. 2008;97:1512–7.
28. Kodama H, Yamaguchi R, Fukuda J, Kasi H, Tanak T. Increased deoxyribonucleic acid damage in the spermatozoa of infertile male patients. Fertil Steril. 1997;65:519–24.
29. De Iuliis GN, Thomson LK, Mitchell LA, Finnie JM, Koppers AJ, Hedges A, et al. DNA damage in human spermatozoa is highly correlated with the efficiency of chromatin remodeling and the formation of 8-hydroxy-2′-deoxyguanosine, a marker of oxidative stress. Biol Reprod. 2009;81:517–24.
30. Kwenang A, Krous MJ, Koster JF, Van Eijk HG. Iron, ferritin and copper in seminal plasma. Hum Reprod. 1987;2:387–8.
31. Aitken RJ, Harkiss D, Buckingham DW. Analysis of lipid peroxidation mechanisms in human spermatozoa. Mol Reprod Dev. 1993;35:302–15.
32. Gomez E, Irvine DS, Aitken RJ. Evaluation of a spectrophotometric assay for the measurement of malondialdehyde and 4-hydroxyalkenals in human spermatozoa: relationships with semen quality and sperm function. Int J Androl. 1998;21:81–94.
33. Virji N, Eliasson R. LDH-C4 in human seminal plasma and testicular function. III. Relationship to other semen variables. Int J Androl. 1985;8:376–84.
34. Suleiman SA, Elamin Ali M, Zaki ZMS, El-Malik EMA, Nasr MA. Lipid peroxidation and human sperm motility: protective role of vitamin E. J Androl. 1996;7:530–7.
35. van Overveld FW, Haenen GR, Rhemrev J, Vermeiden JP, Bast A. Tyrosine as important contributor to the antioxidant capacity of seminal plasma. Chem Biol Interact. 2000;127:151–61.
36. Rhemrev JP, van Overveld FW, Haenen GR, Teerlink T, Bast A, Vermeiden JP. Quantification of the nonenzymatic fast and slow TRAP in a postaddition assay in human seminal plasma and the antioxidant contributions of various seminal compounds. J Androl. 2000; 21:913–20.
37. Aitken RJ, Buckingham D, Harkiss D. Use of a xanthine oxidase free radical generating system to investigate the cytotoxic effects of reactive oxygen species on human spermatozoa. J Reprod Fertil. 1993; 97:441–50.
38. Aitken RJ, Gordon E, Harkiss D, Twigg JP, Milne P, Jennings Z, et al. Relative impact of oxidative stress on the functional competence and genomic integrity of human spermatozoa. Biol Reprod. 1998;59:1037–46.
39. Oehninger S, Blackmore P, Mahony M, Hodgen G. Effects of hydrogen peroxide on human spermatozoa. J Assist Reprod Genet. 1995;12:41–7.
40. Storey BT, Alvarez JG, Thompson KA. Human sperm glutathione reductase activity in situ reveals limitation in the glutathione antioxidant defense system due to supply of NADPH. Mol Reprod Dev. 1998;49:400–7.
41. Vernet P, Rigaudiére N, Ghyselinck N, Dufaure JP, Drevet JR. In vitro expression of a mouse tissue specific glutathione-peroxidase-like protein lacking the selenocysteine can protect stably transfected mammalian cells against oxidative damage. Biochem Cell Biol. 1996;74:125–31.

42. Mennella MRF, Jones R. Properties of spermatozoal superoxide dismutase and lack of involvement of superoxides in metal-ion-catalysed lipid-peroxidation reactions in semen. Biochem J. 1980;191:289–97.
43. Aitken RJ, West KM. Analysis of the relationship between reactive oxygen species production and leucocyte infiltration in fractions of human semen separated on Percoll gradients. Int J Androl. 1990; 13:433–51.
44. Aitken RJ, Buckingham DW, Brindle J, Gomez E, Baker HW, Irvine DS. Analysis of sperm movement in relation to the oxidative stress created by leukocytes in washed sperm preparations and seminal plasma. Hum Reprod. 1995;10:2061–71.
45. Aitken RJ, Clarkson JS. Significance of reactive oxygen species and antioxidants in defining the efficacy of sperm preparation techniques. J Androl. 1988;9: 367–76.
46. Krausz C, Mills C, Rogers S, Tan SL, Aitken RJ. Stimulation of oxidant generation by human sperm suspensions using phorbol esters and formyl peptides: relationships with motility and fertilization in vitro. Fertil Steril. 1994;62:599–605.
47. Baker HW, Brindle J, Irvine DS, Aitken RJ. Protective effect of antioxidants on the impairment of sperm motility by activated polymorphonuclear leukocytes. Fertil Steril. 1996;65:411–9.
48. Aitken RJ, Buckingham DW, West K, Brindle J. On the use of paramagnetic beads and ferrofluids to assess and eliminate the leukocytic contribution to oxygen radical generation by human sperm suspensions. Am J Reprod Immunol. 1996;35:541–51.
49. Twigg J, Irvine DS, Houston P, Fulton N, Michael L, Aitken RJ. Iatrogenic DNA damage induced in human spermatozoa during sperm preparation: protective significance of seminal plasma. Mol Hum Reprod. 1998;4:439–45.
50. Aitken RJ, Buckingham DW, West KM. Reactive oxygen species and human spermatozoa: analysis of the cellular mechanisms involved in luminol- and lucigenin-dependent chemiluminescence. J Cell Physiol. 1992;151:466–77.
51. Aitken RJ, Ryan AL, Curry BJ, Baker MA. Multiple forms of redox activity in populations of human spermatozoa. Mol Hum Reprod. 2003;9:645–61.
52. Baker MA, Krutskikh A, Curry BJ, Hetherington L, Aitken RJ. Identification of cytochrome-b5 reductase as the enzyme responsible for NADH-dependent lucigenin chemiluminescence in human spermatozoa. Biol Reprod. 2005;73:334–42.
53. Baker MA, Krutskikh A, Curry BJ, McLaughlin EA, Aitken RJ. Identification of cytochrome P450-reductase as the enzyme responsible for NADPH-dependent lucigenin and tetrazolium salt reduction in rat epididymal sperm preparations. Biol Reprod. 2004;71:307–18.
54. Aitken RJ, Baker MA, O'Bryan M. Shedding light on chemiluminescence: the application of chemiluminescence in diagnostic andrology. J Androl. 2004;25: 455–65.
55. De Iuliis GN, Wingate JK, Koppers AJ, McLaughlin EA, Aitken RJ. Definitive evidence for the nonmitochondrial production of superoxide anion by human spermatozoa. J Clin Endocrinol Metab. 2006;91: 1968–75.
56. Koppers AJ, De Iuliis GN, Finnie JM, McLaughlin EA, Aitken RJ. Significance of mitochondrial reactive oxygen species in the generation of oxidative stress in spermatozoa. J Clin Endocrinol Metab. 2008; 93:3199–207.
57. Muriel L, Garrido N, Fernández JL, Remohí J, Pellicer A, de los Santos MJ, et al. Value of the sperm deoxyribonucleic acid fragmentation level, as measured by the sperm chromatin dispersion test, in the outcome of in vitro fertilization and intracytoplasmic sperm injection. Fertil Steril. 2006;85:371–83.
58. Evenson DP, Kasperson K, Wixon RL. Analysis of sperm DNA fragmentation using flow cytometry and other techniques. Soc Reprod Fertil Suppl. 2007;65:93–113.
59. Lewis SE, Agbaje IM. Using the alkaline comet assay in prognostic tests for male infertility and assisted reproductive technology outcomes. Mutagenesis. 2008;23:163–70.
60. Mitchell LA, De Iuliis GN, Aitken RJ. The TUNEL assay consistently underestimates DNA damage in human spermatozoa and is influenced by DNA compaction and cell vitality: development of an improved methodology. Int J Androl. 2011;34(1):2–13.
61. Aitken RJ, De Iuliis GN, Finnie JM, Hedges A, McLachlan RI. Analysis of the relationships between oxidative stress, DNA damage and sperm vitality in a patient population: development of diagnostic criteria. Hum Reprod. 2010;25:2415–26.
62. Cavalieri EL, Rogan EG. A unifying mechanism in the initiation of cancer and other diseases by catechol quinones. Ann N Y Acad Sci. 2004;1028:247–57.
63. Meyer-Ficca ML, Lonchar J, Credidio C, Ihara M, Li Y, Wang ZQ, et al. Disruption of poly(ADP-ribose) homeostasis affects spermiogenesis and sperm chromatin integrity in mice. Biol Reprod. 2009;81: 46–55.
64. Chabory E, Damon C, Lenoir A, Kauselmann G, Kern H, Zevnik B, et al. Epididymis seleno-independent glutathione peroxidase 5 maintains sperm DNA integrity in mice. J Clin Invest. 2009;119:2074–85.
65. Fraga CG, Motchnik PA, Wyrobek AJ, Rempel DM, Ames BN. Smoking and low antioxidant levels increase oxidative damage to DNA. Mutat Res. 1996;351:199–203.
66. Ji BT, Shu XO, Linet MS, Zheng W, Wacholder S, Gao YT, et al. Paternal cigarette smoking and the risk of childhood cancer among offspring of non-smoking mothers. J Natl Cancer Inst. 1997;89:238–44.
67. Imai H, Hakkaku N, Iwamoto R, Suzuki J, Suzuki T, Tajima Y, et al. Depletion of selenoprotein GPx4 in spermatocytes causes male infertility in mice. J Biol Chem. 2009;284:32522–32.
68. Zubkova EV, Robaire B. Effect of glutathione depletion on antioxidant enzymes in the epididymis, seminal

vesicles, and liver and on spermatozoa motility in the aging brown Norway rat. Biol Reprod. 2004; 71:1002–8.
69. Imai H, Suzuki K, Ishizaka K, Ichinose S, Oshima H, Okayasu I, et al. Failure of the expression of phospholipid hydroperoxide glutathione peroxidase in the spermatozoa of human infertile males. Biol Reprod. 2001;64:674–83.
70. O WS, Chen H, Chow PH. Male genital tract antioxidant enzymes – their ability to preserve sperm DNA integrity. Mol Cell Endocrinol. 2006;250:80–3.
71. Shamsi MB, Venkatesh S, Tanwar M, Talwar P, Sharma RK, Dhawan A, et al. DNA integrity and semen quality in men with low seminal antioxidant levels. Mutat Res. 2009;665:29–36.
72. Song GJ, Norkus EP, Lewis V. Relationship between seminal ascorbic acid and sperm DNA integrity in infertile men. Int J Androl. 2006;29:569–75.
73. Abd-Elmoaty MA, Saleh R, Sharma R, Agarwal A. Increased levels of oxidants and reduced antioxidants in semen of infertile men with varicocele. Fertil Steril. 2010;94:1531–4.
74. Smith R, Kaune H, Parodi D, Madariaga M, Rios R, Morales I, et al. Increased sperm DNA damage in patients with varicocele: relationship with seminal oxidative stress. Hum Reprod. 2006;21:986–93.
75. Brackett NL, Ibrahim E, Grotas JA, Aballa TC, Lynne CM. Higher sperm DNA damage in semen from men with spinal cord injuries compared with controls. J Androl. 2008;29:93–9.
76. Moskovtsev SI, Willis J, White J, Mullen JB. Leukocytospermia: relationship to sperm deoxyribonucleic acid integrity in patients evaluated for male factor infertility. Fertil Steril. 2007;88:737–40.
77. Erenpreiss J, Hlevicka S, Zalkalns J, Erenpreisa J. Effect of leukocytospermia on sperm DNA integrity: a negative effect in abnormal semen samples. J Androl. 2002;23:717–23.
78. Fariello RM, Del Giudice PT, Spaine DM, Fraietta R, Bertolla RP, Cedenho AP. Effect of leukocytospermia and processing by discontinuous density gradient on sperm nuclear DNA fragmentation and mitochondrial activity. J Assist Reprod Genet. 2009; 26:151–7.
79. Sakkas D, Mariethoz E, St. John JC. Abnormal sperm parameters in humans are indicative of an abortive apoptotic mechanism linked to the Fas-mediated pathway. Exp Cell Res. 1999;251:350–5.
80. Gorczyca W, Traganos F, Jesionowska H, Darzynkiewicz Z. Presence of DNA strand breaks and increased sensitivity of DNA in situ to denaturation in abnormal human sperm cells: analogy to apoptosis of somatic cells. Exp Cell Res. 1993;207(1):202–5.
81. Barroso G, Morshedi M, Oehninger S. Analysis of DNA fragmentation, plasma membrane translocation of phosphatidylserine and oxidative stress in human spermatozoa. Hum Reprod. 2000;15:1338–44.
82. Paasch U, Grunewald S, Agarwal A, Glandera HJ. Activation pattern of caspases in human spermatozoa. Fertil Steril. 2004;81 Suppl 1:802–9.
83. Grunewald S, Sharma R, Paasch U, Glander HJ, Agarwal A. Impact of caspase activation in human spermatozoa. Microsc Res Tech. 2009;72:878–88.
84. Lozano GM, Bejarano I, Espino J, González D, Ortiz A, García JF, et al. Relationship between caspase activity and apoptotic markers in human sperm in response to hydrogen peroxide and progesterone. J Reprod Dev. 2009;55:615–21.
85. Libman J, Gabriel MS, Sairam MR, Zini A. Catalase can protect spermatozoa of FSH receptor knock-out mice against oxidant-induced DNA damage in vitro. Int J Androl. 2010;33:818–22.
86. Espino J, Bejarano I, Ortiz A, Lozano GM, García JF, Pariente JA, et al. Melatonin as a potential tool against oxidative damage and apoptosis in ejaculated human spermatozoa. Fertil Steril. 2010;94:1915–7.
87. De Iuliis GN, Newey RJ, King BV, Aitken RJ. Mobile phone radiation induces reactive oxygen species production and DNA damage in human spermatozoa in vitro. PLoS One. 2009;4:e6446.
88. Aitken RJ, Wingate JK, De Iuliis GN, Koppers AJ, McLaughlin EA. Cis-unsaturated fatty acids stimulate reactive oxygen species generation and lipid peroxidation in human spermatozoa. J Clin Endocrinol Metab. 2006;91:4154–63.
89. Sotolongo B, Huang TT, Isenberger E, Ward WS. An endogenous nuclease in hamster, mouse, and human spermatozoa cleaves DNA into loop-sized fragments. J Androl. 2005;26:272–80.
90. Greco E, Romano S, Iacobelli M, Ferrero S, Baroni E, Minasi MG, et al. ICSI in cases of sperm DNA damage: beneficial effect of oral antioxidant treatment. Hum Reprod. 2005;20:2590–4.
91. Bianchi PG, Manicardi GC, Bizzaro D, Bianchi U, Sakkas D. Effect of deoxyribonucleic acid protamination on fluorochrome staining and in situ nick-translation of murine and human mature spermatozoa. Biol Reprod. 1993;49:1083–8.
92. Lolis D, Georgiou I, Syrrou M, Zikopoulos K, Konstantelli M, Messinis I. Chromomycin A3-staining as an indicator of protamine deficiency and fertilization. Int J Androl. 1996;19:23–7.
93. Iranpour FG, Nasr-Esfahani MH, Valojerdi MR, al-Taraihi TM. Chromomycin A3 staining as a useful tool for evaluation of male fertility. J Assist Reprod Genet. 2000;17:60–6.
94. Bennetts LE, Aitken RJ. A comparative study of oxidative DNA damage in mammalian spermatozoa. Mol Reprod Dev. 2005;71:77–87.
95. Den Boer PJ, Poot M, Verkerk A, Jansen R, Mackenbach P, Grootegoed JA. Glutathione-dependent defence mechanisms in isolated round spermatids from the rat. Int J Androl. 1990;13:26–38.
96. Ling J, Söll D. Severe oxidative stress induces protein mistranslation through impairment of an aminoacyl-tRNA synthetase editing site. Proc Natl Acad Sci USA. 2010;107:4028–33.
97. Aitken RJ, Roman SD. Antioxidant systems and oxidative stress in the testes. Adv Exp Med Biol. 2008;636:154–71.

Abortive Apoptosis and Sperm Chromatin Damage

20

Hasan M. El-Fakahany and Denny Sakkas

Abstract

The term *apoptosis* refers to a morphologically distinct form of cell death that plays a major role during the normal development and homeostasis of multicellular organisms. This mode of cell death is a tightly regulated series of energy-dependent molecular and biochemical events orchestrated by a genetic program. Apoptosis is either developmentally regulated (launched in response to specific stimuli, such as deprivation of survival factors, exposure to ionizing radiation and chemotherapeutic drugs, or activation by various death factors and their ligands) or induced in response to cell injury or stress. Apoptosis is also being recognized in the pathogenesis of many diverse human diseases including cancer, acquired immune deficiency syndrome, neurodegenerative disorders, atherosclerosis, and cardiomyopathy. Maintaining the homeostatic relationship between apoptosis and cell proliferation is important for tissue development and degeneration. Decreased apoptosis may lead to neoplasia, whereas increased apoptosis may lead to a dystrophic condition.

Keywords

Abortive apoptosis • Sperm chromatin damage • Programmed cell death • Apoptosis

The term *programmed cell death* was originally used to describe the coordinated series of events leading to cell demise during development. The term *apoptosis* refers to a morphologically distinct form of cell death that plays a major role during the normal development and homeostasis of multicellular organisms. This mode of cell death is a tightly regulated series of energy-dependent molecular and biochemical events orchestrated by a genetic program [1].

Apoptosis is either developmentally regulated (launched in response to specific stimuli, such as deprivation of survival factors, exposure to ionizing radiation and chemotherapeutic drugs, or activation by various death factors and their

H.M. El-Fakahany (✉)
Department of Dermatology, STD's and Andrology, Al-Minya University, PO Box 61519, Al-Minya, Egypt
e-mail: fakahany@hotmail.com

A. Zini and A. Agarwal (eds.), *Sperm Chromatin: Biological and Clinical Applications in Male Infertility and Assisted Reproduction*, DOI 10.1007/978-1-4419-6857-9_20,

ligands) or induced in response to cell injury or stress. It is now widely accepted that apoptosis serves as a prominent force in sculpting body parts, in deleting unneeded structures, in maintaining tissue homeostasis, and it also serves as a defense mechanism to remove unwanted and potentially dangerous cells, such as self-reactive lymphocytes, virus-infected cells, and tumor cells. Apoptosis is also being recognized in the pathogenesis of many diverse human diseases including cancer, acquired immune deficiency syndrome, neurodegenerative disorders, atherosclerosis, and cardiomyopathy. Maintaining the homeostatic relationship between apoptosis and cell proliferation is important for tissue development and degeneration. Decreased apoptosis may lead to neoplasia, whereas increased apoptosis may lead to a dystrophic condition [1].

Cellular Characteristics of Apoptosis vs. Necrosis

The process of apoptosis is associated with well-defined morphological and biochemical changes, including a reduction in cell volume, blebbing of the cell membrane, chromatin condensation and margination, and formation of apoptotic bodies. In contrast to physiological cell death or apoptosis, necrosis is a passive process that does not require energy expenditure by the cell and occurs in response to a wide variety of noxious agents. Necrosis does not occur in a developmental context, usually affects a group of contiguous cells, and is characterized by swelling of the cell and its organelles (as a result of ion pump failure) and results ultimately in membrane rupture and cell lysis [1].

A unique biochemical event in apoptosis is the activation of calcium–magnesium-dependent endonuclease activity, which specifically cleaves cellular DNA between regularly spaced nucleosomal units. Such fragments possess a characteristic DNA pattern, which is considered the hallmark of apoptosis. In necrosis, as opposed to apoptosis, the genomic DNA is degraded randomly by a host of cytosolic and lysosomal endonucleases, producing a continuous spectrum of sizes [2].

Another important distinguishing feature of apoptosis is the rapid clearance of dead cells by "professional" phagocytes (such as macrophages) before they can lyse, spill their noxious contents, and cause an inflammatory reaction. This clearance mechanism is efficient and rapid. By contrast, during the pathological or accidental cell death that results from overwhelming cellular injury, cells swell and lyse, releasing noxious contents that often trigger an inflammatory response. An additional change associated with cells during the early phases of apoptosis is the alteration of plasma membrane phosphatidylserine asymmetry. In normal cells, the phosphatidylserine is located on the cytoplasmic side or on the inner leaflet of the plasma membrane. Early in apoptosis, phosphatidylserine is translocated from the inner to the outer surface of the plasma membrane and, consequently, is exposed to the external cellular environment. Surface exposure of phosphatidylserine occurs along with chromatin condensation, and it precedes the increase in membrane permeability and constitutes one of the principal targets of phagocyte recognition [3].

A disruption in the mitochondrial transmembrane potential occurring before nuclear changes has been observed in many cells undergoing apoptosis. This permeability transition involves the opening of a large channel in the inner membrane of the mitochondrion that leads to the release of apoptosis-inducing factors (AIF) from mitochondria to the cytosol. In addition, permeability transition causes the mitochondrial generation of ROS, and rapid expression of phosphatidylserine residues in the outer plasma membrane leaflet [4].

Moreover, during apoptosis, mitochondrial inner membrane proteins, such as cytochrome c, leak out into the cytosol. At least two other cytosolic proteins, apoptotic protease activating factor 1 (Apaf-1) and Apaf-3, have been identified that collaborate with cytochrome c (also known as Apaf-2) to induce proteolytic processing and caspase activation and, in turn, kill cells by apoptosis [5–7].

Programmed Cell Death Cascade

Broadly, the programmed cell death cascade can be divided into at least three to four phases: signal activation, control, execution, and structural alterations. Multiple signaling pathways lead from death-triggering extrinsic signals to a central control and execution stage [8].

Three major pathways are involved in the process of caspase activation and apoptosis in mammalian cells. The intrinsic pathway for apoptosis involves the release of cytochrome c into the cytosol where it binds to Apaf-1. Once activated by the cytochrome c, Apaf-1 then binds to procaspase 9, resulting in the activation of the initiator caspase 9 and the subsequent proteolytic activation of the executioner caspase 3, 6, and 7. The active executioners are then involved in the cleavage of a set of proteins as poly (ADP-ribose) polymerase (PARP) and cause morphological changes to the cell and nucleus typical of apoptosis. Members of the Bcl-2 family of proteins play a major role in governing this mitochondria-dependent apoptotic pathway, with proteins such as Bax functioning as inducers and proteins such as Bcl-2 as suppressors of cell death [5].

The extrinsic pathway for apoptosis involves ligation of a death receptor (e.g., Fas) to its ligand [e.g., Fas ligand (FasL)]. For the Fas pathway, binding of FasL to Fas activates Fas receptors, which recruit the Fas-associated death domain, which in turn binds to the initiator caspase 8 or 10 [9].

A third subcellular compartment, the endoplasmic reticulum, has also shown to be involved in apoptotic execution. Cross talk between these pathways does occur at numerous levels. In certain cells, caspase 8 through cleavage of Bid, a proapoptotic Bcl-2 family member, can induce cytochrome c release from mitochondria in Fas-mediated death signaling. All these pathways converge on caspase 3 and other executioner caspases and nucleases that drive the terminal events of programmed cell death [9].

Testicular Germ Cells Apoptosis in Normal Spermatogenesis

Spermatogenesis is a dynamic process of germ cell proliferation and differentiation. Sertoli cells and germ cells, the only cell types within the seminiferous epithelium, are in close contact. Sertoli cells, lining the seminiferous epithelium, supervise spermatogenesis by providing structural and nutritional support to germ cells. The seminiferous epithelium of the testis is a rapidly proliferating tissue in which germ cells degenerate spontaneously. Up to 75% of the spermatogonia die in the process of programmed cell death before reaching maturity. The testes of normal men produce 10^8 spermatozoa daily. This output depends on proliferative activity in the basal compartment of the seminiferous epithelium where the spermatogonial cells are found and differentiate toward the lumen where meiosis and spermatogenesis occur. During regular spermatogenesis testicular germ cells degenerate by an apoptotic process. The significance of regulating the cell population by apoptosis is more apparent when sperm production is halted. A number of factors can trigger regression of the epithelium and render the testis sterile [10].

In mammals, germ cell death is conspicuous during spermatogenesis and occurs spontaneously at various phases of germ cell development such that seminiferous epithelium yields fewer spermatozoa than might be anticipated from spermatogonial proliferations [11].

In normal newborns, apoptotic cells in the seminiferous cords were identified as being mostly spermatogonia, even though Sertoli cells were also detected. The extent of testicular cell proliferation during fetal and neonatal development determines the final adult testis size and potential for sperm output in the human with subsequent stabilization during the first years of prepuberty. Even though gonadotropins start to increase during the first month of life, it is remarkable that the peak of the activation of the gonadotropin testicular axis that takes place during the second and third month of life was not associated with a lower rate of apoptosis, or with increase in testis weight. Hormonal or growth factors present in the fetoplacental unit might influence testicular cell growth for a few weeks after birth. The newborn period is characterized by increased cell mass in the two compartments of the testis. This cell growth seems to be mainly mediated by decreased apoptosis. The main mechanism for modulation of cell number in the prepubertal testis is the regulation of apoptotic cell death relative to cell proliferation [12].

Apoptosis is the underlying mechanism of germ cell death during normal spermatogenesis in humans. Human testes exhibit spontaneous occurrence of germ cell apoptosis involving all three classes of germ cells, including spermatogonia, spermatocytes, and spermatids. The incidence of spontaneous germ cell apoptosis in humans varies with ethnic background. For example, the incidence of spermatogonial and spermatid apoptosis was higher in Chinese men than in Caucasian men. The triggering factors for spontaneous germ cell apoptosis during normal spermatogenesis are not known, and it is uncertain why there are ethnic differences in the inherent susceptibility of germ cells to programmed cell death. However, it should be noted that, in testes, as in many other tissues, the contribution of spontaneous germ cell apoptosis has been grossly underestimated due to the rapid and efficient clearance of apoptotic cells by professional phagocytes (Sertoli cells) [13].

The survival of conjoined spermatogonial cell progenies depends in part on maintaining structural and functional relationships with both neighboring Sertoli cells and with the basal lamina of the seminiferous tubular wall. Spermatocytes are less dependent on the basal lamina relationship and more dependent on Sertoli cell support. When apoptosis signaling is activated, caspases initiate a cell disassembling procedure, generating apoptotic bodies and leading to the final demise of entire spermatogonial and spermatocyte progenies [14].

During spermatogenesis, spermatogonia and round spermatids almost certainly die by apoptosis [15]. Peak germ cell loss has been observed during the stages of mitosis of type A spermatogonia, during meiotic division of spermatocytes, and during spermiogenesis [16]. Apoptotic germ cells are either sloughed into the tubule lumen or phagocytosed by Sertoli cells. Spermatozoa also demonstrate changes consistent with apoptosis. The percentage of germ cells undergoing apoptosis in normal subjects is significantly lower than that seen in men with oligoasthenoteratozoospermia, Hodgkin's disease, and testicular cancer [17].

Five possible functional roles have been proposed in the literature for the presence of apoptosis during normal spermatogenesis:

1. *Maintenance of an optimal germ cell/Sertoli cell ratio*. It has been established that each Sertoli cell can support only a definite number of germ cells throughout their development into spermatozoa. Therefore, supraoptimal numbers of spermatogonia may undergo apoptosis to maintain an optimal ratio [18].
2. *Elimination of abnormal germ cells*. There may be a selective process in which abnormal germ cells, especially chromosomally abnormal germ cells, are eliminated from the population by apoptosis [11].
3. *The formation of the blood–testis barrier by tight junctions between Sertoli cells requires the elimination of excessive germ cells*. Suppression of germ cell apoptosis by means of inactivating Bax, an apoptosis-inducing gene, prevents the formation of these tight junctions [19].
4. *Creation of a prepubertal apoptotic wave facilitates the eventual functional development of mature spermatogenesis*. A massive wave of germ cell apoptosis normally takes place as mammalian species approach puberty. This wave serves as a regulator of the ratio between germinal cells in various stages and Sertoli cells. There is evidence that preventing this wave of apoptosis by expression of apoptosis-inhibitory proteins, such as Bcl-xL or Bcl-2, results in highly abnormal adult spermatogenesis accompanied by sterility [20].
5. *Selective removal of unneeded portions of sperm cytoplasm*. Apoptosis plays an important role in the spermatogenesis such as removing abnormal sperm. For example, spermatids display many of the histological and molecular fingerprints of apoptosis. Maturing spermatids form darkly staining basophilic bodies and express multiple caspases within these "residual bodies." In addition, these bodies contain proteins linked to the regulation of cell death such as Fas and p53. The cytoplasm of maturing spermatids is collected and removed by residual bodies. This is probably done by

neighboring Sertoli cells, which recognize and phagocytose them as they are shed. All of this has led to the idea that developing spermatozoa use the apoptotic machinery to selectively dissipate unneeded portions of their cytoplasm. In this view, apoptotic factors are somehow segregated to the cytoplasm – away from the nucleus – and this segregation permits the emerging sperm to utilize the apoptotic machinery without dying [21].

Regulators of Testicular Apoptosis

Apoptotic cell death seems to be strictly regulated by extrinsic and intrinsic factors and can be triggered by a wide variety of stimuli. Examples of extrinsic stimuli potentially important in testicular apoptosis are irradiation, trauma, viral infection, toxin exposure, and the withdrawal of hormonal support. It has been widely assumed that certain hormones, growth factors, or cytokines are necessary for cell survival and cell cycle progression and that their absence leads to apoptosis of their target cells. Moreover, genetic control plays a prominent role in apoptosis through molecular regulatory factors, which act as intrinsic mediators [22].

Intrinsic Regulators

Genes Regulating Germ Cell Apoptosis

Disruption of a number of genes can result in infertility through accelerated germ cell apoptosis in mice. These findings give a first glimpse of the regulatory mechanisms involved in the regulation of germ cell apoptosis and may help in defining important genetic principles that may apply to genes important for human fertility. Male mice deficient in Bax were infertile and displayed accumulation of premeiotic germ cells with complete loss of advanced spermatids. In addition, mice misexpressing Bcl-2 in spermatogonia displayed an accumulation of spermatogonia before puberty but during adulthood exhibited a loss of germ cells in the majority of the tubules [23].

Fas-FasL. The cell surface receptor Fas is a transmembrane glycoprotein that belongs to the tumor necrosis factor/nerve growth factor family. The Fas–FasL interaction triggers the death of cells expressing Fas. Expression of Fas and FasL is detected not only in Sertoli cells but also in germ cells and Leydig cells [24].

In testis, the Fas system has been implicated in maintaining immune privilege. According to this hypothesis, FasL-expressing Sertoli cells eliminate Fas-positive activated T-cells, providing general protection against rejection in the testicular environment. Moreover, if Sertoli cells are injured, they increase the expression of FasL to eliminate Fas-positive germ cells, which cannot be supported adequately. These findings strongly implicate the Sertoli cell in the paracrine control of germ cell output during spermatogenesis by a Fas-mediated pathway [25].

Although Fas may contribute to germ cell homeostasis, it is not essential. Mice with complete lack of Fas are fertile without any overt defects in germ cell apoptosis [26].

Bcl-2 Family. Bcl-2 is the first member identified of a growing family of genes that regulates cell death in either a positive or a negative fashion. The Bcl-2 family of proteins, which contains both proapoptotic (Bax, Bak, Bcl-xs, Bad) and antiapoptotic (Bcl-2, Bcl-xL, Mcl, A1) proteins, constitutes a critical, intracellular checkpoint within a common cell-death pathway that determines the susceptibility of a cell to apoptosis. It is generally believed that the ratio of proapoptotic to antiapoptotic Bcl-2 family proteins is the critical determinant of cell fate, with an excess of Bcl-2 resulting in cell survival but an excess of Bax resulting in cell death. Although these molecules compete, it has not been established firmly yet whether antiapoptotic or proapoptotic members are dominant in determining the key survival-promoting decision point. Paradoxically, a given family member may perform either function, depending on the cell systems used [8].

Bcl-2 protects cells from apoptosis by its capacity to reduce production of ROS. Other members of the Bcl-2 family, including Bax, Bak, and Bad, can block the ability of Bcl-2 to inhibit apoptosis and subsequently to promote cell death. Bax, for example, functions to increase the sensitivity of cells to apoptotic stimuli [27]. Disruption of Bax, an apoptosis-inducing gene, prevented the process of apoptosis in the testis and resulted in an accumulation of immature germ cells (mainly spermatocytes) in the tubules [19].

p53. p53 suppresses oncogenic transformation by promoting apoptosis. p53 is found in high concentration in the testis and plays a significant role in temperature-induced germ cell apoptosis. This cell-cycle regulator also seems to be required for radiation-induced apoptosis of spermatogonia, as evidenced by de novo induction of p53 expression in spermatogonia and degenerating giant cells in the testes following irradiation [22].

p53-induced testicular apoptosis involves the following:

1. Activation of redox-related genes also known as p53-induced genes.
2. Generation of ROS.
3. Oxidative degradation of mitochondrial components permitting the release of apoptosis-inducing factors, including AIF, cytochrome c, Apaf-1, Apaf-3, into the cytosol to activate the Caspases [28].

Caspases. Caspases are cystein proteases that promote apoptosis in mammals. Evidence for the role of caspases in cell death is based on findings that their inhibition can prevent apoptosis, whereas their overexpression and activation cause apoptosis. Caspases mediate apoptosis by cleaving selected intracellular proteins, including PARP, lamin and actin, and cause morphological changes to the cell and nuclei [29].

In vitro, apoptosis of human male germ cells can be prevented by caspase inhibition [30]. On the contrary, caspase activity could not be detected in human adult germ cells obtained from men with normal spermatogenesis and cultured in vitro under conditions that led to massive DNA fragmentation, suggesting the implication of an alternative, caspase-independent mechanism [31].

c-Myc. c-Myc is a nuclear phosphoprotein, encoded by a proto-oncogene, c-Myc. It plays a key role in the control of cell proliferation by acting as a transcription factor. Overexpression of the c-Myc gene in transgenic rats induces germ cell apoptosis at the meiotic prophase of primary spermatocytes. Depletion of sperm and seminiferous tubule atrophy, causing sterility, have been observed in the male transgenic rats [32].

Cyclic adenosine monophosphate responsive element modulator (CREM). The transcriptional activator, cyclic adenosine monophosphate (cAMP) responsive element modulator (CREM), which is highly expressed in postmeiotic cells, may be responsible for the activation of haploid germ-cell-specific genes involved in the structuring of the spermatozoa. CREM is responsive to the cAMP signal pathway and is required for expression of postmeiotic germ-cell-specific genes. Mice that are CREM-deficient are phenotypically normal but have a maturation arrest at the early spermatid stage associated with a marked increase in apoptosis [33].

CREM is expressed in the nuclei of round spermatids, but not in elongated spermatids. CREM may be important for spermatid development and as a stage-specific regulator of human spermatogenesis. Absence of CREM may play a causative role in testicular failure associated with various types of human male infertility [34].

c-kit. c-kit has been identified as a germ cell apoptosis preventing gene. Blockade or loss of the c-kit receptor results in the inability of the mature spermatozoa to undergo the acrosome reaction. Decreased expression of the c-kit receptor and its ligand, stem cell factor, may alter the balance between cell proliferation/differentiation and cell death, resulting in increased apoptosis in the testes [35].

In mice, c-kit is involved in the migration of primordial germ cells and is expressed early in spermatogenesis. It is expressed in type A, intermediate, and type B spermatogonia, and its ligand is expressed in Sertoli cells [36].

Genetic Regulators of DNA Repair

DNA damage is one of the most potent triggers of apoptosis. DNA damage (e.g., chromosomal abnormalities, failure of DNA repair or genetic recombination, ionizing radiation, chemotherapy) leads to the elimination of damaged cells scattered within the epithelium via apoptosis [37].

PARP is a chromatin-associated enzyme with a presumptive role in DNA repair during replication and recovery from strand breaks caused by genotoxic agents. It is particularly active in the testis, where its expression varies according to the stage of germ cell differentiation. The degradation of PARP is also one of the classic indicators of apoptosis [38].

Extrinsic Regulation (Hormonal Regulation)

Withdrawal of gonadotropins or testosterone can markedly accelerate germ cell apoptosis. In rodents, spermatogenesis and apoptosis have been shown to be hormonally dependent. As in other hormonally sensitive reproductive organs, such as the prostate, endometrium, and ovary, the withdrawal of hormonal stimulation results in the selective degeneration of specific cell types [22].

Assessing the relationship between hormonal deprivation and the induction of germ cell apoptosis in adult rats following the withdrawal of testosterone demonstrated a significant rise in testicular cells with a low DNA content in combination with a decrease in haploid cells after testosterone deprivation [39].

Glucocorticoids act at the level of the pituitary and testis to suppress testosterone secretion and as a result may generate testicular apoptosis. Also, administration of exogenous glucocorticoid resulted in testicular germ cell apoptosis in rats. Severe stress may provoke the release of endogenous glucocorticoids in men, resulting in decreased serum testosterone and possibly triggering apoptosis [40].

There is an increase in DNA fragmentation in seminiferous tubules after hypophysectomy, further supporting the concept that androgen deprivation increases programmed cell death in the seminiferous epithelium. GnRH antagonist-induced germ cell apoptosis is most prominent among meiotic spermatocytes. Administration of a GnRH antagonist resulted in morphologic signs of germ cell degeneration in spermatocytes and spermatids [41].

Gonadotropin-dependent germ cell apoptosis seems to be age-related. A marked increase in apoptotic DNA fragmentation was seen in aging rats treated with a potent GnRH antagonist to suppress circulating levels of FSH, LH, and testosterone. Testicular apoptosis may, therefore, be enhanced in the aging male given the decline in free testosterone levels that occur with advancing age [42].

Testicular Germ Cells Apoptosis During Testicular Dysfunction Conditions

Aging

With aging, both potential daily sperm production and Leydig cell function decline. As for spermatogenesis, histopathological examination reveals that there is a significant decline in the number of Sertoli cells per seminiferous tubule and the number of spermatids and primary spermatocytes per Sertoli cell [43].

Germ cell loss associated with aging occurs by apoptosis, probably because of a combination of a primary testicular defect and secondary hypothalamic pituitary dysfunction. Reproductive aging in the rat is characterized by decreased Leydig cell steroidogenesis associated with seminiferous tubule dysfunction. Accelerated germ cell apoptosis involving spermatogonia, spermatocytes, and spermatids is greater in the testes of aging rats than in the testes of younger animals [44].

Downregulated apoptosis of spermatogonia was detected with aging. Diminished spermatogonial proliferation was also found concomitant with low spermatogonial apoptosis. The decline of spermatogonial apoptosis might reflect a compensatory role of apoptosis in spermatogonia for the diminished proliferation that occurred during aging. Accelerated apoptosis of primary spermatocytes was detected in the testis of elderly men. It was speculated that apoptosis of primary spermatocytes might be the most relevant cause of impaired spermatogenesis in the aged testis. Apoptotic rates of round spermatids and elongated spermatids showed no significant elevations, whereas quantitative analysis revealed a reduction in their number. Sertoli cells might already have digested many apoptotic spermatids at the time of the detection of DNA fragmentation because those cells are phagocytosed in the early phase of the apoptotic process in the rat testis [45].

Varicocele

Several varicocele-associated factors, including heat stress, androgen deprivation, and exposure to toxic elements, may induce pathways, which result in apoptosis [46].

Apoptosis in the ejaculate of men with varicocele. Varicocele induces apoptosis, which is initiated in the testicular tissue and is then expressed in the semen. Up to 10% of sperm cells in the ejaculate of men with a varicocele were apoptotic, as compared with 0.1% in fertile controls [47]. Saleh et al. [48] showed that infertile men with varicoceles had significantly greater DNA damage in spermatozoa than had normal men. Bertolla et al. [49] also evaluated DNA fragmentation in adolescents with clinically diagnosed varicoceles, and determined that these boys had a higher percentage of cells with DNA fragmentation than did adolescents with no varicocele.

The expression of Fas protein was upregulated in semen samples obtained from patients with varicocele when compared to a control group, whereas little or no changes in FasL expression were detected in both groups. The relationship between varicoceles and apoptosis was explored by monitoring the concentrations of the soluble form of Fas (s-Fas) in seminal plasma, to characterize the Fas signaling system with regard to hypospermatogenesis as a result of varicocele. By screening the seminal plasma of oligospermic men with varicoceles, oligospermic men with no varicocele, and normal controls, for the levels of s-Fas and the s-Fas ligand, s-Fas ligand was not detected in any of the cases, whereas s-Fas levels were specifically lower only in cases of varicocele. These reduced s-Fas levels were reversed by varicocelectomy. However, although higher temperatures may inhibit s-Fas production in patients with varicocele, the reason for this decrease in s-Fas levels remains unknown [50].

By contrast, Chen et al. [51] identified no relationship between semen quality and apoptosis. Although the varicocele patients had a significantly higher apoptotic index (AI) than fertile controls, semen quality and sperm motion characteristics were not significantly different between the groups.

Seminal ROS may result in sperm DNA damage in patients with varicoceles. At the molecular level, ROS affect DNA directly and alter the levels of intracellular Ca^{+2}, which is known to be one of the most effective means of inducing apoptosis. Morphological alterations in testicular tissues have been reported as "stress patterns" in patients with varicoceles. This stress pattern is reminiscent of, although not identical to, the cytomorphological changes in apoptosis [46].

High levels of seminal ROS and reduced total antioxidant capacity were detected in both fertile and infertile men with a clinical diagnosis of varicocele. Therefore, it was hypothesized that spermatozoal dysfunction in association with varicoceles may be related, at least in part, to elevated levels of sperm DNA damage induced by the high levels of ROS, which are common in such patients [52].

Infertile men with varicoceles had significant increase in spermatozoal DNA damage, which appeared to be associated with high ROS levels in the semen. This finding of high seminal ROS levels in patients with varicoceles might indicate that ROS plays a role in the pathogenesis of sperm DNA damage in such patients [48].

Apoptosis in the testicular tissue in men with varicocele. Simsek et al. [53] evaluated the presence of apoptosis in testicular tissue, using the TUNEL assay. Apoptosis was very rare in the testicular tissues of the control group compared to the varicocele group. The mean percentage of apoptotic cells per total germ cell was 2% in the control and 14.7% in the varicocele group.

Hurley et al. [54] also reported that there were far more apoptotic nuclei in the seminiferous tubules of men with varicocele than in normal controls. Recently, Benoff et al. [55] have reported that the percentage of apoptotic nuclei was noticeably higher in some men with varicoceles.

On the contrary, Fujisawa et al. [56] reported fewer apoptotic germ cells in testicular biopsy material obtained from subfertile men with varicoceles than in biopsies of normal men. There were also fewer apoptotic cells per Sertoli cell in the testes of men with varicocele than in those of normal men.

Although Bcl-2 was not expressed in the germ cells in infertile patients with varicocele, these cells expressed low levels of Bax, with no significant differences to the specimens from fertile men. In the testes from infertile patients with varicoceles stained for caspase 3, significantly fewer germ cells were detected than those in the testes of normal controls. It was suggested that apoptosis might be suppressed as the result of reduced expression of caspase 3 and that the mitochondrial pathway involving Bcl-2 and Bax may not be involved in apoptotic regulation in germ cells [57].

Spermatogenesis Failure

The causes of complete spermiogenesis failure are not completely known. These include the withdrawal of some developmentally important ligands, such as testosterone [58] or vitamin A [59], mutations of the receptors with which these ligands and their metabolites can act, such as the retinoic acid receptor A [60] or the retinoid X receptor B [61], alterations of molecules involved in signal transduction pathways, downstream of receptors, such as CREM protein [33], or mutations of components of cell DNA repair enzyme systems [62]. Such conditions are often associated with germ cell apoptosis [63].

Reduced expression of CREM was also detected in patients with predominant round spermatid maturation arrest in comparison with men with normal spermatogenesis or with mixed testicular atrophy [34], and increased apoptosis of testicular cells has been demonstrated in patients with abnormal spermatogenesis [64]. It can, thus, be postulated that the low efficacy of round spermatid sperm injection in cases of complete spermiogenesis failure is due to the activation of apoptosis-promoting mechanisms similar to those operating in the experimental models of spermiogenesis arrest [65].

Apoptosis is involved in the removal of arrested germ cells from the testis of patients with spermatogenic disorders. The degree of spermatocyte and spermatid DNA fragmentation in the group of patients with incomplete spermiogenesis failure appears higher as compared to men with normal spermatogenesis [13].

In addition to DNA fragmentation, apoptotic cells also undergo a rearrangement of plasma membrane lipids, leading to translocation of phosphatidylserine from the inner side of the plasma membrane to the outer layer, probably as a result of disintegration of plasma membrane cytoskeleton that, in healthy cells, stabilizes membrane structure by connecting plasma membrane components to the cellular interior. It was suggested that this plasma membrane modification may serve to mark apoptotic cells for subsequent recognition and removal by the phagocytotic machinery [66].

Tesarik et al. [67], using double labeling with TUNEL and Annexin-V, concluded that patients with complete spermiogenesis failure (round spermatids is the latest stage detected histolgically in the testicular biopsy in azoospermic patients) had significantly higher frequencies of primary spermatocytes and round spermatids carrying the apoptosis-specific DNA damage in comparison with patients with incomplete spermiogenesis failure (elongated spermatids is the latest stage detected histolgically in the testicular biopsy in azoospermic patients). Apoptosis-related phosphatidylserine externalization occurs rarely until the advanced stages of spermiogenesis. Since externalized phosphatidylserine is expected to be involved in the recognition of apoptotic cells by phagocytes, apoptotic spermatocytes and round spermatids may not be removed easily by phagocytosis. The high frequency of DNA damage in round spermatids from patients with complete spermiogenesis failure explains the low success rates of spermatid conception in these cases. They also recommended that the evaluation of apoptosis could help to predict success rates of spermatid conception.

Caspase activation and DNA fragmentation are frequent phenomena in germ cells from men with nonobstructive azoospermia, especially in cases of meiotic and postmeiotic maturation arrest. The incidence of Caspase activation and DNA fragmentation is somewhat lower in samples from patients with hypospermatogenesis, in which some germ cells achieve the late elongated spermatid stage [68].

Obstructive Azoospermia

The mechanism inducing apoptosis after obstruction remains unknown. Since the obstruction of the vas deferens would also induce an increase of pressure in the seminal tract, it may cause apoptosis. Increased pressure occurring prior to testicular development might have a more adverse effect than that occurring in adulthood. The difference in apoptotic change between prepubertal and adult cases might, thus, relate to the susceptibility to pressure. However, these pressure increases also seem to be reduced by epididymal development [69].

Flickinger et al. [70] reported that obstruction of the seminal tract in immature rats caused epididymal granulomas, which might in turn have caused fairly high pressure to the seminal tract. In the case of prepubertal obstruction, when epididymis is not well developed, the increased pressure may directly affect the testes to cause increased germ cell apoptosis.

Patients with congenital absence of the vas deferens who generally have good spermatogenesis are somewhat different from acquired obstructions. They have life-long history of seminal tract obstruction; however, the increase or the fluctuation of the pressure may not occur. This could be supported by the report that the vasectomized men showed significantly greater seminiferous tubular wall thickness than the patients who had congenital absence of the vas deferens [71].

References

1. Sinha HA, Swerdloff RS. Hormonal and genetic control of germ cell apoptosis in the testis. Rev Reprod. 1999;4(1):38–47.
2. Tilly J, Ratts V. Biological and clinical importance of ovarian cell death. Contemp Obstet Gynecol. 1996;41:49–86.
3. Kroemer G, Petit P, Zamzami N, Vayssiere J, Mignotte B. The biochemistry of programmed cell death. FASEB J. 1995;9:1277–87.
4. Kroemer G, Zamzami N, Susin SA. Mitochondrial control of apoptosis. Immunol Today. 1997;18:44–51.
5. Reed JC. Mechanisms of apoptosis. Am J Pathol. 2000;157(5):1415–30.
6. Reed JC. Double identity for proteins of the Bcl-2 family. Nature. 1997;387(6635):773–6.
7. Reed J. Cytochrome C: can't live with it – can't live without it. Cell. 1997;91:559–62.
8. Sinha-Hikim A, Swerdloff R. Hormonal and genetic control of germ cell apoptosis in the testis. Rev Reprod. 1999;4:38–47.
9. Sinha-Hikim A, Lue Y, Diaz-Romero M, Yen P, Wang C, Swerdloff R. Deciphering the pathways of germ cell apoptosis in the testis. J Steroid Biochem Mol Biol. 2003;85:175–82.
10. Milligan C, Schwartz L. Programmed cell death during animal development. Br Med Bull. 1997;52(3):570–90.
11. Sharpe R. Regulation of spermatogenesis. In: Knobil E, Neill J, editors. The physiology of reproduction. New York: Raven; 1994. p. 1363–434.
12. Berensztein E, Sciara M, Rivarola M, Belgorosky A. Apoptosis and proliferation of human testicular somatic and germ cells during prepuberty: high rate of testicular growth in newborns mediated by decreased apoptosis. J Clin Endocrinol Metab. 2002;87:5113–8.
13. Sinha-Hikim A, Wang C, Lue Y, Johnson L, Wang X, Swerdloff R. Spontaneous germ cell apoptosis in humans: evidence for ethnic differences in the susceptibility of germ cells to programmed cell death. J Clin Endocrinol Metab. 1998;83:152–6.
14. Tres L, Kierszenbaum A. Cell death patterns of the rat spermatogonial cell progeny induced by Sertoli cell geometric changes and Fas (CD95) agonist. Dev Dyn. 1999;214:361–71.
15. Henriksen K, Hakovirta H, Parvinen M. In-situ quantification of stage-specific apoptosis in the rat seminiferous epithelium: effects of short-term experimental cryptorchidism. Int J Androl. 1995;18:256–62.
16. de Rooij D, Janssen J. Regulation of the density of spermatogonia in the seminiferous epithelium of the Chinese hamster. Anat Rec. 1987;217:124–30.
17. Gandini L, Lombardo F, Paoli D, Caponecchia L, Familiari G, Verlengia C, et al. Study of apoptotic DNA fragmentation in human spermatozoa. Hum Reprod. 2000;15:830–9.
18. Orth J, Gunsalus G, Lamperti A. Evidence from Sertoli cell-depleted rats indicates that spermatid number in adults depends on numbers of Sertoli cells produced during perinatal development. Endocrinology. 1988; 122:787–94.
19. Knudson C, Tung K, Tourtellotte W, Brown G, Korsmeyer S. Bax-deficient mice with lymphoid hyperplasia and male germ cell death. Science. 1995;270:96–9.
20. Rodriguez I, Ody C, Araki K, Garcia I, Vasalli P. An early and massive wave of germ cell apoptosis is required for the development of functional spermatogenesis. EMBO J. 1997;16:2262–70.
21. Blanco-Rodriguez J, Martinez-Garcia C. Apoptosis is physiologically restricted to a specialized cytoplasmic

compartment in rat spermatids. Biol Reprod. 1999;61: 1541–7.
22. Kim E, Barqawi A, Seo J, Meacham R. Apoptosis: its importance in spermatogenic dysfunction. Urol Clin North Am. 2002;29(4):755–65.
23. Furuchi T, Masuko K, Nishimune Y, Obinata M, Matsui Y. Inhibition of testicular germ cell apoptosis and differentiation in mice misexpressing Bcl-2 in spermatogonia. Development. 1996;122:1703–9.
24. Sugihara A, Saiki S, Tsuji M, Tsujimura T, Nakata Y, Kubota A, et al. Expression of Fas and Fas ligand in the testes and testicular germ cell tumors: an immunohistochemical study. Anticancer Res. 1997;17:3861–5.
25. Lee J, Richburg J, Shipp E, Meistrich M, Boekelheide K. The Fas system, a regulator of testicular germ cell apoptosis, is differentially upregulated in Sertoli cell versus germ cell injury of the testis. Endocrinology. 1999;140:852–8.
26. Adachi M, Suematsu S, Kondo T, Ogasawara J, Tanaka T, Yoshida N, et al. Targeted mutation in the Fas gene causes hyperplasia in peripheral lymphoid organs and liver. Nat Genet. 1995;11:294–300.
27. Kane D, Sarafian T, Anton R, Hahn H, Gralla E, Valentine J, et al. Bcl-2 inhibition of neural death: decreased generation of reactive oxygen species. Science. 1993;262:1274–7.
28. Polyak K, Xia Y, Zweier JL, Kinzler K, Vogelstein B. A model for p53-induced apoptosis. Nature. 1997;389:300–5.
29. Salvesen G, Dixit V. Caspases: intracellular signaling by proteolysis. Cell. 1997;91:443–6.
30. Pentikainen V, Erkkila K, Dunkel L. Fas regulates germ cell apoptosis in the human testis in vitro. Am J Physiol. 1999;276:310–6.
31. Tesarik J, Martinez F, Rienzi L, Iacobelli M, Ubaldi F, Mendoza C, et al. In-vitro effects of FSH and testosterone withdrawal on caspase activation and DNA fragmentation in different cell types of human seminiferous epithelium. Hum Reprod. 2002;17:1811–9.
32. Kodaira K, Takahashi R, Hirabayashi M, Suzuki T, Obinata M, Ueda M. Overexpression of c-myc induces apoptosis at the prophase of meiosis of rat primary spermatocytes. Mol Reprod Dev. 1996;45:403–10.
33. Nantel F, Monaco L, Foulkes N, Masquilier D, LeMeur M, Henriksen K, et al. Spermiogenesis deficiency and germ cell apoptosis in CREM-mutant mice. Nature. 1996;380:159–62.
34. Weinbauer G, Nieschlag E. The role of testosterone in spermatogenesis. In: Nieschlag E, Behre H, editors. Testosterone: action, deficiency, substitution. 2nd ed. New York: Springer; 1998. p. 143–68.
35. Sandlow J, Feng H, Zheng L, Sandra A. Migration and ultrastructural localization of the c-kit receptor protein in spermatogenic cells and spermatozoa of the mouse. J Urol. 1999;161:1676–80.
36. Feng H, Sandlow J, Sparks A, Sandra A, Zheng L. Decreased expression of the c-kit receptor is associated with increased apoptosis in subfertile human testes. Fertil Steril. 1999;71:85–9.
37. Blanco-Rodriguez J. A matter of death and life: the significance of germ cell death during spermatogenesis. Int J Androl. 1998;21:236–48.
38. Tramontano F, Malanga M, Farina B, Jones R, Quesada P. Heat stress reduces poly(ADPR)polymerase expression in rat testis. Mol Hum Reprod. 2000;6:575–81.
39. Henriksen K, Hakovirta H, Parvinen M. Testosterone inhibits and induces apoptosis in rat seminiferous tubules in a stage-specific manner: in situ quantification in squash preparations after administration of ethane dimethane sulfonate. Endocrinology. 1995;136: 3285–9321.
40. Yazawa H, Sasagawa I, Nakada T. Apoptosis of testicular germ cells induced by exogenous glucocorticoid in rats. Hum Reprod. 2000;15:1917–20.
41. Sinha-Hikim A, Wang C, Leung A, Swerdloff R. Involvement of apoptosis in the induction of germ cell degeneration in adult rats after gonadotropin-releasing hormone antagonist treatment. Endocrinology. 1995;136:2770–5.
42. Billig H, Furuta I, Rivier C, Tapanainen J, Parvinen M, Hsueh A. Apoptosis in testis germ cells: developmental changes in gonadotropin dependence and localization to selective tubule stages. Endocrinology. 1995;136:5–12.
43. Johnson L. Spermatogenesis and aging in the human. J Androl. 1986;7:331–54.
44. Wang C, Sinha-Hikim A, Lue Y, Baravarian S, Swerdloff R. Reproductive ageing in the Brown Norway rat is characterized by accelerated germ cell apoptosis and is not altered by luteinizing hormone replacement. J Androl. 1999;20:509–18.
45. Kimura M, Itoh N, Takagi S, Sasao T, Takahashi A, Masumori N, et al. Balance of apoptosis and proliferation of germ cells related to spermatogenesis in aged men. J Androl. 2003;24:185–91.
46. Ku J, Shim H, Kim S, et al. The role of apoptosis in the pathogenesis of varicocele. BJU Int. 2005;96:1092–9.
47. Baccetti B, Collodel G, Piomboni P. Apoptosis in human ejaculated sperm cells (notulae seminologicae 9). J Submicrosc Cytol Pathol. 1996;28:587–96.
48. Saleh R, Agarwal A, Sharma R, Said T, Sikka S, Thomas A. Evaluation of nuclear DNA damage in spermatozoa from infertile men with varicocele. Fertil Steril. 2003;80:1431–6.
49. Bertolla R, Cedenho A, Hassun Filho P, Lima S, Ortiz V, Srougi M. Sperm nuclear DNA fragmentation in adolescents with varicocele. Fertil Steril. 2006;85:625–8.
50. Fujisawa M, Ishikawa T. Soluble forms of Fas and Fas ligand concentrations in the seminal plasma of infertile men with varicocele. J Urol. 2003;170:2363–5.
51. Chen C, Lee S, Chen D, Chien H, Chen I, Chu Y, et al. Apoptosis and kinematics of ejaculated spermatozoa in patients with varicocele. J Androl. 2004;25:348–53.
52. Hendin B, Kolettis P, Sharma R, Thomas A, Agarwal A. Varicocele is associated with elevated spermatozoal reactive oxygen species production and

diminished seminal plasma antioxidant capacity. J Urol. 1999;161:1831–4.
53. Simsek F, Turkeri L, Cevik I, Bircan K, Akdas A. Role of apoptosis in testicular tissue damage caused by varicocele. Arch Esp Urol. 1998;51:947–50.
54. Hurley I, Cooper G, Napolitano B, Gilbert B, Marmar J, Benoff S. High testicular cadmium (Cd^{2+}) levels in varicocele-associated infertility (VAI). Andrologia. 2000;32:190–6.
55. Benoff S, Millan C, Hurley I, Napolitano B, Marmar J. Bilateral increased apoptosis and bilateral accumulation of cadmium in infertile men with left varicocele. Hum Reprod. 2004;19:616–27.
56. Fujisawa M, Hiramine C, Tanaka H, Okada H, Arakawa S, Kamidono S. Decrease in apoptosis of germ cells in the testes of infertile men with varicocele. World J Urol. 1999;17:296–300.
57. Tanaka H, Fujisawa M, Tanaka H, Okada H, Kamidono S. Apoptosis-related proteins in the testes of infertile men with varicocele. BJU Int. 2002;89:905–9.
58. O'Donnell L, McLachlan R, Wreford N, de Kretser D, Robertson D. Testosterone withdrawal promotes stage-specific detachment of round spermatids from the rat seminiferous epithelium. Biol Reprod. 1996;55: 895–901.
59. Eskild W, Hansson V. Vitamin A functions in the reproductive organs. In: Blomhoff R, editor. Vitamin A in health and disease. New York: Marcel Dekker; 1994. p. 531–59.
60. Akmal K, Dufour J, Kim K. Retinoic acid receptor gene expression in the rat testis: potential role during the prophase of meiosis and in the transition from round to elongating spermatids. Biol Reprod. 1997;56:549–56.
61. Kastner P, Mark M, Leid M, Gansmuller A, Chin W, Grondona J, et al. Abnormal spermatogenesis in RXR mutant mice. Genes Dev. 1996;10:80–92.
62. Roest H, van Klaveren J, de Wit J, van Grup C, Kohen M, Vermey M, et al. Inactivation of the HR6B ubiquitin-conjugating DNA repair enzyme in mice causes male sterility associated with chromatin modification. Cell. 1996;86:799–810.
63. Sassone-Corsi P. Transcriptional checkpoints determining the fate of male germ cells. Cell. 1997;88(2):163–6.
64. Lin W, Lamb D, Wheeler T, Abrams J, Lipshultz L, Kim E. Apoptotic frequency is increased in spermatogenic maturation arrest and the hypospermatogenic states. J Urol. 1997;158(5):1791–3.
65. Amer M, Soliman E, El-Sadek M, Mendoza C, Tesarik J. Is complete spermiogenesis failure a good indication for spermatid conception? Lancet. 1997;350: 116–22.
66. van Engeland M, Kuijpers H, Ramaekers F, Reutelingsperger C, Schutte B. Plasma membrane alterations and cytoskeletal changes in apoptosis. Exp Cell Res. 1997;235:421–30.
67. Tesarik J, Greco E, Cohen-Bacrie P, Mendoza C. Germ cell apoptosis in men with complete and incomplete spermiogenesis failure. Mol Hum Reprod. 1998;4(8):757–62.
68. Tesarik J, Ubaldi F, Rienzi L, Martinez F, Jacobelli M, Mendoza C, et al. Caspase-dependent and independent DNA fragmentation in Sertoli and germ cells from men with primary testicular failure: relationship with histological diagnosis. Hum Reprod. 2004;19(2): 254–61.
69. Inaba Y, Fujisawa M, Okada H, Arakawa S, Kamidod S. The apoptotic changes of testicular germ cells in the obstructive azoospermia models of prepubertal and adult rats. Invest Urol. 1998;160(2):540–4.
70. Flickinger C, Herr J, Baran M, Howards S. Testicular development and the formation of spermatic granulomas of the epididymis after obstruction of the vas deferens in immature rats. J Urol. 1995;154:1539–44.
71. Hirsch I, Choi H. Quantitative testicular biopsy in congenital and acquired genital obstruction. J Urol. 1990;143:311–9.

Spermiogenesis in Sperm Genetic Integrity

21

Marie-Chantal Grégoire, Frédéric Leduc, and Guylain Boissonneault

Abstract

This review suggests that spermiogenesis has probably been overlooked as an important source of genetic instability that can provide a repertoire of mutations distributed through millions of spermatozoa, each having the potential to transfer genetic alterations to the next generation. Further investigation will be needed to establish whether this could be considered as a new component of evolution.

Keywords

Chromatin remodeling • DNA breaks • Genomic integrity • Topoisomerase • Spermiogenesis

Spermiogenesis is the haploid phase of male germ cell differentiation, spanning from postmeiotic spermatids to their release as spermatozoa into the lumen of the seminiferous tubules. This differentiation is one of the most radical programs found in the eukaryotic world associated with nuclear events never observed in somatic cells. First, the acrosome forms throughout the spermiogenesis by a process depending on the Golgi apparatus. It undergoes several changes from proacrosomal granules to fully developed acrosome, which contains several proteolytic enzymes essential for fertilization. At mid-spermiogenesis, the flagellum starts to develop arising from the centriole pair, which migrate to the nucleus membrane to implant the flagellum on the opposite side of the acrosome, providing the typical polarity of the nucleus [1]. To achieve the highly compacted elongated nucleus, the chromatin is remodeled by a set of abundant transition proteins (TPs) subsequently replaced by the protamines (PRMs). The PRMs bind DNA, neutralizing the phosphodiester backbone of the double helix [2] and allowing a tight compaction of the DNA as torroids [3]. Round spermatids massively synthesize mRNAs under the rigorous control of several cell-specific transcription factors. These mRNAs are stored to be translated at later steps when chromatin remodeling no longer supports transcription.

M.-C. Grégoire (✉)
Department of Biochemistry, University of Sherbrooke, Sherbrooke, QC, Canada
e-mail: marie-chantal.gregoire@usherbrooke.ca

A. Zini and A. Agarwal (eds.), *Sperm Chromatin: Biological and Clinical Applications in Male Infertility and Assisted Reproduction*, DOI 10.1007/978-1-4419-6857-9_21,

Chromatin Remodeling in Spermatids

Specific Histones and Histone Variants Present During Spermiogenesis

To achieve the tightly compacted structure of the nucleus, several differentiation steps are needed from the somatic-like histone-bound chromatin structure to the large-scale genome compaction provided by PRMs late during spermiogenesis. In different species, several histone variants are exclusively expressed in male germ cells [4]. Interestingly, incorporation of one of the many testis-specific histone variant is thought to form nucleosomes with lower stability than those containing canonical histones [5–7]. These testis-specific histones include H1 variants [8–10] (H1T, H1T2, HILS1), H2A variants [11, 12] (mouse: H2AL1, H2AL2, H2AL3; human: H2A. Bbd), H2B variants [11, 13–15] (mouse: H2BL1, H2BL2, TH2B ; human: hTSH2B, H2BFWT), and H3 variants [16, 17] (H3T). Apart from these testis-specific structural histones, other noncanonical variants shared by other tissues also exist. For instance, H2AFX plays a role in the DNA damage response [18, 19], while H3F3A and H3F3B are involved in histone replacement and chromatin regulation [20, 21]. Also, CENPA and H2AZ are also present during spermatogenesis, being involved in centromeric structure and gene activation, respectively [22]. The majority of these variants may participate in the progressive inhibition of transcription and in the correct DNA compaction, as well as morphological changes of the spermatid nuclei.

Posttranslational Modifications and Their Contribution to the Remodeling Program

In addition to the incorporation of histone variants, posttranslational modifications (PTM) of histones, either alone or in combination, are important for the successful completion of spermiogenesis. PTM such as acetylation, ubiquitination, phosphorylation, methylation, and sumoylation may add to the remarkable plasticity of the spermatidal chromatin. It has been shown that massive H3 (unpublished data, Leduc and Boissonneault) and H4 hyperacetylation is observed at chromatin remodeling steps in spermatids, which would provide a better context for histone withdrawal by lowering their affinity for DNA and establish a more open chromatin structure [23–28]. For somatic cells, it has been shown that histone ubiquitination is also associated with destabilization of nucleosomes, in relation to active gene transcription [29]. In elongating spermatids, ubiquitinated forms of H2A and H3 were shown [30, 31] while the absence of the ubiquitin ligase RNF8 has been shown to impair the removal of histones leading to infertility [32]. While the phosphorylation of H2AFX on serine 139 (γH2AFX, previously known as γH2AX) has been observed throughout spermatogenesis [33], elongating spermatids seems to be particularly enriched in this histone variant, at steps associated with detection of DNA strand breaks [18, 19]. Moreover, Krishnamoorthy and colleagues [34] reported that phosphorylation of histone H4 at serine 1 is essential for chromatin compaction in yeast. Interestingly, they also reported that this modification is present during mouse spermiogenesis and disappears in elongating spermatids when TP2 is translated. Finally, lysine methylation, known to be involved in transcriptional regulation and the propagation of chromosome stability [35], was reported in elongating spermatids [35]. More specifically, the onset of spermatid elongation is characterized by mono-, di-, and tri-methylation of lysine 4 on histone H3 (H3K4) accompanied by an increase in the lysine-specific histone demethylase AOF2, also coincident with the chromatin remodeling process [36]. In addition, trimethylation of lysine 9 on histone H3 (H4K9me3) and lysine 20 on histone H4 (H4K20me3) were reported to occur at chromocenters following the onset of nuclear elongation in spermatids [37]. These observations suggest that the timely methylation of histone lysines plays a key role in the chromatin remodeling process. Furthermore, sumoylation pathway is also regulated and expressed in the elongating spermatids, but its contribution remains unclear [38, 39]. Hence, PTM of histones seem to be essential to orchestrate the nucleosome-to-PRM transition leading to efficient compaction of the male haploid genome.

Nuclear Proteins Transition

While the histone variants incorporation in nucleosomes and the posttranslational histone modification are known to destabilize the nucleosome–DNA interactions, the mechanism controlling the transition from a nucleosome-based chromatin to such a densely packed nucleus is yet unknown. In most mammals, nucleosomes are first replaced by TPs and then PRMs [40]. In vitro studies showed that when the DNA–nucleosome interactions are being disrupted by either histone PTM or histone variants, both the TPs or PRMs are able to replace DNA-bound nucleosome, since they have a higher affinity for DNA [41, 42]. By contrast, in vivo studies have recently shown that histone exchange occur normally in mice lacking both TPs, suggesting that the latter proteins may be accessory to the process [43]. To efficiently pack the genome, haploid cells are expressing positively charged PRMs, which efficiently neutralize the DNA phosphate backbone, allowing to bring adjacent DNA molecules in close juxtaposition. Protamination is, however, necessary, as alteration in the PRM level such as those resulting from haploinsufficiency induced in mice may lead to infertility [44]. Normal protamination of the spermatid nucleus provides both chemical and mechanical stability to the haploid genome [45] throughout their transit to fertilization [46, 47].

Endogenous DNA Breaks as Part of the Normal Differentiation Program of Spermatids

A topological transition occurs between a nucleosome-based supercoiled chromatin to a PRM-based tightly compacted linear structure, as the cell must remove most of the negative DNA supercoiling in the process [48, 49]. Since DNA is bound to the nuclear matrix and wrapped around nucleosomes, DNA breaks could provide the necessary swivel to relieve torsional stress [50].

Detection and Characterization of DNA Breaks in Elongating Spermatids

As early as 1981, reports suggested that some DNA damage in the form of strand breaks was associated with the massive chromatin remodeling in elongating spermatids, since endogenous DNA polymerases activity was detected [51–56]. More recently, our group has established that DNA breaks are present in the whole population of fertile mouse and human spermatids and are, therefore, part of the normal differentiation program of these cells [26]. Both nick translation and terminal deoxynucleotidyl transferase-mediated dUTP nick-end labeling (TUNEL) were used to demonstrate the presence of free 3'OH groups. As both techniques can potentially label single- and double-strand breaks, earlier reports could not distinguish between these types of DNA damage. However, single-cell gel electrophoresis, also known as the comet assay, performed in either neutral or alkaline conditions suggested that transient double-stranded breaks are created in elongating spermatids [57].

Possible Origins of DNA Breaks

As stated above, DNA breaks either single- or double-stranded would be expected to relieve the torsional stress induced by the withdrawal of histones leaving free supercoils [58]. One possibility is that the mechanical stress itself could induce the breaks as the chromatin remodeling is extensive and takes place within a few differentiation steps. Enzymatic induction of DNA strand breaks is most likely as they can be end-labeled with enzymes using 3′OH as substrate. Topoisomerases have long been considered as likely candidates to support chromatin remodeling because of their ubiquitous role in chromosome dynamics during the somatic cell cycle.

Type II Topoisomerases as Likely Candidates

Change in DNA topology can be achieved by single-stranded breaks changing the linking number in steps of one. Single-stranded breaks induced by type I topoisomerases, would be considered a much smaller threat on the genome's integrity than a DSB generated by type II topoisomerases. However, Roca and Mezquita demonstrated more than 30 years ago that type I topoisomerase activity was largely associated with transcription, whereas type II topoisomerase activity was observed throughout spermatogenesis and particularly present at stages of spermatidal DNA compaction in chicken [59–62]. Similar conclusions were drawn from the study of rat spermatogenesis [52, 53, 63]. The presence of topoisomerases II in rat elongating spermatids was confirmed by immunoblots and its expected nuclear localization by immunofluorescence. Interestingly, it was also demonstrated that elongating spermatids had topoisomerase II of lower molecular weight (142 and 148 kDa), whereas bands of 170 and 177 kDa were observed in round spermatids, which correspond to the α and β isoforms, respectively. Although this observation has not yet been confirmed in other species, it raises the possibility of an atypical topoisomerase activity in elongating spermatids (see below).

Using purified elongating spermatids nuclei, we also demonstrated that type II topoisomerase inhibitors, such as suramin and etoposide, abolished TUNEL positivity, suggesting that most DNA strand breaks originate from type II topoisomerase activity [57]. Topoisomerase IIβ foci were observed in elongating spermatids, whereas topoisomerase IIα remained undetected [18]. In mammal somatic cells, the topoisomerase α and β are differentially expressed; topoisomerase IIα is mostly found in replicating cells, whereas topoisomerase IIβ predominates in quiescent cells [64, 65]. Hence, detection of topoisomerase IIβ in elongating spermatids is not surprising, as spermatids are nonreplicative cells. Topoisomerase IIβ was also found in spermatozoa and is considered to be part of the nuclear matrix, supporting a role in the chromatin remodeling of spermatids [66].

Alternatively, one interesting possibility is that DSB could be induced by retrotransposon nucleases that are expressed throughout spermatogenesis and also detected in the nucleus of spermatids [59–61]. The open chromatin induced by the PTM of histones may present an ideal opportunity for such nucleases and retrotransposition in general.

DNA Breaks and DNA Packaging: The Chicken or the Egg?

Observations in infertile men and transgenic mice models demonstrated that low PRM content in sperm or altered PRM1–PRM2 ratio is associated with infertility [44, 67–71]. In addition, altered sperm chromatin correlates with high level of DNA strand breaks. Since sperm chromatin is preferentially established in elongating spermatids steps, this suggests a link between this important transition and the final genetic integrity of the mature gamete. A less compacted sperm nucleus would be more vulnerable to any chemical or physical insults, such as those resulting from reactive oxygen species [72]. Using a double knockout mouse model, Zhao and colleagues demonstrated that the absence of both TP1 and TP2 seriously compromises chromatin condensation, leading to infertility [43]. Interestingly, DNA breaks were found to persist beyond the normal chromatin remodeling steps. DNA strand breaks were observed primarily in less condensed nuclei of an atypical heterogeneous population of spermatids therefore supporting the link between condensation and DNA integrity. It is noteworthy that mice lacking only one of the TPs were fertile as one TP partially compensates for the absence of the other.

Transition proteins are known to enhance DNA ligation activity in vitro [73]. They may act as a linker and provide the proper scaffold for DNA repair processes in a histone-depleted chromatin environment. So, condensing proteins such as TPs and PRMs may serve a dual purpose by condensing the nucleus and improving DNA repair. Moreover, in vitro interaction assay has recently been used to demonstrate that PARP2, a poly(ADP-ribosyl) polymerase involved in DNA repair and apoptosis, interacts with TP2, whereas PARP1 was found to poly(ADP-ribosyl)ate HSPA2, a newly identified transition protein chaperone of the Hsp70 family. Similarly, PARP family members may also play a dual role in DNA repair and chromatin remodeling. PARPs may facilitate transition proteins incorporation in the spermatidal chromatin by poly(ADP-ribosyl) ation of histones, inducing both chromatin relaxation and modulation of TP chaperones. Although normally present at later steps, it is likely that PRMs play a similar role as the TPs in preserving genetic integrity, as they share the same overall DNA-binding properties.

DNA Damage Response and DNA Repair Processes in Spermatids

DNA Damage Response

In higher eukaryotes, γH2AFX is a universal biomarker of double-strand breaks and is considered one of the most reliable signatures of an active DNA damage response [74, 75]. This PTM appears less than 3 min after the occurrence of a DSB and may serve as a recognition pattern to help recruit DNA repair proteins at the break site [75, 76]. γH2AFX foci were initially reported to be detected during the chromatin remodeling steps of rat spermiogenesis [19], and we later confirmed the presence of similar foci during spermiogenesis of both mice [18] and humans (unpublished observations, Leduc and Boissonneault). Based on our recent immunofluorescence data in mouse, γH2AFX immunolabeling is found distributed throughout the nuclei of steps 10 and 11 spermatids [18]. In somatic and germinal cells, this modification spreads up to a megabase surrounding the DSB site [33, 77]. The global distribution of γH2AFX appears not surprising, since to sustain a global change in DNA topology, one would assume that DNA breaks must be distributed throughout most of the genome and that the phosphorylation of H2AFX will follow accordingly. One hypothesis is that such DSB could localize at the bases of matrix attachment regions (MARs) known to be rich in topoisomerase IIβ [78]. In sperm cells, the loops circumscribed by MARs are thought to range between 40 and 50 kb [79]. If a DNA break occurs every 40–50 kb, it is most probable that a majority of the genome would be covered by γH2AFX in elongating spermatids.

Although members of the phosphatidylinositol 3-kinase family, such as ATM, ATR, and DNA-PKcs, are known to phosphorylate H2AFX, other kinases could also spread this PTM in such a unique chromatin context. For example, SSTK (small serine/threonine protein kinase) can phosphorylate in vitro H2AFX amongst other histones [80]. Furthermore, SSTK null mutant mice display a condensation defect during spermiogenesis supporting its role in the chromatin remodeling of spermatids. More research is needed to identify the apical kinase involved.

Do Topoisomerases Trigger DNA Damage Response?

Topoisomerase II activity should not normally trigger the activation of H2AFX because the enzyme catalytic cycle involves cleavage and ligation with an intermediate where both 5′ termini are covalently attached to the enzyme [81], therefore never really leaving a recognizable DSB. As type II topoisomerases are considered as potential inducers of DSBs in elongating spermatids, there is an interesting possibility that a faulty enzyme variant, unable to carry out the full catalytic cycle, leaves unrepaired DSBs. Such a variant would be generated by (1) alternative splicing or specific proteolytic cleavage leading to lower molecular topoisomerases, (2) PTM, (3) separation of the homodimer due to extended unwinding, or (4) incomplete catalysis because of the chromatin context. Indeed, the presence of the tyrosyl phosphodiesterase (TDP1) distributed as foci in the nuclei of elongating spermatids suggests an atypical topoisomerase activity, as TDP1 is known to remove topoisomerase adducts by efficient cleavage of the 3′-phosphotyrosyl bonds (type I topoisomerase adducts) as well as 5′-phosphotyrosyl bonds of stalled type II topoisomerases albeit to a lower extent [82–84].

We then proposed that TDP1 could remove stalled topoisomerase IIβ, leaving a DSB that can be signaled by the phosphorylation of H2AFX [18]. Recently, TTRAP (TRAF and TNF receptor-associated protein) has been identified in humans as a 5′ tyrosyl phosphodiesterase [85], which may represent a more likely candidate to remove topoisomerase IIβ adducts. The status of spermatidal topoisomerases is clearly in need of further investigations.

DNA Repair Mechanisms in Spermatids

As spermatids are haploid cells and cannot rely on HR due to the lack of sister chromatid, DSB repair processes must involve error-prone pathways. These pathways include nonhomologous end joining either DNA-PKcs-dependent (NHEJ-D) or its backup pathway (NHEJ-B), single-strand annealing (SSA), or microhomology-mediated end joining (MMEJ) (Table 21.1). The pathways involved in the repair of endogenous DSB in spermatids are still unknown. If a typical end-joining process is identified, this may reveal a new source of genetic instability in these cells, as such processes can induce deletions and insertions. Alternatively, because of the potential for these cells to generate progeny, it is conceivable that they evolved a more reliable end-joining mechanism that would prevent subtle mutations to be transmitted to the next generation.

Table 21.1 Factors associated with known DNA double-strand break repair pathways

Double-strand break repair pathways	Proteins involved
Homologous recombination [86]	RPA, RAD51, RAD52, RAD54, BRCA1, BRCA2
Nonhomologous end joining, DNA-PKcs-dependent pathway [86, 87]	KU70, KU80, DNA-PKcs, XRCC4, LIGIV, XFL
Nonhomologous end joining, backup pathway [88]	PARP1, XRCC1, LigIII
Single-strand annealing [89]	RPA, RAD52, ERCC1/XPF
Microhomology-mediated end joining	Unknown

The participation of TPs and PRMs in these pathways may enhance reliability of the DNA repair mechanism to be identified.

Nonhomologous End joining

The end-joining repair processes are repressed throughout the meiotic stages of spermatogenesis to promote HR. Such a repression is no longer present during spermiogenesis [86–88]. Although much remain to be known about the repair of endogenous DSB, round spermatids apparently rely on the NHEJ-B pathway to repair radiation-induced DSBs but with slower kinetics than in somatic cells [88, 89]. DNA-PKcs is an important kinase of the NHEJ-D pathway. The later pathways also seem to be involved in spermatidal DSB repair, as DNA-PKcs-deficient SCID mice demonstrated lower repair rates of γH2AFX foci following irradiation.

Evidence of NHEJ-D was reported during spermiogenesis of several grasshopper species as established by the immunofluorescence detection of KU70 and γH2AFX nuclear foci [90]. Further confirmation of this pathway will be needed, as KU proteins also play a role in telomere maintenance [91, 92]. Specialized DNA polymerases, such as polymerases of the X family, polymerase μ and polymerase λ, are also involved in the repair of DSBs by NHEJ, as they process incompatible ends, fill gaps, and remove unwanted flaps [86]. The sole detection of DNA repair factors by immunological techniques do not implicate that they are functional. However, using in situ incorporation of biotinylated dUTP, we have confirmed an endogenous DNA polymerase activity in elongating spermatids of mice, leading to the conclusion of an active repair process [18].

Polymerases of the PARP family, PARP1 and PARP2, are often referred to as guardians of genome integrity [93, 94]. PARPs are chromatin-associated proteins activated by DNA strand breaks. Upon activation, they catalyze the covalent attachment of ADP-ribose from NAD+ substrate to a number of proteins, such as histones, TP53, topoisomerases, and even themselves. This automodification releases PARPs from DNA and can be reversed by the poly(ADP-ribose) glycohydrolase (PARG). PARP1 participates in the base excision repair (BER) and also in the NHEJ backup pathways.

Considering that PARP1 and PARP2 have overlapping functions and that a double-knockout of these proteins is embryonic lethal, it is difficult to study their individual role during spermiogenesis. Inactivation of PARP2 in mice leads to hypofertility, as pachytene spermatocytes display defective meiotic sex chromosome inactivation. Compromised differentiation of spermatids can also be observed [95]. Knockout mice for PARP1, PARG (110 kDa isoform) or both displayed abnormal sperm with varying degrees of residual DNA breaks [96]. Not as striking as one could have expected, the perturbation of the poly(ADP-ribose) metabolism clearly impacts the differentiation program of spermatids.

DNA Repair by Homology in a Haploid Cell

The two other end-joining pathways, SSA and MMEJ, use repetitive DNA and microhomology, respectively, as a template to repair DSB. Although very different from one another, these two systems inevitably introduce errors in the DNA sequence mostly by deletions. The SSA pathway shares several proteins with HR, and the two pathways usually compete against each other in somatic cells [97], a situation that should not prevail in spermatids. Repair of a DSB by the SSA pathway proceeds from long homologous sequences (>30 nucleotides) and the one copy of the repeat sequence and the intervening sequence serving as a template are destroyed upon completion of the repair [98]. In MMEJ, the KU-independent end joining is mediated by a 5–25 nucleotides homology resulting in deletions of sequences, and sometimes insertions, close to the break site [99]. Although MMEJ deletions are smaller than those usually created by SSA, this will, nonetheless, lead to an alteration of the genome's integrity.

Highly Conserved Process Among Higher Eukaryotes

A rapid survey of the recent literature points to the highly conserved nature of the DNA damage response to endogenous breaks in spermatids. Evidence of DNA damage response was presented in mammalian models, such as mice and rats, but it can also be extended to human as we have shown. Rahtke and colleagues observed DNA breaks during spermiogenesis of *Drosophila* [100], whereas others demonstrated that spermatids of several grasshopper species displayed KU70 and γH2AFX foci [90]. Most interestingly, γH2AFX foci was also reported in spermatids of the algae *Chara vulgaris* [101], suggesting that a related process extends to plants. Hence, this process could very well be used throughout the eukaryotic world where gametogenesis requires condensation of the genetic material.

Possible Consequences and Clinical Relevance

Impairment of Genetic Integrity in the Male Gamete

The generation of a transient more "open" chromatin structure during spermiogenesis and the presence of DSBs in such a striking chromatin-remodeling context make it possible that more important genomic alterations could be observed. Interestingly, many studies reported that more than 80% of the structural de novo chromosome aberrations are of paternal origin [102–104]. In healthy men's sperm, the spontaneous frequencies of structural chromosomal abnormalities was shown to be higher than those of numerical abnormalities, and chromosomal breaks are more prevalent than partial duplications and deletions [105].

It is well known that lifestyle factors such as smoking, alcohol, and caffeine consumption have been associated with chromosomal aberration and genomic alterations in somatic cells [106–111]. While several studies showed a deleterious effect of lifestyle factors on the male fertility, only a few studies focused on the effect of tobacco smoking and alcohol consumption on male germ cells' genetic integrity and showed unclear correlations with sperm aneuploidy and DNA fragmentation [112–115]. However, Schmid and colleagues showed that caffeine consumption is associated with increased DSBs in sperm [116]. Interestingly, caffeine might lead to inactivation of H2AFX through the inhibition of kinases related to DNA repair, such as ATM, ATR, and DNA-PKcs [117–119].

Aging was associated with increased genetic alterations and chromosomal aberrations in sperm, suggesting a less efficient DNA packaging process [116, 120–122]. Altogether, these studies suggest that some environmental and lifestyle factors may likely result in chromosomal aberrations, persistent DNA breaks, and genetic impairments in mature spermatozoa, leading to dramatic consequences on the reproductive outcome.

Given the peculiar chromatin of spermatids, one can assume that such a context may favor chromosomal translocation due to the proximity of the breaks if generated by a nuclear matrix associated type II topoisomerases and the DNA repair pathways available. Interestingly, the natural rate of chromosomal aberrations as seen in untreated controls and reported by some studies monitoring the effects of some toxicants is quite high ranging from 0.7 to 5% [123–125].

Retrotransposition is another interesting mechanism of genetic instability potentially occurring in spermiogenesis. Testicular expression of the ORF1 and ORF2 proteins encoded by LINE1 sequence has been demonstrated, particularly in the early steps of spermiogenesis [60]. Knowing that ORF2 protein has an endonuclease activity [126, 127], Gasior and colleagues showed that LINE1 ORFs expression leads to a high level of DSB formation and activation of H2AFX [61]. In addition, it was shown that LINE retrotransposition in transformed human cells can lead to a variety of genomic rearrangements [128]. Together, these findings makes it tempting to speculate that the spermatidal chromatin remodeling would offer a suitable context for retrotransposition, increasing the repertoire of possible mutations distributed throughout the millions of sperm cells.

Finally, as the human genome is composed of nearly 50% of repeated DNA, microhomology-based DNA repair pathways such the SSA and MMEJ described above may prove to play an important role in spermatids and be the cause of several genetic diseases and cancers, as mutagenic deletions often share homology at breakpoint junctions, such as Alu and LINE repeats [129, 130]. For instance, microdeletions in the highly repetitive Y chromosome seem important in the etiology of infertility [131, 132] and may bear also the signature of these alternative mutagenic DNA repair systems.

Impact of This Transient Window of Genetic Instability on Clinical Practices

In contrast to spermatocytes, spermatids are apparently devoid of cell cycle checkpoints. Their differentiation program can be compared to an assembly line where defective products will be discarded through their lack of fitness for fertilization. Moreover, spermatids have a scheduled differentiation program most probably synchronized by Sertoli cells. Any delay in the process is likely to have consequences for the gamete's integrity. Therefore, procedures that bypass the natural selection of gametes, such as ICSI, ROSI, or IVF to a lower extent, bear the risk of selecting unfit gametes.

Although they possess a haploid genome, round spermatids are less compatible with artificial reproduction techniques (ART) as demonstrated by the low successful birth rate following ROSI in mouse (1.7–28.2%) [133, 134]. Recently, it has been shown that 77.5% of the ROSI-generated embryos exhibited abnormal chromosome segregation at the first mitosis, originating from double-strand breakage of the male-derived genomic DNA. ICSI and ROSI procedures resulted in no embryonic development when chromosome segregation was abnormal at the first mitotic division [135]. Therefore, residual DNA breaks in the male gamete may lead to abnormal chromosome segregation and genetic impairment in the developing zygote. Moreover, taking into account that the remodeling process is accompanied with DNA breaks, one can assume that selecting spermatids undergoing this transition should lead to unsuccessful reproductive outcomes. Unfortunately, when ROSI technique is performed in humans, one cannot avoid selecting spermatids undergoing chromatin remodeling, as they have the same apparent morphology that of those immediately preceding or following these crucial steps.

Potential Recovery by the Oocyte After Fertilization

Autosomal aneuploidies are more frequently of maternal origin, whereas point mutations and chromosomal rearrangements are of paternal origin [136, 137]. Moreover, it was shown that the DNA repair capacity of spermatids declines drastically after the nuclear remodeling and continues to decline until spermiation [138]. On the contrary, the repair capacities of the oocyte are quite stable throughout oogenesis and persist after fertilization and may repair DNA damages from both parental genomes [139, 140]. Using first-cleavage metaphases, it was shown that both NHEJ and HR are used by the oocyte to rescue the genetic integrity of the paternal genome after fertilization. However, not all DNA breaks were efficiently repaired, as many residual chromosomal aberrations were found in controls. Thus, even if the oocyte can repair some paternal DNA lesions, chromosomal aberrations can persist after the first zygotic cell cycle [123]. Moreover, if the DNA repair systems of the spermatid create point mutations or chromosomic rearrangements, these will likely escape the oocyte's damage response and will be transmitted to the next generation.

Summary

Altogether, this review suggests that spermiogenesis has probably been overlooked as an important source of genetic instability that can provide a repertoire of mutations distributed through millions of spermatozoa, each having the potential to transfer genetic alterations to the next generation. Further investigation will be needed to establish whether this could be considered as a new component of evolution.

Acknowledgments Funded by the Natural Sciences and Engineering Research Council of Canada (grant # 155182) to G.B.

References

1. Loonie D. Russell APSH, Robert Ettlin. Histological and histopathological evaluation of the testis: Cache River Press; 1990.
2. Balhorn R. A model for the structure of chromatin in mammalian sperm. J Cell Biol. 1982 May 1;93(2):298–305.
3. Ward W. Deoxyribonucleic acid loop domain tertiary structure in mammalian spermatozoa. Biol Reprod. 1993;48(6):1193–201.
4. Talbert PB, Henikoff S. Histone variants–ancient wrap artists of the epigenome. Nat Rev Mol Cell Biol. 2010 Apr 1;11(4):264–75.
5. Li A, Maffey AH, Abbott WD, Conde e Silva N, Prunell A, Siino J, et al. Characterization of nucleosomes consisting of the human testis/sperm-specific histone H2B variant (hTSH2B). Biochemistry. 2005 Feb 22;44(7):2529–35.
6. Syed SH, Boulard M, Shukla MS, Gautier T, Travers A, Bednar J, et al. The incorporation of the novel histone variant H2AL2 confers unusual structural and functional properties of the nucleosome. Nucleic Acids Res. 2009 Aug 1;37(14):4684–95.
7. González-Romero R, Méndez J, Ausió J, Eirín-López JM. Quickly evolving histones, nucleosome stability and chromatin folding: all about histone H2A.Bbd. Gene. 2008 Apr 30;413(1–2):1–7.
8. Seyedin SM, Kistler WS. Isolation and characterization of rat testis H1t. An H1 histone variant associated with spermatogenesis. J Biol Chem. 1980 Jun 25;255(12):5949–54.
9. Tanaka H, Iguchi N, Isotani A, Kitamura K, Toyama Y, Matsuoka Y, et al. HANP1/H1T2, a novel histone H1-like protein involved in nuclear formation and sperm fertility. Mol Cell Biol. 2005;25(16):7107–19.
10. Yan W, Ma L, Burns KH, Matzuk MM. HILS1 is a spermatid-specific linker histone H1-like protein implicated in chromatin remodeling during mammalian spermiogenesis. Proc Natl Acad Sci USA. 2003 Sep 2;100(18):10546–51.
11. Govin J, Escoffier E, Rousseaux S, Kuhn L, Ferro M, Thévenon J, et al. Pericentric heterochromatin reprogramming by new histone variants during mouse spermiogenesis. J Cell Biol. 2007 Jan 29;176(3):283–94.
12. Chadwick BP, Willard HF. A novel chromatin protein, distantly related to histone H2A, is largely excluded from the inactive X chromosome. J Cell Biol. 2001 Jan 22;152(2):375–84.

13. Shires A, Carpenter MP, Chalkley R. A cysteine-containing H2B-like histone found in mature mammalian testis. J Biol Chem. 1976 Jul 10;251(13):4155–8.
14. Zalensky AO, Siino JS, Gineitis AA, Zalenskaya IA, Tomilin NV, Yau P, et al. Human testis/sperm-specific histone H2B (hTSH2B). Molecular cloning and characterization. J Biol Chem. 2002 Nov 8;277(45):43474–80.
15. Churikov D, Siino J, Svetlova M, Zhang K, Gineitis A, Morton Bradbury E, et al. Novel human testis-specific histone H2B encoded by the interrupted gene on the X chromosome. Genomics. 2004 Oct 1;84(4):745–56.
16. Franklin SG, Zweidler A. Non-allelic variants of histones 2a, 2b and 3 in mammals. Nature. 1977 Mar 17;266(5599):273–5.
17. Witt O, Albig W, Doenecke D. Testis-specific expression of a novel human H3 histone gene. Exp Cell Res. 1996 Dec 15;229(2):301–6.
18. Leduc F, Maquennehan V, Nkoma GB, Boissonneault G. DNA damage response during chromatin remodeling in elongating spermatids of mice. Biol Reprod. 2008 Feb;78(2):324–32.
19. Meyer-Ficca M, Scherthan H, Burkle A, Meyer R. Poly(ADP-ribosyl)ation during chromatin remodeling steps in rat spermiogenesis. Chromosoma. 2005;114(1):67–74.
20. Bramlage B, Kosciessa U, Doenecke D. Differential expression of the murine histone genes H3.3A and H3.3B. Differentiation. 1997 Oct 1;62(1):13–20.
21. Elsaesser SJ, Goldberg AD, Allis CD. New functions for an old variant: no substitute for histone H3.3. Curr Opin Genet Dev. 2010 Feb 11.
22. Zalensky AO, Breneman JW, Zalenskaya IA, Brinkley BR, Bradbury EM. Organization of centromeres in the decondensed nuclei of mature human sperm. Chromosoma. 1993 Sep 1;102(8):509–18.
23. Oliva R, Mezquita C. Histone H4 hyperacetylation and rapid turnover of its acetyl groups in transcriptionally inactive rooster testis spermatids. Nucleic Acids Res. 1982 Dec 20;10(24):8049–59.
24. Christensen M, Rattner J, Dixon G. Hyperacetylation of histone H4 promotes chromatin decondensation prior to histone replacement by protamines during spermatogenesis in rainbow trout. Nucleic Acids Res. 1984;12(11):4575–92.
25. Grimes S, Henderson N. Hyperacetylation of histone H4 in rat testis spermatids. Exp Cell Res. 1984;152(1):91–7.
26. Marcon L, Boissonneault G. Transient DNA strand breaks during mouse and human spermiogenesis new insights in stage specificity and link to chromatin remodeling. Biol Reprod. 2004 Apr;70(4): 910–8.
27. Meistrich M, Trostle-Weige P, Lin R, Bhatnagar Y, Allis C. Highly acetylated H4 is associated with histone displacement in rat spermatids. Mol Reprod Dev. 1992;31(3):170–81.
28. Hazzouri M, Pivot-Pajot C, Faure A, Usson Y, Pelletier R, Sele B, et al. Regulated hyperacetylation of core histones during mouse spermatogenesis: involvement of histone deacetylases. Eur J Cell Biol. 2000;79(12):950–60.
29. Li W, Nagaraja S, Delcuve GP, Hendzel MJ, Davie JR. Effects of histone acetylation, ubiquitination and variants on nucleosome stability. Biochem J. 1993 Dec 15;296 (Pt 3):737–44.
30. Baarends W, Hoogerbrugge J, Roest H, Ooms M, Vreeburg J, Hoeijmakers J, et al. Histone ubiquitination and chromatin remodeling in mouse spermatogenesis. Dev Biol. 1999;207(2):322–33.
31. Chen HY, Sun JM, Zhang Y, Davie JR, Meistrich ML. Ubiquitination of histone H3 in elongating spermatids of rat testes. J Biol Chem. 1998 May 22;273(21):13165–9.
32. Lu L-Y, Wu J, Ye L, Gavrilina GB, Saunders TL, Yu X. RNF8-dependent histone modifications regulate nucleosome removal during spermatogenesis. Dev cell. 2010 Feb 10.
33. Blanco-Rodríguez J. GammaH2AX marks the main events of the spermatogenic process. Microsc Res Tech. 2009 Apr 29.
34. Krishnamoorthy T, Chen X, Govin J, Cheung W, Dorsey J, Schindler K, et al. Phosphorylation of histone H4 Ser1 regulates sporulation in yeast and is conserved in fly and mouse spermatogenesis. Genes Dev. 2006;20(18):2580.
35. Sims RJ, Nishioka K, Reinberg D. Histone lysine methylation: a signature for chromatin function. Trends Genet. 2003 Nov 1;19(11):629–39.
36. Godmann, Auger, Ferraroni-Aguiar, Sauro D, Sette, Behr, et al. Dynamic regulation of histone H3 methylation at lysine 4 in mammalian spermatogenesis. Biol Reprod. 2007 Jul 18.
37. van der Heijden G, Derijck A, Ramos L, Giele M, van der Vlag J, de Boer P. Transmission of modified nucleosomes from the mouse male germline to the zygote and subsequent remodeling of paternal chromatin. Dev Biol. 2006 Aug 5;298(2):458–69.
38. Vigodner M, Morris P. Testicular expression of small ubiquitin-related modifier-1 (SUMO-1) supports multiple roles in spermatogenesis: silencing of sex chromosomes in spermatocytes, spermatid microtubule nucleation, and nuclear reshaping. Dev Biol. 2005;282(2):480–92.
39. La Salle S, Sun F, Zhang X-D, Matunis MJ, Handel MA. Developmental control of sumoylation pathway proteins in mouse male germ cells. Dev Biol. 2008 Sep 1;321(1):227–37.
40. Balhorn R, Weston S, Thomas C, Wyrobek A. DNA packaging in mouse spermatids. Synthesis of protamine variants and four transition proteins. Exp Cell Res. 1984;150(2):298–308.
41. Marushige K, Marushige Y, Wong TK. Complete displacement of somatic histones during transformation of spermatid chromatin: a model experiment. Biochemistry. 1976 May 18;15(10):2047–53.

42. Oliva R, Mezquita C. Marked differences in the ability of distinct protamines to disassemble nucleosomal core particles in vitro. Biochemistry. 1986 Oct 21;25(21):6508–11.
43. Zhao M, Shirley C, Hayashi S, Marcon L, Mohapatra B, Suganuma R, et al. Transition nuclear proteins are required for normal chromatin condensation and functional sperm development. Genesis. 2004 Apr 15;38(4):200–13.
44. Cho C, Willis WD, Goulding EH, Jung-Ha H, Choi YC, Hecht NB, et al. Haploinsufficiency of protamine-1 or −2 causes infertility in mice. Nat Genet. 2001 May 1;28(1):82–6.
45. Sotolongo B, Lino E, Ward W. Ability of hamster spermatozoa to digest their own DNA. Biol Reprod. 2003;69(6):2029–35.
46. Kuretake S, Kimura Y, Hoshi K, Yanagimachi R. Fertilization and development of mouse oocytes injected with isolated sperm heads. Biol Reprod. 1996 Oct 1;55(4):789–95.
47. Tateno H, Kimura Y, Yanagimachi R. Sonication per se is not as deleterious to sperm chromosomes as previously inferred. Biol Reprod. 2000 Jul 1;63(1):341–6.
48. Risley MS, Einheber S, Bumcrot DA. Changes in DNA topology during spermatogenesis. Chromosoma. 1986 Jan 1;94(3):217–27.
49. Ward WS. The structure of the sleeping genome: implications of sperm DNA organization for somatic cells. J Cell Biochem. 1994 May;55(1):77–82.
50. Laberge RM, Boissonneault G. Chromatin remodeling in spermatids: a sensitive step for the genetic integrity of the male gamete. Arch Androl. 2005 Mar-Apr;51(2):125–33.
51. Hecht N, Parvinen M. DNA synthesis catalysed by endogenous templates and DNA-dependent DNA polymerases in spermatogenic cells from rat. Exp Cell Res. 1981;135(1):103–14.
52. McPherson S, Longo F. Localization of DNase I-hypersensitive regions during rat spermatogenesis: stage-dependent patterns and unique sensitivity of elongating spermatids. Mol Reprod Dev. 1992 Apr 1;31(4):268–79.
53. McPherson S, Longo F. Nicking of rat spermatid and spermatozoa DNA: possible involvement of DNA topoisomerase II. Dev Biol. 1993;158(1):122–30.
54. McPherson S, Longo F. Chromatin structure-function alterations during mammalian spermatogenesis: DNA nicking and repair in elongating spermatids. European journal of histochemistry : EJH. 1993 Jan 1;37(2):109–28.
55. Sakkas D, Manicardi G, Bianchi P, Bizzaro D, Bianchi U. Relationship between the presence of endogenous nicks and sperm chromatin packaging in maturing and fertilizing mouse spermatozoa. Biol Reprod. 1995;52(5):1149–55.
56. Iseki S. DNA strand breaks in rat tissues as detected by in situ nick translation. Exp Cell Res. 1986 Dec 1;167(2):311–26.
57. Laberge R, Boissonneault G. On the nature and origin of DNA strand breaks in elongating spermatids. Biol Reprod. 2005;73(2):289–96.
58. Boissonneault G. Chromatin remodeling during spermiogenesis: a possible role for the transition proteins in DNA strand break repair. FEBS Lett. 2002 Mar 13;514(2–3):111–4.
59. Branciforte D, Martin SL. Developmental and cell type specificity of LINE-1 expression in mouse testis: implications for transposition. Mol Cell Biol. 1994 Apr 1;14(4):2584–92.
60. Ergün S, Buschmann C, Heukeshoven J, Dammann K, Schnieders F, Lauke H, et al. Cell type-specific expression of LINE-1 open reading frames 1 and 2 in fetal and adult human tissues. J Biol Chem. 2004 Jun 25;279(26):27753–63.
61. Gasior SL, Wakeman TP, Xu B, Deininger PL. The human LINE-1 retrotransposon creates DNA double-strand breaks. J Mol Biol. 2006 Apr 14;357(5):1383–93.
62. Roca J, Mezquita C. DNA topoisomerase II activity in nonreplicating, transcriptionally inactive, chicken late spermatids. EMBO J. 1989 Jun;8(6):1855–60.
63. Chen J, Longo F. Expression and localization of DNA topoisomerase II during rat spermatogenesis. Mol Reprod Dev. 1996 Sep 1;45(1):61–71.
64. Morse-Gaudio M, Risley MS. Topoisomerase II expression and VM-26 induction of DNA breaks during spermatogenesis in *Xenopus laevis*. J Cell Sci. 1994 Oct 1;107 (Pt 10):2887–98.
65. Turley H, Comley M, Houlbrook S, Nozaki N, Kikuchi A, Hickson I, et al. The distribution and expression of the two isoforms of DNA topoisomerase II in normal and neoplastic human tissues. Br J Cancer. 1997;75(9):1340–6.
66. Shaman J, Prisztoka R, Ward W. Topoisomerase IIB and an extracellular nuclease interact to digest sperm DNA in an apoptotic-like manner. Biol Reprod. 2006;75(5):741–8.
67. Balhorn R, Reed S, Tanphaichitr N. Aberrant protamine 1/protamine 2 ratios in sperm of infertile human males. Experientia. 1988 Jan 15;44(1):52–5.
68. Belokopytova IA, Kostyleva EI, Tomilin AN, Vorob'ev VI. Human male infertility may be due to a decrease of the protamine P2 content in sperm chromatin. Mol Reprod Dev. 1993 Jan 1;34(1):53–7.
69. Aoki V, Emery B, Liu L, Carrell D. Protamine levels vary between individual sperm cells of infertile human males and correlate with viability and DNA integrity. J Androl. 2006.
70. Ravel C, Chantot-Bastaraud S, El Houate B, Berthaut I, Verstraete L, De Larouziere V, et al. Mutations in the protamine 1 gene associated with male infertility. Mol Hum Reprod. 2007 Jul 1;13(7):461–4.
71. Carrell D, Emery B, Hammoud S. Altered protamine expression and diminished spermatogenesis: what is the link? Hum Reprod Update. 2007 Jan 1;13(3):313–27.

72. Aitken R, De Iuliis G. On the possible origins of DNA damage in human spermatozoa. Mol Hum Reprod. 2009 Jul 31.
73. Caron N, Veilleux S, Boissonneault G. Stimulation of DNA repair by the spermatidal TP1 protein. Mol Reprod Dev. 2001 Apr;58(4):437–43.
74. Lowndes NF, Toh GW-L. DNA repair: the importance of phosphorylating histone H2AX. Curr Biol. 2005 Feb 8;15(3):R99-R102.
75. Rogakou EP, Pilch DR, Orr AH, Ivanova VS, Bonner WM. DNA double-stranded breaks induce histone H2AX phosphorylation on serine 139. J Biol Chem. 1998 Mar 6;273(10):5858–68.
76. Pilch DR, Sedelnikova OA, Redon C, Celeste A, Nussenzweig A, Bonner WM. Characteristics of gamma-H2AX foci at DNA double-strand breaks sites. Biochem Cell Biol. 2003 Jun 1;81(3):123–9.
77. Srivastava N, Raman M. Homologous recombination-mediated double-strand break repair in mouse testicular extracts and comparison with different germ cell stages. Cell Biochem Funct. 2007 Jan 1;25(1):75–86.
78. Martins RP, Krawetz SA. Decondensing the protamine domain for transcription. Proc Natl Acad Sci USA. 2007 May 15;104(20):8340–5.
79. Ward WS. Function of sperm chromatin structural elements in fertilization and development. Mol Hum Reprod. 2010 Jan 1;16(1):30–6.
80. Spiridonov NA, Wong L, Zerfas PM, Starost MF, Pack SD, Paweletz CP, et al. Identification and characterization of SSTK, a serine/threonine protein kinase essential for male fertility. Mol Cell Biol. 2005 May 1;25(10):4250–61.
81. Deweese J, Osheroff N. The DNA cleavage reaction of topoisomerase II: wolf in sheep's clothing. Nucleic Acids Research. 2008 Nov 28:gkn937v1.
82. Barthelmes H, Habermeyer M, Christensen M, Mielke C, Interthal H, Pouliot J, et al. TDP1 overexpression in human cells counteracts DNA damage mediated by topoisomerases I and II. J Biol Chem. 2004 Oct 21;279(53):55618–25.
83. Nitiss K, Malik M, He X, White S, Nitiss J. Tyrosyl-DNA phosphodiesterase (Tdp1) participates in the repair of Top2-mediated DNA damage. Proc Natl Acad Sci USA. 2006 Jun 13;103(24):8953–8.
84. Interthal H, Chen H, Champoux J. Human Tdp1 cleaves a broad spectrum of substrates, including phosphoamide linkages. J Biol Chem. 2005 Oct 28;280(43):36518–28.
85. Cortes Ledesma F, El Khamisy SF, Zuma MC, Osborn K, Caldecott KW. A human 5'-tyrosyl DNA phosphodiesterase that repairs topoisomerase-mediated DNA damage. Nature. 2009 Oct 1;461(7264): 674–8.
86. Weterings E, Chen DJ. The endless tale of non-homologous end-joining. Cell Res. 2008 Jan 1;18(1): 114–24.
87. Pastwa E, Somiari R, Malinowski M, Somiari S, Winters T. In vitro non-homologous DNA end joining assays-The 20th anniversary. Int J Biochem Cell Biol. 2008 Dec 6.
88. Ahmed EA, de Boer P, Philippens MEP, Kal HB, de Rooij DG. Parp1–XRCC1 and the repair of DNA double strand breaks in mouse round spermatids. Mutat Res. 2010 Jan 5;683(1–2):84–90.
89. Valerie K, Povirk LF. Regulation and mechanisms of mammalian double-strand break repair. Oncogene. 2003 Sep 1;22(37):5792–812.
90. Cabrero J, Palomino-Morales RJ, Camacho JPM. The DNA-repair Ku70 protein is located in the nucleus and tail of elongating spermatids in grasshoppers. Chromosome Res. 2007 Jan 1;15(8): 1093–100.
91. Celli GB, Denchi EL, de Lange T. Ku70 stimulates fusion of dysfunctional telomeres yet protects chromosome ends from homologous recombination. Nat Cell Biol. 2006 Aug 1;8(8):885–90.
92. Boulton SJ, Jackson SP. Components of the Ku-dependent non-homologous end-joining pathway are involved in telomeric length maintenance and telomeric silencing. EMBO J. 1998 Mar 16;17(6): 1819–28.
93. Maymon B, Cohenarmon M, Yavetz H, Yogev L, Lifschitzmercer B, Kleiman S, et al. Role of poly(ADP-ribosyl)ation during human spermatogenesis. Fertil Steril. 2006 Nov 1;86(5):1402–7.
94. Di Meglio S, Denegri M, Vallefuoco S, Tramontano F, Scovassi AI, Quesada P. Poly(ADPR) polymerase-1 and poly(ADPR) glycohydrolase level and distribution in differentiating rat germinal cells. Mol Cell Biochem. 2003 Jun 1;248(1–2):85–91.
95. Dantzer F, Mark M, Quenet D, Scherthan H, Huber A, Liebe B, et al. Poly(ADP-ribose) polymerase-2 contributes to the fidelity of male meiosis I and spermiogenesis. Proc Natl Acad Sci USA. 2006 Oct 3;103(40):14854–9.
96. Meyer-Ficca ML, Lonchar J, Credidio C, Ihara M, Li Y, Wang Z-Q, et al. Disruption of poly(ADP-ribose) homeostasis affects spermiogenesis and sperm chromatin integrity in mice. Biol Reprod. 2009 Jul 1;81(1):46–55.
97. Stark JM, Pierce AJ, Oh J, Pastink A, Jasin M. Genetic steps of mammalian homologous repair with distinct mutagenic consequences. Mol Cell Biol. 2004 Nov 1;24(21):9305–16.
98. Weinstock D, Richardson C, Elliott B, Jasin M. Modeling oncogenic translocations: distinct roles for double-strand break repair pathways in translocation formation in mammalian cells. DNA Repair (Amst). 2006 Jul 4;5(9–10):1065–74.
99. McVey M, Lee SE. MMEJ repair of double-strand breaks (director's cut): deleted sequences and alternative endings. Trends Genet. 2008 Nov 1;24(11): 529–38.
100. Rathke C, Baarends WM, Jayaramaiah-Raja S, Bartkuhn M, Renkawitz R, Renkawitz-Pohl R. Transition from a nucleosome-based to a protamine-based chromatin configuration during spermiogenesis in Drosophila. J Cell Sci. 2007 May 1;120(Pt 9): 1689–700.
101. Wojtczak A, Popłoska K, Kwiatkowska M. Phosphorylation of H2AX histone as indirect evidence for

double-stranded DNA breaks related to the exchange of nuclear proteins and chromatin remodeling in Chara vulgaris spermiogenesis. Protoplasma. 2008 Nov 1;233(3–4):263–7.

102. Thomas NS, Durkie M, Van Zyl B, Sanford R, Potts G, Youings S, et al. Parental and chromosomal origin of unbalanced de novo structural chromosome abnormalities in man. Hum Genet. 2006 May 1;119(4):444–50.
103. Chandley AC. On the parental origin of de novo mutation in man. J Med Genet. 1991 Apr 1;28(4):217–23.
104. Olson SB, Magenis, R.E. Preferential paternal origin of de novo structural chromosome rearrangements. In: Daniel A, editor. The cytogenetics of mammalian autosomal rearrangements. New York: Liss; 1988. p. 583–99.
105. Sloter ED, Lowe X, Moore II DH, Nath J, Wyrobek AJ. Multicolor FISH analysis of chromosomal breaks, duplications, deletions, and numerical abnormalities in the sperm of healthy men. Am J Hum Genet. 2000 Oct 1;67(4):862–72.
106. Maffei F, Forti GC, Castelli E, Stefanini GF, Mattioli S, Hrelia P. Biomarkers to assess the genetic damage induced by alcohol abuse in human lymphocytes. Mutat Res. 2002 Feb 15;514(1–2):49–58.
107. Obe G, Herha J. Chromosomal aberrations in heavy smokers. Hum Genet. 1978 Apr 24;41(3):259–63.
108. Hopkins JM, Evans HJ. Cigarette smoke-induced DNA damage and lung cancer risks. Nature. 1980 Jan 24;283(5745):388–90.
109. Glei M, Habermann N, Osswald K, Seidel C, Persin C, Jahreis G, et al. Assessment of DNA damage and its modulation by dietary and genetic factors in smokers using the Comet assay: a biomarker model. Biomarkers. 2005 Jan 1;10(2–3):203–17.
110. Fedeli D, Fedeli A, Luciani F, Massi M, Falcioni G, Polidori C. Lymphocyte DNA alteration by subchronic ethanol intake in alcohol-preferring rats. Clin Chim Acta. 2003 Nov 1;337(1–2):43–8.
111. Katsuki Y, Nakada S, Yokoyama T, Imoto I, Inazawa J, Nagasawa M, et al. Caffeine yields aneuploidy through asymmetrical cell division caused by misalignment of chromosomes. Cancer Sci. 2008 Aug 1;99(8):1539–45.
112. Rubes J, Lowe X, Moore D, Perreault S, Slott V, Evenson D, et al. Smoking cigarettes is associated with increased sperm disomy in teenage men. Fertil Steril. 1998 Oct 1;70(4):715–23.
113. Belcheva A, Ivanova-Kicheva M, Tzvetkova P, Marinov M. Effects of cigarette smoking on sperm plasma membrane integrity and DNA fragmentation. Int J Androl. 2004 Oct;27(5):296–300.
114. Martin QS, Evelyn Ko, Leona Barclay, Tina Hoang, Alfred Rademaker, RenÉe. Cigarette smoking and aneuploidy in human sperm. Mol Reprod Dev. 2001 Jan 1;59(4):417–21.
115. Sepaniak S, Forges T, Gerard H, Foliguet B, Bene M-C, Monnier-Barbarino P. The influence of cigarette smoking on human sperm quality and DNA fragmentation. Toxicology. 2006 Jun 1;223(1–2):54–60.
116. Schmid T, Eskenazi B, Baumgartner A, Marchetti F, Young S, Weldon R, et al. The effects of male age on sperm DNA damage in healthy non-smokers. Hum Reprod. 2007 Jan 1;22(1):180.
117. Rybaczek, Bodys, Maszewski. H2AX foci in late S/G2- and M-phase cells after hydroxyurea- and aphidicolin-induced DNA replication stress in Vicia. Histochem Cell Biol. 2007 Jul 18.
118. Block W, Yu Y, Merkle D, Gifford J, Ding Q, Meek K, et al. Autophosphorylation-dependent remodeling of the DNA-dependent protein kinase catalytic subunit regulates ligation of DNA ends. Nucleic Acids Res. 2004 Aug 16;32(14):4351.
119. Sarkaria JN, Busby EC, Tibbetts RS, Roos P, Taya Y, Karnitz LM, et al. Inhibition of ATM and ATR kinase activities by the radiosensitizing agent, caffeine. Cancer Res. 1999 Sep 1;59(17):4375–82.
120. Bosch M, Rajmil O, Egozcue J, Templado C. Linear increase of structural and numerical chromosome 9 abnormalities in human sperm regarding age. Eur J Hum Genet. 2003 Oct 1;11(10):754–9.
121. Sloter E, Nath J, Eskenazi B, Wyrobek AJ. Effects of male age on the frequencies of germinal and heritable chromosomal abnormalities in humans and rodents. Fertil Steril. 2004 Apr 1;81(4):925–43.
122. Tiemann-Boege I, Navidi W, Grewal R, Cohn D, Eskenazi B, Wyrobek AJ, et al. The observed human sperm mutation frequency cannot explain the achondroplasia paternal age effect. Proc Natl Acad Sci USA. 2002 Nov 12;99(23):14952–7.
123. Marchetti F, Essers J, Kanaar R, Wyrobek AJ. Disruption of maternal DNA repair increases sperm-derived chromosomal aberrations. Proc Natl Acad Sci USA. 2007 Nov 6;104(45):17725–9.
124. Marchetti F, Wyrobek AJ. DNA repair decline during mouse spermiogenesis results in the accumulation of heritable DNA damage. DNA Repair (Amst). 2008 Apr 2;7(4):572–81.
125. Kusakabe H, Kamiguchi Y. Chromosome analysis of mouse zygotes after injecting oocytes with spermatozoa treated in vitro with green tea catechin, (−)-epigallocatechin gallate (EGCG). Mutat Res. 2004 Dec 12;564(2):195–200.
126. Feng Q, Moran JV, Kazazian HH, Boeke JD. Human L1 retrotransposon encodes a conserved endonuclease required for retrotransposition. Cell. 1996 Nov 29;87(5):905–16.
127. Goodier JL, Ostertag EM, Engleka KA, Seleme MC, Kazazian HH. A potential role for the nucleolus in L1 retrotransposition. Hum Mol Genet. 2004 May 15;13(10):1041–8.
128. Gilbert N, Lutz S, Morrish TA, Moran JV. Multiple fates of L1 retrotransposition intermediates in cultured human cells. Mol Cell Biol. 2005 Sep 1;25(17):7780–95.
129. Deininger PL, Batzer MA. Alu repeats and human disease. Mol Genet Metab. 1999 Jul 1;67(3):183–93.
130. Wei Y, Sun M, Nilsson G, Dwight T, Xie Y, Wang J, et al. Characteristic sequence motifs located at the genomic breakpoints of the translocation t(X;18)

in synovial sarcomas. Oncogene. 2003 Apr 10;22(14): 2215–22.
131. Minor A, Wong E, Harmer K, Ma S. Molecular and cytogenetic investigation of Y chromosome deletions over three generations facilitated by intracytoplasmic sperm injection. Prenat Diagn. 2007 May 29.
132. Aitken R, Krausz C. Oxidative stress, DNA damage and the Y chromosome. Reproduction. 2001 Oct 1;122(4):497–506.
133. Ogura A, Matsuda J, Yanagimachi R. Birth of normal young after electrofusion of mouse oocytes with round spermatids. Proc Natl Acad Sci USA. 1994 Aug 2;91(16):7460–2.
134. Kimura Y, Yanagimachi R. Mouse oocytes injected with testicular spermatozoa or round spermatids can develop into normal offspring. Development. 1995 Aug 1;121(8):2397–405.
135. Yamagata K, Suetsugu R, Wakayama T. Assessment of chromosomal integrity using a novel live-cell imaging technique in mouse embryos produced by intracytoplasmic sperm injection. Hum Reprod. 2009 Oct 1;24(10):2490–9.
136. Hassold T, Hunt P. To err (meiotically) is human: the genesis of human aneuploidy. Nat Rev Genet. 2001 Apr 1;2(4):280–91.
137. Crow JF. The origins, patterns and implications of human spontaneous mutation. Nat Rev Genet. 2000 Oct 1;1(1):40–7.
138. Olsen A, Lindeman B, Wiger R, Duale N, Brunborg G. How do male germ cells handle DNA damage? Toxicol Appl Pharmacol. 2005;207(2 suppl.): 521–31.
139. Brandriff B, Pedersen RA. Repair of the ultraviolet-irradiated male genome in fertilized mouse eggs. Science. 1981 Mar 27;211(4489):1431–3.
140. Ashwood-Smith MJ, Edwards RG. DNA repair by oocytes. Mol Hum Reprod. 1996 Jan 1;2(1): 46–51.

22 Male Subfertility and Sperm Chromatin Damage

Mona Bungum, Aleksander Giwercman, and Marcello Spanò

Abstract

Male subfertility contributing to about half of all involuntary childlessness is affected by a variety of genetic, behavioural and environmental factors. The value of the conventional semen quality analysis used to diagnose male infertility/subfertility is limited. The parameters recommended by the World Health Organization (WHO), sperm concentration, motility and morphology, are not sufficient to fully describe all the sperm biological properties needed to achieve fertilization and pregnancy, in spontaneous or assisted conception. During the last ten years, the assessment of sperm DNA integrity has emerged as a strong candidate as a new biomarker of semen quality that may help in the discrimination between infertile and fertile men and in predicting pregnancy outcome and risk of adverse reproductive events.

Keywords

Male subfertility • Sperm chromatin damage • DNA damage in men • Infertility in men

Infertility is defined as a state in which a couple desiring a child is unable to conceive following 12 months of unprotected intercourse. Infertility represents one of the most common diseases and affects between 17 and 25% of couples [1, 2]. For long, female factors have been regarded as the primary causes of failure to conceive. However, male causes are involved in about half of the cases [3]. Male infertility is a multifactorial disease that can be due to a variety of genetic and acquired factors. However, in about half of the men the aetiology of impaired semen quality remains unexplained [3]. In a high proportion of the cases, no cause-related treatment is possible [4].

The traditional semen analysis where the World Health Organization (WHO) has set criteria in regard to sperm concentration, motility and morphology is the cornerstone procedure used to diagnose male infertility [5]. However, the WHO parameters only address few aspects of sperm

M. Bungum (✉)
Reproductive Medicine Centre, Skåne University Hospital Malmö, Lund University, Sodra Forstadsgatan, Malmö 20502, Sweden
e-mail: mona.bungumed.lu.se

A. Zini and A. Agarwal (eds.), *Sperm Chromatin: Biological and Clinical Applications in Male Infertility and Assisted Reproduction*, DOI 10.1007/978-1-4419-6857-9_22,

quality and function, and thus, the discriminative power in relation to fertility is quite low [6, 7]. Finding better markers of male fertility may have important clinical and biological implications [8–10]. It may improve understanding of mechanisms underlying subfertility and facilitate development of new specific therapies. Moreover, better markers could help in deciding for which couple assisted reproductive technology (ART) is needed and to identify the most effective type of ART treatment for a given couple [11].

Fertility requires fusion of the genomes of an oocyte and a sperm, and the completion of this process and subsequent embryo development depends, in addition to the repair capacity of the oocyte, on the inherent integrity of sperm DNA [12–14]. Animal studies have shown that a male gamete with damaged DNA can transmit genetic defects and in worse cases can lead to pregnancy loss, infant mortality, birth defects and genetic diseases in offspring [15, 16]. Extensive laboratory animal literature unequivocally reports that the genetic integrity of the male gamete is pivotal to ensure normal embryo development [17]. In support to animal studies are the findings relating paternal smoking and sperm DNA damage passed from the father to the offspring following ART [18] and the evidence of an association between paternal smoking and an increased risk of childhood cancer in the offspring [19].

Today, subfertile couples can be helped through ART. However, concerns have been raised about the increasing use of ART and in particular ICSI that bypasses natural biological barriers preventing against fertilization by defective sperm and, as a consequence, chromatin/DNA alterations can be transmitted to the embryo and the offspring. Therefore, during the last decades, a growing attention has gained the assessment of sperm chromatin integrity in the pathophysiology of infertility [20, 21]. In this chapter, we review how sperm chromatin/DNA integrity can impact male fertility.

Male Infertility/Subfertility

In 20% of involuntary childlessness couple, the predominant cause is solely male related, and in another 27%, anomalies in both partners contribute to the childlessness [3]. Reduced male fertility can be the result of congenital and acquired urogenital abnormalities, infections of the genital tract, varicocele, endocrine disturbances and genetic or immunological factors [3]. Environmental, occupational, lifestyle and therapeutic exposures have also been invoked as possible cofactors hampering male fertility [22–25].

However, the underlying cause of infertility remains unexplained in at least 50% of the infertile men. Genetic abnormalities [26–28] are thought to account for 15–30% of male factor infertility. Approximately 5% of infertile men have chromosomal abnormalities, a prevalence that increases up to 15% in the population of azoospermic males [29, 156].

Recent studies have shown that epigenetic modifications in sperm may also cause infertility [26, 30–34]. One of the main epigenetic mechanisms in sperm appears to be DNA methylation [156]. Several studies indicate that DNA methylation is altered, in at least some imprinted genes, in oligozoospermic men and men with improper histone to protamine replacement [35–37]. Furthermore, methylation defects as well as other epigenetic defects may play an important role in the development and growth of ART offspring [38–41].

Another important cause of male infertility is considered, and this is the main topic of this chapter, the occurrence of chromatin and nuclear abnormalities manifesting themselves as breaks in sperm nuclear DNA [16, 20, 42–45].

Diagnosis of Male Infertility/Subfertility

Mostly, the diagnosis of male infertility/subfertility is based solely on the presence of an abnormal semen analysis of sperm concentration, motility and morphology [5]. The standard sperm parameters vary significantly between individuals, seasons, countries and regions and even between consecutive samples within the same man [5, 46–50]. As the analysis is mainly performed by standard light microscopy of 100–200 spermatozoa, the analysis implies a high level of subjectivity resulting in a high grade of intra- and interlaboratory variation [51, 52], and thus, a low predictive power of the analysis is seen. Another problem when assessing predictivity of semen parameters is that female factors only rarely are taken into account. Whilst mostly the term infertility are used, most patients are actually subfertile, rather than sterile (infertile), but the degree of subfertility is difficult to predict [53]. A fertile partner may compensate for a less fertile spouse, and thus, in most cases the term subfertility better covers the condition.

During the last decades, several other laboratory tests of sperm function have been developed, such as antisperm antibody test, vital staining, biochemical analysis of semen, hypoosmotic swelling test, sperm penetration assay, hemizona assay, creatine kinase test, reactive oxygen species (ROS) tests and computer-assisted sperm analysis (CASA), to mention the most commonly used [8]. However, the clinical value of these tests has been questioned, and few are implemented in clinical routine [54].

Although the origin and the mechanisms responsible for sperm DNA damage are not yet fully clarified, a bulk of data have accumulated, demonstrating an association between genetic damage and fertility or progeny outcome [16, 20, 43, 45, 55–59, 156]. It has been proposed that sperm DNA integrity could be a possible fertility predictor to be used as a supplement to the traditional sperm parameters [11, 42].

Assessment of Sperm Chromatin Damage

During the past decades, a variety of new techniques to assess sperm nuclear integrity have been developed [42, 60–62]. This issue is reviewed in depth in other chapters of this book. Briefly, such techniques, using microscopy-based and flow-cytometry-based analyses, can evaluate sperm DNA and chromatin integrity in situ on cell-by-cell basis. Each test uses a different strategy to detect DNA/chromatin damages. Unspecific DNA strand breaks can be detected by the single-cell gel electrophoresis assay (comet assay) in its alkaline, neutral and two-tailed versions [63–65], the terminal deoxynucleotidyl transferase dUTP nick-end labelling (TUNEL) assay [66, 67], in situ nick translation (ISNT) [68] or DNA breakage detection fluorescence in situ hybridization (DBD-FISH) [69]. In addition, DNA breaks can be evaluated indirectly through the DNA denaturability by the sperm chromatin structure assay (SCSA) [42, 70], the sperm chromatin dispersion (SCD) test [71, 72] and the toluidine blue assay [73]. Chromatin integrity can also be assessed with respect to the degree of protamination by the CMA3 assay [68, 74], which relies on the detection of lysine residues as a measure of an excess of histones remaining bound to the sperm DNA, and by measuring the level of compaction due to the formation of inter- and intraprotamine disulphide bridges [75]. More recently, techniques have been developed to assess the epigenetic components of sperm such as the global DNA methylation level [76, 77]. These tests can measure a parameter generally known as sperm DNA fragmentation. For sake of brevity, we use the abbreviation DFI (DNA Fragmentation Index) to identify the fraction of DNA defective sperm independently from the various DNA fragmentation assays used. From studies carried out both in normal and infertile men, it turned out that, with few exceptions, sperm DNA integrity tests generally correlate well with each other, even though the level of correlation between the same techniques can vary across different studies. In Table 22.1, the studies correlating different DNA fragmentation assays, together with their correlation levels, and involving more than 50 individuals are reported.

Genesis of Sperm DNA Damage

The most common types of sperm DNA damage include single- or double-strand breaks, base modifications and adducts, DNA intra-/inter-strand and DNA–protein cross links [16]. Even though the mechanisms leading to the formation of DNA damage in sperm are only partially elucidated, it has generally been proposed that DNA damage in sperm can be produced by unrepaired DNA breaks during the spermiogenetical chromatin packaging [78], by partial or complete protamine deficiency [26, 79, 80], abortive apoptosis during spermatogenesis [81] and the

Table 22.1 Studies (with>50 individuals) correlating different DNA fragmentation assays

Technique	N Normal men	Infertile men	Results	References
SCSA vs. AOT		185	N.S.	Apedaile et al. [110]
SCSA vs. AOT	7	60	N.S.	Chohan et al. [157]
SCSA vs. comet (neutral pH 8)	80		N.S.	Schmid et al. [156]
SCSA vs. comet (alkaline pH 13)	80		N.S.	Schmid et al. [156]
SCSA vs. comet (alkaline pH 12.1)		55	$r=0.3$	O'Flaherty et al. [158]
SCSA vs. FIM-TUNEL	7	60	$r=0.9$	Chohan et al. [157]
SCSA vs. FIM-TUNEL	25	55	$r=0.50$	Smith et al. [131]
SCSA vs. FCM-TUNEL	24	96	$r=0.41$	Ståhl et al. [159]
SCSA vs. FCM-TUNEL	666		$r=0.56$	Toft (personal communication, 2006)
SCSA vs. FCM-TUNEL		58	$r=0.27$	O'Flaherty et al. [158]
SCSA vs. SCD	7	60	$r=0.9$	Chohan et al. [157]
SCSA vs. Toluidine Blue	63	79	$r=0.47$	Tsarev et al. [145]
M-TUNEL vs. SCD	30	60	$r=0.6–0.9$	Zhang et al. [160]
FIM-TUNEL vs. FCM-TUNEL		66	$r=0.72$	Domínguez-Fandos et al. [161]
FIM-TUNEL vs. SCD	7	60	$r=0.9$	Chohan et al. [157]
FIM-TUNEL vs. CMA3		61	$r=0.76$	Plastira et al. [162]
FIM-TUNEL vs. CMA3		132	$r=0.53$	Tarozzi et al. [57]
FCM-TUNEL vs. FIM-TUNEL		68	$r=0.94$	Cohen-Bacrie et al. [112]
FCM TUNEL vs. comet (alkaline pH 10)	42	21	$r=0.56$	Bian et al. [163]
FCM TUNEL vs. CMA3 (FCM)	39	28	$r=0.83–0.96$	De Iuliis et al. [164]
FCM TUNEL vs. 8-OHdG (FCM)		94	$r=0.25$ ($r=0.76$ in the high-density Percoll fraction)	De Iuliis et al. [164]
SCD vs. CMA3		78	$r=0.29$	Tavalaee et al. [165]

SCSA sperm chromatin structure assay; *AOT* acridine orange test; *TUNEL* terminal deoxynucleotidyl transferase dUTP nick-end labelling; *M-TUNEL* TUNEL assay, bright field microscopy; *FIM-TUNEL* TUNEL assay, fluorescence microscopy; *FCM-TUNEL* flow cytometry TUNEL; *SCD* sperm chromatin dispersion test; *CMA3* chromomycin A3; *ISNT* in situ nick translation; *8-OHdG* 8-hydroxydeoxyguanosine level evaluated by high-performance liquid chromatography; *N.S.* not statistically significant

action of (sperm- or leucocyte-derived) oxidative damage [44]. Furthermore, a variety of external factors such as genotoxic agents due to therapeutical, occupational and environmental exposures [44, 45, 82] may cause sperm DNA breaks by some of the mechanisms mentioned above. At least, some of these exposures directly target DNA, whereas others induce oxidative stress. ROS can damage sperm DNA [44, 83], and a reliable biomarker of the oxidative attack on the DNA molecule is the formation of 8-hydroxy-2′-deoxyguanosine (8-OHdG). Higher levels of this adduct have been found in the sperm DNA of infertile men [84, 85]. 8-OHdG adducts representing a modified DNA structure potentially leading to a DNA break and strong relationships between the two forms of damage have been reported [44]. These issues are reviewed in detail elsewhere in this book. The proposed mechanisms are obviously not mutually exclusive, and recently, a two-step hypothesis has been put forward where faulty spermatogenesis can lead to defective chromatin remodelling with the DNA more susceptible and vulnerable to a variety of stressors [16, 44].

Factors Reported to Impact Sperm Chromatin Integrity

The personal burden of sperm DNA fragmentation can increase under the influence of variety of stressors [20, 21, 86]. Exposures to physical agents and chemicals including therapeutic drugs, pesticides, metals, air pollutants and tobacco smoking are known to target male germ cells. At least, some of these exposures directly target DNA, whereas others induce oxidative stress. Smokers have an increased level of oxidative damage in their sperm DNA compared to non-smokers [87], and several studies have reported a negative effect of cigarette smoking on sperm DNA [88].

Moreover, recent studies have indicated an association between high body mass index (BMI) and reduced semen quality [89–91]. A higher fraction of sperm with high DNA damage was reported in obese men than in normal-weight men [92]. However, results are conflicting, and so far, no such prospective or intervention studies have been published.

Fever can have marked effects on both the conventional semen parameters and sperm DNA integrity [21, 93, 94]. Also, several studies have reported that the higher is the abstinence period, the higher is the fraction of DNA defective sperm. It seems likely that this correlation stems from a longer exposure of sperm to ROS attacks. A weak positive correlation has been found both in the general population and in infertility patients [43, 95–100]. However, this relation did not emerge in other studies [101, 102], including a study designed to specifically address this issue [103].

Another important source of deterioration of the DNA integrity of spermatozoa is aging. It is known that a major proportion of abnormal reproductive outcomes are associated with paternally transmitted numerical and structural chromosomal abnormalities and advancing paternal age has been implicated in a broad range of abnormal reproductive and genetic outcomes [104]. Older men are reported to have sperm with more DNA fragmentation than younger men [95, 100, 105, 106, 156], a finding well in accordance with the age-dependent decline in standard semen parameters [107]. Increased life expectancies, changes in family-planning practices and advances in ART in industrialized countries are resulting in an increasing number of births in couples aged 34–54 years. The age-related increase of damage in male germ cells raises substantial health concerns regarding the possible long-term consequences of increasing paternal ages on the viability and genetic health of the offspring.

Male Infertility and Sperm Chromatin Damage

The Association Between the Traditional Sperm Parameters and Sperm DNA Integrity

Although the level will vary, in all men, sperm cells with DNA breaks are always present in the ejaculate. Whilst unselected men planning for their first pregnancy had a mean DFI of 14% [96], infertile men have a mean DFI of 23% as compared to a DFI of 12% observed for fertile men [108].

In Table 22.2, the studies involving more than 100 men reporting prevalence of DNA-defective sperm in infertile men as compared to normal controls are quoted.

Several studies have demonstrated a weak-to-moderate inverse correlation, if any, between sperm DNA fragmentation measured by the various sperm integrity assays and the traditional semen parameters [95, 98, 109–114]. The correlation levels among different studies comparing the same techniques can vary, probably because minor variations in the protocols and in the semen samples can influence the final figures. However, by and large, sperm DNA fragmentation assessment is quite independent from the WHO standard parameters. Motility has generally been the parameter with the highest degree of association to sperm DNA defects, probably because both sperm chromatin compaction and acquisition of

Table 22.2 Studies (with >100 individuals) reporting prevalence of DNA defective sperm in infertile men as compared to normal controls

Technique	Controls, *n*	Infertile, *n*	References
M-TUNEL	20	236	Høst et al. [166]
FlM-TUNEL	23	87	Gandini et al. [167]
FlM-TUNEL	49	61	Plastira et al. [162]
FCM-TUNEL	47	66	Sergerie et al. [144]
CMA3	49	61	Plastira et al. [162]
SCSA	165	115	Evenson et al. [122]
SCSA	13	88	Zini et al. [125]
SCSA	16	92	Saleh et al. [126]
SCSA	13	101	
SCSA	100	200	Pant et al. [168]
SCSA	137	127	Giwercman et al. [108]
Aniline Blue	75	90	Hammadeh et al. [169]
Toluidine Blue	63	79	Tsarev et al. [145]
8-OHdG	54	60	Shen et al. [85]

SCSA sperm chromatin structure assay; *TUNEL* terminal deoxynucleotidyl transferase dUTP nick-end labelling; *M-TUNEL* TUNEL assay, bright field microscopy; *FlM-TUNEL* TUNEL assay, fluorescence microscopy; *FCM-TUNEL* flow cytometry TUNEL; *CMA3* Chromomycin A3; *8-OHdG* 8-hydroxydeoxyguanosine level evaluated by high-performance liquid chromatography

motility are parallel differentiation processes culminating during the passage of the maturing male gamete in the epididymal tract.

The relation between the fractions of DNA defective sperm and blood concentration of sexual hormones and other biomarkers of the sexual accessory glands has also been studied. In a study involving 278 young men with no knowledge of their fertility status [115], the DFI as evaluated by the SCSA was weakly correlated, negatively with estradiol and free testosterone and positively correlated with the seminal concentration of zinc and fructose. In another study involving 362 male partners of infertile couples [116], the fraction of sperm with high DNA damage, evaluated by the neutral comet assay, resulted positively associated with free thyroxine (T4) and total triiodothyronine (T3).

Intraindividual Variation of Sperm Chromatin Parameters

Traditional semen parameters usually exhibit a high intraindividual variability [51, 52] and coefficient of variations (CVs) as high as 54% has been reported [117]. The issue of possible intraindividual changes of sperm chromatin parameters with time has specifically been addressed by some groups, and DFI is demonstrated to be a sperm parameter characterized by a lower level of variability. In a study of 45 men who delivered eight monthly semen samples, the average within-donor CV of DFI as measured by SCSA was around 23% [101]. These results were confirmed by other SCSA studies. Zini et al. [118] measured the DFI in 21 men who provided two semen samples, 2–6 weeks apart and observed a within-subject CV of 21%. In another study, involving 277 men, semen was measured twice during 6 months and a within-subject CV of 23% for DFI was obtained [95]. Altogether, these data point to a lower level of intraindividual variation for SCSA measurements as compared to the standard sperm parameters.

Time stability of sperm DNA integrity was assessed both by the SCSA and the TUNEL assays in a healthy non-smoking fertile volunteer, characterized by a low DFI, over a 10-year period. Compared with TUNEL data, SCSA measurements showed less variation over the data collection period with a DFI within-subject CV of 47.4 and 22.3%, respectively. DFI remained normal, and no trend was observed over the period of observation [119]. Finally, the stability over time of the flow cytometry TUNEL assay, during a 6-month period, was tested in a longitudinal study using 15 men donors who provided monthly multiple semen samples. A good reproducibility of the TUNEL assay was obtained: individual CVs for sperm DFI ranged from 12.9 to 43.9%, whereas parallel measurements on cell counts showed within-donor CVs ranging from 16.7 up to 63.2% [120].

In a study of 282 patients undergoing ART with repeated (between 2 and 5) SCSA measurements, CV of DFI was a bit higher than previously reported, about 29%, showing that intraindividual variability in DFI could be of significance. Repeated measurements were recommended in men having a DFI >20% [121], since a switch of DFI to a higher level may have implications for the selection of the ART treatment [11].

Impact of Sperm DNA Damage on Fecundity in General Population

Whether sperm chromatin integrity parameters, independently from the WHO parameters, could predict the chances of spontaneous pregnancy was a question addressed by two almost concomitant SCSA studies, one carried out in USA (the Georgetown study, 165 couples) and the other carried out in Europe (the Danish first pregnancy planners study, 215 couples). Both demonstrated that in couples from the general population, the chance of spontaneous pregnancy, measured by the time-to-pregnancy (TTP), decreases when DFI exceeded 20–30% [96, 122]. If the DFI was more than 30%, TTP tended to become infinite and the chances of spontaneous pregnancies were quite negligible [96, 122]. Stratifying the population into two groups, below and above a DFI threshold at 30%, the probability of pregnancy for the group with DFI <30% was significantly higher than that for the group with DFI >30%. These two in vivo studies showed that the pregnancy rates are significantly higher for the group with DFI below the thresholds of 30% [123]. In the same population of Danish first-pregnancy planners [96], the likelihood of pregnancy occurring in a single menstrual cycle was inversely associated with the level of 8-OHdG [124], corroborating the result of the previous SCSA analysis and reinforcing the notion that oxidative DNA damage can play a major role in the genesis of DNA breaks [44].

Impact of Sperm DNA Damage on Fecundity in Subfertile Men

There are few studies addressing the issue of prevalence of high levels of sperm DNA damage among infertile men. Such prevalence was reported to be 17% when the 30% SCSA derived DFI threshold was used [125], and 58% using a 24% SCSA-DFI threshold [126]. On the other hand, Verit and coworkers did not find increased DFI levels among infertile men with normal conventional semen parameters, as compared to fertile donors [127].

The prevalence of sperm DNA damage in 350 men from infertile couples with both normal and abnormal semen parameters was studied to investigate whether sperm DNA fragmentation, assessed by the SCSA, could add to the information obtained by routine semen analysis when explaining the causes of infertility [128]. In this study, 28% of men had a DFI >20%, while 12% had a DFI >30%. In the subgroup of 224 men with abnormal semen parameters, 35% had a DFI >20% and 16% had a DFI >30%, whereas these figures were 15 and 5%, respectively, in the subgroup of men with normal semen parameters. Men with low sperm motility and abnormal morphology had significantly higher odds ratios (OR) for having a DFI >20% (4.0 for motility and 1.9 for morphology) and DFI >30% (6.2 for motility and 2.8 for morphology) compared with men with normal sperm motility and morphology.

In a more recent study, 127 men from infertile couples where female factors contributing to the infertility problem were excluded, and 137 men with proven fertility were considered. Also in this work, DFI was assessed using SCSA. The risk of being infertile was increased when DFI >20% (OR 5.1) in men with normal standard semen parameters, whereas if one of the WHO parameters were abnormal, the OR for infertility was increased already at DFI above 10% (OR 16). DFI above 20% was found in 40% of men with otherwise normal standard parameters. Thus, the DFI as measured by SCSA adds to the value of semen analysis in prediction of the chance of natural conception [108]. Furthermore, in almost 50% of so-called "unexplained" cases of infertility, sperm DNA defects seem to be at least a contributing factor to the problem.

Moreover, a SCSA derived DFI threshold at 30% has been observed in two large ART studies where ORs of 8–14 were observed in the probability of delivery after intrauterine insemination [11, 129]. Thus, DFI was found to be an independent predictor of fertility in vivo. The role of sperm DNA fragmentation in ART is reviewed in another chapter of this book.

In addition to the relationship between sperm DNA fragmentation and pregnancy outcome, sperm DNA integrity has also been started to be used in sperm quality assessment in other andrological pathologies, such as varicocele, cancer and infections, providing valuable information on disease severity and therapeutic efficacy. Usually, a higher percentage of defective sperm is found in varicocele patients, probably attributable to oxidative stress [65, 130–132]. In many cases, sperm chromatin quality was improved after surgery [133–135] or by antioxidant therapy [136].

Patients with genitourinary infection by *Chlamydia trachomatis* or *Mycoplasma* showed an increased DFI in comparison with fertile controls and antibiotic therapy resulted important in providing a remedy for infection-induced high DNA fragmentation levels [137]. This is consistent with the results from another study where patients with bacteriospermia had improvement in DFI results after antibiotic treatment [136].

Among pathologies unrelated with andrology disturbances where the fraction of sperm with fragmented DNA was higher in patients than in suitable controls, thalassemia major [138], spinal cord injury [139] and type I diabetes mellitus [140] should be mentioned. In the latter, DFI measured by the alkaline comet assay and levels of oxidative DNA modification were evaluated in spermatozoa of diabetic and non-diabetic men. A positive correlation was observed between DFI and concentrations of 8-OHdG, again supporting the hypothesis that oxidative stress may play a major role in the genesis of DNA breaks.

Studies on patients with testicular cancer have shown that sperm DNA might be damaged already before irradiation or chemotherapy [141]. However, cancer therapy was shown to further contribute to increased DNA damage [142].

Future Perspectives

There is an urgent call for better methods of assessing male fertility potential. A more precise diagnosing would enable clinicians to better counsel the infertile couple and may also result in improvement and further development of cause-related therapy, which is very little used in today's clinical practice.

Unfortunately, few of the assays used to assess sperm DNA integrity have been standardized sufficiently, and there remain wide variations in results obtained from different laboratories [58, 62]. Statistically validated threshold values could be of help in the future clinical applications of sperm DNA integrity tests. So far, SCSA has been the only method providing thresholds of clinical relevance for in vivo and in vitro pregnancy [11, 42, 96, 122, 123, 129, 143]. Other tests in which clinical thresholds for infertility have been suggested are the comet [58], the TUNEL [144] assays, the SCD [72] and the toluidine blue tests [145]. However, differently from the SCSA, none of these other assays has provided stable thresholds based on large study populations. Each of the techniques seems to have its own specificity and limitations and it is still not clear what is unequivocally measured by each test. Thus, we can choose from a variety of assays, often proposing some variations of the application protocol, but so far, we cannot comfortably decide which is the best and the most robust and why. It should be noted that all these tests can detect only a subgroup of the possible alterations of the DNA molecule [16]. Likely, only a "tip of the iceberg" of the overall DNA damage is measured [93]. An array of methods measuring "potential" DNA damage, in terms of precursors to actual strand breaks, combined with DNA fragmentation assays, may give a more complete picture of the extent of total DNA damage [146]. The predictive value of sperm DNA fragmentation assays could also depend on a variety of other unknown factors such as the extent of DNA damage per sperm, the location of DNA damage in coding or non-coding regions, the association of DNA breaks to other type of DNA lesions and how much sperm DNA damage an oocyte can deal with [16, 45, 62, 64, 146]. Further fundamental research is mandatory to solve these key questions.

Cause-related therapy is almost non-existent in male subfertility. As oxidative damage is considered one of the main, if not the most important, factors underlying the induction of sperm DNA damage, the effects of antioxidant therapy (generally based on antioxidants such as vitamin C, vitamin E, carotene, selenium, zinc, etc.) on sperm DNA quality has been attempted to verify the theoretical possibility of an amelioration of the DNA damage caused by oxidative stress [64, 83, 136, 147–151]. However, the studies have been small and conflicting. Further trials are needed to investigate whether such therapy and other types of causal treatment are effective.

Conclusions and Clinical Recommendations

It can be concluded that infertile men generally have more sperm DNA damage than fertile men. Studies of both fertile and infertile men have shown that DFI as measured by SCSA is an independent predictor of male subfertility. Despite this, there is no worldwide consensus if sperm chromatin integrity testing should be implemented as a routine in infertility work-up and in ART [59, 62, 64]. Although there seem to be insufficient data to recommend an indiscriminate application of sperm DNA testing, there are specific conditions where men would certainly benefit from this analysis. These could be male partners of couples planning to undergo ART to evaluate the impact of sperm DNA damage on reproductive outcomes (fertilization, embryo development, pregnancy, miscarriage, post-natal development) and to select which type of ART should be preferable. Also, male partners of couples with unexplained infertility or recurrent pregnancy loss could benefit from SCSA analysis [59].

So far, SCSA is the only method which has demonstrated clear and clinically useful cut-off levels for inferring male infertility potential [96, 108, 122], and its prognostic value in ART has also been shown [11, 129]. The SCSA can be used as a tool to discriminate among the different treatment options, IUI, IVF and ICSI. In men having WHO sperm parameters that indicate ICSI, there are no therapeutic advantages of performing SCSA [11, 129, 151–153]. However, in the group of men with unexplained infertility causes, 40% may have a DFI level that exceeds 20%, and thus the chances of in vivo fertility are reduced [11, 108]. In these men, the SCSA analysis, in adjunct to the standard semen quality parameters, can be valuable to disclose the causes of their infertility. In cases where a traditional semen quality analysis shows one or no abnormality, a SCSA check should be performed as the chance of spontaneous pregnancy is significantly reduced for DFI above 10% [108]. In these "unexplained" infertile couples, provided that female age is <35 years and the duration of infertility is short, IUI should be the treatment of first choice. In long-standing unexplained infertility (>5 years) and a female partner above 35 years, where DFI is above 20%, a direct referral to IVF may be the best alternative [154]. It is worth to stress again the combined impact of both female and male reproductive capability determining the cumulative fertility of a couple. A highly fertile partner can often compensate for a less fertile one. Female factors, not at least age [155] and duration of infertility [154], should always be taken into consideration when evaluating sperm DNA integrity in counselling a couple seeking ART.

Despite the fact that DFI, as other sperm parameters, but to a lesser extent, is subject of intraindividual variation, SCSA analysis was found to be a strong predictor of infertility in vivo [108]. However, in men seeking ART having a

DFI above 20%, there is a 27% risk of having a DFI above 30% in the next semen sample, and this could influence the result of the ART-treatment negatively [11]. Thus, these men having a DFI above 20% should have their SCSA analysis repeated close to their ART treatment [121].

Lastly, human studies relating sperm DNA defects to health of the offspring is not yet published. However, successful mammalian reproduction depends partly on the inherent integrity of the sperm DNA, as sperm DNA damage may adversely impact reproductive outcomes. From a clinical long-term perspective, we cannot overlook the message from extensive animal experiments providing unequivocal links between DNA damage in spermatozoa and defects in embryonic development or in the health of the offspring. Therefore, a sort of "precautionary principle" should be adopted deploying all possible strategies aiming at reducing the involvement of defective sperm in the fertilization process. Sperm DNA integrity testing has also demonstrated to be potentially useful beyond the framework of fertility assessment and should be used as an adjunct tool for the sperm quality assessment in andrological pathologies, such as varicocele, cancer and infections, providing precious information on disease severity and therapeutic efficacy.

Acknowledgements We wish to thank all the authors whose papers have been quoted and apologize to those whose contributions could not be cited because of lack of space.

References

1. de Kretser DM. Male infertility. Lancet. 1997;349: 787–90.
2. Dunson DB, Baird DD, Colombo B. Increased infertility with age in men and women. Obstet Gynecol. 2004;103:51–6.
3. WHO. WHO manual for the standardised investigation, diagnosis and management of the infertile male. Cambridge: Cambridge University Press; 2000.
4. Skakkebæk NE, Giwercman A, de Kretser D. Pathogenesis and management of male infertility. Lancet. 1994;343:1473–9.
5. World Health Organisation. WHO laboratory manual for examination of human semen and semen-cervical mucus interaction. Cambridge: Cambridge University Press; 1999. p. 1–20.
6. Bonde JP, Ernst E, Jensen TK, Hjøllund NHI, Kolstad H, Henriksen TB, et al. Relation between semen quality and fertility: a population-based study of 430 first-pregnancy planners. Lancet. 1998;352:1172–7.
7. Guzick DS, Overstreet JW, Factor-Litvak P, Brazil CK, Nakajima ST, Coutifaris C, et al. Sperm morphology, motility, and concentration in fertile and infertile men. New Engl J Med. 2001;345:1388–93.
8. Aitken RJ. Sperm function tests and fertility. Int J Androl. 2006;29:69–75.
9. Lewis SE. Is sperm evaluation useful in predicting human fertility? Reproduction. 2007;134:31–40.
10. Agarwal A, Bragais FM, Sabanegh E. Assessing sperm function. Urol Clin North Am. 2008;35:157–71.
11. Bungum M, Humaidan P, Axmon A, Spano M, Bungum L, Erenpreiss J, et al. Sperm DNA integrity assessment in prediction of assisted reproduction technology outcome. Hum Reprod. 2007;22:174–9.
12. Twigg JP, Irvine DS, Aitken RJ. Oxidative damage to DNA in human spermatozoa does not preclude pronucleus formation at intracytoplasmic sperm injection. Hum Reprod. 1998;13:1864–71.
13. Ahmadi A, Ng SC. Fertilizing ability of DNA-damaged spermatozoa. J Exp Zool. 1999;284:696–704.
14. Marchetti F, Essers J, Kanaar R, Wyrobek AJ. Disruption of maternal DNA repair increases sperm-derived chromosomal aberrations. Proc Natl Acad Sci USA. 2007;104:17725–9.
15. Fernández-Gonzalez R, Moreira PN, Pérez-Crespo M, Sánchez-Martín M, Ramirez MA, Pericuesta E, et al. Long-term effects of mouse intracytoplasmic sperm injection with DNA-fragmented sperm on health and behavior of adult offspring. Biol Reprod. 2008;78:761–72.
16. Aitken RJ, De Iuliis GN, McLachlan RI. Biological and clinical significance of DNA damage in the male germ line. Int J Androl. 2009;32:46–56.
17. Marchetti F, Wyrobek AJ. Mechanisms and consequences of paternally-transmitted chromosomal abnormalities. Birth Defects Res C Embryo Today. 2005;75:112–29.
18. Zenzes MT. Smoking and reproduction: gene damage to human gametes and embryos. Hum Reprod Update. 2000;6:122–31.
19. Chang JS. Parental smoking and childhood leukemia. Methods Mol Biol. 2009;472:103–37.
20. Erenpreiss J, Spano M, Erenpreisa J, Bungum M, Giwercman A. Sperm chromatin structure and male fertility: biological and clinical aspects. Asian J Androl. 2006;8:11–29.
21. Evenson DP, Jost LK, Corzett M, Balhorn R. Characteristics of human sperm chromatin structure following an episode of influenza and high fever: a case study. J Androl. 2000;21:739–46.
22. Perry MJ. Effects of environmental and occupational pesticide exposure on human sperm: a systematic review. Hum Reprod Update. 2008;14:233–42.
23. Varghese AC, du Plessis SS, Agarwal A. Male gamete survival at stake: causes and solutions. Reprod Biomed Online. 2008;17:866–80.

24. Wigle DT, Arbuckle TE, Turner MC, Bérubé A, Yang Q, Liu S, et al. Epidemiologic evidence of relationships between reproductive and child health outcomes and environmental chemical contaminants. J Toxicol Environ Health B Crit Rev. 2008;11:373–517.
25. Bonde JP. Male reproductive organs are at risk from environmental hazards. Asian J Androl. 2010; 12:152–6.
26. Carrell DT, Emery BR, Hammoud S. The aetiology of sperm protamine abnormalities and their potential impact on the sperm epigenome. Int J Androl. 2008;31:537–45.
27. Matzuk MM, Lamb DJ. The biology of infertility: research advances and clinical challenges. Nat Med. 2008;14:1197–213.
28. O'Flynn O'Brien KL, Varghese AC, Agarwal A. The genetic cause of male factor infertility: a review. Fertil Steril 2010;93:1–12.
29. Ferlin A, Raicu F, Gatta V, Zuccarello D, Palka G, Foresta C. Male infertility: role of genetic background. Reprod Biomed Online. 2007;14:734–45.
30. Jirtle RL, Skinner MK. Environmental epigenomics and disease susceptibility. Nat Rev Genet. 2007; 8:253–62.
31. Sasaki H, Matsui Y. Epigenetic events in mammalian germ-cell development: reprogramming and beyond. Nat Rev Genet. 2008;9:129–40.
32. Nanassy L, Carrell DT. Paternal effects on early embryogenesis. J Exp Clin Assist Reprod. 2008;5:2.
33. Carrell DT, Hammoud SS. The human sperm epigenome and its potential role in embryonic development. Mol Hum Reprod. 2010;16:37–47.
34. Miller D, Brinkworth M, Iles D. Paternal DNA packaging in spermatozoa: more than the sum of its parts? Reproduction. 2010;139:287–301.
35. Houshdaran S, Cortessis VK, Siegmund K, Yang A, Laird PW, Sokol RZ. Widespread epigenetic abnormalities suggest a broad DNA methylation erasure defect in abnormal human sperm. PLoS One. 2007;2:e1289.
36. Kobayashi H, Sato A, Otsu E, Hiura H, Tomatsu C, Utsunomiya T, et al. Aberrant DNA methylation of imprinted loci in sperm from oligospermic patients. Hum Mol Genet. 2007;16:2542–51.
37. Filipponi D, Feil R. Perturbation of genomic imprinting in oligozoospermia. Epigenetics. 2009;4:27–30.
38. Bowdin S, Allen C, Kirby G, Brueton L, Afnan M, Barratt C, et al. A survey of assisted reproductive technology births and imprinting disorders. Hum Reprod. 2007;22:3237–40.
39. Doornbos ME, Maas SM, McDonnell J, Vermeiden JP, Hennekam RC. Infertility, assisted reproduction technologies and imprinting disturbances: a Dutch study. Hum Reprod. 2007;22:2476–80.
40. Kobayashi H, Hiura H, John RM, Sato A, Otsu E, Kobayashi N, et al. DNA methylation errors at imprinted loci after assisted conception originate in the parental sperm. Eur J Hum Genet. 2009;17:1582–91.
41. Manipalviratn S, DeCherney A, Segars J. Imprinting disorders and assisted reproductive technology. Fertil Steril. 2009;91:305–15.
42. Evenson DP, Larson KL, Jost LK. Sperm chromatin structure assay: its clinical use for detecting sperm DNA fragmentation in male infertility and comparisons with other techniques. J Androl. 2002;23:25–43.
43. Spanò M, Seli E, Bizzaro D, Manicardi GC, Sakkas D. The significance of sperm nuclear DNA strand breaks on reproductive outcome. Curr Opin Obstet Gynecol. 2005;17:255–60.
44. Aitken RJ, De Iuliis GN. On the possible origins of DNA damage in human spermatozoa. Mol Hum Reprod. 2010;16:3–13.
45. Sakkas D, Alvarez JG. Sperm DNA fragmentation: mechanisms of origin, impact on reproductive outcome, and analysis. Fertil Steril. 2010;93:1027–36.
46. Chia SE, Tay SK, Lim ST. What constitutes a normal seminal analysis? Semen parameters of 243 fertile men. Hum Reprod. 1998;13:3394–8.
47. Auger J, Eustache F, Andersen AG, Irvine DS, Jorgensen N, Skakkebaek NE, et al. Sperm morphological defects related to environment, lifestyle and medical history of 1001 male partners of pregnant women from four European cities. Hum Reprod. 2001;16:2710–7.
48. Jorgensen N, Andersen AG, Eustache F, Irvine DS, Suominen J, Petersen JH, et al. Regional differences in semen quality in Europe. Hum Reprod. 2001; 16:1012–9.
49. Chen Z, Toth T, Godfrey-Bailey L, Mercedat N, Schiff I, Hauser R. Seasonal variation and age-related changes in human semen parameters. J Androl. 2003;24:226–31.
50. Jorgensen N, Asklund C, Carlsen E, Skakkebaek NE. Coordinated European investigations of semen quality: results from studies of Scandinavian young men is a matter of concern. Int J Androl. 2006;29:54–61; discussion 105–8.
51. Neuwinger J, Behre HM, Nieschlag E. External quality control in the andrology laboratory: an experimental multicenter trial. Fertil Steril. 1990; 54:308–14.
52. Cooper TG, Neuwinger J, Bahrs S, Nieschlag E. Internal quality control of semen analysis. Fertil Steril. 1992;58:172–8.
53. Baker HWG. Male infertility. Philadelphia: Saunders; 2001.
54. Muller CH. Rationale, interpretation, validation, and uses of sperm function tests. J Androl. 2000;21:10–30.
55. Evenson DP, Wixon R. Clinical aspects of sperm DNA fragmentation detection and male infertility. Theriogenology. 2006;65:979–91.
56. Zini A, Libman J. Sperm DNA damage: importance in the era of assisted reproduction. Curr Opin Urol. 2006;16:428–34.
57. Tarozzi N, Bizzaro D, Flamigni C, Borini A. Clinical relevance of sperm DNA damage in assisted reproduction. Reprod Biomed Online. 2007;14:746–57.
58. Lewis SE, Agbaje IM. Using the alkaline comet assay in prognostic tests for male infertility and assisted reproductive technology outcomes. Mutagenesis. 2008;23:163–70.

59. Zini A, Sigman M. Are tests of sperm DNA damage clinically useful? Pros and cons. J Androl. 2009; 30:219–29.
60. Perreault SD, Aitken RJ, Baker HW, Evenson DP, Huszar G, Irvine DS, et al. Integrating new tests of sperm genetic integrity into semen analysis: breakout group discussion. Adv Exp Med Biol. 2003; 518:253–68.
61. Aitken RJ, De Iuliis GN. Value of DNA integrity assays for fertility evaluation. Soc Reprod Fertil Suppl. 2007;65:81–92.
62. Barratt CL, Aitken RJ, Björndahl L, Carrell DT, de Boer P, Kvist U, et al. Sperm DNA: organization, protection and vulnerability: from basic science to clinical applications–a position report. Hum Reprod. 2010;25:824–38.
63. Singh NP, Danner DB, Tice RR, McCoy MT, Collins GD, Schneider EL. Abundant alkali-sensitive sites in DNA of human and mouse sperm. Exp Cell Res. 1989;184:461–70.
64. Lewis SE, Agbaje I, Alvarez J. Sperm DNA tests as useful adjuncts to semen analysis. Syst Biol Reprod Med. 2008;54:111–25.
65. Enciso M, Muriel L, Fernández JL, Goyanes V, Segrelles E, Marcos M, et al. Infertile men with varicocele show a high relative proportion of sperm cells with intense nuclear damage level, evidenced by the sperm chromatin dispersion test. J Androl. 2006;27:106–11.
66. Gorczyca W, Traganos F, Jesionowska H, Darzynkiewicz Z. Presence of DNA strand breaks and increased sensitivity of DNA in situ to denaturation in abnormal human sperm cells: analogy to apoptosis of somatic cells. Exp Cell Res. 1993;207:202–5.
67. Mitchell LA, De Iuliis GN, Aitken RJ. The TUNEL assay consistently underestimates DNA damage in human spermatozoa and is influenced by DNA compaction and cell vitality: development of an improved methodology. Int J Androl. 2011;3:2–13.
68. Bianchi PG, Manicardi GC, Bizzaro D, Bianchi U, Sakkas D. Effect of deoxyribonucleic acid protamination on fluorochrome staining and in situ nick-translation of murine and human mature spermatozoa. Biol Reprod. 1993;49:1083–8.
69. Fernández JL, Vázquez-Gundín F, Delgado A, Goyanes VJ, Ramiro-Díaz J, de la Torre J, et al. DNA breakage detection-FISH (DBD-FISH) in human spermatozoa: technical variants evidence different structural features. Mutat Res. 2000;453:77–82.
70. Evenson DP, Darzynkiewicz Z, Melamed MR. Relation of mammalian sperm chromatin heterogeneity to fertility. Science. 1980;210:1131–3.
71. Fernandez JL, Muriel L, Rivero MT, Goyanes V, Vazquez R, Alvarez JG. The sperm chromatin dispersion test: a simple method for the determination of sperm DNA fragmentation. J Androl. 2003;24:59–66.
72. Fernández JL, Muriel L, Goyanes V, Segrelles E, Gosálvez J, Enciso M, et al. Simple determination of human sperm DNA fragmentation with an improved sperm chromatin dispersion test. Fertil Steril. 2005;84:833–42.
73. Erenpreiss J, Bars J, Lipatnikova V, Erenpreisa J, Zalkalns J. Comparative study of cytochemical tests for sperm chromatin integrity. J Androl. 2001; 22:45–53.
74. Manicardi GC, Bianchi PG, Pantano S, Azzoni P, Bizzaro D, Bianchi U, et al. Presence of endogenous nicks in DNA of ejaculated human spermatozoa and its relationship to chromomycin A3 accessibility. Biol Reprod. 1995;52:864–7.
75. Zubkova EV, Wade M, Robaire B. Changes in spermatozoal chromatin packaging and susceptibility to oxidative challenge during aging. Fertil Steril. 2005;84:1191–8.
76. Benchaib M, Ajina M, Lornage J, Niveleau A, Durand P, Guerin JF. Quantitation by image analysis of global DNA methylation in human spermatozoa and its prognostic value in in vitro fertilization: a preliminary study. Fertil Steril. 2003;80:947–53.
77. Benchaib M, Braun V, Ressnikof D, Lornage J, Durand P, Niveleau A, et al. Influence of global sperm DNA methylation on IVF results. Hum Reprod. 2005;20:768–73.
78. Leduc F, Nkoma GB, Boissonneault G. Spermiogenesis and DNA repair: a possible etiology of human infertility and genetic disorders. Syst Biol Reprod Med. 2008;54:3–10.
79. Oliva R. Protamines and male infertility. Hum Reprod Update. 2006;12:417–35.
80. Carrell DT, Emery BR, Hammoud S. Altered protamine expression and diminished spermatogenesis: what is the link? Hum Reprod Update. 2007;13:313–27.
81. Sakkas D, Seli E, Bizzaro D, Tarozzi N, Manicardi GC. Abnormal spermatozoa in the ejaculate: abortive apoptosis and faulty nuclear remodelling during spermatogenesis. Reprod Biomed Online. 2003; 7:428–32.
82. Delbès G, Hales BF, Robaire B. Toxicants and human sperm chromatin integrity. Mol Hum Reprod. 2010;16:14–22.
83. Kefer JC, Agarwal A, Sabanegh E. Role of antioxidants in the treatment of male infertility. Int J Urol. 2009;16:449–57.
84. Kodama H, Yamaguchi R, Fukuda J, Kasai H, Tanaka T. Increased oxidative deoxyribonucleic acid damage in the spermatozoa of infertile male patients. Fertil Steril. 1997;68:519–24.
85. Shen HM, Chia SE, Ong CN. Evaluation of oxidative DNA damage in human sperm and its association with male infertility. J Androl. 1999;20:718–23.
86. Aitken RJ, De Iuliis GN. Origins and consequences of DNA damage in male germ cells. Reprod Biomed Online. 2007;14:727–33.
87. Fraga CG, Motchnik PA, Wyrobek AJ, Rempel DM, Ames BN. Smoking and low antioxidant levels increase oxidative damage to DNA. Mutat Res. 1996;351:199–203.
88. Sepaniak S, Forges T, Gerard H, Foliguet B, Bene MC, Monnier-Barbarino P. The influence of cigarette smoking on human sperm quality and DNA fragmentation. Toxicology. 2006;223:54–60.

89. Pauli EM, Legro RS, Demers LM, Kunselman AR, Dodson WC, Lee PA. Diminished paternity and gonodal function with increasing obesity in men. Fertil Steril. 2008;90:346–51.
90. Aggerholm AS, Thulsttrup AM, Toft G, Ramlau-Hansen CH, Bonde JP. Is overweight a risk factor for reduced semen quality and altered serum sex hormone profile? Fertil Steril. 2008;90:619–26.
91. Hammoud AO, Wilde N, Gibson M, Parks A, Carrell DT, Meikle W. Male obesity and alterations in sperm parameters. Fertil Steril. 2008;90:2222–5.
92. Chavarro JE, Toth TL, Wright DL, Meeker JD, Hauser R. Body mass index in relation to semen quality, sperm DNA integrity, and serum reproductive hormone levels among men attending an infertility clinic. Fertil Steril. 2009;93:2222–31.
93. Evenson D, Jost L. Sperm chromatin structure assay is useful for fertility assessment. Methods Cell Sci. 2000;22:169–89.
94. Sergerie M, Mieusset R, Croute F, Daudin M, Bujan L. High risk of temporary alteration of semen parameters after recent acute febrile illness. Fertil Steril. 2007;88:970.e1–7.
95. Spanò M, Kolstad AH, Larsen SB, Cordelli E, Leter G, Giwercman A, et al. The applicability of the flow cytometric sperm chromatin structure assay in epidemiological studies. Asclepios. Hum Reprod. 1998;13:2495–505.
96. Spanò M, Bonde JP, Hjøllund HI, Kolstad HA, Cordelli E, Leter G. Sperm chromatin damage impairs human fertility. The Danish First Pregnancy Planner Study Team. Fertil Steril. 2000;73:43–50.
97. Bonde JP, Joffe M, Apostoli P, Dale A, Kiss P, Spano M, et al. Sperm count and chromatin structure in men exposed to inorganic lead: lowest adverse effect levels. Occup Environ Med. 2002;59:234–42.
98. Giwercman A, Richthoff J, Hjøllund H, Bonde JP, Jepson K, Frohm B, et al. Correlation between sperm motility and sperm chromatin structure assay parameters. Fertil Steril. 2003;80:1404–12.
99. Rignell-Hydbom A, Rylander L, Giwercman A, Jönsson BA, Lindh C, Eleuteri P, et al. Exposure to PCBs and p, p′-DDE and human sperm chromatin integrity. Environ Health Perspect. 2005;113:175–9.
100. Wyrobek AJ, Eskenazi B, Young S, Arnheim N, Tiemann-Boege I, Jabs EW, et al. Advancing age has differential effects on DNA damage, chromatin integrity, gene mutations, and aneuploidies in sperm. Proc Natl Acad Sci USA. 2006;103:9601–6.
101. Evenson DP, Jost LK, Baer RK, Turner TW, Schrader SM. Individuality of DNA denaturation patterns in human sperm as measured by the sperm chromatin structure assay. Reprod Toxicol. 1991;5:115–25.
102. Sun JG, Jurisicova A, Casper RF. Detection of deoxyribonucleic acid fragmentation in human sperm: correlation with fertilization in vitro. Biol Reprod. 1997;56:602–7.
103. De Jonge C, LaFromboise M, Bosmans E, Ombelet W, Cox A, Nijs M. Influence of the abstinence period on human sperm quality. Fertil Steril. 2004;82:57–65.
104. Kühnert B, Nieschlag E. Reproductive functions of the ageing male. Hum Reprod Update. 2004;10:327–39.
105. Moskovtsev SI, Willis J, Mullen JB. Age-related decline in sperm deoxyribonucleic acid integrity in patients evaluated for male infertility. Fertil Steril. 2006;85:496–9.
106. Belloc S, Benkhalifa M, Junca AM, Dumont M, Bacrie PC, Ménézo Y. Paternal age and sperm DNA decay: discrepancy between chromomycin and aniline blue staining. Reprod Biomed Online. 2009;19:264–9.
107. Sartorius GA, Nieschlag E. Paternal age and reproduction. Hum Reprod Update. 2010;16:65–79.
108. Giwercman A, Lindstedt L, Larsson M, Bungum M, Spano M, Levine RJ, et al. Sperm chromatin structure assay as an independent predictor of fertility in vivo: a case-control study. Int J Androl. 2010;33:e221–7.
109. Zini A, Bielecki R, Phang D, Zenzes MT. Correlations between two markers of sperm DNA integrity, DNA denaturation and DNA fragmentation, in fertile and infertile men. Fertil Steril. 2001;75:674–7.
110. Apedaile AE, Garrett C, Liu DY, Clarke GN, Johnston SA, Baker HW. Flow cytometry and microscopic acridine orange test: relationship with standard semen analysis. Reprod Biomed Online. 2004;8:398–407.
111. Trisini AT, Singh NP, Duty SM, Hauser R. Relationship between human semen parameters and deoxyribonucleic acid damage assessed by the neutral comet assay. Fertil Steril. 2004;82:1623–32.
112. Cohen-Bacrie P, Belloc S, Ménézo YJ, Clement P, Hamidi J, Benkhalifa M. Correlation between DNA damage and sperm parameters: a prospective study of 1,633 patients. Fertil Steril. 2009;91:1801–5.
113. de Jager C, Aneck-Hahn NH, Bornman MS, Farias P, Leter G, Eleuteri P, et al. Sperm chromatin integrity in DDT-exposed young men living in a malaria area in the Limpopo Province, South Africa. Hum Reprod. 2009;24:2429–38.
114. Moskovtsev SI, Willis J, White J, Mullen JB. Sperm DNA damage: correlation to severity of semen abnormalities. Urology. 2009;74:789–93.
115. Richthoff J, Spano M, Giwercman YL, Frohm B, Jepson K, Malm J, et al. The impact of testicular and accessory sex gland function on sperm chromatin integrity as assessed by the sperm chromatin structure assay (SCSA). Hum Reprod. 2002;17:3162–9.
116. Meeker JD, Singh NP, Hauser R. Serum concentrations of estradiol and free T4 are inversely correlated with sperm DNA damage in men from an infertility clinic. J Androl. 2008;29:379–88.
117. Keel BA. Within- and between-subject variation in semen parameters in infertile men and normal semen donors. Fertil Steril. 2006;85:128–34.
118. Zini A, Kamal K, Phang D, Willis J, Jarvi K. Biologic variability of sperm DNA denaturation in infertile men. Urology. 2001;58:258–61.
119. Sergerie M, Mieusset R, Daudin M, Thonneau P, Bujan L. Ten-year variation in semen parameters and

sperm deoxyribonucleic acid integrity in a healthy fertile man. Fertil Steril. 2006;86:1513.e11–8.
120. Sergerie M, Laforest G, Boulanger K, Bissonnette F, Bleau G. Longitudinal study of sperm DNA fragmentation as measured by terminal uridine nick end-labelling assay. Hum Reprod. 2005;20:1921–7.
121. Erenpreiss J, Bungum M, Spano M, Elzanaty S, Orbidans J, Giwercman A. Intra-individual variation in sperm chromatin structure assay parameters in men from infertile couples: clinical implications. Hum Reprod. 2006;21:2061–4.
122. Evenson DP, Jost LK, Marshall D, Zinaman MJ, Clegg E, Purvis K, et al. Utility of the sperm chromatin structure assay as a diagnostic and prognostic tool in the human fertility clinic. Hum Reprod. 1999;14:1039–49.
123. Evenson DP, Wixon R. Data analysis of two in vivo fertility studies using sperm chromatin structure assay-derived DNA fragmentation index vs. pregnancy outcome. Fertil Steril. 2008;90:1229–31.
124. Loft S, Kold-Jensen T, Hjollund NH, Giwercman A, Gyllemborg J, Ernst E, et al. Oxidative DNA damage in human sperm influences time to pregnancy. Hum Reprod. 2003;18:1265–72.
125. Zini A, Fischer MA, Sharir S, Shayegan B, Phang D, Jarvi K. Prevalence of abnormal sperm DNA denaturation in fertile and infertile men. Urology. 2002;60:1069–72.
126. Saleh RA, Agarwal A, Nelson DR, Nada EA, El-Tonsy MH, Alvarez JG, et al. Increased sperm nuclear DNA damage in normozoospermic infertile men: a prospective study. Fertil Steril. 2002;78:313–8.
127. Verit FF, Verit A, Kocyigit A, Ciftci H, Celik H, Koksal M. No increase in sperm DNA damage and seminal oxidative stress in patients with idiopathic infertility. Arch Gynecol Obstet. 2006;274:339–44.
128. Erenpreiss J, Elzanaty S, Giwercman A. Sperm DNA damage in men from infertile couples. Asian J Androl. 2008;10:786–90.
129. Bungum M, Humaidan P, Spano M, Jepson K, Bungum L, Giwercman A. The predictive value of sperm chromatin structure assay (SCSA) parameters for the outcome of intrauterine insemination, IVF and ICSI. Hum Reprod. 2004;19:1401–8.
130. Saleh RA, Agarwal A, Sharma RK, Said TM, Sikka SC, Thomas Jr AJ. Evaluation of nuclear DNA damage in spermatozoa from infertile men with varicocele. Fertil Steril. 2003;80:1431–6.
131. Smith R, Kaune H, Parodi D, Madariaga M, Rios R, Morales I, et al. Increased sperm DNA damage in patients with varicocele: relationship with seminal oxidative stress. Hum Reprod. 2006;21:986–93.
132. Blumer CG, Fariello RM, Restelli AE, Spaine DM, Bertolla RP, Cedenho AP. Sperm nuclear DNA fragmentation and mitochondrial activity in men with varicocele. Fertil Steril. 2008;90:1716–22.
133. Zini A, Blumenfeld A, Libman J, Willis J. Beneficial effect of microsurgical varicocelectomy on human sperm DNA integrity. Hum Reprod. 2005;20:1018–21.
134. Werthman P, Wixon R, Kasperson K, Evenson DP. Significant decrease in sperm deoxyribonucleic acid fragmentation after varicocelectomy. Fertil Steril. 2008;90:1800–4.
135. Nasr-Esfahani MH, Abasi H, Razavi S, Ashrafi S, Tavalaee M. Varicocelectomy: semen parameters and protamine deficiency. Int J Androl. 2009;32:115–22.
136. Moskovtsev SI, Lecker I, Mullen JB, Jarvi K, Willis J, White J, et al. Cause-specific treatment in patients with high sperm DNA damage resulted in significant DNA improvement. Syst Biol Reprod Med. 2009; 55:109–15.
137. Gallegos G, Ramos B, Santiso R, Goyanes V, Gosálvez J, Fernández JL. Sperm DNA fragmentation in infertile men with genitourinary infection by *Chlamydia trachomatis* and *Mycoplasma*. Fertil Steril. 2008;90:328–34.
138. Perera D, Pizzey A, Campbell A, Katz M, Porter J, Petrou M, et al. Sperm DNA damage in potentially fertile homozygous beta-thalassaemia patients with iron overload. Hum Reprod. 2002;17:1820–5.
139. Brackett NL, Ibrahim E, Grotas JA, Aballa TC, Lynne CM. Higher sperm DNA damage in semen from men with spinal cord injuries compared with controls. J Androl. 2008;29:93–9.
140. Agbaje IM, McVicar CM, Schock BC, McClure N, Atkinson AB, Rogers D, et al. Increased concentrations of the oxidative DNA adduct 7,8-dihydro-8-oxo-2-deoxyguanosine in the germ-line of men with type 1 diabetes. Reprod Biomed Online. 2008; 16:401–9.
141. Kobayashi H, Larson K, Sharma RK, Nelson DR, Evenson DP, Toma H, et al. DNA damage in patients with untreated cancer as measured by the sperm chromatin structure assay. Fertil Steril. 2001;75:469–75.
142. Ståhl O, Eberhard J, Cavallin-Ståhl E, Jepson K, Friberg B, Tingsmark C, et al. Sperm DNA integrity in cancer patients: the effect of disease and treatment. Int J Androl. 2009;32:695–703.
143. Evenson D, Wixon R. Meta-analysis of sperm DNA fragmentation using the sperm chromatin structure assay. Reprod Biomed Online. 2006;12:466–72.
144. Sergerie M, Laforest G, Bujan L, Bissonnette F, Bleau G. Sperm DNA fragmentation: threshold value in male fertility. Hum Reprod. 2005;20:3446–51.
145. Tsarev I, Bungum M, Giwercman A, Erenpreisa J, Ebessen T, Ernst E, et al. Evaluation of male fertility potential by Toluidine Blue test for sperm chromatin structure assessment. Hum Reprod. 2009;24:1569–74.
146. Alvarez JG. The predictive value of sperm chromatin structure assay. Hum Reprod. 2005;20:2365–7.
147. Greco E, Iacobelli M, Rienzi L, Ubaldi F, Ferrero S, Tesarik J. Reduction of the incidence of sperm DNA fragmentation by oral antioxidant treatment. J Androl. 2005;26:349–53.
148. Greco E, Romano S, Iacobelli M, Ferrero S, Baroni E, Minasi MG, et al. ICSI in cases of sperm DNA damage: beneficial effect of oral antioxidant treatment. Hum Reprod. 2005;20:2590–4.

149. Silver EW, Eskenazi B, Evenson DP, Block G, Young S, Wyrobek AJ. Effect of antioxidant intake on sperm chromatin stability in healthy nonsmoking men. J Androl. 2005;26:550–6.
150. Song GJ, Norkus EP, Lewis V. Relationship between seminal ascorbic acid and sperm DNA integrity in infertile men. Int J Androl. 2006;29:569–75.
151. Gandini L, Lombardo F, Paoli D, Caruso F, Eleuteri P, Leter G, et al. Full-term pregnancies achieved with ICSI despite high levels of sperm chromatin damage. Hum Reprod. 2004;19:1409–17.
152. Virro MR, Larson-Cook KL, Evenson DP. Sperm chromatin structure assay (SCSA) parameters are related to fertilization, blastocyst development, and ongoing pregnancy in in vitro fertilization and intracytoplasmic sperm injection cycles. Fertil Steril. 2004;81:1289–95.
153. Ménézo YJ, Hazout A, Panteix G, Robert F, Rollet J, Cohen-Bacrie P, et al. Antioxidants to reduce sperm DNA fragmentation: an unexpected adverse effect. Reprod Biomed Online. 2007;14:418–21.
154. Hull MG, Glazener CMA, Kelly NJ, Conway DI, Foster PA, Hinton RA, et al. Population study of causes, treatment, and outcome of infertility. BMJ. 1985;291:1693–7.
155. Hull MG, Fleming CF, Hughes AO, McDermott A. The age-related decline in female fecundity: a quantitative controlled study of implanting capacity and survival of individual embryos after in vitro fertilization. Fertil Steril. 1996;65:783–90.
156. Schmid TE, Eskenazi B, Baumgartner A, Marchetti F, Young S, Weldon R, et al. The effects of male age on sperm DNA damage in healthy non-smokers. Hum Reprod. 2007;22:180–7.
157. Chohan KR, Griffin JT, Lafromboise M, De Jonge CJ, Carrell DT. Comparison of chromatin assays for DNA fragmentation evaluation in human sperm. J Androl. 2006;27:53–9.
158. O'Flaherty C, Vaisheva F, Hales BF, Chan P, Robaire B. Characterization of sperm chromatin quality in testicular cancer and Hodgkin's lymphoma patients prior to chemotherapy. Hum Reprod. 2008;23:1044–52.
159. Ståhl O, Eberhard J, Jepson K, Spano M, Cwikiel M, Cavallin-Ståhl E, et al. Sperm DNA integrity in testicular cancer patients. Hum Reprod. 2006;21:3199–205.
160. Zhang LH, Qiu Y, Wang KH, Wang Q, Tao G, Wang LG. Measurement of sperm DNA fragmentation using bright-field microscopy: comparison between sperm chromatin dispersion test and terminal uridine nick-end labeling assay. Fertil Steril. 2010;94:1027–32.
161. Domínguez-Fandos D, Camejo MI, Ballescà JL, Oliva R. Human sperm DNA fragmentation: correlation of TUNEL results as assessed by flow cytometry and optical microscopy. Cytom A. 2007;71:1011–8.
162. Plastira K, Msaouel P, Angelopoulou R, Zanioti K, Plastiras A, Pothos A, et al. The effects of age on DNA fragmentation, chromatin packaging and conventional semen parameters in spermatozoa of oligoasthenoteratozoospermic patients. J Assist Reprod Genet. 2007;24:437–43.
163. Bian Q, Xu LC, Wang SL, Xia YK, Tan LF, Chen JF, et al. Study on the relation between occupational fenvalerate exposure and spermatozoa DNA damage of pesticide factory workers. Occup Environ Med. 2004;61:999–1005.
164. De Iuliis GN, Thomson LK, Mitchell LA, Finnie JM, Koppers AJ, Hedges A, et al. DNA damage in human spermatozoa is highly correlated with the efficiency of chromatin remodeling and the formation of 8-hydroxy-2′-deoxyguanosine, a marker of oxidative stress. Biol Reprod. 2009;81:517–24.
165. Tavalaee M, Razavi S, Nasr-Esfahani MH. Influence of sperm chromatin anomalies on assisted reproductive technology outcome. Fertil Steril. 2009;91:1119–26.
166. Høst E, Lindenberg S, Smidt-Jensen S. DNA strand breaks in human spermatozoa: correlation with fertilization in vitro in oligozoospermic men and in men with unexplained infertility. Acta Obstet Gynecol Scand. 2000;79:189–93.
167. Gandini L, Lombardo F, Paoli D, Caponecchia L, Familiari G, Verlengia C, et al. Study of apoptotic DNA fragmentation in human spermatozoa. Hum Reprod. 2000;15:830–9.
168. Pant N, Shukla M, Kumar Patel D, Shukla Y, Mathur N, Kumar Gupta Y, et al. Correlation of phthalate exposures with semen quality. Toxicol Appl Pharmacol. 2008; 231:112–6.
169. Hammadeh ME, Zeginiadov T, Rosenbaum P, Georg T, Schmidt W, Strehler E. Predictive value of sperm chromatin condensation (aniline blue staining) in the assessment of male fertility. Arch Androl 2001; 46:99–104.

23 Aging and Sperm DNA Damage

Fábio F. Pasqualotto and Eleonora B. Pasqualotto

Abstract

As life expectancy increases and our lives become busier every day, many couples are waiting longer to establish their families. Female fecundity declines slowly after age 30 and more rapidly after 40 and is considered the main limiting factor in the treatment of infertility. Also, the effects of paternal age on a couple's fertility are real and may be greater than have previously been thought. After adjustments for other factors, it has been demonstrated that the probability that a fertile couple will take >12 months to conceive nearly doubles from 8% when the man is <25 years to 15% when he is >35 years; thus, paternal age is a further factor to be taken into account when deciding the prognosis of infertile couples. Also, the increased male age is associated with a significant decline in fertility (fivefold longer time to pregnancy at the age of 45 years), which is independent of the woman's age, coital frequency, and lifestyle effect, as well as the effect of other subfertility risk factors. Furthermore, fathering at older ages may have significant effects on the viability and genetic health of human pregnancies and offspring, primarily as a result of structural chromosomal aberrations in sperm. The evidence for sex chromosomal aneuploidy suggests that there may be about a twofold increase in risk at the age of 50. In fact, the risk for a father over 40 years old to have a child with an autosomal dominant mutation equals the risk of Down syndrome for a child whose mother is 35–40 years old. Recent reports have raised concern about decreasing male fertility caused by genomic abnormalities. There are reports of increased congenital anomalies and testicular cancer in children. Sperm DNA is known to contribute one half of the genomic material to offspring. Thus, normal sperm genetic material is required for

F.F. Pasqualotto (✉)
Department of Urology, University of Caxias do Sul, Rua Pinheiro Machado, 2569, sl 23/24, Bairro São Pelegrino, Caxias do Sul, RS, Brazil
e-mail: fabio@conception-rs.com.br

A. Zini and A. Agarwal (eds.), *Sperm Chromatin: Biological and Clinical Applications in Male Infertility and Assisted Reproduction*, DOI 10.1007/978-1-4419-6857-9_23,

fertilization, embryo and fetal development, and postnatal child well-being. Abnormal DNA can lead to derangements in any of these processes. The abnormality or defect in the genomic material may take the form of condensation or nuclear maturity defects, DNA breaks or DNA integrity defects, and sperm chromosomal aneuploidy. Evidence for the decline in men's fertility with increasing age and its quantification are provided.

Keywords
Aging and sperm • Sperm DNA damage • DNA damage in sperm • Oxidative stress in sperm

It has become more socially acceptable to delay fatherhood, but the heritable consequences of this trend remain poorly understood. Approximately 15% of couples of reproductive age experience infertility, and approximately 1/3 to half of infertility cases may be attributed to male factors [1]. It is well known that maternal age is a significant contributor to human infertility [2], primarily due to the precipitous loss of functional oocytes in women by their late 30s [3]. Human spermatogenesis, on the contrary, continues well into advanced ages, allowing men to reproduce during senescence. Although very little is known about the topic, paternal age may also contribute to human infertility.

It is well known that practically no children are born to mothers aged >50 years and it is common to all older fathers that they have younger partners. The discrepancy in the reproductive arena between males and females is astonishing, and reduced fertility and higher reproductive risks associated with advancing maternal age raise the question whether advanced paternal age is also associated with compromised fertility and increasing risks. In addition, it is well documented that because of a progressive decrease of fertility due to both quantitative and qualitative loss of oocytes, eventually ending in menopause, women experience an age-dependent increase of miscarriages, obstetric morbidities, and chromosomal anomalies of the fetus [4]. This question should be discussed with younger age groups, since increasing numbers of couples postpone parenthood into their fourth or fifth decade of life.

In contrast to the female, male reproductive functions do not cease abruptly, but androgen production and spermatogenesis continue lifelong. However, evaluating a possible decline in the semen quality is a little bit difficult. Some men are reluctant to provide semen samples unless actively concerned about their fertility. For instance, population-based studies typically recruit at least 20% of young men willing to provide semen samples [5], constituting an inevitable participation bias in such studies [6, 7]. In addition, most of the published studies on sperm output in older men are largely restricted to patients attending infertility clinics, where few are older than 50 years [8]. An uncertain, but probably high, proportion of such men have unrecognized defects in sperm production and/or function. Furthermore, access to such specialized medical services may be strongly influenced by nonbiological factors, and the results from infertility clinics may not be reliably extrapolated to the general male population.

Anyway, the effects of paternal age on a couple's fertility are real and may be greater than have previously been thought. Ford et al. [9] stated that, after adjustments for other factors, the probability that a fertile couple will take >12 months to conceive nearly doubles from 8% when the man is <25 years to 15% when he is >35 years; thus, paternal age is a further factor to be taken into account when deciding the prognosis of infertile couples. It has been demonstrated that in men between 26 and 59 years of age and who undergo IVF or ICSI treatment, the rising age is detrimental to sperm DNA integrity and ejaculate volume [10]. Furthermore, Dutch men showed a significantly poorer sperm quality based on a higher DFI compared to migrants. In fact, the age-related decrease in sperm quality below 59 years of age – based on increased sperm DNA damage and decreased ejaculate volume – suggests that delaying childbearing, not only in women but also in men, contributes to a reduced reproductive capacity. The significantly higher DFI in Dutch men compared to migrants could not be explained by differences in age and the most prominent unhealthy lifestyles.

To explain the age-dependent changes observed in semen quality, two issues should be considered [8–12]. First, cellular or physiological changes due to aging have been described in testicles, seminal vesicles, prostate, and epididymis. Age-related narrowing and sclerosis of the testicular tubular lumen, decreases in spermatogenic activity, increased degeneration of germ cells, and decreased numbers and function of Leydig cells have been found in autopsies of men who died from accidental causes [13]. Smooth muscle atrophy and a decrease in protein and water content, which occur in the prostate with aging, may contribute to decreased semen volume and sperm motility. Also, the epididymis, a hormonally sensitive tissue, may undergo age-related changes. The hormonal or epididymal senescence may lead to decreased motility in older men. Second, increasing age implies more frequent exposure to exogenous damage or disease [8]. In addition to age per se, factors such as urogenital infections, vascular diseases, or an accumulation of toxic substances (cigarettes) may be responsible for worsening semen parameters. Indeed, a retrospective cross-sectional study in 3,698 infertile men showed an infection rate of the accessory glands in 6.1% in patients aged <25 years but in 13.6% of patients >40 years, and total sperm counts were significantly lower in patients with an infection of the accessory glands [14]. In addition, an age-dependent increase of polychlorinated biphenyls (PBC) in men has been described, and in men with normal semen parameters, the PBC concentration is inversely correlated with sperm count and progressive motility [15]. The concentration of cadmium also increases with age in the human testis, epididymis, and prostate, although lead and selenium remain constant over the whole age range in the reproductive organs [16, 17].

Handelsman and Staraj [18] demonstrated that after exclusion of men with different diseases associated with diminishing testicular size, the specific effects of age on testicular volume appears only in the eighth decade of life. In healthy men of this age group, the testis volume is 31% lower than in 18–40-year-old men [19]. However, recently a study has shown a decline in testicular volume over time, specifically, after the age of 45 [20].

Morphological characteristics of aging testes varies from Sertoli cells accumulating cytoplasmic lipid droplets to the cells with reduced number of the droplets [21], as are the Leydig cells [22], which may also be multinucleated [23]. Tubule involution is associated with an enlargement of the tunica propria, leading to progressive sclerosis parallel to a reduction of the seminiferous epithelium with complete tubular sclerosis as an end point [24]. Testicular sclerosis is associated with defective vascularization of the testicular parenchyma and with systemic arteriosclerosis of affected men [25]. Arteriographic patterns of the epididymis and the testes support these findings and are correlated with the degree of systemic arteriosclerosis [25]. In addition, age-dependent alterations of the prostate are well known [26] and are detectable histologically in 50% of 50 year-old men, but in 90% of men aged >90 years [27].

Semen Analysis

Considering the age-dependent changes in reproductive organs of men, variations in semen parameters over time are not surprising; however, only few studies are controlled for abstinence time and other possible factors that may influence semen quality such as hypertension or smoking habits. Most studies are retrospective and rarely include males with more than 60 or 70 years old. Pasqualotto et al. [20] have recently described a decrease in semen volume across the groups evaluated in the study. In fact, reports in the literature have shown a decrease in semen volume with aging [8, 28, 29]. The higher number of days' abstinence in men over 50 years old could explain these results. In the studies where the analyses were adjusted for abstinence days, a decrease in semen volume of 3–22% was observed [9].

Regarding sperm motility, many studies adjusted for time of abstinence found a significant decrease in sperm motility associated with age and a yearly decrease ranging between 0.17 [30] and 0.7% [31]. However, these studies were performed in sperm donors [30–33] as well as infertile patients [34, 35]. Pasqualotto et al. are on the same page as others showing that sperm motility tends to decrease over time. Those studies that have been adjusted for duration of abstinence have reported statistically significant effects, such as negative linear relationships and decreases in motility ranging from 0.17 to 0.6% for each year of age [8, 30, 36, 37].

Computer-assisted semen analysis (CASA) has been developed as a specific tool to make the assessment of semen quality more objective and detailed [38]. Several specific motility parameters describing the movements of spermatozoa in a more detailed manner can be obtained with CASA. In addition, the classification into motile and immotile spermatozoa can be based on well-defined velocity thresholds. However, no correlations are detected between specific motion parameters as evaluated with CASA and the aging effect in the study by Pasqualotto et al. [20].

When focusing on sperm concentration, abstinence-adjusted studies do not provide a uniform picture. Even though some studies have reported a decrease in sperm concentration with increased age, several other studies have reported an increase in sperm concentration with age or found little or no association between age and sperm concentration [9, 12, 36, 39]. In fact, there are two different populations that we have to be considered before evaluating the results: fertile vs. infertile men. A significant age-dependent decrease [31, 33] as well as constant values over the age range [32] or even a nonsignificant age-dependent increase with age [30] has been detected in healthy men. Regarding the infertile population, sperm concentration increases [34, 35] or remains unaltered [14], as indicated in abstinence-adjusted studies.

One of the good indicators of the germinal epithelium status is the sperm morphology. Degenerative changes in the germinal epithelium because of aging may affect spermatogenesis and, thus, sperm morphology. Pasqualotto et al. [20], based on a linear regression analysis, stated that normal sperm morphology tends to decrease by 0.039% each year. Auger et al. [33], in a linear regression model, have shown that the normal sperm morphology decreases 0.9% yearly. Thus, as compared to an average 30-year-old man, an average 50-year-old man had a 18% decrease in normally shaped sperm [34]. Ng et al. [29] showed that older men had more abnormal sperm morphology with decreasing numbers of normal forms and reduced vitality, as well as increased numbers of cytoplasmic droplets and sperm tail abnormalities (30% vs. 17%) compared to younger men. The aberrant sperm morphology in older men was most evident in defects of tail morphology, possibly reflecting the complex cellular structural assembly process of the axoneme. Such increasing proportion of defects may reflect degenerative changes with aging in the germinal epithelium and/or in the intrinsic program directing spermiogenesis. In fact, the decrease per year varies from 0.2 [37] to 0.9% [33].

All reported changes of histological and seminal parameters develop gradually without a sudden age threshold. The alterations in semen parameters fall within normal ranges. Nevertheless, the age-dependent alterations of testicular histology and semen parameters are accompanied by a significant increase in FSH [20, 40] and a slight but significant decrease in inhibin B [19, 41], which are also found in men with apparently normal semen parameters.

Fertility of Aging Men

Without any type of doubt, male fertility is basically maintained until very late in life, and it has been documented scientifically up to more than

90 years of age [42]. Besides female age, further confounders, such as reduced coital frequency, an increasing incidence of erectile dysfunction, and smoking habits have to be considered in studies analyzing male fertility. All studies focused on a nonclinical population found a significant negative relationship between male age and couples' fertility.

A retrospective study of a large sample of European couples analyzed the risk of difficulties (due to adverse pregnancy outcome, such as ectopic pregnancy, miscarriage, or stillbirth, or due to delayed conception) and the risk of delay in pregnancy onset [43]. Age-related changes were also found in a prospective study that estimated day-specific probabilities for pregnancy relative to ovulation [44]. Frequency of sexual intercourse was monitored by sexual diaries, and ovulation was based on basal body temperature measurements. According to this study, fertility for men aged >35 years is significantly reduced and the age effect of men aged 35–40 years is about the same as when intercourse frequency drops from twice per week to once per week [45]. In studies dealing with subfertile couples, a significant decrease in pregnancy rates [34] or an increase in TTP [46] was observed with female, but not with male, age, possibly indicating that male age-dependent alterations are masked by the infertility as such.

With methods of assisted reproduction, prerequisites for natural conception such as motility or fertilizing capacity are circumvented. In fact, the more invasive the treatment, the less important male age appears. Therefore, the success rates of ISCI [47] or IVF [48–50] are not associated with male age. On the contrary, the success rate of intrauterine insemination (IUI), a method requiring much higher quality and capability of sperm, is without question related to male age [51, 52].

Genetic Risks of the Aging Male

A maternal age effect has been found for all trisomy conditions but varies among chromosomes, with an exponential increase in chromosome 21 and a linear increase, for instance, in chromosome 16 [53]. Early observations also associate paternal age with certain syndromes [54]. Meanwhile, it has become evident that some mutations, consisting of single-base substitutions in three different genes, namely, RET, FGFR2 (fibroblast growth factor receptor 2), and FGFR3 (fibroblast growth factor receptor 3), are exclusively of paternal origin and may increase with male age [55].

A possible explanation for this male-specific age effect is the much higher number of germ cell divisions in males than in females: in the fetal ovary, germ cells undergo 22 mitotic divisions before they enter the meiotic prophase [56]. They remain in meiotic arrest and continue meiosis in adulthood when ovulation has taken place. Thus, while it was formerly believed that in women germ cell divisions are completed before birth, a recent publication has suggested that adult mouse ovaries still possess mitotically active germ cells [57].

On the contrary, male germ cells divide continuously. It has been estimated that 30 spermatogonial stem cell divisions take place before puberty, when they begin to undergo meiotic divisions. From then on, 23 mitotic divisions per year occur, resulting in 150 replications by the age of 20 years and 840 replications by the age of 50 years [55]. Therefore, due to these numerous divisions of stem cells, older men may have an increased risk of errors in DNA transcription. Consequently, the association between elevated paternal age and serious birth defects is the reason why the age of semen donors is limited to 40 years in certain countries [58, 59]. On the contrary, male age is not an indicator for prenatal diagnosis.

Numerical Chromosome Disorders

Aneuploidy, the presence of an extra or missing chromosome, is the leading genetic cause of pregnancy loss. Aneuploidies are detected in 35% of spontaneous abortions, in 4% of stillbirths, and in 0.3% of live births [60]. Among spontaneous abortions, Turner's syndrome (45,X) and trisomy 16, 21, and 22 are the most prevalent aneuploidies. In general, aneuploidies arise by the process of nondisjunction, for instance, the failure of paired chromosomes to separate in the first meiotic division of maternal meiosis [61, 62]. Sperm reveal an aneuploidy incidence of 2% with a high variability of disomy frequency of individual sperm from different fluorescence in situ hybridization (FISH) studies [61]. The disomy frequency was calculated to be 0.26% for the sex chromosomes and 0.15% for the autosomes with an exception for chromosomes 14, 21, and 22, which display higher disomy frequencies [63].

Studies analyzing the age-dependent alteration of aneuploidy frequency in chromosomes are highly limited due to low case numbers. Interestingly, the age-dependent increase of XY disomy was also detected in sperm from fathers of boys with Klinefelter's syndrome [64], irrespective of paternal or maternal inheritance of the extra X chromosome [31]. Fifty percent of Klinefelter's syndrome cases are of paternal origin, and other gonosomal aneuploidies are even more often paternally inherited in live births, as are 80% of Turner's syndrome cases (45,X) and 100% of XYY karyotypes [65, 66]. However, none of these syndromes are related to paternal age [67]. Similarly, the incidence of autosomal aneuploidies, such as trisomy 13, 16, and 18, is independent of paternal age [66, 68]. Therefore, the paternal age effect for trisomy 21 remains to be elucidated.

Early studies with small sample sizes reflect different results in the same study population depending on the method of statistical analysis [68, 69]. In spontaneous abortions, a nonsignificant paternal age effect was detected [66], and in live births, no age effect [70, 71] or a significant paternal age effect [72, 73] was evident. It should be kept in mind that only 10% of Down's syndrome patients receive the excess chromosome from their father [74], so an age effect could be confined to this small category of cases and subtle age effects might go undetected unless those derived paternally are considered separately. However, with respect to paternally inherited Down's syndrome cases, no paternal age effect became evident [75]. Paternal age effect was seen in association with a maternal age lower than 35 years, so a paternal age effect in aged couples can no longer be neglected concerning trisomy 21, whereas other autosomal or sex chromosomal aneuploidies are not associated with increased paternal age [72].

Structural Chromosomal Anomalies

Structural chromosomal anomalies result from chromosomal breakage and the following abnormal rearrangement within the same or within different chromosomes. In 84% of cases, de novo structural aberrations are of paternal origin [76], and they are found in 2% of spontaneous abortions and in 0.6% of live births [77]. Cytogenetic studies on structural chromosomal anomalies in sperm are rare but consistently describe an increase of mutations with age [78].

FISH was used for the structural analysis of individual chromosomes: duplications and deletions for the centromeric and subtelomeric regions of chromosome 9 increase significantly with age [79]. In spite of these age-dependent structural alterations in sperm, no increase of de novo structural chromosomal anomalies has been detected in newborns of older fathers [74].

Autosomal Dominant Diseases

There is a direct relationship between paternal aging and offspring development. Considerable evidence shows a connection between aging and offspring learning and cognition.

Achondroplasia, the most common form of dwarfism, is the first genetic disorder that was hypothesized to have a paternal age component [55]. Apert's syndrome and achondroplasia have been amenable to direct sperm DNA mutation analysis [80, 81], and both are characterized by an age-dependent increase of mutations in sperm, but there are some peculiarities. For sporadic cases of Crouzon's or Pfeiffer's syndrome, 11 different mutations of the FGFR2 gene are responsible, indicating that, unlike Apert's syndrome or achondroplasia, these are genetically heterogeneous conditions [82]. These mutations also arise in the male germ line, and advanced paternal age was noted for fathers of those patients.

The relationship between mutation frequency and paternal age is heterogeneous among autosomal dominantly inherited diseases [83]. In contrast to the above-mentioned diseases, osteogenesis imperfecta, neurofibromatosis, or bilateral retinoblastoma shows a weak paternal age effect [84]. This may be due to the fact that a significant fraction of new mutations is not base substitutions [55]. Many of the mutations of the neurofibromatosis gene are intragenic deletions. These deletions are not age-dependent because they occur by mechanisms other than the base substitutions and are maternally derived in 16 of 21 cases [85].

Owing to this heterogeneity of the paternal age effect in autosomal dominant diseases, the risk estimates proposed by Friedman for paternal age and autosomal dominant mutations may be overestimated [86]. Friedman calculated a risk for autosomal dominant diseases of 0.3–0.5% among offspring of fathers aged >40 years. This risk is comparable with the risk of Down's syndrome for 35–40 year old women. However, the calculation was based on the assumption that the paternal age effect found in achondroplasia is typical of all autosomal dominant diseases.

There are conflicting are the data for Alzheimer's disease. Few studies conclude that paternal age is a risk factor [87]. However, the inconsistent results may be due to small sample sizes of the studies or due to the genetic heterogeneity of the disease.

Regarding schizophrenia, there are more conclusive data. In fact, the studies identified an increased risk of schizophrenia with paternal age [88]. Patients without a family history of schizophrenia had significantly older fathers than familial patients, so de novo mutations were considered responsible [89]. Preeclampsia, which is considered to be a risk factor for schizophrenia, is also associated with paternal age [90].

A recent study, by using 1997–2004 data from the National Birth Defects Prevention Study, has performed a logistic regression models with paternal and maternal age as continuous variables while adjusting for demographic and other factors. They demonstrated elevated odds ratios (ORs) for each year increase in paternal age in the following congenital malformation: cleft palate (OR, 1.02, 95% confidence interval [95% CI], 1.00–1.04), diaphragmatic hernia (OR, 1.04; 95% CI, 1.02–1.06), right ventricular outflow tract obstruction (OR, 1.03; 95% CI, 1.01–1.04), and pulmonary valve stenosis (OR, 1.02, 95% CI, 1.01–1.04). At younger paternal ages, each year increase in paternal age correlated with increased OR of having offspring with encephalocele, cataract, esophageal atresia, anomalous pulmonary venous return, and coarctation of the aorta, but these increased ORs were not observed at older paternal ages. The effect of paternal age was modified by maternal age for gastroschisis, omphalocele, spina bifida, all orofacial clefts, and septal heart defects. This study suggested that paternal age may be a risk factor for some multifactorial birth defects [91].

One very important point we should never forget is that advanced paternal age increases the risk of other cancers in offspring. According to the Swedish Family-Cancer Database, there is an effect of paternal age on the incidence of sporadic breast and sporadic nervous system cancer in offspring [92]. Interestingly, an association between paternal age and the son's risk of prostate cancer was found [93]. The association of paternal age with early-onset prostate cancer (<65 years) was greater than that with late-onset prostate cancer. In fact, older men are having children, but the reality of a male biological clock makes this trend worrisome [94].

Sperm DNA Damage

Sperm chromatin and DNA integrity is essential to ensure that the fertilizing sperm can support normal embryonic development of the zygote. To better inform treatment pathways and, more importantly, to ensure a generation of healthy children from assisted reproductive technologies (ART), we urgently require tests of sperm function, including the normalcy of sperm DNA, that provide high quality and robust diagnostic and prognostic information.

Understanding the effects of male age on sperm DNA damage is especially relevant for men attending reproductive clinics because of the increasing reliance on modern technologies, especially among marginally fertile older men. ICSI and IVF enhance the probability of achieving fatherhood, yet they also circumvent the natural barriers against fertilization by damaged sperm.

Schmid et al. [95] demonstrated an association between male age and sperm DNA strand damage in a nonclinical sample of active healthy nonsmoking workers and retirees. Sperm of older men had significantly higher frequencies of sperm with DNA damage measured under alkaline conditions, which is thought to represent alkali labile DNA sites and single-strand DNA breaks. However, age was not associated with sperm DNA damage under neutral conditions, which is thought to represent double-strand DNA breaks. The observations of differential effects of age on genomic damage is consistent with the recent finding of Wyrobek et al. [96] who reported age-related effects on DNA fragmentation and achondroplasia mutations, but not on aneuploidy, Apert syndrome mutations, or sex ratio.

The finding of age-related increases in DNA strand damage under alkaline conditions is consistent with the findings of Morris [97] who studied 60 men participating in an IVF program. They reported that sperm DNA damage was positively correlated with donor age and with impairment of postfertilization embryo cleavage following ICSI, indicating an overall decline in the integrity of sperm DNA in older men. The findings by Schmid et al. of no association between age and sperm DNA damage under neutral conditions are in contrast with the study of Singh et al. [98] who studied 66 men, aged 20–57 years, from an infertility clinic and a nonclinical group. However, Singh et al. [98] did not investigate sperm DNA damage under alkaline conditions in sperm, and Morris [97] did not investigate sperm damage under neutral conditions. Using a different assay for measuring DNA strand damage in sperm, the SCSA, Spano et al. [99] found a strong association of DFI with age among men 18–55 year olds, a finding confirmed by Wyrobek et al. [96] using a larger group of men that spanned 20–80 years of age.

Older men may produce more sperm with DNA damage as a consequence of age-associated increased oxidative stress in their reproductive tracts [100, 101]. Oxidative stress can damage sperm DNA as well as mitochondrial and nuclear membranes [102, 103]. Kodama et al. [103] reported an association between oxidative DNA damage in sperm and male infertility. Alternatively, apoptotic functions of spermatogenesis may be less effective in older males resulting in the release of more sperm with DNA damage [104, 105]. While apoptosis has been identified in the testes of elderly men [104], there have been no comparisons on rates of apoptosis among men of different ages. Increased sperm DNA damage has been associated with chromosomal abnormalities, developmental loss, and birth defects in mouse model systems [106, 107] and with increases in the percentage of human embryos that failed to develop after ICSI [97].

Increasing oxidative stress levels associated with aging might be responsible for this increase in DNA damage with age. Oxidative stress-mediated DNA damage may be an etiology for repeated ART failures in older men. Increasing male age may have an influence on DNA fragmentation in the form of single-strand breaks. This may not have any effect on fertilization because the oocyte can repair single-strand breaks. However, if the oocyte repair mechanisms are dysfunctional, this may result in poor, if not failed, blastocyst formation. Thus, oxidative stress-induced DNA damage can lead to various genomic defects [108].

Oxidative Stress and Aging

Mitochondria play an important role in cellular energy generation, apoptosis regulation, and calcium homeostasis [109]. Coupled to the tricarboxylic acid cycle, the electron transport chain (ETC) and adenosine triphosphate (ATP) synthase in the mitochondria generate ATP, a source of most cellular energy. Since reactive oxygen species (ROS) are continually produced in the mitochondria of spermatozoa, they play an important role in age-related male reproductive pathophysiology.

Mitochondria are more susceptible to oxidative damage because of the active production of ROS, as mtDNA is not protected by histones. Consequently, mitochondrial ROS production damages the mitochondria themselves. Mutations or deletions in mtDNA lead to defects in oxidative phosphorylations, defective cellular calcium dyshomeostasis, and other related mtDNA diseases. A dysfunctional mitochondrial respiratory chain would lead to more ROS production. Oxidative damage is more prevalent in mtDNA and protein in vivo than in other cell components.

The increased ROS level in semen observed with aging is associated with a possible decrease in antioxidant enzyme activity. This imbalance between prooxidants and antioxidants induces oxidative damage, resulting in abnormalities in telomeres and telomerase in sperm cells [110–112]. This sequence of events may explain the decrease in sperm concentration seen with aging. ROS-induced telomere shortening may be due to direct injury to guanine repeat telomere DNA by ROS. The addition of an antioxidant suppresses the rate of telomere shortening in somatic cells. The telomere shortening rate slowed after enrichment by ascorbic acid, a strong antioxidant. The rate of telomere shortening in sheep and humans is directly related to the cellular oxidative stress levels [109].

Oxidative stress in aging male reproductive system may inhibit sperm axonemal phosphorylation and increase lipid peroxidation, which can decrease sperm motility. This oxidative stress can also lead to lipofuscin and amyloid accumulation in the male reproductive tract, potentially the cause of decreased Leydig cell function and a subsequent decrease in blood testosterone levels. A higher rate of lipofuscin accumulation in turn may increase the amount of dysfunctional mitochondria in spermatozoa, thus increasing ROS formation. Along with its negative effect on the fertilizing potential of spermatozoa, ROS also leads to offspring malformation (if fertilization is successful). Oxidative stress-induced mtDNA damage and nuclear DNA damage in aging men may put them at a higher risk for transmitting multiple genetic and chromosomal defects [109].

Therefore, ROS might play a central role in decreased male fertility with aging. This hypothesis provides guidance for future study and experiments, focusing on specific biomarkers of aging in men (telomere function, lipofuscin, amyloid) and their comparison with semen parameters and male fertility.

Conclusion

Couples are waiting longer to have children, and advances in reproductive technology are allowing older men and women to consider having children. The lack of appreciation among both medical professionals and the lay public for the reality of a male biological clock makes these trends worrisome. The age-related changes associated with the male biological clock affect sperm quality, fertility, hormone levels, and a lot of nonreproductive physiological issues. Although based on a small number of cases, the data presented for testicular morphology, semen parameters, and fertility in aging males are conclusive and reflect a gradual deterioration with age within a broad individual spectrum. Most studies suggest that reduced fertility begins to become evident in the late 30s in men. Increased male age is associated with an increased risk of miscarriages, and both the risk of infertility and the risk of miscarriage strongly depend on female age. Advancing paternal age is associated with an increased risk for trisomy 21 and with diseases of complex etiology such as schizophrenia.

Advanced paternal age increases the risk for spontaneous abortion as well as genetic abnormalities in offspring due to multiple factors, including DNA damage from abnormal apoptosis and ROS. Older men considering parenthood should have a thorough history and physical examination focused on their sexual and reproductive capacity. Such examination should entail disclosure of any sexual dysfunction and the use of medications, drugs, or lifestyle factors that might impair fertility or sexual response.

Couples should be aware of these age-dependent alterations in fertility and predisposition to genetic risks. Although at the moment increased paternal age is not an indication for prenatal diagnosis, there may be further developments in the future.

References

1. Templeton A. Infertility – epidemiology, aetiology and effective management. Health Bull (Edinb). 1995;53(5):294–8.
2. Joffe M, Li Z. Male and female factors in fertility. Am J Epidemiol. 1995;141(11):1107–8.
3. Lansac J. Delayed parenting. Is delayed childbearing a good thing? Hum Reprod. 1995;10(5):1033–5.
4. Lubna P, Santoro N. Age-related decline in fertility. Endocrinol Metab Clin N Am. 2003;32:669–88.
5. Jensen TK, Jorgensen N, Punab M, Haugen TB, Suominen J, Zilaitiene B, et al. Association of in utero exposure to maternal smoking with reduced semen quality and testis size in adulthood: a cross-sectional study of 1770 young men from the general population in five European countries. Am J Epidemiol. 2004;159:49–58.
6. Handelsman DJ. Sperm output of healthy men in Australia: magnitude of bias due to self-selected volunteers. Hum Reprod. 1997;12:2701–5.
7. Cohn BA, Overstreet JW, Fogel RJ, Brazil CK, Baird DD, Cirillo PM. Epidemiologic studies of human semen quality: considerations for study design. Am J Epidemiol. 2002;155:664–71.
8. Kidd SA, Eskenazi B, Wyrobek AJ. Effects of male age on semen quality and fertility: a review of the literature. Fertil Steril. 2001;75:237–48.
9. Ford WCL, North K, Taylor H, Farrow A, Hull MGR, Golding J and the ALSPAC Study Team. Increasing paternal age is associated with delayed conception in a large population of fertile couples: evidence for declining fecundity in older men. Hum Reprod. 2000;15:1703–8.
10. Hammiche F, Laven JSE, Boxmeer JC, Dohle GR, Steegers EAP, Steegers-Theunissen RPM. Sperm quality decline among men below 60 years of age undergoing IVF or ICSI treatment. J Androl. 2011; 32(1):70–6.
11. Eskenazi B, Wyrobek AJ, Sloter E, Kidd SA, Moore L, Young S, et al. The association of age and semen quality in healthy men. Hum Reprod. 2003;18:447–54.
12. Hassan MAM, Killick SR. Effect of male age on fertility: evidence for the decline in male fertility with increasing age. Fertil Steril. 2003;79:1520–7.
13. Neaves WB, Johnson L, Porter JC, Parker CR, Petty S. Leydig cell numbers, daily sperm production and serum gonadotropin levels in aging men. J Clin Endocrinol Metab. 1984;59:756–63.
14. Rolf C, Kenkel S, Nieschlag E. Age-related disease pattern in infertile men: increasing incidence of infections in older patients. Andrologia. 2002;34: 209–17.
15. Dallinga JW, Moonen EJ, Dumoulin JC, Evers JL, Geraedts JP, Kleinjans JC. Decreased human semen quality and organochlorine compounds in blood. Hum Reprod. 2002;17:1973–99.
16. Oldereid NB, Thomassen Y, Attramadal A, Olaisen B, Purvis K. Concentrations of lead, cadmium and zinc in the tissues of reproductive organs of men. J Reprod Fertil. 1993;99:421–55.
17. Oldereid NB, Thomassen Y, Purvis K. Selenium in human male reproductive organs. Hum Reprod. 1998;13:2172–6.
18. Handelsman DJ, Staraj S. Testicular size: the effects of aging, malnutrition, and illness. J Androl. 1985; 6:144–51.
19. Mahmoud AM, Goemaere S, El-Garem Y, Van Pottelbergh I, Comhaire FH, Kaufman JM. Testicular volume in relation to hormonal indices of gonadal function in community-dwelling elderly men. J Clin Endocrinol Metab. 2003;88:179–84.
20. Pasqualotto FF, Sobreiro BP, Hallak J, Pasqualotto EB, Lucon AM. Sperm concentration and normal sperm morphology decrease and follicle-stimulating hormone level increases with age. BJU Int. 2005; 96:1087–91.
21. Harbitz TB. Morphometric studies of the Sertoli cells in elderly men with special reference to the histology of the prostate. Acta Pathol Microbiol Scand A Pathol. 1973;81:703–17.
22. Johnson L. Spermatogenesis and aging in the human. J Androl. 1986;7:331–54.
23. Paniagua R, Amat P, Nistal M, Martin A. Ultrastructure of Leydig cells in human ageing testes. J Anat. 1986;146:173–83.
24. Paniagua R, Nistal M, Amat P, Rodriguez MC, Martin A. Seminiferous tubule involution in elderly men. Biol Reprod. 1987;36:939–47.
25. Regadera J, Nistal M, Paniagua R. Testis, epididymis, and spermatic cord in elderly men. Correlation of angiographic and histologic studies with systemic arteriosclerosis. Arch Pathol Lab Med. 1985;109: 663–7.
26. Hermann M, Untergasser G, Rumpold H, Berger P. Aging of the male reproductive system. Exp Gerontol. 2000;35:1267–79.
27. Coffey DS, Berry SJ, Ewing LL. An overview of current concepts in the study of benign prostate hyperplasia. In: Rodgers CH, Coffey DS, Cunha G, Grayhack JT, Hinman F, Horton R, editors. Benign prostatic hyperplasia, vol II. NIH publication No. 87-2881; 1987. p. 1–13.
28. Kühnert B, Nieschlag E. Reproductive functions of the ageing male. Hum Reprod Update. 2004;10: 327–39.
29. Ng KK, Donat R, Chan L, Lalak A, Di Pierro I, Handelsman DJ. Sperm output of older men. Hum Reprod. 2004;8:1811–5.
30. Fisch H, Goluboff ET, Olson JH, Feldshuh J, Broder SJ, Barad DH. Semen analysis in 1,283 men from the United States over a 25-year period: no decline in quality. Fertil Steril. 1996;65:1009–14.

31. Eskenazi B, Wyrobek AJ, Kidd SA, Lowe X, Moore II D, Weisiger K, et al. Sperm aneuploidy in fathers of children with paternally and maternally inherited Klinefelter syndrome. Hum Reprod. 2002;17: 576–83.
32. Schwartz D, Mayaux MJ, Spira A, Moscato ML, Jouannet P, Czyglik F, et al. Semen characteristics as a function of age in 833 fertile men. Fertil Steril. 1983;39:530–5.
33. Auger J, Kunstmann JM, Czyglik F, Jouannet P. Decline in semen quality among fertile men in Paris during the past 20 years. N Engl J Med. 1995; 332:281–5.
34. Rolf C, Behre HM, Nieschlag E. Reproductive parameters of older couples compared to younger men of infertile couples. Int J Androl. 1996; 19:135–42.
35. Andolz P, Bielsa MA, Vila J. Evolution of semen quality in North-eastern Spain: a study in 22 759 infertile men over a 36 year period. Hum Reprod. 1999;14:731–5.
36. Berling S, Wolner-Hanssen P. No evidence of deteriorating semen quality among men in infertile relationships during the last decade: a study of males from southern Sweden. Hum Reprod. 1997;12: 1002–5.
37. Carlsen E, Giwercman A, Keiding N, Skakkebaek NE. Evidence for decreasing quality of semen during past 50 years. BMJ. 1992;305:609–13.
38. Agarwal A, Ozturk E, Loughlin KR. Comparison of semen analysis between the two Hamilton-Thorn semen analysers. Andrologia. 1992;24:327–9.
39. Hommonai ZT, Fainman N, David MP, Paz GF. Semen quality and sex hormone pattern of 29 middle aged men. Andrologia. 1982;14:164–70.
40. Nieschlag E, Lammers U, Freischem CW, Langer K, Wickings EJ. Reproductive functions in young fathers and grandfathers. J Clin Endocrinol Metab. 1982;55:676–81.
41. Baccarelli A, Morpurgo PS, Corsi A, Vaghi I, Fanelli M, Cremonesi G, et al. Activin A serum levels and aging of the pituitary–gonadal axis: a cross-sectional study in middle-aged and elderly healthy subjects. Exp Gerontol. 2001;36:1403–12.
42. Seymour FI, Duffy C, Koerner A. A case of authenticated fertility in a man, aged 94. J Am Med Assoc. 1935;105:1423–4.
43. de la Rochebrochard E, Thonneau P. Paternal age and maternal age are risk factors for miscarriage; results of a multicentre European study. Hum Reprod. 2002;17:1649–56.
44. Dunson DB, Colombo B, Baird D. Changes with age in the level and duration of fertility in the menstrual cycle. Hum Reprod. 2002;17:1399–403.
45. Dunson DB, Baird DD, Colombo B. Increased infertility with age in men and women. Obstet Gynecol. 2004;103:51–6.
46. Olson J. Subfecundity according to the age of the mother and the father. Dan Med Bull. 1990;37: 281–2.
47. Spandorfer SD, Avrech OM, Colombero LT, Palermo GD, Rosenwaks Z. Effect of parental age on fertilization and pregnancy characteristics in couples treated by intracytoplasmic sperm injection. Hum Reprod. 1998;13:334–8.
48. Piette C, de Mouzon J, Bachelot A, Spira A. In-vitro fertilization: influence of women's age on pregnancy rates. Hum Reprod. 1990;5:56–9.
49. Gallardo E, Simón C, Levy M. Effect of age on sperm fertility potential: oocyte donation as a model. Fertil Steril. 1996;66:260–4.
50. Paulson RJ, Milligan RC, Sokol RZ. The lack of influence of age on male fertility. Am J Obstet Gynecol. 2001;184:818–22.
51. Mathieu C, Ecochard R, Bied V, Lornage J, Czyba JC. Cumulative conception rate following intrauterine artificial insemination with husband's spermatozoa: influence of husband's age. Hum Reprod. 1995;10:1090–7.
52. Brzechffa PR, Buyalos RP. Female and male partner age and menotrophin requirements influence pregnancy rates with human menopausal gonadotrophin therapy in combination with intrauterine insemination. Hum Reprod. 1997;12:29–33.
53. Wyrobek AJ, Aardema M, Eichenlaub-Ritter U, Ferguson L, Marchetti F. Mechanisms and targets involved in maternal and paternal age effects on numerical aneuploidy. Environ Mol Mutagen. 1996; 28:254–64.
54. Penrose LS. Parental age and mutation. Lancet. 1955;269:312–3.
55. Crow JF. The origins, patterns and implications of human spontaneous mutation. Nat Rev Genet. 2000;1:40–7.
56. Drost JB, Lee WR. Biological basis of germline mutation: comparisons of spontaneous germline mutation rates among drosophila, mouse, and human. Environ Mol Mutagen. 1995;25 Suppl 26:48–64.
57. Johnson J, Canning J, Kaneko T, Pru JK, Tilly JL. Germline stem cells and follicular renewal in the postnatal mammalian ovary. Nature. 2004;428: 145–50.
58. American Society for Reproductive Medicine. Guidelines for therapeutic donor insemination: sperm. Fertil Steril. 1998;70 Suppl 3:1S–3.
59. British Andrology Society. British Andrology Society guidelines for the screening of semen donors for donor insemination. Hum Reprod. 1999;14: 1823–6.
60. Hassold T, Hunt P. To err (meiotically) is human: the genesis of human aneuploidy. Nat Rev Genet. 2001; 2:280–91.
61. Griffin DK. The incidence, origin, and etiology of aneuploidy. Int Rev Cytol. 1996;167:263–96.
62. Eichenlaub-Ritter U. Parental age-related aneuploidy in human germ cells and offspring: a story of past and present. Environ Mol Mutagen. 1996;28: 211–36.
63. Shi Q, Martin RH. Aneuploidy in human sperm: a review of the frequency and distribution of aneuploidy,

effects of donor age and lifestyle factors. Cytogenet Cell Genet. 2000;90:219–26.
64. Lowe X, Eskenazi B, Nelson DO, Kidd S, Alme A, Wyrobek AJ. Frequency of XY sperm increases with age in fathers of boys with Klinefelter syndrome. Am J Hum Genet. 2001;69:1046–54.
65. Lorda-Sanchez I, Binkert F, Maechler M, Robinson WP, Schinzel AA. Reduced recombination and paternal age effect in Klinefelter syndrome. Hum Genet. 1992;89:524–30.
66. Hatch M, Kline J, Levine B, Hutzler M, Wartburton D. Paternal age and trisomy among spontaneous abortions. Hum Genet. 1990;85:355–61.
67. Bordson BL, Leonardo VS. The appropriate upper age limit for semen donors: a review of the genetic effects of paternal age. Fertil Steril. 1991;56:397–401.
68. Stene E, Stene J, Stengel-Rutkowski S. A reanalysis of the New York State prenatal diagnosis data on Down's syndrome and paternal age effects. Hum Genet. 1987;77:299–302.
69. Hook EB, Cross PK. Paternal age and Down's syndrome genotypes diagnosed prenatally: no association in New York state data. Hum Genet. 1982; 62:167–74.
70. de Michelena MI, Burstein E, Lama JR, Vasquez JC. Paternal age as a risk factor for Down syndrome. Am J Med Genet. 1993;45:679–82.
71. Stoll C, Alembik Y, Dott B, Roth MP. Study of Down syndrome in 238 942 consecutive births. Ann Génét. 1998;41:44–51.
72. Fisch H, Hyun G, Golden R, Hensle TW, Olsson CA, Liberson GL. The influence of paternal age on Down syndrome. J Urol. 2003;169:2275–8.
73. McIntosh GC, Olshan AF, Baird PA. Paternal age and the risk of birth defects in offspring. Epidemiology. 1995;6:282–8.
74. Hassold T, Sherman S. Down syndrome: genetic recombination and the origin of the extra chromosome 21. Clin Genet. 2000;57:95–100.
75. Hook EB, Regal RR. A search for a paternal-age effect upon cases of 47, þ 21 in which the extra chromosome is of paternal origin. Am J Hum Genet. 1984;36:413–21.
76. Olsen SD, Magenis RE. Preferential paternal origin of de novo structural chromosome rearrangements. In: Daniel A, editor. The cytogenetics of mammalian autosomal rearrangements. New York: Alan R. Liss; 1988. p. 583–99.
77. Jacobs PA. The chromosome complement of human gametes. Oxf Rev Reprod Biol. 1992;14:47–72.
78. Sartorelli EM, Mazzucatto LF, de Pina-Neto JM. Effect of paternal age on human sperm chromosomes. Fertil Steril. 2001;76:1119–23.
79. Bosch M, Rajmil O, Martinez-Pasarell O, Egozcue J, Templado C. Linear increase of diploidy in human sperm with age: a four-colour FISH study. Eur J Hum Genet. 2001;9:533–8.
80. Glaser RL, Broman KW, Schulman RL, Eskenazi B, Wyrobek AJ, Jabs EW. The paternal-age effect in Apert syndrome is due, in part, to the increased frequency of mutations in sperm. Am J Hum Genet. 2003;73:939–47.
81. Goriely A, McVean GA, Rojmyr M, Ingemarsson B, Wilkie AO. Evidence for selective advantage of pathogenic FGFR2 mutations in the male germ line. Science. 2003;301:643–6.
82. Glaser RL, Jiang W, Boyadjiev SA, Tran AK, Zachary AA, Van Maldergem L, et al. Paternal origin of FGFR2 mutations in sporadic cases of Crouzon syndrome and Pfeiffer syndrome. Am J Hum Genet. 2000;66:768–77.
83. Risch N, Reich EW, Wishnick MM, McCarthy JG. Spontaneous mutation and parental age in humans. Am J Hum Genet. 1987;41:218–48.
84. Sivakumaran TA, Ghose S, Kumar HAS, Kucheria K. Parental age in Indian patients with sporadic hereditary retinoblastoma. Ophthalmic Epidemiol. 2000;7:285–91.
85. Lazaro C, Gaona A, Ainsworth P, Tenconi R, Vidaud D, Kruyer H, et al. Sex differences in the mutation rate and mutational mechanism in the NF gene in neurofibromatosis type 1 patients. Hum Genet. 1996; 98:696–9.
86. Friedman JM. Genetic disease in the offspring of older fathers. Obstet Gynecol. 1981;57:745–9.
87. Bertram L, Busch R, Spiegl M, Lautenschlager NT, Müller U, Kurz A. Paternal age is a risk factor for Alzheimer disease in the absence of a major gene. Neurogenetics. 1998;1:277–80.
88. Dalman C, Allebeck P. Paternal age and schizophrenia: further support for an association. Am J Psychiatry. 2002;159:1591–2.
89. Malaspina D, Corcoran C, Fahim C, Berman A, Harkavy-Friedman J, Yale S, et al. Paternal age and sporadic schizophrenia: evidence for de novo mutations. Am J Med Genet. 2002;114:299–303.
90. Green RF, Devine O, Crider KS, Olney RS, Archer N, Olshan AF, Shapira SK, National Birth Defects prevention study. Association of paternal age and risk for major congenital anomalies from the National Birth Defects Prevention Study, 1997 to 2004. Ann Epidemiol. 2010;20(3):241–9.
91. Harlap S, Paltiel O, Deutsch L, Knaanie A, Masalha S, Tiram E, et al. Paternal age and preeclampsia. Epidemiology. 2002;13:660–7.
92. Hemminki K, Kyyronen P. Parental age and risk of sporadic and familial cancer in offspring: implications for germ cell mutagenesis. Epidemiology. 1999;10:747–51.
93. Zhang Y, Kreger BE, Dorgan JF, Cupples LA, Myers RH, Splansky GL, et al. Parental age at child's birth and son's risk of prostate cancer. The Framingham Study. Am J Epidemiol. 1999;150:1208–12.
94. Fisch H. Older men are having children, but the reality of a male biological clock makes this trend worrisome. Geriatrics. 2009;64(1):14–7.
95. Schmid TE, Eskenazi B, Baumgartner A, Marchetti F, Young S, Weldon R, et al. The effects of male age on sperm DNA damage in healthy non-smokers. Hum Reprod. 2007;22(1):180–7.

96. Wyrobek AJ, Evenson D, Arnheim N, Jabs EW, Young S, Pearson F, et al. Advancing male age increase the frequencies of sperm with DNA fragmentation and certain gene mutations, but not aneuploidies or diploidies. Proc Natl Acad Sci USA. 2006;103(25):9601–6.
97. Morris ID. Sperm DNA damage and cancer treatment. Int J Androl. 2002;25(5):255–61.
98. Singh NP, Muller CH, Berger RE. Effects of age on DNA doublestrand breaks and apoptosis in human sperm. Fertil Steril. 2003;80(6):1420–30.
99. Spano M, Kolstad AH, Larsen SB, Cordelli E, Leter G, Giwercman A, et al. The applicability of the flow cytometric sperm chromatin structure assay in epidemiological studies. Hum Reprod. 1998;13(9): 2495–505.
100. Barnes CJ, Hardman WE, Maze GL, Lee M, Cameron IL. Age-dependent sensitization to oxidative stress by dietary fatty acids. Aging (Milano). 1998;10(6):455–62.
101. Barroso G, Morshedi M, Oehninger S. Analysis of DNA fragmentation, plasma membrane translocation of phosphatidylserine and oxidative stress in human spermatozoa. Hum Reprod. 2000;15(6): 1338–44.
102. Aitken RJ, Baker MA, Sawyer D. Oxidative stress in the male germ line and its role in the aetiology of male infertility and genetic disease. Reprod Biomed Online. 2003;7(1):65–70.
103. Kodama H, Yamaguchi R, Fukuda J, Kasai H, Tanaka T. Increased oxidative deoxyribonucleic acid damage in the spermatozoa of infertile male patients. Fertil Steril. 1997;68(3):519–24.
104. Brinkworth MH, Weinbauer GF, Bergmann M, Nieschlag E. Apoptosis as a mechanism of germ cell loss in elderly men. Int J Androl. 1997;20(4): 222–8.
105. Print CG, Loveland KL. Germ cell suicide: new insights into apoptosis during spermatogenesis. Bioessays. 2000;22(5):423–30.
106. Marchetti F, Lowe X, Bishop J, Wyrobek J. Induction of chromosomal aberrations in mouse zygotes by acrylamide treatment of male germ cells and their correlation with dominant lethality and heritable translocations. Environ Mol Mutagen. 1997;30: 410–7.
107. Marchetti F, Bishop JB, Cosentino L, Moore II D, Wyrobek AJ. Paternally transmitted chromosomal aberrations in mouse zygotes determine their embryonic fate. Biol Reprod. 2004;70:616–24.
108. Barratt CLR, Aitken J, Bjorndahl L, Carrell DT, de Boer P, Kvist U, et al. Sperm DNA: organization, protection and vulnerability: from basic science to clinical applications – a position report. Hum Reprod. 2010;25(4):824–38.
109. Desai N, Sabanegh Jr E, Kim T, Agarwal A. Free radical theory of aging: implications in male infertility. Urology. 2010;75(1):14–9.
110. Pasqualotto FF, Sharma RK, Nelson DR, Thomas Jr AJ, Agarwal A. Relationship between oxidative stress, semen characteristics and clinical diagnosis in men undergoing infertility investigation. Fertil Steril. 2000;73:459–64.
111. Pasqualotto FF, Sharma RK, Kobayashi H, Nelson DR, Thomas Jr AJ, Agarwal A. Oxidative stress in normospermic men undergoing fertility evaluation. J Androl. 2001;22:316–22.
112. Sharma RK, Pasqualotto FF, Nelson DR, Thomas Jr AJ, Agarwal Jr A. ROS-TAC is a novel marker of oxidative stress. Hum Reprod. 1999;14:2801–7.

Cancer in Males: Implications for Sperm Quality, Fertility, and Progeny Outcome

24

Peter Chan and Bernard Robaire

Abstract

There is an expanding population of boys and young men who are surviving cancer. Surveys clearly indicate their strong desire to father children. However, cancer therapies such as surgery, radiation and chemotherapy can impair their potential for normal reproductive health. Even before treatment, there appears to be pre-existing impairment in semen quality, as reflected by a significant level of damage in sperm chromatin. In a prospective longitudinal study on young men with testis cancer, Hodgkin's and non-Hodgkin's lymphoma, we demonstrated that chemotherapy can increase the risks of sperm aneuploidy and chromatin damage at two years post cancer treatment. These results highlight the importance of fertility preservation for cancer survivors. In addition to traditional semen cryopreservation, new strategies aimed at preserving spermatogonial stem cells are currently in development; these include cryopreservation of testicular tissue prior to exposure to testis toxicants with the expectation of restoring fertility when these children become adults and desire a family. A multidisciplinary approach through integration of new knowledge from basic science and clinical studies is the key to allow health-care professionals to optimize the fertility care for young cancer survivors.

Keywords

Cancer and fertility in men • Fertility and cancer in men • Cancer therapy and fertility • Sperm chromatin and chemotoxic therapy

Recent epidemiological studies indicate that there is a worldwide rise in the incidence of many cancers that affect boys and young men [1–3]. Simultaneously, with the advances in medical technology for early detection of cancer and the improvement in the efficacy of cancer therapies, the survival rates of many of these cancer patients have improved dramatically in the past decades. Many of young cancer survivors have

P. Chan (✉)
Department of Surgery, McGill University and the Research Institute of the MUHC, Montréal, QC, Canada
e-mail: mcgillsperminator@yahoo.com

A. Zini and A. Agarwal (eds.), *Sperm Chromatin: Biological and Clinical Applications in Male Infertility and Assisted Reproduction*, DOI 10.1007/978-1-4419-6857-9_24,

not started or completed forming a family. Thus, the impact of cancer and cancer therapies on male reproductive health and the options for fertility preservation are important issues in survivorship for young cancer patients.

In this chapter, we first provide an overview of the epidemiology of cancers in boys and young men of reproductive age with an emphasis on the various issues that arise in the management and counseling of these patients. The biological data on the impact of chemotoxic cancer therapies on sperm chromatin structures are reviewed. Finally, we present fertility preservation and restoration strategies that are currently clinically available and under development for patients in the near future.

Epidemiology of Cancer in Boys and Young Men

Common cancers in men that receive most attention in the public media include lung, colon, prostate, skin, and liver cancers. These cancers, however, tend to affect men who have passed the reproductive age. For boys and young men, the common cancers include testis cancer, lymphomas, leukemia, sarcoma, and brain cancers. The incidence of childhood cancer worldwide has been steadily increasing over the past 50 years [1–3]. With an estimated cumulative incidence of 1,720 per million, equivalent to a risk of 1 in 581, childhood cancer is indeed one of the leading causes of death among children younger than 15 years of age [4, 5]. According to recent data from the National Cancer Institute of USA [4], 10,400 children were newly diagnosed with cancer in 2007. Interestingly, boys were affected 1.2 times more frequently than girls [6].

Fortunately, thanks to tremendous strides in cancer management, including early detection strategies and advances in various treatment modalities such as surgeries, radiation, and combination chemotherapy regimens, the survival rates of many childhood cancers have increased dramatically over the past 40 years [7, 8]. In particular, testis cancer, which is the most commonly diagnosed solitary cancer in young men between the ages of 18 and 35 years [9], has a 5-year survival rate of over 90%, even in cases with metastasis, making testis cancer one of the most curable malignancies.

Approximately half of childhood cancers are hematologic malignancies (leukemia and lymphoma) with an anticipated long-term survival greater than 75%. Improvements in prognosis and survival rates have also been observed for many other childhood malignancies, including Wilm's tumor, malignant bone tumors, and rhabdomyosarcomas. The latest statistics in Canada indicate that the relative 5-year survival rate for all childhood cancers combined is approximately 82% [3]. It is estimated that today in North America approximately 1 in 900 of the population aged 20–45 years is a childhood cancer survivor [10]. In Canada, this translates to approximately 10,000 people who are survivors of childhood cancer and are expected to have 70 years or more of life after successful treatment [11].

Fertility After Cancer Therapy

While many of these young cancer survivors can expect a good quality of life, they may also face a series of undesired consequences related to their cancer and cancer therapies. Impairment in reproductive health is a well-known complication of cancer therapy; it occurs in a significant proportion of cancer survivors due to the spermatotoxicity of cancer treatments such as chemotherapy and radiation therapy. Many young cancer survivors have not initiated or completed forming a family. Interestingly, surveys indicated that almost 80% of childless cancer survivors report the desire to have children and believe that their experience of surviving cancer will make them better parents [12–14].

For the majority of cancer survivors who desire to have children but have poor sperm quantity and quality, assisted reproductive technologies (ART), including in vitro fertilization (IVF) and intracytoplasmic sperm injection (ICSI), are sought to help them to father their own children [15]. While ART is becoming more popular and available, and our knowledge and experience in its efficacy and safety have expanded tremendously in recent years, it does carry significant risks, including an increased risk of congenital malformations, genetic anomalies, low birth weight, and multiple pregnancies [16–19]. Health-care professionals counseling cancer patients and survivors must be prepared to provide them with precise and up-to-date options on postcancer fertility.

Cancer Management Strategies

Generally, cancer management involves three major modalities, namely, surgery, radiation, and chemotherapy. The choice of treatment depends on the nature and stage of the cancer and the comorbidity of the subject. Not uncommonly, a combination of these modalities in various orders may be required to achieve optimal cancer control. Complications of each modality also vary. Mechanisms of how each treatment modality may potentially compromise male reproductive health are discussed in this section.

Reproductive Health Before Cancer Treatment

It should be pointed out that the reproductive health of many cancer patients may be suboptimal even before receiving specific cancer therapies, as revealed by studies on the sperm density and morphology of prechemotherapy sperm banked samples and on case–control studies of their natural fecundity [20–23]. The reason for the impaired fertility status may, in part, be due to the decline in the physical state (poor nutrition, fever, cachexia, pain, etc.) of the patients due to cancer. The psychosocial stress attributed to the cancer diagnosis may play a role in the well-being of the subject. Prolonged periods of sexual abstinence may also contribute to the poor sperm quality before chemotherapy. In testis cancer, poor sperm profile may be explained by the fact that there is only one remaining contralateral noncancerous testis to produce sperm. Indeed, some studies have shown that the contralateral noncancerous testis may have compromised reproductive function due to a higher risk of coexisting intraepithelial germ cell tumors and abnormal spermatogenesis, both quantitatively and qualitatively [24].

Using a complementary panel of molecular genetic assays, including the AO/SCSA®, TUNEL, and comet assays to determine sperm DNA damage and mBBr-SH labeling and the CMA3 assay to assess chromatin packaging, we have recently reported that, prior to chemotherapy, 37% of men with testis cancer and 81% of men with Hodgkin's lymphoma demonstrated abnormal sperm chromatin structure despite having normal sperm density and motility [23]. Our findings suggest that with subsequent cytotoxic cancer therapy, their sperms are at risk for further genetic damage.

Impact of Surgical Management for Cancer on Male Reproductive Status

The purpose of surgical resection of tumor is to remove the tumor with adequate surgical margins to aim for cure or to debulk the volume of tumor to facilitate the effect of adjuvant therapy with radiation or chemotherapy and thus control the cancer. A common surgical management for testicular cancer in young males is radical orchiectomy. Removal of one testis may affect the total spermatogenic activity in an individual. Indeed, men with testis cancer are at risk of having decreased spermatogenic activity in the contralateral testis.

Other surgical managements for cancers in young males may result in damage to the autonomic nervous system required for semen emission. Pelvic and lower intestine surgeries, retroperitoneal lymph node dissection for advanced testicular cancer, or any procedures involving the spine and other parts of the central nervous system may result in postoperative anejaculation. Despite the fact that spermatogenic function is generally not affected in these patients, they are at risk of having impaired fertility due to the absence of semen emission.

Although it is well established in several animal models, including the monkey, that compensatory hypertrophy of the remaining testis occurs in the adult when one testis is removed prior to puberty [25–27], in human subjects, clinical studies indicate that this occurs only to a limited extent [28, 29] and is insufficient to compensate for the loss of one testis.

Radiation Therapy

Germ cells and somatic cells in testes are prone to damage post radiation. The usual clinical dosage of radiation therapy for cancer ranges from 0.2 to 70 Gy, depending on the nature, stage, and anatomical location of the tumor. A cumulative dosage of 2.5–6 Gy directly to the testes may permanently damage germ cells, leading to prolonged or permanent azoospermia [30]. Even for radiation therapy outside the pelvic areas (e.g., paraaortic lymph nodes) with gonadal shielding to reduce the extent of gonadal toxicity, the scattering effects of radiation may still contribute to impaired fertility post irradiation. Such damage to sperm production may be further attributed to damage to cells in the somatic compartment of the testis. Using spermatogonial stem cell (SCC) transplantation in rat, Zhang et al. [31] demonstrated that transplantation of SCCs from irradiated animals into testes of irradiated nude mice (which had normal differentiation of their own spermatogonia) permitted differentiation of the donor spermatogonia to spermatozoa. Conversely, transplantation of SCCs from untreated prepubertal rats into irradiated rat testes showed that the donor spermatogonia were able to colonize along the seminiferous tubules, but could not differentiate. Their findings suggest that the defect caused by radiation in the rat testes that hinder spermatogonial differentiation is due to damage to the somatic compartment [31].

Chemotherapy

Chemotherapy is generally indicated in advanced and metastatic cancer, although its use in certain cancers, such as germ cell tumors at an early, localized stage, may help to lower the risks of subsequent metastasis. In addition to malignant cells, any rapidly dividing cells, including germ cells at various phases of spermatogenesis, are targets of chemotherapy. Gonadotoxicity of chemotherapy to an individual depends on at least three factors: (1) the nature of the malignancy, which dictates the type of chemotherapeutic agents to be used, (2) the stage of the disease, which dictates the duration and dosages of chemotherapy, (3) host factors, such as the baseline reproductive health of the individual. The impact of chemotherapy on male reproductive health is discussed in the next section.

Impact of Chemotherapy on Male Reproductive Health

Animal Studies

Using rodents (rats and mice) as models, a large body of evidence has emerged demonstrating that treatment with chemotherapeutic agents that act by blocking cell division usually have dramatic effects on the production of male germ cells [32, 33]. Depending on the mechanism by which such agents act on the different phases of spermatogenesis (spermatogonial mitotic cell division, meiosis, or spermiogenesis), consequences can range from complete elimination of germ cells from the testis, resulting in Sertoli-cell-only syndrome, to no apparent histological effects on spermatogenesis, but functional effects on germ cells (their motility, fertilizing ability, or capacity to produce normal viable offspring).

Over the past 20 years, studies on male mediated adverse effects of chemotherapeutic drugs, such as cyclophosphamide (CPA), bleomycin, cisplatin, or procarbazine, on fertility and progeny outcome have clearly established some of the underlying molecular mechanisms that result in loss of fertility and altered progeny outcome [34–37]. Using CPA or the combination of drugs used for treating testicular cancer (bleomycin, etoposide, and cisplatin, BEP) as model drugs and the rat as the model animal, it has been demonstrated that paternal exposures result in adverse reproductive outcomes that range from increased preimplantation and postimplantation loss or early postnatal death, to growth retardation and congenital malformation; significantly, some of these outcomes are transmitted to subsequent generations [38–40]. It is particularly noteworthy that the action of such drugs on germ cells not only affects the number of germ cells that the testis can produce but also alters markers of chromatin structure (Comet, acridine orange, TUNEL, MBBr, and CMA3 assays, nuclear proteome) in spermatozoa [41, 42]. It is clear from animal studies that spermatozoa that have damaged chromatin as a result of paternal drug treatment are capable of fertilizing oocytes [38, 43–45].

Animal studies have also revealed that the effects of paternal exposure on progeny can be wide ranging. While treatment with BEP caused a decrease in both sperm production and sperm motility, no apparent effects were observed on progeny at the end of gestation, yet postnatal death rates were dramatically increased [43]. By contrast, chronic CPA treatment had minimal effects on sperm number and motility, yet a wide range of effects were observed in progeny, ranging from abnormalities at birth to learning deficits as adults and in subsequent generations as well as abnormal reproductive capacity [35, 38, 46].

Clinical Studies

The assessment of the consequences on progeny outcome of exposure of men to chemotherapeutic drugs presents remarkable challenges. Chemotherapy often results in transient or permanent azoospermia or oligozoospermia in cancer patients [47]. Large epidemiological studies, discussed above, have revealed that there is clearly an effect on fertility and time to pregnancy [14]. In addition, the standard semen parameters (sperm number, motility, and morphology, as established by the World Health Organization (WHO) [48]) are not sufficiently reliable predictors of male fertility [23, 49–51]. Consequently, the focus has shifted in recent years to assessing the nature and quality of chromatin in spermatozoa. In a recent comprehensive review and position paper, Barratt et al. [52] have outlined our current clinical understanding and uncertainties related to the many assays used to ascertain sperm chromatin quality.

Aneuploidy, an abnormal number of chromosomes, is one of the more striking consequences of anticancer drugs on sperm chromatin quality. Using multicolor fluorescent in situ hybridization to detect sperm aneuploidy for chromosomes 13, 21, X, and Y in testicular cancer and Hodgkin's lymphoma patients before and up to 24 months after the initiation of chemotherapy, Tempest et al. [53] found that at 6 months, all cancer patients showed significantly increased frequencies of XY disomy and nullisomy for chromosomes 13 and 21. Although frequencies of aneuploidy generally declined over time after termination of treatment, increased aneuploidy frequencies persisted in some chromosomes for up to 24 months.

Using a series of assays that provide complementary information on sperm chromatin structure, e.g., extent of single- and double-strand breaks, degree of protamination, cross-linking of sulfhydryl bonds, O'Flaherty et al. [23] have shown that, prior to initiation of chemotherapy, sperm chromatin integrity was poorer in cancer patients than in a control population. After treatment with chemotherapeutics, not only was there the expected decline in sperm production and chromatin quality but also, up to 2 years later, a reduction in spermatozoal chromatin integrity in over 40% of the patients who had a return of spermatogenesis [54, 55].

Based on the limited studies to date, it is clear that the presence of several cancers in young men results, to varying degrees, in sperm chromatin with reduced integrity. Furthermore, treatment of cancer may cause transient partial or complete loss of spermatozoa. Under some conditions, it is clear that the germ cells that eventually return to repopulate the seminiferous epithelium are still damaged, while under others, they appear to be normal. Whether SCCs are able repair all the damage caused by radiation or chemotherapy or not remains to be established.

Male Fertility Preservation and Restoration Strategies

Fertility preservation has become recognized as part of the important global care of cancer patients at the time of cancer diagnosis. This has come about because of the potential long-term negative impact of cytotoxic cancer therapies on male reproductive health and the express desire of many young cancer survivors to have children. Continuing research efforts are being made to contribute to the development of multidisciplinary counseling strategies to best advise cancer patients and survivors regarding their potential risks for adverse pregnancy and progeny outcomes.

Sperm Cryopreservation

Sperm cryopreservation or sperm "banking" is currently the only available strategy to preserve male fertility. Ideally, sperm samples should be collected before any cytotoxic cancer therapies, through ejaculation by masturbation after 2–4 days of sexual abstinence. Then, sperm samples should be analyzed, frozen, and stored in aliquots in liquid nitrogen for future use. With the advances in and increased access to ART, such as IVF/ICSI, a very low number of living spermatozoa are required to achieve fertilization; therefore, even sperm samples that are far from meeting the semen parameters set by the WHO may still be used to achieve fertilization.

Sperm cryopreservation does have its limitations as a fertility preservation strategy. First, only subjects beyond the state of physical maturity of adolescence, when "spermarche" begins within the testes, can have spermatozoa in the semen for cryopreservation. One study of 62 attempts by adolescents to bank sperm before cancer therapy resulted in totally normal semen in only four subjects [56]. Semen procurement by masturbation may not always be feasible among adolescents, even for those who have spermatogenesis. In fact, for cultural and religious reasons, the act of masturbation may be viewed as inappropriate by parents of young adolescent cancer patients [57]. Alternative methods to obtain mature sperms in adolescents using high-frequency penile vibratory stimulus, electroejaculation, or surgical testicular sperm extraction will require sedation/anesthesia and are deemed too invasive for youngsters. Thus, sperm banking is not universally practiced in pediatric-oncology centers, and few adolescent friendly facilities exist.

For preadolescent boys with cancer, there is currently no feasible option for fertility preservation. Early investigators held the view that being prepubertal during anticancer therapy conferred protection against gonadal damage. However, a study evaluating 12 men who survived childhood malignancy revealed that although puberty had progressed apparently normally in all 12, 8 patients were azoospermic, and only 1 had normal semen analysis 2–16.5 years post chemotherapy [58]. In addition, following treatment of Hodgkin's lymphoma in childhood, severe germ cell damage was observed in the majority of patients, even 17 years after chemotherapy [55, 59]. Evidently, there is no gonadal protection in the prepubertal male against chemotherapy-induced damage [60, 61]. In fact, some investigators believe that prepubertal testes are more vulnerable to the cytotoxic effects of chemotherapy than adult testes [62].

Pharmacological Strategies

The hypothesis that blocking the hypothalamic-pituitary-gonadal axis prior to the initiation of chemotherapy to preserve the nondividing germ cell population was first proposed by Glode et al. [63]. Hormonal manipulation, including the use of exogenous GnRH (gonadotropin-releasing hormone) analogs and steroids (testosterone) to suppress the gonadotropin release, has been investigated as a potential fertility preservation strategy. Since cytotoxic treatment acts mainly on rapidly dividing cells, germ cells have been postulated to be less susceptible to cytotoxic effects if hormone treatments are used to render the testes quiescent. This technique has been successful in some rodents (rats, but not mice) [64, 65]; in addition, in rats the extent of the damage of chemotherapeutic agents has been shown to extend beyond the germ cells to the somatic cells surrounding them [66]. There is no evidence of a similar spermatogonial block in monkeys [67]. Thus far, clinical trials have not shown any benefit of this method [60, 68]. Furthermore, this approach would be ineffective for prepubertal children as the proliferation of germ cells in prepubertal primates appears to be gonadotropin-independent [69]. Clearly, there is an urgent need for novel strategies that are effective and minimally invasive for fertility preservation in young male cancer patients.

Fertility Restoration with Germ-Cell Transplantation

Stem cells of the male germ line, termed SSCs, exist in the testis prior to birth. Harvesting either SSCs or tissue blocks from testes for cryopreservation before anticancer therapies offers the hope for prepubertal boys with cancer to preserve fertility and form their family in the future [70]. After the patient is cured and is at an appropriate state of maturity, preserved SSCs, or SSCs derived from frozen tissue blocks, could be autotransplanted back to the seminiferous tubules to regenerate complete spermatogenesis. Cryopreservation of testis tissue from prepubertal boys has revealed that germ cells can be preserved [71]. An important feature of this strategy is that instead of just preserving fertility, it aims to "restore" fertility. This fertility restoration scheme, based on germ cell or tissue transplantation, has been established with mice and other species [66, 70–78] and is currently under investigation to extend its application to humans.

Looking to the Future

While the risks of impaired fertility after cancer therapy have long been recognized, the biological mechanisms and the nature and extent of sperm damage at the molecular level have only been revealed recently. The importance of fertility after cancer is gradually being accepted as an essential survivorship issue for young cancer survivors. The establishment of effective fertility preservation protocols and counseling strategies represents the ongoing efforts of researchers and clinicians.

A multidisciplinary approach, including input from oncologists, reproductive biologists, social workers, ethicists, geneticists, and embryologists, is the essence of successful development and implementation of any fertility management plan for young cancer survivors. Many questions remain to be answered: What is the potential of further recovery of sperm quality in long-term post chemotherapy? How long should a patient wait post chemotherapy before he can safely use his fresh sperm for procreation? What is the nature and extent of risk of adverse reproductive outcomes using sperm with impaired sperm chromatin post chemotherapy? What are the transgenerational risks? To what extent would such risks be reduced by using sperm cryopreserved prior to chemotherapy for procreation? What sperm biological markers and what assays provide the best clinical prediction of the risks of adverse reproductive outcomes when using sperm with impaired chromatin quality? Further research to address these and other related questions is clearly needed to help health-care professionals and health policy makers to enhance the quality of counseling and to establish practice guidelines on the subject of fertility after cancer.

References

1. Burns KC, Boudreau C, Panepinto JA. Attitudes regarding fertility preservation in female adolescent cancer patients. J Pediatr Hematol Oncol. 2006;28:350–4.
2. Richiardi L, Bellocco R, Adami HO, Torrang A, Barlow L, Hakulinen T, et al. Testicular cancer incidence in eight northern European countries: secular and recent trends. Cancer Epidemiol Biomark Prev. 2004;13:2157–66.
3. Canadian Cancer Society, National Cancer Institute of Canada. Canadian cancer statistics 2010. www.cancer.ca (2010). Accessed 23 July 2010.
4. National Cancer Institute. National Cancer Institute research on childhood cancers: fact sheet. http://www.cancer.gov/cancertopics/factsheet/NCI-childhood-cancers-research (2009). Accessed 2 January 2009.
5. Stiller CA. Epidemiology and genetics of childhood cancer. Oncogene. 2004;23:6429–44.
6. Stiller CA, Allen MB, Eatock EM. Childhood cancer in Britain: the National Registry of Childhood Tumours and incidence rates 1978–1987. Eur J Cancer. 1995;31A:2028–34.
7. Waring AB, Wallace WH. Subfertility following treatment for childhood cancer. Hosp Med. 2000;61:550–7.
8. National Cancer Institute. Surveillance epidemiology and end results program of the National Cancer Institute research: fact sheet. http://SEER.Cancer.gov (2009). Accessed 2 January 2009.
9. Garner MJ, Turner MC, Ghadirian P, Krewski D. Epidemiology of testicular cancer: an overview. Int J Cancer. 2005;116:331–9.
10. Bhatia S. Cancer survivorship–pediatric issues. Hematology Am Soc Hematol Educ Program. 2005:507–15.
11. Meadows AT. Pediatric cancer survivorship: research and clinical care. J Clin Oncol. 2006;24:5160–5.
12. Schover LR, Brey K, Lichtin A, Lipshultz LI, Jeha S. Oncologists' attitudes and practices regarding banking sperm before cancer treatment. J Clin Oncol. 2002;20:1890–7.
13. Schover LR, Rybicki LA, Martin BA, Bringelsen KA. Having children after cancer. A pilot survey of survivors' attitudes and experiences. Cancer. 1999;86:697–709.

14. Green DM, Lange JM, Peabody EM, Grigorieva NN, Peterson SM, Kalapurakal JA, et al. Pregnancy outcome after treatment for Wilms tumor: a report from the national Wilms tumor long-term follow-up study. J Clin Oncol. 2010;28:2824–30.
15. Chan PT, Palermo GD, Veeck LL, Rosenwaks Z, Schlegel PN. Testicular sperm extraction combined with intracytoplasmic sperm injection in the treatment of men with persistent azoospermia postchemotherapy. Cancer. 2001;92:1632–7.
16. Hansen M, Bower C, Milne E, de Klerk N, Kurinczuk JJ. Assisted reproductive technologies and the risk of birth defects–a systematic review. Hum Reprod. 2005;20:328–38.
17. Schieve LA, Cohen B, Nannini A, Ferre C, Reynolds MA, Zhang Z, Jeng G, Macaluso M, Wright VC; Massachusetts Consortium for Assisted Reproductive Technology Epidemiologic Research (MCARTER). A population-based study of maternal and perinatal outcomes associated with assisted reproductive technology in Massachusetts. Matern Child Health J. 2007;11:517–25.
18. Schieve LA, Meikle SF, Ferre C, Peterson HB, Jeng G, Wilcox LS. Low and very low birth weight in infants conceived with use of assisted reproductive technology. N Engl J Med. 2002;346:731–7.
19. Steel AJ, Sutcliffe A. Long-term health implications for children conceived by IVF/ICSI. Hum Fertil (Camb). 2009;12:21–7.
20. Meirow D, Schenker JG. Cancer and male infertility. Hum Reprod. 1995;10:2017–22.
21. Meseguer M, Molina N, García-Velasco JA, Remohí J, Pellicer A, Garrido N. Sperm cryopreservation in oncological patients: a 14-year follow-up study. Fertil Steril. 2006;85:640–5.
22. Baker JA, Buck GM, Vena JE, Moysich KB. Fertility patterns prior to testicular cancer diagnosis. Cancer Causes Control. 2005;16:295–9.
23. O'Flaherty C, Vaisheva F, Hales BF, Chan P, Robaire B. Characterization of sperm chromatin quality in testicular cancer and Hodgkin's lymphoma patients prior to chemotherapy. Hum Reprod. 2008;23:1044–52.
24. Dieckmann KP, Linke J, Pichlmeier U, Kulejewski M, German Testicular Cancer Study Group. Spermatogenesis in the contralateral testis of patients with testicular germ cell cancer: histological evaluation of testicular biopsies and a comparison with healthy males. BJU Int. 2007;99:1079–85.
25. Cunningham GR, Tindall DJ, Huckins C, Means AR. Mechanisms for the testicular hypertrophy which follows hemicastration. Endocrinology. 1978;102: 16–23.
26. Medhamurthy R, Aravindan GR, Moudgal NR. Hemiorchidectomy leads to dramatic and immediate alterations in pituitary follicle-stimulating hormone secretion and the functional activity of the remaining testis in the adult male bonnet monkey (*Macaca radiata*). Biol Reprod. 1993;49:743–9.
27. Tsutsui T, Kurita A, Kirihara N, Hori T, Kawakami E. Testicular compensatory hypertrophy related to hemicastration in prepubertal dogs. J Vet Med Sci. 2004;66:1021–5.
28. Laron Z, Dickerman Z, Ritterman I, Kaufman H. Follow-up of boys with unilateral compensatory testicular hypertrophy. Fertil Steril. 1980;33:297–301.
29. Lin WW, Kim ED, Quesada ET, Lipshultz LI, Coburn M. Unilateral testicular injury from external trauma: evaluation of semen quality and endocrine parameters. J Urol. 1998;159:841–3.
30. Clifton DK, Bremner WJ. The effect of testicular x-irradiation on spermatogenesis in man. A comparison with the mouse. J Androl. 1983;4:387–92.
31. Zhang Z, Shao S, Meistrich ML. The radiation-induced block in spermatogonial differentiation is due to damage to the somatic environment, not the germ cells. J Cell Physiol. 2007;211:149–58.
32. Robaire B, Hales BF, editors. Advances in male-mediated reproductive toxicology. New York: Kluwer Academic/Plenum; 2003.
33. Anderson D, Brinkworth MH, editors. Male mediated developmental toxicity. London: Royal Society of Chemistry; 2006.
34. Hales BF, Barton TS, Robaire B. Impact of paternal exposure to chemotherapy on offspring in the rat. J Natl Cancer Inst Monogr. 2005;34:28–31.
35. Hales BF, Crosman K, Robaire B. Increased postimplantation loss and malformations among the F2 progeny of male rats chronically treated with cyclophosphamide. Teratology. 1992;45:671–8.
36. Anderson D. Male-mediated developmental toxicity. Toxicol Appl Pharmacol. 2005;207(2 Suppl): 506–13.
37. Meistrich ML. Male gonadal toxicity. Pediatr Blood Cancer. 2009;53:261–6.
38. Trasler JM, Hales BF, Robaire B. Paternal cyclophosphamide treatment of rats causes fetal loss and malformations without affecting male fertility. Nature. 1985;316:144–6.
39. Qiu J, Hales BF, Robaire B. Effects of chronic low-dose cyclophosphamide exposure on the nuclei of rat spermatozoa. Biol Reprod. 1995;52:33–40.
40. Harrouk W, Codrington A, Vinson R, Robaire B, Hales BF. Paternal exposure to cyclophosphamide induces DNA damage and alters the expression of DNA repair genes in the rat preimplantation embryo. Mutat Res. 2000;461:229–41.
41. Codrington AM, Hales BF, Robaire B. Chronic cyclophosphamide exposure alters the profile of rat sperm nuclear matrix proteins. Biol Reprod. 2007;77: 303–11.
42. Delbès G, Hales BF, Robaire B. Toxicants and human sperm chromatin integrity. Mol Hum Reprod. 2010;16:14–22.
43. Bieber AM, Marcon L, Hales BF, Robaire B. Effects of chemotherapeutic agents for testicular cancer on the male rat reproductive system, spermatozoa, and fertility. J Androl. 2006;27:189–200.
44. Ménézo Y, Dale B, Cohen M. DNA damage and repair in human oocytes and embryos: a review. Zygote. 2010;21:1–9.

45. Sakkas D, Alvarez JG. Sperm DNA fragmentation: mechanisms of origin, impact on reproductive outcome, and analysis. Fertil Steril. 2010;93:1027–36.
46. Auroux M, Dulioust E, Selva J, Rince P. Cyclophosphamide in the F0 male rat: physical and behavioral changes in three successive adult generations. Mutat Res. 1990;229:189–200.
47. Gandini L, Sgro P, Lombardo F, Paoli D, Culasso F, Toselli L, et al. Effect of chemo- or radiotherapy on sperm parameters of testicular cancer patients. Hum Reprod. 2006;21:2882–9.
48. World Health Organization. WHO laboratory manual for examination and processing of human semen. 5th ed. Cambridge: University Press Cambridge; 2010.
49. Spano M, Bonde JP, Hjollund HI, Kolstad HA, Cordelli E, Leter G. Sperm chromatin damage impairs human fertility. The Danish First Pregnancy Planner Study Team. Fertil Steril. 2000;73:43–50.
50. Virro MR, Larson-Cook KL, Evenson DP. Sperm chromatin structure assay (SCSA) parameters are related to fertilization, blastocyst development, and ongoing pregnancy in in vitro fertilization and intracytoplasmic sperm injection cycles. Fertil Steril. 2004;81:1289–95.
51. Payne JF, Raburn DJ, Couchman GM, Price TM, Jamison MG, Walmer DK. Redefining the relationship between sperm deoxyribonucleic acid fragmentation as measured by the sperm chromatin structure assay and outcomes of assisted reproductive techniques. Fertil Steril. 2005;84:356–64.
52. Barratt CL, Aitken RJ, Björndahl L, Carrell DT, de Boer P, Kvist U, et al. Sperm DNA: organization, protection and vulnerability: from basic science to clinical applications–a position report. Hum Reprod. 2010;25:824–38.
53. Tempest HG, Ko E, Chan P, Robaire B, Rademaker A, Martin RH. Sperm aneuploidy frequencies analysed before and after chemotherapy in testicular cancer and Hodgkin's lymphoma patients. Hum Reprod. 2008;23:251–8.
54. O'Flaherty C, Hales BF, Chan P, Robaire B. Impact of chemotherapeutics and advanced testicular cancer or Hodgkin lymphoma on sperm deoxyribonucleic acid integrity. Fertil Steril. 2010;94:1374–9.
55. Heikens J, Behrendt H, Adriaanse R, Berghout A. Irreversible gonadal damage in male survivors of pediatric Hodgkin's disease. Cancer. 1996;78:2020–4.
56. Postovsky S, Lightman A, Aminpour D, Elhasid R, Peretz M, Arush MW. Sperm cryopreservation in adolescents with newly diagnosed cancer. Med Pediatr Oncol. 2003;40:355–9.
57. Rosoff PM, Katsur ML. Preserving fertility in young cancer patients: a medical, ethical and legal challenge. J Philos Sci Law. 2003;3. http://www6.miami.edu/ethics/jpsl/archives/papers/preservingFert.html. Accessed 4 September 2010.
58. Mustieles C, Muñoz A, Alonso M, Ros P, Yturriaga R, Maldonado S, et al. Male gonadal function after chemotherapy in survivors of childhood malignancy. Med Pediatr Oncol. 1995;24:347–51.
59. Shafford EA, Kingston JE, Malpas JS, Plowman PN, Pritchard J, Savage MO, et al. Testicular function following the treatment of Hodgkin's disease in childhood. Br J Cancer. 1993;68:1199–204.
60. Thomson AB, Anderson RA, Irvine DS, Kelnar CJ, Sharpe RM, Wallace WH. Investigation of suppression of the hypothalamic-pituitary-gonadal axis to restore spermatogenesis in azoospermic men treated for childhood cancer. Hum Reprod. 2002;17:1715–23.
61. van Casteren NJ, van der Linden GH, Hakvoort-Cammel FG, Hählen K, Dohle GR, van den Heuvel-Eibrink MM. Effect of childhood cancer treatment on fertility markers in adult male long-term survivors. Pediatr Blood Cancer. 2009;52:108–12.
62. Revel A, Revel-Vilk S. Pediatric fertility preservation: is it time to offer testicular tissue cryopreservation? Mol Cell Endocrinol. 2008;282:143–9.
63. Glode LM, Robinson J, Gould SF. Protection from cyclophosphamide-induced testicular damage with an analogue of gonadotropin-releasing hormone. Lancet. 1981;1(8230):1132–4.
64. Kurdoglu B, Wilson G, Parchuri N, Ye WS, Meistrich ML. Protection from radiation-induced damage to spermatogenesis by hormone treatment. Radiat Res. 1994;139:97–102.
65. Meistrich ML, Shetty G. Hormonal suppression for fertility preservation in males and females. Reproduction. 2008;136:691–701.
66. Zhang X, Ebata KT, Nagano MC. Genetic analysis of the clonal origin of regenerating mouse spermatogenesis following transplantation. Biol Reprod. 2003;69:1872–8.
67. Boekelheide K, Schoenfeld H, Hall SJ, Weng CCY, Shetty G, Leith J, et al. Gonadotropin-releasing hormone antagonist (cetrorelix) therapy fails to protect non human primates (*Macaca arctoides*) from radiation-induced spermatogenic failure. J Androl. 2009; 26:222–34.
68. Waxman JH, Ahmed R, Smith D, Wrigley PF, Gregory W, Shalet S, et al. Failure to preserve fertility in patients with Hodgkin's disease. Cancer Chemother Pharmacol. 1987;19:159–62.
69. Kelnar CJ, McKinnell C, Walker M, Morris KD, Wallace WH, Saunders PT, et al. Testicular changes during infantile 'quiescence' in the marmoset and their gonadotrophin dependence: a model for investigating susceptibility of the prepubertal human testis to cancer therapy? Hum Reprod. 2002;17:1367–78.
70. Nagano MC. A surgical strategy using spermatogonial stem cells for restoring male fertility. In: Tulandi T, Gosden RG, editors. Preservation of fertility. London: Taylor & Francis; 2004. p. 125–40.
71. Keros V, Hultenby K, Borgström B, Fridström M, Jahnukainen K, Hovatta O. Methods of cryopreservation of testicular tissue with viable spermatogonia in pre-pubertal boys undergoing gonadotoxic cancer treatment. Hum Reprod. 2007;22:1384–95.
72. Oatley JM, Brinster RL. Regulation of spermatogonial stem cell self-renewal in mammals. Annu Rev Cell Dev Biol. 2008;24:263–86.

73. Nagano MC. Homing efficiency and proliferation kinetics of male germ line stem cells following transplantation in mice. Biol Reprod. 2003;69:701–7.
74. Ebata KT, Yeh JR, Zhang X, Nagano MC. The application of biomarkers of spermatogonial stem cells for restoring male fertility. Dis Markers. 2008;24: 267–76.
75. Ebata KT, Zhang X, Nagano MC. Expression patterns of cell-surface molecules on male germ line stem cells during postnatal mouse development. Mol Reprod Dev. 2005;72:171–81.
76. Ebata KT, Zhang X, Nagano MC. Male germ line stem cells have an altered potential to proliferate and differentiate during postnatal development in mice. Biol Reprod. 2007;76:841–7.
77. Yeh JR, Zhang X, Nagano MC. Establishment of a short-term in vitro assay for mouse spermatogonial stem cells. Biol Reprod. 2007;77:897–904.
78. Schlatt S, Ehmcke J, Jahnukainen K. Testicular stem cells for fertility preservation: preclinical studies on male germ cell transplantation and testicular grafting. Pediatr Blood Cancer. 2009;53:274–80.

Sperm Chromatin and Environmental Factors

25

Aleksander Giwercman

Abstract

Available experimental and human data show that a number of lifestyle- and environment-related exposures may have negative effects on sperm DNA integrity. The extent of the sperm DNA damage seems to differ between different studies. Such a discrepancy may partly be due to the use of several methods for assessment of sperm DNA integrity, techniques that do not measure exactly the same characteristics of sperm DNA. Another contributing factor may be genetically determined variation in the individual susceptibility. Sperm DNA damage due to environmental and lifestyle factors may have a negative impact on fertility, and there is a potential risk of transmission to the offspring. Therefore, it is of importance to focus on the association between environment and sperm DNA integrity to prevent male subfertility and to avoid potentially serious health effects in the future generation(s).

Keywords

Sperm chromatin and environmental factors • Environment in sperm chromatin • Lifestyle factors in sperm chromatin • DNA and the environment

During the past 20 years, a lot of attention has been given to possible time-related deterioration in the function of male reproductive organs [1]. This debate has been initiated by alarming reports on declining sperm counts – the issue that is still widely debated [2, 3]. It has also been suggested that the incidence of congenital malformations, as cryptorchidism and hypospadias, has increased, although even here some uncertainty exists [4]. On the contrary, there is no doubt that testicular cancer, has become significantly more common in the past 4–5 decades [5]. The rapidity by which this rise in the incidence of testicular cancer has occurred points towards a negative impact of environment- or lifestyle-related factors on male reproductive function.

A. Giwercman (✉)
Reproductive Medicine Centre, Skåne University Hospital Malmö, Lund University, Malmö, Sweden
e-mail: Aleksander.giwercman@med.lu.se

A. Zini and A. Agarwal (eds.), *Sperm Chromatin: Biological and Clinical Applications in Male Infertility and Assisted Reproduction*, DOI 10.1007/978-1-4419-6857-9_25,

However, the deleterious effect of environment and/or lifestyle on semen parameters might not only result in declining sperm counts but even affect the quality of the spermatozoa, including the integrity of the DNA. Such an effect might not only have an impact on the fertility potential of the subject but also introduce genetic aberrations that might be transmitted to the next generation [6]. This chapter focuses on the available evidence regarding environment- and lifestyle-induced changes in the sperm DNA, and the biological and clinical implications of such effects.

Biological and Clinical Relevance

The issue of the effect of environment and lifestyle has some interesting clinical and biological implications. Fist of all, it is now well established that, at least certain types of, sperm DNA damage may have a negative impact on the fertility of the subject *in vivo*, and possibly even *in vitro*. Although in many of the epidemiological studies the increase in percentage of sperms in exposed subjects is rather discrete, even such an effect may be deleterious to the fertility potential of the subject. Thus, we have recently reported that in subjects with normal standard sperm parameters the odds ratio for spontaneous pregnancy significantly decreases when the DNA Fragmentation Index (DFI), as determined by the sperm chromatin structure assay (SCSA), exceeds the level of 20% [7]. However, this decrease in fertility *in vivo* is already seen at DFI above 10%, if one of standard sperm parameters is abnormal. Since many of the environmental toxicants may affect not only sperm DNA integrity but also concentration, motility and/or morphology [8], even slight increase in percentage of sperms with abnormal DNA, combined with deterioration of some other semen characteristics, may lead to a decrease in fertility.

An intriguing, but yet unresolved, issue is the question to which degree these DNA defects are becoming repaired following the process of fertilization. Unrepaired damaged sperm DNA introduced into the embryo might, in theory, lead to poor fertilization, early or late abortion, impaired foetal growth, congenital malformations and/or diseases arising during different phases of the post-natal life. These problems might not only occur in the offspring of the man exposed to such factors but also become manifest in the subsequent generation(s) [9]. The relevance of asking the question whether sperm DNA defects are transmitted to the offspring has become even more relevant in view of the increasing use of advanced techniques of assisted reproduction, *in vitro* fertilization (IVF) and intracytoplasmatic sperm injection (ICSI). Use of this technology has made it possible to achieve pregnancies in couples in whom this has been hampered by factors as sperm DNA breaks [10]. The complexity of the question, the rather recent access to techniques for evaluation of sperm DNA integrity and the relatively short follow-up of IVF and ICSI children do not allow to answer the question, but this issue should have a high priority on the future agenda of evaluating the potential risks of assisted reproduction.

Identification of environmental- and lifestyle-related factors deleterious to sperm DNA does also have implications in relation to the possibility of prevention and treatment of male-related infertility problems. Thus, once the implications of environment and lifestyle on sperm DNA integrity are understood, proper measures aiming to prevent such effects can be taken. Furthermore, studying the mechanisms of environment/lifestyle-related changes in the genome of the male gamete will also increase our level of understanding of the mechanisms involved in impairment of testicular function. Such knowledge is crucial, not only for prevention of infertility but also for development of specific drugs for treatment of fertility problems.

Therefore, studying and understanding the phenomenon involved in the effects of environmental and lifestyle factors on sperm DNA may be an important step in preventing and treating infertility problems as well as other important diseases, not only in relation to the generation actually being exposed but also in their offspring and, possibly, even in the subsequent generations.

How Can Environment/Lifestyle Affect Sperm Chromatin?

Apart from numerical and structural chromosomal changes, environmental exposure may, in principle, affect sperm DNA by introducing DNA fragmentation (or DNA strand breaks) and/or epigenetic changes in the genome of the male gamete.

Abnormal sperm chromatin/DNA structure is thought to arise from four potential sources: (1) deficiencies in recombination during spermatogenesis, (2) abnormal spermatid maturation (protamination disturbancies), (3) abortive apoptosis and (4) oxidative stress [9, 11] (Fig. 25.1).

Meiotic crossing-over is associated with the genetically programmed introduction of DNA double-strand breaks (DSBs) by specific nucleases of SPO11 family [12]. These DNA DSBs should be ligated until the end of meiosis I. Defective repair may interrupt spermatogenesis or lead to persistent sperm DNA fragmentation in ejaculated spermatozoa. Stage-specific occurrence of transient DNA strand breaks during spermiogenesis has been observed [13–15]. Both single-strand breaks (SSBs) and DSBs have been found in round and elongating spermatids. DNA breaks are necessary for transient relief of torsional stress, favouring casting off of the nucleosome histone cores, and aiding their replacement with transitional proteins and protamines during maturation of elongating spermatids [14–16]. Thus, chromatin packaging necessitates endogenous nuclease activity to both create and ligate breaks to reassemble DNA around the new protamine core. Chromatin packaging is completed and DNA integrity is restored during epididymal transit [17]. Although there is little evidence that spermatid maturation-associated DNA breaks are fully ligated, biologically this must be the case [18]. Ligation of DNA breaks is necessary not only to preserve the integrity of the primary DNA structure but also for reassembly of the important unit of genome expression – the DNA loop-domain.

Enzymatic activity involved in the creation of DNA breaks in spermatids has only been proven (by decatenating activity and specific inhibition) for topoisomerase II (Topo II) generating and ligating DSBs [14, 19]. Remodelling of chromatin by histone H4 hyperacetylation weakens the ionic interactions between the DNA and histone cores and is needed for Topo II activity to be introduced in spermatids [19]. Interestingly, Topo II activity seems to be androgen dependent [20], and since many of the environmental toxicants act as endocrine disrupters, they may, in principle, have an impact on sperm DNA integrity.

An alternative aetiology for the DNA DSBs in the spermatozoa of infertile patients can arise through an abortive apoptotic pathway.

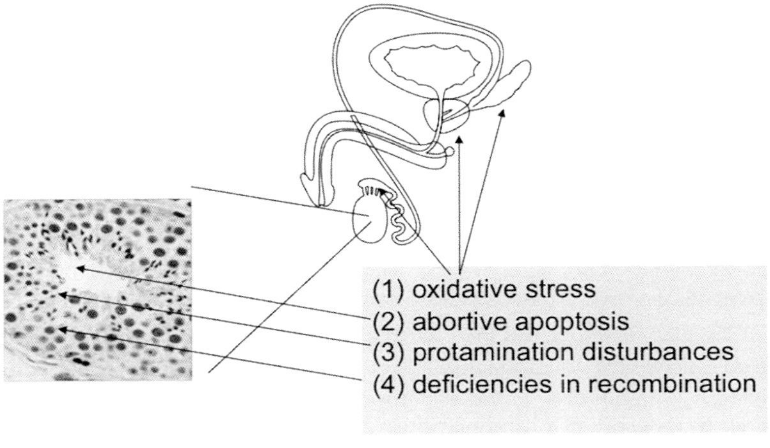

Fig. 25.1 Examples of major mechanisms of inducing DNA damage in spermatozoa related to lifestyle and/or environmental exposure

Apoptosis of testicular germ cells occurs normally throughout life, preventing their overproliferation [21, 22]. It has been suggested that an early apoptotic pathway, initiated in spermatogonia and spermatocytes, is mediated by Fas protein. Fas is a type I membrane protein that belongs to the tumour necrosis factor–nerve growth factor receptor family [23, 24]. It has been shown that Sertoli cells express Fas ligand, which by binding to Fas leads to cell death via apoptosis [23], limiting the size of germ cell population to numbers Sertoli cells can support [22]. Ligation of Fas ligand to Fas in the cellular membrane triggers the activation of caspases; therefore, this pathway is also characterized as a caspase-induced apoptosis [25]. Men exhibiting deficiencies in their semen profile often possess a large number of spermatozoa bearing Fas. This fact prompts the suggestion that these dysfunctional cells are the product of an incomplete apoptotic cascade [26]. Also the testicular process of apoptosis seems to be under the influence of reproductive hormones, thus being a potential target for an adverse effect of chemicals interfering with the endocrine function.

Reactive oxygen species (ROS) play an important physiological role, modulating gene and protein activities vital for sperm proliferation, differentiation and function. In the semen of fertile men, the amount of ROS generation is properly controlled by seminal antioxidants. The pathogenic effects of ROS occur when they are produced in excess of the antioxidant capabilities of the male reproductive tract or seminal plasma [27]. Abnormal spermatozoa and leukocytes are the main source of excess ROS generation [27]. It seems that sperm DNA is more prone to leukocyte induced ROS damage in infertile men with abnormal semen parameters likely possessing "masked" DNA damage and/or more fragile chromatin structure, which are under the sensitivity threshold of the assays used for the sperm DNA damage assessment [28].

Processes leading to DNA damage in ejaculated sperm are interrelated. For example, a defective maturation process during spermiogenesis, resulting in diminished sperm chromatin packaging, makes sperm cells more vulnerable for ROS-induced DNA fragmentation.

Another type of potentially environmentally induced sperm chromatin alterations are *epigenetic* changes in the genome. Epigenetics refers to changes in gene expression caused by mechanisms other than changes in the underlying DNA sequence. Examples of epigentics include gene methylation or demethylation leading to their inactivation or deactivation, respectively [29].

Epidemiological Indications of Environmentally Induced Changes in Sperm DNA

Genetic or epigenetic changes in the sperm genome introduced by environment and/or lifestyle related factors may have a serious impact on the reproductive function of an individual. Thus, such alterations may not only lead to impaired male fertility but once established, these changes may be paternally passed on to subsequent generations [30, 31].

Although there is no direct evidence of sperm DNA alterations induced by environment/lifestyle and then subsequently passed on to the offspring, there are some examples that can be considered as indirect evidence of existence of such mechanisms. The Y chromosome has been argued to be particularly vulnerable to DNA damage because it cannot correct double-stranded DNA deletions by homologous recombination [6].

Thus, paternal smoking, reported to introduce sperm DNA damage [32], has been reported to lead to an increased risk of childhood cancer in offspring [33–35], although others could not find the association [36]. Another possible consequence of sperm DNA damage might be microdeletions in the Y chromosome, which will lead to infertility in the male offspring [37].

It has been suggested, although the data seem somewhat contradictory, that increasing paternal age is associated to a higher frequency of aneuploidies, point mutations, sperm DNA breaks, loss of apoptosis, genetic imprinting and other chromosomal abnormalities, and it has even been considered as the major cause of new mutations in human populations [38]. Apart from age, paternal occupation has been linked to certain birth defects or diseases in the offspring which supposedly would act through genetic or epigenetic mechanisms [39]. Epigenetic abnormalities have been associated to imprinting diseases, for which a paternal role has been reported [40], and have been suggested to be increased in babies following conception by assisted reproduction.

A yet unresolved question being of great importance for evaluation of the risk of transmission of sperm DNA changes to the offspring is the ability of the fertilized oocyte to repair such changes. However, animal experiments might indicate that (1) sensitivity of induction of transmissible genetic damage is germ cell-stage dependent, the male post-meiotic cells being the most sensitive; (2) cytogenetic abnormalities at first metaphase after fertilization are critical intermediates between paternal exposure and abnormal reproductive outcomes and (3) the amount of sperm DNA damage that is converted into chromosomal aberrations in the zygote and that directly affect the risk for abnormal reproductive outcomes is regulated by maternal susceptibility factors [41].

Sperm Chromatin and Environment

Animal Experience

Animal experiments have clearly demonstrated that exposure of laboratory animals to environmental toxicants, irradiation and cytotoxic drugs may induce chromosome structural aberrations to be transmitted into the zygote [42]. However, even less dramatic negative impact of environment on sperm chromatin integrity has been reported. Such an effect has been found in different set-ups, and some examples are summarized below.

Yauk et al. [43] exposed mice in situ to ambient air near two integrated steel mills and a major highway, while the control animal breathed high-efficiency particulate air (HEPA) filtered ambient air. The animals exposed to unfiltered air presented with 1.6-fold increased rate in sperm mutation frequency and even higher levels of DNA strand breaks and hypermethylation. These results confirm findings reported earlier by Somers et al. [44].

In mice, exposure to mono-(2-ethylhexyl) phthalate (MEHP) was found to increase the germ cell apoptosis. Abortive apoptosis was suggested as one of the mechanisms leading to appearance of DNA strand breaks in ejaculated spermatozoa [45] (see above).

Another environmental toxicant, methyl *tert*-butyl ether, was found to exert reproductive system toxicity by increasing the oxidative stress [46], the latter also being one of the suggested causes of sperm DNA strand breaks.

Rather spectacular results were reported by Anway et al. [30]. They found that exposure of a gestating female rat during the period of gonadal sex determination to the endocrine disruptors vinclozolin (an antiandrogenic compound) or methoxychlor (an estrogenic compound) induced an adult phenotype in the F1 generation of decreased spermatogenic capacity (cell number and viability) and increased incidence of male infertility. These effects were transferred through the male germ line to nearly all males of all subsequent (F1–F4) generations examined and correlated with altered DNA methylation patterns in the germ line. These effects were also, partly, found by other groups [47], but some could not verify these findings [48, 49].

Human Data

Several studies have addressed the issue of association between certain lifestyle and environment-related exposures and sperm DNA integrity. Generally, the results are somewhat conflicting. This may to some degree be due to use of different methods for assessment of DNA damage, large variation in sample size as well as variations in recruitment of study subjects, including men from general population, infertility patients or occupationally exposed men. Below, these results, in relation to the most extensively studied exposures, are summarized.

Tobacco and Other Lifestyle Factors

For obvious reasons, the impact of cigarette smoking on sperm DNA integrity has been extensively studied. Some studies have shown increased level of sperm DNA damage in smoking men. Thus, Shen et al. [50] reported on a positive correlation between 8-OHdG amount and blood cotinine levels. The same was true for three studies based on the use of TUNEL [51–53] and one study using SCSA [52]. All these reports were based on relatively small sample sizes, none of them including more than 60 exposed men.

On the contrary, a number of reports could not confirm the association between tobacco smoking and sperm DNA damage [54–58]. This list includes studies based on the use of COMET, TUNEL, SCSA and 8-oxodG analyses. Interestingly, Saleh et al. [59] reported higher levels of ROS but not sperm DNA strand breaks in smokers as compared to non-smokers. Similarly, Viloria et al. [55] found lower level of sperm antioxidative enzymes in smokers as compared to non-smokers, however, without any difference in the degree of sperm DNA damage between the two groups.

One study [60] focused on the effects of prenatal exposure to cigarette smoking, and although sons of mothers smoking during pregnancy presented with lower sperm counts, no difference in regard to sperm DNA integrity was seen.

Thus, although some studies might indicate a negative effect of cigarette smoking on sperm DNA integrity, the results are rather conflicting, the largest of them finding no such effect. A list of studies dealing with effect of cigarette smoking on sperm DNA is given in Table 25.1.

Among other lifestyle related factors, one study addressed the issue of coffee drinking in relation to the sperm DNA integrity. The major finding [61] was that, using the COMET assay, men who consumed more than three cups coffee per day having approximately 20% higher percentage tail DNA under neutral but not alkaline conditions compared with men who consumed no caffeine ($P=0.005$).

Although animal experiments have indicated that cocaine may induce increased apoptosis [62]

Table 25.1 A list of studies dealing with the impact of cigarette smoking on sperm DNA integrity

Smoking as a main exposure or confounding factor	Assay used	No. participants	Effect	References
Main exposure	8-OHdG	60	[a]	[50]
Confounding factor	TUNEL	113	[a]	[51]
Main exposure	SCSA	25	[b]	[100]
Main exposure	SCSA	277	[b]	[101]
Main exposure	TUNEL	70	[a]	[52]
Main exposure	SCSA	70	[a]	[52]
Main exposure	TUNEL	97	[b]	[102]
Main exposure	SCSA	65	[b]	[59]
Confounding factor	COMET (alkaline)	71	[b]	[66]
Confounding factor	8-OHdG	225	[b]	[58]
Main exposure	COMET (alkaline)	40	[b]	[103]
Main exposure	COMET (neutral)	257	[b]	[54]
Confounding factor	SCSA	176	[b]	[73]
Main exposure	TUNEL	108	[a]	[53]
Confounding factor	COMET (neutral)	379	[b]	[56]
Main exposure	OxyDNA assay	55	[b]	[55]
Confounding factor	SCSA	279	[b]	[104]
Main exposure (mother)	SCSA	265	[b]	[60]

Adapted from M Spanó, unpublished data
[a] Exposure related increase in percentage of spermatozoa with DNA damage
[b] No effect found

as well as alteration of gene imprinting in germ cells [63], similar data in humans are lacking.

Occupational Exposure

Surprisingly, few epidemiological studies have addressed the issue of occupational exposure in relation to sperm DNA integrity. Three studies dealing with impact of styrene exposure, two of them using COMET assay and one applying SCSA all found a statistically significant increase in the indices of impairment of DNA integrity in exposed workers as compared to unexposed subjects [64–66].

One study focused on occupational boron exposure reporting no significant correlations between blood and urine boron and adverse semen parameters including sperm DNA breaks and percentage apoptotic cells [67].

Hsu et al. [68] reported on the effect of lead exposure on SCSA parameters in a group of battery factory workers in Taiwan and found a positive correlation between the blood levels of the metal and the percentage of sperms with DNA fragmentation. An earlier study performed by Bonde et al. [69] only partly supported these results reporting some indications of deterioration of sperm chromatin found in men with the highest concentrations of lead within spermatozoa. These data are summarized in Table 25.2. The issue of pesticide exposure is covered in a separate section below.

Air Pollution

Animal studies (se above) have linked air pollution to the level of sperm DNA damage. Similar

Table 25.2 A list of studies dealing with impact of environmental/occupational (except pesticide and PCB) exposure on sperm DNA integrity

Exposure	Assay used	No. participants	Effect	References
Air pollution	SCSA	266	[a]	[70]
Air pollution	SCSA	36	[a]	[71]
Air pollution	SCSA and CMA	228	[b]	[72]
Styrene (mandelic acid urinary concentration)	SCSA	44	[a]	[64]
Styrene (mandelic acid urinary concentration)	COMET (alkaline)	73	[a]	[66]
Styrene (mandelic and phenylglyoxylic acid urinary concentration)	COMET (alkaline)	77	[a]	[65]
Boron (blood and urine)	COMET and TUNEL	103	[b]	[67]
Phthalate esters	Sperm nuclear chromatin decondensation (NCD) test	53	[a]	[76]
Phthalate and phthalate metabolites	COMET (neutral)	168	[a]	[105]
Phthalate and phthalate metabolites	COMET (neutral)	379	[a]	[56]
Phthalate and phthalate metabolites	SCSA	234	[b]	[84]
Phthalate and phthalate metabolites	SCSA	300	[a]	[82]
Acrylonitrile	COMET (alkaline)	60	[a]	[106]
Lead	SCSA	503	At blood Pb concentration <45 μg/dl[b]	[69]
Lead	NCD test	68	[a]	[107]
Lead	SCSA	80	[a]	[68]
Mercury	SCSA	195	No synergism with PCB exposure[b]	[108]

Adapted from M Spanó, unpublished data
[a] Exposure-related increase in percentage of spermatozoa with DNA damage
[b] No effect found

findings have also been done in Czech men, both in a cross-sectional [70] and a longitudinal [71] set-up. However, a recent study by Hansen et al. [72] based on a cohort of 228 fertile men could not find any association between the level of exposure to ozone and particulate matter <2.5 μm in aerodynamic diameter on sperm DNA indices as assessed by SCSA and by chromomycin A3 staining (Table 25.2).

Persistent Organohalogen Pollutants

A number of studies have addressed the issue of the impact of exposure to persistent organohalogen pollutants (POPs) in relation to the sperm chromatin integrity. In a multi-centre European Union funded study (http://www.inuendo.dk) focus was given to association between serum levels of CB-153, a marker of exposure to polychlorinated biphenyls (PCBs) as well as concentrations of *p,p′*-DDE, a metabolite of dichlorodiphenyltrichloroethane (DDT) and sperm parameters, including DNA integrity (Tables 25.3 and 25.4). Using both SCSA and TUNEL, high levels of PCB exposure were found to be associated with increased percentage of spermatozoa with DNA damage. However, interestingly, these associations were found in Caucasian populations (Sweden, Ukraine, and Poland), but not in Greenlandic Inuits, the latter – despite very high levels of CB-153, presenting with significantly lower DFI as compared to the European men [73–75]. This finding might indicate an interaction between POP exposure and genetic (see below) and/or other lifestyle or environmental factors in relation to the integrity of sperm DNA. The Inuendo findings seem to be in agreement with an earlier study by Rozati et al. [76] showing

Table 25.3 A list of studies dealing with the impact of pesticide exposure on sperm DNA integrity

Exposure	Assay used	No. participants	Effect	References
p,p′-DDE	COMET (neutral)	212	[a]	[57]
p,p′-DDE	SCSA	176	[a]	[73]
p,p′-DDE	SCSA	707	[a]	[74]
p,p′-DDE	SCSA	680	Only in subjects with androgen receptor CAG repeat length f 21 or less [b]	[96]
p,p′-DDE/DDT	SCSA	209	[b]	[78]
p,p′-DDE	TUNEL	652	[a]	[75]
p,p′-DDE	Aniline Blue	116	[b]	[79]
Pesticides (occupation exposure)	SCSA	251	[a]	[85]
Pesticides (dietary intake)	SCSA	256	[a]	[86]
Pesticides	SCSA	256	[a]	[109]
Organophosphoric pesticides	SCSA	66	[b]	[88]
Organophosphoric pesticides	ISNT	54	Paraoxonase: 192RR genotype more susceptible [b]	[87]
Hexachlorobenzene	COMET (neutral)	212	[a]	[57]
Insecticides (fenvalerate)	COMET (alkaline) TUNEL	63	[b]	[90]
Insecticides (chlorpyrifos, carbaryl)	COMET (neutral)	260	[b]	[92]
Insecticides Pyrethroids	COMET (neutral)	207	[b]	[91]
Insecticides (carbaryl)	TUNEL	46	[b]	[89]

Adapted from M Spanó, unpublished data

[a] No effect found

[b] Exposure-related increase in percentage of spermatozoa with DNA damage

Table 25.4 List of studies dealing with the impact of PCB exposure on sperm DNA integrity

Exposure	Assay used	No. participants	Effect	References
PCB	NCD	53	[a]	[76]
PCB	COMET (neutral)	212	[b]	[57]
PCB	SCSA	176	[a]	[73]
PCB	SCSA	707	In Caucasians but not in Inuits[a]	[74]
PCB	TUNEL	652	In Caucasians but not in Inuits[a]	[75]

Adapted from M Spanó, unpublished data
[a] Exposure-related increase in percentage of spermatozoa with DNA damage
[b] No effect found

positive correlation between seminal PCB levels and percentage of spermatozoa with single-stranded DNA.

Considering the impact of DDT exposure on sperm chromatin integrity, the results are more diverging. No correlation between *p,p'*-DDE and TUNEL as well as SCSA parameters was found in the Inuendo study – if the impact of genetic polymorphsms was not taken into consideration (see below) [74]. However, it should be kept in mind that there was a high level of correlation between serum levels of CB-153 and the *p,p'*-DDE concentration, in an epidemiological set-up [77], making it impossible to detangle the biological effects of these to compounds. Thus, in a men living in areas with endemic malaria, where due to use of DDT the plasma levels of its metabolite can reach levels 1,000-fold higher than in other populations, there was a positive correlation between SCSA DFI as well as Aniline Blue test assessing the most severe category of incomplete DNA condensation and the concentration of *p,p'*-DDE [78, 79].

Phthalates

During the past few years, a lot of attention has been given to the potential endocrine disrupting effect of phthalate exposure; these chemicals are supposed to interfere with the Leydig cell function [80, 81], thereby affecting the levels of intratesticular testosterone. A recent study has shown positive correlation between the level of phthalate exposure and ROS production [82]. Therefore, these chemicals may exert a negative effect on sperm DNA integrity both by inducing high ROS levels and, through hormonal deregulation, by interfering with normal intratesticular function of DNA repair enzymes.

Three studies, based on men attending infertility clinics, found a positive association between at least some of the phthalate metabolites and indices of sperm DNA damage, assessed by COMET [56, 83] or SCSA [82]. However, no such association was found in a younger group of Swedish military conscripts [84] (Table 25.2).

Insecticides and Pesticides Other than DDT

Within this quite heterogeneous category of environmental toxicants, the studies have focused on either occupational exposure or the one related to consumption of food containing traces of such compounds.

Generally, even within this category the results are somewhat conflicting, mainly as regards the pesticides, with no association in Danish agricultural workers and the opposite findings in Mexico [85–88]. A number of reports related to exposure to insecticides have shown positive association between the levels of these chemicals and markers of sperm DNA damage. These findings have been rather consistent when the exposure has been related to the occupation [89, 90] and when it was rather environmentally related [91, 92].

Also, the exposure to organophosphoric pesticides seems to have a negative impact on sperm DNA integrity [87, 88], whereas in two studies comparing organic and non-organic farmers [85, 86], no such effect was found (Table 25.3).

Gene–Environment Interaction and Sperm Chromatin

Impairment of sperm chromatin integrity due to lifestyle- or environment-related factors represents a unique form of "gene–environment interaction" – namely, environmental stress having a negative impact on the genome of the gamete, those changes being potentially transmittable to the following generation(s). The results of the study by Anway et al. [30], although focusing on epigenetic changes rather than direct DNA damage, illustrate that such scenario is not unlikely. However, in traditional terms, the term "gene–environment interaction" usually refers to interindividual variation in susceptibility to environmental/lifestyle factors based on genetic differences between the subjects. There are several indications of such mechanisms operating even in relation to impairment of sperm DNA integrity.

Thus, as already mentioned, in the Inuendo study, Inuits were found to have significantly lower DFI as compared to Caucasian men [74], the association between levels of PCB exposure being seen among the latter but not in the former ethnic group. Although nutritional or other environmental factors might be the cause of such difference, genetic diversity as a causal factor is also likely.

Our research group has been focusing on polymorphisms in the androgen receptor gene (AR) as modifiers of the effect of endocrine disrupting chemicals, including POPs. One of the polymorphic regions in the *AR* is the glutamine encoding CAG repeats in the exon I of this gene [93]. It has been shown that the number of these repeats, which in a Caucasian normal population varies between 10 and 30 with a mean number of 22, has an impact on the receptor activity [93–95]. In the Inuendo study, we found that an association between *p,p'*-DDE, but not CB-153, levels and the DFI dependent on the CAG number [96]. For CAG lengths of 21 or less, those with high levels of the *p,p'*-DDE presented with 40% higher percentage of spermatozoa with impaired DNA integrity than those with low concentrations of this POP. Such an association between exposure and sperm DNA damage was not seen for other CAG lengths. These findings might, at least partly, explain the robustness of Inuits to the deleterious effects of POP exposure. The mean CAG number is on average 1.5–2 higher in Inuits as compared to Caucasians, thus a less proportion of men having the genotype encoding for higher level of susceptibility [97].

In the study of the impact of sperm pollution on sperm DNA integrity, this effect was shown to be modified by the polymorphisms in the glutathione-*S*-transferase M1 gene [98] as well as by variations in several DNA repair genes [99].

Therefore, it is to be expected that even for the other environmental and lifestyle factors shown to have an impact on sperm DNA integrity, the genetically determined susceptibility may vary between the individuals. Such gene–environment interaction might, at least partly, explain the mechanisms between the above-mentioned somewhat diverging results when different population cohorts are investigated and/or when several techniques are used for detection of sperm DNA damage.

Conclusions

Available experimental and human data show that a number of lifestyle- and environment-related exposures may have negative effects on sperm DNA integrity. The extent of the sperm DNA damage seems to differ between different studies. Such a discrepancy may partly be due to use of several methods for assessment of sperm DNA integrity, techniques that do not measure exactly the same characteristics of sperm DNA. Another contributing factor may be the genetically determined variation in individual susceptibility.

Sperm DNA damage due to environmental and lifestyle factors may have a negative impact on fertility, and there is a potential risk of transmission to the offspring. Therefore, it is of importance to focus on the association between environment and sperm DNA integrity to prevent male subfertility and to avoid potentially serious health effects in the future generation(s).

References

1. Giwercman A, Skakkebaek NE. The human testis–an organ at risk? Int J Androl. 1992;15(5):373–5.
2. Carlsen E, Giwercman A, Keiding N, Skakkeb'k NE. Evidence for decreasing quality of semen during past 50 years. Br Med J. 1992;305:609–13.
3. Carlsen E, Giwercman A, Skakkeb'k NE, Keiding N. Decreasing quality of semen (letter). Br Med J. 1993;306:461.
4. Toppari J, Larsen JC, Christiansen P, et al. Male reproductive health and environmental xenoestrogens. Environ Health Perspect. 1996;104:741–803.
5. Richiardi L, Bellocco R, Adami HO, et al. Testicular cancer incidence in eight northern European countries: secular and recent trends. Cancer Epidemiol Biomark Prev. 2004;13(12):2157–66.
6. Aitken RJ, Krausz C. Oxidative stress, DNA damage and the Y chromosome. Reproduction. 2001;122(4):497–506.
7. Giwercman A, Lindstedt L, Larsson M, et al. Sperm chromatin structure assay as an independent predictor of fertility in vivo: a case-control study. Int J Androl. 2009;33:e221–7.
8. Toft G, Rignell-Hydbom A, Tyrkiel E, et al. Semen quality and exposure to persistent organochlorine pollutants. Epidemiology. 2006;17(4):450–8.
9. Agarwal A, Agarwal A, Said TM. Role of sperm chromatin abnormalities and DNA damage in male infertility. Hum Reprod Update. 2003;9(4):331–45.
10. Bungum M, Humaidan P, Axmon A, et al. Sperm DNA integrity assessment in prediction of assisted reproduction technology outcome. Hum Reprod. 2007;22(1):174–9.
11. Sakkas D, Mariethoz E, Manicardi G, Bizzaro D, Bianchi PG, Bianchi U. Origin of DNA damage in ejaculated human spermatozoa. Rev Reprod. 1999;4(1):31–7.
12. Bannister LA, Schimenti JC. Homologous recombinational repair proteins in mouse meiosis. Cytogenet Genome Res. 2004;107(3–4):191–200.
13. Bench G, Corzett MH, De Yebra L, Oliva R, Balhorn R. Protein and DNA contents in sperm from an infertile human male possessing protamine defects that vary over time. Mol Reprod Dev. 1998;50(3):345–53.
14. de Yebra L, Ballesca JL, Vanrell JA, Bassas L, Oliva R. Complete selective absence of protamine P2 in humans. J Biol Chem. 1993;268(14):10553–7.
15. de Yebra L, Ballesca JL, Vanrell JA, Corzett M, Balhorn R, Oliva R. Detection of P2 precursors in the sperm cells of infertile patients who have reduced protamine P2 levels. Fertil Steril. 1998;69(4):755–9.
16. Carrell DT, Liu L. Altered protamine 2 expression is uncommon in donors of known fertility, but common among men with poor fertilizing capacity, and may reflect other abnormalities of spermiogenesis. J Androl. 2001;22(4):604–10.
17. Mengual L, Ballesca JL, Ascaso C, Oliva R. Marked differences in protamine content and P1/P2 ratios in sperm cells from percoll fractions between patients and controls. J Androl. 2003;24(3):438–47.
18. Nasr-Esfahani MH, Salehi M, Razavi S, et al. Effect of protamine-2 deficiency on ICSI outcome. Reprod Biomed Online. 2004;9(6):652–8.
19. Kierszenbaum AL. Transition nuclear proteins during spermiogenesis: unrepaired DNA breaks not allowed. Mol Reprod Dev. 2001;58(4):357–8.
20. Bakshi RP, Galande S, Bali P, Dighe R, Muniyappa K. Developmental and hormonal regulation of type II DNA topoisomerase in rat testis. J Mol Endocrinol. 2001;26(3):193–206.
21. Fernandez JL, Goyanes VJ, Ramiro-Diaz J, Gosalvez J. Application of FISH for in situ detection and quantification of DNA breakage. Cytogenet Cell Genet. 1998;82(3–4):251–6.
22. Yu YE, Zhang Y, Unni E, et al. Abnormal spermatogenesis and reduced fertility in transition nuclear protein 1-deficient mice. Proc Natl Acad Sci USA. 2000;97(9):4683–8.
23. Zhao M, Shirley CR, Yu YE, et al. Targeted disruption of the transition protein 2 gene affects sperm chromatin structure and reduces fertility in mice. Mol Cell Biol. 2001;21(21):7243–55.
24. Boissonneault G. Chromatin remodeling during spermiogenesis: a possible role for the transition proteins in DNA strand break repair. FEBS Lett. 2002;514(2–3):111–4.
25. Cho C, Jung-Ha H, Willis WD, et al. Protamine 2 deficiency leads to sperm DNA damage and embryo death in mice. Biol Reprod. 2003;69(1):211–7.
26. Bench GS, Friz AM, Corzett MH, Morse DH, Balhorn R. DNA and total protamine masses in individual sperm from fertile mammalian subjects. Cytometry. 1996;23(4):263–71.
27. Said TM, Paasch U, Glander HJ, Agarwal A. Role of caspases in male infertility. Hum Reprod Update. 2004;10(1):39–51.
28. Erenpreiss J, Hlevicka S, Zalkalns J, Erenpreisa J. Effect of leukocytospermia on sperm DNA integrity: a negative effect in abnormal semen samples. J Androl. 2002;23(5):717–23.
29. Trasler JM. Epigenetics in spermatogenesis. Mol Cell Endocrinol. 2009;306(1–2):33–6.
30. Anway MD, Cupp AS, Uzumcu M, Skinner MK. Epigenetic transgenerational actions of endocrine disruptors and male fertility. Science. 2005;308(5727):1466–9.
31. Cheng RY, Hockman T, Crawford E, Anderson LM, Shiao YH. Epigenetic and gene expression changes related to transgenerational carcinogenesis. Mol Carcinog. 2004;40(1):1–11.
32. Elshal MF, El-Sayed IH, Elsaied MA, El-Masry SA, Kumosani TA. Sperm head defects and disturbances in spermatozoal chromatin and DNA integrities in idiopathic infertile subjects: association with cigarette smoking. Clin Biochem. 2009;42(7–8):589–94.

33. Sorahan T, Lancashire RJ, Hulten MA, Peck I, Stewart AM. Childhood cancer and parental use of tobacco: deaths from 1953 to 1955. Br J Cancer. 1997;75(1): 134–8.
34. Sorahan T, McKinney PA, Mann JR, et al. Childhood cancer and parental use of tobacco: findings from the inter-regional epidemiological study of childhood cancer (IRESCC). Br J Cancer. 2001;84(1):141–6.
35. Ji BT, Shu XO, Linet MS, et al. Paternal cigarette smoking and the risk of childhood cancer among offspring of nonsmoking mothers. J Natl Cancer Inst. 1997;89(3):238–44.
36. Schuz J, Kaatsch P, Kaletsch U, Meinert R, Michaelis J. Association of childhood cancer with factors related to pregnancy and birth. Int J Epidemiol. 1999;28(4):631–9.
37. Aitken RJ. The Amoroso Lecture. The human spermatozoon–a cell in crisis? J Reprod Fertil. 1999; 115(1):1–7.
38. Sartorius GA, Nieschlag E. Paternal age and reproduction. Hum Reprod Update. 2010;16(1):65–79.
39. Whalley LJ, Thomas BM, McGonigal G, McQuade CA, Swingler R, Black R. Epidemiology of presenile Alzheimer's disease in Scotland (1974-88) I. Non-random geographical variation. Br J Psychiatry. 1995; 167(6):728–31.
40. Marques CJ, Carvalho F, Sousa M, Barros A. Genomic imprinting in disruptive spermatogenesis. Lancet. 2004;363(9422):1700–2.
41. Marchetti F, Wyrobek AJ. Mechanisms and consequences of paternally-transmitted chromosomal abnormalities. Birth Defects Res C Embryo Today. 2005;75(2):112–29.
42. Pacchierotti F, Tiveron C, Ranaldi R, et al. Reproductive toxicity of 1,3-butadiene in the mouse: cytogenetic analysis of chromosome aberrations in first-cleavage embryos and flow cytometric evaluation of spermatogonial cell killing. Mutat Res. 1998; 397(1):55–66.
43. Yauk C, Polyzos A, Rowan-Carroll A, et al. Germ-line mutations, DNA damage, and global hypermethylation in mice exposed to particulate air pollution in an urban/industrial location. Proc Natl Acad Sci USA. 2008;105(2):605–10.
44. Somers CM, McCarry BE, Malek F, Quinn JS. Reduction of particulate air pollution lowers the risk of heritable mutations in mice. Science. 2004; 304(5673):1008–10.
45. Giammona CJ, Sawhney P, Chandrasekaran Y, Richburg JH. Death receptor response in rodent testis after mono-(2-ethylhexyl) phthalate exposure. Toxicol Appl Pharmacol. 2002;185(2):119–27.
46. Li D, Yuan C, Gong Y, Huang Y, Han X. The effects of methyl tert-butyl ether (MTBE) on the male rat reproductive system. Food Chem Toxicol. 2008; 46(7):2402–8.
47. Stouder C, Paoloni-Giacobino A. Transgenerational effects of the endocrine disruptor vinclozolin on the methylation pattern of imprinted genes in the mouse sperm. Reproduction. 2010;139(2):373–9.
48. Schneider S, Kaufmann W, Buesen R, van Ravenzwaay B. Vinclozolin–the lack of a transgenerational effect after oral maternal exposure during organogenesis. Reprod Toxicol. 2008;25(3):352–60.
49. Inawaka K, Kawabe M, Takahashi S, et al. Maternal exposure to anti-androgenic compounds, vinclozolin, flutamide and procymidone, has no effects on spermatogenesis and DNA methylation in male rats of subsequent generations. Toxicol Appl Pharmacol. 2009;237(2):178–87.
50. Shen HM, Chia SE, Ni ZY, New AL, Lee BL, Ong CN. Detection of oxidative DNA damage in human sperm and the association with cigarette smoking. Reprod Toxicol. 1997;11(5):675–80.
51. Sun JG et al. Detection of deoxyribomucleic acid fragmentation in human sperm: correlation with fertilization in vitro. Biol Reprod. 1997;56:602–7.
52. Potts RJ, Newbury CJ, Smith G, Notarianni LJ, Jefferies TM. Sperm chromatin damage associated with male smoking. Mutat Res. 1999;423(1–2): 103–11.
53. Sepaniak S, Forges T, Gerard H, Foliguet B, Bene MC, Monnier-Barbarino P. The influence of cigarette smoking on human sperm quality and DNA fragmentation. Toxicology. 2006;223(1–2):54–60.
54. Trisini AT, Singh NP, Duty SM, Hauser R. Relationship between human semen parameters and deoxyribonucleic acid damage assessed by the neutral comet assay. Fertil Steril. 2004;82(6):1623–32.
55. Viloria T, Meseguer M, Martinez-Conejero JA, et al. Cigarette smoking affects specific sperm oxidative defense but does not cause oxidative DNA damage in infertile men. Fertil Steril. 2009;94:631–7.
56. Hauser R, Meeker JD, Singh NP, et al. DNA damage in human sperm is related to urinary levels of phthalate monoester and oxidative metabolites. Hum Reprod. 2007;22(3):688–95.
57. Hauser R, Singh NP, Chen Z, Pothier L, Altshul L. Lack of an association between environmental exposure to polychlorinated biphenyls and p, p′-DDE and DNA damage in human sperm measured using the neutral comet assay. Hum Reprod. 2003;18(12): 2525–33.
58. Loft S, Kold-Jensen T, Hjollund NH, et al. Oxidative DNA damage in human sperm influences time to pregnancy. Hum Reprod. 2003;18(6):1265–72.
59. Saleh RA, Agarwal A, Sharma RK, Nelson DR, Thomas Jr AJ. Effect of cigarette smoking on levels of seminal oxidative stress in infertile men: a prospective study. Fertil Steril. 2002;78(3):491–9.
60. Storgaard L, Bonde JP, Ernst E, et al. Does smoking during pregnancy affect sons' sperm counts? Epidemiology. 2003;14(3):278–86.
61. Schmid TE, Eskenazi B, Baumgartner A, et al. The effects of male age on sperm DNA damage in healthy non-smokers. Hum Reprod. 2007;22(1):180–7.
62. Yang GS, Wang W, Wang YM, Chen ZD, Wang S, Fang JJ. Effect of cocaine on germ cell apoptosis in rats at different ages. Asian J Androl. 2006;8(5): 569–75.

63. He F, Lidow IA, Lidow MS. Consequences of paternal cocaine exposure in mice. Neurotoxicol Teratol. 2006;28(2):198–209.
64. Kolstad HA, Bonde JP, Spano M, et al. Sperm chromatin structure and semen quality following occupational styrene exposure. Scand J Work Environ Health. 1999;25:70–3.
65. Migliore L, Colognato R, Naccarati A, Bergamaschi E. Relationship between genotoxicity biomarkers in somatic and germ cells: findings from a biomonitoring study. Mutagenesis. 2006;21(2):149–52.
66. Migliore L, Naccarati A, Zanello A, Scarpato R, Bramanti L, Mariani M. Assessment of sperm DNA integrity in workers exposed to styrene. Hum Reprod. 2002;17(11):2912–8.
67. Robbins WA, Xun L, Jia J, Kennedy N, Elashoff DA, Ping L. Chronic boron exposure and human semen parameters. Reprod Toxicol. 2010;29(2):184–90.
68. Hsu PC, Chang HY, Guo YL, Liu YC, Shih TS. Effect of smoking on blood lead levels in workers and role of reactive oxygen species in lead-induced sperm chromatin DNA damage. Fertil Steril. 2009;91(4):1096–103.
69. Bonde JP, Joffe M, Apostoli P, et al. Sperm count and chromatin structure in men exposed to inorganic lead: lowest adverse effect levels. Occup Environ Med. 2002;59(4):234–42.
70. Selevan SG, Borkovec L, Slott VL, et al. Semen quality and reproductive health of young Czech men exposed to seasonal air pollution. Environ Health Perspect. 2000;108(9):887–94.
71. Rubes J, Selevan SG, Evenson DP, et al. Episodic air pollution is associated with increased DNA fragmentation in human sperm without other changes in semen quality. Hum Reprod. 2005;20(10):2776–83.
72. Hansen C, Luben TJ, Sacks JD, et al. The effect of ambient air pollution on sperm quality. Environ Health Perspect. 2010;118(2):203–9.
73. Rignell-Hydbom A, Rylander L, Giwercman A, et al. Exposure to PCBs and p, p′-DDE and human sperm chromatin integrity. Environ Health Perspect. 2005; 113(2):175–9.
74. Spano M, Toft G, Hagmar L, et al. Exposure to PCB and p, p′-DDE in European and Inuit populations: impact on human sperm chromatin integrity. Hum Reprod. 2005;20(12):3488–99.
75. Stronati A, Manicardi GC, Cecati M, et al. Relationships between sperm DNA fragmentation, sperm apoptotic markers and serum levels of CB-153 and p, p′-DDE in European and Inuit populations. Reproduction. 2006;132(6):949–58.
76. Rozati R, Reddy PP, Reddanna P, Mujtaba R. Role of environmental estrogens in the deterioration of male factor fertility. Fertil Steril. 2002;78(6):1187–94.
77. Jonsson BA, Rylander L, Lindh C, et al. Inter-population variations in concentrations, determinants of and correlations between 2,2′,4,4′,5,5′-hexachlorobiphenyl (CB-153) and 1,1-dichloro-2,2-bis (p-chlorophenyl)-ethylene (p, p′-DDE): a cross-sectional study of 3161 men and women from Inuit and European populations. Environ Health. 2005;4:27.
78. de Jager C, Aneck-Hahn NH, Bornman MS, et al. Sperm chromatin integrity in DDT-exposed young men living in a malaria area in the Limpopo Province, South Africa. Hum Reprod. 2009;24(10):2429–38.
79. De Jager C, Farias P, Barraza-Villarreal A, et al. Reduced seminal parameters associated with environmental DDT exposure and p, p′-DDE concentrations in men in Chiapas, Mexico: a cross-sectional study. J Androl. 2006;27(1):16–27.
80. Scott HM, Mason JI, Sharpe RM. Steroidogenesis in the fetal testis and its susceptibility to disruption by exogenous compounds. Endocr Rev. 2009;30(7): 883–925.
81. Svechnikov K, Svechnikova I, Soder O. Inhibitory effects of mono-ethylhexyl phthalate on steroidogenesis in immature and adult rat Leydig cells in vitro. Reprod Toxicol. 2008;25(4):485–90.
82. Pant N, Shukla M, Kumar Patel D, et al. Correlation of phthalate exposures with semen quality. Toxicol Appl Pharmacol. 2008;231(1):112–6.
83. Duty SM, Singh NP, Silva MJ, et al. The relationship between environmental exposures to phthalates and DNA damage in human sperm using the neutral comet assay. Environ Health Perspect. 2003;111(9):1164–9.
84. Jonsson BA, Richthoff J, Rylander L, Giwercman A, Hagmar L. Urinary phthalate metabolites and biomarkers of reproductive function in young men. Epidemiology. 2005;16(4):487–93.
85. Larsen SB, Giwercman A, Spano M, Bonde JP. A longitudinal study of semen quality in pesticide spraying Danish farmers. The ASCLEPIOS Study Group. Reprod Toxicol. 1998;12(6):581–9.
86. Juhler RK, Larsen SB, Meyer O, et al. Human semen quality in relation to dietary pesticide exposure and organic food. Arch Environ Contam Toxicol. 1999;37:415–23.
87. Perez-Herrera N, Polanco-Minaya H, Salazar-Arredondo E, et al. PON1Q192R genetic polymorphism modifies organophosphorous pesticide effects on semen quality and DNA integrity in agricultural workers from southern Mexico. Toxicol Appl Pharmacol. 2008;230(2):261–8.
88. Sanchez-Pena LC, Reyes BE, Lopez-Carrillo L, et al. Organophosphorous pesticide exposure alters sperm chromatin structure in Mexican agricultural workers. Toxicol Appl Pharmacol. 2004;196(1):108–13.
89. Xia Y, Cheng S, Bian Q, et al. Genotoxic effects on spermatozoa of carbaryl-exposed workers. Toxicol Sci. 2005;85(1):615–23.
90. Bian Q, Xu LC, Wang SL, et al. Study on the relation between occupational fenvalerate exposure and spermatozoa DNA damage of pesticide factory workers. Occup Environ Med. 2004;61(12):999–1005.
91. Meeker JD, Barr DB, Hauser R. Human semen quality and sperm DNA damage in relation to urinary metabolites of pyrethroid insecticides. Hum Reprod. 2008;23(8):1932–40.
92. Meeker JD, Singh NP, Ryan L, et al. Urinary levels of insecticide metabolites and DNA damage in human sperm. Hum Reprod. 2004;19(11):2573–80.

93. Tut TG, Ghadessy FJ, Trifiro MA, Pinsky L, Young EL. Long polyglutamine tracts in the androgen receptor are associated with reduced trans-activation, impaired sperm production, and male infertility. J Clin Endocrinol Metab. 1997;82:3777–82.
94. Eckardstein SV, Schmidt A, Kamischke A, Simoni M, Gromoll J, Nieschlag E. CAG repeat length in the androgen receptor gene and gonadotrophin suppression influence the effectiveness of hormonal male contraception. Clin Endocrinol (Oxf). 2002;57(5):647–55.
95. Nenonen H, Bjork C, Skjaerpe PA, et al. CAG repeat number is not inversely associated with androgen receptor activity in vitro. Mol Hum Reprod. 2010;16(3):153–7.
96. Giwercman A, Rylander L, Rignell-Hydbom A, et al. Androgen receptor gene CAG repeat length as modifier of the association between persistent organohalogen pollutant exposure markers and semen characteristics. Pharmacogenet Genomics. 2007;17:391–401.
97. Giwercman C, Giwercman A, Pedersen HS, et al. Polymorphisms in genes regulating androgen activity among prostate cancer low-risk Inuit men and high-risk Scandinavians. Int J Androl. 2008;31(1): 25–30.
98. Rubes J, Selevan SG, Sram RJ, Evenson DP, Perreault SD. GSTM1 genotype influences the susceptibility of men to sperm DNA damage associated with exposure to air pollution. Mutat Res. 2007; 625(1–2):20–8.
99. Rubes J, Rybar R, Prinosilova P, et al. Genetic polymorphisms influence the susceptibility of men to sperm DNA damage associated with exposure to air pollution. Mutat Res. 2010;683(1–2):9–15.
100. Rubes J, Lowe X, Moore II D, et al. Smoking cigarettes is associated with increased sperm disomy in teenage men. Fertil Steril. 1998;70(4):715–23.
101. Spano M, Kolstad AH, Larsen SB, et al. The applicability of the flow cytometric sperm chromatin structure assay as diagnostic and prognostic tool in then human fertility clinic. Hum Reprod. 1998;13:2495–505.
102. Sergerie M, Ouhilal S, Bissonnette F, Brodeur J, Bleau G. Lack of association between smoking and DNA fragmentation in the spermatozoa of normal men. Hum Reprod. 2000;15(6):1314–21.
103. Belcheva A, Ivanova-Kicheva M, Tzvetkova P, Marinov M. Effects of cigarette smoking on sperm plasma membrane integrity and DNA fragmentation. Int J Androl. 2004;27(5):296–300.
104. Smit M, Romijn JC, Wildhagen MF, Weber RF, Dohle GR. Sperm chromatin structure is associated with the quality of spermatogenesis in infertile patients. Fertil Steril. 2009;94:1748–52.
105. Duty SM, Calafat AM, Silva MJ, Ryan L, Hauser R. Phthalate exposure and reproductive hormones in adult men. Hum Reprod. 2005;20:604–10.
106. Xu DX, Zhu QX, Zheng LK, et al. Exposure to acrylonitrile induced DNA strand breakage and sex chromosome aneuploidy in human spermatozoa. Mutat Res. 2003;537(1):93–100.
107. Hernandez-Ochoa I, Garcia-Vargas G, Lopez-Carrillo L, et al. Low lead environmental exposure alters semen quality and sperm chromatin condensation in northern Mexico. Reprod Toxicol. 2005; 20(2):221–8.
108. Rignell-Hydbom A, Axmon A, Lundh T, Jonsson BA, Tiido T, Spano M. Dietary exposure to methyl mercury and PCB and the associations with semen parameters among Swedish fishermen. Environ Health. 2007;6:14.
109. Larsen SB, Spano M, Giwercman A, Bonde JP. Semen quality and sex hormones among organic and traditional Danish farmers. ASCLEPIOS Study Group. Occup Environ Med. 1999;56(2):139–44.

Effects of Male Accessory Gland Infection on Sperm Parameters

26

Aldo E. Calogero, Sandro La Vignera, Rosita A. Condorelli, Rosario D'Agata, and Enzo Vicari

Abstract

Despite an open debate on pros and cons of the role of male accessory gland infection (MAGI) in male infertility, andrologists should at least consider MAGI as a risk factor of male infertility. In fact, MAGI may impair sperm function and cause male infertility through the multiple pathophysiological mechanisms discussed in this chapter.

Keywords

Male accessory gland infection • Sperm and accessory gland infection • Andrology • Infection, male accessory gland • Male infertility

Male Accessory Gland Infection

Male accessory gland infection (MAGI) has been identified among diagnostic categories having a negative impact on male reproductive function and fertility [1]. According to the WHO [1], MAGI is diagnosed when a patient has oligo-, astheno-, and/or teratozoospermia associated with at least one factor A plus one factor B, one factor A plus one factor C, one factor B plus one factor C or two factors C (Table 26.1).

MAGI is an umbrella term that includes the following different clinical categories: prostatitis, prostatovesiculitis, and prostatovesiculoepididymitis (PVE). They share some characteristics: they are common diseases, have mainly a chronic course, rarely cause obstruction of the seminal pathways, can have an unpredictable intracanicular spread to one or more sexual accessory glands of the reproductive tract, as well as to one or both sides. Therefore, ultrasound evaluation of epididymis, prostate, and seminal vesicles is an important diagnostic tool that helps to define MAGI extension to the various accessory glands. Thus, we have developed the following ultrasonographic criteria to evaluate the inflammatory involvement of each male accessory gland [2–4] (Table 26.2).

A.E. Calogero (✉)
Section of Endocrinology, Andrology and Internal Medicine, and Master in Andrological, Human Reproduction and Biotechnology Sciences, Department of Internal Medicine and Systemic Diseases, University of Catania, Policlinico "G. Rodolico", Via S. Sofia, 78, Building 4, Catania 95123, Italy
e-mail: acaloger@unict.it

A. Zini and A. Agarwal (eds.), *Sperm Chromatin: Biological and Clinical Applications in Male Infertility and Assisted Reproduction*, DOI 10.1007/978-1-4419-6857-9_26,

Table 26.1 WHO diagnosis of male accessory gland infection

Factors	Description
A	*History*: positive for urinary infection, epididymitis, and/or sexually transmitted disease
	Physical signs: thickened or tender epididymis, tender vas deferens, and/or abnormal digital rectal examination
B	*Prostatic fluid*: abnormal prostate fluid expression and/or abnormal urine after prostatic massage
C	*Ejaculate signs*: leucocyte >1 mil/mL, culture with significant growth of pathogenic bacteria, abnormal appearance, increased viscosity, increased pH, and/or abnormal biochemistry of the seminal plasma

According to the WHO [1], MAGI is diagnosed when a patient has oligo-, astheno-, and/or teratozoospermia associated with at least one factor A plus one factor B, one factor A plus one factor C, one factor B plus one factor C or two factors C
From World Health Organization [1], with permission of Cambridge University Press

Table 26.2 Ultrasonographic criteria to evaluate the inflammatory involvement of each male accessory gland

Gland	Ultrasonograhic abnormalities (presence of at least two of the following)
Prostatitis	Asymmetry of the gland volume
	Areas of ipoechogenicity
	Areas of iperechogenicity
	Dilatation of periprostatic venous plexus
Vesiculitis	Mono- or bilateral increased (>14 mm) anteroposterior diameter
	Asymmetry >2.5 mm (normal 7–14 mm) compared with the contralateral vesicle
	Mono- or bilateral reduced (<7 mm) anteroposterior diameter
	Thickened and/or calcified glandular epithelium
	Polycyclic areas separated by hyperechoic septa in one or both vesicles
Epididymitis	Increased size of the head (craniocaudal diameter >12 mm) and/or of the tail (craniocaudal diameter >6 mm) present mono- or bilaterally
	Presence of multiple microcystis in the head and/or tail present mono- or bilaterally
	Mono or bilateral hypo- or hyperechogenicity
	Large mono- or bilateral hydrocele

Using scrotal and transrectal prostate-vesicular ultrasound scans, MAGI may be classified into (a) uncomplicated form, which includes prostatitis alone and (b) complicated forms, which encounter the inflammatory involvement of both prostate and seminal vesicles (prostatovesiculitis) or the involvement of all the three glands (i.e., PVE). This categorization of MAGI is of clinical importance because of the different impact it has on male fertility. Indeed, the negative impact of the inflammatory process on the sperm quality and, consequently, on fertility is more profound in patients with PVE compared to patients with prostatovesiculitis or prostatitis alone [2, 5]. More recently, we have also reported that the ultrasonographic evaluation of patients with PVE allows for discriminating whether there is a unilateral or bilateral involvement of the accessory glands. As expected, patients with bilateral PVE have poorer sperm parameters compared to those with a unilateral involvement [4].

The presence of a significant number of the above-reported ultrasound abnormalities found in one or more male accessory glands, found associated with elevated bacteriospermia and radical oxygen species (ROS) production [2], likely depicts the following peculiar anatomopathology hallmarks of chronic inflammation occurring simultaneously: (1) inflammation processes primarily mediated by monocytes, long-lived macrophages, and lymphocytes. Macrophages engulf and digest microorganisms, foreign invaders, and senescent cells. Macrophages release several different chemical mediators, including

IL-1, TNFα, and prostaglandins, that perpetuate the proinflammatory response; (2) destruction of the inflamed tissue through macrophages and other leukocytes release of ROS, resulting in an "oxidative burst" from neutrophils/macrophages as a first-line defense mechanism, and of proteases; (3) repair of the damaged tissue by replacement with cells of the same type or with fibrous connective tissue. An important part of the inflammatory process involves local angiogenesis, resulting in the development of new blood vessels. In some instances, the host is unable to repair the damaged tissue and the (chronic) inflammatory cascade continues.

In this chapter, we show that all components of the inflammatory response (from the agents that first trigger it to each component of the inflammatory response dynamic) can deteriorate conventional and/or nonconventional sperm parameters arising from one or more of the following mechanisms: (a) altered secretory function of the epididymis, seminal vesicles, and prostate, which reduces the antioxidant properties or scavenging role of the seminal plasma, (b) deterioration of spermatogenesis, and (c) (unilateral or bilateral) organic or functional subobstruction of the seminal tract.

Effects of MAGI on Sperm Parameters

Over the years, a debate has been going on to establish the effects, if any, of MAGI on sperm parameters. A large body of literature suggests that MAGI may negatively interfere with sperm quality in many ways. Indeed, sperm output and quality is the final product of (a) microorganisms/viruses intrinsic properties (degree of virulence, bacterial/viral load, etc.), (b) time of interaction between the microorganism and the germ cells, and (c) the involvement of one or more male accessory glands.

The inflammatory response of one or more glands contributes to the negative impact on sperm function, since many inflammatory mediators released in higher amounts during MAGI have a detrimental effect on germ cells [6, 7]. These include ROS and cytokines [3, 8–12]. These bioactive substances may persist even after successful treatment with antimicrobials because the antioxidant capacity of the seminal plasma is progressively exhausted and cannot be restored because of often dysfunctional male accessory glands. Indeed, ultrasound abnormalities have been found in the accessory glands (prostate, seminal vesicles, and/or epididymis) of infertile patients with MAGI and elevated bacteriospermia ($\geq 10^5$ CFU/mL) or with *Chlamydia trachomatis* or *Ureaplasma urealyticum* infection (in urethral swabs after prostate massage) [2, 4]. These patients have also an increased inflammatory response and an impaired semen quality directly related to the extension of MAGI being progressively worst in patients with prostatitis alone, prostatovesiculitis, or PVE [2, 5].

Conventional sperm parameters, biofunctional markers, and chromatin/DNA integrity have been reported to be altered in patients with MAGI. Three main different mechanisms have been hypothesized as a cause of sperm DNA damage. These include abortive apoptosis, abnormal chromatin packaging, or increased oxidative stress, which is often present in patients with MAGI [13]. Infections acting at the testicular level cause sperm death very likely due to necrosis by itself or necrosis that occurs as a final step of apoptosis [14]. Interestingly, these authors reported that the recovery from infections does not seem to coincide with improved sperm quality, probably because of a persistent inflammatory state, suggested by a high percentage of sperm necrosis sometimes associated with leukocytospermia. The effects of inflammation could progress even in the absence of germs due to the hyperproduction of proinflammatory mediators. Therefore, the results of this study suggest that the presence of necrosis, sometimes associated with apoptosis, may be regarded as an indicator of male genital tract inflammation [15].

In summary, germ cells are the target of many possible pathophysiological mechanisms that may contribute to the onset of infertility in the course of MAGI. We briefly review the effects of (a) microorganisms and viruses, (b) ROS hyperproduction, and (c) the main proinflammatory cytokines.

Effects of Various Microorganisms and Viruses

Some Gram-negative Enterobacteriaceae, such as *Escherichia coli*, *Klebsiella* sp., *Proteus*, *Serratia*, *Pseudomonas* sp., etc., have been recognized as known prostate pathogens (category II, NIH classification), since they have a strong association with a clear positive clinical history (prior and/or recurrent urinary tract infection, sexually transmitted disease, congenital urogenital abnormalities) and some urogenital abnormalities during physical examination. On the contrary, the only presence of some microorganisms is interpreted by some investigators as "probable" (when Gram-positive pathogens, such as *Enterococcus* sp. and *Staphylococcus aureus*, are present) or "possible" (when coagulase-negative anaerobic pathogens, such as *Staphylococcus*, *C. trachomatis*, *U. urealyticum*, are present) prostate infection. The major difficulty in interpreting microbiological findings is the presence of contaminating, indigenous microbiota, or of inhibitory substances known to be present in the prostatic secretions, as well as previous courses of antibiotics. Thus, the diagnosis of bacterial prostatitis may be confirmed by quantitative bacteriological cultures in the semen (growth of $>10^3$ pathogenic bacteria or $>10^4$ nonpathogenic bacteria in seminal plasma diluted 1:2 with saline solution) [16] or segmented cultures, i.e., four [17] and/or two [18] glass test.

Various germs and viruses have been shown able to alter sperm function. Indeed, they may damage conventional sperm parameters, particularly motility, as well as sperm mitochondrial function and/or chromatin/DNA integrity. Paradoxically, literature has focused more toward experimental infection mediated by "possible" microbiota responsible for urogenital infection, reporting in vitro models and impaired effects on conventional and nonconventional sperm parameters, whereas a lower attention has been devoted to evaluate the effects of "known prostate pathogens."

Escherichia coli

Many studies have explored the effects of *E. coli* on sperm function mainly using an in vitro approach. Diemer et al. evaluated the effect of the uropathogenic *E. coli* serotype 06 on normal spermatozoa separated by swim-up and found a significant inhibition of sperm progressive motility. The inhibitory effect was achieved at a sperm–bacteria ratio of 1, and it was prevented by chloramphenicol. On the contrary, no effect on sperm motility was observed after incubation with *E. coli* culture filtrates. Electron microscopy analysis revealed multiple adhesions of *E. coli* to spermatozoa [19]. An inhibitory effect of *E. coli*, but not of the enterococcus, on sperm motility was subsequently confirmed by the same group of researchers [20]. The coincubation of normal spermatozoa with *E. coli* and polymorphonuclear (PMN) has been reported to reduce sperm motility evaluated by computer-assisted sperm analysis (CASA) more profoundly than when spermatozoa were incubated with PMN alone, suggesting that *E. coli* is the primary agent that interferes with sperm motility [21].

Normal spermatozoa incubated with *E. coli* resulted in an increased percentage of spermatozoa with phosphatidylserine (PS) externalization (early apoptosis event) and with apoptosis/necrosis (annexin V-FITC-positive/propidium iodide-positive), whereas the incubation with PMN activated by phorbol-12-myristate-13-acetate showed only a small increase in apoptosis/necrosis [22]. These results suggest that *E. coli* is directly able to alter ejaculated sperm function without involving any of the molecular mechanisms that alter their motility, vitality, and DNA integrity. Accordingly, incubation with *E. coli* decreased the percentage of spermatozoa with elevated mitochondrial membrane potential (MMP); this was found associated with decreased sperm motility and viability. Reactive oxygen species (ROS) production and PS externalization did not change significantly. Interestingly, a similar effect was observed incubating spermatozoa with the supernatant from *E. coli* culture, suggesting the soluble factors damage sperm function [23]. Very recently, in an attempt to understand the mechanism by which *E. coli* inhibits sperm motility, Prabha et al. have isolated and purified the factor responsible for such an effect which they named sperm immobilization factor (SIF). SIF is a 56-kDa molecule that causes instant immobilization without agglutination of human spermatozoa at a concentration of about 1 mg/mL and death at a concentration of about 2 mg/mL. Spermatozoa incubated with SIF revealed multiple and profound alterations involving all superficial structures of spermatozoa as observed by electron microscopy [24].

Neisseria gonorrhoeae

Few studies have explored the effects of *Neisseria gonorrhoeae* on sperm parameters. Liu et al. evaluated the effects of this microorganism on the motility parameters of normal spermatozoa by CASA at a ratio of 1:50. They did not found any effect after 2 and 4 h of incubation, whereas using the same experimental model, *S. aureus* significantly decreased sperm motility and viability [25]. Interestingly, it has been shown that *N. gonorrhoeae* is able to upregulate several host antiapoptotic mechanisms on urethral epithelium and that the gonococcal infection protects host cells from subsequent in vitro staurosporine exposure-induced death. The upregulation of antiapoptotic mechanisms in the urethral epithelium by the gonococcus may represent a mechanism employed by this pathogen to survive and proliferate in host epithelium [26]. It is not known whether a similar mechanism is also exerted on germ cells.

Chlamydia trachomatis

C. trachomatis infection may cause sperm apoptosis because the rate of cells with fragmented DNA has been reported to be higher in patients with chlamydial infection compared with controls [27]. Asymptomatic men with ejaculates positive for chlamydial infection, diagnosed by nested plasmid polymerase chain reaction (PCR), have a significantly higher number of leukocytes and a higher ejaculate volume than those whose ejaculates resulted PCR negative for chlamydial infection. No significant differences were observed for all the other parameters [28]. By contrast, sperm concentration, motility, and morphology were significantly worse in men with both chlamydial and/or mycoplasma infection, whereas sperm viability was not significantly affected. Interestingly, these patients had also an increased percentage of spermatozoa with DNA fragmentation, which decreased after antibiotic administration [29]. Ultrastructural examination suggested that the presence of abnormal spermatozoa during chlamydial infection may relate to the microorganism per se or to the host immune/inflammatory response. In addition, bacteria were detected within the leukocytes of these semen samples. This intracellular persistence of germs may be responsible for the establishment of a latent or chronic infection that may circumvent bactericidal immune mechanisms, impair the efficacy of the antimicrobial treatment, and favor the spreading of the infection in the female genital tract [30]. More recently, chlamydial infection has been found associated with significantly higher pH and seminal leukocyte number as well as a significantly lower percentage of progressive motile spermatozoa in infertile patients compared to fertile men with chlamydial infection. This was associated with higher semen plasma IL-8 and IL-6 levels [31].

Some studies have tried to elucidate the mechanism(s) by which *C. trachomatis* alters sperm function. An in vitro model showed that elementary bodies (EB) of *C. trachomatis* serovar E, incubated with spermatozoa of normal men for 1–6 h, reduced significantly sperm motility and viability, whereas serovar LGV reduced only sperm viability. No effect was reported on the rate of the acrosome reaction. The coincubation with dead EB did not have any effect, suggesting that the detrimental effects on sperm motility and viability are due to live microorganisms and not due to their soluble components [32]. A subsequent study showed that the lipopolysaccharide (LPS) extracted from *C. trachomatis* EB decreased sperm motility and increased the number of dead spermatozoa by the same extent as serovar E EBs, suggesting that LPS mediate the spermicidal effects of *C. trachomatis* [33]. In addition, LPS has been shown to cause sperm apoptosis when incubated in vitro with normal spermatozoa. This effect is caspase 3-mediated, as shown by the inhibition of DNA fragmentation in presence of a pancaspase or caspase-3 inhibitor. These data suggest that sperm death is, at least in part, due to apoptosis [34]. We have investigated the effects of *C. trachomatis* on sperm apoptosis by incubating spermatozoa from normozoospermic healthy men with increasing concentrations of *C. trachomatis* serovar E EBs for 6 and 24 h. After 6 h of incubation, *C. trachomatis* did not have any effect on the percentage of spermatozoa with PS externalization, whereas a significant effect on this parameter was observed after 24 h of incubation. Sperm DNA fragmentation increased significantly after 6 and 24 h of incubation. These findings support the contention that *C. trachomatis* alters directly sperm fertilizing capability [35]. To further evaluate the role of the various LPS molecules on sperm function, Hakimi et al. showed that the lipid A and the 3-deoxy-d-manno-octulosonic acid, toxic components of the *C. trachomatis* LPS, have spermicidal effects similar to LPS. In addition, both molecules were shown to induce sperm apoptosis with a mechanism caspase-mediated [36].

Ureaplasma urealyticum

U. urealyticum is the most common microorganism found in infertile men with a prevalence ranging between 10 and 40% (for review see Dieterle [37]). The presence of *U. urealyticum* in the human male genital tract has been found associated with a significantly lower sperm concentration, whereas no effect has been reported on semen volume and sperm motility, viability, or morphology. Seminal biochemical parameters (zinc, magnesium, acid phosphatase, and fructose) were not affected by *U. urealyticum* [38]. A more profound effect on sperm parameters was subsequently reported. Indeed, infertile men with genital tract infection caused by various microorganisms including *U. urealyticum* have decreased semen volume, sperm concentration, motility, morphology, and viability. However, this study does not allow for identifying specifically the effects of *U. urealyticum* on sperm parameters because these end points have been reported regardless of the etiology of the infection [39]. In patients with isolated *U. urealyticum* infection, Wang et al. found an altered semen viscosity, pH value and sperm concentration, whereas all the other parameters were not affected significantly [40]. Altogether these findings suggest that *U. urealyticum* affects negatively sperm concentration, but does not seem to have a relevant effect on the other conventional sperm parameters. However, a study conducted in Chinese infertile men who have an elevated prevalence of *U. urealyticum* infection (about 34%), showed that, in addition to sperm concentration, sperm motility and viability were also significantly lower compared with patients without *U. urealyticum* infection. Computerized sperm analysis showed that several sperm motility parameters were significantly lower in the patients with the infection. These effects on motility were associated to a decreased seminal plasma α-glucosidase levels, whereas seminal plasma acid phosphatase and fructose were unchanged, suggesting a possible epididymal site of action [41].

In vitro overnight incubation with *U. urealyticum*, as well as with *Mycoplasma hominis*, decreased significantly sperm motility and the percentage of spermatozoa with normal form, hyperactivation, and calcium ionophore-induced acrosome reaction [42]. A reduction of sperm acrosome reaction inducibility has also been reported in vivo in men with *U. urealyticum* infection. This alteration normalized after antimicrobial treatment in about two thirds of the patients treated. The effect on the acrosome reaction seems specific to *U. urealyticum*, since *M. hominis* affected sperm functions in vitro, but had no effects in vivo [43]. *U. urealyticum* has been shown to bind spermatozoa, to reduce sperm motility, and to alter sperm membrane after a long-term incubation (4 h or overnight) in vitro, whereas it increases sperm velocity after a short time (45 min) [44]. To explain these opposite effects of *U. urealyticum*, the authors hypothesized that when sperm activity depends on mitochondrial oxidative phosphorylation, usually at low pH, *U. urealyticum* competes with mitochondrial energy production with a consequent decline of sperm motility and viability, whereas when sperm energy metabolism depends on glycolysis, usually at higher pH, *U. urealyticum* stimulates glycolysis and, therefore, sperm activity [45].

U. urealyticum serotype 4 was most effective in reducing the Hamster's oocyte sperm penetration rate compared with other mycoplasmas. Since the number of spermatozoa adsorbed to Hamster's oocytes was not influenced by mycoplasma preincubation. This suggests that the inhibition of penetration is not due to a masking of sperm membrane sites [46, 47].

Pyospermia was reported in patients with the simultaneous presence of *U. urealyticum* and *Gardnerella vaginalis* [48]. Examination of specimens from infertile patients and fertile men showed the adhesion of *U. urealyticum* to the membrane of spermatozoa, mainly in the midpiece and the postacrosomial region, and exfoliated germ cells. To further study the effects of *U. urealyticum* on fertility, the authors infected artificially male rats with *U. urealyticum*, serotype 8 (T960). A drastic spermatogenesis impairment was found in about a quarter of the rats and infertility in a similar percentage of animals after mating experiments. In addition, the offspring of the infected rats were significantly smaller than those of controls in terms of prenatal and birth weights, suggesting a profound impact on the reproductive function [49].

Shi et al. showed that *U. urealyticum* has antigens (UreG) which cross-react with human sperm membrane proteins and in particular with the nuclear autoantigenic sperm protein [50]. Because of the cross-reaction between NASP and UreG, some men infected with *U. urealyticum* display positive antisperm antibodies in their serum and/or semen, which may cause infertility with an autoimmune mechanism, as reported in an experimental mouse model.

Interestingly, the infection with *U. urealyticum* has also been reported to be able to alter the concentration of microelements in the seminal fluid of infertile patients. In fact, patients with *U. urealyticum* infection had an increased ratios Cu/Zn and Cd/Zn and of the concentrations of As and Mg in the seminal fluid [51]. These abnormalities may contribute to the sperm quality decline found by some authors.

It is noteworthy to recall that mycoplasma infection may alter glycolipid metabolism in the early primary spermatocytes. Particularly, these microorganisms may desulfate sulfogalactosylglycerolipid (SGG), an important molecule for the sperm–egg binding. Therefore, this mechanism may contribute to the negative impact of *U. urealyticum* infection on human fertility [52]. Furthermore, the presence of *U. urealyticum* may affect negatively the implantation of the embryo [37].

To gain further insight into the effects of *U. urealyticum* on sperm function, nonconventional sperm parameters have also been studied. Shang et al. found that patients with *U. urealyticum* infection have an increased number of spermatozoa with fragmented DNA, evaluated by TUNEL assay, compared to controls [53]. This has been confirmed by a subsequent study, which also reported an increased percentage of spermatozoa with less stable chromatin. After treatment with doxycyclin, a significant improvement of both parameters was observed. The authors replicated these in vivo findings in an in vitro model. Spermatozoa incubated with *U. urealyticum* showed a significant dose- and time-dependent chromatin decondensation and DNA damage. The percentage of human spermatozoa with denatured DNA increased by almost 50% after 30 min of incubation with the serotypes 3 and 8, at a concentration of 100 ureaplasmas/spermatozoon compared with uninfected control spermatozoa [54]. A study in male rats experimentally infected with *U. urealyticum* (serotype 8) showed an increased number of TUNEL-positive cells and areas in the testis and a Fas-FasL overexpression in germinal and Sertoli cells. These findings suggest that *U. urealyticum* increases germ cell apoptosis [55].

Despite these evidences, other studies have reported no effect of *U. urealyticum* infection on sperm parameters. *U. urealyticum* infection had no effect on sperm function as assessed by seminal fluid analysis, in vitro sperm penetration of bovine cervical mucus, and the Hamster's oocyte sperm penetration assay [56]. In vitro, *U. urealyticum* experimental infection did not alter sperm motility or penetration capability when spermatozoa were incubated with the germ for 45 min at very high *U. urealyticum*–spermatozoa ratios (up to 100:1) [57]. In vivo studies showed no statistically significant difference between sperm parameters in subfertile patients with or without *U. urealyticum* infection [58], and no correlation was found between abnormal sperm parameters and the presence of *U. urealyticum* in 86 unselected asymptomatic men [59]. Similarly, infertile patients with *U. urealyticum* infection, diagnosed by PCR analysis in their semen sample, did not have any significant difference in seminal volume, sperm concentration, viability, motility, morphology, and leukocyte count [60]. The same authors confirmed these findings in a group of asymptomatic male partners of infertile Tunisian couples who had the concomitant presence of *Mycoplasma* and *U. urealyticum* DNA in their semen samples [61].

Mycoplasma hominis and Others

The effects of *M. hominis* on sperm parameters have often been evaluated in the presence of other germs [29, 61–64]. These studies reported a detrimental effect on sperm motility [29, 62, 64], morphology [29, 61, 63], and concentration [29, 61]. Agbakoba et al. reported that many patients infected with various strains of mycoplasmas were oligozoospermics [65]. The presence of *M. hominis* DNA in semen samples is associated with low sperm concentration and abnormal sperm morphology; a negative correlation between sperm concentration and the detection of *Mycoplasma genitalium* in semen samples of infertile men has also been reported [60].

A direct in vitro interaction between *M. hominis* and spermatozoa has also been evaluated. An overnight incubation with mycoplasma species decreased significantly sperm motility and the percentage of normally shaped and the proportion of acrosome-reacted spermatozoa after incubation with the calcium ionophore [42]. Confocal microscopy showed that *M. hominis* binds sperm heads, tails and, to a lower extent, the midpiece 10 min after coincubation. Moreover, infected spermatozoa had the germ within the head and the midpiece in cytosolic space. Only a subtle sperm damage was observed after a short-term *M. hominis* interaction with spermatozoa [66]. Interestingly, experimentally *M. genitalium* attaches to motile spermatozoa, and thus, the microorganism may be carried along with the spermatozoa to the female genital tract [67].

Spermatozoa preincubated with various strains of mycoplasmas had lower penetration rate using the sperm–Hamster egg fertilization test compared to controls. A lower penetration rate has been reported in Percoll-washed spermatozoa, which resulted positive for the presence of mycoplasma DNA compared to those without infection. The similarities of hypoosmotic swelling and kinematic parameters between the two groups suggest that the reduced sperm–oocyte penetration rate is not due to the latter two parameters [68].

By contrast, a number of studies failed to show any effect of mycoplasmas on sperm parameters both in vivo and in vitro. The presence of *M. hominis* and/or *U. urealyticum* in semen was not associated with any significant difference in sperm parameters in men attending an IVF unit [69]. Eggert-Kruse et al. reported no difference on conventional sperm parameters following antimicrobial treatment in patients with *C. trachomatis*, *M. hominis*, *U. urealyticum*, and *N. gonorrhoeae* infections [70]. Similar results were reported examining semen samples for routine analysis. Despite the high prevalence of mycoplasmas in these samples, conventional sperm parameters of the men infected resulted similar to those of the uninfected men [71]. On this account, a systematic search for mycoplasmas infection has not been suggested [72].

Candida albicans

Studies suggest that *Candida albicans* infection has a negative effect on sperm function and spermatozoon fertilizing ability. Experimentally induced *C. albicans* infection has been reported to inhibit sperm motility in a time-dependent manner [73]. A significant inhibitory effect of *C. albicans* was only detected in the samples with the initial bacterial concentration of 20 million microorganisms/mL [20]. A significant degree of sperm nonspecific agglutination, detected after 2 and 4 h of incubation, was also reported, as well as a clear head-to-head sperm agglutination with *C. albicans* interposition [74], suggesting the formation of a mechanical barrier that hampers sperm motility [20]. Subsequent studies showed, however, that mitochondrial and tail alterations may contribute to the sperm motility decline. In addition, spermatozoa in contact with *C. albicans* undergo acrosomal swelling, vesiculation (outer membrane), and rupture [74], which may impair sperm fertilization capability. In this regard, a case report showed that in the presence of *C. albicans* no fertilization occurred after IVF and ICSI [75]. Subsequently, we reported that spermatozoa isolated from normozoospermic healthy men and incubated with increasing concentrations *C. albicans* had a significantly sperm motility decline associated with an increased percentage of spermatozoa with low MMP or PS externalization. *C. albicans* did not seem to have any significant effect on sperm DNA fragmentation or chromatin integrity, at least under these experimental conditions [76]. Indeed, we found an increased sperm chromatin packaging damage and apoptosis in a patient with *C. albicans* infection [75]. This suggests that the adverse effects of *C. albicans* on sperm chromatin/DNA integrity require the presence of other factors (leukocyte, etc.) that are present in vivo. Recently, it has been shown that farnesol, a sesquiterpene alcohol produced by many organisms, which acts as a quorum sensing molecule and as a virulence factor of *C. albicans*, reduces sperm motility and causes sperm apoptosis and necrosis. Moreover, sublethal doses of this signaling molecule induce premature acrosome loss [77].

Trichomonas vaginalis

Trichomonas vaginalis is a flagellated parasite often found as an occult resident of the genital tract of sexually active women and men. Its presence in the seminal samples of asymptomatic men resulted in a significant increase of viscosity and number of particulate debris, decreased sperm motility, number of normal forms, and viability (evaluated by the hypoosmotic swelling test). After a single course of treatment with metronidazole (400 mg × 3/day for 10 days), a significant improvement of the semen characteristics was observed in about half of the patients treated [78]. These findings suggest that *T. vaginalis* may cause infertility.

In vitro, this protozoan has been shown to be capable of reducing sperm motility after 2, 4, and 6 h of incubation without causing any sperm agglutination [79]. Subsequent studies confirmed a detrimental effect of *T. vaginalis* on sperm motility and have attempted to establish the mechanism(s) by which this happens [80–83]. Jarecki-Black et al. reported that spent medium of *T. vaginalis* culture caused complete cessation of sperm motility after 15 min of incubation. Trophozoite soluble fraction or formalin-killed trophozoites caused a 50% reduction in sperm motility, compared to 25% reduction caused by the trophozoite particulate fraction or the sterile medium and 3% by saline (control). The *T. vaginalis* spermicidal activity was heat-stable, trypsin-sensitive, and had a molecular weight of 12–15 kDa by gel filtration. This proteinaceous substance was present in and secreted by *T. vaginalis* trophozoites during normal growth in axenic culture [80]. An inhibitory role of *T. vaginalis* metabolites [81] or of a soluble extract [82] of this protozoan on sperm motility was further reported. The incubation with a *T. vaginalis* soluble factor resulted also in an increased viscosity, number of debris, and sperm membrane damage in vitro [82]. Benchimol et al. reported that *T. vaginalis* is also able to bind sperm head and flagella and that the reduction of sperm motility was associated with an intense agglutination. In this regard, *T. vaginalis* appeared to be much more virulent than *T. foetus* whose effects were evaluated in the same study on bull spermatozoa [83].

By contrast, Daly et al. did not report any effect of *T. vaginalis* on sperm motility up to 24 h of incubation, though protozoa survived well in the semen samples [84]. The lack of effect may relate to low number of *T. vaginalis* (about 2,500/mL semen) used in this study compared to the higher range used (10^4–10^7 protozoa/mL) in other studies [79].

Hepatitis B Virus and Hepatitis C Virus

Various studies have explored the effects of Hepatitis B virus (HBV) or Hepatitis C virus (HCV) infection on sperm parameters. Garrido et al., in an attempt to determine the predictive value of sperm parameters, sperm washing procedure, and the infection status for the postwash viral positivity, found that sperm parameters of HCV-affected patients did not differ from those of noninfected men [85]. We evaluated the sperm parameters of infertile patients in Child–Pugh classification A with HBV or HCV infection, compared with those of a group of 30 patients with primary infertility due to causes different from liver diseases. HBV patients (median HBV-DNA load of 6×10^5 copies/mL, range: 1×10^5–1×10^7 copies/mL) had sperm density, total number, forward motility, morphology, and viability significantly worse than those found in patients with HCV (median HCV-RNA load of 2.3×10^6 copies/mL, range: 2×10^5–1.2×10^7 copies/mL). No significant correlation between sperm parameters and the duration of viral infection or the viral HBV-DNA load was found with the exception of sperm morphology, which exhibited a trend for a negative correlation with the viral HBV-DNA load [86]. HCV-infected patients had a significantly lower sperm motility and percentage of normal forms than controls. Combined antiviral treatment with interferon and ribavirin worsened sperm morphology, while it did not have any effect on the other sperm parameters [87]. A negative effect on sperm motility [88, 89] and morphology [89] has been confirmed in HCV- and HBV-positive patients. However, Moretti et al. did not find any significant effect on sperm concentration [88], whereas Lorusso et al. found lower sperm concentration and viability in both HBV and HCV seropositive men compared with controls [89].

Very little is known about the mechanism by which HBV affects sperm function. A recent study, evaluating the role of the HBV S protein (HBs), the main component of HBV envelop protein, has reported that HBs reduces sperm motility in a dose- and time-dependent fashion and increases the number of spermatozoa with low MMP. The fertilization rate in HBs-treated group was significantly lower than that of the control group [90].

Electronic microscopy revealed significantly higher values of sperm apoptosis and necrosis in patients with HBV- or HCV-infection compared with controls, whereas the disomy and diploidy rates for chromosomes 18, X, and Y did not differ significantly from controls [88]. By contrast, a significantly higher total sperm chromosome abnormalities, evaluated after zona-free Hamster oocyte penetration, were found in patients with HBV infection compared with healthy men. In addition, sperm chromosomes in HBV patients present stickiness, clumping, failure to staining, etc. These findings suggest that HBV infection may cause sperm chromosome aberrations [91].

The possibility that the HBV may integrate into sperm chromosomes has been evaluated in patients with HBV infection. Specific fluorescent spots for HBV DNA have been detected in sperm chromosomes, although with a different intensity. These results suggest the possibility of vertical transmission of HBV via the germ line to the next generation [91, 92].

Human Immunodeficiency Virus Type 1

The effect of human immunodeficiency virus (HIV) type 1 infection on sperm parameters was evaluated in asymptomatic or minimally symptomatic HIV-seropositive men and in men with AIDS. All the men with AIDS had leukocytospermia and grossly abnormal spermatozoa. By contrast, sperm parameters of seropositive men did not differ significantly from those of healthy seronegative donors. Zidovudine therapy did not affect sperm morphology or seminal characteristics [93]. No sperm parameters alteration was subsequently confirmed in HIV seropositive men [94]. However, this study showed that HIV seropositive men had a significantly higher percentage of (a) spermatozoa with cytoplasmic droplet, (b) immature germ cells, and (c) spermiophages. In addition, HIV seropositive men showed a significant positive correlation between blood CD4+ and sperm motility, as well as a significant inverse correlation between CD4+ and sperm abnormalities [94].

In contrast to seropositive men, HIV type 1 men have a significantly lower ejaculate volume, sperm concentration, total count, progressive motility, and normal morphology compared with controls. A significant positive correlation was observed between CD4 count and sperm concentration, total count, motility, progressive motility (type a and b) [95]. These data demonstrate that sperm parameters are significantly impaired by the presence of HIV infection. Men with HIV have been reported to have low sperm motility compared to HIV negative men and leucocytospermia irrespective of a previous history of sexual transmitted diseases. These findings suggest that sperm motility impairment in HIV positive men may relate to an increase oxidative stress leukocyte-mediated [96]. However, Garrido et al. did not find any significant alteration of the sperm parameters in HIV-affected patients compared with non-infected men [85]. Because of this inconsistency in the results on sperm parameters in HIV-infected men, Bujan et al. investigated sperm parameters in 190 HIV type 1-infected patients and compared them with those of a control group of fertile, non-infected men ($n=218$). They found that semen volume, percentages of progressive motile spermatozoa, total sperm counts, and seminal leukocytes were lower, while pH values and spermatozoa multiple anomaly indices were higher in HIV-infected patients [97]. Abnormal sperm parameters have been found in the 83% of HIV-infected and in 42% HIV-uninfected ($n=83$) male partners of 130 HIV-infected women seeking fertility with an Odds ratio of 7 (95% CI=2.1–23) [98]. Principal component analysis method showed that HIV-positive men have worst sperm parameters, whereas the distribution of mannose receptors and cytokine levels in HIV-1-positive men were similar to uninfected individuals. The similar distribution of mannose receptors suggests that spermatozoa from infected individuals interact normally with oocytes [99]. Recently, a study conducted in HCV-HIV seropositive men has showed that the only sperm parameter affected was progressive motility (grade a+b), which was significantly lower compared to that of controls [89]. TUNEL analysis revealed an increased percentage of DNA-fragmented ejaculated spermatozoa in semen of HIV-infected men [100].

A prolonged exposure to asymptomatic, untreated HIV-1 infection does not seem to affect sperm parameters. Indeed, no significant variation was observed in 55 men with HIV-1 infection whose sperm parameters were evaluated biannually for a mean follow-up period of 77 weeks. These findings should be reassuring for untreated men infected with HIV-1 who wish to father a child [101].

Aside HIV, many drugs used for the treatment of HIV-infected men have profound spermotoxic effect. Nucleoside analogs reverse transcriptase inhibitors (NRTI), used for the treatment of HIV-infected patients, have important adverse effects that are linked with a common mechanism: alteration of mitochondrial activity. Given the relevant role played by these organelles on sperm function, the effects of these drugs have been evaluated on sperm function. Studies suggest that NRTI exposure alters mitochondrial energy-generating ability in spermatozoa. NRTI are known to increase ROS production, which results in a decreased MMP. The reduced MMP leads to the release some specific apoptotic factors, such as cytochrome C, that initiate programmed cell death [102]. The effects of antiretroviral therapy on semen quality were longitudinally evaluated in a cohort of male patients with different estimated duration of HIV-1 infection. The median period of follow-up was 48 weeks. Five patients underwent thymidine analog-containing treatment, 23 used tenofovir-based treatment, and 6 used other regimens. At all time points, the percentage of progressively motile spermatozoa was low, and it decreased significantly from 28 to 17% during follow-up. All other semen parameters were in the normal range and remained stable [103].

Papillomavirus

Over the years, the role of papillomavirures (HPV) on sperm parameters and/or function has been examined with contrasting results. The presence of HPV gene sequences have been shown in the 64% of Percoll-separated spermatozoa. The HPV type 16 was detected about twice as often as the type 18 [104]. Lai et al. reported that not only HPV types 16 and 18 are able to infect human spermatozoa but also some of their genes are actively transcribed in the infected germ cells [105]. Following experimental infection, the viral DNA appears tenaciously bound to spermatozoa, suggesting an internalization into the sperm. Indeed, sperm washing (centrifuge, two-layer Isolate colloid wash, or test-yolk buffer procedures) was not capable of removing exogenous HPV DNA [106]. In an attempt to clarify the mechanism(s) by which HPV binds to spermatozoa, Pèrez-Andino et al. reported that the capsids of HPV type 16 specifically interact with spermatozoa. Purified HPV16 virions directly adsorb to live spermatozoa in native semen and in conditions that resemble the female genital tract. In particular, the authors found that HPV16 capsids bind to two distinct sites at the equatorial region of the sperm head surface [107]. More recently, the presence of HPV DNA has been shown in about the 25% of the sperm head of infected young (18 years old) adults who had unprotected sexual intercourse. However, the authors could not clarify whether the virus was integrated in the nucleus or not [108]. The presence of the virus makes spermatozoa carriers for the sexual transmission of HPV to sexual partners.

The evaluation of the effects of the in vivo HPV infection on sperm parameters suggests a detrimental role of HPV on sperm motility. Indeed, the incidence of asthenozoospermia has been reported to be higher among patients HPV (type 16 and 18)-positive compared with those without infection (75 vs. 8%). Nevertheless, many sperm kinematic parameters did not differ significantly between the two groups [109]. A reduction of sperm motility has been, recently, reported in infertile patients and subjects with risk factors, in particular when the infection was present in spermatozoa [110], and in young adults [108]. By contrast, no effects on semen quality and assisted reproductive technique (ART) variables (pregnancy and abortion rates) have been reported in men and women HPV type 16-positive [111]. The lack of effect on the HPV infection on sperm parameters has also been confirmed by Rintala et al. Indeed, the presence of HPV DNA did not affect semen volume, sperm concentration, motility, and vitality. Neither oligozoospermia nor asthenozoospermia was associated with the presence of seminal HPV DNA [112].

Using an experimental in vitro model of infection, HPV DNA seems to increase sperm motility. HPV DNA increased sperm total motility and progression, evaluated by computer-assisted sperm analysis. This suggests that HPV DNA increases sperm metabolism or enhances the calcium-regulated motility mechanism. Although an artifact of PCR products cannot be ruled out [106], Connelly et al. confirmed that normal spermatozoa had higher motility after incubation with HPV types 16, 18, 31, and 33, but not 6/11, and increased linearity after incubation with all HPV types tested with the exception of the type 18 [113]. An opposite effect of HPV types 6b/11, 16, 18, 31, and 33 exposure has been reported on motility (decreased) and hyperactivation (increased), which suggests that HPV-exposed spermatozoa retain some fertilizing capacity [114].

Normal motile spermatozoa incubated with E6-E7 HPV DNA fragments had increased DNA fragmentation after exposure to DNA of the HPV types 16 and 31, whereas the types 18, 33, and 6/11 did not alter sperm DNA integrity [113]. Lee et al., in the attempt to further evaluate the role, if any, of HPV on sperm DNA of specific gene regions, examined the effects of HPV exposure on the integrity of exons 5 and 8 of the p53 gene. Fragmentation of exon 5 occurred after exposure to HPV DNA type 18. By contrast, only exon 8 was affected by HPV type 16. HPV DNA from type 31 or 33 was without effect on the p53 exons [114].

Effects of Oxidative Stress

An increased production of ROS and/or a decrease of the antioxidant defenses cause sperm abnormalities. These include decreased sperm motility, acrosine activity, and sperm–oocyte fusion capability (see Lanzafame et al. [115] for review). Indeed, a sperm–oocyte penetration rate <25% is associated with an increased ROS production in an elevated number of oligozoospermic patients with this abnormality of sperm function [116]. Sperm motility inhibition caused by ROS has been reported to negatively correlate with MDA seminal plasma levels [117], whereas a decrement of MDA is associated with an increased pregnancy rate [118]. An increased oxidative stress has been suggested to cause seminal plasma hyperviscosity in infertile males [119].

An increased oxidative stress damages sperm chromatin/DNA integrity also. Indeed, ROS exposure increases DNA fragmentation in normal spermatozoa [116], causes DNA protein cross-linking in chromatin [120], increases the frequency of DNA single and double-strand breaks [121], and oxidates DNA base changes in asthenozoospermic and normozoospermic infertile patients compared with fertile men [122]. Sperm DNA fragmentation does not correlate with the fertilization rate, but is associated with a significant reduction of the pregnancy rate in ART programs when TUNEL-positive spermatozoa are used [123]. Therefore, spermatozoa with damaged DNA are able to fertilize oocytes, but at the time when the paternal genome is switched on, further development stops [124]. DNA damage seems to lead to an amplified risk of miscarriage and chromosomal abnormalities [125].

Effects of Proinflammatory Cytokines

Cytokines are a group of soluble mediators produced by lymphoid and nonlymphoid cells that play a key role in the afferent and efferent phases of immune responses of both the innate and acquired immune systems. In the dynamic of the inflammatory response, cytokines have pleiotropic and redundant effects, being the same cytokines present in more moments of the inflammatory response. For example, tumor necrosis factor-α (TNFα) is present in the initial inflammatory trigger, but it is also an inductor of chemokines, contributes to the neutrophil chemotaxis, enhances the toxic final effect, and induces apoptosis; IL-6 contributes to the initial inflammatory trigger, but it also causes activation and differentiation of leukocytes, as well as it contributes to the toxic final effect through ROS hyperproduction; IL-8 contributess to the phase of chemoattraction of neutrophils to the site of inflammation, and to the activation of neutrophils toward phagocytosis. Thus, cytokines have a multitasking role that reverberates negatively on male accessory gland function.

Interleukin 1

The seminal plasma concentration of interleukin 1 (IL-1) has been reported to be higher in infertile patients than in normal controls. However, no difference was found different subgroups of patients divided on the basis of progressive motility or percentage of sperm with abnormal forms [126]. IL-1 has been reported to have no effect on both spontaneous or calcium ionophore-induced acrosome reaction in normal spermatozoa [127] as well as sperm MDA production in vitro when used alone or in combination with leukocytes [128].

Interleukin 6

Interleukin 6 (IL-6) seminal plasma levels have been reported to be higher in infertile patients than in normal fertile men and to negatively correlate with sperm MDA formation, suggesting a ROS-mediated lipoperoxidation process [129]. An inhibitory dose- and time-dependent effect of IL-6 on sperm motility has been reported in vitro, which seems to relate to hyperproduction of nitric oxide [130]. In addition, IL-6 has been shown to decrease both spontaneous and calcium ionophore- or progesterone-induced acrosome reaction of normal spermatozoa separated by swim-up procedure. This inhibitory effect was, however, of lower intensity compared with that obtained by incubating spermatozoa with TNF-α in the same experimental model [131].

Interleukin 8

Fedder and Ellerman-Eriksen showed that interleukin 8 (IL-8) had no effect on sperm motility and on the ionophore-induced acrosome reaction in vitro [132]. By contrast, in subfertile patients, IL-8 seminal plasma concentrations have been shown to negatively correlate with the total number of motile spermatozoa or with the number of motile spermatozoa harvested after swim-up technique. A significant positive correlation was found between seminal plasma IL-8 concentration and leukocyte counts [133]. An increasing effect of IL-8 has also been reported on normal spermatozoa in vitro, both after physiological or infection–inflammation concentrations [134].

Interferon Gamma

A significant inhibitory effect of interferon-γ (IFNγ) on sperm motility has been reported in vitro [69, 132]. Such an effect has been confirmed in experiments using both TNFα and IFNγ [135]. Sperm motility inhibition was associated with a significantly reduced capacity of spermatozoa to penetrate Hamster oocytes [136].

At physiological concentration, IFNγ increased sperm membrane lipoperoxidation, but no further increment of MDA production was observed when this cytokine was used at higher concentrations, such as those measured during infection/inflammation [134]. IFNγ has been reported to have both no significant effect on calcium ionophore-induced acrosome reaction [132] and a suppressive effect on spontaneous acrosome reaction and acrosine activity [137]. A marked reduction of Na^+/K^+-ATPase, Ca^{2+}-ATPase and superoxide dismutase activities and an increased production of nitric oxide have been reported in normal spermatozoa incubated with IFNγ [137]. These latter effects may explain the detrimental effects of IFNγ on sperm acrosine activity and acrosome reaction. It is noteworthy that IFNγ did not alter motility and viability of normal spermatozoa following incubation with this cytokine for up to 3 h [138].

Macrophage Migration Inhibitory Factor

Macrophage migration inhibitory factor (MIF), a proinflammatory cytokine, is a constituent of the seminal plasma. It is expressed in the epididymis and has been shown to be an important factor in sperm maturation [139]. Sperm-associated, but not seminal plasma, MIF negatively correlates with sperm motility [140]. We have shown a negative correlation between MIF levels in human seminal fluid and fertility status. In addition, MIF added to normal spermatozoa decreased sperm total and progressive motility and increased the percentage of spermatozoa with PS externalization or with DNA fragmentation [141]. A deleterious effect on sperm motility was also reported by Carli et al. but only at high concentrations, whereas MIF may play a physiological role in sperm capacitation process at lower concentrations [142].

Tumor Necrosis Factor-α

Several studies have shown that TNFα is present in the seminal plasma of normal men at a concentration similar to that found in the seminal plasma of patients with bacterial infection [143]. Other studies have instead shown that the seminal plasma concentrations of TNFα are higher in patients with bacterial or mycoplasma infections than in normal controls [144]. In addition, it has been shown that leukocytospermia [145, 146] and/or bacteriospermia [145] are associated with a higher release of TNFα.

Though several studies have explored the effect of TNFα on sperm parameters, no clear conclusion can be drawn. Wincek et al. showed that sperm motility and Hamster oocyte penetration were not affected by the incubation with TNFα [147]. Haney et al. reported that motile spermatozoa obtained from fertile men and separated by the swim-up technique did not show any decreased motility after of exposure to TNFα, IL-1α, and IFNγ alone or in combination even at doses higher than those observed in vivo [148]. Accordingly, no relationship between seminal plasma TNFα concentration and sperm parameters has been reported in normal men [143]. Fedder and Ellerman-Eriksen showed that TNFα had no effect on sperm motility and on the ionophore-induced acrosome reaction [132]. Lewis et al. did not report any effect of TNF-α on sperm viability [149].

On the contrary, a significant in vitro negative effects of TNFα on sperm motility and sperm fertilizing ability of Hamster oocytes have been reported [69, 136]. Similarly, Gruschwitz et al. showed that seminal plasma TNFα concentrations in patients with bacterial or mycoplasma infections correlated negatively with the number of progressively motile spermatozoa [144]. Kocak et al. reported that TNFα levels correlate negatively with sperm motility and morphology, but not with total sperm counts [150]. Estrada et al. showed that although the inflammatory cytokines TNFα plus IFNγ have only partial detrimental effects on sperm motility, viability, membrane integrity, and lateral head displacement, they may contribute to the poor fertilizing potential of human spermatozoa during inflammatory conditions [135]. Accordingly, the peritoneal fluid of women with endometriosis which contains elevated concentrations of TNFα caused a significant reduction in both total and progressive sperm motility after 4 and 21 h incubation compared with spermatozoa incubated with peritoneal fluid which did contain TNFα. The ability of TNFα to hamper sperm motility in vitro suggests that this may be a mechanism for the infertility observed in women with minimal endometriosis [151]. We found that TNFα inhibits total and progressive sperm motility in a concentration- and time-dependent manner [152]. This detrimental effect may relate to a reduced sperm mitochondrial function, as shown by an increased number of spermatozoa with low MMP [152, 153], as well as an increased nitric oxide production [130].

Divergent results have been reported about the effects of TNFα on lipid sperm membrane peroxidation, evaluated by the production of malondialdehyde. In fact, TNFα has been reported both to increase MDA production at physiological concentrations and, to a greater extent, at infection–inflammation concentrations [134] and to have no effect on MDA production from spermatozoa isolated by swim-up technique [128].

TNFα has also been reported to inhibit spontaneous and induced (by calcium ionophore or progesterone) acrosome reaction in normal spermatozoa [127, 131, 137].

In keeping with previous observation showing TNFα capable of inducing apoptosis, we found that this proinflammatory cytokine causes sperm apoptosis also. Indeed, TNFα increased both the percentage the PS externalization, an early molecular event of apoptosis, and DNA fragmentation, a late sign of apoptosis. Similar TNFα toxic effects were reported on sperm motility, functional integrity of the sperm membrane, and DNA fragmentation. These effects were reversed by coincubation with infliximab, a selective TNF-α antibody [154]. More recently, a positive correlation has been reported between seminal plasma TNFα levels and apoptotic spermatozoa as shown by an increased percentage of spermatozoa with PS externalization [155].

Conclusions

Though an open debate on pros and cons of the role of MAGI in male infertility is going on, andrologists should at least consider MAGI as a risk factor of male infertility [5]. In fact, MAGI may impair sperm function and cause male infertility through the above-reported multiple pathophysiological mechanisms.

References

1. World Health Organization. In: Rowe P, Comhaire F, Hargreave TB, Mellows HJ, editors. World Health Organization manual for the standardised investigation and diagnosis of the infertile couple. Cambridge: Cambridge University Press; 1993.
2. Vicari E. Seminal leukocyte concentration and related specific reactive oxygen species production in patients with male accessory gland infections. Hum Reprod. 1999;14:2025–30.
3. Vicari E. Effectiveness and limits of antimicrobial treatment on seminal leukocyte concentration and related reactive oxygen species production in patients with male accessory gland infection. Hum Reprod. 2000;15:2536–44.
4. Vicari E, La Vignera S, Castiglione R, et al. Sperm parameters abnormalities, low seminal fructose and reactive oxygen species overproduction do not discriminate patients with unilateral or bilateral post-infectious inflammatory prostato-vesiculo-epididymitis. J Endocrinol Invest. 2006;29:18–25.
5. Bayasgalan G, Naranbat D, Radnaabazar J, et al. Male infertility: risk factors in Mongolian men. Asian J Androl. 2004;6:305–11.
6. Agarwal A, Saleh RA, Bedaiwy MA. Role of reactive oxygen species in the pathophysiology of human reproduction. Fertil Steril. 2003;79:829–43.
7. Sanocka D, Jedrzejczak P, Szumała-Kaekol A, et al. Male genital tract inflammation: the role of selected interleukins in regulation of pro-oxidant and anti-oxidant enzymatic substances in seminal plasma. J Androl. 2003;24:448–55.
8. Ochsendorf FR. Infections in the male genital tract and reactive oxygen species. Hum Reprod Update. 1999;5:399–420.
9. Vicari E, Calogero AE. Effects of treatment with carnitines in infertile patients with prostato-vesiculo-epididymitis. Hum Reprod. 2001;16:2338–42.
10. Vicari E, La Vignera S, Calogero AE. Antioxidant treatment with carnitines is effective in infertile patients with prostatovesiculoepididymitis and elevated seminal leukocyte concentrations after treatment with nonsteroidal anti-inflammatory compounds. Fertil Steril. 2002;78:1203–8.
11. Weidner W, Colpi GM, Hargreave TB, et al. EAU guidelines on male infertility. Eur Urol. 2002; 42:313–22.
12. Diemer T, Hales DB, Weidner W. Immune-endocrine interactions and Leydig cell function: the role of cytokines. Andrologia. 2003;35:55–63.
13. Aitken RJ, De Iuliis GN. Origins and consequences of DNA damage in male germ cells. Reprod Biomed Online. 2007;14:727–33.

14. Collodel G, Baccetti B, Capitani S. Necrosis in human spermatozoa. I. Ultrastructural features and FISH study in semen from patients with uro-genital infections. J Submicrosc Cytol Pathol. 2005;37:93–8.
15. Moretti E, Baccetti B, Capitani S, et al. Necrosis in human spermatozoa. II. Ultrastructural features and FISH study in semen from patients with uro-genital infections. J Submicrosc Cytol Pathol. 2005;37:93–8.
16. Comhaire F, Verschraegen G, Vermeulen L. Diagnosis of accessory gland infection and its possible role in male infertility. Int J Androl. 1980; 3:32–45.
17. Meares EM, Stamey TA. Bacteriologic localization patterns in bacterial prostatitis and urethritis. Invest Urol. 1968;5:492–518.
18. Nickel JC. The pre and post massage test (PPMT): a simple screen for prostatitis. Tech Urol. 1997;3: 38–43.
19. Diemer T, Weidner W, Michelmann HW, et al. Influence of *Escherichia coli* on motility parameters of human spermatozoa in vitro. Int J Androl. 1996;19:271–7.
20. Huwe P, Diemer T, Ludwig M, et al. Influence of different uropathogenic microorganisms on human sperm motility parameters in an in vitro experiment. Andrologia. 1998;30 Suppl 1:55–9.
21. Diemer T, Huwe P, Ludwig M, et al. Influence of autogenous leucocytes and *Escherichia coli* on sperm motility parameters in vitro. Andrologia. 2003;35:100–5.
22. Villegas J, Schulz M, Soto L, et al. Bacteria induce expression of apoptosis in human spermatozoa. Apoptosis. 2005;10:105–10.
23. Schulz M, Sànchez R, Soto L, et al. Effect of *Escherichia coli* and its soluble factors on mitochondrial membrane potential, phosphatidylserine translocation, viability and motility of human spermatozoa. Fertil Steril. 2010;94:619–23.
24. Prabha V, Sandhu R, Kaur S, et al. Mechanism of sperm immobilization by *Escherichia coli*. Adv Urol. 2010;2010:240–68.
25. Liu JH, Li HY, Cao ZG, et al. Influence of several uropathogenic microorganisms on human sperm motility parameters in vitro. Asian J Androl. 2002;4:179–82.
26. Binnicker MJ, Williams RD, Apicella MA. Infection of human urethral epithelium with *Neisseria gonorrhoeae* factors and protects cells from staurosporine-induced apoptosis. Cell Microbiol. 2003;5:549–60.
27. Wan CC, Wang H, Hao BJ, et al. Infection of *Chlamydia trachomatis* and apoptosis of spermatogenic cells. Zhonghua Nan Ke Xue. 2003;9:350–1.
28. Hosseinzadeh S, Eley A, Pacey AA. Semen quality of men with asymptomatic chlamydial infection. J Androl. 2004;25:104–9.
29. Gallegos G, Ramos B, Santiso R, et al. Sperm DNA fragmentation in infertile men with genitourinary infection by *Chlamydia trachomatis* and mycoplasma. Fertil Steril. 2008;90:328–34.
30. Gallegos-Avila G, Ortega-Martínez M, Ramos-González B, et al. Ultrastructural findings in semen samples of infertile men infected with *Chlamydia trachomatis* and mycoplasmas. Fertil Steril. 2009; 91:915–9.
31. Kokab A, Akhondi MM, Sadeghi MR, et al. Raised inflammatory markers in semen from men with asymptomatic chlamydial infection. J Androl. 2010; 31:114–20.
32. Hosseinzadeh S, Brewis IA, Eley A, et al. Co-incubation of human spermatozoa with *Chlamydia trachomatis* serovar E causes premature sperm death. Hum Reprod. 2001;16:293–9.
33. Hosseinzadeh S, Pacey AA, Eley A. *Chlamydia trachomatis*-induced death of human spermatozoa is caused primarily by lipopolysaccharide. J Med Microbiol. 2003;52(Pt 3):193–200.
34. Eley A, Pacey AA, Galdiero M, et al. Can *Chlamydia trachomatis* directly damage your sperm? Lancet Infect Dis. 2005;5:53–7.
35. Satta A, Stivala A, Garozzo A, et al. Experimental *Chlamydia trachomatis* infection causes apoptosis in human sperm. Hum Reprod. 2006;21:134–7.
36. Hakimi H, Geary I, Pacey A, et al. Spermicidal activity of bacterial lipopolysaccharide is only partly due to lipid A. J Androl. 2006;27:774–9.
37. Dieterle S. Urogenital infections in reproductive medicine. Andrologia. 2008;40:117–9.
38. Upadhyaya M, Hibbard BM, Walker SM. The effect of *Ureaplasma urealyticum* on semen characteristics. Fertil Steril. 1984;41:304–8.
39. Sanocka-Maciejewska D, Ciupińska M, Kurpisz M. Bacterial infection and semen quality. J Reprod Immunol. 2005;67:51–6.
40. Wang Y, Liang CL, Wu JQ, et al. Do *Ureaplasma urealyticum* infections in the genital tract affect semen quality? Asian J Androl. 2006;8:562–8.
41. Zheng J, Yu SY, Jia DS, et al. *Ureaplasma urealyticum* infection in the genital tract reduces seminal quality in infertile men. Zhonghua Nan Ke Xue. 2008;14:507–12.
42. Rose BI, Scott B. Sperm motility, morphology, hyperactivation, and ionophore-induced acrosome reactions after overnight incubation with mycoplasmas. Fertil Steril. 1994;61:341–8.
43. Köhn FM, Erdmann I, Oeda T, et al. Influence of urogenital infections on sperm functions. Andrologia. 1998;30 Suppl 1:73–80.
44. Núñez-Calonge R, Caballero P, Redondo C, et al. *Ureaplasma urealyticum* reduces motility and induces membrane alterations in human spermatozoa. Hum Reprod. 1998;13:2756–61.
45. Reichart M, Levi H, Kahane I, et al. Dual energy metabolism-dependent effect of *Ureaplasma urealyticum* infection on sperm activity. J Androl. 2001;22:404–12.
46. Busolo F, Zanchetta R. Do mycoplasmas inhibit the human sperm fertilizing ability in vitro? Isr J Med Sci. 1984;20:902–4.
47. Soffer Y, Ron-El R, Golan A, et al. Male genital mycoplasmas and *Chlamydia trachomatis* culture: its relationship with accessory gland function, sperm quality, and autoimmunity. Fertil Steril. 1990;53:331–6.

48. Kjaergaard N, Kristensen B, Hansen ES, et al. Microbiology of semen specimens from males attending a fertility clinic. APMIS. 1997;105:566–70.
49. Xu C, Sun GF, Zhu YF, et al. The correlation of *Ureaplasma urealyticum* infection with infertility. Andrologia. 1997;29:219–26.
50. Shi J, Yang Z, Wang M, et al. Screening of an antigen target for immunocontraceptives from cross-reactive antigens between human sperm and *Ureaplasma urealyticum*. Infect Immun. 2007;75:2004–11.
51. Wang Y, Kang L, Hou Y, et al. Microelements in seminal plasma of infertile men infected with *Ureaplasma urealyticum*. Biol Trace Elem Res. 2005;105:11–8.
52. Ma J, Xu C. Relationship between mycoplasma infection and germ cell sulfogalactosylglycerolipid. Zhonghua Nan Ke Xue. 2004;10:215–7.
53. Shang XJ, Huang YF, Xiong CL, et al. *Ureaplasma urealyticum* infection and apoptosis of spermatogenic cells. Asian J Androl. 1999;1:127–9.
54. Reichart M, Kahane I, Bartoov B. In vivo and in vitro impairment of human and ram sperm nuclear chromatin integrity by sexually transmitted *Ureaplasma urealyticum* infection. Biol Reprod. 2000;63:1041–8.
55. Xu C, Lu MG, Feng JS, et al. Germ cell apoptosis induced by *Ureaplasma urealyticum* infection. Asian J Androl. 2001;3:199–204.
56. Shalhoub D, Abdel-Latif A, Fredericks CM, et al. Physiological integrity of human sperm in the presence of *Ureaplasma urealyticum*. Arch Androl. 1986;16:75–80.
57. Talkington DF, Davis JK, Canupp KC, et al. The effects of three serotypes of *Ureaplasma urealyticum* on spermatozoal motility and penetration in vitro. Fertil Steril. 1991;55:170–6.
58. Cintron RD, Wortham Jr JW, et al. The association of semen factors with the recovery of *Ureaplasma urealyticum*. Fertil Steril. 1981;36:648–52.
59. Gregoriou O, Botsis D, Papadias K, et al. Culture of seminal fluid in infertile men and relationship to semen evaluation. Int J Gynaecol Obstet. 1989;28:149–53.
60. Gdoura R, Kchaou W, Chaari C, et al. *Ureaplasma urealyticum*, *Ureaplasma parvum*, *Mycoplasma hominis* and *Mycoplasma genitalium* infections and semen quality of infertile men. BMC Infect Dis. 2007;7:129.
61. Gdoura R, Kchaou W, Ammar-Keskes L, et al. Assessment of *Chlamydia trachomatis*, *Ureaplasma urealyticum*, *Ureaplasma parvum*, *Mycoplasma hominis*, and *Mycoplasma genitalium* in semen and first void urine specimens of asymptomatic male partners of infertile couples. J Androl. 2008;29:198–206.
62. Hofstetter A, Schmiedt E, Schill WB, et al. Genital mycoplasma strains as a cause of male infertility. Helv Chir Acta. 1978;45:329–33.
63. Bornman MS, Mahomed MF, Boomker D, et al. Microbial flora in semen of infertile African men at Garankuwa hospital. Andrologia. 1990;22:118–21.
64. Corradi G, Molnàr G, Pànovics J. Andrologic significance of genital mycoplasma. Orv Hetil. 1992;133:3085–8.
65. Agbakoba NR, Adetosoye AI, Ikechebelu JI. Genital mycoplasmas in semen samples of males attending a tertiary care hospital in Nigeria: any role in sperm count reduction? Niger J Clin Pract. 2007;10:169–73.
66. Dìaz-Garcìa FJ, Herrera-Mendoza AP, Giono-Cerezo S, et al. *Mycoplasma hominis* attaches to and locates intracellularly in human spermatozoa. Hum Reprod. 2006;21:1591–8.
67. Svenstrup HF, Fedder J, Abraham-Peskir J, et al. *Mycoplasma genitalium* attaches to human spermatozoa. Hum Reprod. 2003;18:2103–9.
68. Kalugdan T, Chan PJ, Seraj IM, et al. Polymerase chain reaction enzyme-linked immunosorbent assay detection of mycoplasma consensus gene in sperm with low oocyte penetration capacity. Fertil Steril. 1996;66:793–7.
69. Hill JA, Haimovici F, Politch JA, et al. Effects of soluble products of activated lymphocytes and macrophages (lymphokines and monokines) on human sperm motion parameters. Fertil Steril. 1987;47:460–5.
70. Eggert-Kruse W, Hofmann H, Gerhard I, et al. Effects of antimicrobial therapy on sperm-mucus interaction. Hum Reprod. 1988;3:861–9.
71. Andrade-Rocha FT. *Ureaplasma urealyticum* and *Mycoplasma hominis* in men attending for routine semen analysis. Prevalence, incidence by age and clinical settings, influence on sperm characteristics, relationship with the leukocyte count and clinical value. Urol Int. 2003;71:377–81.
72. Rosemond A, Lanotte P, Watt S, et al. Systematic screening tests for *Chlamydia trachomatis*, *Mycoplasma hominis* and *Ureaplasma urealyticum* in urogenital specimens of infertile couples. Pathol Biol (Paris). 2006;54:125–9.
73. Tuttle Jr JP, Bannister ER, Derrick FC. Interference of human spermatozoal motility and spermatozoal agglutination by *Candida albicans*. J Urol. 1997;118:797–9.
74. Tian Y-H, Xiong JW, Hu L, et al. *Candida albicans* and filtrates interfere with human spermatozoal motility and alter the ultrastructure of spermatozoa: an in vitro study. Int J Androl. 2007;30:421–9.
75. Burrello N, Calogero AE, Perdichizzi A, et al. Inhibition of oocyte fertilization by assisted reproductive techniques and increased sperm DNA fragmentation in the presence of *Candida albicans*: a case report. Reprod Biomed Online. 2004;8:569–73.
76. Burrello N, Salmeri M, Perdichizzi A, et al. *Candida albicans* experimental infection: effects on human sperm motility, mitochondrial membrane potential and apoptosis. Reprod Biomed Online. 2009;18:496–501.
77. Rennemeier C, Frambach T, Hennicke F. Microbial quorum-sensing molecules induce acrosome loss and cell death in human spermatozoa. Infect Immun. 2009;77:4990–7.

78. Gopalkrishnan K, Hinduja IN, Kumar TC. Semen characteristics of asymptomatic males affected by *Trichomonas vaginalis*. J In Vitro Fert Embryo Transf. 1990;7:165–7.
79. Tuttle Jr JP, Holbrook TW, Derrick FC. Interference of human spermatozoal motility by *Trichomonas vaginalis*. J Urol. 1977;118:1024–5.
80. Jarecki-Black JC, Lushbaugh WB, Golosov L. *Trichomonas vaginalis*: preliminary characterization of a sperm motility inhibiting factor. Ann Clin Lab Sci. 1988;18:484–9.
81. Han Q, Liu J, Wang T. Influence of the metabolite produced by *Trichomonas vaginalis* on human sperm motility in vitro. Zhonghua Nan Ke Xue. 2004;10:272–4.
82. Kranjci -Zec I, Dzami A, Mitrovi S, et al. The role of parasites and fungi in secondary infertility. Med Pregl. 2004;57:30–2.
83. Benchimol M, de Andrade Rosa I, da Silva Fontes R, et al. Trichomonas adhere and phagocytose sperm cells: adhesion seems to be a prominent stage during interaction. Parasitol Res. 2008;102:597–604.
84. Daly JJ, Sherman JK, Green L, et al. Survival of *Trichomonas vaginalis* in human semen. Genitourin Med. 1989;65:106–8.
85. Garrido N, Meseguer M, Remohí J, et al. Semen characteristics in human immunodeficiency virus (HIV)- and hepatitis C (HCV)-seropositive males: predictors of the success of viral removal after sperm washing. Hum Reprod. 2005;20:1028–34.
86. Vicari E, Arcoria D, Di Mauro C, et al. Sperm output in patients with primary infertility and hepatitis B or C virus; negative influence of HBV infection during concomitant varicocele. Minerva Med. 2006;97:65–77.
87. Durazzo M, Premoli A, Di Bisceglie C, et al. Alterations of seminal and hormonal parameters: an extrahepatic manifestation of HCV infection? World J Gastroenterol. 2006;12:3073–6.
88. Moretti E, Federico MG, Giannerini V, et al. Sperm ultrastructure and meiotic segregation in a group of patients with chronic hepatitis B and C. Andrologia. 2008;40:286–91.
89. Lorusso F, Palmisano M, Chironna M, et al. Impact of chronic viral diseases on semen parameters. Andrologia. 2010;42:121–6.
90. Zhou XL, Sun PN, Huang TH, et al. Effects of hepatitis B virus S protein on human sperm function. Hum Reprod. 2009;24:1575–83.
91. Huang JM, Huang TH, Qiu HY, et al. Effects of hepatitis B virus infection on human sperm chromosomes. World J Gastroenterol. 2003;9:736–40.
92. Huang JM, Huang TH, Qiu HY, et al. Studies on the integration of hepatitis B virus DNA sequence in human sperm chromosomes. Asian J Androl. 2002;4:209–12.
93. Krieger JN, Coombs RW, Collier AC, et al. Fertility parameters in men infected with human immunodeficiency virus. J Infect Dis. 1991;164:464–9.
94. Dondero F, Rossi T, D'Offizi G, et al. Semen analysis in HIV seropositive men and in subjects at high risk for HIV infection. Hum Reprod. 1996;11:765–8.
95. Nicopoullos JDM, Almeida PA, Ramsay JWA, et al. The effect of human immunodeficiency virus on sperm parameters and the outcome of intrauterine insemination following sperm washing. Hum Reprod. 2004;19:2289–97.
96. Umapathy E. STD/HIV association: effects on semen characteristics. Arch Androl. 2005;51:361–5.
97. Bujan L, Sergerie M, Moinard N, et al. Decreased semen volume and spermatozoa motility in HIV-1-infected patients under antiretroviral treatment. J Androl. 2007;28:444–52.
98. Coll O, Lopez M, Vidal R, et al. Fertility assessment in non-infertile HIV-infected women and their partners. Reprod Biomed Online. 2007;14:488–94.
99. Cardona-Maya W, Velilla P, Montoya CJ, et al. Presence of HIV-1 DNA in spermatozoa from HIV-positive patients: changes in the semen parameters. Curr HIV Res. 2009;7:418–24.
100. Muciaccia B, Corallini S, Vicini E, et al. HIV-1 viral DNA is present in ejaculated abnormal spermatozoa of seropositive subjects. Hum Reprod. 2007;22:2868–78.
101. van Leeuwen E, Wit FW, Prins JM, et al. Semen quality remains stable during 96 weeks of untreated human immunodeficiency virus-1 infection. Fertil Steril. 2008;90:636–41.
102. Sergerie M, Martinet S, Kiffer N, et al. Impact of reverse transcriptase inhibitors on sperm mitochondrial genomic DNA in assisted reproduction techniques. Gynecol Obstet Fertil. 2004;32:841–9.
103. van Leeuwen E, Wit FW, Repping S, et al. Effects of antiretroviral therapy on semen quality. AIDS. 2008;22:637–42.
104. Chan PJ, Su BC, Kalugdan T, et al. Human papillomavirus gene sequences in washed human sperm deoxyribonucleic acid. Fertil Steril. 1994;61:982–5.
105. Lai YM, Yang FP, Pao CC. Human papillomavirus deoxyribonucleic acid and ribonucleic acid in seminal plasma and sperm cells. Fertil Steril. 1996;65:1026–30.
106. Brossfield JE, Chan PJ, Patton WC, et al. Tenacity of exogenous human papillomavirus DNA in sperm washing. Int J STD AIDS. 1999;15:740–3.
107. Pèrez-Andino J, Buck CB, Ribbeck K. Adsorption of human papillomavirus 16 to live human sperm. PLoS ONE. 2009;4:e5847.
108. Foresta C, Garolla A, Zuccarello D, et al. Human papillomavirus found in sperm head of young adult males affects the progressive motility. Fertil Steril. 2010;93:802–6.
109. Lai YM, Lee JF, Huang HY, et al. The effect of human papillomavirus infection on sperm cell motility. Fertil Steril. 1997;67:1152–5.
110. Foresta C, Pizzol D, Moretti A, et al. Clinical and prognostic significance of human papillomavirus DNA in the sperm or exfoliated cells of infertile patients and subject with risk factors. Fertil Steril. 2010;94:1723–7.

111. Tanaka H, Karube A, Kodama H, et al. Mass screening for human papillomavirus type 16 infection in infertile couples. J Reprod Med. 2000;45:907–11.
112. Rintala MA, Grènman SE, Pöllänen PP, et al. Detection of high-risk HPV DNA in semen and its association with the quality of semen. Int J STD AIDS. 2004;15:740–3.
113. Connelly DA, Chan PJ, Patton WC, et al. Human sperm deoxyribonucleic acid fragmentation by specific types of papillomavirus. J Assist Reprod Genet. 2001;184:1068–70.
114. Lee CA, Huang CT, King A, et al. Differential effects of human papillomavirus DNA types on p53 tumor-suppressor gene apoptosis in sperm. Gynecol Oncol. 2002;85:511–6.
115. Lanzafame F, La Vignera S, Vicari E, et al. Oxidative stress and antioxidant medical treatment in male infertility. Reprod Biomed Online. 2009;19:638–59.
116. Aitken RJ, Clarkson JS, Hargreave TB, et al. Analysis of the relationship between defective sperm function and the generation of reactive oxygen species in cases of oligozoospermia. J Androl. 1989;10:214–20.
117. Saraniya A, Koner BC, Doureradjou P, et al. Altered malondialdehyde, protein carbonyl and sialic acid levels in seminal plasma of microscopically abnormal semen. Andrologia. 2008;40:56–7.
118. Suleiman SA, Ali ME, Zaki ZM, et al. Lipid peroxidation and human sperm motility: protective role of vitamin E. J Androl. 1996;17:530–7.
119. Aydemir B, Onaran I, Kiziler AR, et al. The influence of oxidative damage on viscosity of seminal fluid in infertile men. J Androl. 2008;29:41–6.
120. Twigg JP, Irvine DS, Aitken RJ. Oxidative damage to DNA in human spermatozoa does not preclude pronucleus formation at intracytoplasmic sperm injection. Hum Reprod. 1998;13:1864–71.
121. Barroso G, Morshedi M, Oehringer S. Analysis of DNA fragmentation, plasma membrane translocation of phosphatidylserine and oxidative stress in human spermatozoa. Hum Reprod. 2000;15: 1338–44.
122. Kodama H, Yamaguchi R, Fukuda J, et al. Increased deoxyribonucleic acid damage in the spermatozoa of infertile male patients. Fertil Steril. 1997;65:519–24.
123. Henkel R, Kierspel E, Hajimohammad M, et al. DNA fragmentation of spermatozoa and assisted reproduction technology. Reprod Biomed Online. 2003;7:477–84.
124. Evenson DP, Larson KL, Jost LK. Sperm chromatin structure assay: its clinical use for detecting sperm DNA fragmentation in male infertility and comparisons with the other techniques. J Androl. 2002; 23:25–43.
125. Griveau JF, Le Lannou D. Reactive oxygen species and human spermatozoa: physiology and pathology. Int J Androl. 1997;20:61–9.
126. Dousset B, Hussenet F, Daudin M, et al. Seminal cytokine concentrations (IL-1beta, IL-2, IL-6, sR IL-2, sR IL-6), semen parameters and blood hormonal status in male infertility. Hum Reprod. 1997;12:1476–9.
127. Dimitrov DG, Petrovská M. Effects of products of activated immune cells and recombinant cytokines on spontaneous and ionophore-induced acrosome reaction. Am J Reprod Immunol. 1996;36:150–6.
128. Fraczek M, Sanocka D, Kamieniczna M, et al. Proinflammatory cytokines as an intermediate factor enhancing lipid sperm membrane peroxidation in in vitro conditions. J Androl. 2008;29:85–92.
129. Camejo MI, Segnini A, Proverbio F. Interleukin-6 (IL-6) in seminal plasma of infertile men, and lipid peroxidation of their sperm. Arch Androl. 2001;47:97–101.
130. Lampiao F, du Plessis SS. TNF-alpha and IL-6 affect human sperm function by elevating nitric oxide production. Reprod Biomed Online. 2008;17:628–31.
131. Lampiao F, du Plessis SS. Effects of tumour necrosis factor alpha and interleukin-6 on progesterone and calcium ionophore-induced acrosome reaction. Int J Androl. 2009;32:274–7.
132. Fedder J, Ellerman-Eriksen S. Effect of cytokines on sperm motility and ionophore-stimulated acrosome reaction. Arch Androl. 1995;35:173–85.
133. Eggert-Kruse W, Boit R, Rohr G, et al. Relationship of seminal plasma interleukin (IL) -8 and IL-6 with semen quality. Hum Reprod. 2001;16:517–28.
134. Martínez P, Proverbio F, Camejo MI. Sperm lipid peroxidation and pro-inflammatory cytokines. Asian J Androl. 2007;9:102–7.
135. Estrada LS, Champion HC, Wang R, et al. Effect of tumour necrosis factor-alpha (TNF-alpha) and interferon-gamma (IFN-gamma) on human sperm motility, viability and motion parameters. Int J Androl. 1997;20:237–42.
136. Hill JA, Cohen J, Anderson DJ. The effects of lymphokines and monokines on human sperm fertilizing ability in the zona-free hamster egg penetration test. Am J Obstet Gynecol. 1989;160(5 Pt 1):1154–9.
137. Bian SL, Jin HB, Wang SZ, et al. Effects of interferon-gamma and tumor necrosis factor-alpha on the fertilizing capacity of human sperm and their mechanisms. Zhonghua Nan Ke Xue. 2007;13:681–4.
138. Sikka SC, Champion HC, Bivalacqua TJ, et al. Role of genitourinary inflammation in infertility: synergistic effects of lipopolysaccharide and interferon-gamma on human spermatozoa. Int J Androl. 2001;24:136–41.
139. Eickhoff R, Baldauf C, Koyro HW, et al. Influence of macrophage migration inhibitory factor (MIF) on the zinc content and redox state of protein-bound sulphydryl groups in rat sperm: indications for a new role of MIF in sperm maturation. Mol Hum Reprod. 2004;10:605–11.
140. Frenette G, Légaré C, Saez F, et al. Macrophage migration inhibitory factor in the human epididymis and semen. Mol Hum Reprod. 2005;11:575–82.
141. Aljabari B, Calogero AE, Perdichizzi A, et al. Imbalance in seminal fluid MIF indicates male infertility. Mol Med. 2007;13:199–202.

142. Carli C, Leclerc P, Metz CN, et al. Direct effect of macrophage migration inhibitory factor on sperm function: possible involvement in endometriosis-associated infertility. Fertil Steril. 2007;88(4 Suppl): 1240–7.
143. Hussenet F, Dousset B, Cordonnier JL, et al. Tumour necrosis factor alpha and interleukin 2 in normal and infected human seminal fluid. Hum Reprod. 1993;8:409–11.
144. Gruschwitz MS, Brezinschek R, Brezinschek HP. Cytokine levels in the seminal plasma of infertile males. J Androl. 1996;17:158–63.
145. Omu AE, Al-Qattan F, Al-Abdul-Hadi FM, et al. Seminal immune response in infertile men with leukocytospermia: effect on antioxidant activity. Eur J Obstet Gynecol Reprod Biol. 1999;86:195–202.
146. Sikorski R, Kapec E, Krzeminski A, et al. Levels of proinflammatory cytokines (Il-1 alpha, Il-6, TNF-alpha) in the semen plasma of male partners of infertile couples. Ginekol Pol. 2001;72:1325–8.
147. Wincek TJ, Meyer TK, Meyer MR, et al. Absence of a direct effect of recombinant tumor necrosis factor-alpha on human sperm function and murine preimplantation development. Fertil Steril. 1991;56: 332–9.
148. Haney AF, Hughes SF, Weinberg JB. The lack of effect of tumor necrosis factor-alpha, interleukin-1-alpha, and interferon-gamma on human sperm motility in vitro. J Androl. 1992;13:249–53.
149. Lewis SE, Donnelly ET, Sterling ES, et al. Nitric oxide synthase and nitrite production in human spermatozoa: evidence that endogenous nitric oxide is beneficial to sperm motility. Mol Hum Reprod. 1996;2:873–8.
150. Kocak I, Yenisey C, Dundar M, et al. Relationship between seminal plasma interleukin-6 and tumor necrosis factor alpha levels with semen parameters in fertile and infertile men. Urol Res. 2002;30: 263–7.
151. Eisermann J, Register KB, Strickler RC, et al. The effect of tumor necrosis factor on human sperm motility in vitro. J Androl. 1989;10:270–4.
152. Perdichizzi A, Nicoletti F, La Vignera S, et al. Effects of tumour necrosis factor-alpha on human sperm motility and apoptosis. J Clin Immunol. 2007;27: 152–62.
153. Bian J, Guo X, Xiong C, et al. Experimental study of the effect of rhTNF-alpha on human sperm mitochondrial function and motility in vitro. Zhonghua Nan Ke Xue. 2004;10:415–9.
154. Said TM, Agarwal A, Falcone T, et al. Infliximab may reverse the toxic effects induced by tumor necrosis factor alpha in human spermatozoa: an in vitro model. Fertil Steril. 2005;83:1665–73.
155. Allam JP, Fronhoffs F, Fathy A, et al. High percentage of apoptotic spermatozoa in ejaculates from men with chronic genital tract inflammation. Andrologia. 2008;40:329–34.

Part IV

Sperm Chromatin and Assisted Reproductive Technology Outcomes

The Impact of Sperm Processing and Cryopreservation on Sperm DNA Integrity

27

Dan Yu, Luke Simon, and Sheena E.M. Lewis

Abstract

Here, an overview of the various clinical uses of processing of semen samples and cryopreservation is given. The effects of sperm processing on conventional semen parameters are discussed along with the ramifications of removing seminal plasma, oxidative stress and potential benefits of antioxidant addition in laboratory processing of testicular and ejaculated sperm. In the second part of the chapter, the effects of a second laboratory hazard, namely, cryopreservation, are discussed in terms of effects on conventional parameters of sperm structure and function and also on DNA fragmentation of testicular and ejaculated sperm. The greater vulnerability of sperm from infertile men is described, as well as the cryoinjury displayed by those with teratozoospermia. The mechanisms of cryoinjury are set out with special reference to oxidative damage and the process of repeated freezing and thawing. The efficacy of commercially available cryoprotectants is also discussed. Finally, novel freezing–thawing protocols such as freeze-drying and vitrification of sperm are explored.

Keywords

Sperm processing • Cryopreservation of sperm DNA • Sperm DNA integrity • Sperm DNA processing • Density-gradient centrifugation

The Clinical Need for Sperm Processing

To maintain their fertility, sperm must be separated from seminal plasma as soon as possible after ejaculation, as it has been shown that long exposure to seminal plasma results in reduced motility and vitality [1]. There are also a number of seminal plasma components that inhibit

S.E.M. Lewis (✉)Lewis
Centre for Public Health, Queen's University of Belfast, Room 208, Institute of Clinical Science, Belfast, Northern Ireland, UK
e-mail: s.e.lewis@qub.ac.uk

A. Zini and A. Agarwal (eds.), *Sperm Chromatin: Biological and Clinical Applications in Male Infertility and Assisted Reproduction*, DOI 10.1007/978-1-4419-6857-9_27,

acrosome reactions, capacitation and, thus, the fertilization potential of the sperm. For clinical use, sperm should be separated from seminal plasma as soon as possible after liquefaction. The most common method is by discontinuous two-step density-gradient centrifugation (DCG) that isolates the subpopulation of sperm with the best motility, morphology [2], superior nuclear and mitochondrial DNA quality [3] and without endogenous nicks [4]. The importance and efficacy of DCG in selecting out a population of sperm where most are of high quality are reflected in terms of higher assisted conception rates [5, 6]. However, centrifugation of a semen sample prior to its use in assisted reproductive techniques (ART) can exacerbate sperm oxidative stress. Since sperm do not have any repair mechanisms, as they are transcriptionally silent and lack functional repair enzymes [7–9], this can cause irreversible damage. This can be limited by reducing the time of centrifugation in the preparation of sperm for ART [10, 11]. Furthermore, culturing sperm under low oxygen tension (5%O2/95% CO2 vs. 20% atmospheric O2 content) has been shown to significantly improve sperm quality by reducing seminal leukocyte reactive oxygen species (ROS) production [12, 13].

The Hazards of Seminal Plasma Removal During DCG

Although DGC facilitates the isolation of high-quality sperm, most suitable for use in ART, the removal of the seminal plasma's protective antioxidants makes sperm DNA more vulnerable to oxidative insult through generation of ROS by adjacent damaged sperm [14, 15]. Furthermore, the semen from infertile men is often associated with higher levels of ROS than that of fertile men, and numerous studies have shown the association between male infertility and raised ROS in semen [14, 16, 17]. Negative associations have been reported between ROS and quantitative velocity parameters, sperm DNA integrity and also lower total antioxidant levels (TAC) in sperm of men with male infertility attending a tertiary centre [18]. Increased ROS levels and reduced antioxidants have also been reported in semen of infertile men with varicocele [19].

When these men's sperm are then exposed to sperm processing for ART, they are at further risk of oxidative damage. Previous studies from our group have shown that depriving sperm of seminal antioxidant protection during DCG preparation for ART leads to DNA damage [20].

All sperm are particularly vulnerable to damage from ROS because of their high polyunsaturated fatty acid content and limited ability to repair damage. In contrast to somatic cells, which contain protective antioxidants within their cytoplasm, sperm lose most of their cytoplasm during the maturation process and, therefore, lack the endogenous repair mechanisms and enzymatic defences observed in other cell types. This leaves them at a significant disadvantage especially since the absence of RNA transcription and DNA repair mechanisms means that any damage induced to sperm will be permanent. ROS are also among the most powerful instigators of sperm DNA damage [21, 22].

Antioxidants: Physiological and Therapeutic Uses

Antioxidants act to remove damaging ROS such as O_2 and H_2O_2, and scavengers such as albumin and taurine [23]. Metal chelators can also be useful in reducing ROS generation and preventing lipid peroxidation of sperm membranes, thereby protecting sperm nuclear DNA (reviewed by Agarwal and Said [24]). Paradoxically, the addition of combinations of antioxidants such as vitamins C and E can have damaging effects to DNA in vitro [25] and in vivo where DNA decondensation can increase [26] or they can be ineffective [27]. Ascorbate and catalase, which are both found naturally in seminal plasma, reduce the level of ROS that induce sperm nuclear DNA damage, improving the quality of sperm following cryopreservation prior to ART [28]. Mature ejaculated sperm are protected from oxidative insult by the surrounding seminal plasma, which contains enzymes such as superoxide dismutase, catalase and chain breaking antioxidants such as ascorbate, which is ten times more concentrated than in blood plasma [29], emphasizing the physiological importance of antioxidants. Alpha tocopherol and acetyl cysteine have also been found [30, 31] to be of benefit in protecting motility against ROS impairment and enhancing sperm zona binding. Our group has shown that protection from the DNA damage that can occur during DCG can be provided by supplementing media with antioxidants [3, 20]. Ascorbic acid, alpha tocopherol and urate separately significantly decreased the level of sperm DNA fragmentation.

Recently the human endogenous cannabinoid system (ECS) has been strongly implicated in various aspects of female and male fertility (reviewed by Battista et al. [32]). Some members of the ECS (*N*-acylethanolamide, oleoylethanolamide and Cannabidiol) are potent antioxidants in somatic cells, which may also exert protective effects on sperm DNA, if supplemented during assisted conception processes. Further research is urgently needed to find the most effective antioxidant therapy and dosage for sperm nuclear DNA protection during ART.

The Heightened Vulnerability of Testicular Sperm

Testicular sperm, retrieved for ART via testicular biopsy, are especially vulnerable to oxidative assault and resultant DNA damage in comparison to mature ejaculated sperm, as chromatin packaging is not completed until SH bonds are oxidized during transit through the epididymis. All sperm naturally produce low levels of ROS as by-products of the electron transfer chain, which are essential for normal sperm maturation and function. However, testicular sperm retain a significant proportion of cytoplasm that may facilitate excess ROS generation, and unlike ejaculated sperm, testicular sperm have no seminal plasma to confer antioxidant protection. This presents a significant clinical problem, since the use of testicular sperm for assisted conception is an increasingly used avenue of treatment for males with problems such as obstructive azoospermia and ejaculatory dysfunction.

The Clinical Need for Sperm Cryopreservation

A second inadvertent cause of damage in the laboratory is through cryopreservation. Semen cryopreservation is a core technique in the process of preservation and storage of male gametes prior to ART, or before cytotoxic chemotherapy [33], radiotherapy or surgical treatment, whichmay lead to testicular damage or ejaculatory dysfunction. The process of freezing sperm before beginning the treatment, which may affect fertility potential, enables many patients to father their own children post treatment through the use of IVF or intracytoplasmic sperm injection (ICSI). In addition, sperm cryopreservation is mandatory in donor-insemination programmes, as the use of frozen semen allows screening of sperm donors for infections such as HIV and hepatitis B prior to release for insemination [34]. The technique is also widely used for storage of sperm retrieved from azoospermic patients who have undergone testicular sperm biopsy or percutaneous epididymal sperm aspiration, avoiding the need for repeat biopsies or aspiration on the day of ART. Despite many refinements in methodology (reviewed by Anger et al. [35]), the procedure is not without risk and adverse affects. The quality of post-thaw samples remains suboptimal, and IUI and IVF success rates are lower with frozen sperm than with fresh samples [36].

The Impact of Cryopreservation on Conventional Sperm Parameters

Sperm motility is the function most vulnerable to cryoinjury [37]. Post-thaw motilities are routinely only 50% of pre-freeze values (reviewed by Nijs and Ombelet [38]; Anger et al. [35]). Quantitative motility assessments show reductions in straight line and curvilinear velocities of 25–75% [39]. This functional impairment is due to structural damage in the flagella caused by alterations in permeability and membrane fluidity [40], and conformation of phospholipid bilayers [41, 42]. Pentoxifylline and 2-deoxyadenosine have been utilized to optimize flagging energy levels by inhibiting the breakdown of cAMP and cGMP [42–45] but the adverse effects of phosphodiesterase inhibitors on fertilization rates and early embryo cleavage [46, 47] demand caution in their clinical usage. Reduced sperm penetration of the cervical mucus has also been reported [48].

Organelle damage is also observed in mitochondrial distortion [49]. Alterations in plasma and mitochondrial membrane potentials, observed by reduced R123 uptake [47] leads to reductions in $[Ca^{2+}]_i$. This, in turn, impairs the cell's response to progesterone and ability to progress into capacitative motility [50]. Further damage has been reported as a reduction in intact acrosomal caps and in acrosin activity. There is also an increase in gross morphological abnormalities; particularly in amorphous sperm heads, midpiece anomalies and, cytoplasmic vacuolation [49, 51]. The ultimate cryoinjury, which occurs in up to 30% of sperm, is the fatal loss of membrane integrity [35].

The Impact of Cryopreservation on Human Sperm DNA

All previous semen freezing suitability criteria have been based on concentration, motility and morphology [24, 52]. However, these have now been largely rejected as fertility biomarkers. In their place, sperm DNA integrity is recognized as a more robust measure of male fertility potential. As a result the scientific community has been readdressing the impact of numerous clinical procedures, including cryopreservation, on sperm DNA integrity.

For the past decade there was a general belief that sperm DNA was impervious to cryodamage [53]. This was largely based on a small study by Duru et al. [54] comparing sperm DNA fragmentation measured by the TUNEL of freeze-thawed sperm from 5 donors and 10 men undergoing infertility investigations and finding no significant differences. This study was supported by several groups using the sperm chromatin structure assay [55–57] where semen samples are frozen in liquid Nitrogen without cryoprotectant and transported by dry shipper to a central lab for SCSA testing. By contrast, studies from our group reported that sperm DNA was fragmented by cryopreservation. Recently interest in sperm DNA cryoinjury has revived and a quite a number of interesting papers have been published [58–63]. In the study by de Paula et al. [59] of men with oligozoospermia, higher DNA damage, by TUNEL, was observed before and after freezing in comparison with a group of normozoospermic men attending for infertility treatment because of female problems. In this study, the increase in post-thaw damage in both groups was similar.

The Greater Susceptibility of Infertile Mens' Sperm to Cryoinjury

The degree of DNA damage in sperm from infertile men has been reported to be significantly higher than in sperm from fertile donors [39, 64]. This was even true of infertile men with normozoospermic profiles susceptibility to cryoinjury so our groups has suggested that resistance to cryoinjury might be used as an additional diagnostic test to semen analysis. In another study of men with abnormal semen profiles; in this case teratozoospermia, a threefold increase on DNA fragmentation, by Comet and acridine orange binding, was reported in the teratozoospermic samples compared with a normozoospermic group. This adds to the literature confirming greater vulnerability of "infertile" sperm. It also suggests a relationship between abnormal morphology and DNA damage. Teratospermic semen samples have increased levels of ROS [65]. Since many of these abnormal sperm have retained cytoplasm, major source of free radicals, the amount of ROS produced during cryopreservation of such sperm may be higher than that of morphologically normal sperm, which may be the cause of the increased levels of DNA damage in these teratospermic samples [61].

Mechanisms of Cryoinjury

Cryopreservation can result in cryodamage at different levels and functions of the cell, such as thermal shock, formation of intracellular ice crystals, cellular dehydration, increased concentration of salts and osmotic shock [66]. Such processes can lead to alterations of the acrosomal structure, decrease of acrosome activity, swelling or shrinkage of nuclei and cytoplasmic membranes and loss of plasma membrane integrity [40, 41, 67, 68].

Some of the cellular damage that human sperm encounter in cryopreservation has been attributed to the formation of intracellular ice. Clinical cryopreservation usually uses high and very high cooling rates [69, 70]. Supercooling can lead to intracellular ice formation, which can be fatal to the cell. However, no direct evidence of intracellular ice damage in sperm has been presented. Morris et al. [71] carried out a study to examine whether intracellular ice formation during rapid cooling causes the observed damage. Their results suggested that sperm damage at least for cooling rates up to 3,000°C/min is not caused by intracellular ice formation. Further, there was no evidence for intracellular ice, even upon warming and refreezing samples; conditions that would be expected to result in the recrystallization of any ice present within cells.

If intracellular ice formation is not the reason of cell damage at rapid rates of cooling, other physical factors, such as extracellular ice formation in the cryosolution surrounding the spermatozoa, must be responsible. During freezing of cell suspensions, the water outside the cells forms ice first in the extracellular space, which sets up an osmotic gradient between the intracellular isomotic solution and the freeze-concentrated extracellular solution. Morris et al. [72] demonstrated that the viscosity of the freeze-concentrated material can be increased rapidly by freezing an aqueous solution of the glycerol. Following ice nucleation, water from the adjacent solution migrates to the ice crystal, which causes the growth of ice crystal. During the thawing process of rapidly cooled glycerol solutions, a number of recrystallization patterns have been observed [73]. It is believed that either the crystallization during freezing or recrystallization during thawing could be a major cause of sperm cryoinjury.

The efflux of water from the cell can cause extracellular ice formation and, more damagingly, cellular dehydration. Various transport pathways in cell membranes for substrates, fluids, ions and gases preserve optimal osmotic balance between the intracellular and the extracellular environments [74, 75]. Cell transport machinery associated with cryopreservation involves water and permeable solutes [76, 77]. At the freezing temperature, the extracellular solution almost always freezes first. The extracellular solutes are concentrated in the remaining unfrozen extracellular water, so all solutes and suspended materials, including the cells, get localized in freeze-concentrated compartments [78]. During the further reduction in temperature, the cells are exposed to increasingly concentrated solutions. The hypertonic conditions that the cells encounter lead to an osmotic loss of water, which dehydrates the cells by osmosis as water diffuses from the cytoplasm into the more concentrated external solution [72].

The exposure of sperm to hypotonic solutions and the subsequent changes are termed osmotic shock. Osmotic damage caused by the exposure of frozen–thawed spermatozoa to isotonic conditions after a period of hypertonic exposure, is lethal due to extensive cell shrinkage. Subsequent rewarming and thawing of the cells can further deteriorate their viability through possible excessive osmotic swelling [53, 79, 80].

Cryosurvival of human sperm is also associated with cryoprotectants, as they were confirmed to lower the water freezing point and prevent the formation of ice crystals during freezing and, therefore, avoid structural damage and motility loss after cryopreservation [81].

Is Apoptosis a Cause of DNA Cryodamage?

In studies by Baumber et al. [82] the percentage of apoptotic sperm significantly increased after cryopreservation. Apoptosis is physiologically programmed cell death and an underlying mechanism for normal spermatogenesis [83, 84]. "Abortive apoptosis" is a theory proposed by Sakkas et al. [85, 86], in which the correct clearance of sperm via apoptosis is failed. Therefore, spermatozoa showing abnormal morphological forms, irregular biochemical function [87] or DNA damage fails to be eliminated. Abortive apoptosis may play a role in cryoinjury to sperm DNA because cryopreservation of spermatozoa resulted in activation of caspase, which has been reported in both human [88, 89] and bull sperm [90]. Caspases are particular aspartic acid-directed cysteine proteases, which are shown to play a key role in the cellular apoptotic and eventual cell death [91]. Although a correlation between the presence of activated caspases and sperm DNA fragmentation are reported [92, 93], the results are far from compelling. The total amount of DNA damage in spermatozoa cannot be explained by apoptosis alone. DNA damage can also occur due to oxidative stress [94, 95]. Furthermore, there is no any strong evidence to suggest a caspase/apoptosis-related increase in sperm DNA fragmentation during cryopreservation [90, 96]. Thomson et al. [63] carried out a study on the mechanisms of sperm DNA fragmentation increase following cryopreservation and measured caspase activation as an indicator of apoptosis, but this did not affect damage levels; so, this led them to conclude that cryopreservation causes damage via oxidative stress and not by apoptosis.

Is Sperm DNA Damage a Result of Oxidation?

The effects of cryopreservation on sperm DNA have recently been assessed using novel tests. Zribi et al. [62] determined DNA fragmentation by TUNEL supplemented with a measure of sperm DNA oxidation b using oxy-DNA test. They found an increase in fragmentation after thawing but just an insignificant trend towards increased DNA oxidation and therefore no relationship between DNA fragmentation and oxidation. However, this study was small ($n=15$) and perhaps larger numbers would show differences. In contrast to this study, in a larger group ($n=60$), Thomson et al. [97] cryopreservation caused a marked increase in sperm DNA fragmentation, by TUNEL and DNA oxidation by 8-OHdG using the oxyDNA test with a positive correlation before ($r=0.756$, $p<0.001$) and after treatment ($r=0.528$, $p<0.017$).

Why Freeze Neat Semen?

To prevent the damage to healthy sperm by weaker ROS releasing sperm during cryopreservation, the solution may be to prepare sperm before freezing and freeze only the DCG population. Perez-Sanchez et al. [98] reported an improvement in post-thaw sperm quality if sperm were prepared beforehand. Freezing prepared sperm has been shown to have no adverse effects on fertilization as indicated by sperm zona binding [99]. However, in studies by Donnelly et al. [39] and Thomson et al. [97], DCG sperm frozen without seminal plasma protection showed marked damage. The removal of seminal plasma protection, evidently necessary to resist cryoinjury, was probably from antioxidants, as cryoprotectants without antioxidants were not sufficient. Thomson et al. [97] demonstrated the percentage of sperm DNA fragmentation post-DCG significantly increase after cryopreservation both with and without the addition of cryoprotectant. The observed increase of DNA fragmentation in DGC-prepared spermatozoa might due to the removal of seminal fluid via DCG and the stresses of centrifugation. Centrifugation is also known to exacerbate oxidative stress within a semen sample [10, 22], which could lead to further damage to sperm.

When DGC sperm were frozen with seminal plasma and cryoprotectant added, post-thaw DNA fragmentation was the same as pre-freeze levels showing the efficacy of this combination of removing damaging sperm together with adding physiological protection [39]. The disadvantage of this method is the reduced numbers of sperm, but this may be outweighed by their quality and preservation of structure and function.

Recent Advances in Cryopreservation

The Efficacy of Different Cryoprotectants

In an attempt to reduce chilling injury and improve the optimal survival and fertility capacity of human sperm following cryopreservation, many different cryoprotectant media have been developed. Currently, the most widely used cryoprotectant is the permeating agent glycerol, as it has been confirmed the most effective in lowering the freezing point of intracellular water [100]. Other compounds are added to glycerol-containing media as buffers to yield optimal cryosurvival rates [101], In spite of these advances, a gold-standard method of cryopreserving human semen with an optimal cryoprotectant is yet to be determined.

Cryoprotectants themselves can pose a threat to cellular survival and cellular structures by causing the cell to shrink and swell beyond viable limits, thereby inducing osmotic shock and spermolysis [102, 103]. Commonly used cryoprotectants include glycerol and buffers. Higher glycerol concentrations have also been linked to increased activation of caspases via direct toxic effects to mitochondria during cryopreservation of spermatozoa [89].

This third paper by Thomson et al. [97] has contributed appreciably to our knowledge of the usefulness of cryoprotectants. In it, seven of the most commonly used cryoprotectant media were compared in how they protected sperm from 320 men from DNA damage. Neither the presence nor the type of cryoprotectant protected sperm DNA from cryoinjury followed by DCG. This conclusion was disappointing, but not surprising. The medium called SpermCryo gave least protection against DNA damage compared to Medicult Sperm Freezing Medium and FertiPro Sperm Freeze. This is the medium with the highest proportion of glycerol (68% glycerol compared with the others <25%). The authors explain this by highlighting a study showing high levels of glycerol cause cell death [42]. With their high proportion of membrane lipids, sperm, in particular, are sensitive to osmotic changes and vulnerable to lethal injury in hyper-osmotic conditions [104] through inappropriate re aggregation post-thawing [105]. How these processes impair DNA specifically needs elucidation.

There were, however, surprising results for men with low levels of DNA damage before freezing in that their sperm appeared to undergo a higher degree of cryoinjury than those who had higher levels of damage in their fresh samples. One explanation for this may be that these samples with high DNA damage were suboptimal in other parameters too as seen by O'Connell et al. [49] and so they also lose their motility through cryoinjury. So that only the sperm with better DNA are isolated in the DGC fraction.

Effects of Repeated Freezing and Thawing

There are many reasons for freezing sperm in small numbers. Oligozoospermia is a very common problem in men attending for ART. Further, 10% of male infertility cases are azoospermic and of those more than half have obstructive azoospermia so spermatogenesis may be relatively normal and testicular sperm may be extracted. As this is an invasive procedure, it would be useful to store sperm surplus to one ART cycle so that repeated biopsies are unnecessary. Third, cancer patients often wish to have sperm stored before treatment with spermotoxic drugs. Often when they present to have sperm stored, they are unwell and sperm concentration and/or quality is reduced [106–108]. However, techniques for cryopreservation of individual or small numbers of sperm have not been optimized and very small numbers of pregnancies have been reported using any of the techniques available. This being the case, it is of interest to examine the effects of repeated freezing and thawing on sperm DNA. In ART clinics, as donor sperm is expensive and increasingly scarce, it is routine to offer patients a repeated cryocycle to maximize the use of sperm. For example, couples may wish to have siblings from the same donor's sample.

There is overall agreement that, following each freeze–thaw cycle, the number of recovered motile and viable sperm decrease steadily [109–111] and the standard semen parameters of overall motility and vitality drop steadily [109–111]. A study carried by Thomson et al. [63] shows that the percentage of motile sperm and vital sperm dropped by half following the first cycle of freezing and thawing, and continued to drop by half following each subsequent cycle.

In the same study, the effects of repeated freeze thawing on sperm DNA fragmentation by TUNEL were also assessed. They found that repeated freezing and thawing increases the percentage of sperm exhibiting DNA fragmentation in raw, non-separated semen samples. Furthermore, when the samples were washed and fresh cryoprotectant added after each thaw, the percentage of sperm DNA fragmentation increased significantly. However, when the sample was refrozen in the original cryoprotectant without and further treatment or wash, the percentage of sperm DNA fragmentation only increased slightly after the second and third thaw. Therefore, Thomson et al. [97] recommended to avoid washing steps and the addition of fresh cryoprotectant in between each freeze–thaw cycle. Samples are refrozen in their original cryoprotectant and not washed or altered in any way in between, and separated by DGC or swim-up before use in ART. By this protocol, the increase of percentage of sperm DNA fragmentation will be terminated in up to three cycles of freezing and thawing, even though it is still higher than the fresh sample.

Thus, preparing sperm by DCG for ART with the concomitant removal of seminal protection before freezing increased DNA damage and reduced vitality. If samples are frozen in their original cryoprotectant without further processing but are subjected to DGC after thawing, the "risk" (described as the relative chance of fertilization with a cell containing fragmented DNA) of three F–T cycles is equivalent to one cycle. This study again highlights the necessity of seminal plasma and the adverse effects of laboratory processing.

The Benefits of Antioxidant Supplements to Cryoprotectant Media

As described previously, sperm DNA damage has been associated with high levels of ROS in fresh and cryopreserved semen [28, 112, 113]. A study by Li et al. [28] shows that the addition of ascorbate or catalase in human semen samples reduces ROS levels and sperm nuclear DNA damage, and improves the human sperm quality in the process of freezing and thawing. In addition, Gadea et al. [114] reported that the addition of glutathione (L-gamma-glutamyl-L-cysteinylglycine) to the thawing medium resulted in a similar result in frozen bull spermatozoa.

The work of Bilodeau et al. [115] and Peris et al. [116] have confirmed the belief that DNA instability is increased during cryopreservation as a result of reduced sperm antioxidant defence mechanisms. Specific factors such as alpha – tocopherol and ascorbate have been shown to increase post-thaw viability. Further studies have reported the addition of superoxide dismutase and catalase an increase in hamster egg penetration, increased embryo numbers [117] and increased implantation in bovine studies [118].

Freezing has been shown to reduce gluthathione (GSH) and superoxide dismutase levels [119], whereas post-thaw addition of thiols (GSH, Cysteine, N acetyl cysteine) and pyruvate, metal chelators or oviductal catalase prevented H_2O_2 prevented a H_2O_2 mediated reduction in sperm motility [115, 119].

In the Thomson study [97], all of the cryoprotectants, except SpermCryo and Medicult Cryosperm, included human serum albumin at an unknown concentration. Albumin has traditionally been viewed as a useful antioxidant as well as a plasma protein [120], but it is also known to be ineffective against NADPH [120], so oxidative stress by this route may still be the cause of the DNA fragmentation observed in this study.

Recent Advances in Freeze–Thaw Protocols

Freeze-Drying of Sperm

Long-term preservation of mouse sperm has been achieved with freeze-drying without cryoprotectants by Yanagimachi's group [121]. The sperm were plunged into liquid nitrogen for 20 s and then freeze-dried for 4 h. After storage at 4°C for periods of 1–12 months, the sperm were thawed by bringing to room temperature and hydrating with sterile water. Chromosomal stability was maintained in these sperm and embryos generated by ICSI. In another study using freeze-dried mouse sperm (new) [122], 96% of resultant zygotes had normal chromosomes and 58% developed into normal viable foetuses. Live offspring were obtained after storage of 1.5 years. Major advantages of this technique are the convenience of its short protocol and reduced need for storage space and sophisticated cryofacilities and expensive shipping procedures.

Vitrification of Sperm

Conventional freezing techniques have been shown to cause physical–chemical damage to human sperm. Vitrification is an alternative method that can eliminate ice crystallization and, thus, decrease the cryodamage. The earliest information on vitrification comes from as far back as 1937 [123, 124]. However, vitrification lost its appeal for many years because critical speeds of cooling (~700,000°K/min) were unachievable at that time. Recently, the technique has been revisited, as rapid cooling is now possible. Two conditions must be fulfilled for vitrification to occur: an increase in the viscosity and a depression of the freezing temperature. The cryoprotectants used here have this purpose: to act like antifreeze, lowering the freezing temperature and increasing the viscosity so that instead of crystallizing, the syrupy solution turns into an amorphous ice – i.e. it vitrifies.

One of the difficult compromises faced in vitrifying cryopreservation is limiting the damage produced by the cryoprotectant. The protocol of vitrification currently used for the sperm cryopreservation involves the use of very high concentrations (3.5–8 M) of permeating cryoprotectants and relatively high cooling rates (up to 10^{4}°K/min) [81]. It is known that high concentrations of cryoprotectants have a marked toxic effect [89, 102, 103]. It is possible to decrease cryoprotectant toxicity by reducing the amount of cryoprotectant and increasing freezing and thawing rates [125]. Cryoprotectant-free vitrification has been reported [81] with promising results in that no DNA damage was observed with either vitrification or conventional slow cooling. However, this study was performed on healthy volunteers. If it were used with infertile men with more vulnerable sperm, the damage might be greater (see previous section).

Conclusions and Future Recommendations

There is still much progress to be made in the field of semen cryopreservation. The most promising areas for research appear to be addition of antioxidants to cryomedia and sperm vitrification.

References

1. Mortimer D. Sperm recovery techniques to maximize fertilizing capacity. Reprod Fertil Dev. 1994; 6:25–31.
2. Donnelly ET, Lewis SE, McNally JA, et al. In vitro fertilization and pregnancy rates: the influence of sperm motility and morphology on IVF outcome. Fertil Steril. 1998;70:305–14.
3. Donnelly ET, McClure N, Lewis SE. Glutathione and hypotaurine in vitro: effects on human sperm motility, DNA integrity and production of reactive oxygen species. Mutagenesis. 2000;15:61–8.
4. Sakkas D, Manicardi GC, Tomlinson M, et al. The use of two density gradient centrifugation technique and the swim up method to separate spermatozoa with chromatin and nuclear DNA anomalies. Hum Reprod. 2000;15(5):1112–6.
5. Kanwar KC, Yanagimachi R, Lopata A. Effects of human seminal plasma on fertilizing capacity of human spermatozoa. Fertil Steril. 1979;31:321–7.
6. Rogers BJ, Perreault SD, Bentwood BJ, et al. Variability in the human–hamster in vitro assay for fertility evaluation. Fertil Steril. 1983;39:204–11.

7. van Loon AA, Den Boer PJ, van der Schans GP, et al. Immunochemical detection of DNA damage induction and repair at different cellular stages of spermatogenesis of the hamster after in vitro or in vivo exposure to ionizing radiation. Exp Cell Res. 1991;193:303–9.
8. van Loon AA, Sonneveld E, Hoogerbrugge J, et al. Induction and repair of DNA single-strand breaks and DNA base damage at different cellular stages of spermatogenesis of the hamster upon in vitro exposure to ionizing radiation. Mutat Res. 1993;294:139–48.
9. Drost JB, Lee WR. Biological basis of germline mutation: comparisons of spontaneous germline mutation rates among drosophila, mouse, and human. Environ Mol Mutagen. 1995;25:48–64.
10. Shekarriz M, DeWire DM, Thomas AJ, et al. A method of human semen centrifugation to minimize the iatrogenic sperm injuries caused by reactive oxygen species. Eur Urol. 1995;28:31–5.
11. Shekarriz M, Thomas AJ, Agarwal A. Incidence and level of seminal reactive oxygen species in normal men. Urology. 1995;45:103–7.
12. Griveau JF, Le Lannou D. Influence of oxygen tension on reactive oxygen species production and human sperm function. Int J Androl. 1997;20: 195–200.
13. Whittington K, Ford WC. The effect of incubation periods under 95% oxygen on the stimulated acrosome reaction and motility of human spermatozoa. Mol Hum Reprod. 1998;4:1053–7.
14. Aitken RJ, Buckingham DW, West KM. Reactive oxygen species and human spermatozoa: analysis of the cellular mechanisms involved in luminol- and lucigenin dependent chemiluminescence. J Cell Physiol. 1992;151:466–77.
15. Aitken RJ, Clarkson JS. Significance of reactive oxygen species and antioxidants in defining the efficiency of sperm preparation techniques. J Androl. 1998;9(6):367–76.
16. Sharma RK, Agarwal A. Role of reactive oxygen species in male infertility. Urology. 1996;48:835–50.
17. Agarwal A, Makker K, Sharma R. Clinical relevance of oxidative stress in male factor infertility: an update. Am J Reprod Immunol. 2008;59:2–11.
18. Mahfouz R, Sharma R, Thiyagarajan A, et al. Semen characteristics and sperm DNA fragmentation in infertile men with low and high levels of seminal reactive oxygen species. Fertil Steril. 2010;94:2141–6.
19. Abd-Elmoaty MA, Saleh R, Sharma R, et al. Increased levels of oxidants and reduced antioxidants in semen of infertile men with varicocele. Fertil Steril. 2009;94:1531–4.
20. Hughes CM, Lewis SE, McKelvey-Martin VJ, et al. The effects of antioxidant supplementation during Percoll preparation on human sperm DNA integrity. Hum Reprod. 1998;13:1240–7.
21. Twigg JP, Irvine DS, Aitken RJ. Oxidative damage to DNA in human spermatozoa does not preclude pronucleus formation at intracytoplasmic sperm injection. Hum Reprod. 1998;13:1864–71.
22. Tremellen K. Oxidative stress and male infertility–a clinical perspective. Hum Reprod Update. 2008; 14:243–58.
23. Halliwell B, Gutteridge JMC. Free radicals in biology and medicine. 2nd ed. Oxford: Clarendon; 1989.
24. Agarwal A, Said TM. Oxidative stress, DNA damage and apoptosis in male infertility: a clinical approach. BJU Int. 2004;95:503–7.
25. Donnelly ET, McClure N, Lewis SE. Antioxidant supplementation in vitro does not improve human sperm motility. Fertil Steril. 1999;72:484–95.
26. Menezo YJR, Hazout A, Panteix G, et al. Antioxidants to reduce sperm DNA fragmentation: an unexpected adverse effect. Reprod Biomed Online. 2007; 14(4):418–21.
27. Rolf C, Cooper TG, Yeung CH, et al. Antioxidant treatment of patients with asthenozoospermia or moderate oligoasthenozoospermia with high-dose vitamin C and vitamin E: a randomized, placebo-controlled, double-blind study. Hum Reprod. 1999;14:1028–33.
28. Li ZL, Lin QL, Liu RJ, et al. Reducing oxidative DNA damage by adding antioxidants in human semen samples undergoing cryopreservation procedure. Zhonghua Yi Xue Za Zhi. 2007;87:3174–7.
29. Lewis SE, Boyle PM, McKinney KA, et al. Total antioxidant capacity of seminal plasma is different in fertile and infertile men. Fertil Steril. 1995;64: 868–70.
30. Kessopoulou E, Powers HJ, Sharma KK, et al. A double-blind randomized placebo cross-over controlled trial using the antioxidant vitamin E to treat reactive oxygen species associated male infertility. Fertil Steril. 1995;64:825–31.
31. Baker HW, Brindle J, Irvine DS, et al. Protective effect of antioxidants on the impairment of sperm motility by activated polymorphonuclear leukocytes. Fertil Steril. 1996;65:411–9.
32. Battista N, Rapino C, Di Tommaso M, et al. Regulation of male fertility by the endocannabinoid system. Mol Cell Endocrinol. 2008;286:S17–23.
33. Sanger WG, Olson JH, Sherman JK. Semen cryobanking for men with cancer – criteria change. Fertil Steril. 1992;58:1024–7.
34. Sherman JK. Current status of clinical cryobanking of human semen. In: Paulson JD, Negro-Vlar A, Lucena E, Martini L, editors. Andrology: male fertility and sterility. Orlando: Academic; 1986. p. 517–47.
35. Anger JT, Gilbert BR, Goldstein M. Cryopreservation of sperm: indications, methods and results. J Urol. 2003;170:1079–84.
36. Borges E, Rossi LM, de Freitas CVL, et al. Fertilization and pregnancy outcome after intracytoplasmic injection with fresh or cryopreserved ejaculated spermatozoa. Fertil Steril. 2007;87:316–20.
37. Centola GM, Raubertas RF, Mattox JH. Cryopreservation of human semen. Comparison of cryopreservatives, sources of variability, and prediction of post-thaw survival. J Androl. 1992;13:283–8.

38. Nijs M, Ombelet W. Cryopreservation of human sperm. Hum Fertil. 2001;4(3):158–63.
39. Donnelly ET, McClure N, Lewis SE. Cryopreservation of human semen and prepared sperm: effects on motility parameters and DNA integrity. Fertil Steril. 2001;76:892–900.
40. Esteves SC, Sharma RK, Thomas AJ, et al. Suitability of the hypo-osmotic swelling test for assessing the viability of cryopreserved sperm. Fertil Steril. 1996;66:798–804.
41. Cross NL, Hanks SE. Effects of cryopreservation on human sperm acrosomes. Hum Reprod. 1991;6: 1279–83.
42. Mazur P, Rall WF, Leibo SP. Kinetics of water loss and the likelihood of intracellular freezing in mouse ova. Influence of the method of calculating the temperature dependence of water permeability. Cell Biophys. 1984;6:197–213.
43. Lewis SEM, McKinney KA, Thompson W. Influence of pentoxifylline on human sperm motility in asthenozoospermic individuals using computer-assisted analysis. Arch Androl. 1994;32(3):175–83.
44. Glenn DRJ, McVicar CM, McClure N, et al. Sidenafil citrate improves sperm motility but causes a premature acrosome reaction in vitro. Fertil Steril. 2007;87:1064–70.
45. Sharma RK, Wang JH, Wu Z. Mechanisms of inhibition of calmodulin-stimulated cyclic nucleotide phosphodiesterase by dihydropyridine calcium antagonists. J Neurochem. 1997;69(2):845–50.
46. Tournaye H, van der Linden M, van den Abbeel E, et al. Mouse in vitro fertilization using sperm treated with pentoxifylline and 2-deoxyadenosine. Fertil Steril. 1994;62:644–7.
47. Glenn DR, McClure N, Cosby SL, et al. Sildenafil citrate (Viagra) impairs fertilization and early embryo development in mice. Fertil Steril. 2009;91(3):893–9.
48. Fjällbrant B, Ackerman DR. Cervical mucus penetration in vitro by fresh and frozen-preserved human semen specimens. J Reprod Fertil. 1969;20:515–7.
49. O'Connell M, McClure N, Lewis SEM. The effects of cryopreservation on sperm morphology, motility and mitochondrial function. Hum Reprod. 2002;17:704–9.
50. Rossato M, La Sala GB, Balasini M, et al. Sperm treatment with extracellular ATP increases fertilization rates in in-vitro fertilization for male factor infertility. Hum Reprod. 1999;14:694–7.
51. Verheyen G, Nagy Z, Joris H, et al. Quality of frozen thawed testicular sperm and its preclinical use for intracytoplasmic sperm injection into in vitro-matured germinal-vesicle stage oocytes. Fertil Steril. 1997;67:74–80.
52. Sabanegh ES, Ragheb AM. Male fertility after cancer. Urology. 2009;73(2):225–31.
53. Watson PF. Recent developments and concepts in the cryopreservation of spermatozoa and the assessment of their post thawing function. Reprod Fertil Dev. 1995;7:871–91.
54. Duru NK, Morshedi MS, Schuffner A, et al. Cryopreservation-thawing of fractionated human spermatozoa is associated with membrane phosphatidylserine externalization and not DNA fragmentation. J Androl. 2001;22(4):646–51.
55. Virro MR, Larson-Cook KL, Evenson DP. Sperm chromatin structure assay (SCSA) parameters are related to fertilization, blastocyst development, and ongoing pregnancy in in vitro fertilization and intracytoplasmic sperm injection cycles. Fertil Steril. 2004;81:1289–95.
56. Smit M, Dohle GR, Hop WC, et al. Clinical correlates of the biological variation of sperm DNA fragmentation in infertile men attending an andrology outpatient clinic. Int J Androl. 2007;30(1):48–55.
57. Lin MH, Lee KKR, Li SH, et al. Sperm chromatin structure assay parameters are not related to fertilization rates, embryo quality, and pregnancy rates in in vitro fertilization and intracytoplasmic sperm injection, but might be related to spontaneous abortion rates. Fertil Steril. 2008;90:352–9.
58. Gandini L, Lombardo F, Lenzi A, et al. Cryopreservation and sperm DNA integrity. Cell Tissue Bank. 2006;7:91–8.
59. de Paula TS, Bertolla RP, Spaine DM, et al. Effect of cryopreservation on sperm apoptotic deoxyribonucleic acid fragmentation in patients with oligozoospermia. Fertil Steril. 2006;86:597–600.
60. Calamera JC, Buffone MG, Doncel GF, et al. Effect of thawing temperature on the motility recovery of cryopreserved human spermatozoa. Fertil Steril. 2010;93:789–94.
61. Kalthur G, Adiga SK, Upadhya D, et al. Effect of cryopreservation on sperm DNA integrity in patients with teratospermia. Fertil Steril. 2008;89:1723–7.
62. Zribi N, Chakroun FN, Euch EH, et al. Effects of cryopreservation on human sperm deoxyribonucleic acid integrity. Fertil Steril. 2010;93:159–66.
63. Thomson LK, Fleming SD, Schulke L, et al. The DNA integrity of cryopreserved spermatozoa separated for use in assisted reproductive technology is unaffected by the type of cryoprotectant used but is related to the DNA integrity of the fresh separated preparation. Fertil Steril. 2009;92:991–1001.
64. Dalzell LH, Thompson-Cree MEM, McClure N, et al. Effects of 24-hour incubation after freeze-thawing on DNA fragmentation of testicular sperm from infertile and fertile men. Fertil Steril. 2003;79(3):1670–2.
65. Pasqualotto FF, Sharma RK, Kobayashi H, et al. Oxidative stress in normospermic men undergoing infertility evaluation. J Androl. 2001;22:316–22.
66. Stanic P, Tandara M, Sonicki Z, et al. Comparison of protective media and freezing techniques for cryopreservation of human semen. Eur J Obstet Gynecol Reprod Biol. 2000;91:65–70.
67. Esteves SC, Sharma RK, Thomas AJ, et al. Evaluation of acrosomal status and sperm viability in fresh and cryopreserved specimens by the use of fluorescent peanut agglutinin lectin in conjunction with hypo-osmotic swelling test. Int Braz J Urol. 2007;33:364–74.

68. Nallella KP, Sharma RK, Allamaneni SS, et al. Cryopreservation of human spermatozoa: comparison of two cryopreservation methods and three cryoprotectants. Fertil Steril. 2004;82:913–8.
69. Sherman JK. Synopsis of the use of frozen human semen since 1964: state of the art of human semen banking. Fertil Steril. 1973;24:397–412.
70. Verheyen G, Pletincx I, van Steirteghem A. Effect of freezing method, thawing temperature and post-thaw dilution/washing on motility (CASA) and morphology characteristics of high-quality human sperm. Hum Reprod. 1993;8:1678–84.
71. Morris GJ. Rapidly cooled human sperm: no evidence of intracellular ice formation. Hum Reprod. 2006;21:2075–83.
72. Morris GJ, Goodrich M, Acton E, et al. The high viscosity encountered during freezing in glycerol solutions: effects on cryopreservation. Cryobiology. 2006;52:323–34.
73. Ablett S, Izzard MJ, Lillford PJ. Differential scanning calorimetric study of frozen sucrose and glycerol solutions. J Chem Soc Faraday Trans. 1992;88:789–94.
74. Denda M, Hosoi J, Asida Y. Visual imaging of ion distribution in human epidermis. Biochem Biophys Res Commun. 2000;272:134–7.
75. Hernandez JA, Cristina E. Modeling cell volume regulation in nonexcitable cells: the roles of the Na+ pump and of cotransport systems. Am J Physiol. 1998;275:1067–80.
76. Li LY, Tighe BJ, Ruberti JW. Mathematical modeling of corneal swelling. Biomech Model Mechanobiol. 2004;3:114–23.
77. Chen HH, Purtteman JJ, Heimfeld S, et al. Development of a microfluidic device for determination of cell osmotic behavior and membrane transport properties. Cryobiology. 2007;55:200–9.
78. Mazur P. Principles of cryobiology. In: Fuller BJ, Lane N, Benson EE, editors. Life in the frozen state. Boca Raton: CRC Press; 2004. p. 3–65.
79. Gao DY, Liu C, McGann LE, et al. Prevention of osmotic injury to human spermatozoa during addition and removal of glycerol. Hum Reprod. 1995;10:1109–22.
80. Gao D, Mazur P, Critser J. Fundamental cryobiology of mammalian spermatozoa. In: Karow AM, Critser JK, editors. Reproductive tissue banking. London: Academic; 1997. p. 263–328.
81. Isachenko E, Isachenko V, Katkov II, et al. DNA integrity and motility of human spermatozoa after standard slow freezing versus cryoprotectant-free vitrification. Hum Reprod. 2004;19:932–9.
82. Baumber J, Ball BA, Linfor JJ, et al. Reactive oxygen species and cryopreservation promote DNA fragmentation in equine spermatozoa. J Androl. 2003;24:621–8.
83. Print CG, Loveland KL. Germ cell suicide: new insights into apoptosis during spermatogenesis. Bioessays. 2000;22(5):423–30.
84. Kierszenbaum AL. Transition nuclear proteins during spermiogenesis: unrepaired DNA breaks not allowed. Mol Reprod Dev. 2001;58:357–8.
85. Sakkas D, Mariethoz E, John JC. Abnormal sperm parameters in humans are indicative of an abortive apoptotic mechanism linked to the Fas-mediated pathway. Exp Cell Res. 1999;251:350–5.
86. Sakkas D, Mariethoz E, Manicardi G, et al. Origin of DNA damage in ejaculated human spermatozoa. Rev Reprod. 1999;4:431–7.
87. Huszar G, Sbracia M, Vigue L, et al. Sperm plasma membrane remodeling during spermiogenic maturation in men: relationship among plasma membrane beat 1,4 galactosyltransferase, cytoplasmic creatine phosphokinase and creatine phosphokinase isoform ratios. Biol Reprod. 1997;56:1020–4.
88. Paasch U, Sharma RK, Gupta AK, et al. Cryopreservation and thawing is associated with varying extent of activation of apoptotic machinery in subsets of ejaculated human spermatozoa. Biol Reprod. 2004;71:1828–37.
89. Wundrich K, Paasch U, Leicht M, et al. Activation of caspases in human spermatozoa during cryopreservation–an immunoblot study. Cell Tissue Bank. 2006;7:81–90.
90. Martin G, Sabido O, Durand P, et al. Cryopreservation induces an apoptosis-like mechanism in bull sperm. Biol Reprod. 2004;71:28–37.
91. Thornberry NA, Lazebnik Y. Caspases: enemies within. Science. 1998;281:1312–6.
92. Weng SL, Taylor SL, Morshedi M, et al. Caspase activity and apoptotic markers in ejaculated human sperm. Mol Hum Reprod. 2002;8:984–91.
93. Marchetti C, Gallego MA, Defossez A, et al. Staining of human sperm with fluorochrome-labeled inhibitor of caspases to detect activated caspases: correlation with apoptosis and sperm parameters. Hum Reprod. 2004;19:1127–34.
94. Agarwal A, Ranganathan P, Kattal N, et al. Fertility after cancer: a prospective review of assisted reproductive outcome with banked semen specimens. Fertil Steril. 2004;81:342–8.
95. Sharma RK, Said T, Agarwal A. Sperm DNA damage and its clinical relevance in assessing reproductive outcome. Asian J Androl. 2004;6:139–48.
96. Paasch U, Grunewald S, Agarwal A, et al. Activation pattern of caspases in human spermatozoa. Fertil Steril. 2004;81:802–9.
97. Thomson LK, Fleming SD, Aitken RJ, et al. Cryopreservation-induced human sperm DNA damage is predominantly mediated by oxidative stress rather than apoptosis. Hum Reprod. 2009;24:2061–70.
98. Perez-Sanchez F, Cooper TG, Yeung CH, et al. Improvement in quality of cryopreserved spermatozoa by swim-up before freezing. Int J Androl. 1994;17:115–20.
99. Yogev L, Gamzu R, Paz G, Kleiman S, Botchan A, Hauser R, et al. Pre-freezing sperm preparation does

not impair thawed spermatozoa binding to the zona pellucida. Hum Reprod. 1999;14:114–7.
100. Royere D, Barthelemy C, Hamamah S, et al. Cryopreservation of spermatozoa: a 1996 review. Hum Reprod Update. 1996;2:553–9.
101. Hallak J, Sharma RK, Wellstead C, et al. Cryopreservation of human spermatozoa: comparison of TEST-yolk buffer and glycerol. Int J Fertil Womens Med. 2000;45:38–42.
102. Gilmore JA, Liu J, Gao DY, et al. Determination of optimal cryoprotectants and procedures for their addition and removal from human spermatozoa. Hum Reprod. 1997;12:112–8.
103. Critser JK, Huse-Benda AR, Aaker DV, et al. Cryopreservation of human spermatozoa. III. The effect of cryopreservation on motility. Fertil Steril. 1988;50:314–20.
104. Gao DY, Ashworth E, Watson PF, et al. Hyperosmotic tolerance of human spermatozoa: separate effects of glycerol, sodium chloride, and sucrose on spermolysis. Biol Reprod. 1993;49:112–23.
105. Hammerstedt RH, Graham JK, Nolan JP. Cryopreservation of mammalian sperm: what we ask them to survive. J Androl. 1990;11:73–88.
106. Padron OF, Brackett NL, Sharma RK, et al. Seminal reactive oxygen species, sperm motility and morphology in men with spinal cord injury. Fertil Steril. 1997;67:115–1120.
107. Lass A, Akagbosu F, Abusheikha N, et al. A programme of semen cryopreservation for patients with malignant disease in a tertiary infertility centre: lessons from 8 years' experience. Hum Reprod. 1998;13(11):3256–61.
108. O'Flaherty C, Hales BF, Chan P, et al. Impact of chemotherapeutics and advanced testicular cancer or Hodgkin lymphoma on sperm deoxyribonucleic acid integrity. Fertil Steril. 2010;94:1374–9.
109. Polcz TE, Stronk J, Xiong C, et al. Optimal utilization of cryopreserved human semen for assisted reproduction: recovery and maintenance of sperm motility and viability. J Assist Reprod Genet. 1998;15:504–12.
110. Rofeim O, Brown TA, Gilbert BR. Effects of serial thaw-refreeze cycles on human sperm motility and viability. Fertil Steril. 2001;75:1242–3.
111. Bandularatne E, Bongso A. Evaluation of human sperm function after repeated freezing and thawing. J Androl. 2002;23:242–9.
112. Zini A, Libman J. Sperm DNA damage: clinical significance in the era of assisted reproduction. CMAJ. 2006;175:495–500.
113. Wang X, Sharma RK, Sikka SC, et al. Oxidative stress is associated with increased apoptosis leading to spermatozoa DNA damage in patients with male factor infertility. Fertil Steril. 2003;80:531–5.
114. Gadea J, Gumbao D, Canovas S, et al. Supplementation of the dilution medium after thawing with reduced glutathione improves function and the in vitro fertilizing ability of frozen-thawed bull spermatozoa. Int J Androl. 2008;31:40–9.
115. Bilodeau JF, Chatterjee S, Sirard MA, et al. Levels of antioxidant defenses are decreased in bovine spermatozoa after a cycle of freezing and thawing. Mol Reprod Dev. 2000;55:282–8.
116. Peris SI, Bilodeau JF, Dufour M, et al. Impact of cryopreservation and reactive oxygen species on DNA integrity, lipid peroxidation, and functional parameters in ram sperm. Mol Reprod Dev. 2007;74:878–92.
117. Roca J, Rodriguez MJ, Gil MA, et al. Survival and in vitro fertility of boar spermatozoa frozen in the presence of superoxide dismutase and/or catalase. J Androl. 2005;26:15–24.
118. Beconi MT, Francia CR, Mora NG, et al. Effect of natural antioxidants on frozen bovine semen preservation. Theriogenology. 1993;40:841–51.
119. Chatterjee S, de Lamirande E, Gagnon C. Cryopreservation alters membrane sulfhydryl status of bull spermatozoa: protection by oxidized glutathione. Mol Reprod Dev. 2001;60(4):498–506.
120. Twigg J, Fulton N, Gomez E, et al. Analysis of the impact of intracellular reactive oxygen species generation on the structural and functional integrity of human spermatozoa: lipid peroxidation, DNA fragmentation and effectiveness of antioxidants. Hum Reprod. 1998;13(6):1429–36.
121. Kusakabe H, Szczygiel MA, Whittingham DG, et al. Maintenance of genetic integrity in frozen and freeze-dried mouse spermatozoa. Proc Natl Acad Sci USA. 2001;98:13501–6.
122. Ward MA, Kaneko T, Kusakabe H, et al. Long-term preservation of mouse spermatozoa after freeze-drying and freezing without cryoprotection. Biol Reprod. 2003;69(6):2100–8.
123. Luyet BJ, Hodapp R. Revival of frog spermatozoa vitrified in liquid air. Proc Meet Soc Exp Biol. 1938;39:433–4.
124. Luyet BJ. Differential staining for living and dead cells. Science. 1937;85:106.
125. Liebermann J, Tucker M, Graham J, et al. Blastocyst development after vitrification of multipronucleate zygotes using the flexipet denuding pipette (FDP). Reprod Biomed Online. 2002;4:148–52.

28 Structure of Chromatin in Human Sperm Bound to Hyaluronic Acid: The Benefits of PICSI Dish Mediated Sperm Selection

Gabor Huszar and Leyla Sati

Abstract

In the past two decades, the Huszar laboratory focused upon the objective biochemical markers of sperm cellular maturity and development, and the characteristics of sperm in which cellular development is arrested during spermatogenesis and spermiogenesis. In assays directed to sperm cellular maturity/development, several sperm attributes were studied, including (a) excess cytoplasm (arrested cytoplasmic extrusion), (b) low expression of the HspA2 chaperone protein, a protein of the synaptonemal complex that supports meiosis, as well as the delivery of essential components in developing sperm, such as DNA repair enzymes or chromatin elements, (c) sperm shape (affected by cytoplasmic extrusion and insertion of the tail) according to the criteria detected by the Metamorph computer-assisted program or by the Tygerberg strict morphology, and (d) the relationship between sperm nuclear events (such as presence of chromosomal aneuploidies and excess residual histones). The primary focus of this chapter is related to recent reports that studied human sperm chromatin complexity and sperm HA-binding capacity. The studies suggest that the numerical chromosomal aberrations and persistent histones, i.e., errors in the spermatogenetic and spermiogenetic phases of male germ cell development, are associated. This chromatin deficiency associated with high persistent histones and relative lack of protamines affects DNA folding and renders the sperm DNA more vulnerable. It appears that these two phenomena correspond with each other and, thus, is responsible for defining sperm DNA integrity, fertility, and sterility.

Keywords

Sperm chromatin • Hyaluronic acid and sperm • PICSI dish-mediated sperm selection • Sperm selection with PICSI

In the past two decades, the Huszar laboratory focused upon the objective biochemical markers of sperm cellular maturity and development, and the characteristics of sperm in which cellular development is arrested during spermatogenesis

G. Huszar (✉)
Director, Sperm Physiology Laboratory and Male Fertility Program, Department of Obstetrics, Gynecology and Reproductive Sciences, Yale University, School of Medicine, New Haven, CT, USA
e-mail: Gabor.Huszar@Yale.edu

A. Zini and A. Agarwal (eds.), *Sperm Chromatin: Biological and Clinical Applications in Male Infertility and Assisted Reproduction*, DOI 10.1007/978-1-4419-6857-9_28,

and spermiogenesis. In assays directed to sperm cellular development, several sperm attributes were studied, including (a) excess cytoplasm (arrested cytoplasmic extrusion), (b) low expression of the HspA2 chaperon protein, a protein of the synaptonemal complex that supports meiosis, as well as the delivery of essential component in developing sperm, such as DNA repair enzymes or chromatin elements, (c) sperm shape (affected by cytoplasmic extrusion and insertion of the tail) according to the criteria detected by the Metamorph computer-assisted program or by the Tygerberg strict morphology, and (d) the relationship between sperm nuclear events (such as presence of chromosomal aneuploidies and excess residual histones). This last approach was made possible by the discovery that the morphology is maintained in sperm that was decondensed and denatured for DNA testing with fluorescence in situ hybridization (FISH) or for DNA breaks according to the DNA nick translation method [1].

In line with these studies, we have shown that spermatozoa of arrested development, which also contain surplus cytoplasm and exhibit consequential abnormal morphology, have diminished ability to bind to zona pellucida or to hyaluronic acid (HA) [2, 3]. Conversely, spermatozoa that are able to bind to solid-state HA are fully developed and are devoid of cytoplasmic retention, excess persistent histones, apoptotic processes, DNA chain fragmentation, the apoptotic marker of Caspase 3, and show a normal frequency of chromosomal aneuploidies [4, 5]. Thus, we formulated the hypothesis that HA binding is also related to sperm development.

The primary focus of this chapter is related to recent reports from our laboratory that studied human sperm chromatin complexity and sperm HA-binding capacity. The studies suggest that the numerical chromosomal aberrations and persistent histones, i.e., errors in the spermatogenetic and spermiogenetic phases of male germ cell development, are related [6]. We observed that sperm with high levels of retained histone retention (darkly stained with aniline blue) also exhibited a high degree of DNA fragmentation (most likely due to the vulnerability of DNA chains as a consequence of high histone content, thus lower protamine levels). This chromatin deficiency with high persistent histones and relative lack of protamines affects DNA folding and renders the sperm DNA more vulnerable. It appears that these two phenomena correspond with each other and, thus, is responsible for defining sperm DNA integrity, fertility, and sterility. Sperm with persistent histones that stained darkly with aniline blue showed no FISH signal. Our explanation is as follows: owing to the high histone content, the sperm DNA is more vulnerable, and thus, the DNA fragments are of limited size and are unable to attach\retain the FISH chromosome probes [7]. In another recent study, we have found an inverse correlation between the proportion of sperm with dark aniline blue staining and curvilinear sperm velocity [7].

Sperm Chromatin Maturation and Its Importance

The formation of mature spermatozoa is a unique process involving a series of meiotic and mitotic changes in both the nuclear and cytoplasmic compartments including the histone–transition protein–protamine replacement. In this process, first somatic histones are replaced by testis-specific histone variants, which are then replaced by transition proteins (TP1 and TP2) in a process that involves extensive DNA rearrangement and remodeling [8]. During the final postmeiotic phases of spermatogenesis, more than tenfold compaction of sperm chromatin structure is achieved when almost 85% of the histones are replaced by protamines (protamine 1 and protamine 2) [9–11]. Finally, sperm chromatin becomes a highly organized compact structure consisting of DNA and heterogeneous nucleoproteins. This occurs in a specific manner that keeps the chromatin in the nucleus stable and packed with a special type of small, basic protein in a tight, almost crystalline status [12].

The entire sperm genome is organized into DNA loop domains that have an average length of 27 kDa; and are attached at their bases to a structural element within the sperm nucleus known as the nuclear matrix. Therefore, the DNA remains anchored to the base when the human sperm undergoes decondensation process. Such DNA organization is important not only for the transfer of highly packaged genetic information to the oocyte but also to ensure that the DNA is delivered in a physical and chemical form that allows the developing embryo to access the genetic information [13].

The most abundant nucleoproteins in mature sperm are the positively charged protamines. Protamines are highly basic proteins about half the size of a typical histone [12]. Arginines are represented as 55–79% of the amino-acid residues in protamines, permitting a strong DNA binding [14]. In mammalian spermatozoa, it is suggested that the DNA packaging process in human spermatozoa is not completed during sperm production in the testis. Further condensation occurs in the majority of the nuclei as they pass through the epididymal lumen [15]. Thus, sperm chromatin occupies almost the entire nuclear volume, whereas somatic cell DNA only partially fills the nucleus.

Earlier studies showed an association between diminished histone–transition protein–protamine exchange that may be detected by aniline blue staining of the excess persistent lysine-rich histones [6, 16–20]. Accordingly, based on the variations in sperm maturity, a polymorphic pattern of the sperm staining intensity was found with aniline blue staining, depicted as light, intermediate, and dark patterns that represent sperm with mature, moderately immature, and severely immature developmental and maturation states, respectively [4, 21].

It is clear that sperm chromatin structure and the accurate transition of the histone–transition protein–protamine sequence is essential for sperm function and subsequent embryonic development because defects in sperm chromatin are linked to natural reproductive malfunctions, including spontaneous abortion as failure in assisted reproduction attempts [22–24]. However, the requirement for TPs in during this process is not clear, since spermatogenesis in TP null mice is shown to be almost normal, fertile with testis weights and epididymal sperm counts being unaffected, albeit with smaller litter sizes. The studies suggest that each TP performs some exclusive function during spermiogenesis, even though sperm phenotypes strongly indicate that defects occur largely because of a selective deficiency of gene products [25–27]. It has been suggested that DNA damage is the main cause of implantation failure in embryos derived from healthy eggs fertilized by sperm with chromatin defects [28, 29].

This idea is further supported by experimental findings in our laboratory. As described above, there was a deficiency of FISH probe binding in spermatozoa that stained dark with aniline blue, which reflects high levels of persistent histones in diminished maturity ejaculated spermatozoa [7]. Thus, the arrest of chromatin development and improper DNA folding causes consequential increased DNA chain fragmentation. The lack of FISH signal in these spermatozoa, which are frequent in semen of oligozoospermic men, casts doubt on some of the data related to disomies and diploidies reported in men with male factor infertility. In addition, environmental stress, gene defects, and chromosomal abnormalities can disturb critical biochemical compaction processes that occur during spermatogenesis, and may also cause an abnormal chromatin structure that would finally interfere with fertility [30–33].

Impact of Sperm Chromatin Maturation and Imprinting

Haploid male germ cells contain only 10% volume of DNA compared to somatic cells. This is due to the supercompaction of the sperm chromatin along with the replacement of histones with protamines. Protamine replacement may also be necessary for silencing of the paternal genome and reprogramming of the imprinting pattern of the gamete [34, 35]. Imprinted genes play important roles in embryo development, placental function, neurological processes, individual behavior patterns, and cancer. However, the suggested reports about the imprinting defects in cases of disrupted spermatogenesis raised the possibility that they could be directly associated with infertility [36–38].

The process of histone replacement with transition proteins and protamines include side chain modification of the histones, for instance, with methylation and acetylation. The latter addition provides additional charges that reduce the histone's affinity to DNA. Several studies have assessed how amino-acid side chain methylation levels in chromatin may affect IVF success in terms of both fertilization and pregnancy rates [39–42]. Some of these data support the notion that children conceived by ART do not show a higher degree of imprinting variability and do not have a higher risk for imprinting disorders. Others suggested that the extent of sperm DNA methylation does not influence the fertilization rates but does adversely influence embryo development if the aggregate DNA methylation level is below a threshold value [43, 44]. However, the details of how methylation is regulated during the process of chromatin evolution and sequential nucleoprotein replacement are not clearly understood. Thus, in mammalian spermatozoa chromatin exhibits a special compaction pattern that may provide a novel epigenetic signature that contributes to the developmental competence of the embryo and enhances the normal embryo development.

Another functional advantage of the highly compacted DNA package that it reduces the size of the sperm nucleus and head, and the smaller head size is related to more efficient sperm motility and velocity. The above-mentioned recent data from our laboratory supports that related factors of the lack of persistent histones, compaction of chromatin, and more optimal head shape contribute to the movement of the sperm and to the efficiency of fertilization by increasing the frequency of sperm–oocyte encounters [7].

Relationship Between the Nuclear and Cytoplasmic Aspects of Sperm Mismaturation

A further dimension related to chromatin maturation/remodeling has emerged with recent studies [6]. Double-stained human spermatozoa, first with aniline blue and, after recording the sperm, with a separate second probe for the same sperm, provided evidence for relationships between the various biochemical markers of maturity/mismaturity. The methods used and the experimental process are summarized in Fig. 28.1. The biochemical attributes within the data pairs studied in the same sperm included (a) aniline blue staining (residual histones) and creatine kinase immunocytochemistry (cytoplasmic retention, Fig. 28.2), (b) aniline blue staining and Caspase 3 immunostaining (apoptotic process in the sperm, Fig. 28.3), (c) aniline blue staining and DNA nick translation (DNA chain fragmentation/integrity, Fig. 28.4), and finally (d) aniline blue staining and Tygerberg normal sperm morphology (Fig. 28.5). When the sperm were scored for various nuclear and cytoplasmic attributes staining, there was an >70% agreement between the patterns of marker pair staining within the same spermatozoa. This is a reverse aspect of the notion that sperm cells with no persistent histone retention, and thus no arrest of chromatin development, will also fail to display other attributes of developmental/maturation arrest, such as cytoplasmic retention or abnormal Tygerberg morphology.

These data indicate that the development/dysmaturity biochemical markers are related within the same sperm, and thus, the regulation of chromatin remodeling or the disturbed process of chromatin remodeling such as evolution from DNA–histone complexes to DNA–transition protein or DNA–protamine complexes is related to other attributes of sperm development or developmental arrest. In a global view, disordered evolution of chromatin remodeling and associated problems are likely related to upstream defects of spermatogenesis or spermiogenesis, for instance, the low expression of the functionally important HspA2 chaperone, which supports meiosis via the synaptonemal complex, the delivery of cellular building elements and DNA repair enzymes, and several other important functions [45, 46]. Low levels of HspA2 have predicted the lack of IVF pregnancies in two previous studies, independently from normal sperm concentration and motility in the husband's semen [46, 47].

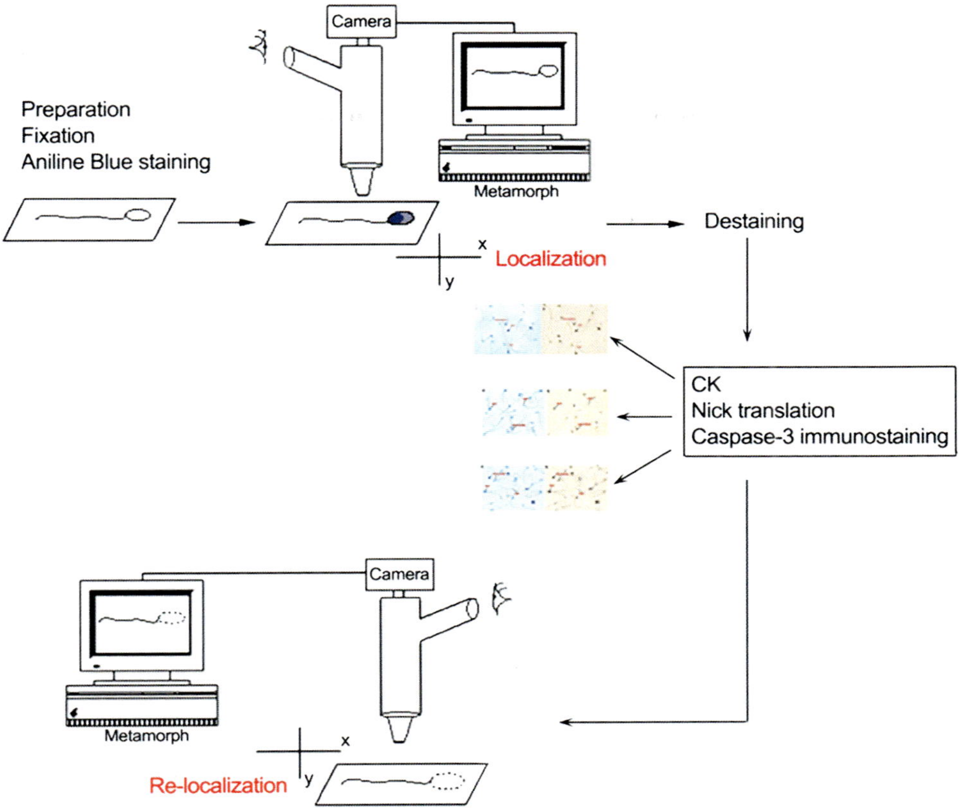

Fig. 28.1 Flowchart of the experimental design. Sperm are treated with aniline blue, stained fields are recorded, aniline blue is destained, second cytoplasmic or nuclear probes are applied anew (CK – creatine kinase, nick translation – DNA chain integrity, Caspase 3 – apoptosis, etc.), sperm field is localized and recorded with the second probe

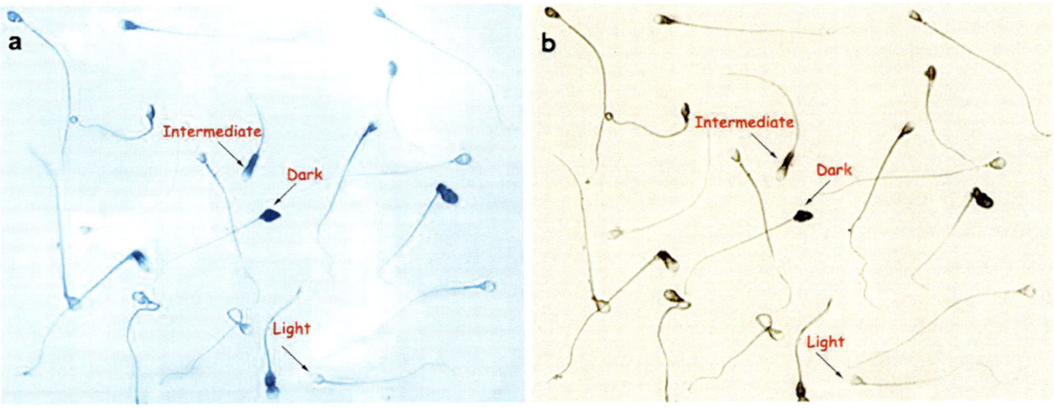

Fig. 28.2 (**a**, **b**) Aniline blue staining and creatine kinase (CK)-immunostaining of the same spermatozoa field. Note the substantial degree of similarity in the light-, intermediate-, and dark-staining patterns with aniline blue and CK. Based on evaluation of 1,284 sperm image pairs (samples of four men), we established 82% conforming staining patterns (light, intermediate, or dark) with the two biochemical probes

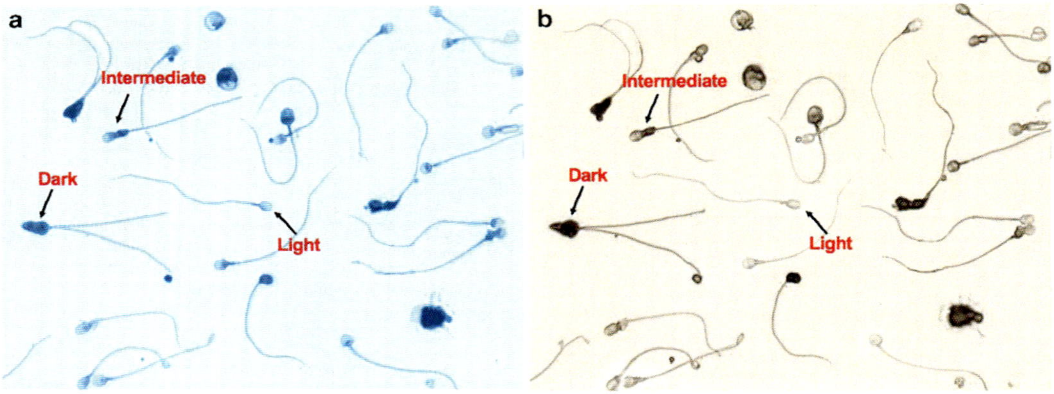

Fig. 28.3 (**a**, **b**) Aniline blue staining and Caspase 3 immunostaining of the same spermatozoa field. Based on the evaluation of a total of 2,101 spermatoza pairs (originating in four samples), there were approximately 85% conforming staining patterns with aniline blue and the apoptotic marker

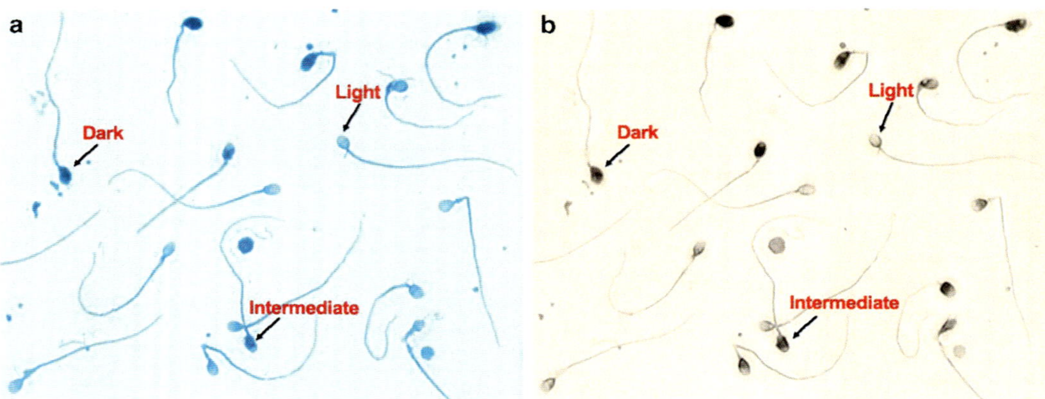

Fig. 28.4 (**a**, **b**) Aniline blue staining and DNA nick translation, probing for DNA chain integrity, of the same spermatozoa. Data on 2,446 spermatozoa indicate an 84% agreement in staining pattern. The light spermatozoa with no persistent histones showed high DNA chain integrity with no staining, whereas the spermatozoa with dark aniline blue staining, reflecting increased levels of persistent histones, exhibited substantial degree of DNA chain fragmentation

Confirming results from other laboratories [28, 29] indicated that protamine insufficiency (which is corollary to our finding of persistent histones in ejaculated sperm) is related to DNA chain breaks and irregular DNA repair (which is likely to be related to our finding of low HspA2 chaperone activity). In the Aioki and Ramos papers, the authors indicate that the sperm with the biochemical signs of immaturity (complementary to our findings) are deficient in the process of fertilization and the paternal contributions/biochemical markers support of the embryo [28, 29].

These independent but related findings are very important from the point of view of this chapter in which we discuss the advantages of HA-mediated sperm selection for ICSI. The HA-binding-mediated selection of spermatozoa yields sperm that are clearly devoid of attributes of chromatin or cytoplasmic developmental arrest as well as DNA chain fragmentation, which are otherwise related to arrested sperm development/maturation.

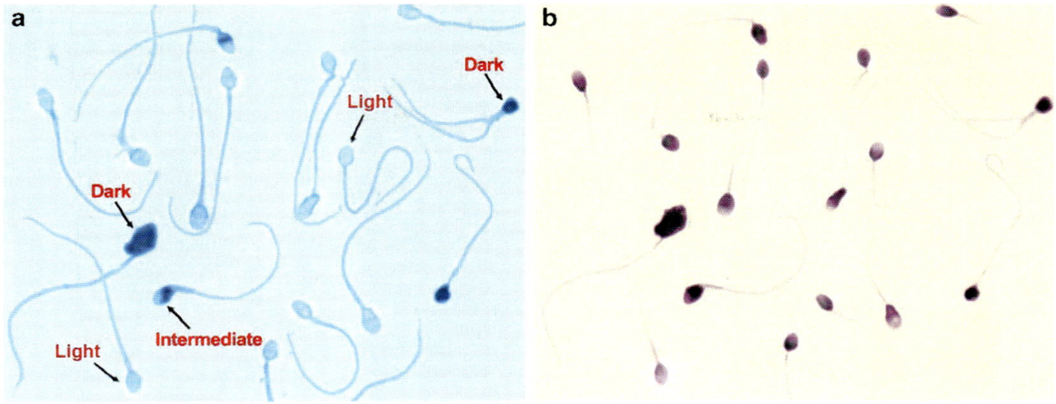

Fig. 28.5 (**a**, **b**) Aniline blue staining and sperm morphology evaluated according to the Tygerberg strict criteria. Sperm dysmaturity and abnormal morphology are related because cytoplasmic retention is an underlying factor of abnormal sperm head and midpiece shape, and abaxial insertion of the tail. Also, the shorter sperm tail, a characteristic of dysmature sperm, is an adverse component of the Tygerberg classification. This figure, based on 3,882 sperm image pairs, showed significant differences in the proportion of normal sperm between the light and intermediate staining groups ($p<0.05$), and the intermediate vs. dark staining groups ($p<0.01$). Thus, there is a relationship between persistent histones and sperm shape as detected by the Tygerberg criteria

Sperm–Hyaluronic Acid (HA) Binding: Spermatozoa with Full Cellular Development Selectively Binds to Solid-State HA

Another important marker of normal sperm development is the sperm plasma membrane remodeling during spermiogenesis, which facilitates the expression of the receptors for zona pellucida and HA [48]. Thus, sperm that did not go through the remodeling process do not recognize zona pellucida or HA. Conversely, all the data that have been developed so far indicate that the HA-bound spermatozoa exhibit nuclear, cytoplasmic, and shape properties identical to those bound to the zona pellucida of oocytes. Indeed, a study of about 60 semen samples bound to both hemizonae and HA showed a significant correlation between the binding to the two entities at $r=0.76$, $p<0.001$ [2, 49]. We feel that the correlation would even be closer if some of the hemizonae would not originate in unfertilized oocytes, thus with uncertain sperm binding properties. The slides used in the sperm-HA binding tests are of uniform quality, with a very low (<5%) intraassay variation.

When we applied several biochemical sperm probes, the data have also revealed that the sperm membrane remodeling process is inherently related to upstream spermatogenetic and spermiogenetic events, For instance, sperm with cytoplasmic retention (attribute of arrested cellular maturation) and persistent histones with dark aniline blue staining, showed staining with biochemical and apoptotic markers, such as LDH-C_4 Caspase 3, DNA fragmentation, and abnormal sperm shape [4].

Furthermore, in an advanced version of these experiments we have attached semen aliquots to glass slides and fixed the sperm for various markers. Another semen aliquot was incubated on HA-coated slides, the unbound sperm were gently rinsed off, and both the whole semen fraction on glass slides and the HA-bound sperm fraction on the HA-coated slides were stained with various sperm biochemical markers. In the semen sperm fraction, we have found sperm with cytoplasmic retention, DNA degradation (detected in individual spermatozoa with DNA-nick translation), aberrant sperm shape, persistent histones with aniline blue staining, whereas in the HA-bound sperm fraction, there was no presence of sperm with any of the cytoplasmic or nuclear defects [4, 49].

Another interesting line of research revealed that the HA-bound sperm fractions (evaluated by three blinded investigators) were enriched in sperm with normal Tygerberg morphology. Furthermore, the degree of enrichment was comparable to that of the rate reported by the Tygerberg group with respect to the improved proportion of normally shaped sperm in the zona-pellucida-bound sperm fractions vs. that in the respective semen [50, 51].

Does Sperm HA-Binding Test Predict DNA Chain Integrity in Bound Sperm?

Another recent example for similarity in zona-pellucida-bound and HA-bound sperm was developed by acridine orange staining probe, which provides green fluorescence for DNA with high chain integrity and orange fluorescence for sperm with damaged DNA. It was reported that zona-pellucida-bound sperm has mostly green fluorescence [52]. We performed this assay with sperm bound to the HA-spot of the PICSI dish (Origio-Midatlantic Inc., Mt. Laurel, NJ), which is used for ICSI sperm selection. Our finding indicated, using the very fine Polyscience Inc. acridine orange reagent, that virtually all of the HA-bound sperm exhibited green fluorescence [53]. Thus, whether probing sperm DNA with nick translation or with acridine orange, the DNA of HA-bound sperm had high DNA integrity, and no attributes of arrested sperm cellular maturation, such as cytoplasmic retention, persistent histones, or apoptotic markers, were detectable.

In addition to the DNA integrity, there is now focus on the increase in chromosomal anuploidies, which within the ICSI offspring are 3–4 times elevated, if the sperm used for ICSI are eye-selected by the embryologist. Pointing out the relationship between meiotic and late-spermiogenetic events, we have shown that sperm with consequential cytoplasmic retention defects also have increased frequencies of chromosomal aneuploidies with a substantial correlation and significance, e.g., sperm with cytoplasmic retention and frequency of Y chromosome disomy: $r=0.78$, $p<0.001$ [54]. Further filtering effect of the zona pellucida has been reconstructed and tested by HA binding. No matter how high the aneuploidy frequency was in the semen sperm fraction, sperm bound and removed from HA had 4–6× lower disomy and diploidy frequencies within the 0.1–0.2% normal range, which is customary in babies conceived with natural conception or with conventional IVF conception [5]. Thus, the PICSI dish seems to be an ideal platform for ICSI sperm selection [5].

Regarding the relationship between excess persistent histones and improperly packed vulnerable DNA in sperm, we have published recently that in sperm with solid aniline blue staining, indicating the failure in the histone–intermediate protein–protamine cycle and high levels of persistent histones, there were no signals after testing the sperm for aneuploidies using FISH. This indicated that in sperm with arrested chromatin maturation and fragmented DNA, there were no DNA sites left with chain length sufficient for the binding of the FISH chromosomal markers [7].

Sperm Chromatin and Sperm Cellular Development

In considering the key questions of sperm chromatin and cellular development/maturity studies, we should turn our attention to two basic questions: (1) Is there an interrelationship between arrest of sperm cellular development at the level of excess histone retention and other attributes of nuclear and cytoplasmic defects in human spermatozoa? (2) Considering the various phases of spermatogenesis and spermiogenesis, is there a relationship between the nuclear and cytoplasmic biochemical markers that would support the idea that some spermiogenetic or other late-appearing defects have an upstream spermatogenetic origin?

(1) Regarding the relationship between maturational arrest and a defect of the histone–transition protein–protamine replacement (which also has consequences in DNA folding and DNA chain vulnerability) that manifests in strong aniline blue staining, due to excess histone retention, the

answer is yes. There are several sperm attributes associated with arrested cellular maturation.

These include the following: (a) Cytoplasmic retention, which is measured as sperm creatine kinase activity. High sperm CK activity in semen of oligozoospermic men treated with intrauterine insemination has diminished occurrence of pregnancies, independently from sperm concentration and motility in their semen [55]; (b) The low expression of the HspA2 chaperone protein. Indeed, men with low sperm HspA2 levels failed to achieve pregnancy in couples treated with conventional IVF in two studies, one in a blinded Yale-Norfolk collaboration (84 couples) [46], and one in a Yale IVF study (119 couples) [47]; (c) Men with sperm cytoplasmic retention and low HspA2 expression had a higher incidence of sperm with aniline blue staining, indicating elevated content of histones; (d) In semen samples with decreased frequency of normal Tygerberg shaped spermatozoa, there were also increased levels of sperm creatine kinase and aniline blue staining; (e) Sperm with cytoplasmic retention have a higher rate of aneuploidy, which actually-showed a statistically significant relationship (i.e., sperm with cytoplasmic retention vs. Y disomy: $r=0.78$, $p=<0.001$) [54]. Further, with the establishment that sperm after the decondensation step, necessary for FISH or DNA integrity studies, maintain their initial shape as it was in semen, we could also demonstrate the relationship between the association of sperm shape and aneuploidies within the same spermatozoa [1].

Conversely, study of HA-bound spermatozoa (signifying that the bound spermatozoa completed spermiogenetic plasma membrane remodeling) has indicated that the HA-bound sperm lacked any attributes of arrested sperm development, such as cytoplasmic retention, persistent histones, DNA fragmentation (detected by two methods: in situ DNA nick translation, and acridine orange fluorescence), increased frequency of chromosomal aneuploidies, regardless how elevated the aneuploidy and diploidy levels were in the semen sperm fraction. Further, testing sperm in semen and in the respective HA-bound sperm fraction indicted that within the HA-bound sperm fraction there was an improved Tygerberg strict morphology (approximately 3× enrichment of normal sperm vs. the semen sperm population). This improvement has corresponded with the improvement in zona-pellucida-bound spermatozoa when compared to the respective semen samples [51].

Conclusions and Overview

First, regarding the validation of the attributes discussed above, there are two points of interest. First, in a study comparing sperm binding to hemizonae and HA, there was a high correlation and a significant relationship ($r=0.76$, $p<0.001$), which validates the HA-binding assay and reinforces the idea that the formation of the zona pellucida and HA receptors are related during the spermiogenetic plasma membrane remodeling. The location of these receptors is also common in the acrosomal region, as sperm binding to the zona pellucida and HA follows a common head-first pattern. Second, it has recently been published that sperm with dark aniline blue staining show no DNA staining with probes for in situ fluorescence hybridization, or with the DNA probe [7]. These data indicate that sperm with high levels of persistent histones and diminished protamine content suffers major DNA chain fragmentations, and thus, the FISH probes are unable to bind, and the fragmented DNA dissipates from sperm during the multiple steps of the FISH process.

Second, with respect to the interrelationship between the biochemical markers of arrested sperm development and chromatin maturation, there is a major connection [6]. The double-stained sperm method in which sperm cells were stained with aniline blue for probing persistent histones, and after recording the fields, the same sperm were probed for sperm shape, for cytoplasmic retention with creatine kinase immunostaining, for apoptotic process with Caspase 3 immunostaining, and for DNA chain fragmentation with in situ DNA nick translation (Figs. 28.1–28.5).

The evaluation of the same double-stained sperm with the first and second probes showed an approximately 75% agreement between the staining patterns whether light (no probe presence), intermediate (some probe detection), and dark staining (heavy probe presence). This indicated that, indeed, there is a relationship between the attributes of incomplete development in the same sperm. Also, the experiment demonstrated that the nuclear and cytoplasmic as well as the early and late events of spermatogenesis and spermiogenesis are related. The data strongly support the idea that the later manifestations of arrested cellular maturation of spermatozoa are related and may originate from common upstream events of early spermatogenesis [6].

Third, the data and ideas presented in this review support the validity of the HA-mediated sperm selection for ICSI by the HA-coated PICSI dish. This method is unparalleled when compared to other sperm selection methods, for three reasons: (a) The research base of the sperm plasma membrane remodeling during terminal spermiogenesis and the common origin of the formation of the receptors for zona pellucida and HA are novel ideas and are well supported by solid work of various lines. (b) The detailed characterization of HA-bound spermatozoa with respect to the lack of cytoplasmic retention, lack of persistent histones, lack of Tygerberg normal morphology, lack of apoptotic processes, and the close correlation of the rate of binding to zona pellucida or HA by sperm in the same semen sample, all point to the high level of similarity between zona pellucida- and HA-selected spermatozoa [3, 5, 6, 21, 49, 51].

(c) The genetic properties of the HA-bound sperm, with respect to the lack of DNA fragmentation tested with the methods of DNA nick translation and acridine orange fluorescence methods, and the data indicating the frequencies of chromosomal aneuploidies in the normal range, no matter how high the rates were in the original semen sperm population, all support the notion that the HA-selected spermatozoa are equivalent to those sperm bound to and selected by zona pellucida. Thus, HA-mediated sperm selection provides a method for performing ICSI and initiating fertilization with the selected sperm that have no DNA fragmentation or chromatin aberrations, and comparable to those fertilizing sperm selected by the zona pellucida under physiological or conventional IVF conditions [5, 53, 54].

References

1. Celik-Ozenci C, Catalanotti J, Jakab A, Aksu C, Ward D, Bray-Ward P, et al. Human sperm maintain their shape following decondensation and denaturation for fluorescent in situ hybridization: shape analysis and objective morphometry. Biol Reprod. 2003;69:1347–55.
2. Cayli S, Jakab A, Ovari L, Delpiano E, Celik-Ozenci C, Sakkas D, et al. Biochemical markers of sperm function: male fertility and sperm selection for ICSI. Reprod Biomed Online. 2003;7:462–8.
3. Huszar G, Vigue L, Oehninger S. Creatine kinase immunocytochemistry of human sperm-hemizona complexes: selective binding of sperm with mature creatine kinase-staining pattern. Fertil Steril. 1994;61:136–42.
4. Cayli S, Sakkas D, Vigue L, Demir R, Huszar G. Cellular maturity and apoptosis in human sperm: creatine kinase, caspase-3 and Bcl-XL levels in mature and diminished maturity sperm. Mol Hum Reprod. 2004;10:365–72.
5. Jakab A, Sakkas D, Delpiano E, Cayli S, Kovanci E, Ward D, et al. Intracytoplasmic sperm injection: a novel selection method for sperm with normal frequency of chromosomal aneuploidies. Fertil Steril. 2005;84:1665–73.
6. Sati L, Ovari L, Bennett D, Simon SD, Demir R, Huszar G. Double probing of human spermatozoa for persistent histones, surplus cytoplasm, apoptosis and DNA fragmentation. Reprod Biomed Online. 2008;16:570–9.
7. Ovari L, Sati L, Stronk J, Borsos A, Ward DC, Huszar G. Double probing individual human spermatozoa: aniline blue staining for persistent histones and fluorescence in situ hybridization for aneuploidies. Fertil Steril. 2010;93:2255–61.
8. Ward WS, Coffey DS. DNA packaging and organization in mammalian spermatozoa: comparison with somatic cells. Biol Reprod. 1991;44:569–74.
9. Braun RE. Packaging paternal chromosomes with protamine. Nat Genet. 2001;28:10–2.
10. Dadoune JP. The nuclear status of human sperm cells. Micron. 1995;26:323–45.
11. Steger K. Transcriptional and translational regulation of gene expression in haploid spermatids. Anat Embryol (Berl). 1999;199:471–87.
12. Fuentes-Mascorro G, Serrano H, Rosado A. Sperm chromatin. Arch Androl. 2000;45:215–25.
13. Sakkas D, Mariethoz E, Manicardi G, Bizzaro D, Bianchi PG, Bianchi U. Origin of DNA damage in

ejaculated human spermatozoa. Rev Reprod. 1999; 4:31–7.
14. Balhorn R, Brewer L, Corzett M. DNA condensation by protamine and arginine-rich peptides: analysis of toroid stability using single DNA molecules. Mol Reprod Dev. 2000;56:230–4.
15. Golan R, Cooper TG, Oschry Y, Oberpenning F, Schulze H, Shochat L, et al. Changes in chromatin condensation of human spermatozoa during epididymal transit as determined by flow cytometry. Hum Reprod. 1996;11:1457–62.
16. Auger J, Mesbah M, Huber C, Dadoune JP. Aniline blue staining as a marker of sperm chromatin defects associated with different semen characteristics discriminates between proven fertile and suspected infertile men. Int J Androl. 1990;13:452–62.
17. Dadoune JP, Mayaux MJ, Guihard-Moscato ML. Correlation between defects in chromatin condensation of human spermatozoa stained by aniline blue and semen characteristics. Andrologia. 1988;20:211–7.
18. Foresta C, Zorzi M, Rossato M, Varotto A. Sperm nuclear instability and staining with aniline blue: abnormal persistence of histones in spermatozoa in infertile men. Int J Androl. 1992;15:330–7.
19. Morel F, Mercier S, Roux C, Elmrini T, Clavequin MC, Bresson JL. Interindividual variations in the disomy frequencies of human spermatozoa and their correlation with nuclear maturity as evaluated by aniline blue staining. Fertil Steril. 1998;69:1122–7.
20. Morel F, Roux C, Bresson JL. Disomy frequency estimated by multicolour fluorescence in situ hybridization, degree of nuclear maturity and teratozoospermia in human spermatozoa. Reproduction. 2001;121: 783–9.
21. Huszar G, Ozenci CC, Cayli S, Zavaczki Z, Hansch E, Vigue L. Hyaluronic acid binding by human sperm indicates cellular maturity, viability, and unreacted acrosomal status. Fertil Steril. 2003;79 Suppl 3:1616–24.
22. Bungum M, Humaidan P, Axmon A, Spano M, Bungum L, Erenpreiss J, et al. Sperm DNA integrity assessment in prediction of assisted reproduction technology outcome. Hum Reprod. 2007;22:174–9.
23. Cebesoy FB, Aydos K, Unlu C. Effect of sperm chromatin damage on fertilization ratio and embryo quality post-ICSI. Arch Androl. 2006;52:397–402.
24. Filatov MV, Semenova EV, Vorob'eva OA, Leont'eva OA, Drobchenko EA. Relationship between abnormal sperm chromatin packing and IVF results. Mol Hum Reprod. 1999;5:825–30.
25. Shirley CR, Hayashi S, Mounsey S, Yanagimachi R, Meistrich ML. Abnormalities and reduced reproductive potential of sperm from Tnp1- and Tnp2-null double mutant mice. Biol Reprod. 2004;71:1220–9.
26. Zhao M, Shirley CR, Hayashi S, Marcon L, Mohapatra B, Suganuma R, et al. Transition nuclear proteins are required for normal chromatin condensation and functional sperm development. Genesis. 2004;38:200–13.
27. Zhao M, Shirley CR, Mounsey S, Meistrich ML. Nucleoprotein transitions during spermiogenesis in mice with transition nuclear protein Tnp1 and Tnp2 mutations. Biol Reprod. 2004;71:1016–25.
28. Aoki VW, Moskovtsev SI, Willis J, Liu L, Mullen JB, Carrell DT. DNA integrity is compromised in protamine-deficient human sperm. J Androl. 2005;26: 741–8.
29. Ramos L, van der Heijden GW, Derijck A, Berden JH, Kremer JA, van der Vlag J, et al. Incomplete nuclear transformation of human spermatozoa in oligo-astheno-teratospermia: characterization by indirect immunofluorescence of chromatin and thiol status. Hum Reprod. 2008;23:259–70.
30. Aoki VW, Liu L, Jones KP, Hatasaka HH, Gibson M, Peterson CM, et al. Sperm protamine 1/protamine 2 ratios are related to in vitro fertilization pregnancy rates and predictive of fertilization ability. Fertil Steril. 2006;86:1408–15.
31. Carrell DT, Emery BR, Hammoud S. Altered protamine expression and diminished spermatogenesis: what is the link? Hum Reprod Update. 2007; 13:313–27.
32. Evenson DP, Larson KL, Jost LK. Sperm chromatin structure assay: its clinical use for detecting sperm DNA fragmentation in male infertility and comparisons with other techniques. J Androl. 2002;23:25–43.
33. Hammadeh ME, al-Hasani S, Stieber M, Rosenbaum P, Kupker D, Diedrich K, et al. The effect of chromatin condensation (aniline blue staining) and morphology (strict criteria) of human spermatozoa on fertilization, cleavage and pregnancy rates in an intracytoplasmic sperm injection programme. Hum Reprod. 1996;11:2468–71.
34. Aoki VW, Carrell DT. Human protamines and the developing spermatid: their structure, function, expression and relationship with male infertility. Asian J Androl. 2003;5:315–24.
35. Miller D, Brinkworth M, Iles D. Paternal DNA packaging in spermatozoa: more than the sum of its parts? DNA, histones, protamines and epigenetics. Reproduction. 2010;139:287–301.
36. Marques CJ, Carvalho F, Sousa M, Barros A. Genomic imprinting in disruptive spermatogenesis. Lancet. 2004;363:1700–2.
37. Marques CJ, Costa P, Vaz B, Carvalho F, Fernandes S, Barros A, et al. Abnormal methylation of imprinted genes in human sperm is associated with oligozoospermia. Mol Hum Reprod. 2008;14:67–74.
38. Schaefer CB, Ooi SK, Bestor TH, Bourc'his D. Epigenetic decisions in mammalian germ cells. Science. 2007;316:398–9.
39. Fauque P, Jouannet P, Lesaffre C, Ripoche MA, Dandolo L, Vaiman D, et al. Assisted Reproductive Technology affects developmental kinetics, H19 Imprinting Control Region methylation and H19 gene expression in individual mouse embryos. BMC Dev Biol. 2007;7:116.
40. Le Bouc Y, Rossignol S, Azzi S, Steunou V, Netchine I, Gicquel C. Epigenetics, genomic imprinting and assisted reproductive technology. Ann Endocrinol (Paris). 2010;71:237–8.

41. Li T, Vu TH, Ulaner GA, Littman E, Ling JQ, Chen HL, et al. IVF results in de novo DNA methylation and histone methylation at an Igf2-H19 imprinting epigenetic switch. Mol Hum Reprod. 2005;11: 631–40.
42. Lim D, Bowdin SC, Tee L, Kirby GA, Blair E, Fryer A, et al. Clinical and molecular genetic features of Beckwith-Wiedemann syndrome associated with assisted reproductive technologies. Hum Reprod. 2009;24:741–7.
43. Benchaib M, Braun V, Ressnikof D, Lornage J, Durand P, Niveleau A, et al. Influence of global sperm DNA methylation on IVF results. Hum Reprod. 2005;20:768–73.
44. Tierling S, Souren NY, Gries J, Loporto C, Groth M, Lutsik P, et al. Assisted reproductive technologies do not enhance the variability of DNA methylation imprints in human. J Med Genet. 2010;47:371–6.
45. Huszar G, Stone K, Dix D, Vigue L. Putative creatine kinase M-isoform in human sperm is identified as the 70-kilodalton heat shock protein HspA2. Biol Reprod. 2000;63(3):925–32.
46. Huszar G, Vigue L, Morshedi M. Sperm creatine phosphokinase M-isoform ratios and fertilizing potential of men: a blinded study of 84 couples treated with in vitro fertilization. Fertil Steril. 1992;57:882–8.
47. Ergur AR, Dokras A, Giraldo JL, Habana A, Kovanci E, Huszar G. Sperm maturity and treatment choice of in vitro fertilization (IVF) or intracytoplasmic sperm injection: diminished sperm HspA2 chaperone levels predict IVF failure. Fertil Steril. 2002;77:910–8.
48. Huszar G, Sbracia M, Vigue L, Miller DJ, Shur BD. Sperm plasma membrane remodeling during spermiogenetic maturation in men: relationship among plasma membrane beta 1,4-galactosyltransferase, cytoplasmic creatine phosphokinase, and creatine phosphokinase isoform ratios. Biol Reprod. 1997;56: 1020–4.
49. Huszar G, Jakab A, Sakkas D, Ozenci CC, Cayli S, Delpiano E, et al. Fertility testing and ICSI sperm selection by hyaluronic acid binding: clinical and genetic aspects. Reprod Biomed Online. 2007;14: 650–63.
50. Menkveld R, Franken DR, Kruger TF, Oehninger S, Hodgen GD. Sperm selection capacity of the human zona pellucida. Mol Reprod Dev. 1991;30(4): 346–52.
51. Prinosilova P, Kruger T, Sati L, Ozkavukcu S, Vigue L, Kovanci E, et al. Selectivity of hyaluronic acid binding for spermatozoa with normal Tygerberg strict morphology. Reprod Biomed Online. 2009;18:177–83.
52. Liu DY, Baker HW. Human sperm bound to the zona pellucida have normal nuclear chromatin as assessed by acridine orange fluorescence. Hum Reprod. 2007;22:1597–602.
53. Yagci A, Murk W, Stronk J, Huszar G. Spermatozoa bound to solid state hyaluronic acid show chromatin structure with high DNA chain integrity: an acridine orange fluorescence study. J Androl. 2010;31(6): 566–72.
54. Kovanci E, Kovacs T, Moretti E, Vigue L, Bray-Ward P, Ward DC, et al. FISH assessment of aneuploidy frequencies in mature and immature human spermatozoa classified by the absence or presence of cytoplasmic retention. Hum Reprod. 2001;16:1209–17.
55. Huszar G, Vigue L, Corrales M. Sperm creatine kinase activity in fertile and infertile oligospermic men. J Androl. 1990;11:40–6.

Electrophoretic Sperm Separation

29

Steven Fleming and R. John Aitken

Abstract

By virtue of its ability to rapidly isolate spermatozoa with good morphology and low levels of DNA fragmentation, electrophoretic sperm separation promises to be an extremely versatile, time-saving and efficient method for preparing spermatozoa for a wide variety of applications in assisted reproduction.

Keywords

Electrophoretic sperm separation • Sperm DNA • Sperm separation • Assisted reproduction

Principles of Electrophoresis

Electrophoresis is a term used to define the motion of a particle within a liquid medium, the electrolyte, in response to a spatially uniform electric field. This electrokinetic phenomenon occurs as a result of the particles displaying a net positive or negative surface charge against which an external electric field can exert an electrostatic force. In fact, a surface charge may not even be necessary for electrokinesis, as it is theoretically possible that even neutral particles could migrate in response to an electric field by virtue of the molecular structure of water at their interface. This concept relates to the so-called double layer theory, whereby a diffuse layer of ions having the same but opposite charge to the particle surface screens them from the surrounding medium. Consequently, the electric field exerts an electrostatic force on the ions within the diffuse layer in the opposite direction to that exerted upon the particles, resulting in viscous stress, termed the electrophoretic retardation force. This hydrodynamic friction applied to the particles depends also upon the viscosity of the liquid medium in which they are dispersed, ultimately determining their electrophoretic mobility. Hence, it is necessary to carefully consider the molecular weight and charge of the particles relative to the conductivity and viscosity of the electrolyte to achieve the electrophoretic mobility required.

S. Fleming (✉)
Greenslopes Private Hospital, Suite 9A, Administration Building, Greenslopes, QLD 4120, Australia
e-mail: Steven.fleming@acaivf.com.au

A. Zini and A. Agarwal (eds.), *Sperm Chromatin: Biological and Clinical Applications in Male Infertility and Assisted Reproduction*, DOI 10.1007/978-1-4419-6857-9_29,

Electrophoretic Properties of Spermatozoa

Normal, mature spermatozoa carry a net negative charge that is imparted by the sperm glycocalyx, which is rich in sialic acid residues [1, 2]. One of these residues, called CD52, is a highly sialated glycosylphosphatidylinositol (GPI) anchored protein that is acquired during epididymal transit and located on the sperm plasmalemma [3–5]. During spermatogenesis, there is a massive cell–cell transfer of GPI-anchored CD52 that occurs at the sperm surface, the magnitude of which may be dependent upon the negative charge associated with the sperm plasmalemma [6]. Therefore, the presence of a negative charge may reflect normal spermatogenesis, especially since CD52 expression appears to be significantly correlated with capacitation and normal sperm morphology [5]. Consequently, this differential negative charge imparted by the sperm plasmalemma has been exploited as a means for sperm separation using either simple electrostatic [7, 8] or sophisticated electrophoretic techniques [9–11].

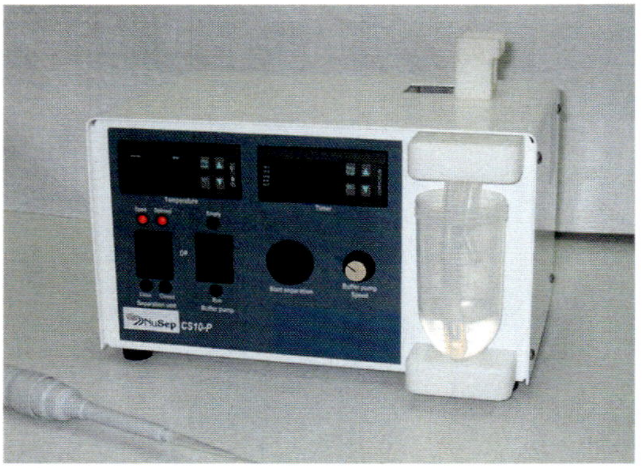

Fig. 29.1 The prototype CS10 instrument

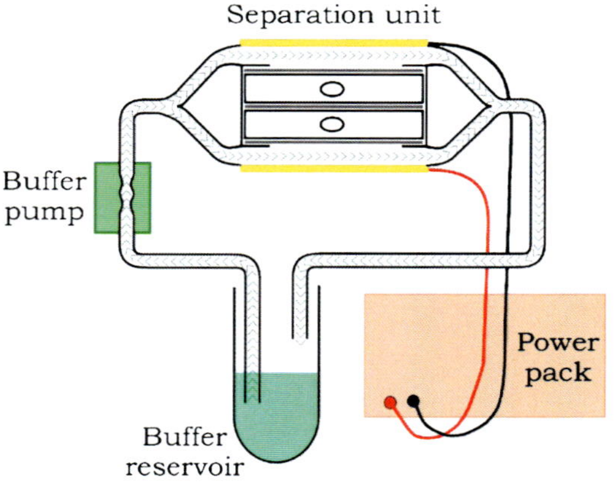

Fig. 29.2 Diagrammatic representation of the CS10 design

Development of Electrophoretic Technology for Sperm Sorting

The life separations company, NuSep, has been concerned with the development of bioseparations products for the past 30 years. In collaboration with Prof. John Aitken at the University of Newcastle, NuSep further developed their laboratory protein separations instrument, the ProteomeSep MF110, to create a prototype instrument designed for sperm separation, called the cell sorter 10 (CS10; Fig. 29.1). The CS10 was based upon preparative isolation by membrane electrophoresis (PrIME) technology, a patented technique that is capable of purifying most macromolecules from complex biological samples. The principle of this mode of separation was developed from the hypothesis that the CS10 preferentially selects cells on the basis of charge differences between human spermatozoa due to the differential presence of sialated proteins on the sperm plasmalemma [9]. A subsidiary commercial entity of NuSep, called SpermGen, is developing the CS10 into a regulatory compliant production unit, known as the SpermSep CS10.

The CS10 applies an electric potential via platinum-coated titanium mesh electrodes to move spermatozoa across a 5-µm polycarbonate separation membrane, the pore size of which allows the passage of morphologically normal spermatozoa while restricting larger cells within semen, such as immature germ cells and leukocytes (Figs. 29.2 and 29.3). Spermatozoa, which are negatively charged when suspended in a physiological buffer, are attracted towards the positive electrode, or anode. Consequently, spermatozoa not possessing a normal negative charge have less electrophoretic mobility and do not manage to

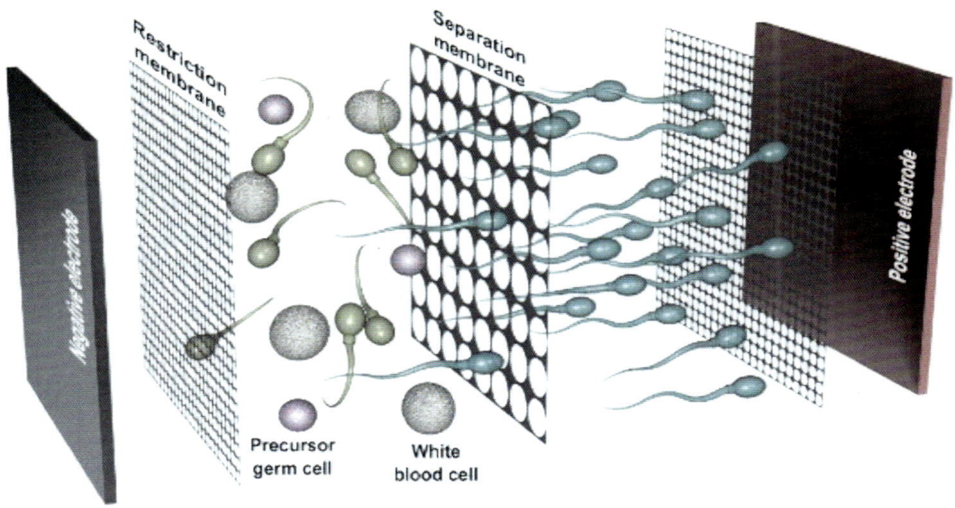

Fig. 29.3 Schematic diagram showing sperm electrophoretic mobility

pass through the separation membrane during the relatively short period (5 min) of electrophoresis. The exploitation of this concept has been found to yield a high percentage of morphologically normal, motile spermatozoa with intact DNA following electrophoretic sperm separation [9].

Equipment Set-Up and Separation Parameters

Separation Cartridges and Sample Handling

The separation cartridge of the prototype CS10 is a self-assembled device that has either a symmetric or an asymmetric format. In the asymmetric design, the inoculation or loading chamber has a volume of 2 mL and a collection or separation chamber has a volume of 400 μL (Fig. 29.4). Conveniently, 400 μL is also the estimated mean volume of the human uterine cavity and is, therefore, often the volume of sperm preparation inserted during intrauterine insemination (IUI) procedures. Consequently, the potential exists for electrophoretic sperm separation to be followed immediately by IUI of the entire volume of the sperm preparation retrieved, providing that the prostaglandins present within seminal plasma have been removed or reduced to clinically insignificant levels. However, for the purposes of in vitro fertilisation (IVF) and intracytoplasmic sperm injection (ICSI), the spermatozoa could be used directly, provided that they are separated into an appropriate medium within the separation chamber. The component parts of the separation cartridge can be autoclaved to ensure sterility. A 5-μm polycarbonate membrane, with an active membrane area of 20×15 mm, separates the loading and separation chambers, which are bound by polyacrylamide restriction membranes with a pore size of 15 kDa that prevent cross-contamination between the semen sample and electrophoresis buffer while permitting free transit of electrolytes (Figs. 29.3 and 29.4).

The separation cartridge is inserted into the cartridge housing on top of the SpermSep CS10 (Fig. 29.1), the housing being designed to ensure the cartridge can only be inserted in the correct orientation. Once the separation unit sealing mechanism is activated, the cartridge components are made watertight by the machine sealing pressure applied by the SpermSep CS10. Semen samples are simply pipetted into the loading chamber of the cartridge using a sterile, non-toxic, disposable plastic pipette, left for 5 min to

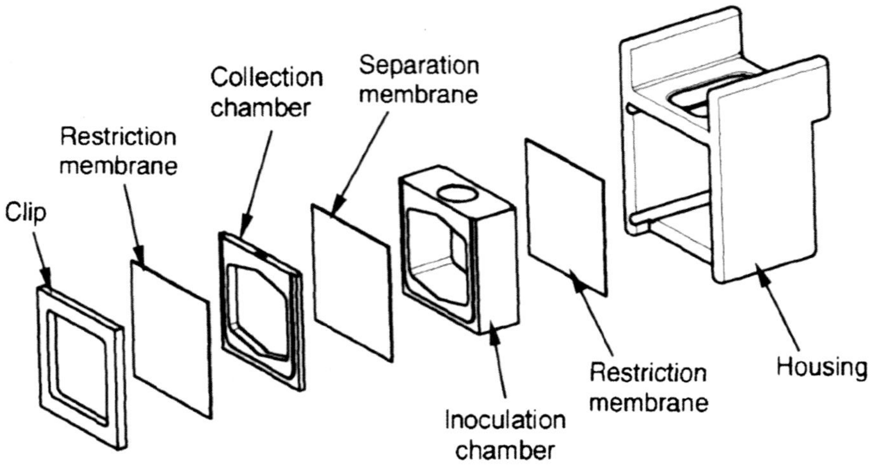

Fig. 29.4 Exploded diagram of the asymmetric separation cartridge

equilibrate and then subjected to electrophoresis. Once separated, the sperm preparation is aspirated from the separation chamber of the cartridge using an elongated, sterile, non-toxic, disposable micropipette tip, as typically used in standard gel electrophoresis.

Electrophoresis Buffers and Temperature Settings

The electrophoresis buffer contains 10 mM Hepes, 30 mM NaCl and 0.2 mM sucrose, having an osmolarity of 310 mOsm kg^{-1} and a pH of 7.4, following adjustment using 2 M KOH. It is filter-sterilised prior to use with a 0.22-μm filter (Millipore Corp., Bedford, USA). In order to provide a physiological medium in which to maintain sperm viability, 400 μL of electrophoresis buffer is placed into the separation chamber prior to running a sperm separation. A sterile, disposable buffer reservoir is filled with 80 mL electrophoresis buffer and placed into the reservoir housing on the front of the SpermSep CS10 (Fig. 29.1). In order to prevent overheating during operation of the instrument, the buffer is maintained at 25°C and is circulated around the instrument by means of a buffer pump (Fig. 29.2). In order to complete the electrical circuit, the buffer pump is run for at least 1 min prior to performing any sperm separations.

Current and Voltage Settings

The input power specifications of the SpermSep CS10 are 115–240 V at 50–60 Hz. Electrophoresis is achieved via a constant current of 75 mA at a variable voltage of 18–21 V applied over a 5-min period. No electrical potential is applied until the separation run is initiated.

Cleaning of Equipment

At the conclusion of each sperm separation, any electrophoresis buffer remaining in the buffer reservoir is replaced with sterile distilled water, and the buffer pump is actuated to rinse the buffer lines. If no more separations are to be performed that day, the water is replaced with a 0.1 M NaOH cleaning solution and the buffer pump is run for 30 s to circulate it through the lines of the SpermSep CS10, and the cleaning solution is left in place overnight. The following morning, the cleaning solution is thoroughly rinsed out with a minimum of three washes of sterile distilled water.

Method Validation

Initial validation of the SpermSep CS10 system was performed using semen samples from normozoospermic sperm donors and a separation cartridge with a symmetrical design, the loading and separation chambers both having a capacity of 400 μL [9].

Sample Recovery and Purity

The mean sample concentration loaded into the system was $52 \pm 5.2 \times 10^6$ mL^{-1}. During an initial 5-min equilibration period, the starting concentration of spermatozoa in the separation chamber was $1.67 \pm 0.58 \times 10^6$ mL^{-1} (3.2% recovery), presumably as a consequence of the inherent motility of spermatozoa. Following just 30 s of electrophoresis, the sperm concentration increased to $3.55 \pm 0.42 \times 10^6$ mL^{-1} (6.8% recovery), reaching a peak concentration of $22.31 \pm 5.85 \times 10^6$ mL^{-1} (42.9% recovery) after 15 min. The purity of the electrophoretically separated sperm preparations was extremely high, with contamination by round cells proving undetectable using phase-contrast microscopy [9].

Sperm Vitality and Motility

Sperm vitality, assessed using the eosin dye (0.05% eosin in phosphate-buffered saline) exclusion test, was $83 \pm 1.5\%$ in the original semen samples prior to electrophoresis. The percentage of viable spermatozoa in the electrophoretically separated sperm preparations was found to be consistent with that of the original samples and there was no significant change in vitality observed over the entire period (15 min) of electrophoresis [9].

Sperm motility, assessed using computer-assisted semen analysis (CASA), was $72 \pm 2.1\%$ in the original semen samples prior to electrophoresis. Similar to sperm vitality, percentage sperm motility was found to be consistent with that of the original samples and not significantly affected by the duration of electrophoresis, though a slight reduction was observed after 15 min [9]. Similarly, kinematic analysis by CASA demonstrated that the duration of electrophoresis had no significant effect upon the quality of sperm motility observed.

Sperm Morphology and DNA Integrity

The percentage of normal spermatozoa observed following staining by a modification of the Papanicolaou method [12] and assessed using the sperm deformity index (SDI) [13] was significantly increased ($P<0.001$) by electrophoresis [9]. A higher percentage of morphologically normal spermatozoa within the separated sperm preparation was observed regardless of the duration of electrophoresis, with no significant variation between different time periods. Furthermore, SDI values for the separated spermatozoa were significantly below ($P<0.001$) the threshold SDI value of 0.93 for all electrophoretic time points, indicating their normal fertilisation potential [13].

DNA damage, assessed using the terminal deoxynucleotidyl transferase-mediated deoxyuridinetriphosphate nick-end labelling (TUNEL) assay, was significantly reduced ($P<0.05$) in the sperm preparation separated by electrophoresis [9]. This reduction was only observed at all time points up to 10 min of electrophoresis, beyond which there was no significant difference in the percentage of DNA-damaged spermatozoa.

Clinical Applications

The first successful clinical application of electrophoretic sperm separation was published as a case report following ICSI [10]. This provided proof-of-principle, that electrophoresis could be used to prepare spermatozoa for use in assisted reproduction. However, since ICSI had been used to fertilise the oocytes in this instance, it was still unknown whether electrophoresis might compromise aspects of sperm function necessary for normal fertilisation. This uncertainty was resolved following a prospective, split-sample, split-cohort controlled clinical trial, involving patients having both ICSI and IVF, with sperm prepared by either standard density-gradient centrifugation (DGC) or by electrophoresis [11]. The design of this trial ensured that any differences in gamete quality between semen samples and cohorts of oocytes were controlled for. Approximately 400 oocytes were inseminated by either DGC or electrophoretically prepared spermatozoa, resulting in comparable rates of fertilisation (63.6% vs. 62.4%, respectively), cleavage (88.5% vs. 99.0%, respectively), and embryo quality (26.1% vs. 27.4% top-grade embryos, respectively), regardless of whether ICSI or IVF was employed as the method of insemination [11]. Furthermore, six pregnancies resulted from the use of electrophoretically prepared spermatozoa, two of them from patients receiving ICSI and four from patients receiving IVF [11].

Previous work has demonstrated that spermatozoa can be efficiently isolated from a variety of sources [10]. Separation of frozen-thawed, cryostored semen ($39.6 \pm 11.1 \times 10^6$ mL^{-1}) resulted in 27% recovery of separated spermatozoa ($10.8 \pm 3.8 \times 10^6$ mL^{-1}) after just 5 min electrophoresis [10]. These sperm preparations were devoid of detectable contaminating cells, the separated spermatozoa displaying significantly greater viability ($P<0.01$), motility ($P<0.05$) and normal morphology ($P<0.001$) than the cryostored semen [10]. Therefore, electrophoresis may prove an advantageous method for preparing cryostored semen, especially since it has recently been shown that such material is particularly vulnerable to oxidative stress and subsequent DNA damage during processing by standard DGC [14–16].

A particularly promising potential application of electrophoretic sperm separation is the isolation of spermatozoa exhibiting low levels of sperm DNA damage from more complex mixtures of cells such as those found in surgically recovered aspirates and biopsies of the epididymis and testis. Testicular biopsy material, containing a range of mature and immature spermatozoa, has been shown to rapidly yield cells with greater residual motility, vitality and normal morphology than those in the original biopsy following electrophoretic sperm preparation [10]. Importantly, the recovery of spermatozoa from the biopsy material was good ($28.4 \pm 7.1\%$).

Closing Remarks

Combined, the basic scientific and clinical data suggest that electrophoretic sperm separation is particularly suitable for those patients requiring ICSI or IVF where the cause of infertility is due to poor sperm morphology and/or significantly damaged sperm DNA. Though electrophoresis has previously been demonstrated to be detrimental to sperm motility in a free-flow electrophoretic system [17], such impacts on sperm quality do not appear to be a problem with the SpermSep CS10. The latter would, therefore, seem to offer some promise as a fast, efficient method for isolating spermatozoa exhibiting low levels of DNA damage for assisted conception applications, ranging from IUI to ICSI [11].

References

1. Kallajoki M, Virtanen I, Suominen J. Surface glycoproteins of human spermatozoa. J Cell Sci. 1986;82: 11–22.
2. Calzada L, Salazar EL, Pedron N. Presence and chemical composition of glycoproteic layer on human spermatozoa. Arch Androl. 1994;33:87–92.
3. Schroter S, Kirchhoff C, Yeung CH, Cooper T, Meyer B. Purification and structural analysis of sperm CD52, a GPI-anchored membrane protein. Adv Exp Med Biol. 1997;424:233–4.
4. Kirchhoff C, Schroter S. New insights into the origin, structure and role of CD52: a major component of the mammalian sperm glycocalyx. Cells Tissues Organs. 2001;168:93–104.
5. Giuliani V, Pandolfi C, Santucci R, Pelliccione F, Macerola B, Focarelli R, et al. Expression of gp20, a human sperm antigen of epididymal origin, is reduced in spermatozoa from subfertile men. Mol Reprod Dev. 2004;69:235–40.
6. Schroter S, Derr P, Conradt HS, Nimtz M, Hale G, Kirchhoff C. Male-specific modification of human CD52. J Biol Chem. 1999;274:29862–73.
7. Chan PJ, Jacobson JD, Corselli JU, Patton WC. A simple zeta method for sperm selection based on membrane charge. Fertil Steril. 2006;85:481–6.
8. Kam TL, Jacobson JD, Patton WC, Corselli JU, Chan PJ. Retention of membrane charge attributes by cryopreserved-thawed sperm and zeta selection. J Assist Reprod Genet. 2007;24:429–34.
9. Ainsworth C, Nixon B, Aitken RJ. Development of a novel electrophoretic system for the isolation of human spermatozoa. Hum Reprod. 2005;20:2261–70.
10. Ainsworth C, Nixon B, Jansen RP, Aitken RJ. First recorded pregnancy and normal birth after ICSI using electrophoretically isolated spermatozoa. Hum Reprod. 2007;22:197–200.
11. Fleming SD, Ilad RS, Griffin AM, Wu Y, Ong KJ, Smith HC, et al. Prospective controlled trial of an electrophoretic method of sperm preparation for assisted reproduction: comparison with density gradient centrifugation. Hum Reprod. 2008;23:2646–51.

12. WHO. WHO laboratory manual for the examination of human semen and sperm-cervical mucus interaction. Cambridge: Cambridge University Press; 1999.
13. Panidis D, Matalliotakis I, Papathanasiou K, Roussos C, Koumantakis E. The sperm deformity and the sperm multiple anomalies indexes in patients who underwent unilateral orchectomy and preventive radiotherapy. Eur J Obstet Gynecol Reprod Biol. 1998;80:247–50.
14. Thomson LK, Fleming SD, Aitken RJ, de Iuliis GN, Zieschang JA, Clark AM. Cryopreservation-induced human sperm DNA damage is predominantly mediated by oxidative stress rather than apoptosis. Hum Reprod. 2009;24:2061–70.
15. Thomson LK, Fleming SD, Barone K, Zieschang JA, Clark AM. The effect of repeated freezing and thawing on human sperm DNA fragmentation. Fertil Steril. 2010;93:1147–56.
16. Thomson LK, Fleming SD, Schulke L, Barone K, Zieschang JA, Clark AM. The DNA integrity of cryopreserved spermatozoa separated for use in assisted reproductive technology is unaffected by the type of cryoprotectant used but is related to the DNA integrity of the fresh separated preparation. Fertil Steril. 2009;92:991–1001.
17. Engelmann U, Krassnigg F, Schatz H, Schill WB. Separation of human X and Y spermatozoa by free-flow electrophoresis. Gamete Res. 1988;19:151–60.

30 Antioxidants and Sperm DNA Damage

Armand Zini and Maria San Gabriel

Abstract

Infertile men have higher levels of sperm DNA damage than do fertile men, and this damage may reduce male fertility potential and may impact on reproductive capacity. This is particularly important in the context of assisted reproductive technologies, as there is a mounting concern regarding the safety of utilizing DNA-damaged spermatozoa in this setting. A better understanding of the etiology of sperm DNA damage may help identify strategies to reduce sperm DNA damage. In this chapter, we discuss the rationale for antioxidant therapy, examine the relationship between oxidative stress and sperm DNA damage, and evaluate the studies on dietary and in vitro antioxidants on sperm DNA damage. The review focuses primarily on clinical (human) studies with some examples taken from experimental (animal) data.

Keywords

Sperm DNA fragmentation • Oxidative stress • Vitamins • Sperm washing • Male infertility

Infertile men have higher levels of sperm DNA damage than do fertile men, and this damage may reduce male fertility potential and may impact on reproductive capacity. This is particularly important in the context of assisted reproductive technologies (ARTs), as there is a mounting concern regarding the safety of utilizing DNA-damaged spermatozoa in this setting. A better understanding of the etiology of sperm DNA damage may help identify strategies to reduce sperm DNA damage. In this chapter, we will discuss the rationale for antioxidant therapy, examine the relationship between oxidative stress and sperm DNA damage, and evaluate the studies on dietary and in vitro antioxidants on sperm DNA damage. The review focuses primarily on clinical (human) studies with some examples taken from experimental (animal) data.

A. Zini (✉)
Department of Surgery, Division of Urology, McGill University, St. Mary's Hospital Center, Montreal, QC, Canada
e-mail: ziniarmand@yahoo.com

A. Zini and A. Agarwal (eds.), *Sperm Chromatin: Biological and Clinical Applications in Male Infertility and Assisted Reproduction*, DOI 10.1007/978-1-4419-6857-9_30,

Etiology of Sperm DNA Damage

The etiology of sperm DNA damage in humans is multifactorial. Several clinical conditions have been associated with sperm DNA damage (e.g., chemotherapy, smoking, genital tract infection, varicocele) [1–9]. These conditions can be categorized as primary defects in spermatogenesis (e.g., genetic or developmental abnormalities) and secondary or extrinsic factors (e.g., gonadotoxins, hyperthermia, oxidants, endocrine disruption).

A number of theories have been proposed to explain the DNA damage in human spermatozoa at the cellular level. Studies have suggested that protamine deficiency (with aberrant chromatin remodeling), reactive oxygen species (ROS), and abortive apoptosis may be responsible for sperm DNA damage [10–13]. Recently, De Iuliis et al. [12] have proposed a two-step hypothesis to explain the generation of sperm DNA damage. Based on their model, sperm DNA damage is sustained as a result of an oxidative injury (second step) to poorly protaminated cells (i.e., cells with incomplete replacement of histones by protamines) that are generated by defective spermiogenesis (first step).

Relationship Between Oxidative Stress and Sperm DNA Damage

Several studies have reported that sperm DNA damage is associated with oxidative stress, and this represents the basis for the use of antioxidants in the treatment of sperm DNA damage [14–23]. Moreover, both exogenous and endogenous ROS can induce sperm DNA damage in vitro, indicating that ROS can cause sperm DNA damage [15, 23, 24]. Approximately 25% of infertile men have high levels of semen ROS [25, 26], and the levels of sperm DNA oxidation are higher in infertile men compared to fertile men [27, 28]. Semen ROS are generated by spermatozoa (especially, defective or immature) and semen leukocytes [29–33]. While the controlled release of low levels of ROS is necessary for normal sperm function, high levels of ROS can cause sperm dysfunction [29]. The levels of sperm-derived ROS have been associated with sperm DNA damage, although there is no established ROS threshold level above which sperm DNA damage is detected [8, 17, 30].

The susceptibility of human spermatozoa to oxidative stress stems primarily from the characteristics of the sperm plasma membrane. The human sperm plasma membrane contains an abundance of unsaturated fatty acids, and these fatty acids provide fluidity that is necessary for sperm motility and membrane fusion events, such as the acrosome reaction and sperm–egg interaction. However, this characteristic of the membrane predisposes spermatozoa to free radical attack and peroxidation of the plasma membrane lipids. Once this process has been initiated, accumulation of lipid peroxides occurs on the sperm surface and oxidative damage to DNA can ensue [23, 34]. It has been shown that ROS can cause damage to the sperm DNA directly or indirectly via production and subsequent translocation of lipid peroxides [35–38].

Seminal Antioxidant Capacity and Sperm DNA Damage

Seminal fluid is a rich source of enzymatic and nonenzymatic antioxidants (ROS scavengers), and this fluid protects spermatozoa from oxidative injury [20, 26, 39–41]. The antioxidant properties of seminal plasma are vital to the survival of spermatozoa because these cells have minimal antioxidant capacity (spermatozoa have little cytoplasmic fluid and no capacity for protein synthesis) [26]. The endogenous ROS scavenging enzymes in the male reproductive tract include superoxide dismutase (SOD), catalase, and glutathione peroxidase (GPX) [26, 40, 42–46]. Experimental studies have shown that a deficiency in any of these enzymes can increase oxidative stress and lead to male infertility [47, 48]. These antioxidant enzymes (SOD, catalase, and GPX) are also found in semen [35]. Additionally, there are several nonenzymatic antioxidants (e.g., vitamins C and E, hypotaurine, taurine, L-carnitine, lycopene) that are found in semen, and these nonenzymatic antioxidants are believed to account for much of the total seminal antioxidant activity [26, 49].

A number of investigators have proposed that oxidative sperm DNA damage may be secondary to reduced semen antioxidant capacity; however, clinical studies have reported conflicting results in this respect. Several studies have, indeed, demonstrated that a deficiency in semen antioxidants is associated with sperm DNA damage, whereas other studies have not observed the same relationship [50–53]. Similarly, several studies have shown that seminal antioxidant activity is lower in infertile men with high levels of seminal ROS (relative to those with normal levels of ROS), whereas others have not shown this to be the case [26, 54–56].

Although a relationship between male infertility and systemic antioxidant deficiency has not been reported to date, it is possible that a subset of infertile men may be at risk for antioxidant deficiency, particularly, vitamin C deficiency [57]. We suspect that infertile men with specific lifestyles (e.g., smoking, increased alcohol intake, dieting) may be at high risk for antioxidant or vitamin deficiency but this remains to be tested [58, 59]. Recently, investigators have evaluated dietary antioxidant intake (vitamins C, E, or β-carotene) and sperm DNA damage in a cohort of fertile men but failed to identify any relationships between these parameters [60].

Clinical Studies

Effect of Dietary Antioxidants on Sperm DNA Damage

An effective dietary antioxidant should be readily absorbed and concentrated in reproductive tract tissues. Ideally, the antioxidant preparation must also replete a deficiency (in the testis, epididymis or semen) and be a vital element of reproductive function. The antioxidant must either improve testicular function and spermatogenesis and/or epididymal function, resulting in improved sperm function and chromatin compaction and integrity. Additionally, the antioxidant preparation should enhance semen antioxidant capacity and reduce seminal oxidative stress.

There are few reports on the role of dietary antioxidant supplements and sperm DNA integrity. Most of the studies are small with no evaluation of the mechanism of action of antioxidants and the only outcomes measured are the integrity of the sperm DNA and/or the pregnancy rate. Moreover, all of the studies evaluate the effects of a short treatment course (with no long-term follow-up), and most are not randomized and fail to include a placebo-control group. Additionally, there may be an inherent bias because many studies select men with high levels of sperm DNA damage or oxidative stress at baseline, and in these studies, treatment is generally associated with an improvement in sperm DNA integrity and fertility potential (Table 30.1) [51, 61–67].

Fraga et al. [51] provided the most convincing demonstration that antioxidants can protect sperm DNA from oxidative damage. In their experiments, they demonstrated that oral vitamin C intake increases semen vitamin C levels and improves sperm DNA integrity (lowers DNA oxidation levels) in men on a vitamin C-depleted diet (with vitamin C deficiency). As stated earlier, several studies of infertile men with high levels of sperm DNA damage or oxidative stress (two were randomized controlled studies and four uncontrolled trials) have shown that antioxidant therapy is effective in improving sperm DNA integrity or pregnancy rates (Table 30.1). In men with idiopathic infertility, the effect of dietary antioxidants on sperm DNA integrity is equivocal with one of two controlled trials showing a benefit of antioxidants on sperm DNA integrity (Table 30.1) [27, 68]. However, in these eight recent studies of antioxidants and sperm DNA damage, there has been no evaluation of systemic or semen vitamin levels and no estimation of seminal oxidative stress. As such, the precise mechanism of action of these antioxidant supplements on sperm DNA quality is unknown.

Table 30.1 Effect of dietary antioxidant supplements on sperm DNA integrity

Study	Patients/test	Treatment(s)	*n*	Results
Infertile men with high sperm DNA fragmentation levels or oxidative stress				
Greco et al. [63]	1 Failed ICSI TUNEL >15%	Vits C 1 g, E 1 g	38	Rx (2 months): ↓DD in 76%, 48% ICSI pregnancy No control group
Greco et al. [62]	Infertility TUNEL >15%	Vits C 1 g, E 1 g	32 32	Rx (2 months): ↓DD (22→9%) Placebo group: no effect on DD (22→22%)
Menezo et al. [64]	2 Failed ICSI DFI >15% Decond >15%	Vits C, E (400 mg), zinc, Se, β-carotene	57	Rx (90 days):↓sperm %DFI (32→26%: by 19%), but ↑ sperm %HDS (17.5→25.5%: by 23%) No control group
Tremellen et al. [65]	Male Infert TUNEL >25%	Menevit (lycopene, vits C, E, zinc, Se, folate, garlic)	36 16	Rx (3 months): 39% ICSI pregnancy rate, but no ↑ in embryo quality, no post-Rx DD Placebo group: 16% ICSI pregnancy rate
Gil-Villa et al. [61]	Pregnancy loss ↑LPO or DFI	Vits C, E, zinc, β-carotene	9	Rx (3 months): six (of nine) couples got pregnancy No control group
Tunc et al. [66]	Male Infert ↑Semen OS	Menevit (lycopene, vits C, E, zinc, Se, folate, garlic)	45	Rx (3 months): ↓DD (22→18%) ↓ROS production and ↑sperm protamination No control group
Unselected infertile men				
Piomboni et al. [68]	Asthenosp. AO stain	Vits C, E, β-glucan, papaya, lactoferrin	36 15	Rx (90 days): ↑motility and morph but not DD Control group: no effect
Kodama et al. [27]	Male infert 8-OHdG	Vits C, E (200 mg) Glutathione (400 mg)	14 7	Rx (2 months): ↓in 8-OHdG (1.5→1.1/10^5 dG) Control group: no change in 8-OHdG levels

8-OHdG 8-hydroxy-2-deoxyguanosine; *AO* acridine orange; *DD* DNA damage; *Decond* decondensation; *DFI* DNA fragmentation index; *LPO* lipid peroxidation; *OS* oxidative stress; *Rx* Treatment; *ROS* reactive oxygen species; *Se* selenium; *TUNEL* terminal nucleotidyl transferase dUTP nick-end labeling; *vit* vitamin

Effect of In Vitro Antioxidants on Sperm DNA Damage

Several studies have evaluated the potential benefit of adding antioxidants to in vitro preparations so as to protect the sperm DNA from oxidative damage. This has important clinical relevance because sperm collection and subsequent in vitro processing is routinely performed prior to the application of ARTs (e.g., intrauterine insemination and in vitro fertilization). Oxidative injury to the sperm DNA may result particularly from sperm processing techniques (e.g., centrifugation, aerobic incubation), as spermatozoa are vulnerable to oxidants because seminal plasma (rich in antioxidants) has been removed in the process [41].

There is good evidence to show that subpopulations of spermatozoa will exhibit differing susceptibility to oxidative stress: the DNA of normal spermatozoa is less susceptible to gentle processing techniques than is the DNA of abnormal or immature spermatozoa [33, 69]. Experimental studies suggest that the susceptibility of the sperm DNA to oxidative injury is related to the degree of sperm chromatin compaction (i.e., level of protamination) [12, 70]. We have recently shown that the spermatozoa of FSH-receptor knock-out mice are more susceptible to oxidative DNA injury but also benefit more so from antioxidant

Table 30.2 Role of in vitro antioxidant supplements in protecting sperm DNA from exogenous ROS.

Study	Assay	Exogenous ROS	Antioxidant supplement and results
Lopes et al. [18]	TUNEL	X+XO	GSH+hypotaurine protect spz from X+XO-induced DD Catalase protects spz from X+XO-induced DD *n*-Acetylcysteine protects spz from X+XO-induced DD
Potts et al. [20]	TUNEL	H_2O_2+Fe+ADP	S. plasma (>60%v/v) lowers oxidative spz damage (↓DD, LPO)
Sierens et al. [73]	Comet	H_2O_2	Isoflavones, vitamin C and E protect spz from H_2O_2-induced DD (Isoflavones: genistein, equol). Dose effect noted
Russo et al. [72]	Comet	H_2O_2 Benzopyrene H_2O_2+Fe+ADP	Propolis lowers oxidative spz damage (↓LPO, DD, LDH) (Propolis – a natural resinous hive product)

ADP adenosine diphosphate; *Comet* single-cell gel electrophoresis; *DD* DNA damage; *Fe* iron; *GSH* glutathione; *LDH* lactate dehydrogenase; *LPO* lipid peroxidation; *S. plasma* seminal plasma; *spz* sperm; *TUNEL* terminal nucleotidyl transferase dUTP nick-end labeling; *X* xanthine; *XO* xanthine oxidase

Table 30.3 Role of in vitro antioxidant supplements in protecting sperm DNA from stimulated endogenous reactive oxygen species (ROS) generation

Study	Assay	ROS stimulant	Antioxidant supplement and results
Twigg et al. [23]	ISNTL	NADPH	Vit E, SOD, catalase, hypotaurine, albumin all ineffective in protecting spz DNA from endogenous ROS
Cemeli et al. [76]	Comet	Estrogens (1 h 37°C)	Flavonoid (Kaempferol) protects sperm from estrogen-induced oxidative DD
Dobrzynska et al. [77]	Comet	DES, T3, T4, NA (1 h 37°C)	Flavonoids and catalase protect spz from stimulant-induced oxidative DD (Flavonoids: Kaempferol, Quercetin)
Anderson et al. [75]	Comet	Estrogens	Catalase protects spz from estrogen-induced oxidative DD, SOD and vit C less effective (Estrogens: equol, daidzein, genistein, DES, E2)

Comet alkaline single-cell gel electrophoresis; *DD* DNA damage; *ISNTL* in situ nick translation assay; *NA* noradrenaline; *ROS* reactive oxygen species; *SOD* superoxide dismutase; *spz* sperm; *T3* triiodothyronine; *T4* thyroxine; *vit* vitamin

treatment than do spermatozoa of wild-type animals [71]. These data suggest that the spermatozoa of infertile men may be more susceptible to oxidative injury yet may be afforded greater protection by antioxidants.

Studies on in vitro antioxidant supplementation have evaluated the capacity of antioxidants to protect spermatozoa from exogenous and endogenous ROS and from the effects of semen processing and cryopreservation. It is quite clear from several studies that antioxidants (e.g., vitamins C and E, catalase, glutathione) can effectively protect sperm DNA from the effects of exogenous ROS (see Table 30.2) [18, 20, 72, 73]. This is clinically relevant because many semen samples contain leukocytes and the sperm processing (with removal of seminal plasma) can cause these cells to generate high levels of unchecked exogenous ROS (e.g., centrifugation) [74]. By contrast, a number of studies have shown that antioxidants are of limited value in protecting the DNA of normal spermatozoa (with normal chromatin compaction) from endogenous ROS production (for example, ROS may be generated by incubating spermatozoa with NADPH or by centrifugation) (Table 30.3) [23, 75–77].

Table 30.4 Role of in vitro antioxidant supplements in protecting sperm DNA from semen processing

Study	Assay	Semen processing	Antioxidant supplement and results
Hughes et al. [81]	Comet	Percoll DGC	Vitamins C, E or urate lower sperm DD after DGC Vitamins C + E or AC increase sperm DD after DGC
Donnelly et al. [79]	Comet	Percoll DGC	Vit C or E do not lower baseline sperm ROS and DD Vit C or E protect sperm from H_2O_2 induced ROS and DD Vits C + E induce sperm DD and increase H_2O_2-induced DD
Donnelly et al. [80]	Comet	Percoll DGC $\pm H_2O_2$	GSH, hypotaurine or both do not alter baseline sperm DD GSH, hypotaurine or both do not alter sperm motility at 4 h GSH and/or hypotaurine lower H_2O_2-induced sperm DD
Chi et al. [78]	Comet	Centrifugation (1,000 rpm ×2) + 1 h incubation	EDTA or catalase lower centrifugation-induced sperm ROS EDTA or catalase lower centrifugation-induced sperm DD EDTA or datalase have no protective effect on LPO

AC Acetyl cysteine; *Comet alkaline* single-cell gel electrophoresis; *DD* DNA damage; *DGC* density-gradient centrifugation; *GSH* glutathione; *LPO* lipid peroxidation; *ROS* reactive oxygen species; *vit* vitamin

Although there are few data to support this, some studies suggest that in those samples with poor morphology and poor sperm chromatin compaction, antioxidants may protect the sperm DNA from endogenous ROS production, as these samples are inherently more vulnerable to oxidative stress [33, 69].

Antioxidants appear to be of minimal value in terms of protecting sperm DNA from gentle semen processing (e.g., incubation or density-gradient centrifugation) (Table 30.4) [78–81], and in some cases, these antioxidants (e.g., combination of vitamins C and E) may increase the levels of sperm DNA damage [80, 81]. Five clinical studies have evaluated the potential protective effect(s) of antioxidants on sperm DNA integrity during cryopreservation. Although Taylor et al. [82] reported that the antioxidant vitamin E does not protect sperm DNA during cryopreservation, four other studies have shown that antioxidants (vitamin C, catalase, resveratrol, genistein) can protect the sperm DNA from oxidative injury during cryopreservation and subsequent thawing [82–86] (Table 30.5).

Summary

In vitro studies have demonstrated a beneficial effect of antioxidant supplements in protecting normal sperm DNA from exogenous oxidants, but the effect of these antioxidants in protecting normal spermatozoa from endogenous ROS and gentle sperm processing has not been established. By contrast, when evaluating spermatozoa from infertile men, clinical and experimental studies indicate a beneficial effect of antioxidant supplements in protecting the DNA from exogenous and endogenous oxidants and from gentle sperm processing. The limited data on the protective effect(s) of antioxidants on sperm DNA integrity during cryopreservation and thawing suggest that antioxidants are useful in this context. Dietary antioxidants may be beneficial in reducing sperm DNA damage, particularly in men with high levels of DNA fragmentation. However, the exact mechanism of action of dietary antioxidants has not been established and most of studies on this subject are small.

Table 30.5 The role of in vitro antioxidants in protecting human sperm DNA from injury caused by cryopreservation and thawing

Study	Assay	Antioxidant	Effect of antioxidant on cryopreservation and thawing
Taylor et al. [82]	TUNEL	Vitamin E	No effect on sperm DNA integrity Improved post-thaw motility
Li et al. [84]	Comet	Catalase Ascorbic acid	Improved sperm DNA integrity Reduced ROS production
Branco et al. [83]	Comet	Resveratol or Ascorbic acid	Improved sperm DNA integrity
Martinez-Soto et al. [85]	TUNEL	Genistein	Improved sperm DNA integrity Reduced ROS production, improved post-thaw motility
Thomson et al. [86]	8-OHdG TUNEL	Genistein	Improved sperm DNA integrity (reduced oxidative damage)

8-OHdG 8-Hydroxy-2-deoxyguanosine; *COMET* alkaline single-cell gel electrophoresis; *ROS* reactive oxygen species; *TUNEL* terminal nucleotidyl transferase dUTP nick-end labeling

Expert Commentary

The biological basis for the use of antioxidants in male infertility is sound and is based on the body of literature showing that sperm dysfunction (including DNA damage) is strongly related to oxidative stress. Clinical studies of dietary antioxidants demonstrate a promising positive effect of these antioxidants on the integrity of the sperm DNA; however, most studies are small and mechanistic studies are lacking. Moreover, the optimal antioxidant complement has not been defined, but most studies report on one or more of the following: vitamins C and E, folic acid, and zinc. Clinical studies of in vitro antioxidants support the use of antioxidants in protecting spermatozoa (particularly abnormal spermatozoa) from oxidative stress. However, the optimal antioxidant and its concentration have not been established yet.

Five-Year View

In order to see a real advance in the field of dietary antioxidants for male infertility, we need to undertake larger studies with a longer treatment course and some evaluation of the mechanism of action of these agents. Additional in vitro antioxidant studies are needed to better define the differences in treatment response between normal (fertile) and subnormal (infertile) semen samples and identify the optimal protocol (type and concentration of antioxidant).

References

1. Banks S, King SA, Irvine DS, Saunders PT. Impact of a mild scrotal heat stress on DNA integrity in murine spermatozoa. Reproduction. 2005;129:505–14.
2. Bungum M, Humaidan P, Axmon A, Spano M, Bungum L, Erenpreiss J, et al. Sperm DNA integrity assessment in prediction of assisted reproduction technology outcome. Hum Reprod. 2007;22:174–9.
3. Erenpreiss J, Hlevicka S, Zalkalns J, Erenpreisa J. Effect of leukocytospermia on sperm DNA integrity: a negative effect in abnormal semen samples. J Androl. 2002;23:717–23.
4. Fossa SD, De Angelis P, Kraggerud SM, Evenson D, Theodorsen L, Clausen OP. Prediction of posttreatment spermatogenesis in patients with testicular cancer by flow cytometric sperm chromatin structure assay. Cytometry. 1997;30:192–6.
5. O'Flaherty C, Vaisheva F, Hales BF, Chan P, Robaire B. Characterization of sperm chromatin quality in testicular cancer and Hodgkin's lymphoma patients prior to chemotherapy. Hum Reprod. 2008;23:1044–52.
6. Potts RJ, Newbury CJ, Smith G, Notarianni LJ, Jefferies TM. Sperm chromatin damage associated with male smoking. Mutat Res. 1999;423:103–11.
7. Sailer BL, Sarkar LJ, Bjordahl JA, Jost LK, Evenson DP. Effects of heat stress on mouse testicular cells and sperm chromatin structure. J Androl. 1997;18:294–301.
8. Saleh RA, Agarwal A, Sharma RK, Said TM, Sikka SC, Thomas Jr AJ. Evaluation of nuclear DNA

damage in spermatozoa from infertile men with varicocele. Fertil Steril. 2003;80:1431–6.
9. Zini A, Sigman M. Are tests of sperm DNA damage clinically useful? Pros and cons. J Androl. 2009; 30:219–29.
10. Aoki VW, Emery BR, Liu L, Carrell DT. Protamine levels vary between individual sperm cells of infertile human males and correlate with viability and DNA integrity. J Androl. 2006;27:890–8.
11. Aoki VW, Moskovtsev SI, Willis J, Liu L, Mullen JB, Carrell DT. DNA integrity is compromised in protamine-deficient human sperm. J Androl. 2005;26: 741–8.
12. De Iuliis GN, Thomson LK, Mitchell LA, Finnie JM, Koppers AJ, Hedges A, et al. DNA damage in human spermatozoa is highly correlated with the efficiency of chromatin remodeling and the formation of 8-hydroxy-2'-deoxyguanosine, a marker of oxidative stress. Biol Reprod. 2009;81:517–24.
13. Sakkas D, Seli E, Bizzaro D, Tarozzi N, Manicardi GC. Abnormal spermatozoa in the ejaculate: abortive apoptosis and faulty nuclear remodelling during spermatogenesis. Reprod Biomed Online. 2003;7: 428–32.
14. Agarwal A, Said TM. Oxidative stress, DNA damage and apoptosis in male infertility: a clinical approach. BJU Int. 2005;95:503–7.
15. Aitken RJ, Gordon E, Harkiss D, Twigg JP, Milne P, Jennings Z, et al. Relative impact of oxidative stress on the functional competence and genomic integrity of human spermatozoa. Biol Reprod. 1998;59: 1037–46.
16. Aitken RJ, Krausz C. Oxidative stress, DNA damage and the Y chromosome. Reproduction. 2001;122: 497–506.
17. Irvine DS, Twigg JP, Gordon EL, Fulton N, Milne PA, Aitken RJ. DNA integrity in human spermatozoa: relationships with semen quality. J Androl. 2000;21: 33–44.
18. Lopes S, Jurisicova A, Sun JG, Casper RF. Reactive oxygen species: potential cause for DNA fragmentation in human spermatozoa. Hum Reprod. 1998; 13:896–900.
19. Oger I, Da Cruz C, Panteix G, Menezo Y. Evaluating human sperm DNA integrity: relationship between 8-hydroxydeoxyguanosine quantification and the sperm chromatin structure assay. Zygote. 2003;11: 367–71.
20. Potts RJ, Notarianni LJ, Jefferies TM. Seminal plasma reduces exogenous oxidative damage to human sperm, determined by the measurement of DNA strand breaks and lipid peroxidation. Mutat Res. 2000;447:249–56.
21. Said TM, Aziz N, Sharma RK, Lewis-Jones I, Thomas Jr AJ, Agarwal A. Novel association between sperm deformity index and oxidative stress-induced DNA damage in infertile male patients. Asian J Androl. 2005;7:121–6.
22. Saleh RA, Agarwal A, Nada EA, El-Tonsy MH, Sharma RK, Meyer A, et al. Negative effects of increased sperm DNA damage in relation to seminal oxidative stress in men with idiopathic and male factor infertility. Fertil Steril. 2003;79 Suppl 3:1597–605.
23. Twigg J, Fulton N, Gomez E, Irvine DS, Aitken RJ. Analysis of the impact of intracellular reactive oxygen species generation on the structural and functional integrity of human spermatozoa: lipid peroxidation, DNA fragmentation and effectiveness of antioxidants. Hum Reprod. 1998;13:1429–36.
24. Sawyer DE, Mercer BG, Wiklendt AM, Aitken RJ. Quantitative analysis of gene-specific DNA damage in human spermatozoa. Mutat Res. 2003;529:21–34.
25. Iwasaki A, Gagnon C. Formation of reactive oxygen species in spermatozoa of infertile patients. Fertil Steril. 1992;57:409–16.
26. Zini A, de Lamirande E, Gagnon C. Reactive oxygen species in semen of infertile patients: levels of superoxide dismutase- and catalase-like activities in seminal plasma and spermatozoa. Int J Androl. 1993;16: 183–8.
27. Kodama H, Yamaguchi R, Fukuda J, Kasai H, Tanaka T. Increased oxidative deoxyribonucleic acid damage in the spermatozoa of infertile male patients. Fertil Steril. 1997;68:519–24.
28. Shen HM, Chia SE, Ong CN. Evaluation of oxidative DNA damage in human sperm and its association with male infertility. J Androl. 1999;20:718–23.
29. Aitken RJ, West K, Buckingham D. Leukocytic infiltration into the human ejaculate and its association with semen quality, oxidative stress, and sperm function. J Androl. 1994;15:343–52.
30. Barroso G, Morshedi M, Oehninger S. Analysis of DNA fragmentation, plasma membrane translocation of phosphatidylserine and oxidative stress in human spermatozoa. Hum Reprod. 2000;15:1338–44.
31. de Lamirande E, Jiang H, Zini A, Kodama H, Gagnon C. Reactive oxygen species and sperm physiology. Rev Reprod. 1997;2:48–54.
32. Gomez E, Buckingham DW, Brindle J, Lanzafame F, Irvine DS, Aitken RJ. Development of an image analysis system to monitor the retention of residual cytoplasm by human spermatozoa: correlation with biochemical markers of the cytoplasmic space, oxidative stress, and sperm function. J Androl. 1996;17: 276–87.
33. Muratori M, Piomboni P, Baldi E, Filimberti E, Pecchioli P, Moretti E, et al. Functional and ultrastructural features of DNA-fragmented human sperm. J Androl. 2000;21:903–12.
34. Alvarez JG, Touchstone JC, Blasco L, Storey BT. Spontaneous lipid peroxidation and production of hydrogen peroxide and superoxide in human spermatozoa. Superoxide dismutase as major enzyme protectant against oxygen toxicity. J Androl. 1987;8: 338–48.
35. Aitken RJ, Clarkson JS. Cellular basis of defective sperm function and its association with the genesis of reactive oxygen species by human spermatozoa. J Reprod Fertil. 1987;81:459–69.

36. de Lamirande E, Gagnon C. Reactive oxygen species and human spermatozoa. I. Effects on the motility of intact spermatozoa and on sperm axonemes. J Androl. 1992;13:368–78.
37. Lewis SE, Aitken RJ. DNA damage to spermatozoa has impacts on fertilization and pregnancy. Cell Tissue Res. 2005;322:33–41.
38. Yang MH, Schaich KM. Factors affecting DNA damage caused by lipid hydroperoxides and aldehydes. Free Radic Biol Med. 1996;20:225–36.
39. Gagnon C, Iwasaki A, De Lamirande E, Kovalski N. Reactive oxygen species and human spermatozoa. Ann N Y Acad Sci. 1991;637:436–44.
40. Jeulin C, Soufir JC, Weber P, Laval-Martin D, Calvayrac R. Catalase activity in human spermatozoa and seminal plasma. Gamete Res. 1989;24:185–96.
41. Twigg J, Irvine DS, Houston P, Fulton N, Michael L, Aitken RJ. Iatrogenic DNA damage induced in human spermatozoa during sperm preparation: protective significance of seminal plasma. Mol Hum Reprod. 1998;4:439–45.
42. Jow WW, Schlegel PN, Cichon Z, Phillips D, Goldstein M, Bardin CW. Identification and localization of copper-zinc superoxide dismutase gene expression in rat testicular development. J Androl. 1993;14:439–47.
43. Watanabe S. Frequent structural chromosome aberrations in immotile human sperm exposed to culture media. Hum Reprod. 2004;19:940–7.
44. Zini A, Schlegel PN. Catalase mRNA expression in the male rat reproductive tract. J Androl. 1996;17:473–80.
45. Zini A, Schlegel PN. Expression of glutathione peroxidases in the adult male rat reproductive tract. Fertil Steril. 1997;68:689–95.
46. Zini A, Schlegel PN. Identification and characterization of antioxidant enzyme mRNAs in the rat epididymis. Int J Androl. 1997;20:86–91.
47. Chabory E, Damon C, Lenoir A, Kauselmann G, Kern H, Zevnik B, et al. Epididymis seleno-independent glutathione peroxidase 5 maintains sperm DNA integrity in mice. J Clin Invest. 2009;119:2074–85.
48. Weir CP, Robaire B. Spermatozoa have decreased antioxidant enzymatic capacity and increased reactive oxygen species production during aging in the Brown Norway rat. J Androl. 2007;28:229–40.
49. Holmes RP, Goodman HO, Shihabi ZK, Jarow JP. The taurine and hypotaurine content of human semen. J Androl. 1992;13:289–92.
50. Appasamy M, Muttukrishna S, Pizzey AR, Ozturk O, Groome NP, Serhal P, et al. Relationship between male reproductive hormones, sperm DNA damage and markers of oxidative stress in infertility. Reprod Biomed Online. 2007;14:159–65.
51. Fraga CG, Motchnik PA, Shigenaga MK, Helbock HJ, Jacob RA, Ames BN. Ascorbic acid protects against endogenous oxidative DNA damage in human sperm. Proc Natl Acad Sci USA. 1991;88:11003–6.
52. Song GJ, Lewis V. Mitochondrial DNA integrity and copy number in sperm from infertile men. Fertil Steril. 2008;90:2238–44.
53. Verit FF, Verit A, Kocyigit A, Ciftci H, Celik H, Koksal M. No increase in sperm DNA damage and seminal oxidative stress in patients with idiopathic infertility. Arch Gynecol Obstet. 2006;274:339–44.
54. Lewis SE, Boyle PM, McKinney KA, Young IS, Thompson W. Total antioxidant capacity of seminal plasma is different in fertile and infertile men. Fertil Steril. 1995;64:868–70.
55. Sanocka D, Miesel R, Jedrzejczak P, Kurpisz MK. Oxidative stress and male infertility. J Androl. 1996; 17:449–54.
56. Smith R, Vantman D, Ponce J, Escobar J, Lissi E. Total antioxidant capacity of human seminal plasma. Hum Reprod. 1996;11:1655–60.
57. Hampl JS, Taylor CA, Johnston CS. Vitamin C deficiency and depletion in the United States: the Third National Health and Nutrition Examination Survey, 1988 to 1994. Am J Public Health. 2004;94:870–5.
58. Jacob RA. Assessment of human vitamin C status. J Nutr. 1990;120 Suppl 11:1480–5.
59. Ryle PR, Thomson AD. Nutrition and vitamins in alcoholism. Contemp Issues Clin Biochem. 1984;1: 188–224.
60. Silver EW, Eskenazi B, Evenson DP, Block G, Young S, Wyrobek AJ. Effect of antioxidant intake on sperm chromatin stability in healthy nonsmoking men. J Androl. 2005;26:550–6.
61. Gil-Villa AM, Cardona-Maya W, Agarwal A, Sharma R, Cadavid A. Role of male factor in early recurrent embryo loss: do antioxidants have any effect? Fertil Steril. 2009;92:565–71.
62. Greco E, Iacobelli M, Rienzi L, Ubaldi F, Ferrero S, Tesarik J. Reduction of the incidence of sperm DNA fragmentation by oral antioxidant treatment. J Androl. 2005;26:349–53.
63. Greco E, Romano S, Iacobelli M, Ferrero S, Baroni E, Minasi MG, et al. ICSI in cases of sperm DNA damage: beneficial effect of oral antioxidant treatment. Hum Reprod. 2005;20:2590–4.
64. Menezo YJ, Hazout A, Panteix G, Robert F, Rollet J, Cohen-Bacrie P, et al. Antioxidants to reduce sperm DNA fragmentation: an unexpected adverse effect. Reprod Biomed Online. 2007;14:418–21.
65. Tremellen K, Miari G, Froiland D, Thompson J. A randomised control trial examining the effect of an antioxidant (Menevit) on pregnancy outcome during IVF-ICSI treatment. Aust N Z J Obstet Gynaecol. 2007;47:216–21.
66. Tunc O, Thompson J, Tremellen K. Improvement in sperm DNA quality using an oral antioxidant therapy. Reprod Biomed Online. 2009;18:761–8.
67. Tunc O, Tremellen K. Oxidative DNA damage impairs global sperm DNA methylation in infertile men. J Assist Reprod Genet. 2009;26:537–44.
68. Piomboni P, Gambera L, Serafini F, Campanella G, Morgante G, De Leo V. Sperm quality improvement after natural anti-oxidant treatment of asthenoteratospermic men with leukocytospermia. Asian J Androl. 2008;10:201–6.

69. Said TM, Agarwal A, Sharma RK, Thomas Jr AJ, Sikka SC. Impact of sperm morphology on DNA damage caused by oxidative stress induced by beta-nicotinamide adenine dinucleotide phosphate. Fertil Steril. 2005;83:95–103.
70. Cho C, Willis WD, Goulding EH, Jung-Ha H, Choi YC, Hecht NB, et al. Haploinsufficiency of protamine-1 or -2 causes infertility in mice. Nat Genet. 2001;28:82–6.
71. Libman J, Gabriel MS, Sairam MR, Zini A. Catalase can protect spermatozoa of FSH receptor knock-out mice against oxidant-induced DNA damage in vitro. Int J Androl. 2010;33(6):818–22.
72. Russo A, Troncoso N, Sanchez F, Garbarino JA, Vanella A. Propolis protects human spermatozoa from DNA damage caused by benzo[a]pyrene and exogenous reactive oxygen species. Life Sci. 2006;78: 1401–6.
73. Sierens J, Hartley JA, Campbell MJ, Leathem AJ, Woodside JV. In vitro isoflavone supplementation reduces hydrogen peroxide-induced DNA damage in sperm. Teratog Carcinog Mutagen. 2002;22:227–34.
74. Aitken RJ, Buckingham DW, Brindle J, Gomez E, Baker HW, Irvine DS. Analysis of sperm movement in relation to the oxidative stress created by leukocytes in washed sperm preparations and seminal plasma. Hum Reprod. 1995;10:2061–71.
75. Anderson D, Schmid TE, Baumgartner A, Cemeli-Carratala E, Brinkworth MH, Wood JM. Oestrogenic compounds and oxidative stress (in human sperm and lymphocytes in the Comet assay). Mutat Res. 2003; 544:173–8.
76. Cemeli E, Schmid TE, Anderson D. Modulation by flavonoids of DNA damage induced by estrogen-like compounds. Environ Mol Mutagen. 2004;44:420–6.
77. Dobrzynska MM, Baumgartner A, Anderson D. Antioxidants modulate thyroid hormone- and noradrenaline-induced DNA damage in human sperm. Mutagenesis. 2004;19:325–30.
78. Chi HJ, Kim JH, Ryu CS, Lee JY, Park JS, Chung DY, et al. Protective effect of antioxidant supplementation in sperm-preparation medium against oxidative stress in human spermatozoa. Hum Reprod. 2008;23: 1023–8.
79. Donnelly ET, McClure N, Lewis SE. The effect of ascorbate and alpha-tocopherol supplementation in vitro on DNA integrity and hydrogen peroxide-induced DNA damage in human spermatozoa. Mutagenesis. 1999;14:505–12.
80. Donnelly ET, McClure N, Lewis SE. Glutathione and hypotaurine in vitro: effects on human sperm motility, DNA integrity and production of reactive oxygen species. Mutagenesis. 2000;15:61–8.
81. Hughes CM, Lewis SE, McKelvey-Martin VJ, Thompson W. The effects of antioxidant supplementation during Percoll preparation on human sperm DNA integrity. Hum Reprod. 1998;13:1240–7.
82. Taylor K, Roberts P, Sanders K, Burton P. Effect of antioxidant supplementation of cryopreservation medium on post-thaw integrity of human spermatozoa. Reprod Biomed Online. 2009;18:184–9.
83. Branco CS, Garcez ME, Pasqualotto FF, Erdtman B, Salvador M. Resveratrol and ascorbic acid prevent DNA damage induced by cryopreservation in human semen. Cryobiology. 2010;60(2):235–7.
84. Li Z, Lin Q, Liu R, Xiao W, Liu W. Protective effects of ascorbate and catalase on human spermatozoa during cryopreservation. J Androl. 2010;31(5):437–44.
85. Martinez-Soto JC, de Dioshourcade J, Gutierrez-Adan A, Landeras JL, Gadea J. Effect of genistein supplementation of thawing medium on characteristics of frozen human spermatozoa. Asian J Androl. 2010; 12(3):431–41.
86. Thomson LK, Fleming SD, Aitken RJ, De Iuliis GN, Zieschang JA, Clark AM. Cryopreservation-induced human sperm DNA damage is predominantly mediated by oxidative stress rather than apoptosis. Hum Reprod. 2009;24:2061–70.

Sperm Chromatin and ART (IUI, IVF and ICSI) Pregnancy

31

Mona Bungum

Abstract

Diagnosis of male infertility has been based mainly on the traditional semen parameters, namely, concentration, motility and morphology. In assisted reproductive technologies (ART), sperm samples are prepared by methods such as swim-up or density-gradient centrifugation to sort out sperm populations believed to have the highest fertilization potential. Traditionally, results of semen analysis and sperm preparation have been the fundaments on which clinicians could base their decision of what ART method should be used for a given couple. It has, however, become apparent that none of these procedures are sufficient for the determination of male fertility capacity. Owing to the lack of adequate methods to evaluate the fertility potential of a couple, the choice of ART method is made more or less blindly. A continuous search for better markers of male fertility has led to an increased focus on sperm chromatin integrity testing in fertility workup and ART. Numerous sperm DNA integrity tests have been developed. In the context of fertility, the comet, TUNEL, and Sperm Chromatin Structure assays are the most frequently used. Sperm DNA fragmentation has shown to be an independent predictor of success in couples undergoing intrauterine insemination. More contrasting data exist regarding the role of sperm DNA fragmentation in relation to fertilization, pre-embryo development and pregnancy outcome in in vitro fertilization and intracytoplasmic sperm injection (ICSI).

Keywords

Male infertility • Assisted reproductive technology • Intracytoplasmic sperm injection • Intrauterine insemination • In vitro fertilization

In Western countries, 17–25% of couples in reproductive age are seeking medical care for problems of conception [1, 2]. Thanks to the introduction of assisted reproductive technologies (ART), now, almost every involuntarily

M. Bungum (✉)
Reproductive Medicine Centre, Skåne University Hospital Malmö, Lund University, Sodra Forstadsgatan, Malmö 20502, Sweden
e-mail: mona.bungum@med.lu.se

A. Zini and A. Agarwal (eds.), *Sperm Chromatin: Biological and Clinical Applications in Male Infertility and Assisted Reproduction*, DOI 10.1007/978-1-4419-6857-9_31,

childless couple has a realistic hope of parenting. In particular, the introduction of intracytoplasmic sperm injection (ICSI) has revolutionized the area of fertility [3]. The number of ART treatments, in particular ICSI cycles, is steadily increasing [4]. While in the beginning of the era of ICSI the indication for this type of treatment was severe male infertility, now also couples whose male partners are without sperm defects request and are treated with ICSI. However, by ICSI all natural biological barriers that prevent fertilization with defective sperm are bypassed, and its increasing use has led to a growing concern of transmission of genetic and epigenetic diseases.

Although the development of ART has brought us further and led to a vast increase in our understanding of early reproductive function, ART performances have been stable and we have witnessed no net improvement in healthy term pregnancy rate during the last two decades [5]. One reason for this can be a lack of adequate methods to evaluate the fertility potential of a couple and also a lack of methods to identify the most effective type of ART treatment for a given couple.

So far, the traditional semen analysis has been a cornerstone in the diagnosis of male fertility and also used as a tool to decide which ART method to use. The sperm parameters, namely, concentration, motility and morphology, are, however, claimed to be poorly standardized, subjective [6] and not powerful predictors of fertility [7, 8]. A search for better predictors of fertility has contributed to a growing focus on the genomic integrity of the male gametes used for ART [9, 10]. During the last few decades, several methods to assess sperm DNA damage have been developed. Although still many questions remain to be answered, it is evident that sperm DNA integrity is a valuable marker of male fertility, alone or in combination with the conventional semen parameters, in natural conception as well as in ART. This chapter reviews the role of sperm chromatin integrity in ART.

Assisted Reproductive Technologies

The term ART covers all reproductive technologies that involve the handling of gametes outside the body, either sperm alone as in intrauterine insemination (IUI), or oocytes, sperm and embryos as in in vitro fertilization (IVF) and ICSI [11]. ART is primarily used as a treatment of infertility/subfertility and also, to some extent, in establishing pregnancy in couples carrying inherited genetic diseases. The very first documented successful use of ART in humans was in 1978 when the world first IVF-baby was born [12]. Now, about 30 years later, ART is applied worldwide, and it is estimated that more than three million babies have been born as a result of ART since then [13]. The number of ART treatments is rising every year [14].

The first choice of treatment used in ovulatory dysfunction, minimal endometriosis, unexplained subfertility and milder forms of male subfertility is the relatively simple IUI. Following a mild controlled ovarian stimulation, prepared semen is inseminated into the woman's uterus. In tubal factors, IVF is used [11]. In IVF, oocytes are fertilized by sperm in vitro. Two to five days later the pre-embryo is replaced into the woman's uterus. In ICSI, nearly the same principles are followed, but one single spermatozoon is selected and injected directly into the cytoplasm of the oocyte.

Traditional Markers of ART Fertility Potential

Prediction of the fertility potential of a couple has never been more crucial than now. We are facing delayed childbearing and falling sperm counts as possible threats to fertility. Various predictors of fertility have been suggested; however, none is shown to be ideal. While in the female age is the only parameter that has been shown to have the potential to predict ART outcome [15], for long it was thought that the traditional sperm parameters could predict male fertilization capability. In ART, sperm samples are prepared by methods such as swim-up or density gradient centrifugation to sort out populations of sperm believed to have the highest fertilization potential. Traditionally, concentration and motility after sperm preparation have been one of the fundaments driving clinicians decisions about the choice of the specific ART method recommended for a given couple.

It has, however, not shown to be sufficient for assessing of the fertilizing capacity of a sperm.

Several other laboratory tests of sperm function have been suggested, such as antisperm antibody test, vital staining, biochemical analysis of semen, hypoosmotic swelling test, sperm penetration assay, hemizona assay, creatine kinase, reactive oxygen species (ROS) tests and computer-assisted sperm analysis (CASA) [16]; however, the clinical value of these tests has been questioned [17], and only a few of them have been implemented in routine clinical use.

Owing to the lack of tools to predict sperm fertilizating capacity, the criteria for choosing ICSI and, as a consequence, the ratio between IVF and ICSI vary from clinic to clinic. Despite the fact that, in unexplained infertility, fertilization rates are as good in IVF as in ICSI [18], many clinics now perform ICSI as their primary, if not the only, ART technique [4, 19].

Sperm Chromatin Integrity Testing

The evidence that infertile men in general possess substantially more sperm DNA damage than fertile men [20–28] has led to a growing focus on sperm chromatin integrity testing as an adjunct tool to the traditional sperm parameters in prediction of fertility. During the past three decades, a variety of techniques to assess sperm chromatin integrity have been developed. Principles, procedures and other aspects of the different tests are reviewed in detail in other chapters of this book. Briefly, mainly four tests assessing sperm DNA damage are used in ART, namely, the comet assay (single cell gel electrophoresis) [29], the TUNEL (terminal deoxynucleotidyl transferase-mediated dUDP nick-end labelling) assay [30], the Sperm Chromatin Structure assay (SCSA) [31, 32] and the Sperm Chromatin Dispersion (SCD) test [33].

Comet assay is a fluorescence-microscopy-based test. In this assay, spermatozoa are mixed with melted agarose and then placed on a glass slide. Thereafter, the cells are lysed and subjected to horizontal electrophoresis. DNA is visualized with the help of DNA-specific fluorescent dyes, and DNA damage is quantified by measuring the displacement between the nuclear genetic material of the comet head and the broken DNA migrated in the tail.

TUNEL assay can be run using both bright-field/fluorescence microscopy and flow cytometry. In the TUNEL assay, terminal deoxynucleotidyl transferase (TdT) incorporates labelled nucleotides to 3′-OH at single- and double-strand DNA breaks to create a signal, which increases with the number of DNA breaks. On a microscope slide, sperm are scored and classified as positive or negative depending whether they are labelled or not. In flow cytometry, the fraction of positive sperm is represented by the cells above a threshold channel value on a relative fluorescence intensity scale.

SCSA is a flow-cytometric test that measures the susceptibility of sperm DNA to acid-induced DNA denaturation in situ, followed by staining with acridine orange [31, 32]. The level of DNA denaturation is determined by measuring the shift from green fluorescence (double-stranded, native DNA) to red fluorescence (single-stranded, denatured DNA) in a flow cytometer, followed by further analysis by a specific SCSA software. The extent of DNA denaturation is expressed as DNA fragmentation index (DFI) [32]. The fraction of high DNA stainable (HDS) cells, thought to represent immature spermatozoa, is also recorded [32].

Similar to the SCSA, the fluorescence/light microscopic SCD test determines the susceptibility of sperm DNA to acid denaturation [33, 34]. Briefly, intact spermatozoa are immersed in an agarose matrix on a slide, treated with an acid solution to denature DNA that contains breaks and then treated with lysis buffer to remove membranes and proteins. Removal of nuclear proteins results in nucleoids with a central core and a peripheric halo of dispersed DNA loops. Sperm nuclei with elevated DNA fragmentation produce very small or no halos of DNA dispersion, whereas those sperm with low levels of DNA fragmentation release their DNA loops forming large halos. The sperm nucleoids may be visualized using fluorescence microscopy, after staining with a DNA-specific fluorochrome, or bright-field microscopy.

Moderate-to-high correlations between these different tests have been reported [30, 34–36], indicating that, very likely, these tests are not addressing identical aspects of the complex processes underlying sperm nuclear packaging potentially resulting in DNA breaks [37]. The test that has been most extensively tested clinically and found to have the most stable threshold values is the SCSA [20, 24, 31, 38], and this chapter focuses mainly on the results from SCSA-based studies.

Several reports have demonstrated that the association between sperm DNA damage and the traditional semen parameters is only weak-to-moderate [39, 40]. It is also shown that infertile men may have normal standard sperm characteristics according to WHO criteria, but a high number of sperm DNA defects.

In a recent case–control study on infertile vs. fertile men, the risk of being infertile resulted increased when DFI, as measured by SCSA, was above 20% in men with normal standard semen parameters, with an odds ratio (OR) of 5.1 (CI: 1.2–23). If any one of the WHO parameters were abnormal, the OR for infertility was increased already at DFI above 10% (OR 16, CI: 4.2–60). DFI above 20% was found in 40% of men with otherwise normal standard parameters [41]. In another study of 350 Latvian men from infertile couples [42], 20% of the men with otherwise normal WHO sperm parameters had a SCSA-DFI above 20%. This is clinically relevant in counselling couples during the fertility workup, and also in couples seeking ART where the choice of treatment most often is based upon the traditional sperm parameters and where an underlying high DFI can hinder a pregnancy.

Sperm Chromatin Integrity Testing in ART

Intrauterine Insemination

The first study indicating an association between sperm DNA damage and reduced pregnancy chances after IUI was published by Duran et al. [43]. In a retrospective study of 154 IUI cycles, they found that pregnancy could not be achieved when DFI, as measured by the TUNEL assay, was above 12%. Similar findings have been reported by Saleh's group [28] who performed a small study where 12 of 19 couples had a DFI value as measured by SCSA above 28% and none of these couples achieved a pregnancy. Boe-Hansen et al. [44] used SCSA in a study on 48 IUI couples. Only two of the couples had a DFI value above 30%, and none of the couples achieved a pregnancy. Recently, in a study of 387 IUI cycles, it shown that the SCSA parameter DFI can be used as an independent predictor of fertility [38]. While the proportion of children born per cycle was 19.0% when the DFI value was below 30%, those with a DFI value above 30% only had a take-home-baby rate of 1.5%. These IUI results are in good accordance with those results obtained from natural conception. In fact, both Evenson et al. [20] and Spanò et al. [24] demonstrated that, after unprotected intercourse, time-to-pregnancy increased (fertile couples took longer to conceive) as a function of the proportion of sperm with abnormal chromatin measured by the SCSA [20, 24]. By contrast, no correlation was found between SCD results and pregnancy outcome in 100 Spanish IUI patients [45].

Normal sperm DNA integrity seems to be particularly important when the contact between the two gametes occurs in a natural way as in natural conception and IUI. It has been suggested that selective pressures operate to avoid the development of an embryo derived from sperm with a high load of genetic damage in a natural environment [29]. Additionally, spermatozoa with damaged DNA could be more prone to undergo apoptosis during the transport through the genital tract than spermatozoa with normal DNA integrity. For an overview of IUI-papers, see Table 31.1.

In Vitro Fertilization and Intracytoplasmic Sperm Injection

Numerous of retrospective studies have examined the role of sperm chromatin damage in IVF and ICSI. In Table 31.2, an overview of studies using SCSA, TUNEL, comet or SCD assays is presented.

Sperm DNA Damage in Relation to Pregnancy Outcome

Some of the first studies relating outcome of ART to sperm DNA damage suggested that a DFI

Table 31.1 Influence of sperm DNA damage on pregnancy rates in IUI treatment

References	Patients (*n*)	Pregnancy rates impaired	Test applied	DNA fragmentation index (DFI)-threshold suggested (%)
Duran et al. [43]	154	Yes	TUNEL	12
Saleh et al. [28]	19	Yes	SCSA	30
Bungum et al. [48]	131	Yes	SCSA	27
Muriel et al. [45]	100	No	SCD	–
Bungum et al. [38]	387	Yes	SCSA	30

IUI Intrauterine insemination; *SCSA* Sperm Chromatin Structure assay; *TUNEL* terminal deoxynucleotidyl transferase dUTP nick-end labelling; *SCD* Sperm Chromatin Dispersion test

Table 31.2 Influence of sperm DNA damage on fertilization, embryo development and pregnancy rates in IVF and ICSI

References	IVF (*n*)	ICSI (*n*)	Fertilization rates impaired	Embryo development impaired	Pregnancy rates impaired	Test applied
Tomsu et al. [62]	40	0	No	Yes	Yes	Comet
Morris et al. [29]	20	40	No	Yes	NA	Comet
Caglar et al. [116]	0	56	No	No	No	Comet
Lewis et al. [64]	0	77	No	NA	Yes	Comet
Nasr-Esfahani et al. [66]	0	28	No	No	NA	Comet
Larson-Cook et al. [47]	55	34	No	No	Yes	SCSA
Larson et al. [46]	24 IVF/ICSI	NA	No	No	Yes	SCSA
Saleh et al. [28]	10	4	Yes	Yes	Yes	SCSA
Bungum et al. [48]	109	66	No	No	Yes	SCSA
Gandini et al. [49]	12	24	No	Yes (blastocysts)	Yes	SCSA
Virro et al. [50]	249 IVF/ICSI	NA	No	No	Yes	SCSA
Check et al. [117]	0	106	No	No	Yes	SCSA
Payne et al. [52]	46	54	No	No	No	SCSA
Boe-Hansen et al. [44]	139	47	No	No	Yes	SCSA
Bungum et al. [38]	388	223	No	No	Yes	SCSA
Sun et al. [67]	143	0	Yes	Yes	NA	TUNEL
Lopes et al. [68]	0	150	Yes	No	NA	TUNEL
Host et al. [22]	50	61	Yes	NA	NA	TUNEL
Tomlinson et al. [61]	140	0	No	No	Yes	TUNEL
Benchaib et al. [85]	50	54	Yes	No	Yes	TUNEL
Henkel et al. [63]	208	54	No	No	No	TUNEL
Huang et al. [65]	217	86	Yes	No	No	TUNEL
Seli et al. [75]	49	NA	NA	Yes	No	TUNEL
Henkel et al. [118]	208	54	No	No	No	TUNEL
Hammadeh et al. [87]	26	22	NA	NA	No	TUNEL
Borini et al. [88]	82	50	NA	NA	Only for ICSI	TUNEL
Benchaib et al. [86]	88	234	Only for ICSI	Only for ICSI	No	TUNEL
Bakos et al. [119]	45	68	Only for IVF	No	Only for ICSI	TUNEL
Frydman et al. [120]	117	0	NA	NA	Yes	TUNEL

(continued)

Table 31.2 (continued)

References	IVF (*n*)	ICSI (*n*)	Fertilization rates impaired	Embryo development impaired	Pregnancy rates impaired	Test applied
Tarozzi et al. [121]	82	50	NA	NA	Only for ICSI	TUNEL
Muriel et al. [45]	85 IVF/ICSI	NA	NA	NA	No	SCD
Velez de la Calle et al. [122]	622 IVF/ICSI	NA	No	Yes	No	SCD
Tavalaee et al. [123]	92 IVF/ICSI	NA	Only for ICSI	NA	No	SCD

IVF In vitro fertilization; *ICSI* intracytoplasmic sperm injection; *SCSA* Sperm Chromatin Structure assay; *TUNEL* terminal deoxynucleotidyl transferase dUTP nick-end labelling; *SCD* Sperm Chromatin Dispersion test; *NA* not applicable

above 27% as measured by SCSA could be used as a cut-off value for infertility. The authors reported that in couples with a DFI above 27%, no pregnancy could be obtained, regardless of the type of ART applied [46, 47]. However, in 2004 when three independent SCSA reports demonstrated that a DFI level above 27% was indeed compatible with pregnancy and delivery after both IVF and ICSI [48–50], it became evident that ART can compensate poor sperm chromatin quality.

Gandini et al. [49], in a study involving 34 couples (12 IVF and 22 ICSI), did not note any difference between patients initiating pregnancies or not. They reported healthy full-term pregnancies with levels of DFI up to 66.3%. Bungum et al. [48] investigated 109 consecutive couples undergoing IVF and 66 couples undergoing ICSI. No statistically significant difference in the pregnancy outcome was noted by dividing patients according to the DFI level of 27%. However, in the group with a DFI above 27%, the results of ICSI were significantly better than those of IVF, clinical pregnancy (52.9 vs. 22.2%), implantation (37.5 vs. 19.4%) and delivery (47.1 vs. 22.2%). Virro et al. [50] studied 249 couples undergoing IVF/ICSI and noted that men with DFI below 33% had a significantly greater chance of initiating a pregnancy, lower rate of spontaneous abortions and an increased rate of ongoing pregnancies at 12 weeks (47 vs. 28%) than those with a DFI above 33%.

These data were in agreement with other previous smaller reports using TUNEL or comet assays, showing that sperm DNA damage is more predictive in IVF and, less in ICSI [22, 51]. This was later confirmed in a larger data set including nearly 1,000 men in IUI, IVF or ICSI treatment using DFI 30% as threshold level. No statistically significant difference between the outcomes of ICSI vs. IVF in the group with DFI ≤30% was seen. In the DFI >30% group, however, the results of ICSI were significantly better than those of IVF. The odds ratios (ORs) for biochemical pregnancy (BP), clinical pregnancy (CP) and delivery (D) were 3.0 (95% CI: 1.4–6.2), 2.3 (5% CI: 1.1–4.6) and 2.2 (95% CI: 1.0–4.5), respectively. For ICSI, there was even a tendency towards higher rates of BP, CP and D with a DFI >30% vs. a DFI ≤30%, however, not reaching a statistically significant difference. Moreover, the implantation rate in the ICSI group with DFI >30% seemed to be higher than in any other subgroup. The other SCSA parameter, HDS did, however, not predict the outcome of IVF or ICSI, neither alone nor in combination with DFI [38]. By contrast, one single study had, however, reported that DFI and HDS threshold values were not valid [52]. The authors found that the poorer the integrity of sperm nuclear DNA, the better is the pregnancy outcome and suggested to "redefine the relationship between SCSA data and ART outcomes". The study was, however, based on only 100 IVF/ICSI treatments where female factor infertility not was taken into consideration.

Despite convincing data from several authors, some reports have challenged the predictive value of the SCSA test [53]. One example is a position paper from the Practice Committee of the American Society for Reproductive Medicine [54]. Although ASRM, after a meta-analysis on 14 published studies, stated that fragmented sperm DNA is more frequent in infertile than in fertile and may contribute to poor reproductive performance, but concluded that, so far, there was no proven role for routine DNA integrity testing in the evaluation of infertility. Other examples are two meta-analyses including studies using either TUNEL and SCSA assays. Both Collins et al. [55], who considered 13 IVF/ICSI studies (9 carried by SCSA and 4 by the TUNEL assay), and Zini et al. [56], who considered 9 IVF (6 carried out by TUNEL assay and 3 by SCSA) and 11 ICSI studies (6 carried by SCSA and 5 by the TUNEL assay) found only small associations between sperm DNA integrity test results and pregnancy in IVF and ICSI. Two other meta-analysis including only SCSA-studies have been performed. Based on 14 papers, Evenson and Wixon [57] reported that in IVF and ICSI, CP was closely related to DFI as measured by SCSA. By contrast, based on three papers, Li et al. [58] found that neither DFI nor HDS had an effect on the chance of CP after IVF or ICSI treatment.

Sperm DNA Damage in Relation to Fertilization

There is conflicting evidence about the relationship between sperm DNA fragmentation and fertilization rates after IVF and ICSI. Ahmadi and Ng [59] in a mouse model demonstrated that, despite a high DNA damage load, sperm were able to fertilize an oocyte. Also, several studies in the human have shown that men with high number of sperm with damaged DNA can have the same ability to fertilize in vitro as men with a lower fraction of sperm with DNA damage as measured by SCSA [38, 46–50, 58] or by other sperm DNA integrity assays [29, 60–66].

On the contrary, the presence of damaged sperm DNA was shown to have a significant inverse relationship with fertilization in other studies [22, 67] and to contribute to a failure of fertilization even in ICSI [68]. Host et al. [22] found a negative correlation between the proportion of spermatozoa with DNA strand breaks and the fertilization rates in all groups except for those undergoing ICSI.

Also, the SCSA parameter HDS, thought to represent immature spermatozoa with incomplete protamination, was found to be related to IVF fertilization rates, but not in ICSI [50]. Consequently, the authors suggested that men with HDS >15% should be treated with ICSI. This finding has, however, not been confirmed by others, and thus, HDS does not seem to have any clinical impact.

Sperm DNA Damage in Relation to Pre-Embryo Development

Although fertilization may be independent of sperm DNA integrity, the post-fertilization development of the pre-embryo can be impaired by sperm DNA damage.

It has been speculated in if and how sperm DNA damage has impact on human embryo and foetal development as well as on offspring health [69]. Incomplete or aberrant sperm DNA repair by the oocyte is hypothesized to create mutations in the genome of the zygote, which potentially could lead to implantation failure, early miscarriages or, in worst cases, diseases in the offspring [9, 70, 71]. While the mature spermatozoon itself does not have the capability to repair DNA damage, oocytes and early embryos may have this capacity [72] to a certain degree [73].

Among the first reports to indicate that sperm DNA damage is related to poor embryo development was studies in mice by Ahmadi and Ng [73]. The human data regarding pre-embryo development in relation to sperm DNA damage is somewhat conflicting. While some authors have reported similar cleavage stage embryo developmental rates between high and low DFI groups as measured by SCSA [44, 46, 47, 52, 74], others have shown that sperm DNA damage is negatively correlated with embryo quality after IVF and ICSI [28, 29, 67]. Two studies have also reported that men with high levels of DNA fragmentation are at increased risk of low blastocyst formation compared to men with a low DFI [50, 75], and consequently, it has been suggested to practice blastocyst culture as a routine in ART.

Raw vs. Prepared Semen

In a vast majority of cases, spermatozoa used for ART are prepared by density-gradient centrifugation or swim-up methods. Both approaches aim at separating normal sperm from lymphocytes, epithelial cells, abnormal or immature sperm, cell debris, bacteria and seminal fluid. Several previous reports have shown an improvement in the sperm chromatin parameters comparing neat semen samples and samples prepared for ART [39, 49, 61, 76–82]. On the contrary, other reports showed unchanged or worse results [29, 45, 62, 65, 75, 83–87].

One study has analyzed the same semen samples before and after density gradient centrifuged considering 510 ART cycles. In contrast to what has been seen for raw semen, no predictive value of the SCSA parameters DFI and HDS, evaluated on the prepared semen, emerged in relation to pregnancy outcome [82]. These data supported the two first SCSA-ART studies where the SCSA parameters were assessed also on prepared semen, even if on a more limited number of patients, 24 and 34, respectively [46, 49]. Using the TUNEL assay, Borini et al. [88] in ICSI patients found DFI >10% in density-gradient-centrifuged semen to be discriminative for pregnancy. Also, Duran et al. [43], in a study on IUI couples, used washed semen samples and found no pregnancy if DFI, as measured by the TUNEL assay, exceeded the level of 12%. Larson et al. [46] suggested that elevated DFI in neat semen may reflect chromatin or other abnormalities within the entire sperm population interfering with the ability of the sperm to fertilize, but not completely eliminated by the sperm preparation procedure.

Incubation of Sperm

Temperature and pH are known to influence on stability and developmental potential of gametes [89, 90], but as yet there is no developed sufficient good laboratory standards for incubation of sperm during the period between sperm preparation and fertilization. The duration and environment for sperm incubation vary from clinic to clinic. Peer et al. [91] found that a 2-h incubation of density-gradient-prepared ejaculates at 37°C led to increased nuclear degradation in terms of vacuolated nuclei in comparison to that at 21°C. Testicular sperm appear to be more susceptible to damage than ejaculated sperm, yet they are subjected to conditions under the assumption that they have similar resistance to injury. For example, incubation under aerobic conditions for 4 or 24 h at 37°C leads to marked sperm DNA damage [92, 93].

Testicular vs. Ejaculated Sperm in ART

Previous reports have shown that sperm DNA damage is significantly lower in the seminiferous tubules compared with the epididymis [94] or in ejaculated sperm [95]. Use of testicular sperm in couples with repeated pregnancy failure in ART and high sperm DNA fragmentation resulted in significant better pregnancy rates [94, 95]. Although use of testicular sperm may only have a potential of solving ROS-induced sperm DNA damage, these findings should be followed up by larger prospective, randomized studies. In the majority of cases, sperm DNA damage is believed to be ROS-induced [96].

The Use of Cryopreserved Sperm in ART

Some studies of cryopreservation of sperm have demonstrated that freezing–thawing has a negative effect on sperm DNA integrity [74], especially in infertile men [93, 97]. Cryopreservation can induce an increased rate of lipid peroxidation in the sperm plasma membrane, causing an overall increase in the concentration of oxygen radicals in the sample. Exposure to high ROS concentrations can result in the disruption of mitochondrial and plasma membranes, causing DNA fragmentation and a reduction in sperm motility [98]. Adding antioxidants to the cryoprotection media [99] have shown to be a promising ameliorating procedure. Another strategy shown to cause less chromatin damage to sperm is freezing of density-gradient-prepared semen instead of raw semen [97]. However, larger studies are needed to clarify whether these are more effective and gentle methods compared to those in use.

Intraindividual Variation of DFI in Relation to ART

One of the drawbacks by the conventional sperm analysis is the huge intraindividual variation reported for concentration, motility and morphology [100]. By contrast, the first SCSA reports found a lower intraindividual variation for DFI [101]. A more recent study of infertile men in ART treatment has, however, demonstrated a significant day-to-day variation of DFI with a mean coefficient of variation (CV) of 29% [102]. Data from a so far unpublished study has shown that among 616 men who had their semen analyzed by SCSA both in infertility workup and in the actual ART cycle, 85% of the men remained in the same DFI category; ≤30% or >30% from measurement 1 to 2. This implies that only 15% had a clinical effect of repeating the SCSA measurement (Oleszczuk et al., unpublished). Also, data from Giwercman et al. [41] demonstrated that a single SCSA analysis is a strong predictor of infertility.

Future Perspectives

Despite the growing knowledge in the field of sperm chromatin integrity testing in fertility, fundamental questions remain to be answered as part of a more detailed understanding of sperm chromatin and its packaging during spermatogenesis, sperm maturation, ejaculation and unpackaging in the oocyte. For further clinical relevance, we need to know more about the following: (1) the type of DNA damage, (2) the percentage of sperm with DNA damage, (3) the extent of DNA damage per spermatozoon, (4) whether there is combined nucleotide damage and DNA fragmentation, (5) whether DNA damage affects introns or exons and (6) the ability of the oocyte to repair sperm DNA damage in the fertilizing sperm [103]. Developing standardized sperm DNA integrity assays providing such information is of highest value.

We also know too little about the origins of the damage and what can be done to prevent or cure sperm DNA damage. Cause-related therapy in the form of antioxidants has been attempted to reduce DNA damage caused by oxidative stress [95, 104–110]. However, such studies have been rather limited in size and the data are conflicting. Further large-scale studies are needed to investigate the type, role and mode of antioxidant therapy, as well as other types of causal treatment.

Another important issue for the future should be the development of new sperm separation or sorting techniques where individual or populations of sperm with intact DNA are isolated. Currently, a number of new techniques to favour sperm with normal sperm DNA integrity have been suggested and used; however, none of them are implemented into clinical practice. These include the so-called high-magnification ICSI, a method where spermatozoa with surface vacuoles are discarded [111] and the recently introduced confocal light absorption scattering spectroscopy (CLASS) technology, which allows for the non-invasive visualization of subcellular structures [112]. Also, the use of Annexin-V columns has shown to reduce the number of sperm with DNA fragmentation [113].

Another strategy suggested to follow the role of sperm DNA damage on pre-implantation development is to assess whether a quantity of known DNA damage has been repaired by the oocyte or the embryo by analyzing DNA damage in the trophoblast cells obtained by blastocyst biopsy [103].

Data from mice show links between DNA damage in spermatozoa and defects in embryonic development as well as the long-term health of the offspring [114]. However, knowledge on if and how sperm DNA defects may influence the human offspring is lacking, and it is urgent to initiate such studies.

Lastly, the question whether sperm DNA integrity tests can be used as a tool in ART to find the most effective treatment type in a given couple is only partly solved. Although it is clearly shown that men with a SCSA-DFI above 30%

should benefit from being referred directly to IVF/ICSI, it is still questionable whether there is, in these men, a clear difference in efficacy between IVF and ICSI [38]. As all available IVF/ICSI data come from retrospective studies, prospective randomized controlled trials should be conducted.

Conclusions and Clinical Recommendations

ART fertility is a multifactorial issue and involves factors from both partners. Sperm DNA integrity status is only one piece in this puzzle. However, it covers an important aspect of sperm quality and function and should be routinely implemented as an adjunct to the conventional sperm parameters in fertility workup and ART, especially in unexplained subfertility. Among the sperm DNA integrity tests currently available, SCSA has provided the most stabile clinical threshold values in relation to infertility.

Based on existing data, it is evident that the relevance of sperm DNA integrity testing concerns, first of all, in vivo fertilization. In addition to its role as a predictor of natural conception, the SCSA parameter DFI, as measured in raw semen, can be used as an independent predictor of success in couples undergoing IUI. The predictive role of SCSA in IVF and ICSI are, however, more doubtful and needs to be further investigated by prospective randomized studies. In IVF and ICSI, it seems clear that no association between sperm DNA damage and fertilization rates exist. The same seems to be the case for embryo development until day 3. This has, however, indicated that blastocyst development is impaired in patients with high numbers of sperm with DNA damage.

In men having standard sperm parameters that indicate ICSI, there are no therapeutic consequences of performing SCSA [38, 48–50]. Men with high numbers of DNA-fragmented sperm have similar chances of obtaining pregnancy by IVF and ICSI as men with low sperm DNA fragmentation. However, the group of men who, first of all, will benefit from SCSA assessment would be unexplained subfertile men. Roughly, 20–25% of subfertile men, one out of four, with normal WHO sperm parameters have a SCSA-DFI above 20–30%, which is the DFI level where the chance of giving rise to a spontaneous or IUI-induced pregnancy reduces significantly. In order to find men with sperm DNA damage as a hidden cause to their childlessness, where the traditional semen analysis shows one or no abnormality, a SCSA analysis should be offered [41]. In men where all standard parameters are normal, chances of in vivo pregnancy starts to reduce for DFI above 20%. In the presence of one abnormal semen quality parameter, the chance of spontaneous pregnancy is significantly reduced already at DFI above 10%. Thus, in such couples DFI should be taken into consideration and the couples should be referred directly to IVF/ICSI [38].

The SCSA parameter DFI is more stable than the conventional WHO parameters.[115] In most men, a single analysis is enough to be of clinical value for the choice of ART treatment. However, when DFI is above 20% it is recommended to repeat the test prior to the actual ART treatment. Unfortunately, couples seeking ART are only to a limited degree counselled in regard to the impact of lifestyle factors on fertility. Existing knowledge on factors contributing to sperm DNA damage as for instance smoking and obesity should to a higher degree be communicated to the couples.

Laboratory procedures can harm sperm DNA integrity. In order to prevent further sperm DNA damage and to sort out sperm with fragmented DNA, density-gradient preparation is a good choice for sperm preparation. However, one should be aware that repeated centrifugations as well as the speed of centrifugation could have negative effects on sperm chromatin. Also, to prevent further DNA damage, semen samples should be processed as close to the fertilization procedure as possible. Cryopreserved and thawed semen should be tested in regard to sperm DNA damage prior to use in ART, and the fertilization method chosen according to the DFI should be assessed post thawing.

In conclusion, more research is needed to improve our current knowledge on DNA anomalies in spermatozoa. It is necessary to standardize better the methods of DNA damage evaluation and the time to apply them as pregnancy predictors in assisted reproduction. Moreover, a greater insight into the causes of sperm DNA damage is needed to develop appropriate treatment strategies and to enhance the genomic integrity of spermatozoa, thus contributing to optimize assisted reproduction outcome. So far, available data has shown that DFI as measured by SCSA can be used as a valuable tool in ART treatment and adds to the clinical management of subfertile/infertile couples. DFI has also shown to be an independent predictor of fertility in IUI and can be used to decide which type of ART treatment is needed for a couple.

References

1. De Kretser DM. Male infertility. Lancet. 1997;349:787–90.
2. Dunson DB, Baird DD, Colombo B. Increased infertility with age in men and women. Obstet Gynecol. 2004;103:51–6.
3. Palermo G, Joris H, Devroey P, Van Steirteghem AC. Pregnancies after intracytoplasmic injection of single spermatozoon into an oocyte. Lancet. 1992;340:17–8.
4. Andersen AN, Goossens V, Ferraretti AP, Bhattacharya S, Felberbaum R, de Mouzon J, et al. Assisted reproductive technology in Europe, 2004: results generated from European registers by ESHRE. Hum Reprod. 2008;23:756–71.
5. Andersen AN, Gianaroli L, Felberbaum R, de Mouzon J, Nygren KG. Assisted reproductive technology in Europe, 2001. Results generated from European registers by ESHRE. Hum Reprod. 2005;20:1158–76.
6. Auger J, Eustache F, Andersen AG, Irvine DS, Jorgensen N, Skakkebaek NE, et al. Sperm morphological defects related to environment, lifestyle and medical history of 1001 male partners of pregnant women from four European cities. Hum Reprod. 2001;16:2710–7.
7. Bonde JPE, Ernst E, Jensen TK, Hjollund NHI, Kolstad H, Henriksen TB, et al. Relation between semen quality and fertility: a population-based study of 430 first-pregnancy planners. Lancet. 1998;352:1172–7.
8. Guzick DS, Overstreet JW, Factor-Litvak P, Brazil CK, Nakajima ST, Coutifaris C, et al. Sperm morphology, motility, and concentration in fertile and infertile men. N Engl J Med. 2001;345:1388–93.
9. Agarwal A, Said TM. Role of sperm chromatin abnormalities and DNA damage in male infertility. Hum Reprod Update. 2003;9:331–45.
10. Erenpreiss J, Spanó M, Erenpreisa J, Bungum M, Giwercman A. Sperm chromatin structure and male fertility: biological and clinical aspects. Asian J Androl. 2006;8:11–29.
11. Edwards RG, Brody SA. Principles and practice of assisted human reproduction. Philadelphia: W.B Saunders; 1995.
12. Steptoe PC, Edwards RG. Birth after the reimplantation of a human embryo. Lancet. 1978;2:366.
13. Zegers-Hochschild F, Nygren KG, Adamson GD, de Mouzon J, Lancaster P, Mansour R, et al. The International Committee Monitoring Assisted Reproductive Technologies (ICMART) glossary on ART terminology. Fertil Steril. 2006;86:16–9.
14. Dobson R. Number of babies born by assisted reproduction rises by 12%. BMJ. 2009;338:2208.
15. Hull MG, Fleming CF, Hughes AO, McDermott A. The age-related decline in female fecundity: a quantitative controlled study of implanting capacity and survival of individual embryos after in vitro fertilization. Fertil Steril. 1996;65:783–90.
16. Aitken RJ. Sperm function tests and fertility. Int J Androl. 2006;29:69–75; discussion 105–8.
17. Muller CH. Rationale, interpretation, validation, and uses of sperm function tests. J Androl. 2000;21:10–30.
18. Bhattacharya S, Hamilton MP, Shaaban M, Khalaf Y, Seddler M, Ghobara T, et al. Conventional in-vitro fertilisation versus intracytoplasmic sperm injection for the treatment of non-male-factor infertility: a randomised controlled trial. Lancet. 2001;357:2075–9.
19. Jain T, Gupta RS. Trends in the use of intracytoplasmic sperm injection in the United States. N Engl J Med. 2007;357:251–7.
20. Evenson DP, Jost LK, Marshall D, Zinaman MJ, Clegg E, Purvis K, et al. Utility of the sperm chromatin structure assay as a diagnostic and prognostic tool in the human fertility clinic. Hum Reprod. 1999;14:1039–49.
21. Gandini L, Lombardo F, Paoli D, Caponecchia L, Familiari G, Verlengia C, et al. Study of apoptotic DNA fragmentation in human spermatozoa. Hum Reprod. 2000;15:830–9.
22. Host E, Lindenberg S, Smidt-Jensen S. The role of DNA strand breaks in human spermatozoa used for IVF and ICSI. Acta Obstet Gynecol Scand. 2000;79:559–63.
23. Irvine DS, Twigg JP, Gordon EL, Fulton N, Milne PA, Aitken RJ. DNA integrity in human spermatozoa: relationships with semen quality. J Androl. 2000;21:33–44.
24. Spanò M, Bonde JP, Hjøllund HI, Kolstad HA, Cordelli E, Leter G. Sperm chromatin damage impairs human fertility. The Danish First Pregnancy Planner Study Team. Fertil Steril. 2000;73:43–50.
25. Carrell DT, Liu L. Altered protamine 2 expression is uncommon in donors of known fertility, but common among men with poor fertilizing capacity, and may reflect other abnormalities of spermiogenesis. J Androl. 2001;22:604–10.

26. Zini A, Bielecki R, Phang D, Zenzes MT. Correlations between two markers of sperm DNA integrity, DNA denaturation and DNA fragmentation, in fertile and infertile men. Fertil Steril. 2001;75:674–7.
27. Zini A, Fischer MA, Sharir S, Shayegan B, Phang D, Jarvi K. Prevalence of abnormal sperm DNA denaturation in fertile and infertile men. Urology. 2002; 60:1069–72.
28. Saleh RA, Agarwal A, Nada EA, El-Tonsy MH, Sharma RK, Meyer A, et al. Negative effects of increased sperm DNA damage in relation to seminal oxidative stress in men with idiopathic and male factor infertility. Fertil Steril. 2003;79 Suppl 3:1597–605.
29. Morris ID, Ilott S, Dixon L, Brison DR. The spectrum of DNA damage in human sperm assessed by single cell gel electrophoresis (Comet assay) and its relationship to fertilization and embryo development. Hum Reprod. 2002;17:990–8.
30. Gorczyca W, Gong J, Darzynkiewicz Z. Detection of DNA strand breaks in individual apoptotic cells by the in situ terminal deoxynucleotidyl transferase and nick translation assays. Cancer Res. 1993;53:1945–51.
31. Evenson DP, Darzynkiewicz Z, Melamed MR. Relation of mammalian sperm chromatin heterogeneity to fertility. Science. 1980;210:1131–3.
32. Evenson DP, Larson KL, Jost LK. Sperm chromatin structure assay: its clinical use for detecting sperm DNA fragmentation in male infertility and comparisons with other techniques. J Androl. 2002;23:25–43.
33. Fernandez JL, Muriel L, Rivero MT, Goyanes V, Vazquez R, Alvarez JG. The sperm chromatin dispersion test: a simple method for the determination of sperm DNA fragmentation. J Androl. 2003;24:59–66.
34. Fernández JL, Muriel L, Goyanes V, Segrelles E, Gosálvez J, Enciso M, et al. Simple determination of human sperm DNA fragmentation with an improved sperm chromatin dispersion test. Fertil Steril. 2005; 84:833–42.
35. Aravindan GR, Bjordahl J, Jost LK, Evenson DP. Susceptibility of human sperm to in situ DNA denaturation is strongly correlated with DNA strand breaks identified by single-cell electrophoresis. Exp Cell Res. 1997;236:231–7.
36. Erenpreiss J, Jepson K, Giwercman A, Tsarev I, Erenpreisa J, Spanó M. Toluidine blue cytometry test for sperm DNA conformation: comparison with the flow cytometric sperm chromatin structure and TUNEL assays. Hum Reprod. 2004;19:2277–82.
37. Makhlouf AA, Niederberger C. DNA integrity tests in clinical practice: it is not a simple matter of black and white (or red and green). J Androl. 2006;27:316–23.
38. Bungum M, Humaidan P, Axmon A, Spanó M, Bungum L, Erenpreiss J, et al. Sperm DNA integrity assessment in prediction of assisted reproduction technology outcome. Hum Reprod. 2007;22:174–9.
39. Spanò M, Kolstad AH, Larsen SB, Cordelli E, Leter G, Giwercman A, et al. The applicability of the flow cytometric sperm chromatin structure assay in epidemiological studies. Asclepios. Hum Reprod. 1998;13:2495–505.
40. Giwercman A, Richthoff J, Hjollund H, Bonde JP, Jepson K, Frohm B, et al. Correlation between sperm motility and sperm chromatin structure assay parameters. Fertil Steril. 2003;80:1404–12.
41. Giwercman A, Lindstedt L, Larsson M, Bungum M, Spanó M, Levine RJ, et al. Sperm chromatin structure assay as an independent predictor of fertility in vivo: a case-control study. Int J Androl. 2010;33:221–7.
42. Erenpreiss J, Elzanaty S, Giwercman A. Sperm DNA damage in men from infertile couples. Asian J Androl. 2008;10:786–90.
43. Duran EH, Morshedi M, Taylor S, Oehninger S. Sperm DNA quality predicts intrauterine insemination outcome: a prospective cohort study. Hum Reprod. 2002;17:3122–8.
44. Boe-Hansen GB, Fedder J, Ersboll AK, Christensen P. The sperm chromatin structure assay as a diagnostic tool in the human fertility clinic. Hum Reprod. 2006;21:1576–82.
45. Muriel L, Meseguer M, Fernandez JL, Alvarez J, Remohi J, Pellicer A, et al. Value of the sperm chromatin dispersion test in predicting pregnancy outcome in intrauterine insemination: a blind prospective study. Hum Reprod. 2006;21:738–44.
46. Larson KL, DeJonge CJ, Barnes AM, Jost LK, Evenson DP. Sperm chromatin structure assay parameters as predictors of failed pregnancy following assisted reproductive techniques. Hum Reprod. 2000;15:1717–22.
47. Larson-Cook KL, Brannian JD, Hansen KA, Kasperson KM, Aamold ET, Evenson DP. Relationship between the outcomes of assisted reproductive techniques and sperm DNA fragmentation as measured by the sperm chromatin structure assay. Fertil Steril. 2003;80:895–902.
48. Bungum M, Humaidan P, Spanó M, Jepson K, Bungum L, Giwercman A. The predictive value of sperm chromatin structure assay (SCSA) parameters for the outcome of intrauterine insemination, IVF and ICSI. Hum Reprod. 2004;19:1401–8.
49. Gandini L, Lombardo F, Paoli D, Caruso F, Eleuteri P, Leter G, et al. Full-term pregnancies achieved with ICSI despite high levels of sperm chromatin damage. Hum Reprod. 2004;19:1409–17.
50. Virro MR, Larson-Cook KL, Evenson DP. Sperm chromatin structure assay (SCSA) parameters are related to fertilization, blastocyst development, and ongoing pregnancy in in vitro fertilization and intracytoplasmic sperm injection cycles. Fertil Steril. 2004;81:1289–95.
51. Hammadeh ME, Stieber M, Haidl G, Schmidt W. Association between sperm cell chromatin condensation, morphology based on strict criteria, and fertilization, cleavage and pregnancy rates in an IVF program. Andrologia. 1998;30:29–35.
52. Payne JF, Raburn DJ, Couchman GM, Price TM, Jamison MG, Walmer DK. Redefining the relationship between sperm deoxyribonucleic acid fragmentation as measured by the sperm chromatin structure assay and outcomes of assisted reproductive techniques. Fertil Steril. 2005;84:356–64.

53. Aitken RJ, DeIuliis GN. Value of DNA integrity assays for fertility evaluation. Soc Reprod Fertil Suppl. 2007;65:81–92.
54. The Practice Committee of the American Society of Reproductive Medicine. The clinical utility of sperm DNA integrity testing. Fertil Steril. 2006;86 Suppl 4:35–7.
55. Collins JA, Barnhart KT, Schlegel PN. Do sperm DNA integrity tests predict pregnancy with in vitro fertilization? Fertil Steril. 2008;89:823–31.
56. Zini A, Boman JM, Belzile E, Ciampi A. Sperm DNA damage is associated with an increased risk of pregnancy loss after IVF and ICSI: systematic review and meta-analysis. Hum Reprod. 2008;23:2663–8.
57. Evenson DP, Wixon R. Meta-analysis of sperm DNA fragmentation using the sperm chromatin structure assay. Reprod Biomed Online. 2006;12:466–72.
58. Li Z, Wang L, Cai J, Huang H. Correlation of sperm DNA damage with IVF and ICSI outcomes: a systematic review and meta-analysis. J Assist Reprod Genet. 2006;23:367–76.
59. Ahmadi A, Ng SC. Fertilizing ability of DNA-damaged spermatozoa. J Exp Zool. 1999;284: 696–704.
60. Sakkas D, Urner F, Bianchi PG, Bizzaro D, Wagner I, Jaquenoud N, et al. Sperm chromatin anomalies can influence decondensation after intracytoplasmic sperm injection. Hum Reprod. 1996;11:837–43.
61. Tomlinson MJ, Moffatt O, Manicardi GC, Bizzaro D, Afnan M, Sakkas D. Interrelationships between seminal parameters and sperm nuclear DNA damage before and after density gradient centrifugation: implications for assisted conception. Hum Reprod. 2001;16:2160–5.
62. Tomsu M, Sharma V, Miller D. Embryo quality and IVF treatment outcomes may correlate with different sperm comet assay parameters. Hum Reprod. 2002;17:1856–62.
63. Henkel R, Kierspel E, Hajimohammad M, Stalf T, Hoogendijk C, Mehnert C, et al. DNA fragmentation of spermatozoa and assisted reproduction technology. Reprod Biomed Online. 2003;7:477–84.
64. Lewis SE, O'Connell M, Stevenson M, Thompson-Cree L, McClure N. An algorithm to predict pregnancy in assisted reproduction. Hum Reprod. 2004; 19:1385–94.
65. Huang CC, Lin DP, Tsao HM, Cheng TC, Liu CH, Lee MS. Sperm DNA fragmentation negatively correlates with velocity and fertilization rates but might not affect pregnancy rates. Fertil Steril. 2005;84:130–40.
66. Nasr-Esfahani MH, Salehi M, Razavi S, Anjomshoa M, Rozbahani S, Moulavi F, et al. Effect of sperm DNA damage and sperm protamine deficiency on fertilization and embryo development post-ICSI. Reprod Biomed Online. 2005;11:198–205.
67. Sun JG, Jurisicova A, Casper RF. Detection of deoxyribonucleic acid fragmentation in human sperm: correlation with fertilization in vitro. Biol Reprod. 1997;56:602–7.
68. Lopes CH, Mazzini MN, Tortorella H, Konrath RA, Brandelli A. Isolation, partial characterization and biological activity of mannosyl glycopeptides from seminal plasma. Glycoconj J. 1998;15:477–81.
69. Lewis SE, Aitken RJ. DNA damage to spermatozoa has impacts on fertilization and pregnancy. Cell Tissue Res. 2005;322:33–41.
70. Aitken RJ, Baker MA. Oxidative stress and male reproductive biology. Reprod Fertil Dev. 2004;16: 581–8.
71. Liu CH, Tsao HM, Cheng TC, Wu HM, Huang CC, Chen CI, et al. DNA fragmentation, mitochondrial dysfunction and chromosomal aneuploidy in the spermatozoa of oligoasthenoteratozoospermic males. J Assist Reprod Genet. 2004;21:119–26.
72. Matsuda Y, Tobari I. Chromosomal analysis in mouse eggs fertilized in vitro with sperm exposed to ultraviolet light (UV) and methyl and ethyl methanesulfonate (MMS and EMS). Mutat Res. 1988;198:131–44.
73. Ahmadi A, Ng SC. Developmental capacity of damaged spermatozoa. Hum Reprod. 1999;14:2279–85.
74. Gandini L, Lombardo F, Lenzi A, Spano M, Dondero F. Cryopreservation and sperm DNA integrity. Cell Tissue Bank. 2006;7:91–8.
75. Seli E, Gardner DK, Schoolcraft WB, Moffatt O, Sakkas D. Extent of nuclear DNA damage in ejaculated spermatozoa impacts on blastocyst development after in vitro fertilization. Fertil Steril. 2004;82: 378–83.
76. Golan R, Cooper TG, Oschry Y, Oberpenning F, Schulze H, Shocha L, et al. Changes in chromatin condensation of human spermatozoa during epididymal transit as determined by flow cytometry. Hum Reprod. 1996;11:1457–62.
77. Larson KL, Brannian JD, Timm BK, Jost LK, Evenson DP. Density gradient centrifugation and glass wool filtration of semen remove spermatozoa with damaged chromatin structure. Hum Reprod. 1999;14:2015–9.
78. Donnelly ET, O'Connell M, McClure N, Lewis SE. Differences in nuclear DNA fragmentation and mitochondrial integrity of semen and prepared human spermatozoa. Hum Reprod. 2000;15:1552–61.
79. Younglai EV, Holt D, Brown P, Jurisicova A, Casper RF. Sperm swim-up techniques and DNA fragmentation. Hum Reprod. 2001;16:1950–3.
80. McVicar CM, McClure N, Williamson K, Dalzell LH, Lewis SE. Incidence of Fas positivity and deoxyribonucleic acid double-stranded breaks in human ejaculated sperm. Fertil Steril. 2004;81 Suppl 1:767–74.
81. Morrell JM, Moffatt O, Sakkas D, Manicardi GC, Bizzaro D, Tomlinson M, et al. Reduced senescence and retained nuclear DNA integrity in human spermatozoa prepared by density gradient centrifugation. J Assist Reprod Genet. 2004;21:217–22.
82. Bungum M, Spanó M, Humaidan P, Eleuteri P, Rescia M, Giwercman A. Sperm Chromatin Structure Assay (SCSA) parameters measured after density gradient centrifugation are not predictive for the outcome of ART. Hum Reprod. 2008;23:4–10.

83. Zini A, Mak V, Phang D, Jarvi K. Potential adverse effect of semen processing on human sperm deoxyribonucleic acid integrity. Fertil Steril. 1999;72:496–9.
84. Zini A, Finelli A, Phang D, Jarvi K. Influence of semen processing technique on human sperm DNA integrity. Urology. 2000;56:1081–4.
85. Benchaib M, Braun V, Lornage J, Hadj S, Salle B, Lejeune H, et al. Sperm DNA fragmentation decreases the pregnancy rate in an assisted reproductive technique. Hum Reprod. 2003;18:1023–8.
86. Benchaib M, Lornage J, Mazoyer C, Lejeune H, Salle B, Francois Guerin J. Sperm deoxyribonucleic acid fragmentation as a prognostic indicator of assisted reproductive technology outcome. Fertil Steril. 2007;87:93–100.
87. Hammadeh ME, Radwan M, Al-Hasani S, Micu R, Rosenbaum P, Lorenz M, et al. Comparison of reactive oxygen species concentration in seminal plasma and semen parameters in partners of pregnant and non-pregnant patients after IVF/ICSI. Reprod Biomed Online. 2006;13:696–706.
88. Borini A, Tarozzi N, Bizzaro D, Bonu MA, Fava L, Flamigni C, et al. Sperm DNA fragmentation: paternal effect on early post-implantation embryo development in ART. Hum Reprod. 2006;21:2876–81.
89. Hamamah S, Magnoux E, Royere D, Barthelemy C, Dacheux JL, Gatti JL. Internal pH of human spermatozoa: effects of ions, follicular fluid and progesterone. Mol Hum Reprod. 1996;2:219–24.
90. Hamamah S, Gatti JL. Role of the ionic environment and internal pH on sperm activity. Hum Reprod. 1998;13 Suppl 4:20–30.
91. Peer S, Eltes F, Berkovitz A, Yehuda R, Itsykson P, Bartoov B. Is fine morphology of the human sperm nuclei affected by in vitro incubation at 37 degrees C? Fertil Steril. 2007;88:1589–94.
92. Dalzell LH, Thompson-Cree ME, McClure N, Traub AI, Lewis SE. Effects of 24-hour incubation after freeze-thawing on DNA fragmentation of testicular sperm from infertile and fertile men. Fertil Steril. 2003;79 Suppl 3:1670–2.
93. Dalzell JA, McVicar CM, McClure N, Lutton D, Lewis S. Effects of short and long incubations on DNA fragmentation of testicular sperm. Fertil Steril. 2004;82:1443–5.
94. Steele EK, McClure N, Maxwell RJ, Lewis SE. A comparison of DNA damage in testicular and proximal epididymal spermatozoa in obstructive azoospermia. Mol Hum Reprod. 1999;5:831–5.
95. Greco E, Scarselli F, Iacobelli M, Rienzi L, Ubaldi F, Ferrero S, et al. Efficient treatment of infertility due to sperm DNA damage by ICSI with testicular spermatozoa. Hum Reprod. 2005;20:226–30.
96. Aitken RJ, De Iuliis GN, McLachlan RI. Biological and clinical significance of DNA damage in the male germ line. Int J Androl. 2009;32:46–56.
97. Ahmad L, Jalali S, Shami SA, Akram Z, Batool S, Kalsoom O. Effects of cryopreservation on sperm DNA integrity in normospermic and four categories of infertile males. Syst Biol Reprod Med. 2010;56:74–83.
98. Baumber J, Ball BA, Linfor JJ, Meyers SA. Reactive oxygen species and cryopreservation promote DNA fragmentation in equine spermatozoa. J Androl. 2003;24:621–8.
99. Taylor K, Roberts P, Sanders K, Burton P. Effect of antioxidant supplementation of cryopreservation medium on post-thaw integrity of human spermatozoa. Reprod Biomed Online. 2009;18:184–9.
100. Jorgensen N, Auger J, Giwercman A, Irvine DS, Jensen TK, Keiding N, et al. Semen analysis performed by different laboratory teams: an intervariation study. Int J Androl. 1997;20:201–8.
101. Evenson DP, Jost LK, Baer RK, Turner TW, Schrader SM. Individuality of DNA denaturation patterns in human sperm as measured by the sperm chromatin structure assay. Reprod Toxicol. 1991;5:115–25.
102. Erenpreiss J, Bungum M, Spanó M, Elzanaty S, Orbidans J, Giwercman A. Intra-individual variation in Sperm Chromatin Structure Assay parameters in men from infertile couples: clinical implications. Hum Reprod. 2006;21:2061–4.
103. Sakkas D, Alvarez JG. Sperm DNA fragmentation: mechanisms of origin, impact on reproductive outcome and analysis. Fertil Steril. 2010;93:1027–36.
104. Lewis SE, Agbaje IM. Using the alkaline comet assay in prognostic tests for male infertility and assisted reproductive technology outcomes. Mutagenesis. 2008;23:163–70.
105. Kefer JC, Agarwal A, Sabanegh E. Role of antioxidants in the treatment of male infertility. Int J Urol. 2009;16:449–57.
106. Silver EW, Eskenazi B, Evenson DP, Block G, Young S, Wyrobek AJ. Effect of antioxidant intake on sperm chromatin stability in healthy nonsmoking men. J Androl. 2005;26:550–6.
107. Song GJ, Norkus EP, Lewis V. Relationship between seminal ascorbic acid and sperm DNA integrity in infertile men. Int J Androl. 2006;29:569–75.
108. Ménézo YJ, Hazout A, Panteix G, Robert F, Rollet J, Cohen-Bacrie P, et al. Antioxidants to reduce sperm DNA fragmentation: an unexpected adverse effect. Reprod Biomed Online. 2007;14:418–21.
109. Moskovtsev SI, Lecker I, Mullen JB, Jarvi K, Willis J, White J, et al. Cause-specific treatment in patients with high sperm DNA damage resulted in significant DNA improvement. Syst Biol Reprod Med. 2009;55:109–15.
110. Greco E, Iacobelli M, Rienzi L, Ubaldi F, Ferrero S, Tesarik J. Reduction of the incidence of sperm DNA fragmentation by oral antioxidant treatment. J Androl. 2005;26:349–53.
111. Bartoov B, Berkovitz A, Eltes F. Selection of spermatozoa with normal nuclei to improve the pregnancy rate with intracytoplasmic sperm injection. N Engl J Med. 2001;345:1067–8.

112. Itzkan I, Qiu L, Fang H, Zaman MM, Vitkin E, Ghiran IC, et al. Confocal light absorption and scattering spectroscopic microscopy monitors organelles in live cells with no exogenous labels. Proc Natl Acad Sci USA. 2007;104:17255–60.
113. Said TM, Grunewald S, Paasch U, Glander HJ, Baumann T, Kriegel C, et al. Advantage of combining magnetic cell separation with sperm preparation techniques. Reprod Biomed Online. 2005;10:740–6.
114. Fernández-Gonzalez R, Moreira PN, Pérez-Crespo M, Sánchez-Martín M, Ramirez MA, Pericuesta E, et al. Long-term effects of mouse intracytoplasmic sperm injection with DNA-fragmented sperm on health and behavior of adult offspring. Biol Reprod. 2008;78:761–72.
115. World Health Organization. WHO laboratory manual for the examination of human semen and sperm-cervical mucus interaction. Cambridge: Cambridge University Press; 1999.
116. Caglar GS, Köster F, Schöpper B, Asimakopoulos B, Nehls B, Nikolettos N, et al. Semen DNA fragmentation index, evaluated with both TUNEL and Comet assay, and the ICSI outcome. In Vivo. 2007;21:1075–80.
117. Check JH, Graziano V, Cohen R, Krotec J, Check ML. Effect of an abnormal sperm chromatin structural assay (SCSA) on pregnancy outcome following (IVF) with ICSI in previous IVF failures. Arch Androl. 2005;51:121–4.
118. Henkel R, Hajimohammad M, Stalf T, Hoogendijk C, Mehnert C, Menkveld R, et al. Influence of deoxyribonucleic acid damage on fertilization and pregnancy. Fertil Steril. 2004;81:965–72.
119. Bakos HW, Thompson JG, Feil D, Lane M. Sperm DNA damage is associated with assisted reproductive technology pregnancy. Int J Androl. 2008;31:518–26.
120. Frydman N, Prisant N, Hesters L, Frydman R, Tachdjian G, Cohen-Bacrie P, et al. Adequate ovarian follicular status does not prevent the decrease in pregnancy rates associated with high sperm DNA fragmentation. Fertil Steril. 2008;89:92–7.
121. Tarozzi N, Nadalini M, Stronati A, Bizzaro D, Dal Prato L, Coticchio G, et al. Anomalies in sperm chromatin packaging: implications for assisted reproduction techniques. Reprod Biomed Online. 2009;18:486–95.
122. Velez de la Calle JF, Muller A, Walschaerts M, Clavere JL, Jimenez C, Wittemer C, et al. Sperm deoxyribonucleic acid fragmentation as assessed by the sperm chromatin dispersion test in assisted reproductive technology programs: results of a large prospective multicenter study. Fertil Steril. 2008;90:1792–9.
123. Tavalaee M, Razavi S, Nasr-Esfahani MH. Influence of sperm chromatin anomalies on assisted reproductive technology outcome. Fertil Steril. 2009;91:1119–26.

Sperm DNA Damage and Pregnancy Loss After IVF/ICSI

32

Armand Zini and Jason Matthew Boman

Abstract

Sperm DNA damage has been associated with reproductive difficulties in the male. While not yet routine, sperm DNA damage testing appears to be securing a place in the evaluation of the infertile male and infertile couple. The demonstration of a relationship between sperm DNA damage and pregnancy loss after IVF and ICSI provides yet another potential clinical application of this type of testing. At the very least, this relationship highlights the need to find ways to reduce sperm DNA damage in men and provides further rationale for ongoing research in this field. We believe that sperm DNA integrity testing may now be justified in the context of IVF and ICSI to help understand the possible cause of pregnancy loss and to provide prognostic information regarding a couple's potential risk of pregnancy loss following these assisted reproductive technologies (ARTs).

Keywords

Sperm DNA damage • In vitro fertilization • DNA damage testing • Male infertility • Pregnancy loss after IVF

Standard semen parameters that exhibit a high degree of biological variability are only fair measures of fertility potential and are poor predictors of reproductive outcomes [1]. As such, there is a need for better markers that might help distinguish fertile from infertile men and help predict pregnancy outcome and adverse reproductive events. Animal studies have shown that embryo development and implantation depend, at least in part, on the integrity of the sperm DNA and that there may be a threshold of sperm DNA damage beyond which these processes are impaired [2]. While the clinical utility sperm DNA integrity testing has yet to be firmly established, there is now clear evidence that infertile men possess substantially more sperm DNA damage than do fertile men [3–7]. In addition, sperm DNA damage is associated with lower natural, IUI, and IVF pregnancy rates [8–14].

Interestingly, sperm DNA and chromatin defects are not associated with lower ICSI

A. Zini (✉)
Department of Surgery, Division of Urology, McGill University, St. Mary's Hospital Center, Montreal, QC, Canada
e-mail: ziniarmand@yahoo.com

A. Zini and A. Agarwal (eds.), *Sperm Chromatin: Biological and Clinical Applications in Male Infertility and Assisted Reproduction*, DOI 10.1007/978-1-4419-6857-9_32,

pregnancy rates [15–19]. This is clinically relevant because men with severe male-factor infertility are the most likely patients to possess sperm DNA defects and are the most likely to require IVF with ICSI as a means of reproducing. While sperm DNA damage does not seem to impact pregnancy rates with IVF/ICSI, a higher level of DNA defects may infer a greater risk of losing the pregnancy once established.

The risk of pregnancy loss after IVF/ICSI has been reported in a number of studies, and these results have now been reviewed in a systematic fashion. This chapter reviews the etiology of sperm DNA damage, the tests used to measure DNA damage, and relationship between DNA damage and reproductive outcomes. In particular, the impact that sperm DNA has on pregnancy loss after using IVF/ICSI to achieve pregnancy is discussed.

Pregnancy Loss

Definitions

Confirmation of pregnancy can be achieved either by biochemical means (serum hCG elevation) or clinically (presence of a heartbeat, confirmed by ultrasound). Pregnancy loss (spontaneous abortion or miscarriage) refers to a pregnancy that ends spontaneously before the fetus has reached a viable gestational age. The World Health Organization defines it, more specifically, as expulsion or extraction of an embryo or fetus weighing 500 g or less from its mother. This typically corresponds to a gestational age of 20–22 weeks or less.

Etiologies

Female *chromosomal abnormalities* account for approximately 50% of all miscarriages. The most frequently encountered chromosomal abnormalities in decreasing order of frequency are as follows: autosomal trisomies (52%), monosomy X (19%), polyploidies (22%), and others (7%). Trisomy 16 is the most common autosomal trisomy and is always lethal [20].

Congenital anomalies that can result from either genetic abnormalities, extrinsic factors such as amniotic bands, or exposure to teratogens can also lead to pregnancy loss. *Trauma* resulting from invasive diagnostic procedures (e.g., amniocentesis or chorionic villus sampling) or from blunt injuries to the maternal abdomen is yet another potential cause of early loss pregnancy. A myriad of *maternal host factors* that might include anatomic uterine anomalies, acute maternal infections or endocrinopathies, hypercoagulable states, and finally immunologic rejection are all potential causes of pregnancy loss.

While the majority of pregnancy losses result from female factors, increasing evidence suggests that male factors can also play a role in miscarriage. Although, it is not known exactly how sperm defects contribute to pregnancy loss, some studies suggest that abnormal sperm DNA integrity may affect embryo development and increase miscarriage risk [2, 8, 11, 12, 14, 16–18, 21].

Sperm DNA Damage

Human Sperm DNA and Chromatin Structure

Sperm chromatin is very tightly compacted by virtue of the unique associations between the DNA and sperm nuclear proteins (histones and protamines) [22]. During the later stages of spermatogenesis, the haploid spermatid nucleus is remodeled and condensed further as a result of the sequential displacement of histones by transition proteins and then by protamines [22, 23]. The DNA strands are tightly wrapped around the protamine molecules forming tight and highly organized loops [24], and it is thought that this nuclear compaction is important to protect the sperm genome from external stresses such as oxidation or temperature elevation [25]. In humans, up to 15% of the DNA remains packaged by histones at specific DNA sequences (i.e., there is a nonrandom association between histones and DNA sequences) [26]. The histone-bound DNA sequences are less tightly compacted and more available for expression than non-histone-bound

DNA sequences. It is thought that these DNA sequences and/or genes may be involved in fertilization and early embryo development [27].

Infertile men have an increased sperm histone to protamine ratio when compared to fertile controls [28]. The spermatozoa of infertile men can also exhibit incomplete nuclear sulfhydryl group oxidation – the reaction leading to the formation of stabilizing disulfide cross-links [29, 30]. These sperm abnormalities (histone to protamine ratio and sulfhydryl group status) can potentially result in defective chromatin compaction [31] and in an increased susceptibility to DNA damage [7]. Sperm nuclear compaction or condensation may be an important determinant of sperm head morphology. Both animal and human studies have demonstrated a correlation between sperm DNA stainability and head morphology, which may, in part, be due to reduced nuclear compaction [32, 33].

Etiology of Sperm DNA Damage

The etiology of sperm DNA damage is multifactorial. Clinically, several conditions have been associated with sperm DNA and chromatin damage (e.g., chemotherapy, smoking, genital tract infection, varicocele, etc.) [8, 34–41]. Broadly, these conditions can be categorized as primary or intrinsic defects in spermatogenesis (e.g., genetic or developmental abnormalities) and secondary or extrinsic noxious factors (e.g., gonadotoxins, hyperthermia, oxidants, endocrine disruption, etc.).

At the cellular level, a number of theories have been proposed to explain the DNA damage in human spermatozoa. Studies have suggested that protamine deficiency (with aberrant chromatin remodeling), reactive oxygen species (ROS) and abortive apoptosis may be responsible for sperm DNA damage [42–47]. Recently, De Iuliis et al. have proposed a two-step hypothesis to explain the generation of sperm DNA damage. Based on the model, oxidative stress acts on poorly protaminated cells (i.e., cells with incomplete replacement of histones by protamines) generated as a result of defective spermiogenesis [48].

Tests of Sperm DNA Damage

Several tests of sperm DNA and chromatin damage have been described [41, 49, 50]. These tests have been developed in the hope that they may (1) help in the diagnosis of male infertility, (2) predict reproductive outcomes in the context of assisted reproductive technologies (ARTs), and (3) provide some assurance regarding the integrity of the male gamete genome. Several factors must be considered when evaluating studies of sperm DNA and chromatin integrity. First, the different assays measure different aspects of sperm DNA and chromatin. Second, the assay conditions can greatly influence the accessibility of the dye or enzyme to the sites of damaged DNA and, therefore, impact on the final results. Third, current assays are limited because they do not selectively differentiate clinically important DNA fragmentation (e.g., degree or gene specificity) from clinically insignificant damage. Finally, sample preparation and handling prior to sperm DNA and chromatin integrity testing can impact on the final test results.

The Comet (single cell gel electrophoresis) and TUNEL (terminal deoxynucleotidyl transferase-mediated dUTP nick end-labeling) assays are commonly utilized assays that detect DNA strand breaks directly. Some assays measure the susceptibility of DNA to denaturation – that is the formation of single-stranded DNA from native double-stranded DNA (e.g., SCSA-sperm chromatin structure assay) – and depend on the premise that nicked DNA will denature more readily than intact DNA. Other assays rely on the differential binding of dyes or agents to single-stranded and [50] double-stranded DNA (e.g., acridine orange) or to protamine-deficient sites (e.g., aniline blue or CMA3 test). Remarkably, the results of most sperm DNA or chromatin integrity assays correlate highly with each other – with the exception of the manual acridine orange test [50]. In order to provide clinically relevant information, an upper normal level (cutoff) of the percentage of cells with DNA fragmentation or chromatin defect has been defined for most of these assays. Samples with test results above the threshold or cutoff value are considered to have high levels of DNA damage [41].

Relationship Between Sperm DNA Damage and Pregnancy Loss After IVF and ICSI

Several studies have reported on the risk of pregnancy loss after standard IVF and after ICSI, and we have recently carried out a systematic review of the literature and performed a meta-analysis of these studies to further evaluate the impact of sperm DNA damage on pregnancy loss after IVF without and with ICSI [51].

In our review of the literature, we found seven eligible reports (with 11 studies) that involved 1,549 cycles of treatment (808 IVF and 741 IVF/ICSI cycles), 640 pregnancies (345 with IVF and 295 with IVF/ICSI), and 122 pregnancy losses in total. The characteristics of the studies were highly variable in terms of data collection (i.e., prospective vs. retrospective), definition of pregnancy loss (biochemical vs. clinical), population characteristics (unselected vs. repeated IVF failures), female inclusion/exclusion criteria, sperm DNA damage test, and sperm DNA test cutoff. The characteristics of the studies are summarized in Tables 32.1 and 32.2. In all but one study, sperm DNA damage was evaluated on whole

Table 32.1 Selected diagnostic properties of studies on sperm DNA damage and pregnancy loss (PL) after IVF and IVF/ICSI

Study	*n*	ART	Assay	PL	Ab Test*	Sens (%)	Spec (%)	PPV	NPV	OR (95% CI)
Check et al. [17]	104	ICSI	SCSA	47	24	0.31	0.83	0.63	0.58	2.27 (0.45, 11.59)
Zini et al. [18]	60	ICSI	SCSA	16	19	0.40	0.85	0.33	0.88	3.67 (0.46, 29.42)
Borini et al. [11]	82	IVF	TUNEL	6	11	0.91	0.94	0.50	0.99	160 (0.18, 1,44,708)
Borini et al. [11]	50	ICSI	TUNEL	25	25	0.97	0.99	0.97	0.99	2,700 (0.38, 2×10^7)
Benchaib et al. [12]	84	IVF	TUNEL	15	15	0.50	0.91	0.50	0.91	10.0 (0.87, 114.8)
Benchaib et al. [12]	218	ICSI	TUNEL	12	15	0.38	0.88	0.30	0.91	4.54 (0.89, 23.28)
Bungum et al. [8]	388	IVF	SCSA	24	14	0.11	0.85	0.19	0.76	0.73 (0.23, 2.33)
Bungum et al. [8]	223	ICSI	SCSA	19	40	0.50	0.63	0.24	0.84	1.69 (0.63, 4.49)
Frydman et al. [14]	117	IVF	TUNEL	19	32	0.64	0.75	0.37	0.90	5.25 (1.31, 21.11)
Lin et al. [19]	137	IVF	SCSA	10	17	0.29	0.84	0.17	0.92	2.16 (0.37, 12.72)
Lin et al. [19]	86	ICSI	SCSA	18	23	0.50	0.83	0.40	0.88	5.00 (0.97, 25.77)

ART assisted reproductive technology; *Abn Test* proportion of abnormal sperm DNA test among documented pregnancies; *PL* pregnancy loss; *Sens* sensitivity; *Spec* specificity; *PPV* positive predictive value; *NPV* negative predictive value; *OR* odds ratio

Table 32.2 Characteristics of studies on sperm DNA damage and pregnancy loss (PL) after IVF and IVF/ICSI

Study	*n*	ART	Assay	Population	Study design	PL-Def	Female Dx
Check et al. [17]	104	ICSI	SCSA	Failed IVFx2	Unspecified	Per CP	Unspecified
Zini et al. [18]	60	ICSI	SCSA	Unspecified	Prospective	Per CP	<40
Borini et al. [11]	82	IVF	TUNEL	Unspecified	Unspecified	Per CP	Unspecified
Borini et al. [11]	50	ICSI	TUNEL	Unspecified	Unspecified	Per CP	Unspecified
Benchaib et al. [12]	84	IVF	TUNEL	Unspecified	Prospective	Per CP	Unspecified
Benchaib et al. [12]	218	ICSI	TUNEL	Unspecified	Prospective	Per CP	Unspecified
Lin et al. [21]	137	IVF	SCSA	Unspecified	Prospective	Per CP	<40, FSH<15
Lin et al. [21]	86	ICSI	SCSA	Male factor	Prospective	Per CP	<40, FSH<15
Bungum et al. [18]	388	IVF	SCSA	Female factor	Prospective	Per BP	<40, FSH<12
Bungum et al. [8]	223	ICSI	SCSA	Male factor	Prospective	Per BP	<40, FSH<12
Frydman et al. [14]	117	IVF	TUNEL	Unspecified	Prospective	Per CP	<38, FSH<10

n number of IVF or ICSI cycles; *ART* assisted reproductive technology; *PL-Def* pregnancy loss definition; *CP* clinical pregnancy; *BP* biochemical pregnancy; *Female Dx* female diagnosis; <40 or <38 = <40 or <38 year-old; FSH<15 (<12, <10) = day 3 serum FSH < 15 (<12, <10) IU/L (unprocessed) semen. In the Borini et al. [11] study, sperm samples were washed prior to assessing DNA damage. This needs mentioning because there may a difference in sperm DNA damage between whole and prepared semen, and the sperm DNA damage cutoffs may not be reliable when evaluating washed semen in predicting outcome of ART [52].

Our meta-analysis of the evaluable studies demonstrated a combined OR of 2.48 (95% CI; 1.52, 4.04, $p<0.0001$), indicating an important association between sperm DNA damage and the rate of pregnancy loss after IVF and ICSI [51]. Repeating the meta-analysis with the Borini et al. [11] study excluded also demonstrated a significant OR estimate (OR = 2.37), which was not significantly different from the overall meta-analysis.

We found no significant difference in the OR according to the type of ART (IVF or ICSI) However, there was a significant difference in the OR estimates between the TUNEL and the SCSA studies (the combined OR of the studies using TUNEL assay (OR = 7.04) was significantly higher than that of the studies using SCSA (OR = 1.77)).

The finding of an association between sperm DNA damage and pregnancy loss provides a mechanism by which sperm defects may impact pregnancy loss, particularly after IVF and IVF/ICSI, where the barriers to natural selection are bypassed.

Although it is uncertain whether knowledge of one's level of sperm DNA damage will influence a couple's decision to proceed with ARTs, assessing sperm DNA damage may still provide clinically valuable information. In our analysis of the 11 studies discussed previously, there was a positive predictive value (PPV) of 37% and a median negative predictive value (NPV) of 90% (with a median pregnancy loss rate of 18%). In other words, in populations with an overall pregnancy loss rate of 18%, the risk of pregnancy loss is estimated at 37% with an abnormal test result and 10% with a normal one. In this scenario, the clinician might want to discuss the effect of DNA damage with the patients, since testing could discriminate between pregnancy loss rates of 10 and 38%.

Conclusion

Sperm DNA damage has been associated with reproductive difficulties in the male. While not yet routine, sperm DNA damage testing appears to be securing a place in the evaluation of the infertile male and infertile couple. The demonstration of a relationship between sperm DNA damage and pregnancy loss after IVF and ICSI provides yet another potential clinical application of this type of testing. At the very least, this relationship highlights the need to find ways to reduce sperm DNA damage in men and provides further rationale for ongoing research in this field. We believe that sperm DNA integrity testing may now be justified in the context of IVF and ICSI to help understand the possible cause of pregnancy loss and to provide prognostic information regarding a couple's potential risk of pregnancy loss following these ARTs.

References

1. Guzick DS, Overstreet JW, Factor-Litvak P, National Cooperative Reproductive Medicine Network, et al. Sperm morphology, motility, and concentration in fertile and infertile men. N Engl J Med. 2001;345(19): 1388–93.
2. Ahmadi A, Ng SC. Fertilizing ability of DNA-damaged spermatozoa. J Exp Zool. 1999;284(6):696–704.
3. Evenson DP, Darzynkiewicz Z, Melamed MR. Relation of mammalian sperm chromatin heterogeneity to fertility. Science. 1980;210(4474):1131–3.
4. Irvine DS, Twigg JP, Gordon EL, Fulton N, Milne PA, Aitken RJ. DNA integrity in human spermatozoa: relationships with semen quality. J Androl. 2000;21(1): 33–44.
5. Shen H, Ong C. Detection of oxidative DNA damage in human sperm and its association with sperm function and male infertility. Free Radic Biol Med. 2000;28(4):529–36.
6. Spanò M, Bonde JP, Hjøllund HI, Kolstad HA, Cordelli E, Leter G. Sperm chromatin damage impairs human fertility. The Danish First Pregnancy Planner Study Team. Fertil Steril. 2000;73(1):43–50.
7. Zini A, Kamal KM, Phang D. Free thiols in human spermatozoa: correlation with sperm DNA integrity. Urology. 2001;58:80–4.
8. Bungum M, Humaidan P, Axmon A, Spano M, Bungum L, Erenpreiss J, et al. Sperm DNA integrity assessment in prediction of assisted reproduction technology outcome. Hum Reprod. 2007;22:174–9.

9. Larson-Cook KL, Brannian JD, Hansen KA, Kasperson KM, Aamold ET, Evenson DP. Relationship between the outcomes of assisted reproductive techniques and sperm DNA fragmentation as measured by the sperm chromatin structure assay. Fertil Steril. 2003;80(4):895–902.
10. Henkel R, Kierspel E, Hajimohammad M, Stalf T, Hoogendijk C, Mehnert C, et al. DNA fragmentation of spermatozoa and assisted reproduction technology. Reprod Biomed Online. 2003;7(4):477–84.
11. Borini A, Tarozzi N, Bizzaro D, Bonu MA, Fava L, Flamigni C, et al. Sperm DNA fragmentation: paternal effect on early post-implantation embryo development in ART. Hum Reprod. 2006;21:2876–81.
12. Benchaib M, Lornage J, Mazoyer C, Lejeune H, Salle B, Francois Guerin J. Sperm deoxyribonucleic acid fragmentation as a prognostic indicator of assisted reproductive technology outcome. Fertil Steril. 2007;87:93–100.
13. Collins JA, Barnhart KT, Schlegel PN. Do sperm DNA integrity tests predict pregnancy with in vitro fertilization? Fertil Steril. 2008;89(4):823–31.
14. Frydman N, Prisant N, Hesters L, Frydman R, Tachdjian G, Cohen-Bacrie P, et al. Adequate ovarian follicular status does not prevent the decrease in pregnancy rates associated with high sperm DNA fragmentation. Fertil Steril. 2008;89:92–7.
15. Lombardo F, Gandini L, Lenzi A, Dondero F. Antisperm immunity in assisted reproduction. J Reprod Immunol. 2004;62(1–2):101–9.
16. Virro MR, Larson-Cook KL, Evenson DP. Sperm chromatin structure assay (SCSA) parameters are related to fertilization, blastocyst development, and ongoing pregnancy in in vitro fertilization and intracytoplasmic sperm injection cycles. Fertil Steril. 2004;81:1289–95.
17. Check JH, Graziano V, Cohen R, Krotec J, Check ML. Effect of an abnormal sperm chromatin structural assay (SCSA) on pregnancy outcome following (IVF) with ICSI in previous IVF failures. Arch Androl. 2005;51:121–4.
18. Zini A, Meriano J, Kader K, Jarvi K, Laskin CA, Cadesky K. Potential adverse effect of sperm DNA damage on embryo quality after ICSI. Hum Reprod. 2005;20:3476–80.
19. Lin KQ, Fan HG, Xu H, Zhang XM. Diagnosis and treatment of complicated interstitial pregnancy. Zhejiang Da Xue Xue Bao Yi Xue Ban. 2008;37(6): 638–41.
20. Sierra S, Stephenson M. Genetics of recurrent pregnancy loss. Semin Reprod Med. 2006;24:17.
21. Lin MH, Kuo-Kuang Lee R, Li SH, Lu CH, Sun FJ, Hwu YM. Sperm chromatin structure assay parameters are not related to fertilization rates, embryo quality, and pregnancy rates in in vitro fertilization and intracytoplasmic sperm injection, but might be related to spontaneous abortion rates. Fertil Steril. 2008;90:352–9.
22. Ward WS, Coffey DS. DNA packaging and organization in mammalian spermatozoa: comparison with somatic cells. Biol Reprod. 1991;44:569–74.
23. Steger K, Pauls K, Klonisch T, Franke FE, Bergmann M. Expression of protamine-1 and -2 mRNA during human spermiogenesis. Mol Hum Reprod. 2000;6: 219–25.
24. Brewer LR, Corzett M, Balhorn R. Protamine induced condensation and decondensation of the same DNA molecule. Science. 1999;286:120–3.
25. Kosower NS, Katayose H, Yanagimachi R. Thiol-disulfide status and acridine orange fluorescence of mammalian sperm nuclei. J Androl. 1992;13:342–8.
26. Gatewood JM, Cook GR, Balhorn R, Bradbury EM, Schmid CW. Sequence-specific packaging of DNA in human sperm chromatin. Science. 1987;236:962–4.
27. Gineitis AA, Zalenskaya IA, Yau PM, Bradbury EM, Zalensky AO. Human sperm telomere-binding complex involves histone H2B and secures telomere membrane attachment. J Cell Biol. 2000;151:1591–8.
28. Zhang X, San Gabriel M, Zini A. Sperm nuclear histone to protamine ratio in fertile and infertile men: evidence of heterogeneous subpopulations of spermatozoa in the ejaculate. J Androl. 2006;27(3):414–20.
29. Barrera C, Mazzolli AB, Pelling C, Stockert JC. Metachromatic staining of human sperm nuclei after reduction of disulphide bonds. Acta Histochem. 1993;94:141–9.
30. Katayose H, Yanagida K, Hashimoto S, Yamada H, Sato A. Use of diamide–acridine orange fluorescence staining to detect aberrant protamination of human-ejaculated sperm nuclei. Fertil Steril. 2003;79 Suppl 1:670–6.
31. Aravindan GR, Krishnamurthy H, Mougdal NR. Enhanced susceptibility of follicle-stimulating-hormone-deprived infertile bonnet monkey (Macaca radiata) spermatozoa to dithiothreitol-induced DNA decondensation in situ. J Androl. 1997;18:688–97.
32. Beletti ME, Mello ML. Comparison between the toluidine blue stain and the Feulgen reaction for evaluation of rabbit sperm chromatin condensation and their relationship with sperm morphology. Theriogenology. 2004;62:398–402.
33. Zini A, Phillips S, Courchesne A, Boman JM, Baazeem A, Bissonnette F, et al. Sperm head morphology is related to high deoxyribonucleic acid stainability assessed by sperm chromatin structure assay. Fertil Steril. 2009;91(6):2495–500.
34. Fosså SD, De Angelis P, Kraggerud SM, Evenson D, Theodorsen L, Clausen OP. Prediction of posttreatment spermatogenesis in patients with testicular cancer by flow cytometric sperm chromatin structure assay. Cytometry. 1997;30(4):192–6.
35. Sailer BL, Sarkar LJ, Bjordahl JA, Jost LK, Evenson DP. Effects of heat stress on mouse testicular cells and sperm chromatin structure. J Androl. 1997;18(3): 294–301.
36. Potts RJ, Newbury CJ, Smith G, Notarianni LJ, Jefferies TM. Sperm chromatin damage associated with male smoking. Mutat Res. 1999;423(1–2): 103–11.
37. Erenpreiss J, Hlevicka S, Zalkalns J, Erenpreisa J. Effect of leukocytospermia on sperm DNA integrity:

a negative effect in abnormal semen samples. J Androl. 2002;23(5):717–23.
38. Saleh RA, Agarwal A, Sharma RK, Said TM, Sikka SC, Thomas Jr AJ. Evaluation of nuclear DNA damage in spermatozoa from infertile men with varicocele. Fertil Steril. 2003;80(6):1431–6.
39. Banks S, King SA, Irvine DS, Saunders PT. Impact of a mild scrotal heat stress on DNA integrity in murine spermatozoa. Reproduction. 2005;129(4):505–14.
40. O'Flaherty C, Vaisheva F, Hales BF, Chan P, Robaire B. Characterization of sperm chromatin quality in testicular cancer and Hodgkin's lymphoma patients prior to chemotherapy. Hum Reprod. 2008;23(5):1044–52.
41. Zini A, Sigman M. Are tests of sperm DNA damage clinically useful? Pros and cons. J Androl. 2009; 30(3):219–29.
42. Cho C, Willis WD, Goulding EH, Jung-Ha H, Choi YC, Hecht NB, et al. Haploinsufficiency of protamine-1 or -2 causes infertility in mice. Nat Genet. 2001;28(1):82–6.
43. Sakkas D, Seli E, Bizzaro D, Tarozzi N, Manicardi GC. Abnormal spermatozoa in the ejaculate: abortive apoptosis and faulty nuclear remodelling during spermatogenesis. Reprod Biomed Online. 2003;7(4):428–32.
44. Aoki VW, Moskovtsev SI, Willis J, Liu L, Mullen JB, Carrell DT. DNA integrity is compromised in protamine-deficient human sperm. J Androl. 2005; 26(6):741–8.
45. Aoki VW, Emery BR, Liu L, Carrell DT. Protamine levels vary between individual sperm cells of infertile human males and correlate with viability and DNA integrity. J Androl. 2006;27(6):890–8.
46. Aitken RJ, De Iuliis GN. Origins and consequences of DNA damage in male germ cells. Reprod Biomed Online. 2007;14(6):727–33.
47. Leduc F, Maquennehan V, Nkoma GB, Boissonneault G. DNA damage response during chromatin remodeling in elongating spermatids of mice. Biol Reprod. 2008;78(2):324–32.
48. De Iuliis GN, Thomson LK, Mitchell LA, et al. DNA damage in human spermatozoa is highly correlated with the efficiency of chromatin remodeling and the formation of 8-hydroxy-2'-deoxyguanosine, a marker of oxidative stress. Biol Reprod. 2009;81(3):517–24.
49. Evenson DP, Jost LK, Marshall D, et al. Utility of the sperm chromatin structure assay as a diagnostic and prognostic tool in the human fertility clinic. Hum Reprod. 1999;14(4):1039–49.
50. Chohan KR, Griffin JT, Lafromboise M, De Jonge CJ, Carrell DT. Comparison of chromatin assays for DNA fragmentation evaluation in human sperm. J Androl. 2006;27(1):53–9.
51. Zini A, Boman JM, Belzile E, Ciampi A. Sperm DNA damage is associated with an increased risk of pregnancy loss after IVF and ICSI: systematic review and meta-analysis. Hum Reprod. 2008;23(12):2663–8.
52. Bungum M, Spanò M, Humaidan P, et al. Sperm chromatin structure assay parameters measured after density gradient centrifugation are not predictive for the outcome of ART. Hum Reprod. 2008;23(1):4–10.

Postnatal Effects of Sperm Chromatin Damage

33

Miriam Pérez-Crespo, Raúl Fernández-González, Miguel Ángel Ramírez, Eva Pericuesta, Alexandra Calle, and Alfonso Gutiérrez-Adán

Abstract

The use of spermatozoa with fragmented DNA has been linked to developmental and postnatal effects in animal models. Environmental and toxic factors such as radiation, heat stress, air pollution, chemotherapeutic agents, etc. are known to have detrimental effects on sperm chromatin. Sperm chromatin damage has also been observed following sperm manipulation techniques (freeze–thawing without cryoprotectants, freeze-drying, preincubation under different conditions, etc.). The developmental and postnatal consequences observed in the progeny of affected males depend on several factors: the DNA repair capacity of the zygote, the intensity of insult to germ cells, the affected germ cell types, the time interval between insult and mating, or the assisted reproductive technique used. The consequences of using sperm with fragmented DNA can be observed as early as the preimplantation stages of development, through other less obvious alterations may pass unnoticed during embryonic and fetal development and emerge in later life. These alterations have been observed in mouse models after the use of sperm with fragmented DNA in intracytoplasmic sperm injection (ICSI) procedures and include aberrant growth, premature aging, behavioral changes, and mesenchymal tumors. The experimental animal model is effectively the only system available to address the long-term consequences of the use of DNA-damaged sperm in fertilization protocols.

Keywords

Sperm chromatin damage • Postnatal effects of sperm chromatin damage • Environmental factors in sperm chromatin damage • Spermatozoa with fragmented DNA

Over the past 20 years, numerous studies have identified several reproductive consequences of using sperm with damaged DNA both in animals and humans. The integrity of sperm DNA is crucial for the correct transmission of genetic

A. Gutiérrez-Adán (✉)
Department of Animal Reproduction and Animal Genetic Resources Conservation, National Research Institute for Agriculture and Food Technology (INIA), Madrid, Spain
e-mail: agutierr@inia.es

A. Zini and A. Agarwal (eds.), *Sperm Chromatin: Biological and Clinical Applications in Male Infertility and Assisted Reproduction*, DOI 10.1007/978-1-4419-6857-9_33,

information to future generations. There is evidence that sperm DNA fragmentation (SDF) may lead to conception failure, abortion, malformation, and genetic diseases [1, 2]. A significant proportion of infertile men has elevated levels of damaged DNA in their ejaculated sperm [3], and it is still unclear whether assisted reproductive technology (ART) techniques can compensate for poor chromatin packaging and/or DNA damage. Several authors have reported negative effects on pregnancy rates of increased proportions of spermatozoa with damaged DNA in sperm samples used for ART techniques [4, 5]. Developmental abnormalities arising from such chromatin damage may not be observed until postimplantation stages [6–8]. The biological impact of an abnormal sperm chromatin structure depends on the combined effects of the extent of sperm DNA or chromatin damage and the capacity of the oocyte to repair that damage [9]. In a mouse model system, spermatozoa with defective DNA have been observed to fertilize an oocyte and produce high-quality early-stage embryos, but as the extent of the DNA damage increases, the likelihood of a successful pregnancy to term decreases [10]. The authors of this report suggest that the oocyte has the capacity to repair damaged sperm DNA when less than 8% of the DNA is affected [10]. However, many situations have been described in which oocytes show a reduced capacity for repair (e.g., aged oocytes) [11]. Under these circumstances, the consequences of using sperm with fragmented DNA could be critical. Depending on the level of SDF, we would expect three possible scenarios. Thus, in some cases, the oocyte repair machinery will not be capable of repairing the damaged sperm, and the embryo may fail to develop or implant in the uterus, or may be aborted naturally at a later stage (unrecoverable damage). In other cases, the oocyte will be able to repair the breaks in the DNA strand before initiation of the first cleavage division such that the sperm will generate normal offspring (recoverable damage). Finally, the worst scenario arises when the oocyte is able to partially repair the damaged sperm DNA, since deletions or sequence errors may be introduced possibly resulting in abnormal offspring (partially recoverable damage). In effect, some authors claim that the origin of 80% of de novo structural chromosome aberrations in humans is of paternal origin [12]. For instance, if DNA damage involves an oncogene, the result would be an increased risk of cancer in the offspring.

The routine performing of ART techniques such as intracytoplasmic sperm injection (ICSI) for sperm-related infertility problems determines a need to examine the postnatal consequences of SDF. Using this technique, full-term pregnancies are achieved despite high levels of sperm DNA damage [13]. ICSI is able to overcome the natural barriers that impede spermatozoa with a high load of damaged DNA to fertilize the oocyte and initiate a successful pregnancy, when this would hardly be possible through natural conception, intrauterine insemination (IUI), or even to some extent in vitro fertilization (IVF). Knowledge that SDF is common in infertile men, and preliminary reports on genetic and epigenetic abnormalities in children conceived through ICSI, prompted us to further address the issue of sperm DNA damage. Although the efficiency of ART techniques in humans is backed by a large body of data, some experiments designed to elucidate key mechanisms are not feasible in humans. In animals, however, DNA damage to the paternal germ line may be induced, and the long-term consequences of this damage in the offspring can then be assessed. In a mouse model, we identified a strong link between damage to the paternal genome and compromised embryo development, and more importantly, this had negative consequences on the newborns and subsequent generations [14]. Also, in mice, we detected selection mechanisms operating in nature that are able to discriminate the quality of spermatozoa DNA. Thus, the female reproductive tract and the zona pellucida (ZP) binding/penetration process play an important role in selecting sperm that, besides showing normal motility and morphology, feature intact chromatin. If we are able to understand the basis of these naturally imposed selection mechanisms that can distinguish the quality of spermatozoa, this could help clarify which of the many laboratory tests in current use are likely to be the most informative about fertility.

In this chapter, we review the findings of studies conducted in animal models, in which the offspring of males with sperm chromatin damage are characterized. First, we analyze the long-term effects on the offspring of the exposure of males to environmental or other toxic agents that affect the integrity of sperm DNA, and second, we report on various experiments that have explored the reproductive and long-term health consequences of the use of sperm with damaged DNA in natural matings, IVF, and ICSI.

DNA damage to the paternal germ line is experimentally induced by several methods: (a) using different sperm manipulation techniques such as preincubation under different conditions or freeze–thawing without cryoprotectants and (b) using physical agents such as scrotal heat stress or whole-body gamma radiation. These procedures serve to assess the long-term consequence of fertilizing oocytes using spermatozoa with different extents and types of DNA damage. Finally, we review recent experiments that have analyzed the transgenerational consequences of SDF and studies conducted in humans.

Long-Term Effects on Progeny of Paternal Exposure to Harmful Environmental and Medical Agents That Affect Sperm Chromatin Structure in Animals

Male germ cells are targets for a wide variety of physical agents, such as radiation or heat, or chemicals including therapeutic drugs, such as chemotherapy agents and environmental pollutants (pesticides, metals, and tobacco smoke or air pollutants). Exposure to these factors may have adverse effects on sperm production and sperm function, with the risk that a damaged male genome may be transmitted to subsequent generations. Some of these agents directly target DNA, whereas others induce oxidative stress, in which case it is the reactive oxygen species generated that form DNA adducts and damage DNA [15].

The effects we have observed on the progeny of males exposed to these factors could be attributed to sperm chromatin damage, but it should be considered that these environmental agents or drugs can induce other alterations such as epigenetic modifications, which could be associated with the phenotypes observed in the progeny [16]. Table 33.1 shows the effects on the progeny of males exposed to environmental and pharmacological toxic agents that damage sperm chromatin structure. At present, three main (not mutually exclusive) hypotheses have been proposed to explain the manifestations of germ-cell genetic damage such as malformations or cancer susceptibility in the offspring [17]: (1) germ-cell mutations: mutagen–DNA interactions may fix a mutation in a spermatogenic cell line; (2) genomic stability: genetic damage in a germ cell may induce the appearance of mutations in subsequent cell generations, germ cell generations and/or embryonic cell generations [18]; and (3) apoptosis suppression: exposure to a toxic agent could prevent germ cells from entering apoptosis when signaled to do so, leading to the build-up of

Table 33.1 Postnatal consequences of paternal exposure to harmful environmental and medical agents known to affect sperm chromatin structure

Harmful agent	Sperm quality	Species	Consequences in the offspring	Authors
Cyclophosphamide (chemotherapeutic agent)	Abnormalities in sperm chromatin and composition of sperm head basic proteins [29]	Rat	Pre- and postimplantation losses External malformations Altered behavior	[32–34]
Radiation	Abnormal sperm chromatin structure [54] DNA strand breaks [22]	Mouse	Malformations Heritable chromosomal translocations Heritable gene mutations Genomic instability and cancer	[24, 91–93]
Air pollution	DNA strand breaks [37] Mutations [38]	Mouse	Inherited mutations	[38]

genetically damaged cells among the mature spermatozoa [19, 20].

Long-Term Effects on Progeny of Paternal Exposure to Radiation

Radiation is a well-established DNA damaging agent that affects the male germ line [21, 22]. Exposure of mice to X-rays has shown that maximum DNA damage is produced in differentiating spermatogonia [22]. Currently, it is well accepted that preconception whole-body exposure to radiation poses a significant threat to the progeny of the irradiated parents by inducing DNA damage to sperm cells [23]. Sperm cell damage may affect fertilization and embryo development by causing numerous harmful phenotypic and genotypic effects in the offspring [23]. Phenotypic effects include reduced fertility and a variety of teratogenic effects. Genotypic alterations consist of increased mutation rates and elevated frequencies of chromosome aberrations, micronuclei formation, altered gene expression, and many other signs of transgeneration genome instability [24–26]. Such genotypic alterations may confer the progeny of irradiated parents an increased risk of genetic diseases, infertility, and cancer [24, 25, 27]. Most studies on germ-line and transgeneration radiation effects have analyzed the consequences of parental whole-body irradiation. However, this type of exposure is relatively rare compared to the localized body-part exposure that is frequently incurred during radiation diagnostics and therapy.

Long-Term Effects on Progeny of Paternal Exposure to Chemical and Environmental Factors

A number of studies performed in humans and animal models have linked exposure to numerous environmental pollutants to sperm DNA damage (reviewed in [28]). Using animal models, it has been shown that in addition to causing increased aneuploidy rates, treatment with one or a cocktail of chemotherapeutic agents causes sperm chromatin damage (as detected using the sperm chromatin structure assay [SCSA]) and alters the sperm nuclear proteome [29]. Paternal cyclophosphamide exposure before conception induces aberrant epigenetic programming in early embryos sired by these males [30]. Likewise, this exposure has been shown to alter the expression of important DNA repair genes in preimplantation rat embryos [31]. The treatment of male rats with cyclophosphamide has been noted to give rise to preimplantation losses [32], postimplantation losses, malformed and growth-retarded fetuses [20, 33], and even to behavioral changes [34].

Air pollution has been correlated with sperm chromatin damage in humans [35, 36] and with an increase in sperm DNA strand breaks in mice [37]. Air pollution represents a mixture of genotoxic substances. Animal studies have recently provided evidence that air pollution, under ambient conditions, can induce germ-line mutations in vertebrate sperm at minisatellite loci and expanded simple tandem repeat loci [38]. Moreover, inherited mutations have been observed in the progeny of males exposed to air pollution [38].

The effects of many chemical and environmental factors on postnatal development vary according to the time elapsed between exposure and mating. For example, the effects of a well-known mutagen, cyclophosphamide, were observed to progressively diminish, from postimplantation losses to behavioral disorders, as a function of the time between exposure and mating [33, 34]. It, thus, seems that the longer the delay between cessation of cyclophosphamide treatment and mating, the less radical the detrimental effects on the progeny.

Using animal models, it has been possible to correlate sperm DNA damage with the traditional variables of progeny outcome used in developmental toxicity studies, such as litter size, pre- and post-implantation losses, and external or internal malformations [32–34]. Nevertheless, outcome measures, such as these, are insufficiently sensitive to predict the potential impact of exposure to drugs or environmental factors on postnatal and adult end points such as neurodevelopment, immunocompetence, or normal reproductive function, some of which may be the expected consequences of epigenetic modifications.

Long-Term Consequences of Fertilizing Oocytes Using Spermatozoa with Different Extents and Types of Chromatin Damage Determined in Mouse Models

In this section, we review several experiments in which sperm chromatin damage is produced either by manipulating the sperm sample or by disrupting spermatogenesis using physical agents. The most outstanding experiments performed in mice are summarized in Table 33.2.

Sperm Chromatin Damage Induced by Sperm Preincubation Conditions

During the preparation of sperm samples, nuclei can be damaged, and the developmental potential of these spermatozoa may be reduced compared to that of fresh spermatozoa. Several authors have demonstrated that the incubation conditions to which spermatozoa are subjected before ICSI can modify sperm DNA integrity [39–41]. We speculate that spontaneous DNA fragmentation during in vitro sperm incubation involves sperm endonuclease activity. Thus, when ICSI was

Table 33.2 Developmental and postnatal consequences of fertilizing oocytes using spermatozoa with different extents and types of chromatin damage

DNA damage induced by	Method used to assess DNA damage	ART technique	Embryo development	Implantation rate	Effects on fetuses/ newborns/adults	Authors
Sperm preincubation conditions	TUNEL	ICSI	No difference in Bl yields (44 vs. 49% control)	Lower (21 vs. 54% control)	No difference in percentages of fetuses (33 vs. 40% control)	[40]
Subjecting males to γ-radiation	Comet	Natural conception	N.D.	No difference	Lower percentage of fetuses (30 vs. 92% control)	[63]
		IVF	Lower Bl yields (63 vs. 96% control)	N.D	N.D	
		ICSI	Lower Bl yields (43 vs. 67% control)	N.D	N.D	
Scrotal heat stress (42°C, 30 min)	TUNEL	Natural conception	N.D	Lower number of implantation sites (13.3 ± 2.2 vs. 24.7 ± 4 control)	Lower number of fetuses (11.2 ± 2.4 vs. 22.7 ± 4.3 control) Sex ratio distorted	[51]
Scrotal heat stress (42°C, 30 min)	Comet	IVF	No difference in Bl yields (90 vs. 96% control)	N.D	N.D	[63]
		ICSI	Lower Bl yields (6 vs. 67% control)	N.D	N.D	
Freeze–thawing spermatozoa without cryoprotectant	TUNEL and Comet	ICSI	No difference in cleavage rates (71 vs. 86% control)	N.D	Lower % live pups (13 vs. 26% control) Premature symptoms of aging; increased incidence of tumors Behavioral alterations in females	[14]
Freeze–thawing spermatozoa without cryoprotectant	ACS	ICSI	No significant difference in the incidence of ACS (38.1 vs. 26.3 control)	N.D	Lower percentage of fetuses (58.3 vs. 65% control)	[44]

(continued)

Table 33.2 (contined)

DNA damage induced by	Method used to assess DNA damage	ART technique	Embryo development	Implantation rate	Effects on fetuses/ newborns/adults	Authors
Freeze-drying spermatozoa	SCSA	ICSI using sperm with a DFI of 46.72%	47% Bl development	85%	Live pups (19%)	[48]
		ICSI using sperm with a DFI of 2.52%	69% Bl development	78%	Live pups (24%)	
Freeze-drying spermatozoa	N.D	ICSI	57% Bl vs. 73% control	N.D	Lower percentage of live pups (14 vs. 33%) Viable, healthy, and genomically stable	[49]

Studies conducted in mouse models

Bl blastocysts; *ART* assisted reproduction technology; *N.D* no data available; *N.C* natural conception; *ACS* abnormal chromosome segregation; *DFI* DNA fragmentation index

conducted using spermatozoa incubated in a medium containing endonucleases that putatively cause sperm chromatin damage (data shown in Table 33.2), embryo implantation was significantly impaired [40]. Other authors have also shown in the mouse model that certain sperm incubation conditions prior to ICSI can produce structural chromosome aberrations in the resultant one-cell embryos. These authors described that chromosomal damage increased during sperm preincubation and that its incidence depended on the composition of the medium. They also observed that these chromosome aberrations were transmissible to offspring, with some fetuses displaying a structurally abnormal karyotype (containing reciprocal translocations, inversions, and deletions) [42]. Several authors have also suggested that chromosomal aberrations in zygotes are highly predictive of subsequent abnormal embryonic development [43].

Sperm Chromatin Damage Produced by Freeze–Thawing in the Absence of Cryoprotectants

Recently, it has been hypothesized that the presence of double-strand DNA breaks in the spermatozoa used in ICSI procedures could give rise to embryos undergoing abnormal chromosome segregation (ACS) at the first mitotic division, depending on the spermatozoa manipulation protocol performed prior to ICSI [44]. The rate of ACS in ICSI embryos produced using frozen–thawed spermatozoa without cryoprotectants was significantly higher than among those derived from fresh sperm. Embryos showing ACS at first mitotic division appeared normal during preimplantation stages and could develop to the morula or blastocyst stage and become implanted in the uterus, yet died 7.5 days after fertilization. Accordingly, ACS during first mitosis appears to be a major cause of early pregnancy losses in ICSI-generated mouse embryos.

At our laboratory, several experiments have examined the consequences of injecting mouse spermatozoa with DNA damage following a freeze–thaw cycle in the absence of cryoprotectants into mouse oocytes. Effects were assessed in terms of the success of pregnancy and/or the health and well-being of the progeny [14] This extensive study sought to demonstrate the powerful adverse effects of using sperm with damaged DNA on embryo development, postnatal growth, and the behavior and longevity of the offspring, as well as their susceptibility to tumors. The mouse strains used were CD1 and B6D2 and DNA damage was induced by freeze–thawing the sperm in the absence of cryoprotectants prior to ICSI, since we observed higher rates of SDF in these sperm samples compared to fresh semen using the TUNEL and comet assays (data shown in Table 33.2).

We observed that ICSI using the spermatozoa with fragmented DNA produced no effects on the percentage of embryos returned by microinjection or that cleaved to the 2–4 cell stage. However, the proportion of transferred embryos that gave rise to live offspring was twofold when fresh sperm cells were used for fertilization (26 vs. 13%). Immunofluorescence staining with an antibody against 5-methylcytosine (MeC) revealed a 2-h delay in the active demethylation of the male pronucleus in the embryos derived from sperm with fragmented DNA.

Twenty weeks after parturition, ICSI produced animals and in vivo produced controls were subjected to behavioral tests: locomotor activity (open field), exploratory/anxiety behavior (elevated plus maze, open field), and spatial memory (free-choice exploration paradigm in a Y maze). The female CD1 mice produced by ICSI using sperm with fragmented DNA showed general alterations in behavioral responses in both early and later stages of life. These animals suffered increased anxiety, lack of habituation patterns, deficient short-term spatial memory and exhibited age-dependent hypolocomotion in the open field test.

Anatomopathological analysis of the animals at 16 months of age revealed enlarged organs and an increased number of pathologies (33% of the ICSI-produced CD1 females developed solid tumors in the lungs or skin on the back or neck). Postmortem anatomical and histological findings indicated that ICSI using spermatozoa with fragmented DNA led to a significant increase in the number of tumors. Moreover, of the B6D2F1 mice derived from sperm with damaged DNA, 20% died before 5 months of age, 25% of those surviving showed symptoms of premature aging, and 70% died earlier than controls due to different tumors.

We suggest that, depending on the level of SDF, oocytes may be able to repair some of the fragmented DNA to produce blastocysts capable of implanting and producing live offspring. However, incomplete repair could lead to longer-term deficient phenotypes. Of most concern is that our data indicate that the use of spermatozoa with fragmented DNA in ICSI procedures can generate effects that only emerge later in life, including aberrant growth, premature aging, behavioral changes, and mesenchymal tumors. We believe that the increased incidence of tumors observed is related to DNA fragmentation in the sperm used since ICSI conducted with fresh sperm did not produce this effect. The ICSI procedure using DNA-damaged sperm could also be the cause of premature aging and the tumors associated with aging we detected. Aging and cancer are two sides of the same coin: in one case, cells stop dividing and in the other, they cannot stop dividing. DNA damage is thought to contribute to aging [45], and chromatin hypomethylation has also been related to premature aging in mice [46].

Sperm Chromatin Damage Produced by Freeze-Drying

Freeze-drying is a very useful technique for the long-term storage and transport of viable genetic material. It is, thus, essential to check the level of DNA fragmentation in freeze-dried spermatozoa before use. Some authors have reported that mouse sperm can be freeze-dried without damaging their chromosomes [47]. Recently, Kawase et al. [48] have determined the level of DNA fragmentation in freeze-dried spermatozoa using the SCSA before undertaking ICSI. The extent of fragmentation was found to depend on the initial drying pressure and the storage temperature chosen to preserve the sperm samples. These authors observed good correlation between the DNA fragmentation index (DFI) determined by SCSA and the developmental rate to the blastocyst stage (data shown in Table 33.2). However, no correlation was detected between DFI and the percentage of live fetuses, indicating the negative consequences of using spermatozoa with fragmented DNA induced by freeze-drying affect the preimplantation period [48]. Other authors who evaluated the postnatal consequences of using freeze-dried spermatozoa in ICSI have concluded that viable, healthy, and genomically stable mice can be derived from ICSI using freeze-dried mouse sperm stored in the refrigerator for at least 2 months [49].

Sperm Chromatin Damage Produced by Scrotal Heat Stress

Several reports have confirmed that heat applied to the scrotum leads to sperm chromatin damage [50–54]. In effect, paternal heat stress affects most protein patterns in preimplantation embryos [55], which could perhaps explain some of the negative consequences observed in embryo and fetal development. A link between paternal heat stress and embryo survival has been identified in a few studies [56–59]. Several effects on the offspring of heat-stressed males have been reported: a reduction in litter size [50, 51, 60, 61], a reduction in placental and fetus weight [50, 56, 62], and a distortion in the sex ratio toward females, when males are mated to females on the day of heat treatment [51]. We propose this shift in the sex ratio could be attributed to the different functionality of sperm carrying the X or Y chromosome.

All of the above-mentioned studies have assessed the reproductive consequences of naturally mating heat-stressed males to nonheat treated females. These consequences vary significantly depending on the moment after heat treatment when mating takes place. Some authors have observed that the most significant consequences arise when mating takes place 21–28 days after heat stress (Table 33.2). We have examined the reproductive consequences of performing IVF and ICSI using DNA-damaged spermatozoa from scrotal heat-stressed males. Spermatozoa were collected from the cauda epididymis and vas deferens of males 21–25 days after heat treatment to determine the outcome of using sperm that had developed from spermatocytes subjected to heat. In subsequent IVF experiments, a lower percentage of 2-cell embryos was recorded in the heat-stressed compared to the control group. This could be due to the lower motility of the sperm. However, blastocyst development failed to differ between the two groups, indicating a similar postimplantation development potential. However, when natural barriers of fertilization were overcome by ICSI, though the number of surviving oocytes was unaffected by the treatment, the cleavage rate decreased significantly. Moreover, the blastocyst development rate was significantly lower than for the control group, suggesting reduced DNA quality in these spermatozoa [63] (Table 33.2). Other authors conducting IVF using spermatozoa from heat-stressed males observed that the number of embryos moving into the blastocyst stage was greatly reduced (40% that recorded in controls) when sperm was obtained from mice 16 h after heat treatment. However, none of the embryos generated from sperm retrieved from males 23 days after treatment progressed beyond the 4-cell stage.

Sperm Chromatin Damage Produced by Whole-Body Exposure to γ-Rays

The main characteristic of γ-radiation as a factor inducing SDF is the wide range of DNA damages it provokes [64]. The most likely lesion caused by γ-radiation is the presence in the DNA of double-strand breaks (DSBs). In our laboratory, γ-radiation has been used to induce SDF, which is subsequently detected by TUNEL and comet assay. Gamma radiation was applied to mice as described elsewhere [65] using a ^{137}Cs irradiator to deliver a 4-Gy dose at a rate of 1.25 Gy/min. Our objective was to determine the reproductive consequence of mating males, both naturally and using ART techniques (IVF and ICSI), 21–25 days after irradiation treatment. After natural mating, γ-radiation did not affect the implantation rate, yet the number of fetuses conceived by irradiated males was lower than that in the control group. Radiation significantly increased the resorption rate, indicating that spermatozoa from irradiated mice, though capable of in vivo fertilization and producing blastocysts able to be implanted, give rise to an embryo viability that is somehow compromised as reflected by the lower proportion of fetuses obtained (Table 33.2).

When spermatozoa from irradiated males were used for IVF, cleavage and blastocyst rates were lower than those recorded in controls, indicating that some extent of DNA-damage induced by γ-rays allows fertilization, but compromises embryo development. When treated spermatozoa were used for ICSI, the reproductive consequences were similar to those described for the heat-stressed males. Thus, the number of surviving oocytes was unaffected by treatment, but cleavage rates were significantly reduced. In addition, blastocyst development was lower than in the control group, suggesting the reduced DNA quality of the spermatozoa [63] (Table 33.2).

Transgenerational Consequences of the Use of Spermatozoa with Fragmented DNA

There is growing evidence that DNA damage in the fertilizing gamete as a mediator of postfertilization processes contributes to the genomic instability of subsequent generations. Transgenerational genomic instability most likely involves epigenetic mechanisms or error-prone DNA repair processes in the early embryo. Maternal and embryonic DNA repair processes during the early stages of mammalian embryonic development can have far-reaching consequences for the genomic integrity and health of subsequent generations. A series of recent studies have suggested that DNA damage in germ cells can mediate postfertilization processes that lead to an increased risk of genomic instability in the progeny. Paternal exposure to chemical mutagens [66], ionizing radiation [67, 68], and particulate air pollution [37, 38], aside from increasing mutation frequencies in sperm, was more importantly found to induce persistent genomic instability in the F1 and F2 offspring of exposed mice [69–72]. In addition, the introduction of DNA damage by irradiated sperm triggers a genomic instability that can induce mutations in the unirradiated maternal genome [68]. These findings suggest that the mechanisms contributing to transgenerational genomic instability most likely involve epigenetic or error-prone DNA repair processes in the embryo.

There is still little experimental evidence of a transgenerational effect of sperm DNA damage. Adiga et al. have recently examined the transgenerational influence of varying the level of sperm DNA damage in both the somatic and germ-line compartment of F1 offspring in an irradiation model [73]. The data presented reveal increasing levels of genome instability in preimplantation embryos with increasing loads of damaged sperm DNA as evidenced by micronucleus analysis. Interestingly, the genetic instability is transmitted to both somatic and germ-line compartments of the F1 offspring derived from the DNA-damaged sperm. However, the extent of instability observed in embryos, somatic cells, and germ-line cells is dependent on the amount of DNA damage present in the paternal spermatozoa. Previously, Baulch et al. had observed that spermatogonial irradiation causes negative effects on embryonic cell proliferation rates and juvenile offspring protein levels in four generations [65, 74, 75]. In a later investigation, using the sperm comet assay to evaluate SDF, these authors confirmed their previous findings and clearly demonstrated heritable effects of paternal F0 spermatogonial irradiation history on chromatin [65]. This study also clearly demonstrates that the ATM (ataxia telangiectasia mutated) gene heterozygosity of the sire has a significant impact on these heritable effects.

In the same context, Dubrova et al. observed an increment and uniformity in the frequency of germ-line mutations in the F1 generation arising from different parental exposures to ionizing radiation. Dubrova found that this indirect effect leads to the destabilization of ESTR (expanded simple-tandem repeat) loci in the germ-line offspring. A significant level of mosaicism of these mutations is observed in the offspring due to early germ-line development. The uniformity observed in these increased ESTR mutation rates suggests that an epigenetic process is responsible for these alterations [70]. These observations indicate that mutation rates in the offspring of irradiated parents are substantially elevated. Also remarkable is the effect of the parental genotype on transgenerational instability. Thus, oocytes of female *scid* (severe combined immunodeficient) mice are unable to fully support the repair of double-strand breaks induced in paternal sperm which may, in turn, result in the elimination of cells/embryos containing high levels of DNA damage, thus partially preventing the manifestation of genomic instability. The suppression of mutation induced and radiation-induced genomic instability in homozygous *scid* cells can be explained by the DNA-PK-independent activation of p53 and p21, resulting in a high level of apoptosis and cell-cycle arrest in the irradiated DNA-PKcs-deficient cells [76]. Hatch et al. also demonstrated interstrain variation in responses to ionizing radiation, including the manifestation of radiation-induced genomic instability. This variation has been explained by differences in the intensity of apoptosis [77]. According to the results of Hatch et al., cells from radiation-resistant C57BL/6 mice undergo rapid apoptosis after irradiation, which could in turn suppress radiation-induced genomic instability in this strain.

Induced genomic instability can give rise to oncogenic mutations in somatic cells and malignant transformation [78]. Effectively, radiation-induced delayed transgenerational instability may have important health consequences that may become apparent in subsequent generations after the original exposure. It is clear that understanding the DNA repair capacity of the zygote and the mechanisms that contribute to transgenerational genomic instability are areas that will require significant attention in the future. The main studies that have assessed the transgenerational consequences of the use of spermatozoa with fragmented DNA are reviewed in Table 33.3.

Table 33.3 Transgenerational changes in subsequent generations of offspring derived from chromatin-damaged sperm

Treatment	Sperm chromatin damage	Species	Time of mating after treatment	Generations affected	Alterations in the offspring	Authors
Gamma-irradiation of the testicular area	Denatured DNA; increased percentage tail DNA	Mouse	18 h after irradiation	First generation	Increased genomic instability in fetal liver cells and sperm chromatin modifications in F1 males	[73]
WB irradiation	Increased comet tail length and percentage tail DNA assessed by neutral comet assay	Mouse	45 days after irradiation	Third-generation descendants	Heritable chromatin effects in sperm	[65]
WB irradiation	Increased mutation frequencies in sperm	Mouse	6 weeks after irradiation	F1 offspring	Transgenerational destabilization of the F(1) genome; endogenous DNA lesions	[24]
WB ionizing radiation	N.D	Mouse	N.D	F1 and F2 offspring	Elevated mutation rate in somatic and germinal cells. Increased cancer incidence	Reviewed in [25]
WB gamma radiation	N.D	Mouse	6 weeks after irradiation	F1 offspring	Decreased fertilization rates of spermatozoa from the F1 offspring	[94]
Cranial irradiation	Elevated DNA strand breaks	Rat	1 week after irradiation	F1 offspring	Epigenetic dysregulation	[95]
WB gamma radiation	N.D.	Rat	N.D	F1 and F2 generations	Impaired regeneration of liver tissue	[96]
Cyclophosphamide	Sperm DNA damage [31] Sperm chromatin alterations [29]	Rat	Immediately after treatment N.D	F2 progeny F1 and F2 progeny	Postimplantation loss and malformations Behavioral alterations	[97, 98]

N.D no data available; *WB* whole-body

Postnatal Consequences of Sperm Chromatin Damage in Humans

Sperm DNA damage can affect the health of the embryo, fetus, and offspring [17, 79]. A possible consequence of sperm DNA damage is infertility in the offspring [80, 81]. A concern emerging from studies conducted in smokers is an increased risk of childhood cancer observed in the offspring of men with a high proportion of sperm with fragmented DNA in their semen. The study in question revealed that the children of these men, whose ejaculates are under oxidative stress [82] and characterized by a high level of chromatin fragmentation, are 4–5 times more likely to develop cancer in childhood than the children of nonsmoking fathers [83]. Another study has demonstrated that 15% of all childhood cancers are directly attributable to paternal smoking [84]. These studies suggest that there may be a link between sperm DNA damage and the subsequent development of childhood diseases. Moreover, because this particular mutation is fixed in the germ line, it has the potential to impact upon the health and well-being of all the future descendants of a given individual [80]. The link between sperm DNA damage and offspring abnormalities is not confined to smokers. For example, powerful associations exist between childhood disease and paternal occupation [85]. Aitken and Krausz [80] proposed that sperm DNA damage is promutagenic and can give rise to mutations after fertilization. As the oocyte attempts to repair DNA damage prior to the initiation of first cleavage, mutations occurring at this point will be fixed in the germ line and not only may be responsible for the induction of such pathologies as described above (infertility and childhood cancer), but may also confer a higher risk of disease imprinting [86, 87].

Conclusions

The consequences of using sperm with fragmented DNA can be observed as early as at the preimplantational stages of development. Examples of these consequences may be found in epigenetic changes [88], chromosome aberrations observed in ICSI-produced one-cell embryos derived from spermatozoa that have been preincubated [42], or the increased embryo losses incurred after treatment of males with cyclophosphamide [32]. However, other alterations attributable to the consequences of DNA-damaged sperm could go unnoticed during embryo and fetal development and only emerge later in life. These alterations have been identified in the mouse model following the use of DNA-damaged sperm in ICSI and include aberrant growth, premature aging, behavioral changes, and mesenchymal tumors [14]. Another factor that can have long-term consequences is radiation. Radiation induces phenotypic and genotypic alterations in the progeny of treated males. Such genotypic alterations may predispose the progeny of irradiated parents to an increased risk of genetic diseases, infertility, or cancer [24, 25, 27]. Moreover, the consequences of sperm chromatin damage are not limited to the progeny of males exposed to the toxic agent, and several future generations can be affected. Most studies that have reported transgenerational damage have dealt with the harmful effects of radiation or chemotherapeutic agents [67–69, 72–75, 89].

Studies conducted in animal models in which spermatozoa with fragmented DNA are used for ART techniques such as ICSI have confirmed the negative effects on pregnancy rates already reported in humans [4, 40, 87]. However, the consequences observed in the mouse model are not restricted to reproductive failure, since several health and behavioral abnormalities are observed [14]. To avoid such undesirable consequences in the offspring when ICSI is performed, some precautions should be taken. It is advisable to check the preincubation conditions of the sperm prior to ICSI to avoid inducing sperm chromatin damage and, therefore, subsequent negative consequences on embryo development [40, 42, 90]. Further studies are needed in humans to validate the data obtained in animal models on postnatal alterations and the transgenerational genetic risks associated with DNA-damaged sperm. The long-term follow-up of children born through ICSI is also recommended. These studies will have significant implications for the growing use of ART to resolve male infertility problems.

References

1. Sakkas D, Mariethoz E, Manicardi G, Bizzaro D, Bianchi PG, Bianchi U. Origin of DNA damage in ejaculated human spermatozoa. Rev Reprod. 1999;4(1):31–7.
2. Evenson DP. Loss of livestock breeding efficiency due to uncompensable sperm nuclear defects. Reprod Fertil Dev. 1999;11(1):1–15.
3. Irvine DS, Twigg JP, Gordon EL, Fulton N, Milne PA, Aitken RJ. DNA integrity in human spermatozoa: relationships with semen quality. J Androl. 2000; 21(1):33–44.
4. Benchaib M, Braun V, Lornage J, et al. Sperm DNA fragmentation decreases the pregnancy rate in an assisted reproductive technique. Hum Reprod. 2003;18(5):1023–8.
5. Morris ID, Ilott S, Dixon L, Brison DR. The spectrum of DNA damage in human sperm assessed by single cell gel electrophoresis (Comet assay) and its relationship to fertilization and embryo development. Hum Reprod. 2002;17(4):990–8.
6. Hamamah S, Fignon A, Lansac J. The effect of male factors in repeated spontaneous abortion: lesson from in-vitro fertilization and intracytoplasmic sperm injection. Hum Reprod Update. 1997;3(4):393–400.
7. Twigg JP, Irvine DS, Aitken RJ. Oxidative damage to DNA in human spermatozoa does not preclude pronucleus formation at intracytoplasmic sperm injection. Hum Reprod. 1998;13(7):1864–71.
8. Rienzi L, Martinez F, Ubaldi F, et al. Polscope analysis of meiotic spindle changes in living metaphase II human oocytes during the freezing and thawing procedures. Hum Reprod. 2004;19(3):655–9.
9. Genesca A, Caballin MR, Miro R, Benet J, Germa JR, Egozcue J. Repair of human sperm chromosome aberrations in the hamster egg. Hum Genet. 1992;89(2): 181–6.
10. Ahmadi A, Ng SC. Fertilizing ability of DNA-damaged spermatozoa. J Exp Zool. 1999;284(6):696–704.
11. Menezo Y. Oocyte capacity to repair DNA damage induced in sperm. J Gynecol Obstet Biol Reprod (Paris). 2006;35(5 Pt 2):2S19–2S23.
12. Tomar D, Magenis E, Chamberlin J, Allen L, Olson S, Donlon T, et al. Preferential paternal origin of de novo structural chromosome rearrangements. Am J Hum Genet. 1984;36(6):115s.
13. Gandini L, Lombardo F, Paoli D, et al. Full-term pregnancies achieved with ICSI despite high levels of sperm chromatin damage. Hum Reprod. 2004;19(6):1409–17.
14. Fernandez-Gonzalez R, Moreira PN, Perez-Crespo M, et al. Long-term effects of mouse intracytoplasmic sperm injection with DNA-fragmented sperm on health and behavior of adult offspring. Biol Reprod. 2008;78(4):761–72.
15. Aitken RJ, De Iuliis GN, McLachlan RI. Biological and clinical significance of DNA damage in the male germ line. Int J Androl. 2009;32(1):46–56.
16. Wyrobek AJ. Methods and concepts in detecting abnormal reproductive outcomes of paternal origin. Reprod Toxicol. 1993;7 Suppl 1:3–16.
17. Brinkworth MH. Paternal transmission of genetic damage: findings in animals and humans. Int J Androl. 2000;23(3):123–35.
18. Luke GA, Riches AC, Bryant PE. Genomic instability in haematopoietic cells of F1 generation mice of irradiated male parents. Mutagenesis. 1997;12(3):147–52.
19. Barnes CJ, Covington BWt, Cameron IL, Lee M. Effect of aging on spontaneous and induced mouse testicular germ cell apoptosis. Aging (Milano). 1998; 10(6):497–501.
20. Brinkworth MH, Nieschlag E. Association of cyclophosphamide-induced male-mediated, foetal abnormalities with reduced paternal germ-cell apoptosis. Mutat Res. 2000;447(2):149–54.
21. Haines GA, Hendry JH, Daniel CP, Morris ID. Increased levels of comet-detected spermatozoa DNA damage following in vivo isotopic- or X-irradiation of spermatogonia. Mutat Res. 2001;495(1–2):21–32.
22. Haines GA, Hendry JH, Daniel CP, Morris ID. Germ cell and dose-dependent DNA damage measured by the comet assay in murine spermatozoaa after testicular X-irradiation. Biol Reprod. 2002;67(3):854–61.
23. Aitken RJ, De Iuliis GN. Origins and consequences of DNA damage in male germ cells. Reprod Biomed Online. 2007;14(6):727–33.
24. Barber RC, Hickenbotham P, Hatch T, et al. Radiation-induced transgenerational alterations in genome stability and DNA damage. Oncogene. 2006;25(56):7336–42.
25. Dubrova YE. Radiation-induced transgenerational instability. Oncogene. 2003;22(45):7087–93.
26. Nomura T. Transgenerational carcinogenesis: induction and transmission of genetic alterations and mechanisms of carcinogenesis. Mutat Res. 2003;544(2–3):425–32.
27. Mohr U, Dasenbrock C, Tillmann T, et al. Possible carcinogenic effects of X-rays in a transgenerational study with CBA mice. Carcinogenesis. 1999;20(2):325–32.
28. Delbes G, Hales BF, Robaire B. Toxicants and human sperm chromatin integrity. Mol Hum Reprod. 2010;16:14–22.
29. Codrington AM, Hales BF, Robaire B. Exposure of male rats to cyclophosphamide alters the chromatin structure and basic proteome in spermatozoa. Hum Reprod. 2007;22(5):1431–42.
30. Barton TS, Robaire B, Hales BF. Epigenetic programming in the preimplantation rat embryo is disrupted by chronic paternal cyclophosphamide exposure. Proc Natl Acad Sci USA. 2005;102(22):7865–70.
31. Harrouk W, Codrington A, Vinson R, Robaire B, Hales BF. Paternal exposure to cyclophosphamide induces DNA damage and alters the expression of DNA repair genes in the rat preimplantation embryo. Mutat Res. 2000;461(3):229–41.
32. Hales BF, Smith S, Robaire B. Cyclophosphamide in the seminal fluid of treated males: transmission to females by mating and effect on pregnancy outcome. Toxicol Appl Pharmacol. 1986;84(3):423–30.
33. Trasler JM, Hales BF, Robaire B. Chronic low dose cyclophosphamide treatment of adult male rats: effect on fertility, pregnancy outcome and progeny. Biol Reprod. 1986;34(2):275–83.

34. Auroux M, Dulioust E, Selva J, Rince P. Cyclophosphamide in the F0 male rat: physical and behavioral changes in three successive adult generations. Mutat Res. 1990;229(2):189–200.
35. Evenson DP, Wixon R. Environmental toxicants cause sperm DNA fragmentation as detected by the sperm chromatin structure assay (SCSA). Toxicol Appl Pharmacol. 2005;207(2 Suppl):532–7.
36. Rubes J, Selevan SG, Evenson DP, et al. Episodic air pollution is associated with increased DNA fragmentation in human sperm without other changes in semen quality. Hum Reprod. 2005;20(10):2776–83.
37. Yauk C, Polyzos A, Rowan-Carroll A, et al. Germline mutations, DNA damage, and global hypermethylation in mice exposed to particulate air pollution in an urban/industrial location. Proc Natl Acad Sci USA. 2008;105(2):605–10.
38. Somers CM, Yauk CL, White PA, Parfett CL, Quinn JS. Air pollution induces heritable DNA mutations. Proc Natl Acad Sci USA. 2002;99(25):15904–7.
39. Estop AM, Munne S, Jost LK, Evenson DP. Studies on sperm chromatin structure alterations and cytogenetic damage of mouse sperm following in vitro incubation. Studies on in vitro-incubated mouse sperm. J Androl. 1993;14(4):282–8.
40. Perez-Crespo M, Moreira P, Pintado B, Gutierrez-Adan A. Factors from damaged sperm affect its DNA integrity and its ability to promote embryo implantation in mice. J Androl. 2008;29(1):47–54.
41. Ward MA, Ward WS. A model for the function of sperm DNA degradation. Reprod Fertil Dev. 2004;16(5):547–54.
42. Tateno H. Chromosome aberrations in mouse embryos and fetuses produced by assisted reproductive technology. Mutat Res. 2008;657(1):26–31.
43. Marchetti F, Bishop JB, Cosentino L, Moore 2nd D, Wyrobek AJ. Paternally transmitted chromosomal aberrations in mouse zygotes determine their embryonic fate. Biol Reprod. 2004;70(3):616–24.
44. Yamagata K, Suetsugu R, Wakayama T. Assessment of chromosomal integrity using a novel live-cell imaging technique in mouse embryos produced by intracytoplasmic sperm injection. Hum Reprod. 2009; 24(10):2490–9.
45. de Boer J, Andressoo JO, de Wit J, et al. Premature aging in mice deficient in DNA repair and transcription. Science. 2002;296(5571):1276–9.
46. Lopes S, Jurisicova A, Sun JG, Casper RF. Reactive oxygen species: potential cause for DNA fragmentation in human spermatozoa. Hum Reprod. 1998;13(4):896–900.
47. Kusakabe H, Yanagimachi R, Kamiguchi Y. Mouse and human spermatozoa can be freeze-dried without damaging their chromosomes. Hum Reprod. 2008; 23(2):233–9.
48. Kawase Y, Wada NA, Jishage K. Evaluation of DNA fragmentation of freeze-dried mouse sperm using a modified sperm chromatin structure assay. Theriogenology. 2009;72(8):1047–53.
49. Li MW, Willis BJ, Griffey SM, Spearow JL, Lloyd KC. Assessment of three generations of mice derived by ICSI using freeze-dried sperm. Zygote. 2009;17(3):239–51.
50. Paul C, Murray AA, Spears N, Saunders PT. A single, mild, transient scrotal heat stress causes DNA damage, subfertility and impairs formation of blastocysts in mice. Reproduction. 2008;136(1):73–84.
51. Perez-Crespo M, Pintado B, Gutierrez-Adan A. Scrotal heat stress effects on sperm viability, sperm DNA integrity, and the offspring sex ratio in mice. Mol Reprod Dev. 2008;75(1):40–7.
52. Love CC, Kenney RM. Scrotal heat stress induces altered sperm chromatin structure associated with a decrease in protamine disulfide bonding in the stallion. Biol Reprod. 1999;60(3):615–20.
53. Banks S, King SA, Irvine DS, Saunders PT. Impact of a mild scrotal heat stress on DNA integrity in murine spermatozoa. Reproduction. 2005;129(4):505–14.
54. Sailer BL, Sarkar LJ, Bjordahl JA, Jost LK, Evenson DP. Effects of heat stress on mouse testicular cells and sperm chromatin structure. J Androl. 1997;18(3):294–301.
55. Zhu B, Maddocks S. The effect of paternal heat stress on protein profiles of pre-implantation embryos in the mouse. Int J Androl. 2005;28(3):128–36.
56. Jannes P, Spiessens C, Van der Auwera I, D'Hooghe T, Verhoeven G, Vanderschueren D. Male subfertility induced by acute scrotal heating affects embryo quality in normal female mice. Hum Reprod. 1998;13(2):372–5.
57. Paul C, Melton DW, Saunders PT. Do heat stress and deficits in DNA repair pathways have a negative impact on male fertility? Mol Hum Reprod. 2008;14(1):1–8.
58. Setchell BP. The Parkes Lecture. Heat and the testis. J Reprod Fertil. 1998;114(2):179–94.
59. Zhu BK, Setchell BP. Effects of paternal heat stress on the in vivo development of preimplantation embryos in the mouse. Reprod Nutr Dev. 2004;44(6):617–29.
60. Rockett JC, Mapp FL, Garges JB, Luft JC, Mori C, Dix DJ. Effects of hyperthermia on spermatogenesis, apoptosis, gene expression, and fertility in adult male mice. Biol Reprod. 2001;65(1):229–39.
61. Yaeram J, Setchell BP, Maddocks S. Effect of heat stress on the fertility of male mice in vivo and in vitro. Reprod Fertil Dev. 2006;18(6):647–53.
62. Ghasemi N, Babaei H, Azizallahi S, Kheradmand A. Effect of long-term administration of zinc after scrotal heating on mice spermatozoa and subsequent offspring quality. Andrologia. 2009;41(4):222–8.
63. Hourcade JD, Perez-Crespo M, Fernandez-Gonzalez R, Pintado B, Gutierrez-Adan A. Selection against spermatozoa with fragmented DNA after postovulatory mating depends on the type of damage. Reprod Biol Endocrinol. 2010;8(1):9.
64. Frankenberg-Schwager M, Frankenberg D, Harbich R, Adamczyk C. A comparative study of rejoining of DNA double-strand breaks in yeast irradiated with 3.5 MeV alpha-particles or with 30 MeV electrons. Int J Radiat Biol. 1990;57(6):1151–68.
65. Baulch JE, Li MW, Raabe OG. Effect of ATM heterozygosity on heritable DNA damage in mice following paternal F0 germline irradiation. Mutat Res. 2007;616(1–2):34–45.
66. Glen CD, Smith AG, Dubrova YE. Single-molecule PCR analysis of germ line mutation induction by anticancer drugs in mice. Cancer Res. 2008;68(10):3630–6.

67. Dubrova YE. Radiation-induced mutation at tandem repeat DNA Loci in the mouse germline: spectra and doubling doses. Radiat Res. 2005;163(2):200–7.
68. Niwa O, Kominami R. Untargeted mutation of the maternally derived mouse hypervariable minisatellite allele in F1 mice born to irradiated spermatozoa. Proc Natl Acad Sci USA. 2001;98(4):1705–10.
69. Barber R, Plumb MA, Boulton E, Roux I, Dubrova YE. Elevated mutation rates in the germ line of first- and second-generation offspring of irradiated male mice. Proc Natl Acad Sci USA. 2002;99(10):6877–82.
70. Dubrova YE, Plumb M, Brown J, Boulton E, Goodhead D, Jeffreys AJ. Induction of minisatellite mutations in the mouse germline by low-dose chronic exposure to gamma-radiation and fission neutrons. Mutat Res. 2000;453(1):17–24.
71. Hatch T, Derijck AA, Black PD, van der Heijden GW, de Boer P, Dubrova YE. Maternal effects of the scid mutation on radiation-induced transgenerational instability in mice. Oncogene. 2007;26(32):4720–4.
72. Dubrova YE, Hickenbotham P, Glen CD, Monger K, Wong HP, Barber RC. Paternal exposure to ethylnitrosourea results in transgenerational genomic instability in mice. Environ Mol Mutagen. 2008;49(4):308–11.
73. Adiga SK, Upadhya D, Kalthur G, Bola Sadashiva SR, Kumar P. Transgenerational changes in somatic and germ line genetic integrity of first-generation offspring derived from the DNA damaged sperm. Fertil Steril. 2010;93(8):2486–90.
74. Baulch JE, Raabe OG. Gamma irradiation of Type B spermatogonia leads to heritable genomic instability in four generations of mice. Mutagenesis. 2005;20(5): 337–43.
75. Baulch JE, Raabe OG, Wiley LM. Heritable effects of paternal irradiation in mice on signaling protein kinase activities in F3 offspring. Mutagenesis. 2001;16(1): 17–23.
76. Jimenez GS, Bryntesson F, Torres-Arzayus MI, et al. DNA-dependent protein kinase is not required for the p53-dependent response to DNA damage. Nature. 1999;400(6739):81–3.
77. Wallace M, Coates PJ, Wright EG, Ball KL. Differential post-translational modification of the tumour suppressor proteins Rb and p53 modulate the rates of radiation-induced apoptosis in vivo. Oncogene. 2001;20(28):3597–608.
78. Little JB. Radiation carcinogenesis. Carcinogenesis. 2000;21(3):397–404.
79. Savitz DA. Paternal exposure to known mutagens and health of the offspring: ionizing radiation and tobacco smoke. Adv Exp Med Biol. 2003;518:49–57.
80. Aitken RJ, Krausz C. Oxidative stress, DNA damage and the Y chromosome. Reproduction. 2001;122(4): 497–506.
81. Silber SJ, Repping S. Transmission of male infertility to future generations: lessons from the Y chromosome. Hum Reprod Update. 2002;8(3):217–29.
82. Manicardi GC, Bianchi PG, Pantano S, et al. Presence of endogenous nicks in DNA of ejaculated human spermatozoa and its relationship to chromomycin A3 accessibility. Biol Reprod. 1995;52(4):864–7.
83. Ji BT, Shu XO, Linet MS, et al. Paternal cigarette smoking and the risk of childhood cancer among offspring of nonsmoking mothers. J Natl Cancer Inst. 1997;89(3):238–44.
84. Sorahan T, Lancashire RJ, Hulten MA, Peck I, Stewart AM. Childhood cancer and parental use of tobacco: deaths from 1953 to 1955. Br J Cancer. 1997;75(1): 134–8.
85. Aitken RJ, Baker MA, Sawyer D. Oxidative stress in the male germ line and its role in the aetiology of male infertility and genetic disease. Reprod Biomed Online. 2003;7(1):65–70.
86. Cox GF, Burger J, Lip V, et al. Intracytoplasmic sperm injection may increase the risk of imprinting defects. Am J Hum Genet. 2002;71(1):162–4.
87. DeBaun MR, Niemitz EL, Feinberg AP. Association of in vitro fertilization with Beckwith-Wiedemann syndrome and epigenetic alterations of LIT1 and H19. Am J Hum Genet. 2003;72(1):156–60.
88. Barton TS, Robaire B, Hales BF. DNA damage recognition in the rat zygote following chronic paternal cyclophosphamide exposure. Toxicol Sci. 2007;100(2): 495–503.
89. Dubrova YE, Plumb M, Gutierrez B, Boulton E, Jeffreys AJ. Transgenerational mutation by radiation. Nature. 2000;405(6782):37.
90. Tateno H, Kamiguchi Y. Evaluation of chromosomal risk following intracytoplasmic sperm injection in the mouse. Biol Reprod. 2007;77(2):336–42.
91. Muller WU, Streffer C, Wojcik A, Niedereichholz F. Radiation-induced malformations after exposure of murine germ cells in various stages of spermatogenesis. Mutat Res. 1999;425(1):99–106.
92. Muller WU. Radiation and malformations in a murine model. Adv Exp Med Biol. 2003;518:163–8.
93. Marchetti F, Essers J, Kanaar R, Wyrobek AJ. Disruption of maternal DNA repair increases sperm-derived chromosomal aberrations. Proc Natl Acad Sci USA. 2007;104(45):17725–9.
94. Burruel VR, Raabe OG, Wiley LM. In vitro fertilization rate of mouse oocytes with spermatozoa from the F1 offspring of males irradiated with 1.0 Gy 137Cs gamma-rays. Mutat Res. 1997;381(1):59–66.
95. Tamminga J, Koturbash I, Baker M, et al. Paternal cranial irradiation induces distant bystander DNA damage in the germline and leads to epigenetic alterations in the offspring. Cell Cycle. 2008;7(9):1238–45.
96. Slovinska L, Elbertova A, Misurova E. Transmission of genome damage from irradiated male rats to their progeny. Mutat Res. 2004;559(1–2):29–37.
97. Hales BF, Crosman K, Robaire B. Increased postimplantation loss and malformations among the F2 progeny of male rats chronically treated with cyclophosphamide. Teratology. 1992;45(6):671–8.
98. Auroux MR, Dulioust EJ, Nawar NN, et al. Antimitotic drugs in the male rat. Behavioral abnormalities in the second generation. J Androl. 1988;9(3):153–9.

Evaluation of Chromatin and DNA Integrity in Testicular Sperm

34

Armand Zini and Naif Al-Hathal

Abstract

There is a universal agreement that the examination of conventional semen parameters alone only provides the clinician with a general sense of male reproductive health. Recently, sperm DNA fragmentation/damage has been studied extensively in an attempt to improve the diagnostic accuracy of the male evaluation, particularly, in couples with idiopathic infertility. However, the pathophysiology and etiology of sperm DNA damage (DD) in humans are incompletely understood, and to date, there are very few data on the treatment options for infertile men with this sperm defect. There are several tests used to assess chromatin and/or DD in ejaculated spermatozoa. Using these assays, attempts have been made toward establishing threshold values for the percentage of sperm with DD, the values above which fertility would be affected. Nonetheless, these assays need to be standardized, as there is wide variation among the various tests of sperm DD and these assays have not been tailored to evaluate testicular sperm DD. An alternative approach to improve assisted reproductive technology (ART) outcomes in men with high levels of sperm DD is to obtain testicular spermatozoa. This approach is based on the assumption that testicular spermatozoa generally have lower levels of DD than ejaculated spermatozoa because sperm DD may in part be caused by a posttesticular insult.

Keywords

Sperm DNA damage • Fragmentation • Testicular sperm • Male infertility • IVF

A. Zini (✉)
Department of Surgery, Division of Urology,
McGill University, St. Mary's Hospital Center,
Montreal, QC, Canada
e-mail: ziniarmand@yahoo.com

A. Zini and A. Agarwal (eds.), *Sperm Chromatin: Biological and Clinical Applications in Male Infertility and Assisted Reproduction*, DOI 10.1007/978-1-4419-6857-9_34,

There is a universal agreement that the examination of conventional semen parameters alone only provides the clinician with a general sense of male reproductive health. Indeed, not infrequently, normozoospermic patients can have underlying fertilization defects [1]. Recently, sperm DNA fragmentation/damage has been studied extensively in an attempt to improve the diagnostic accuracy of the male evaluation, particularly, in couples with idiopathic infertility. However, the pathophysiology and etiology of sperm DNA damage (DD) in humans are incompletely understood, and to date, there are very few data on the treatment options for infertile men with this sperm defect.

The etiology of sperm DD is multifactorial, with the most commonly reported mechanisms being protamine deficiency, leading to defective sperm chromatin packaging, disordered apoptosis (caspase-dependent and independent pathways), and oxidative stress (secondary to the excessive elaboration of reactive oxygen species – ROS) [2, 3]. Clinically, the potential causes of sperm DNA fragmentation include varicocele, bacteriospermia, air pollution, chemotherapy, radiotherapy, drugs, cigarette smoking, cryopreservation, and advancing age [4–6].

There are several tests used to assess chromatin and/or DD in ejaculated spermatozoa. These tests include the sperm chromatin structure assay (SCSA; [7]), the acridine orange test [8], the single cell gel electrophoresis assay (COMET; [9]), the in situ nick translation assay [10, 11], and the terminal deoxynucleotidyl transferase-mediated dUDP nick-end labeling assay (TUNEL [11]) – all tailored to measure DD in ejaculated sperm. Using these assays, attempts have been made toward establishing threshold values for the percentage of sperm with DD, the values above which fertility would be affected. Nonetheless, these assays need to be standardized, as there is wide variation among the various tests of sperm DD [12]. Using the SCSA, sperm DD threshold values have been established: low (≤15%), moderate (>15 and <30%), and high (≥30%) DNA fragmentation index (%DFI, proportion of cells with DD). These thresholds are associated with excellent, good, and fair to poor natural fertility potential, respectively [13–15].

There is now mounting evidence to indicate that sperm DNA integrity can influence reproductive outcomes after assisted reproductive technologies – ARTs (e.g., IUI, IVF, and IVF/ICSI). Although there are few valid IUI studies, the data suggest that sperm DD is associated with lower IUI pregnancy rates [16]. A systematic review of the literature allows us to conclude that sperm DD is associated with lower IVF pregnancy rates, whereas it is not associated with ICSI pregnancy rates [17, 18]. There is also evidence to show that sperm DD is associated with an increased risk of pregnancy loss after both IVF and ICSI [19]. However, there are very few data on the influence of sperm DD on late reproductive outcomes (e.g., live birth rates, neonatal outcomes) after ARTs.

In general, cause-specific treatment of the clinical and biological factors associated with sperm DD is associated with a decrease in DD. For example, repair of varicocele, treatment of genital infections, and use of oral antioxidants have generally been shown to improve sperm DNA integrity [20–25]. Ultimately, these therapies are aimed at improving male fertility potential and reproductive outcomes after ARTs. An alternative approach to improve ART outcomes in men with high levels of sperm DD is to obtain testicular spermatozoa. This approach is based on the assumption that testicular spermatozoa generally have lower levels of DD than ejaculated spermatozoa because sperm DD may in part be caused by a posttesticular insult [26].

Biological Significance of Testicular Sperm DNA Damage

Evaluation of testicular sperm DD may help us better understand the etiology (ies) of sperm DD. Experimental (animal) models with testicular DD may provide some insight into the cause(s) of the DD and its relationship with the quality of spermatogenesis. In the past two decades, several experimental studies (e.g., gene knockouts) have evaluated a number of putative genes involved in male fertility/infertility. Some of these studies have demonstrated the relationship between male infertility and sperm DD, providing some insight into the etiology of sperm DD. For example, mice with a targeted disruption of the protamine gene produce testicular spermatozoa with poor chromatin compaction and an increased level of DD compared to wild-type animals [27]. These studies have shown that ejaculated (epididymal) sperm DD may in part be due to an underlying genetic defect (e.g., defective protamine expression – resulting in a relative increase in the sperm histone to protamine ratio).

Suganuma et al. conducted studies on DD in testicular and epididymal sperm to gain some insight into the influence of the posttesticular environment on DD in ejaculated spermatozoa [26]. They studied spermatozoa from wild-type mice and mice with a targeted disruption of the transition nuclear protein gene. These studies demonstrate that part of the DD observed in ejaculated spermatozoa results from an injury (e.g., oxidative stress, hyperthermia) sustained during the posttesticular transit (e.g., passage through the epididymis). These studies have shown that testicular spermatozoa with proper chromatin compaction are resistant to posttesticular stresses, whereas testicular spermatozoa with poor chromatin compaction are highly vulnerable to posttesticular insults and can sustain DNA oxidation and fragmentation.

Additional studies have evaluated fertilization rates and embryo health based on the source of surgically retrieved spermatozoa (i.e., epididymal, testicular). Suganuma et al. have observed that when sperm from the testis or caput epididymis of males were injected into enucleated mouse oocytes, the sperm chromosomes from mice with a targeted disruption of the transition nuclear protein gene showed no difference from those of wild-type mice [26]. However, the chromosomes from the sperm taken from the cauda epididymis of mutant males showed increased abnormalities. Furthermore, injection of testicular or caput epididymal sperm from males into intact mouse oocytes resulted in normal embryonic and fetal development and yields of live born equivalent to wild-type, but cauda sperm from mutant mice produced lower implantation rates and yields of live born than did those from wild-type mice [26]. Theoretically, failure to fully protect the DNA during epididymal passage may cause injury to the DNA as a result of the presence of protamine 2 precursors, slightly higher levels of residual histones, less disulfide bond formation, and decreased compaction of the sperm nuclei [28].

The results of experimental animal models allow us to conclude that testicular sperms are well protected by the microenvironment of Sertoli cells. By contrast, spermatozoa recovered from the distal epididymis may harbor DD as a result of the prolonged exposure to oxidants due to long epididymal transit and storage times [29]. Together, animal studies suggest that the primary cause of sperm DD is likely the result of a primary testicular injury (e.g., gene defect) associated with abnormal spermatogenesis and improper compaction of the chromatin. Sperm DD can then occur in the testicular and posttesticular environment as a result of the poor chromatin compaction [30].

Clinical Significance of Testicular Sperm DNA Damage

Evaluation of testicular sperm DNA and chromatin damage may be useful in the diagnosis of male infertility. For example, establishing that a patient has high levels of testicular sperm DNA or chromatin damage would suggest an abnormal spermatogenesis, whereas the absence of such damage would be suggestive of normal spermatogenesis. Concomitant evaluation of epididymal or ejaculated sperm DNA may help identify the source of DD and more broadly the cause of the infertility. Ultimately, these types of observations may provide guidance as to the optimal treatment options.

In 2005, Greco et al. evaluated a cohort of infertile men with high levels of sperm DD in the ejaculate [31]. They performed a testicular sperm extraction in these men and observed that the percentage of testicular spermatozoa harboring DD (4.8%) was much lower than the percentage of spermatozoa with DD in the ejaculate (23.6%). They then proceeded to use the testicular sperm for ICSI (these couples had at least one previous failed ICSI cycle with ejaculated sperm). They reported an improvement in ICSI pregnancy rate with the use of testicular spermatozoa (44% ICSI pregnancy rate using testicular sperm vs. 6% pregnancy rate with ejaculated sperm). Similarly, they observed an improvement in ICSI implantation rates with the use of testicular spermatozoa, whereas fertilization rates and embryo morphology scores were similar for the treatment attempts with ejaculated and testicular spermatozoa. However, the authors do not advocate that all couples with sperm DD proceed to testicular sperm retrieval for ICSI (in view of invasiveness of testicular biopsy), but rather suggest that these cases be individualized (perhaps taking into account female age). The authors also caution that the threshold of (ejaculated) sperm DD beyond which use of testicular sperm extraction (for subsequent ICSI) should be contemplated has not been established.

Recently, Moskovtsev et al. have compared DD in ejaculated and testicular spermatozoa in patients with previously unsuccessful oral antioxidant treatment [32]. In their study, both samples (ejaculated and testicular spermatozoa) were collected on the day of ICSI (unlike the study of Greco et al., where there was a 4-month interval between the collection of the two samples). As in the Greco et al. study, ejaculated spermatozoa showed a threefold higher level of DD when compared with testicular spermatozoa (39.7% ± 14.8 vs. 13.3% ± 7.3). It is unknown whether pretreatment with antioxidant agents and vitamins had an impact on the integrity of the testicular sperm DNA.

Different established methods are available for assessment of sperm DD in the ejaculate. However, these methods have not been designed to assess testicular sperm DNA integrity or damage. Preparations of testicular tissue generally have a lower sperm concentration than semen. Moreover, unlike ejaculated sperm, testicular sperm preparations are contaminated (mixed with other cell types) and frequently testicular spermatozoa are bound to other cells (e.g., Sertoli cells). As a result of these features, testicular tissue is not suitable for sperm DNA tests that require flow-cytometry assessment (e.g., SCSA). Rather, testicular sperm DD is best assessed using slide-based techniques (e.g., terminal deoxynucleotidyl transferase-mediated deoxyuridine triphosphate nick-end labeling – TUNEL), where smears of the testicular tissue are prepared and evaluated. Another important difference between ejaculated and testicular sperm is the compactness of the DNA chromatin. When evaluating ejaculated spermatozoa, a standard nuclear decondensation step is undertaken prior to assessing DD in view of the compact nature of ejaculated sperm chromatin. By contrast, testicular sperm chromatin is less compact, and therefore, it is unclear whether the same nuclear decondensation step is needed prior to assessing DD in these cells (testicular sperms have a lower degree of chromatin compaction and, hence, a more rapid decondensation than ejaculated sperms) [12, 33].

Assessing testicular sperm DD may be useful in the management of male infertility. Establishing that a patient has less DD in testicular compared to ejaculated sperm may provide some guidance in the choice of sperm to be used for ICSI (Tables 34.1 and 34.2). Defective sperm DNA

Table 34.1 Fertilization and embryo development after ICSI with ejaculated and testicular spermatozoa

Sperm source	Attempts	Oocytes injected	Normal zygotes	Fertilization rate (%)	Cleaved embryos	Good-morphology embryos
Ejaculate	18	185	131	70.8[e]	124 (94.7%)[e]	59 (47.6%)[e]
Testis	18	187	140	74.9[e]	133 (95.0%)[e]	68 (51.1%)[e]

[e]The differences in fertilization rates, cleaved embryo rates, and good-morphology embryo rates for the two sperm sources were not significant ($P>0.05$)

Adapted from Greco et al. [31], by permission of Oxford University Press

Table 34.2 Implantation and pregnancy after ICSI with ejaculated and testicular spermatozoa

Sperm source	Attempts	Embryos transferred	Clinical pregnancies	Pregnancy rate (%)	Gestational sacs	Implantation rate (%)
Ejaculate	18	56	1	5.6[e]	1	1.8[f]
Testis	18	58	8	44.4[e]	12	20.7[f]

[e, f]The differences in pregnancy rates and implantation rates for the two sperm sources were significant ($P<0.05$)
Adapted from Greco et al. [31], by permission of Oxford University Press

apoptosis and alterations in the ratio of Sertoli cell to germ cells have been proposed as possible mechanisms to explain the lower levels of sperm DD in the testicular compared to ejaculated sperm. However, randomized, controlled trials are needed to define the clinical utility of testicular sperm extraction in men with high levels of sperm DD in the ejaculated sperms.

Future Directions in the Field

Our basic understanding of the organization of the sperm chromatin and the nature of sperm DD in humans are constantly evolving (12). Nonetheless, there is an urgent need to standardize the laboratory methods for assessing DD, as there is wide variation among the various tests of sperm DD. In order to more accurately assess DD in testicular sperm, the current sperm DNA tests must be reevaluated and modified. These modified assays should also be validated by testing testicular sperm from men with different pathologies (e.g., obstructive azoospermia, nonobstructive azoospermia, oligozoospermia). Additional clinical studies are needed to better define the indications for testicular sperm retrieval in infertile men (e.g., high levels of sperm DD, unexplained ICSI failures). These studies should have parallel assessment of testicular, epididymal, and ejaculated sperm DNA integrity with subsequent assessment of ICSI outcomes in terms of fertilization rates, embryo quality, pregnancy rates, and neonatal outcomes.

References

1. Saleh RA, Agarwal A, Nelson DR, Nada EA, El-Tonsy MH, Alvarez JG, et al. Increased sperm nuclear DNA damage in normozoospermic infertile men: a prospective study. Fertil Steril. 2002;78:313–8.
2. O'Brien J, Zini A. Sperm DNA integrity and male infertility. Urology. 2005;65:16–22.
3. Tesarik J, Ubaldi F, Rienzi L, Martinez F, Iacobelli M, Mendoza C, et al. Caspase-dependent and -independent DNA fragmentation in Sertoli and germ cells from men with primary testicular failure: relationship with histological diagnosis. Hum Reprod. 2004;19:254–61.
4. Moskovtsev SI, Mullen JB, Lecker I, Jarvi K, White J, Roberts M, et al. Frequency and severity of sperm DNA damage in patients with confirmed cases of male infertility of different aetiologies. Reprod Biomed Online. 2010;20:759–63.
5. Rubes J, Rybar R, Prinosilova P, Veznik Z, Chvatalova I, Solansky I, et al. Genetic polymorphisms influence the susceptibility of men to sperm DNA damage associated with exposure to air pollution. Mutat Res. 2010;683:9–15.
6. Sakkas D, Seli E, Bizzaro D, Tarozzi N, Manicardi GC. Abnormal spermatozoa in the ejaculate: abortive apoptosis and faulty nuclear remodelling during spermatogenesis. Reprod Biomed Online. 2003;7:428–32.
7. Evenson DP, Darzynkiewicz Z, Melamed MR. Relation of mammalian sperm chromatin heterogeneity to fertility. Science. 1980;210:1131–3.
8. Tejada RI, Mitchell JC, Norman A, Marik JJ, Friedman S. A test for the practical evaluation of male fertility by acridine orange (AO) fluorescence. Fertil Steril. 1984;42:87–91.
9. Hughes CM, Lewis SE, McKelvey-Martin VJ, Thompson W. Reproducibility of human sperm DNA measurements using the alkaline single cell gel electrophoresis assay. Mutat Res. 1997;374:261–8.
10. Gorczyca W, Traganos F, Jesionowska H, Darzynkiewicz Z. Presence of DNA strand breaks and increased sensitivity of DNA in situ to denaturation in abnormal human sperm cells: analogy to apoptosis of somatic cells. Exp Cell Res. 1993;207:202–5.
11. Sailer BL, Jost LK, Evenson DP. Mammalian sperm DNA susceptibility to in situ denaturation associated with the presence of DNA strand breaks as measured by the terminal deoxynucleotidyl transferase assay. J Androl. 1995;16:80–7.
12. Barratt CL, Aitken RJ, Bjorndahl L, Carrell DT, de Boer P, Kvist U, et al. Sperm DNA: organization, protection and vulnerability: from basic science to clinical applications – a position report. Hum Reprod. 2010;25:824–38.
13. Spano M, Bonde JP, Hjollund HI, Kolstad HA, Cordelli E, Leter G. Sperm chromatin damage impairs human fertility. The Danish First Pregnancy Planner Study Team. Fertil Steril. 2000;73:43–50.

14. Giwercman A, Lindstedt L, Larsson M, Bungum M, Spano M, Levine RJ, et al. Sperm chromatin structure assay as an independent predictor of fertility in vivo: a case-control study. Int J Androl. 2010;33(1):e221–7.
15. Evenson DP, Jost LK, Marshall D, Zinaman MJ, Clegg E, Purvis K, et al. Utility of the sperm chromatin structure assay as a diagnostic and prognostic tool in the human fertility clinic. Hum Reprod. 1999;14: 1039–49.
16. Bungum M, Humaidan P, Axmon A, Spano M, Bungum L, Erenpreiss J, et al. Sperm DNA integrity assessment in prediction of assisted reproduction technology outcome. Hum Reprod. 2007;22:174–9.
17. Collins JA, Barnhart KT, Schlegel PN. Do sperm DNA integrity tests predict pregnancy with in vitro fertilization? Fertil Steril. 2008;89:823–31.
18. Zini A, Sigman M. Are tests of sperm DNA damage clinically useful? Pros and cons. J Androl. 2009;30: 219–29.
19. Zini A, Boman JM, Belzile E, Ciampi A. Sperm DNA damage is associated with an increased risk of pregnancy loss after IVF and ICSI: systematic review and meta-analysis. Hum Reprod. 2008;23:2663–8.
20. Greco E, Iacobelli M, Rienzi L, Ubaldi F, Ferrero S, Tesarik J. Reduction of the incidence of sperm DNA fragmentation by oral antioxidant treatment. J Androl. 2005;26:349–53.
21. Greco E, Romano S, Iacobelli M, Ferrero S, Baroni E, Minasi MG, et al. ICSI in cases of sperm DNA damage: beneficial effect of oral antioxidant treatment. Hum Reprod. 2005;20:2590–4.
22. Moskovtsev SI. Management of patients with high sperm DNA damage. Indian J Med Res. 2008;127:101–3.
23. Moskovtsev SI, Lecker I, Mullen JB, Jarvi K, Willis J, White J, et al. Cause-specific treatment in patients with high sperm DNA damage resulted in significant DNA improvement. Syst Biol Reprod Med. 2009;55:109–15.
24. Zini A, Blumenfeld A, Libman J, Willis J. Beneficial effect of microsurgical varicocelectomy on human sperm DNA integrity. Hum Reprod. 2005;20:1018–21.
25. Zini A, Gabriel M, San Baazeem A. Antioxidants and sperm DNA damage: a clinical perspective. J Assist Reprod Genet. 2009;26:427–32.
26. Suganuma R, Yanagimachi R, Meistrich ML. Decline in fertility of mouse sperm with abnormal chromatin during epididymal passage as revealed by ICSI. Hum Reprod. 2005;20:3101–8.
27. Cho C, Willis WD, Goulding EH, Jung-Ha H, Choi YC, Hecht NB, et al. Haploinsufficiency of protamine-1 or −2 causes infertility in mice. Nat Genet. 2001;28:82–6.
28. Zhao M, Shirley CR, Hayashi S, Marcon L, Mohapatra B, Suganuma R, et al. Transition nuclear proteins are required for normal chromatin condensation and functional sperm development. Genesis. 2004;38:200–13.
29. Rajesh Kumar T, Doreswamy K, Shrilatha B, Muralidhara. Oxidative stress associated DNA damage in testis of mice: induction of abnormal sperms and effects on fertility. Mutat Res. 2002;513: 103–11.
30. De Iuliis GN, Thomson LK, Mitchell LA, Finnie JM, Koppers AJ, Hedges A, et al. DNA damage in human spermatozoa is highly correlated with the efficiency of chromatin remodeling and the formation of 8-hydroxy-2'-deoxyguanosine, a marker of oxidative stress. Biol Reprod. 2009;81:517–24.
31. Greco E, Scarselli F, Iacobelli M, Rienzi L, Ubaldi F, Ferrero S, et al. Efficient treatment of infertility due to sperm DNA damage by ICSI with testicular spermatozoa. Hum Reprod. 2005;20:226–30.
32. Moskovtsev SI, Jarvi K, Mullen JB, Cadesky KI, Hannam T, Lo KC. Testicular spermatozoa have statistically significantly lower DNA damage compared with ejaculated spermatozoa in patients with unsuccessful oral antioxidant treatment. Fertil Steril. 2010;93:1142–6.
33. Kosower NS, Katayose H, Yanagimachi R. Thiol-disulfide status and acridine orange fluorescence of mammalian sperm nuclei. J Androl. 1992;13: 342–8.

Part V

Protocols and Integrity Tests

Sperm Chromatin Structure Assay (SCSA®)

35

Donald P. Evenson

Basic Protocol Steps

Fresh or frozen semen/sperm thawed in a 37°C water bath and diluted to $1–2 \times 10^6$ sperm/ml with TNE buffer:
0.01 M tris buffer
0.15 M NaCl
1 mM EDTA
pH 7.4

200 μl sperm suspension + 400 μl of:
0.15 M NaCl
0.08 N HCl
0.1% Triton–X 100
pH 1.20

After 30 s add 1.20 ml of:
0.20 M Na_2HPO_4
1.0 mM EDTA
0.15 M NaCl
0.10 M citric acid
6.0 μg AO/ml staining buffer
pH 6.0
Measure by flow cytometry

D.P. Evenson (✉)
SCSA Diagnostics, PO Box 107, 219 Kasan Ave, Volga, SD 57071, USA

Emeritus, South Dakota State University, Brookings, SD, USA

Department of Obstetrics and Gynecology, Sanford Medical School, University of South Dakota, Sioux Falls, SD, USA
e-mail: don@scsatest.com

Materials

Acridine Orange: (AO) *chromatographically purified* (Cat. # 04539, Polysciences, Inc., Warrington, PA 18976)
Automated solution dispensers: Oxford adjustable, 0.20–0.80 ml automatic dispenser for the acid-detergent solution with glass amber bottle (CAT # 13 687 65, Fisher Scientific, 800-766-7000) and Oxford adjustable, 0.80–3.0 ml automatic dispenser for the AO staining solution glass amber bottle (CAT # 13 687 66, Fisher Scientific).
Pipetters: adjustable 0–10 μl, 10–100 μl, 100–1,000 μl and a nonadjustable 200 μl
Ice buckets (3) for samples and reagent bottles
Water bath (37°C)
Stopwatch

Staining Solutions and Buffers

For solutions, use double distilled water (dd-H_2O). For sterilization, use a 0.22-mm filter. Use only the purest grade reagents. All solutions and buffers are stored at 4°C.

Acridine Orange (AO) Stock Solution, 1.0 mg/ml

Dissolved *chromatographically purified AO* (Polysciences) in dd-H_2O at 1.0 mg/ml can be stored up to several months. *Our laboratory has used only AO obtained from Polysciences, and*

A. Zini and A. Agarwal (eds.), *Sperm Chromatin: Biological and Clinical Applications in Male Infertility and Assisted Reproduction*, DOI 10.1007/978-1-4419-6857-9_35,

thus, we have full confidence in this source. DO NOT use a more crude preparation of AO; failure will result. AO is a toxic chemical and precautions should be taken when handling it. Tare a 15-ml, flat-bottom scintillation vial on a 5-place electronic balance, carefully remove and transfer 3–6 mg AO powder from the stock bottle with a microspatula into the vial. Add an exact equivalent number of milliliters of water. Wrap the capped vial in aluminum foil to protect from light.

Acid-Detergent Solution, pH 1.20
20.0 ml 2.0 N HCl (0.08 N)
4.39 g NaCl (0.15 M)
0.5 ml Triton X-100 (0.1%)
H_2O to 500 ml
pH to 1.20 with 5 N HCl
Store up to several months

Use purchased 2.0 N HCl (e.g., Sigma Cat # 251–2); do not dilute from a more concentrated HCl solution that is likely less pure and may be of questionable strength. The Triton-X stock solution is very viscous. We use a wide-mouth pipette and carefully draw up the exact amount, wipe the outside of the pipette free of Triton-X, and then expel with force in and out of the pipette until all is dispensed.

0.1 M citric acid buffer
21.01 g/L citric acid monohydrate (F.W. = 210.14; 0.10 M)
H_2O to 1.0 L
Store up to several months at 4 C.

0.2 M Na_2PO_4 buffer
28.4 g sodium phosphate dibasic (F.W. = 141.96; 0.2 M)
H_2O to 1.0 L
Store up to several months at 4 C.

Staining buffer, pH 6.0
370 ml 0.10 M citric acid buffer
630 ml 0.20 M Na_2PO_4 buffer
372 mg EDTA (disodium, FW = 372.24; 1 mM)
8.77 g NaCl (0.15 M)
Mix overnight on a stir plate to insure that the EDTA is entirely in solution.
pH to 6.0 with concentrated NaOH pellets
Store up to several months

Slowly and carefully adjust the pH using very small pieces (cut with a scalpel and handled with a forceps) of concentrated NaOH pellets. Note that when the 0.2 M Na_2PO_4 buffer is removed from the refrigerator, salt crystals will be present. Heat in 37°C water bath until the salts are fully dissolved.

AO staining solution
600 µl AO stock solution is added to each 100 ml of staining buffer. Rinse the pipette tip several times. This solution is kept in a glass amber bottle.
Store up to 2 weeks at 4°C.

AO equilibration buffer
400 µl acid-detergent solution
1.20 ml AO staining solution

This is run through the instrument for ≈15 min prior to sample measurement to insure that AO is equilibrated with the sample tubing. This is also run through the instrument between different samples to maintain the AO equilibrium and help clean the prior sample out of the lines.

TNE buffer, 10×, pH 7.4
9.48 g Tris-HCl (FW = 158; 0.01 M)
52.6 g NaCl (FW = 58.44; 0.15 M)
2.23 g EDTA (disodium, FW = 372.24; 1 mM)
pH to 7.4 with 2 N NaOH
Store up to 1 year at 4 C

TNE buffer, 1×, pH 7.4
60 ml 10× TNE
H_2O to 600 ml
Check pH (7.4)
Store for several months at 4 C

FCM Tubing Cleanser (for unclogging FCM sample lines)
50% ETOH
50% household bleach (contains ~5% sodium hypochlorite)
0.5 M NaCl
Store at room temperature

50% household bleach (for eliminating AO from sample lines)
50 ml household bleach (~5% sodium hypochlorite)
50 ml H_2O

Sheath fluid
2× H_2O 0.45 nm filtered water + 0.1% Triton X-100 (this helps minimize bubbles in the flow channel). It is NOT necessary to use commercially sold sheath fluid unless one FCM sorts the sperm in a jet-in-air sorter.

Major Equipment

Ultracold freezer (−70 to −110°C) or, preferably, a LN_2 tank
Biological safety hood

Flow Cytometer(s)

The flow cytometer must have 488 nm excitation wavelength and an approximate 15–35 mW laser power. Fluorescence of individual cells is collected through red (630–650 nm long pass) and green (515–530 nm band pass) filters.

Orthogonal flow cytometer configuration and related signal artifacts. The highly condensed mammalian sperm nucleus has a much higher index of refraction than sample sheath (water) in a flow cytometer. This differential, coupled with the typical nonspherical shape of sperm nuclei and their orientation in the flow channel, produces an optical artifact consisting of an asymmetric, bimodal emission of DNA dye fluorescence when measured in orthogonal configuration flow cytometers where the collection lenses are situated at right angles to both sample flow and excitation source. Since DFI is a computer calculated ratio of red to total (red + green) fluorescence, the optical artifact of AO-stained sperm measured in the orthogonal instruments does not significantly interfere with results, and the DFI frequency histogram is very narrow for a normal population of sperm. Although each type of flow cytometer with different configurations of lens and fluidics produces different cytogram patterns, the DFI data are essentially the same.

The variables of DFI are useful especially, as discussed above, for toxicology and has been shown for animal fertility studies. Future studies will show its importance for human fertility studies. However, a simple determination of the percent of cells with denatured DNA (%DFI) and the percentage of cells with abnormally high green stainability (%HDS) can be reasonably estimated without the ratio calculations. %DFI is currently the most used variable of this assay for human fertility assessment.

Cell Preparation

Collection and Handling

Human semen samples are typically obtained by masturbation into plastic clinical specimen jars preferably after ~2 days abstinence. Of importance is the length of the previous abstinence period; if days of time have elapsed, then sperm stored in the epididymis can become apoptotic in which case such a sample would not be representative of a fresh semen sample. We suggest that a patient ejaculate, wait for two days, and ejaculate again, then the sample for testing be taken after another two days, e.g., ejaculate on Monday and Wednesday and collect clinical sample on Friday. Freshly collected semen should be quick-frozen as soon as liquefaction has occurred in about a half hour. The majority of semen samples may be kept for up to several hours at room temperature prior to measuring/freezing without significant loss of quality, allowing for collections within a medical institution and transport to the flow cytometry unit. However, we have observed in limited studies that an estimated 10% of samples have an increased DNA fragmentation while setting at room temperature; likely, these samples have very low antioxidant capacity. If transport is required outside of a building complex, the sample may be conveyed in an insulated box or jacket pocket to keep from freezing or on liquid ice if the ambient temperature is hot. Once a sample has been diluted in TNE buffer it should be measured or frozen immediately.

Freezing

After allowing ~30 min for semen liquefaction at room temperature, aliquots of raw or TNE diluted

($1–2 \times 10^6$ sperm/ml) semen can be frozen directly without cryoprotectants in an ultracold freezer (−70 to −110°C; 0.5–1.5 ml snap-cap tubes), a shipping box with dry ice, or can be placed directly into a LN_2 tank (cryovials). Samples should be frozen in vials that are approximately ¼ larger in volume than the semen volume to reduce the air–surface interface, thus minimizing related reactive oxygen damage. Keep the tubes vertical when freezing, since samples frozen at the bottom of a tube could be later thawed in a water bath with greater ease and safety. Cryoprotectants are not needed, since quick-frozen cells and those frozen with a cryoprotectant provide equivalent SCSA data. This feature is unique to mammalian sperm cells due to the highly condensed, crystalline nature of the nucleus.

Flow Cytometer Setup

Workstation

The SCSA procedure requires that samples are thawed and processed in the immediate vicinity of the flow cytometer. The following equipment should be handy for quick and easy use.

- Ice buckets containing wet ice to hold the reagent bottles, sample tubes, and TNE buffer
- Disposable gloves
- Stopwatch
- Automatic pipetters and tips
- Reagent bottles deeply embedded in the ice buckets containing wet ice
- Container with disinfectant for sample disposal

Flow Cytometer Alignment

Prior to measuring experimental samples, the instrument must be checked for alignment using standard fluorescent beads. Very importantly, an AO equilibration buffer (400 μl acid-detergent solution and 1.20 ml AO staining solution) must be passed through the instrument sample lines for ≈15 min prior to establishing instrument settings. This insures that AO is equilibrated with the sample tubing. To save time, this AO buffer can be run through the instrument during its warm-up time prior to alignment and again just before measuring samples. Contrary to existing rumors, using AO in a flow cytometer *does not* ruin it for other purposes. The sample lines DO NOT need to be replaced after using AO in a flow cytometer! However, the system DOES need to be fully equilibrated with AO, as AO does transiently adhere to the sample tubing by electrostatic force, thus reducing the required AO concentration. After finishing SCSA measurements, AO can easily be cleansed from the lines by rinsing the system for about 10 min with a 50% filtered household bleach solution followed by 10 min of filtered H_2O. Our laboratory has utilized many fluorescent dyes and sample types after measuring AO stained sperm without any associated problems.

Reference Samples

Because SCSA variables are very sensitive to small changes in chromatin structure, studies on sperm using this protocol require very precise, repeat instrument settings for all comparative measurements whether done on the same or different days. These settings are obtained by using aliquots of a single semen sample called the "reference sample" (this is not a "control" sperm from a fertile donor). A semen sample that demonstrates heterogeneity of DNA integrity (e.g., 15% DFI) is chosen as a reference sample and then diluted with cold (4°C) TNE buffer to a working concentration of $1–2 \times 10^6$ cells/ml.

CLIA and other licensing agencies, e.g., New York Health, require that for every measurement period that a low %DFI and a high %DFI sample become part of the measurement data.

Several hundred 300-μl aliquots of this dilution are immediately and quickly placed into 0.5-ml snap-cap vials and flash frozen at −70 to −100°C in a freezer or, preferably in a LN_2 tank. These reference samples are used to set the red and green photomultiplier tube (PMT) voltage gains to yield the same mean red and green fluorescence levels from day to day. The mean red and green fluorescence values are set at ≈125/1,000 and

≈475/1,000 channels, respectively. The values established by a laboratory (preferably the same as above) should be used consistently thereafter. Strict adherence to keeping the reference values in this range must be maintained throughout the measurement period. A freshly thawed reference samples is measured after every 5–10 experimental samples to insure that instrument drift has not occurred.

Very few FCM protocols are as demanding as the SCSA for using a reference sample. Obviously, it would be advantageous to prepare a new batch of reference samples from the same individual donor. However, if a new donor is used, then first set the PMTs for the previous reference sample to be in the same position and then measure the new reference sample and note the red and green mean values and use these values for further studies.

Since reference samples can be stored in LN2 for years, a donor could provide enough samples for thousands of reference aliquots.

Sample Measurement

Single frozen samples are immersed in a 37°C water bath, just until the last remnant of ice disappears. When analyzing a series of human samples, it is extremely helpful to obtain the sperm count in advance of SCSA preparation so that time is not lost determining the proper dilution. However, if a sample(s) needs to be measured quickly for a clinical decision, then rather than wait for a sperm count, estimate a dilution, check the flow rate, and if necessary, resample with the proper dilution to attain the required flow rate of ~200 events per second. *A 200-ml aliquot of fresh or frozen/thawed semen sample of known sperm concentration* is placed into a 12 × 75 mm conical plastic test tube. Then, 400 µl of the acid-detergent, low pH buffer is added with an automatic dispenser setting deep in the ice bucket. This dispenser needs to be highly accurate and to have a maximum volume capacity only a small volume more than what is being dispensed. At the beginning of sample measurement and after long breaks in measurement, dispense several volumes from both dispensers before starting with the samples, as AO in the delivery tube may have been damaged by light and solutions in the plastic delivery tubes may be warmer than 4°C. A stopwatch is started immediately after the first buffer is dispensed. Exactly 30 s later, the AO staining solution is added. The sample tube is then placed into the flow cytometer sample chamber – which varies in design by different instruments. The sample flow is started immediately after placing it in the sample holder. Using the stopwatch that was started with the addition of the acid-detergent solution, the acquisition of list mode data to computer disk is started at 3 min. This allows ample time for AO equilibration in the sample and hydrodynamic stabilization of the sample within the fluidics, both very important aspects of AO staining. The sperm flow rate is checked during this time, and if it is too fast, i.e., >250 cells, a new sample is made at the appropriate dilution. This protocol provides approximately equal to two AO molecules/DNA phosphate group. Thus, to initially set up the proper hydrodynamic conditions, measure several sperm samples that have a predetermined cell count of ≈1.5×10^6 sperm/ml (or known concentration of fluorescent beads) and adjust the flow rate settings (if possible) for ≈200 cells/beads per second. On a FACScan, the "low flow" rate setting delivers an approximate correct flow rate. If a sample's flow rate is too high, this same sample cannot be diluted with AO buffer to lower the concentration. Sample and sheath flow valve settings of the instrument are never changed during these measurements so the liquid flow rate is constant. Doing so widens the flow sample stream with consequential loss of resolution. Thus, a change in sperm count rate is a function of sperm cell concentration only. PMT settings should be fairly identical from day to day depending on slight alignment differences between days and sample runs. *All samples are measured at least twice in succession* for statistical considerations and data on ~5,000 sperm cells (total events recorded are higher due to debris) are recorded per measurement. For the second measurement, take the sample from the same thawed aliquot; dilute appropriately, process for the SCSA and measure. After the second measurement of a sample is finished, place a tube of AO equilibration buffer on the instrument to maintain the AO conditions and wash any of the previous sample out of the tubing and start preparing the next sample. There is no need to run this buffer between the duplicate measurements of the same sample; just allow the first one to stay running while preparing the second one.

Gating and Debris Exclusion

A very important, *but* sometimes difficult point, is deciding where to draw the computer gates to exclude cellular debris signals (signals located at the origin in the red (X) vs. green (Y) fluorescence cytograms) from the analysis. This gate is usually best set at a 45° angle, i.e., at the same channel value for both red and green fluorescence values. Resolution of debris and sperm signal is partly instrument dependent.

The real SCSA values of a sample cannot be learned if the fluorescence from debris (i.e., free cellular components and other contaminants) is blended in with the sperm fluorescence signal. This can sometimes be eliminated by washing the sperm or processing though gradients. However, there is a risk of losing cell types and the advantage of using whole semen measurements is then compromised. Bacterial debris appears as a straight line to the left of and parallel with the main sperm population in the cytograms; this usually can be gated out, but not in all samples.

Critical Parameters or Points

Computer gating to determine % DFI and % HDS

The left hand panel of the figure below shows how %DFI and %HDS can be calculated by placing computer gates to the right of the cigar-shaped pattern of sperm without DNA fragmentation (%DFI) as well as the % of sperm with increased green fluorescence (%HDS) characteristic of immature sperm and/or sperm with altered protein composition (Fig. 35.1).

As discussed in the SCSA chapter, it is easy to obtain the %DFI from a semen sample represented

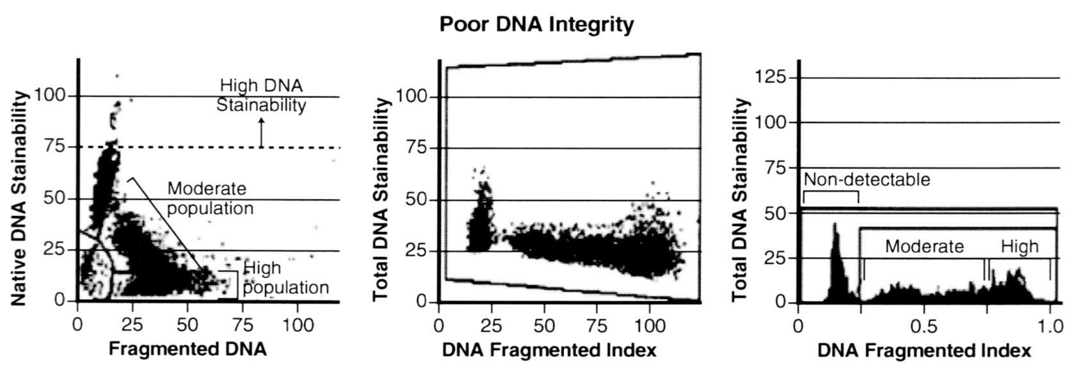

Patient	Date	Measurement	DFI	SD DFI	% DFI	% HDS
7272-113	## ##	1	563.7	307	64.9	6.4
		2	561.4	304.8	64.9	7.2
		mean	562.6	605.9	64.9	6.8
		sd	1.2	1.1	0	0.4

Fig. 35.1 The *middle* and *right hand panels* show the effects of SCSAsoft® calculations without computer gating for %DFI, which is calculated from the DFI frequency histogram as shown in the *right-hand panel*

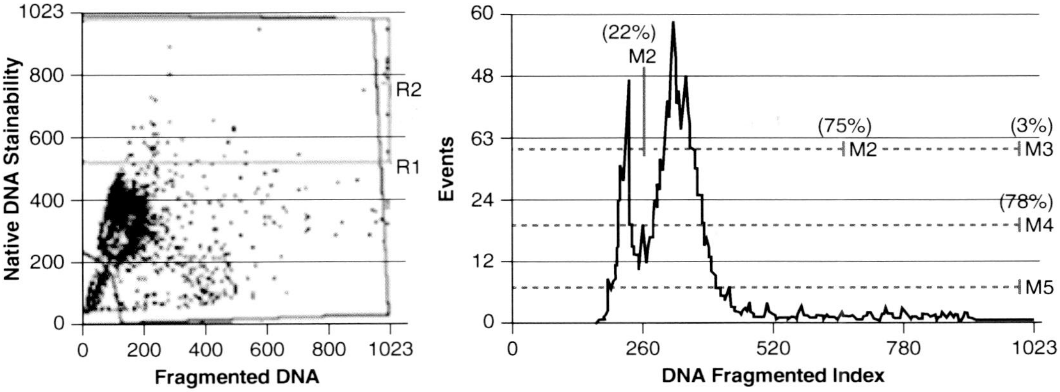

Fig. 35.2 SCSA® data from a sample with a high frequency of sperm with moderate DNA fragmentation. In this case, it is impossible to gate between sperm with no or moderate DNA fragmentation in the FCM dot-plot (*left panel*). With the SCSAsoft®, gating between the two populations is unproblematic (*right panel*, 34)

in the above panel. However, in the semen sample represented in the panel below (Fig. 35.2), it is more difficult to obtain the correct %DFI without the use of SCSAsoft®. The %HDS is equally calculated with or without SCSAsoft®.

In summary, the SCSA protocol appears rather simple offhand; however, there are numerous very critical points that, unless followed exactly, will give very poor data and serious errors in clinical diagnosis and prognosis.

Measurement of DNA Damage in Spermatozoa by TUNEL Assay

36

Rakesh Sharma and Ashok Agarwal

Abstract

Infertile men with severe male-factor infertility have significantly more DNA damage than fertile men. Sperm DNA damage has been shown to affect ART outcomes. DNA damage is increasingly being evaluated as a test in establishing its utility in ART. Various assays have become more common than others. This chapter describes one of the more commonly used assays to measure sperm DNA damage by the terminal deoxynucleotidyltransferase dUTP nick end labeling or the TUNEL assay [1–5].

Keywords

Spermatozoa • Sperm chromatin • Terminal deoxynucleotidyltransferase dUTP nick end labeling • Flow cytometry

DNA fragmentation is a process that results from the activation of endonucleases during apoptosis. These nucleases degrade the higher order sperm chromatin structure into fragments of ~30 kb and subsequently into smaller DNA pieces about ~50 kb in length. This method is used to detect fragmented DNA and utilizes a reaction catalyzed by exogenous terminal deoxynucleotidyltransferase (tdt) and is termed as "end labeling" or "TUNEL" (terminal deoxynucleotidyltransferase dUTP nick end labeling) assay [1–5].

Assay Principle

This single-step staining method labels DNA breaks with FITC-dUTP followed by flow-cytometric analysis. Tdt catalyzes a template-independent addition of bromolated deoxyuridine triphosphatase to the 3'-hydroxyl (OH) termini of double- and single-stranded DNA. After incorporation, these sites are identified by flow-cytometric means by staining the sperm.

R. Sharma (✉)
Andrology Laboratory and Center for Reproductive Medicine, Glickman Urological & Kidney Institute, OB-GYN and Women's Health Institute, Cleveland Clinic, Cleveland, OH, USA
e-mail: Sharmar@ccf.org

A. Zini and A. Agarwal (eds.), *Sperm Chromatin: Biological and Clinical Applications in Male Infertility and Assisted Reproduction*, DOI 10.1007/978-1-4419-6857-9_36,

Specimen Collection

1. Ideally, sample should be collected after a minimum of 48 h and not more than 72 h of sexual abstinence. The name of the patient, period of abstinence, date, and time and place of collection should be recorded on the form accompanying each semen analysis.
2. The sample should be collected in private in a room near the laboratory. If not, it should be delivered to the laboratory within one hour of collection.
3. The sample should be obtained by masturbation and ejaculated into a clean, wide-mouth plastic specimen cup. Lubricants should not be used to facilitate semen collection.
4. Coitus interruptus is not acceptable as a means of collection because it is possible that the first portion of the ejaculate, which usually contains the highest concentration of spermatozoa, will be lost. Moreover, cellular and bacteriological contamination of the sample and the acid pH of the vaginal fluid adversely affect sperm quality.
5. Incomplete samples should be analyzed, but a comment should be entered on the report form.
6. The sample should be protected from extremes of temperature (not less than 20°C and not more than 40°C) during transport to the laboratory.
7. Note down any unusual collection or condition of specimen on the report form.

Equipment and Reagents

- APO-DIRECT™ kit (BD Pharmingen, Catalog % 556381)
- Pipettes
- Pipette tips (200 μL and 1,000 μL)
- Microcell counting chamber
- 3.7% Paraformaldehyde in PBS
- Microfuge ependorf tubes
- Ethanol
- Flow cytometer

Sample Preparation

1. Following liquefaction, evaluate semen specimens for volume, sperm concentration, total cell count, motility, and morphology.
2. Aliquot and load a 5-μL aliquot of the sample on a Microcell slide chamber (Conception Technologies, San Diego, CA) for manual evaluation of concentration and motility. Check the concentration of sperm in the sample. Adjust it to $2–5 \times 10^6$/mL.
3. Using a cryomarker, label one 5-mL tube. Label specimen 1 with the patient name, identification number, and date, i.e., as follows:
   (I) TUNEL
   (II) Smith, John
   (III) No. X-XXX-XXX-X
   (IV) Date
4. Preparation of paraformaldehyde:
   (a) To 10.0 mL of formaldehyde (37%), add 90.0 mL of PBS (pH 7.4).
5. Check the concentration of sperm in the sample. Adjust the volume to give $3–5 \times 10^6$/mL. Spin the sample and remove seminal plasma. Add 1.0 mL of 3.7% paraformaldehyde.
6. Place the cell suspension on ice for 30–60 min/overnight.
7. Store cells in 1 mL of ice-cold 70% (v/v) ethanol at −20°C until use. Cells can be stored at −20°C several days before use.

   Note: The samples can be processed from A-G, batched and shipped.

Staining Protocol

1. Resuspend the positive (6552LZ) and negative (6553LZ) control cells by swirling the vials. Remove 2-mL aliquots of the control cell suspensions (approximately 1×10^6 cells/mL) and place in 12 × 75 mm centrifuge tubes. Centrifuge the control cell suspensions for 5 min at $300 \times g$ and remove the 70% (v/v) ethanol by aspiration, being careful to not disturb the cell pellet.

2. Resuspend each tube of control and sample tubes with 1.0 mL of Wash Buffer (6548AZ) (Blue cap) for each tube. Centrifuge as before and remove the supernatant by aspiration.
3. Repeat the Wash Buffer treatment.
4. Resuspend each tube of the control cell pellets in 50 μL of the *Staining Solution* (prepared as described below).
5. Staining solution (single assay)

Staining solution	1 assay	6 assays	12 assays
Reaction buffer (green cap) (μL)	10.00	60.00	120.00
TdT enzyme (yellow cap) (μL)	0.75	4.50	9.00
FITC-dUTP (orange cap) (μL)	8.00	48.00	96.00
Distilled H_2O (μL)	32.25	193.5	387.00
Total volume (μL)	51.00	306.00	612.00

Note: The appropriate volume of Staining Solution to prepare for a variable number of assays is based upon multiples of the component volumes needed for 1 assay. Mix only sufficient volumes of Staining Solution to complete the number of assays prepared per session. The Staining Solution is active for approximately 24 h at 4°C.

6. Incubate the sperm in the Staining Solution for 60 min at 37°C. The reaction can also be carried out at room temperature overnight for the control cells. For test samples, the 60-min incubation time at 37°C may need to be adjusted to longer periods of time.
7. At the end of the incubation time, add 1.0 mL of Rinse Buffer (6550AZ) (Red cap) to each tube and centrifuge each tube at 300 × *g* for 5 min. Remove the supernatant by aspiration.
   Note: If the cell density is low, decrease the amount of PI/ RNase Staining Buffer to 0.3 mL.
8. Repeat the cell rinsing with 1.0 mL of the Rinse Buffer. Centrifuge and remove the supernatant by aspiration.
9. Resuspend the cell pellet in 0.5 ml of the PI/ RNase Staining Buffer (6551AZ).
10. Incubate the cells in the dark for 30 min at RT.
11. Analyze the cells in PI/ RNase solution by flow cytometry.
    Note: The cells must be analyzed within 3 h of staining, as they may begin to deteriorate if left overnight before the analysis.

Reference range: Percentage of cells showing DNA fragmentation is calculated.

Normal range: ≤19% DNA damage.

Panic values: >19% DNA damage.

References

1. Gorczyca W, Gong J, Darzynkiewicz Z. Detection of DNA strand breaks in individual apoptotic cells by the in situ terminal deoxynucleotidyl transferase and nick translation assays. Cancer Res. 1993;53: 1945–51.
2. Paasch U, Sharma RK, Gupta AK, Grunewald S, Mascha EJ, Thomas Jr AJ, et al. Cryopreservation and thawing is associated with varying extent of activation of apoptotic machinery in subsets of ejaculated human spermatozoa. Biol Reprod. 2004;71: 1828–37.
3. Said TM, Agarwal A, Sharma RK, Thomas Jr AJ, Sikka SC. Impact of sperm morphology on DNA damage caused by oxidative stress induced by beta-nicotinamide adenine dinucleotide phosphate. Fertil Steril. 2005;83: 95–103.
4. Said TM, Aziz N, Sharma RK, Lewis-Jones I, Thomas Jr AJ, Agarwal A. Novel association between sperm deformity index and oxidative stress-induced DNA damage in infertile male patients. Asian J Androl. 2005;7:121–6.
5. Sharma RK, Sabanegh E, Mahfouz R, Gupta S, Thiyagarajan A, Agarwal A. TUNEL as a test for sperm DNA damage in the evaluation of male infertility. Urology. 2010;76:1380–6.

Clinical Utility of Sperm DNA Integrity Tests

37

Armand Zini

Abstract

Tests of sperm DNA integrity are being used increasingly in the evaluation of infertile men with the premise that these tests may better diagnose the infertility and predict reproductive outcomes. Indeed, a systematic review of the literature allows us to conclude that sperm DNA damage is associated with lower natural, IUI, and IVF pregnancy rates. By contrast, studies to date have not shown a clear association between sperm DNA and chromatin defects and pregnancy outcomes after ICSI. In couples undergoing IVF or ICSI, there is also evidence to show that sperm DNA damage is associated with an increased risk of pregnancy loss. A limitation of the systematic reviews and meta-analyses is that they do not address an important feature of the clinical studies on sperm DNA damage, the often marked heterogeneity of the individual study characteristics. Although the clinical utility of tests of sperm DNA damage remains to be established, the data suggest that there is clinical value in testing couples prior to assisted reproductive technologies – ARTs (IUI, IVF, and ICSI) and in those couples with recurrent abortions. Large, well-designed prospective studies are needed before testing becomes a routine part of patient care.

Keywords

Sperm DNA integrity tests • DNA integrity tests • Infertility in men • In vitro fertilization • Sperm chromatin defects

Clinical Utility of Sperm DNA Tests

The relationship between sperm chromatin/DNA damage and pregnancy outcomes has been examined by systematic reviews and meta-analyses [1–3]. The strength of these systematic reviews is

A. Zini (✉)
Department of Surgery, Division of Urology, McGill University, St. Mary's Hospital Center, Montreal, QC, Canada
e-mail: ziniarmand@yahoo.com

A. Zini and A. Agarwal (eds.), *Sperm Chromatin: Biological and Clinical Applications in Male Infertility and Assisted Reproduction*, DOI 10.1007/978-1-4419-6857-9_37,

the improved precision of the summary estimates compared with the individual study estimates of the relationship between sperm DNA defects and pregnancy outcomes. On the contrary, a weakness of meta-analyses (particularly on this topic) is the fact that it combines studies with highly variable study characteristics: data collection (prospective or retrospective), population characteristics (unselected, male factor), female inclusion/exclusion criteria, laboratory expertise in assessment of sperm DNA/chromatin damage, sperm DNA/chromatin test type, and sperm DNA test cutoff.

The recommendations for sperm DNA testing are based on (1) systematic reviews and meta-analyses of the relevant studies, (2) the characteristics of sperm DNA testing (e.g., sensitivity, positivity rate), and (3) disease prevalence (e.g., pregnancy, pregnancy loss).

Screening Test for First Pregnancy Planners

The data from three studies [4–6] show that sperm DNA damage is associated with a reduced probability of natural pregnancy (combined OR 7.01, 95% CI 3.68, 13.36, $p<0.0001$). Remarkably, the three studies [4–6] report very similar associations between sperm DNA damage and natural pregnancy rate (with ORs of 6.54, 6.82, and 7.59, respectively, see Table 37.1). An analysis of the three studies reveals a median pregnancy rate of 53%, with a median positive predictive value (PPV) of 83% and a median negative predictive (NPV) of 58% associated with sperm DNA testing [4–6]. As such, the analysis predicts that in populations with an overall pregnancy rate of 53% (at 6–12 months of follow-up), the pregnancy rate is 17% when there is a positive test for sperm DNA damage and at 58% when the test result is normal. Therefore, testing for sperm DNA damage can discriminate between pregnancy rates of 17% and 58%. However, because the prevalence of a positive test in this context (first pregnancy planners) is low (<10%) and 17% of couples with a positive test will achieve a pregnancy, *indiscriminate sperm DNA testing in this context is not advocated. Clinicians may choose to test first pregnancy planners, but they should understand the predictive value and limitations (*e.g., *sensitivity, specificity) of the sperm DNA test in this context and discuss these issues with the patients.*

Couples with Mild Male-Factor Infertility: IUI Candidates

Data from one valid IUI study show that sperm DNA damage is related to a significantly reduced IUI pregnancy rate (OR 9.9, 95% CI, 2.37, 41.51, $p<0.0001$) [7]. In the Bungum et al. study, the overall IUI pregnancy rate is 20%, the PPV is 97%, and the NPV is 24% [7]. Therefore, in populations with an IUI pregnancy rate of 20%, a positive test for sperm DNA damage predicts the pregnancy rate to be 3% and a normal test result predicts the pregnancy rate to be 24%. Therefore, testing for sperm DNA damage prior to IUI can differentiate between pregnancy rates of 3% and 24%. *According to the Bungum* et al. *study, couples with high levels of sperm DNA damage should proceed to IVF and/or ICSI rather than IUI.* However, it is important to note that the sensitivity and prevalence of a positive test in

Table 37.1 Selected diagnostic properties of studies on sperm DNA damage and natural pregnancy

Study	*n*	%hDFI	Sens	Spec	PPV	NPV	OR	(95% CI)
Evenson et al. [4]	144	7	0.19	0.96	0.60	0.81	6.54	(1.72, 24.92)
Spano et al. [6]	215	13	0.23	0.96	0.86	0.55	7.59	(2.54, 22.67)
Giwercman et al. [9]	257	12	0.21	0.96	0.83	0.58	6.82	(2.52, 18.47)

%hDFI proportion of samples with high sperm DNA fragmentation index (DFI); *Sens* sensitivity; *Spec* specificity; *PPV* positive predictive value; *NPV* negative predictive value; *OR* odds ratio; *CI* confidence interval

this context (couples with mild male-factor infertility) are low (<20%) and these recommendations are derived from only one reliable study [7]. As such, *additional IUI studies are needed before routine testing is recommended prior to initiating IUI treatments.*

Couples with Severe Male-Factor Infertility: IVF or ICSI Candidates

Data from more than 20 studies (11 evaluable – see Table 37.2) demonstrate that sperm DNA damage is associated with a modest but significant reduction in the IVF pregnancy rate (combined OR of 1.70, 95% CI 1.30, 2.23, $p<0.05$) [7–17]. Further analysis of the 11 evaluable IVF studies (with a median pregnancy rate of 33%) reveals a median PPV of 77% and median NPV of 34%. In clinical terms, this means that in populations with an overall IVF pregnancy rate of 33%, a positive test for sperm DNA damage predicts the IVF pregnancy rate to be 23% and 34% if the test is negative. As such, couples with sperm DNA damage may choose to proceed to ICSI, where pregnancy rates are independent of test results (see below). However, *the clinical value of an 11% difference in IVF pregnancy rates (23% vs. 34%, with positive and negative test result, respectively) is modest, and it may be hard to justify routine testing in this setting.* However, clinicians may want to test select couples (e.g., with failed IVF) so as to better counsel these couples in future ART cycles.

Data from more than 20 studies (14 evaluable – see Table 37.3) have evaluated the relationship between sperm DNA integrity and pregnancy rates after IVF/ICSI. As with IVF studies, these ICSI studies are quite heterogeneous. In keeping with a recent analysis [1], the results of this updated meta-analysis on ICSI studies indicate that sperm DNA damage is not related to ICSI pregnancy rates (combined OR of 1.15, 95% 0.90, 1.55, $p=0.65$) [7–10, 13–22]. *These data suggest that sperm DNA testing is not clinically valuable in predicting ICSI outcomes.* Perhaps the most concerning aspect of these findings is the unknown long-term consequence (i.e., postnatal health) of a successful pregnancy with high levels of DNA damage.

Testing couples with severe male-factor infertility enrolled in IVF or ICSI may also be valuable because sperm DNA damage is associated with a significantly higher rate of pregnancy loss after IVF or ICSI (combined OR of 2.48, 95% CI; 1.52, 4.04, $p<0.0001$) [3]. Data derived from these studies (PPV and NPV) indicate that in populations with an overall rate of pregnancy loss of 18%, the rate of pregnancy loss is estimated at

Table 37.2 Selected diagnostic properties of 11 studies on sperm DNA damage and pregnancy after IVF

Study	*n*	Assay	%hDD	Sens	Spec	PPV	NPV	OR	(95% CI)
Filatov et al. [11]	176	CC	41	0.46	0.88	0.96	0.21	6.34	(1.82, 22.08)
Host et al. [14]	175	TUNEL	30	0.34	0.79	0.77	0.37	1.92	(0.92, 4.04)
Henkel et al. [13]	208	TUNEL	69	0.35	0.81	0.81	0.35	2.24	(1.09, 4.58)
Huang et al. [15]	217	TUNEL	19	0.22	0.83	0.50	0.57	1.30	(0.66, 2.56)
Boe-Hansen et al. [9]	139	SCSA	5	0.06	0.97	0.86	0.29	2.43	(0.28, 20.83)
Borini et al. [10]	82	TUNEL	16	0.17	0.89	0.85	0.23	1.66	(0.33, 8.28)
Lin et al. [16]	137	SCSA	16	0.15	0.83	0.45	0.51	0.88	(0.35, 2.19)
Benchaib et al. [8]	84	TUNEL	10	0.07	0.86	0.50	0.32	0.46	(0.11, 2.00)
Bungum et al. [7]	388	SCSA	16	0.17	0.86	0.71	0.34	1.24	(0.69, 2.26)
Frydman et al. [12]	117	TUNEL	44	0.58	0.68	0.64	0.35	2.97	(1.39, 6.32)
Tarozzi et al. [17]	82	CMA3	17	0.22	0.97	0.97	0.28	10.86	(0.62, 191.5)

%hDD proportion of samples with high sperm DNA damage; *Sens* sensitivity; *Spec* specificity; *PPV* positive predictive value; *NPV* negative predictive value; *OR* odds ratio; *CC* chromatin compaction; *TUNEL* terminal deoxynucleotidyl transferase-mediated dUTP nick end-labeling; *SCSA* sperm chromatin structure assay; *CMA3* chromomycin A3

Table 37.3 Selected diagnostic properties of 14 studies on sperm DNA damage and pregnancy after ICSI

Study	*n*	Assay	%hDD	Sens	Spec	PPV	NPV	OR	95% CI
Hammadeh et al. [20]	60	ABlue	44	0.50	0.71	0.82	0.35	2.40	(0.72, 7.96)
Host et al. [14]	61	TUNEL	59	0.57	0.38	0.58	0.36	0.79	(0.28, 2.25)
Henkel et al. [13]	54	TUNEL	48	0.68	0.63	0.79	0.50	3.67	(1.12, 12.0)
Gandini et al. [19]	22	SCSA	41	0.31	0.44	0.44	0.31	0.36	(0.06, 2.08)
Huang et al. [15]	86	TUNEL	57	0.64	0.50	0.55	0.60	1.80	(0.76, 4.27)
Zini et al. [22]	60	SCSA	18	0.17	0.81	0.46	0.51	0.87	(0.23, 3.22)
Check et al. [18]	104	SCSA	28	0.29	0.76	0.72	0.34	1.34	(0.52, 3.43)
Boe-Hansen et al. [9]	47	SCSA	38	0.36	0.57	0.67	0.28	0.76	(0.21, 2.72)
Borini et al. [10]	50	TUNEL	60	0.71	0.75	0.90	0.45	7.36	(1.67, 32.4)
Benchaib et al. [8]	218	TUNEL	17	0.19	0.87	0.72	0.37	1.55	(0.70, 3.41)
Bungum et al. [7]	223	SCSA	33	0.29	0.61	0.52	0.37	0.65	(0.37, 1.14)
Lin et al. [16]	86	SCSA	24	0.26	0.77	0.52	0.52	1.21	(0.45, 3.23)
Micinski et al. [21]	50	SCSA	35	0.40	0.85	0.91	0.28	3.73	(0.74, 18.77)
Tarozzi et al. [17]	50	CMA3	56	0.49	0.27	0.61	0.18	0.34	(0.09, 1.29)

%hDD proportion of samples with high sperm DNA damage; *Sens* sensitivity; *Spec* specificity; *PPV* positive predictive value; *NPV* negative predictive value; *OR* odds ratio; *ABlue* aniline blue; *TUNEL* terminal deoxynucleotidyl transferase-mediated dUTP nick end-labeling; *SCSA* sperm chromatin structure assay; *CMA3* chromomycin A3

37% when the test is positive and 10% when it is negative. The difference between a pregnancy loss rate of 37% and 10% may be valuable to patients and clinicians. Although the effect of DNA damage on pregnancy loss should be discussed with patients prior to undergoing ART, many couples will proceed with these treatments regardless of sperm DNA test results and the impact on pregnancy loss.

Couples with Pregnancy Loss After IVF or IVF/ICSI

The prevalence of a positive test, sensitivity and specificity of sperm DNA testing in the context of pregnancy loss after IVF and ICSI are and 25, 40, and 85%, respectively [3]. This indicates that sperm DNA damage is a minor cause of pregnancy loss after IVF and ICSI (based on the low prevalence and low sensitivity). However, if the test is positive, it suggests that the sperm DNA damage (or male-factor) may be the cause of the pregnancy loss (based on the high specificity). In this setting, it may be advisable to evaluate or reevaluate the male and correct any potential male factor (e.g., varicocele) that may contribute to the DNA damage.

Guidelines on Clinical Value of Sperm DNA Tests

The ASRM (American Society for Reproductive Medicine) has published guidelines on the clinical utility of sperm DNA integrity tests in 2006 and again in 2008 [23, 24]. Based on their evaluation of the existing literature (up to 2006 in both the 2006 and 2008 reports), they conclude the following:

1. Existing data on the relationship between abnormal DNA integrity and reproductive outcomes are limited.
2. Sperm DNA damage is more common in infertile men and may affect reproductive outcomes in selected couples, including those with recurrent spontaneous miscarriage or idiopathic infertility.
3. At present, the results of sperm DNA integrity testing alone do not predict pregnancy rates achieved with intercourse, IUI, or IVF and ICSI.
4. Currently, there is no proven role for routine DNA integrity testing in the evaluation of infertility.
5. Treatments for abnormal DNA integrity have not been shown to have clinical value.

Although these guidelines provide clinicians with a fair assessment of the value of sperm DNA tests (based on literature up to 2006), more recent studies have added to our understanding of this test and the data suggest that there may be value in testing couples prior to ARTs.

Summary

Tests of sperm DNA and chromatin integrity are being used in the evaluation of infertile men. To date, the clinical studies on sperm DNA and chromatin defects allow us to conclude that sperm DNA damage is associated with lower natural, IUI, and IVF pregnancy rates, but not with ICSI pregnancy rates. Moreover, sperm DNA damage is associated with an increased risk of pregnancy loss in those couples undergoing IVF or ICSI. Although the clinical utility of tests of sperm DNA/chromatin damage remains to be firmly established, the data suggest that there is clinical value in testing couples with recurrent abortions or prior to initiating ART cycles.

References

1. Collins JA, Barnhart KT, Schlegel PN. Do sperm DNA integrity tests predict pregnancy with in vitro fertilization? Fertil Steril. 2008;89:823–31.
2. Zini A, Sigman M. Are tests of sperm DNA damage clinically useful? Pros and cons. J Androl. 2009;30: 219–29.
3. Zini A, Boman JM, Belzile E, Ciampi A. Sperm DNA damage is associated with an increased risk of pregnancy loss after IVF and ICSI: systematic review and meta-analysis. Hum Reprod. 2008;23:2663–8.
4. Evenson DP, Jost LK, Marshall D, Zinaman MJ, Clegg E, Purvis K, et al. Utility of the sperm chromatin structure assay as a diagnostic and prognostic tool in the human fertility clinic. Hum Reprod. 1999;14: 1039–49.
5. Giwercman A, Lindstedt L, Larsson M, Bungum M, Spano M, Levine RJ, et al. Sperm chromatin structure assay as an independent predictor of fertility in vivo: a case-control study. Int J Androl. 2010;33: 221–7.
6. Spano M, Bonde JP, Hjollund HI, Kolstad HA, Cordelli E, Leter G. Sperm chromatin damage impairs human fertility. The Danish First Pregnancy Planner Study Team. Fertil Steril. 2000;73:43–50.
7. Bungum M, Humaidan P, Axmon A, Spano M, Bungum L, Erenpreiss J, et al. Sperm DNA integrity assessment in prediction of assisted reproduction technology outcome. Hum Reprod. 2007;22: 174–9.
8. Benchaib M, Lornage J, Mazoyer C, Lejeune H, Salle B, Francois Guerin J. Sperm deoxyribonucleic acid fragmentation as a prognostic indicator of assisted reproductive technology outcome. Fertil Steril. 2007; 87:93–100.
9. Boe-Hansen GB, Fedder J, Ersboll AK, Christensen P. The sperm chromatin structure assay as a diagnostic tool in the human fertility clinic. Hum Reprod. 2006;21:1576–82.
10. Borini A, Tarozzi N, Bizzaro D, Bonu MA, Fava L, Flamigni C, et al. Sperm DNA fragmentation: paternal effect on early post-implantation embryo development in ART. Hum Reprod. 2006;21:2876–81.
11. Filatov MV, Semenova EV, Vorob'eva OA, Leont'eva OA, Drobchenko EA. Relationship between abnormal sperm chromatin packing and IVF results. Mol Hum Reprod. 1999;5:825–30.
12. Frydman N, Prisant N, Hesters L, Frydman R, Tachdjian G, Cohen-Bacrie P, et al. Adequate ovarian follicular status does not prevent the decrease in pregnancy rates associated with high sperm DNA fragmentation. Fertil Steril. 2008;89:92–7.
13. Henkel R, Kierspel E, Hajimohammad M, Stalf T, Hoogendijk C, Mehnert C, et al. DNA fragmentation of spermatozoa and assisted reproduction technology. Reprod Biomed Online. 2003;7:477–84.
14. Host E, Lindenberg S, Smidt-Jensen S. The role of DNA strand breaks in human spermatozoa used for IVF and ICSI. Acta Obstet Gynecol Scand. 2000; 79:559–63.
15. Huang CC, Lin DP, Tsao HM, Cheng TC, Liu CH, Lee MS. Sperm DNA fragmentation negatively correlates with velocity and fertilization rates but might not affect pregnancy rates. Fertil Steril. 2005;84:130–40.
16. Lin MH, Kuo-Kuang Lee R, Li SH, Lu CH, Sun FJ, Hwu YM. Sperm chromatin structure assay parameters are not related to fertilization rates, embryo quality, and pregnancy rates in in vitro fertilization and intracytoplasmic sperm injection, but might be related to spontaneous abortion rates. Fertil Steril. 2008;90: 352–9.
17. Tarozzi N, Nadalini M, Stronati A, Bizzaro D, Dal Prato L, Coticchio G, et al. Anomalies in sperm chromatin packaging: implications for assisted reproduction techniques. Reprod Biomed Online. 2009;18:486–95.
18. Check JH, Graziano V, Cohen R, Krotec J, Check ML. Effect of an abnormal sperm chromatin structural assay (SCSA) on pregnancy outcome following (IVF) with ICSI in previous IVF failures. Arch Androl. 2005;51:121–4.
19. Gandini L, Lombardo F, Paoli D, Caruso F, Eleuteri P, Leter G, et al. Full-term pregnancies achieved with ICSI despite high levels of sperm chromatin damage. Hum Reprod. 2004;19:1409–17.

20. Hammadeh ME, al-Hasani S, Stieber M, Rosenbaum P, Kupker D, Diedrich K, et al. The effect of chromatin condensation (aniline blue staining) and morphology (strict criteria) of human spermatozoa on fertilization, cleavage and pregnancy rates in an intracytoplasmic sperm injection programme. Hum Reprod. 1996;11: 2468–71.
21. Micinski P, Pawlicki K, Wielgus E, Bochenek M, Tworkowska I. The sperm chromatin structure assay (SCSA) as prognostic factor in IVF/ICSI program. Reprod Biol. 2009;9:65–70.
22. Zini A, Meriano J, Kader K, Jarvi K, Laskin CA, Cadesky K. Potential adverse effect of sperm DNA damage on embryo quality after ICSI. Hum Reprod. 2005;20:3476–80.
23. The Practice Committee of the American Society for Reproductive Medicine. The clinical utility of sperm DNA integrity testing. Fertil Steril. 2006;86:S35–7.
24. The Practice Committee of the American Society for Reproductive Medicine. The clinical utility of sperm DNA integrity testing. Fertil Steril. 2008;90: S178–80.

Index

A. Zini and A. Agarwal (eds.), *Sperm Chromatin: Biological and Clinical Applications in Male Infertility and Assisted Reproduction*, DOI 10.1007/978-1-4419-6857-9,